RADIATION TRAPPED IN THE EARTH'S MAGNETIC FIELD

ASTROPHYSICS AND SPACE SCIENCE LIBRARY

A SERIES OF BOOKS ON THE DEVELOPMENTS
OF SPACE SCIENCE AND OF GENERAL GEOPHYSICS AND ASTROPHYSICS
PUBLISHED IN CONNECTION WITH THE JOURNAL
SPACE SCIENCE REVIEWS

VOLUME V

RADIATION TRAPPED IN THE EARTH'S MAGNETIC FIELD

PROCEEDINGS OF THE ADVANCED STUDY INSTITUTE
HELD AT THE CHR. MICHELSEN INSTITUTE, BERGEN, NORWAY
AUGUST 16–SEPTEMBER 3, 1965

Edited by

BILLY M. McCORMAC
Geophysics Division, IIT Research Institute,
Chicago, Ill.

D. REIDEL PUBLISHING COMPANY

DORDRECHT-HOLLAND

ISBN-13: 978-94-010-3555-2 e-ISBN-13: 978-94-010-3553-8
DOI: 10.1007/978-94-010-3553-8

PREFACE

This book contains the lectures presented at the Advanced Study Institute, 'Radiation Trapped in the Earth's Magnetic Field' which was held at the Bergen Tekniske Skole, Bergen, Norway, during the period August 16 through September 3, 1965.

Approximately one-third of the time was devoted to discussion. The various Session Chairmen have summarized the essential points brought out in these discussion periods which were generally quite spirited.

The authors and the publisher have made a special effort to rapidly publish an up-to-date status concerning the various aspects of trapped radiation. Almost all authors turned in their manuscripts prior to the end of the Institute and all prior to September 30, 1965. It was clearly recognized that rapid publication was essential in this rapidly changing research area.

Special thanks are due to the Host, Dr. Odd Dahl, Chr. Michelsen Institute, for making all of the arrangements in Bergen, including excellent living and working facilities.

Direct financial support was provided the Institute by: North Atlantic Treaty Organization, Advanced Research Projects Agency, Air Force Cambridge Research Laboratories, Army Research Office, Defence Atomic Support Agency, and the Office of Naval Research.

BILLY M. MCCORMAC

INTRODUCTION

Some of the scientific problems facing the attendees of this Institute are enumerated below by session. The state of knowledge of the various facets of radiation trapped in the earth's magnetic field is rapidly changing. It is hoped that this Institute will make a significant contribution towards improving our understanding of these phenomena and will identify those areas wherein there is general agreement, as well as those key problems which deserve additional attention.

Carl McIlwain's *B-L* coordinate system with the 1962 Jensen-Cain representation of the earth's magnetic field is currently used by much of the scientific community. To study basic processes and report predictions and experimental results it is desirable to use a common coordinate system. The McIlwain coordinate system breaks down at earth radii greater than about 5, especially because of an insufficient magnetic field representation due to the changing environment near and beyond the magnetopause. Additional data have become available closer to the earth where violent magnetic variations are seldom experienced. When should the 1962 Jensen-Cain representation be revised, before or after POGO? What coordinate system is suitable beyond 5 to 6 earth radii, especially for near to and beyond the magnetopause on out into the solar wind?

The observed particle distributions for electrons and protons in both the inner and outer belts need to be summarized. Such information as the energy spectra, flux, pitch angle distribution, and variations with time are required. Many measurements have been taken with many different detectors; however, it is difficult for most of us to identify those aspects of agreement and disagreement and those aspects wherein little or no data exist. This should be stretched to cover the incident solar wind.

The sources and sinks for the observed protons and electrons in both the inner and outer belts are needed. Are any of the trapped particles directly injected as observed such as from albedo neutron decay protons and electrons or solar wind particles? What are the various acceleration mechanisms? Atmospheric scattering seems to be in pretty good shape as a loss mechanism at low *L*-values. Are there suitable acceptable mechanisms at higher *L*-values?

There have been a large number of theoretical and experimental papers published in the last year on the physics of the outer magnetosphere. What has been observed about the magnetic field, solar wind, and charged particles? What are the possible fundamental processes? The various models need to be explained and suggestions made as to how they can be tested. How does the solar wind magnetosphere interaction tie into aurora phenomena, polar cap events, etc.?

Artificially injected radiation can help improve our understanding of trapped radiation by providing known sources of radiation at known times and locations in space. Previous U.S.A. and Soviet high altitude nuclear detonations have provided much useful information on effective life times, trapping efficiencies, diffusion etc. It is important to review what can be learned from these previous artificial sources as well as to discuss possible future sources.

Synchrotron radiation is an interesting phenomenon which has been especially observed at the lower latitudes from trapped radiation. A detailed knowledge of the spectrum, flux, pitch angle distribution and magnetic field is required to predict the synchrotron radiation to be expected. Unfortunately, measurement of the synchrotron radiation does not uniquely fix any of these parameters. It will still be an interesting phenomenon to measure from other planets for a number of years to come, in particular Jupiter.

Trapped radiation produces detrimental effects on solar cells, transistors, and other exposed and sensitive semiconductor devices. This same radiation could be a hazard to prolonged manned flight in the radiation belt, especially the center of the inner belt. Some suitable means of estimating the radiation dose to human beings and sensitive electronic components are needed. What environment is expected? Particular attention should be given to manned flight through the South Atlantic anomaly.

Illinois Institute of Technology, Chicago, Ill. BILLY M. MCCORMAC

CONTENTS

SESSION V / PHYSICS OF THE OUTER MAGNETOSPHERE

OPENING TALK

I have always been intrigued by the urge for knowledge in human nature. The natural sciences evolved out of this urge, creating and making use of opportunities for systematic researches. That is what you do.

In this quest for knowledge the human mind is conditioned to the prevailing mental climate, and restricted by the level of established knowledge. If it strays far afield from the recognized truth, the Galileos will be in trouble, and if it is unaware of the fundamentals, we may see consuming passions, like wanting to make gold.

Within the range of these extremes, the seekers have found their opportunities and two main outlets; geographical exploration for the adventurous, until now only the geographical fine-structure remains as a challenge on our planet, while the more speculative minds wondered about the universe and the heavenly bodies, became astronomers and astrologers, astronomers again and astro-physicists, rapidly turning into all embracing fundamental physicists.

We have gone a long way and the urge has found many new outlets to which the mind has conditioned itself.

In the process, the researchers, but for the very few, have left their ivory towers to work in large laboratories with vast supporting facilities, and knowledge is pouring out as pursued through group efforts. There is another aspect to this development. In too many cases, when he is good at his work, the individual has become a planner and administrator and often the custodian of vast sums of public money.

The trend was clear in the years before the war, but it was the war which gave the scientist the opportunities or forced him to operate on a really large scale. He had no choice, he had to follow up and do his best with nuclear physics and energy, and rockets.

He has been given the means as never before, but he has lost much of his freedom. The experimenter can not any longer design and build his own experiment, and a library, paper and pencil are not enough for the speculator. They have become entirely dependent upon current technology and industry for their work, but they are also shaping technology and creating new opportunities for industries.

We also note that the international aspects of the sciences have acquired a content far beyond anybody's expectation, and through much closer cooperation the quality of science in all industrial countries has become uniformly good.

We have always said that, when there is good science in a country there is industrial growth, and vice versa. As we now must be said to have good international science, we have a reason to believe in a common industrial growth.

Space activity is an outstanding example of this close interdependence between the sciences, technology and industry across national borders. So much of what is radically new happens in this sector, and with such obvious opportunities for international scientific cooperation, acting as a very strong new driving force in industry and colouring many branches of the conventional industries as well, it can only be a good thing.

In this situation, as always, it is difficult to separate out cause and effect, but the fact remains that the importance of science now has permeated through all levels in society and the various branches of government. More than ever it is realized that we are all in the same boat and that the motive power of the boat is science. This is also because space activities appeal to everybody's imagination and constitute a challenge to the adventurous as well as to the speculative mind. Perhaps back of it all we have the fact that research is becoming increasingly more costly and elaborate, forcing groups and nations to pool resources, mental as well as material.

Thus we have the large nuclear physics laboratories in USA and USSR and the inter-European laboratory in Geneva, and the concerted activities in the field of nuclear energy, and of course now space.

I do not have to tell you that the impact of space research has been terrific, felt with consequences in all the natural sciences.

The new is so simple; it is the given opportunity to put sensing instruments out in free space with the ability to relay information back to earth but the inherent technological challenge is just as great as the scientific, and we have the resulting technique of science with all its built-in redundance, forming modern advance-industries, and coupling back.

Perhaps better than words this Institute expresses what I have tried to convey, a world full of opportunities. In itself it is rather remarkable, not as a conference, but through the fact that a large group of people can come together to deal intimately and in great detail with problems around phenomena which hardly could have been foreseen without access to space.

To be host country for such an important event helps science in Norway too, because our people are aware of a connecting line going back from this Institute to the early work of our countrymen, Birkeland and Störmer. Our own space activity follows their line today.

So, I am glad that you have chosen Bergen as the place for the Institute, and that so many of you from so many countries have come so far to participate, to give and to learn. I can assure you that we feel honoured.

I realize that you are here to work, but I hope not all the time, as we wish for you to find time to see some of this part of Norway. After all, there is a large contingent of Institute wives and children here also, and you are under certain obligations.

We are glad to see you all, not as strangers but as friends. I hope you will have a pleasant stay, and I wish you success.

Chr. Michelsen Institute, Bergen, Norway ODD DAHL

SESSION I

B-L SPACE

Chairman

DR. MARTIN WALT

Lockheed Missile and Space Company

MODELS OF THE EARTH'S MAGNETIC FIELD

JOSEPH C. CAIN
NASA-Goddard Space Flight Center,
Greenbelt, Md.

Introduction

In order to discuss models of the earth's magnetic field it is necessary to understand the use to which they will be put and the expected range of their validity. The evaluation of experimental data both of the field itself and of charged particles trapped in that field requires a detailed and accurate reference field structure throughout the magnetosphere. Theoretical studies are generally not as particular and have frequently subsisted on even a dipole approximation to the main field. However, as our knowledge of the field and particle fluctuations extends outward from the earth to the magnetospheric boundary and into the antisolar tail regions, even the theoretical workers will require a better model than that of the extrapolated dipole.

For most experimental work the dipole approximation is inadequate at *any* altitude since the main field is very irregular near the earth, and beyond a few earth radii where it becomes more dipolar, the distortions due to other effects become significant. Also, as experimental work on the field itself near the earth becomes more precise, even the relatively small time variations in these regions will need to be considered.

In order to better define the scope of magnetic field models that are in the process of being developed we define here four separate field sources as follows:

1. *Core field.* This is the main field from the molten core of the earth as it has historically been plotted on surface magnetic charts of the world and has been analyzed into a potential function using series of spherical harmonic coefficients. Its secular variation is characteristically a fraction of a percent per year in intensity and presumably arises from mass and wave motions in the molten core.

2. *Crustal field.* This additional source of the magnetic field is thought to arise from the magnetized crustal material from the earth's surface to depths of the order of 20 km where the temperatures rise above the Curie point. Its contribution to the ambient field near the earth's surface is frequently large in producing irregularities of a few kilometers in extent and amplitudes of a few

thousands of gamma [1γ (gamma)$=10^{-5}$ Γ (gauss)]. Measurements of the total scalar field **F** by sounding rockets (Davis *et al.*, 1965) indicate that there are measurable contributions of the crustal field at least to 50 km altitude. It is hoped that for most geomagnetic work this source may be smoothed over since its inclusion would require an extensive set of parameters.

3. *Ionospheric currents.*

(a) Quiet daily variation (Sq) including equatorial electrojet;

(b) Disturbance variations (DS) including auroral electrojets.

4. *Plasma pressures.*

(a) Internal – Trapped particle and tail (?);

(b) External – Boundary.

The last two sources, 3 and 4, cannot be discussed without making a distinction between a "forecast" and a "definitive" model of the field. In the past, the world magnetic charts for a given epoch (e.g. 1965) were based on survey data for prior years extrapolated to that epoch. This extrapolation of the secular change in the core field was used over a future period of 5 or 10 years and thus is entirely a "forecast" field model. A "definitive" map of the core field, not generally drawn, would instead be based on an interpolation of data taken over roughly equal intervals about some mean epoch.

To extend these concepts to the sources of field distortions external to the earth, we must understand the features of those distortions that are statistically predictable and might be included in a "forecast" model and those that are patently irregular and, except for probability estimates of occurrence, should better be left to "definitive" models.

An example of geomagnetic field variations as measured at the earth's surface is given in Figure 1. These curves are hourly mean values of the horizontal component measured at a number of magnetic observatories and plotted to the same scale but arbitrary baselines. The vertical lines denote 0^h UT between the days indicated, and the position of the dots gives the times of local noon at each station. The interval chosen is moderately quiet and is thus free from large magnetic disturbance. At most of the low latitude stations one can see the regular peak of the Sq variation near local noon. However, contrary to the impression that might be conveyed by statistical treatments of the subject, one can note many day-to-day changes in the data from the lowest latitude stations and increasing irregularities with increasing latitude. For example, the curve for Huancayo shows a noon minimum instead of the usual maximum on January 12. The slow increase of the level of the curves is typical of that following a magnetic disturbance. The average of these disturbances in the horizontal field over all longitudes (Dst) has been plotted many times in the past (e.g. Sugiura, 1964) and is indicative of the irregular changes in the field with the onset of magnetic disturbance. These disturbances occur several times

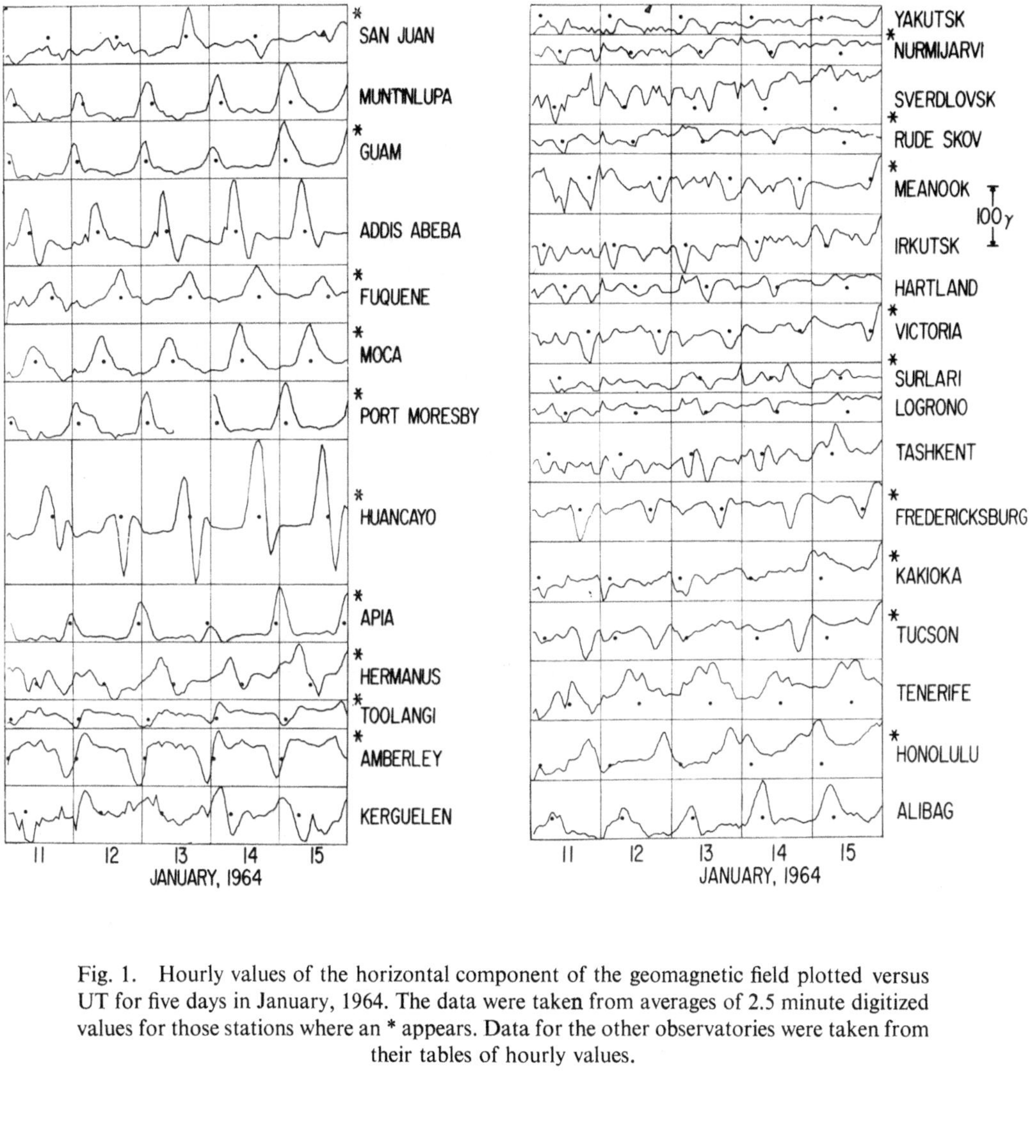

Fig. 1. Hourly values of the horizontal component of the geomagnetic field plotted versus UT for five days in January, 1964. The data were taken from averages of 2.5 minute digitized values for those stations where an * appears. Data for the other observatories were taken from their tables of hourly values.

a year and cause a general lowering of Dst by a few hundred gammas. While the major source of the quiet daily variations Sq are thought to lie in the conducting layer of the ionosphere near 100 km, the Dst variations are speculated as being caused by the expansion of the field by trapped particles. At the present time the distribution of the geoeffective particles is not known nor is it clear that there is a significant steady component (cf. HOFFMAN and BRACKEN, 1965). However, there is evidence (NESS, 1965) that some of the trapped plasma is sufficient to completely distend the earth's field in the antisolar direction.

The external pressures from the solar plasma have now at least been grossly defined and we should see later in the symposium evidence supporting the numerical evaluations by MEAD (1964) of its effect.

Thus in order to effectively model the field of the earth within the magnetosphere even a "forecast" field will need to include these external sources to some extent. For a field reference for some types of experimental data it may eventually be necessary to use a "definitive" model of the field which is based on the interpolated best estimate of the actual field distortions at a given time using whatever surface and satellite measurements of the field that can be obtained. This type of dynamic field modelling may be somewhat unnecessary for present particle work. It would of course enormously complicate the computation of the L parameter used for organizing the data. In order to avoid this complication most of the remainder of this discussion will be confined to the core field and its secular change and thus only cover a fraction of the scope of the subject.

Survey Data

Before discussing the evaluations of the field it is very useful to note the character of the magnetic survey data currently available. These data are characterized more by their heterogeneity than any uniformity. The sources are many and include data from surveys by land and from magnetometers both on board and towed behind ships and aircraft. The quality of the data varies from some where even the component being measured is in doubt, to the high quality observations performed by some survey groups and generally by the magnetic observatories. The data may be comprised of instantaneous measurements or time averages over periods up to a year. They may be presented either with no modifications, modified to attempt to eliminate diurnal variations, or corrected to a given epoch by spatially interpolating the secular change from magnetic observatories. The observatories, though currently numbering some 300, are almost all confined to the land masses of the northern hemisphere and thus cannot give a reliable measure of the secular change in most southern areas.

For the whole interval 1929 to 1950, there were almost no survey observations of the ocean areas. Since that time the U.S. Navy project 'Magnet' has been the main contributor of ocean magnetic data. However, even with the present increased efforts being made during the World Magnetic Survey, there are yet significantly large areas of the earth covered only by a single survey track or not at all. The ocean gaps include the South Pacific centering near the meridian 120 °W, and the Atlantic and Indian Oceans south of $-40°$ latitude. There are also large land areas only sparsely covered such as Asia with only Russian observatory annual mean data since 1940 and the Antarctic with only scattered observations.

The altitude distribution of the data is equally heterogeneous with the only presently available satellite data being those from Vanguard 3 and some less accurate ($\sim 100\gamma$) data from the Alouette satellite. The Vanguard 3 measurements cover only small volumes in areas near the NASA tracking stations and Alouette covers Canada which is already densely covered by aircraft observations.

Clearly, any method of data analysis may produce results that could be greatly affected by this very heterogeneous data distribution.

Technique of Analysis of the Core Field

We have recently completed an analysis of the data available over the period 1940–1962 and reported the results in comparison with most previously available work (CAIN *et al.*, 1965). Our approach to the problem differs from that of most past workers in that we have attempted to fit the secular change of the field simultaneously with the spatial structure. The principal used is to minimize the mean square difference between each *observed* component (H, D, I, Z, F) and a computed component. The field expansion for this work was an internal potential function of the classic form

$$V = a \sum_{n=1}^{\infty} \left(\frac{a}{r}\right)^{n+1} \sum_{m=0}^{n} (g_n^m \cos m\varphi + h_n^m \sin m\varphi) P_n^m(\theta)$$

where r, θ, φ = spherical coordinates corresponding to geocentric radius, geocentric colatitude, and longitude. The coefficients g_n^m, h_n^m are gauss coefficients ($g_1^0 \approx -0.3$) and $P_n^m(\theta)$ are Schmidt's quasi-normalized polynomials. The various field components are evaluated using $\mathbf{F} = -\nabla V$ on the assumption that $\nabla \times \mathbf{F} = 0$ (no current density inside the region of interpolation).

The minimization procedure deviates from a straightforward least-squares only in that each component is expanded into a Taylor's series in the coefficients and, starting with a first approximation to the field, the solution is found for the first order corrections to the coefficients. This procedure is necessary since the fit is made to the components D, I, H and F which vary non-linearly with the coefficients.

In order to include the secular change, the coefficients are expanded in a power series about the mean epoch of the data and the first time derivatives determined along with the coefficients themselves.

In performing the fits it has been found that more consistent results can be obtained by taking into account the earth's oblateness. Thus, whereas the value a in the expression for V is arbitrarily taken to be the mean radius of the earth

6371.2 km, the value of r used for each data point assumes the surface data on the Kaula ellipsoid (equatorial radius of 6378.165 km and a flattening factor of 1/298.3). In addition, the slight ($\sim 0.2°$) rotation of the geocentric *versus* geodetic direction is taken into account. In computing the field from the derived coefficients a precise evaluation needs to allow for both of these factors. Assuming that the geodetic latitude is λ so that the rotation angle is $\delta = \lambda + \theta - 90°$, the measured magnetic components are:

$$X = - F_\theta \cos\delta - F_r \sin\delta$$

$$Y = F_\varphi$$

$$Z = F_\theta \sin\delta - F_r \cos\delta$$

where F_θ, F_φ, F_r are the three geocentric components of **F**. One should also note that although the coefficients are derived using Schmidt normalized polynomials $P_n^m(\theta)$, an internal conversion is normally made in the Fortran computer codes (CAIN *et al.*, 1964) so that the coefficients actually used correspond to the Gauss-Laplace functions $P^{n,m}(\theta)$. This conversion is useful in that the number of computer operations necessary for the evaluation of the field components is reduced by about 10%.

A New Test Model of the Core Field

Before evaluating the relative merits of existing models of the core field we would like to report on an additional model that was recently derived using improved techniques and some added data. This model, which we will here label GSFC (7/65), according to the organization and date of derivation, was intended as a test to see whether higher order spatial coefficients and a simple external field are statistically significant if almost all of the available data are used. Since we have recently published an evaluation of the field (CAIN *et al.*, 1965) we will discuss here only those details which differ from the prior work.

One change in the techniques of computation that allows a significant expansion of the number of coefficients over the previous analysis is the use of single precision (~ 8 digit) arithmetic in the formation and solution of the matrix of values entering the equations for the normal equations. The previous use of double precision arithmetic (~ 17 significant figures) was learned to be helpful in some very special circumstances, but was found not necessary for general use (LEATON, 1964). The error in the final parameters resulting from single precision arithmetic is estimated at less than 10^{-5} of their value.

In order to test the use of higher order coefficients, the number of spatial parameters was expanded to 99 (maximum $n = m = 9$) and first time derivatives to 48 ($\dot{n}_m = 6$), although in the actual analysis to be described here an $\dot{n}_m$ of only 5 was used. In addition to these internal sources, a simple external field was added. In keeping with most theoretical studies that predict that any external source of field would result in a parallel flux across the earth, the form of the external field was taken to be three orthogonal components E_1, E_2, E_3 where E_1 is northward along the geographic axis and E_2 and E_3 are in the equatorial plane directed respectively at geographic longitudes 0 and +90°.

Following the suggestion of KAHLE, KERN and VESTINE (1964) for a checking term, we initially also included the term g_0^0. (The formulation thus assumes $\nabla \times \mathbf{F} = 0$, but $\nabla \cdot \mathbf{F} \neq 0$!).

Since we also wished to test the differences to be obtained by the use of most of the available data, only a small amount of data deletion was done. The data for the period 1945–1964 were grouped into categories of surface, aircraft, observatory, and satellite. The data set used comprised all observatory and satellite (Vanguard 3 and Alouette) data in this interval. The surface and aircraft data were then each "clipped" only in areas of 2° latitude × 2° longitude where the density exceeded about 100 observations per 10^5 km^2. That is, one observation per 2° × 2° block at the poles was allowed and used as a maximum density for the lower latitude blocks. The resulting data distribution of some 197000 component and total field observations is given in Figure 2. The previously discussed "sparse areas" become very evident.

One other difference in this analysis concerns the weighting of the data as entered into the normal equations. Previously, the observatory annual means were weighted by a factor of the order of 6 to 10 over other survey data according to their high estimated accuracy. In this analysis the weights were arbitrarily reduced so that they were only of the order of 2–4 times that of the other survey data since it was felt that the previous weights were too high. It is yet unclear as to the improvements in the derived fields when the data are weighted inversely according to their estimated errors.

Test runs were first made on random samples of the data, skipping to each nth observation, to test the significance levels and ascertain whether there were any spurious correlations between the coefficients. This was the first time that the correlation coefficients have been derived between the various parameters. It was gratifying to find correlation coefficients between all of the parameters of only a few tenths and hence practically insignificant. This result was especially reassuring since it is clear that neither the external terms nor the first time derivatives of the spatial terms have any theoretical reasons for being orthogonal over the earth's surface. As the tests were made first starting with every 200th observation and then increasing the number of observations to only skip to every 100th, and then to every 50th, it was noted that the absolute values of both the g_0^0 and the E terms continuously decreased and were always compa-

Fig. 2. Positions of magnetic survey observations available for the interval 1945–1964 with some thinning as described in text.

rable with their estimated standard error. However, when enough data were added to bring the standard error of the external field E down to about 30γ the components began to stabilize and assume fixed values. However, the g_0^0 term continuously decreased and was deleted when both its absolute value and standard error were of the order of 8γ. It is thus established that if there was any monopole component to the earth's field (indeed, if monopoles existed!) that its level would need to be less than 8γ. This behavior in relation to the estimated standard errors does give some indication that whatever the cause, the external component of the earth's field is a statistically significant aspect of the data set. The resulting set of harmonic coefficients is given in Table 1 to $n=m=9$ in the spatial terms, $n=m=5$ in the first time derivatives, and the three external terms. The estimated standard errors for the external terms and the low order spatial terms range from 4 to 5γ reducing to a 2–3γ level above $n \approx 6$. The standard errors of the time derivatives are of order 1γ/year.

TABLE I

n	m	g_n^m	h_n^m	$\dot{g}_n^m$	$\dot{h}_n^m$	n	m	g_n^m	h_n^m	$\dot{g}_n^m$	$\dot{h}_n^m$
1	0	−30426	−0	−20.4	−0.	7	0	72	−0		
1	1	−2158	5771	4.4	−6.5	7	1	−58	−52		
2	0	−1535	−0	−28.1	−0.	7	2	4	−23		
2	1	3000	−1952	2.0	−11.7	7	3	6	−11		
2	2	1585	205	−0.1	−14.6	7	4	−26	2		
3	0	1304	−0	2.9	−0.	7	5	−2	27		
3	1	−1985	−433	−10.2	4.4	7	6	14	−22		
3	2	1293	236	1.5	2.0	7	7	6	−21		
3	3	871	−128	−1.3	−9.5	8	0	5	−0		
4	0	959	−0	−1.0	−0.	8	1	6	7		
4	1	804	151	2.3	−1.0	8	2	−9	−12		
4	2	503	−269	−0.8	−0.7	8	3	−16	6		
4	3	−393	4	−1.1	4.0	8	4	−4	−18		
4	4	274	−251	−3.8	−5.9	8	5	7	4		
5	0	−227	−0	2.3	−0.	8	6	−6	25		
5	1	359	10	0.8	2.7	8	7	12	−2		
5	2	231	122	1.8	3.4	8	8	7	−16		
5	3	−34	−116	1.2	−1.5	9	0	9	−0		
5	4	−160	−109	0.6	0.7	9	1	3	−22		
5	5	−65	78	2.0	1.4	9	2	7	15		
6	0	50	−0			9	3	−15	4		
6	1	60	−10			9	4	12	−4		
6	2	4	104	$E_1=-19$		9	5	11	−2		
6	3	−244	61	$E_2=9$		9	6	1	10		
6	4	−0	−29	$E_3=16$		9	7	−0	12		
6	5	3	−12			9	8	2	0		
6	6	−108	−11	GSFC(7/65)		9	9	1	−2		

A comparison of the present coefficients with our most recently published set (CAIN *et al.*, 1965) shows good agreement, though for the spatial terms too frequently outside their estimated standard errors. This discrepancy can likely be explained by the fact that although the functions are orthogonal to one another if averaged over the whole sphere, they are not orthogonal over the available data set. As more terms were added for the present analysis it might be predicted that there would be some adjustment of the lower order coefficients as the still neglected higher order harmonics distribute their aliases differently. Some differences are also likely since the data set extends over the period 1945–1964 including some newer data whereas the last fit took an average over the period 1940–1962. As shown in the previous paper the secular changes in the field are not completely linear even over a 20 year interval.

One of the new results of this analysis is the presence of a statistically significant external field. However, even with the estimates of its standard error it is still not conclusive that it is factual due to the poor distribution of data. A plot of its orientation relative to the surface field contours of **F** is shown in Figure 3. Here the + denotes the field *entering* in the northern hemisphere. If we were to speculate on the possible origin of a constant external field we should presently need to turn to one of two effects, both of which would result in almost uniform fields parallel to the earth's axis. The effect of the compression of the earth's field by the solar wind should give a northward flux, whereas the effect of trapped plasma should expand the field to give a net decrease as presumably is enhanced during the main phase of a magnetic storm. The fact that the observed constant external field has such a large equatorial component is puzzling and leads one to believe that perhaps it is a spurious result of the irregular data distribution or some other unknown factor. However, since it is at least in the correct quadrant of the earth to be associated with the eccentric dipole, the result would imply that the net internal plasma pressures more than balance the external to produce a net uniform flux which would weaken the field near the equator.

Spatial Errors of Derived Field

Although it is not possible to know with certainty the absolute errors of the fits, it is possible to obtain some statistical measure of how much the potential function might be varied by moving all of the data through one standard error. Such a spatial function can be derived on the assumptions that the standard errors of all of the data are equal to that of the fit and that all are independent. A resulting spatial map of the standard error in total field F (in gammas) is given in Figure 4. The highs and lows in the field are marked by H and L respectively along with their values in gamma. A comparison between this figure and Figure 2, the data distribution, shows as might be expected, that the fit

Fig. 3. Total field F and plot of external field intercepts using GSFC(7/65) coefficients at earth's surface for 1960.0. External field vector enters at + and exists at ⊙ with ellipses indicating standard errors of position.

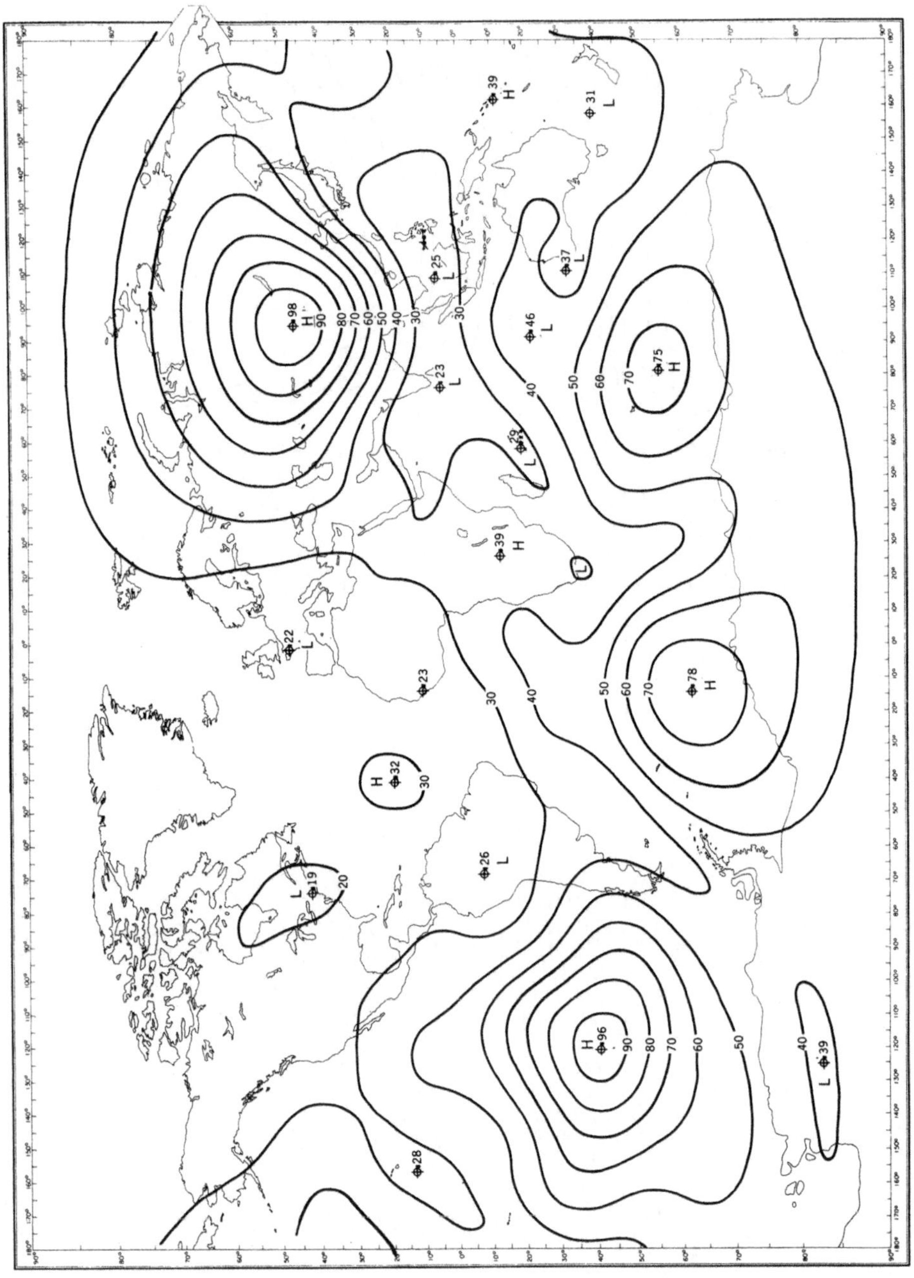

Fig. 4. Standard error of GSFC(7/65) fit computed for total field *F*(gamma) at earth's surface for epoch 1960.0.

could be more easily altered in the regions where the data are sparse. However, although this map gives the possible relative errors in the fitted field, it is likely that the actual errors of the field are several times this. This assertion is based on sample comparisons between measured and computed field values which indicate that there are frequent systematic variations of the order of a few hundred gammas which are not due merely to the crustal anomaly "noise" (cf. Figure 6 in our previous paper (CAIN *et al.*, 1965)). The corresponding standard error map of inclination is given in Figure 5 and shows a range of a factor of four or five between the flexibility of the fit in the sparse data areas to that in the better surveyed regions.

It is useful also to investigate the standard error estimates in regions outside the data volume both with increasing altitude and at years other than those near the mean epoch. Sample calculations of the errors with altitude show that although the ΔF starts at differing values at the surface it tends to decrease to a constant level of the order of 15–20γ by 2000 km altitude and thereafter to remain constant at about 10–15γ. When the errors are computed for epoch 1965 instead of 1960 they increase by about 30–50%. By 1970 the estimated errors are approximately double those computed for 1960. It is thus likely that any definitive field will always have significantly smaller errors than any forecast field where the *latest* data lag by about two years. If future (satellite) data are processed quickly to reduce this time lag, the errors in a forecast field should be correspondingly smaller.

It has been learned by use of fewer coefficients in the fitting process that whereas a smaller coefficient set gives a larger overall standard error in regions of high data density, the error in regions of low density are almost the same as those shown in Figures 4 and 5. Of course if the number of coefficients was increased still more so as to include those with little statistical validity it is likely that the possible variation of the fit in the areas completely void of data could be increased significantly. The conclusion of such studies on the present data set is that whereas the number of coefficients is indeed larger than most derived in the past, this number seems necessary to fit the data and that no significant harm has yet been done in causing large errors in the areas of low data density. If anything, it will be enlightening to expand the coefficient set further to see whether we have reached the optimum number of coefficients necessary to describe the present data. It is likely that there are even shorter wavelengths in the available data. However, before making such a coefficient expansion we intend to investigate independently the spatial spectrum to see whether a better overall result might be better achieved by first filtering the data to eliminate any effect from the neglected harmonics. It is also worthwhile to postpone such evaluations to see whether the POGO satellite will provide sufficient data to relieve some of the problems with the data density.

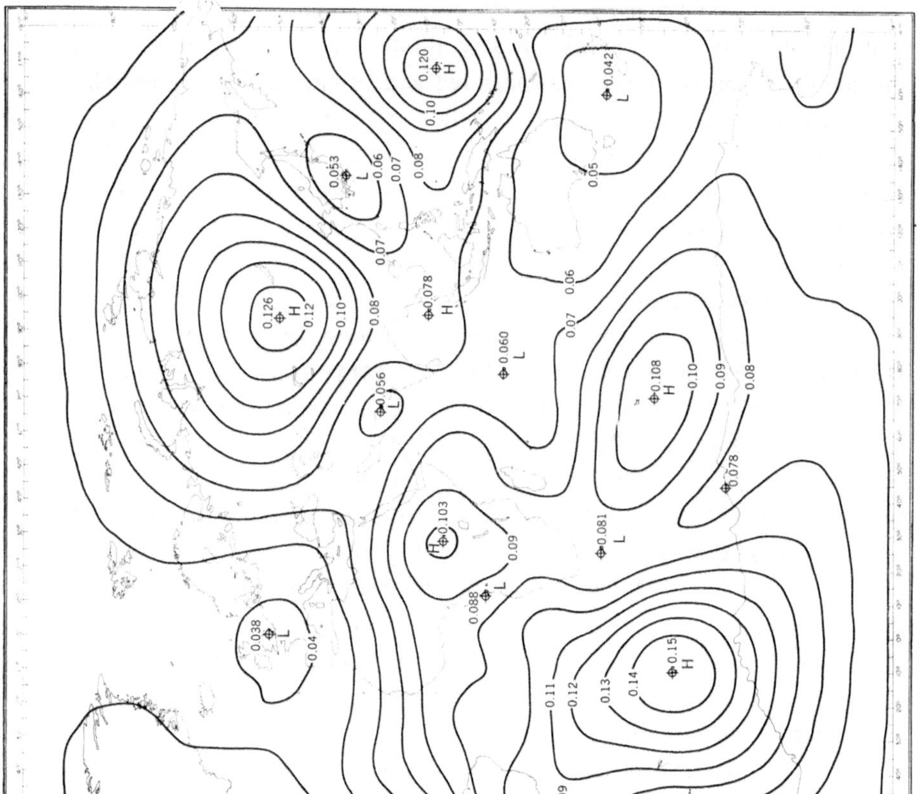

Fig. 5. Standard error of GSFC(7/65) fit computed for inclination I (degrees) at earth's surface for epoch 1960.0.

Comparison of Field Models

The previous paper made comparisons of numerous available core field models with the available survey data. The RMS deviations of all the component observations per year were listed combining such quantities as $(\Delta F)^2$, $(\Delta H)^2$, $(F\Delta I)^2$, $(H\Delta D)^2$ and $(\Delta Z)^2$. We will extend this comparison here and focus on the three best available fields as well as the older J+C model (JENSEN and CAIN, 1962) now in general use for the determination of the L parameter for organizing trapped particle data. A comparison of these four fields (see Table II) is given in Figure 6. Plotted at the top are the RMS deviations from the data for each year and given at the end of each curve the RMS for the whole interval 1945–1965. The distribution of data used per year corresponds to a fifth of that available and was obtained simply by using only every fifth data point from a magnetic tape containing all data sequenced in order of longitude. Such a selection procedure gives almost the same results when compared with deviations calculated using *all* of the available data. One can see that for the twenty years plotted there is about a factor of two between the highest and lowest curves. For the data taken over the years 1960–1963 the J+C field errors are growing to about 2.5 times those of the latest field (GSFC [7/65]). The values for 1964 should not be considered significant since they are based on only 130 observations and are influenced by a few large deviations.

TABLE 2

Characteristics of Four Models of the Core Field

Field	*Number of Coefficients*		*Epoch*	*Derived from*
	spatial	*first derivatives*		
J+C	48	0*	1960	7400 H and F Observations 1940–1960. *First derivatives were actually derived but not published. (JENSEN and CAIN, 1962)
GSFC(64)	63	35	1960	21695 selected observations 1940–1963 (all components) (Table 4 from the paper by CAIN *et al.*, 1965)
LME	80	48	1965	Aircraft and surface observations converted to X and Y components and corrected to epoch 1965 using separate analysis of observatory secular change (LEATON, MALIN and EVANS, 1964)
GSFC(7/65)	99 internal + 3 external	35	1960	197000 selected observations 1945–1964

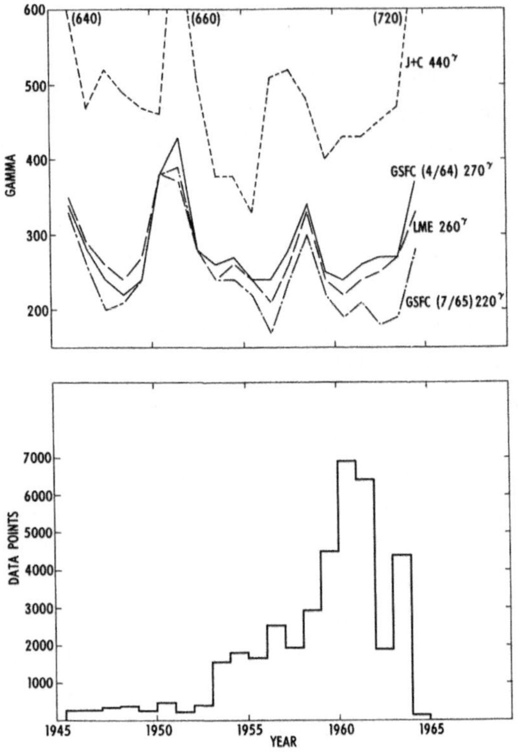

Fig. 6. RMS deviations between survey data by year and computed fields listed in Table II. Numbers at end of each curve give RMS over the whole interval 1945–1964. Histogram on lower diagram gives distribution of data per year.

Figure 6 shows an unusual periodicity that appears to peak both near solar minimum and maximum. However, if one discounts the 1964 points due to a lack of data there is still the large maximum for 1951–1952. On closer inspection of the data it appears that this is also due to the relatively small amount of data (230 and 240 observations respectively) and arises from large deviations of the surface Z measurements. These Z data are frequently found in error with large inaccuracies. The maximum of all the curves near the sunspot maximum years of the IGY (1957–1958) cannot be explained so easily and might indeed reflect the greater incidence of magnetic disturbance in the data.

Of course the use of RMS deviations does not completely characterize the extent to which the fields fit the observed data. As shown in the previous paper, sample detailed deviations of data over various ship and aircraft survey tracks show systematic deviations from the GSFC (64) field with wavelengths up to a few thousand kilometers. We have not yet tested the GSFC (7/65) field to see whether these systematic deviations have been largely eliminated or whether an even larger number of coefficients may be necessary. One indication that there may be a need for still more parameters is that, whereas the Vanguard 3 data taken by themselves are consistent to 17γ (CAIN and HENDRICKS, 1964) the residuals from the GSFC (64) field are 50γ and those from the GSFC (7/65) field are still 28γ.

Conclusions

Some of the conclusions one may draw from this discussion must be qualitative since sufficiently accurate observations of the magnetic field within the magnetosphere are not yet available. In considering the mathematical models necessary to derive the magnetic field we have introduced the concept of a *forecast* versus a *definitive* field and have shown that both must eventually include allowance for plasma-field interactions beyond a few earth radii if the computed fields are to match the ambient field well enough to be useful in helping to understand such experimental data as those from trapped particles. Within a few earth radii, where the effect of the core field predominates, the requirements for increased accuracy may also include making allowance for ionospheric currents and, below a few tens of kilometers, estimates of the crustal field.

Our present estimates of the core field are greatly hampered by the very irregular space-time distribution of magnetic survey data. Nevertheless, the significant spatial wavelengths in this field appear to require at least a factor of two more parameters than the 48 spherical harmonics previously used. Fitting the measured data themselves instead of data read from magnetic charts has proved a useful approach and produced fields whose errors in regions of good data coverage are only a fraction of those previously obtained.

The presence of a simple external field near the earth's surface was determined but more work will need to be done to assure its reality. It is much smaller than any previous estimates but is of the correct order of magnitude to be supported by physical theory.

The problems of determining or minimizing the effect of neglected parameters or of investigating the mathematical significance of the areas void of data have not yet been solved. The present fits to the core field, although containing a large number of parameters, do not appear to be significantly worse in the sparse data areas than fits using a smaller number of coefficients. In the areas of good data coverage the new field can be shown to have much smaller errors. If the satellite magnetometer experiments planned for the next year are successful it is likely that the sparse data area problem can be quickly solved and that work can begin to include the contributions of external sources to the field models.

Acknowledgements

It is a pleasure to acknowledge the significant contributions that are continuing to be made by the personnel at Goddard: S. Hendricks who has organized and processed the survey data and supervised most of the machine computations and W. E. Daniels who has been instrumental in developing some of the statistical procedures. The basic survey data are the result of the international effort of the World Magnetic Survey-IQSY and were supplied on magnetic tape as one of the products of the USC&GS-NASA Cooperative Data Reduction Program.

References

CAIN, Joseph C., and HENDRICKS, Shirley J.: 1964, *J. Geophys. Res.* **69**, 4187–4188.

CAIN, Joseph C., HENDRICKS, Shirley, DANIELS, Walter E., JENSEN, Duane C.: NASA X-611-64-316, 1964.

CAIN, Joseph C., DANIELS, W. E., HENDRICKS, Shirley J., and JENSEN, Duane C.: 1965, *J. Geophys. Res.*, **70**, 3647–3674.

DAVIS, T. N., STOLARIK, J. D., and HEPPNER, J. P.: 1965, *J. Geophys. Res.*, **70**, 5883–5894.

HOFFMAN, R. A., and BRACKEN, P. A.: 1965, *J. Geophys. Res.* **70**, 3541–3556.

JENSEN, D. C., and CAIN, J. C.: *J. Geophys. Res.* **67**, 3568–3569.

KAHLE, A. B., KERN, J. W. and VESTINE, E. H.: *Rand Memorandum* RM-4360-NASA, 1964.

LEATON, B. R.: 1963, *J. Geophys. Res.* **68**, 5902.

LEATON, B. R., MALIN, S. R. C., and EVANS, Margaret J.: 1964, IAGA Symposium on Magnetism of the Earth's Interior, Pittsburg (to be published in *J. Geomag. and Geoelect.*).

MEAD, Gilbert D.: 1964, *J. Geophys. Res.* **69**, 1181–1195.

NESS, Norman F.: 1965, *J. Geophys. Res.* **70**, 2989–3005.

SUGIURA, Masahisa: 1964, *Ann. Intern. Geophys. Yr.* **35**, Pt. 1, 9–45.

Discussion

In the discussion several questions were asked pertaining to the accuracy of the current magnetic field models both as a function of position on the earth and as a function of altitude. Dr. CAIN stated that the field was probably good to between 0.1 and 1.0% at the surface of the earth and was valid to some 10% at a few earth radii. Near the boundary of the magnetosphere the error was a factor of 2. G. MEAD added that at 5 earth radii the external field produced by boundary currents should be about 10% of the total field. However, there was some observational evidence of discrepancies at 3 to 4 earth radii. In regard to errors near the surface Dr. CAIN pointed out that where the complexities are sharp compared to the highest degree of harmonics used, there would be more smoothing than the data would justify. The residuals for the 1965 field indicate that the additional coefficients (a total of 99) are required. However, it is not clear that more coefficients than 99 are needed.

Questions were raised concerning the ability to detect an external ring current from the surface field or to place an upper limit on the monopole moment of nucleons from the integral of B over the earth's surface. Dr. CAIN felt that existing estimates of ring currents were very uncertain due to errors in measured values of the surface field. In regard to the monopole moment of the earth the g_0^0 term was not determined very accurately by these fits, and one can only say that it is less than 8 gammas.

ADIABATIC THEORY OF CHARGED PARTICLE MOTION

T. G. NORTHROP
NASA – Goddard Space Flight Center,
Greenbelt, Md.

1. Introduction

The theory of adiabatic charged particle motion will be described from two points of view. First, from the more physical and intuitive point of view. This method is limited to first order effects in the radius of gyration, since higher order effects become difficult to visualize and one's intuition becomes unreliable as a guide. Second, the formal theory of KRUSKAL (1962) will be reviewed and some results obtained with it described. This and other (see GARDNER, 1959) formal methods are systematic expansions and in principle should require no physical insight into the problem. In practice, as will be seen, the labor of the calculation can be greatly reduced by applying physical insight at the proper points.

Modifications of the theory to particles of relativistic energies will not be given. There are some purely relativistic effects that may be of significance; generally speaking, the inclusion of relativistic effects does not greatly complicate adiabatic theory. For the relativistic theory see NORTHROP and TELLER (1960), VANDERVOORT (1960), and NORTHROP (1963).

2. Adiabatic Theory by Means of Physical Insight

This method was originated by ALFVÉN (1950) and what will be described below is essentially an elaboration and systematizing of his method to bring out all first order effects explicitly. The method is similar to HELLWIG's (1955).

The equation of motion of a charged particle in an electromagnetic field is

$$\ddot{\mathbf{r}} = \frac{e}{mc}\dot{\mathbf{r}} \times \mathbf{B}(\mathbf{r}, t) + \frac{e}{m}\mathbf{E}(\mathbf{r}, t) \tag{1}$$

where $\mathbf{r}$ is the particle position, $\mathbf{B}$ and $\mathbf{E}$ are the magnetic and electric fields, e, m, and c are charge, mass, and light velocity, and dots mean $\mathrm{d}/\mathrm{d}t$. The gravitational force has been omitted in (1), since its effect on trapped particles is too small to be detected by present methods.

In a uniform and constant magnetic field the solution to (1) is a helix; thus slight deviations from uniformity should produce slight perturbations from helicity. Adiabatic theory is just this perturbation theory. To carry out the perturbation theory, let $\mathbf{r}=\mathbf{R}+\boldsymbol{\varrho}$, where the quantities are defined in Figure 1.

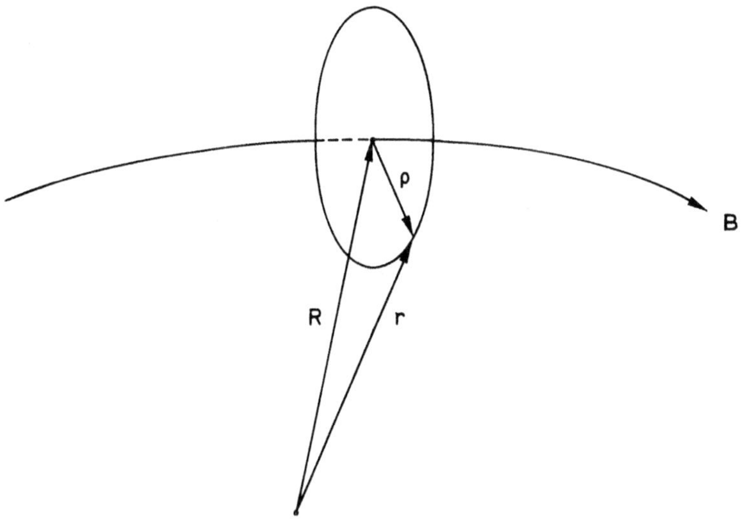

Fig. 1. The charged particle and its gyration circle.

In this Figure, $\boldsymbol{\varrho}$ is defined by

$$\boldsymbol{\varrho} = \frac{mc}{eB^2}\mathbf{B} \times \left(\mathbf{v} - c\frac{\mathbf{E} \times \mathbf{B}}{B^2}\right) \tag{2}$$

so as to give the center of the helix (or the helix with the electric field effect superposed) in a uniform magnetic field ($\mathbf{v}$ is $\dot{\mathbf{r}}$). If the particle position around the axis of the perturbed helix is not of interest, and only the trajectory of the axis is needed, the gyration about the axis should be averaged out in some reasonable fashion. The adiabatic theory of charged particle motion results. To carry out this averaging, let

$$\boldsymbol{\varrho} = \varrho(\mathbf{e}_2 \sin \omega t + \mathbf{e}_3 \cos \omega t) \tag{3}$$

where $\mathbf{e}_2$ and $\mathbf{e}_3$ are two unit vectors locally perpendicular to $\mathbf{B}$ and to each other; ω is eB/mc, the gyration frequency. Substitution of $\mathbf{r}$ into the equation of motion followed by time averaging $\sin \omega t$ and $\cos \omega t$ to zero, but $\sin^2 \omega t$ and $\cos^2 \omega t$ to $\frac{1}{2}$, yields

$$\ddot{\mathbf{R}} = \frac{e}{m}\left[\mathbf{E}(\mathbf{R}, t) + \frac{\dot{\mathbf{R}}}{c} \times \mathbf{B}(\mathbf{R}, t)\right] - \frac{e\varrho^2\omega}{2mc}\nabla B(\mathbf{R}, t) + O(\varepsilon), \tag{4}$$

where ε is m/e, and $O(\varepsilon)$ means terms of order ε and higher. The expansion can be taken in m/e, since if the equation of motion (1) is properly scaled, a dimensionless equation identical in form to (1) can be obtained, with ϱ/L appearing in place of m/e. (L is the characteristic dimension of the system). The gradient of B arises in (4) from a Taylor expansion of $B(\mathbf{r}, t)$ about $B(\mathbf{R}, t)$. The coefficient $e\varrho^2\omega/2mc$ is the magnetic moment M, whose approximate or adiabatic invariance is moved in many places, usually for special geometry. A proof of the invariance of M will not be given here; such a proof is not altogether trivial for the general time-dependent electromagnetic field (NORTHROP, 1963). Special geometries are much easier.

Equation (4) is a differential equation of motion for the "guiding center" $\mathbf{R}$, which is the average particle position. It can be solved by iteration for the "drift velocity" $\dot{\mathbf{R}}_\perp$, the component of $\dot{\mathbf{R}}$ perpendicular to $\mathbf{B}(\mathbf{R})$. The component

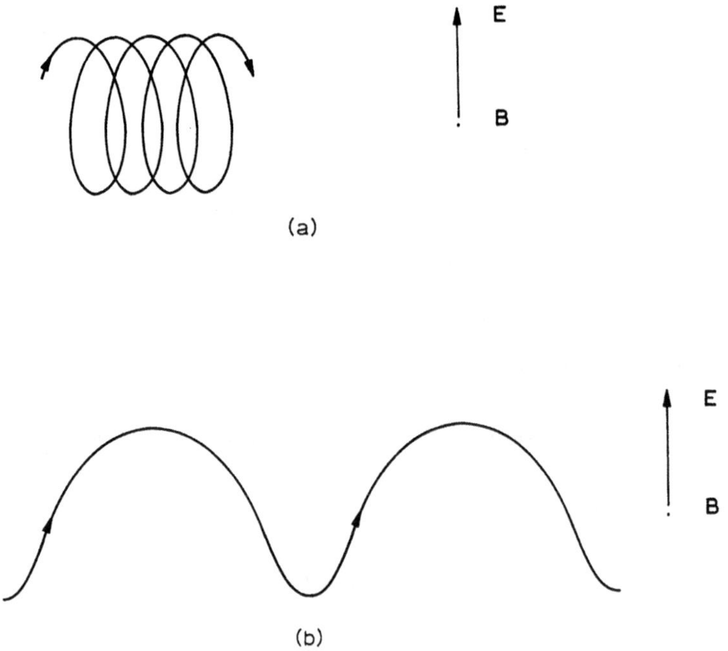

Fig. 2. Particle trajectories when $E_\perp$ is small (a) and large (b).

of (4) parallel to **B** is the equation governing the guiding center motion along a magnetic field line. These two equations are

$$\dot{\mathbf{R}}_{\perp} = \frac{c\mathbf{E} \times \mathbf{B}}{B^2} + \frac{Mc}{e}\frac{\mathbf{B} \times \nabla B}{B^2} + \frac{mc}{e} v_{\parallel}^2 \frac{\mathbf{B}}{B^2} \times \frac{\partial}{\partial s}\left(\frac{\mathbf{B}}{B}\right), \tag{5}$$

and

$$m\frac{\mathrm{d}v_{\parallel}}{\mathrm{d}t} = eE_{\parallel} - M\frac{\partial B}{\partial s}, \tag{6}$$

where s is distance along a field line and $v_{\parallel}$ is $(\mathbf{B}/B)\cdot\dot{\mathbf{R}}$, the component of guiding center velocity parallel to the field line and $E_{\parallel}$ has a similar meaning. Expressions (5) and (6) hold for the case when $\mathbf{E}_{\perp}$ is small, meaning small enough that particle trajectory loops itself as shown in Figure 2a. This will be the case if the thermal velocity of the particle greatly exceeds $c\mathbf{E}\times\mathbf{B}/B^2$, the guiding center drift due to **E**, which will be denoted by $\mathbf{u}_E$. If the thermal velocity is less than u_E, the trajectory is an open cycloid as in Figure 2b and the following more complicated expressions must be used.

$$\begin{aligned}\dot{\mathbf{R}}_{\perp} = \mathbf{u}_E &+ \frac{Mc}{e}\frac{\mathbf{B}\times\nabla B}{B^2} + \frac{mc}{e}v_{\parallel}^2\frac{\mathbf{B}}{B^2}\times\frac{\partial}{\partial s}\left(\frac{\mathbf{B}}{B}\right)\\ &+ \frac{mc}{e}\frac{\mathbf{B}}{B^2}\times\left[v_{\parallel}\frac{\partial}{\partial t}\left(\frac{\mathbf{B}}{B}\right) + v_{\parallel}\mathbf{u}_E\cdot\nabla\left(\frac{\mathbf{B}}{B}\right) + \frac{\partial}{\partial t}\mathbf{u}_E + \right.\\ &\left.+ v_{\parallel}\frac{\partial\mathbf{u}_E}{\partial s} + \mathbf{u}_E\cdot\nabla\mathbf{u}_E\right]\end{aligned} \tag{7}$$

$$\begin{aligned}m\frac{\mathrm{d}v_{\parallel}}{\mathrm{d}t} = eE_{\parallel} - M\frac{\partial B}{\partial s} &+ \mathbf{u}_E\cdot\left[\frac{\partial}{\partial t}\left(\frac{\mathbf{B}}{B}\right) + v_{\parallel}\frac{\partial}{\partial s}\left(\frac{\mathbf{B}}{B}\right)\right.\\ &\left.+ \mathbf{u}_E\cdot\nabla\left(\frac{\mathbf{B}}{B}\right)\right]\end{aligned} \tag{8}$$

When $\mathbf{u}_E$ is comparable to the thermal velocity, it is advisable to examine the extra terms in (7) and (8) so as to see if they are significant for the problem at hand; it is never incorrect to use the more complicated (7) and (8) in place of (5) and (6).

All of the eight drift terms in (7) occur because the radius of curvature of the particle trajectory alternates from large to small at the gyration frequency as in Figure 2. The reason for this alternation differs for each of the eight drifts and examples of each of the drifts are not difficult to find. For example, the $\partial\mathbf{u}_E/\partial t$ term in (7) is responsible for the low frequency dielectric constant of a plasma.

In addition to the magnetic moment, there are two other adiabatic invariants of the motion. Strictly speaking, the second invariant is a constant of the guiding center equations of motion (5) and (6), while the third invariant is conserved by equations of motion to be given below: (15) or (16). As with the magnetic moment, the original discovery of the second and third invariants was not made by rigorous deductive methods, but rather by analogy and intuition. The motion of a guiding center along a magnetic field line as determined by (6) may be periodic if the magnitude of the field increases in both directions from a minimum to form a "magnetic mirror". As the guiding center moves into the increasing magnetic field, its component of velocity parallel to the field decreases because of the $-M(\partial B/\partial s)$ term. If the field becomes sufficiently

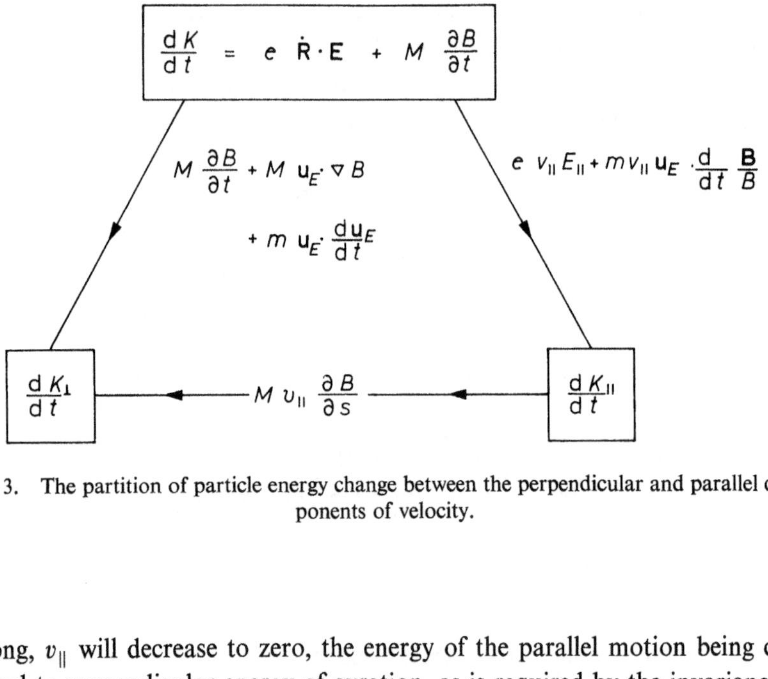

Fig. 3. The partition of particle energy change between the perpendicular and parallel components of velocity.

strong, $v_\parallel$ will decrease to zero, the energy of the parallel motion being converted to perpendicular energy of gyration, as is required by the invariance of the magnetic moment M. The guiding center will then reverse direction and eventually reflect at the opposite mirror. More generally, when there are non-static electric and magnetic fields, the energy flow pattern is more complicated. Figure 3 depicts how the total kinetic energy K changes. In Figure 3, the operator d/dt means $\partial/\partial t + v_\parallel \partial/\partial s + \mathbf{u}_E\cdot\nabla$; $K_\perp$ is $\frac{1}{2}mv_\perp^2$ and $K_\parallel$ is $\frac{1}{2}mv_\parallel^2$.

The expression for dK/dt is intuitively correct: the energy increase comes from motion of the guiding center in the electric field and the betatron effect of the curl of $\mathbf{E}$ acting about the circle of gyration. Figure 3 shows how the energy increase divides between perpendicular and parallel modes. Simultaneously these two modes exchange energy by the mirror effect.

Fermi acceleration (FERMI, 1949, 1954) is a special case of these general energy change expressions, and occurs when there is a frame of reference (other than the observer's) in which the magnetic field is static. In Fermi's type A acceleration, the guiding center velocity is reversed in this static frame by the mirror effect. In type B acceleration, the guiding center moves along a bent field line whose direction reverses, so that the velocity vector of the guiding center is again reversed. The observer sees a net acceleration which for type A comes from the $M(\partial B/\partial t)$ term plus part of the $e\dot{\mathbf{R}}\cdot\mathbf{E}$ term. The Fermi type B comes from the remainder of the latter term. It is thus not correct to distinguish between Fermi acceleration and betatron acceleration; one may distinguish between betatron acceleration and acceleration due to guiding center drift, or when there is a frame of reference in which $\mathbf{B}$ is static, Fermi types A and B may be distinguished.

But to return to the adiabatic invariants – the existence of the second or "longitudinal" invariant was first suggested by Rosenbluth about ten years ago, and used by him and LONGMIRE (1957) to study the interchange instability of a plasma. Because of the drift velocity (5) the motion is not exactly one dimensional (i.e. along a field line) and the guiding center drifts slowly away from any given line. Nevertheless the parallel action

$$J = \oint p_{\parallel}\, \mathrm{d}s \tag{9}$$

is conserved in an average sense as will now be shown, provided the thermal velocity is much greater than $\mathbf{u}_E$, so that (5) is valid and the more general (7) is not needed. In (9), $p_{\parallel}$ is the momentum $mv_{\parallel}$ and $\mathrm{d}s$ is an element of arc length along a field line; the integral is taken over a complete period of oscillation between mirrors. Obviously J cannot even be defined if the parallel motion is not periodic, and in fact no second (or third) invariant exists under these conditions.

The effects of the drifts on J have been studied by KADOMTSEV (1958) for the case of static fields, and by NORTHROP and TELLER (1960) for the general case of relativistic particles in time-dependent fields. The general proof will not be given here, and the proof for the static magnetic field only outlined. Some of the apparatus of the proof will be useful in examples. A very useful coordinate system is the "α, β, s" curvilinear system, in which α and β are two functions of position that are constant along a field line while s is distance along the field line measured from an arbitrary origin on the line; $\alpha(\mathbf{r})$ and $\beta(\mathbf{r})$ can be chosen so that the vector potential $\mathbf{A}$ is $\alpha\nabla\beta$, making $\mathbf{B}$ equal to $\nabla\alpha\times\nabla\beta$. There is an infinite number of α, β pairs to describe a given field and all must be related by a gauge transformation, since $\mathbf{A}=\alpha'\nabla\beta'=\alpha\nabla\beta-\nabla\tau$. The transformation from α, β to α', β' is canonical with τ the generating function. To proceed with the proof, let K be the kinetic energy of the particle as before. Then

$$J(\alpha,\beta,K,M)=\oint\{2m[K-MB(\alpha,\beta,s)]\}^{\frac{1}{2}}\,\mathrm{d}s\,, \tag{10}$$

the integrand being

$$\left[2m\left(K-\frac{mv_\perp^2}{2}\right)\right]^{\frac{1}{2}},$$

which is $p_\parallel$. The instantaneous rate of change of J when the guiding center is at some point s on a field line is caused by the drifts, and is

$$\frac{\mathrm{d}J}{\mathrm{d}t}=\frac{\partial J}{\partial\alpha}\dot{\mathbf{R}}_\perp\cdot\nabla\alpha+\frac{\partial J}{\partial\beta}\dot{\mathbf{R}}_\perp\cdot\nabla\beta \tag{11}$$

A term of the form $(\partial J/\partial M)/(\mathrm{d}M/\mathrm{d}t)$ is omitted because M is constant. The picture here is one where the principal velocity of the guiding center is directed along the direction of $\mathbf{B}$ at point s and the drift velocity at right angles to the line is much slower, but nevertheless is what changes J. After substitution of (5) for $\dot{\mathbf{R}}_\perp$, plus some vector algebra, the following expressions can be derived:

$$\begin{aligned}\dot{\alpha}&=\dot{\mathbf{R}}_\perp\cdot\nabla\alpha=\frac{-Mc}{e}\frac{\partial B(\alpha,\beta,s)}{\partial\beta}-\frac{mc}{e}\frac{\mathrm{d}}{\mathrm{d}t}\left(v_\parallel\frac{\mathbf{B}}{B}\cdot\frac{\partial\mathbf{R}(\alpha,\beta,s)}{\partial\beta}\right)\\ \dot{\beta}&=\dot{\mathbf{R}}_\perp\cdot\nabla\beta=\frac{Mc}{e}\frac{\partial B(\alpha,\beta,s)}{\partial\alpha}+\frac{mc}{e}\frac{\mathrm{d}}{\mathrm{d}t}\left(v_\parallel\frac{\mathbf{B}}{B}\cdot\frac{\partial\mathbf{R}(\alpha,\beta,s)}{\partial\alpha}\right)\end{aligned} \tag{12}$$

By direct differentiation of the integral for J, one obtains

$$\begin{aligned}\frac{\partial J}{\partial\beta}&=\frac{e}{c}\oint\frac{\mathrm{d}s}{v_\parallel}\dot{\alpha}=\frac{e}{c}T\langle\dot{\alpha}\rangle\\ \frac{\partial J}{\partial\alpha}&=-\frac{e}{c}T\langle\dot{\beta}\rangle\,,\end{aligned} \tag{13}$$

where T is the period of the oscillation along the field line $\oint\mathrm{d}s/v_\parallel$, and the angular brackets denote the time average. The fact that

$$\left\langle\frac{\mathrm{d}}{\mathrm{d}t}\left(v_\parallel\frac{\mathbf{B}}{B}\cdot\frac{\partial\mathbf{R}}{\partial\beta}\right)\right\rangle$$

is zero must be used in establishing (13). It vanishes because $v_{\|}$ vanishes at the mirror points. The rate of change of J can now be written

$$\frac{\mathrm{d}J}{\mathrm{d}t} = \frac{eT}{c}\left[\langle\dot{\alpha}\rangle\dot{\beta} - \langle\dot{\beta}\rangle\dot{\alpha}\right]. \tag{14}$$

Clearly $\mathrm{d}J/\mathrm{d}t$ does not vanish unless $\langle\dot{\alpha}\rangle/\langle\dot{\beta}\rangle = \dot{\alpha}/\dot{\beta}$, which does not hold unless the guiding center drifts towards the same adjacent field line at all points s of its motion along the line. Such would be the case for a field with azimuthal symmetry, but in that case the invariance of J does not tell us anything we did not already know. In general the drift $\dot{\mathbf{R}}_{\perp}$ moves the guiding center towards a variety of adjacent field lines during the motion along the line, so that $\mathrm{d}J/\mathrm{d}t$ does not vanish. However, $\langle \mathrm{d}J/\mathrm{d}t\rangle$ does vanish and this property is sufficient to make it useful, because the guiding center stays almost on a field line. Thus a surface composed of field lines on which J is constant is followed closely (to order ε) but not exactly by the guiding center.

Equations (13) constitute new equations of motion when written in the form

$$\begin{aligned}\langle\dot{\alpha}\rangle &= \frac{c}{eT}\frac{\partial J}{\partial\beta}(\alpha,\beta,K,M)\\ \langle\dot{\beta}\rangle &= -\frac{c}{eT}\frac{\partial J}{\partial\alpha},\end{aligned} \tag{15}$$

the longitudinal oscillation now having been averaged over. This doubly averaged motion (once over the gyration and once over the longitudinal oscillation) can be nicely illustrated in Cartesian α, β space as in Figure 4. Equations (15) may appear canonical, with J as Hamiltonian, but they are not because T is a function of α, β, K, and M. This can be rectified, however, by writing $\partial J(\alpha, \beta, K, M)/\partial\beta$ as $-(\partial J/\partial K)(\partial K(\alpha, \beta, J, M)/\partial\beta)$ and similarly for $\partial J/\partial\alpha$. Furthermore, differentiation of (10) yields $\partial J/\partial K = T$. Thus we have

$$\begin{aligned}\langle\dot{\alpha}\rangle &= -\frac{c}{e}\frac{\partial K}{\partial\beta}(\alpha,\beta,J,M)\\ \langle\dot{\beta}\rangle &= \frac{c}{e}\frac{\partial K}{\partial\alpha}\end{aligned} \tag{16}$$

and these are canonical, with energy K as Hamiltonian. Because of the canonical form, a Liouville theorem can be written in α, β space and long-term equilibria under the average drifts characterized. This approach has led to the so-called "minimum B" equilibria of a plasma, in which the components of the pressure tensor are constant on closed surfaces of constant magnitude of field B.

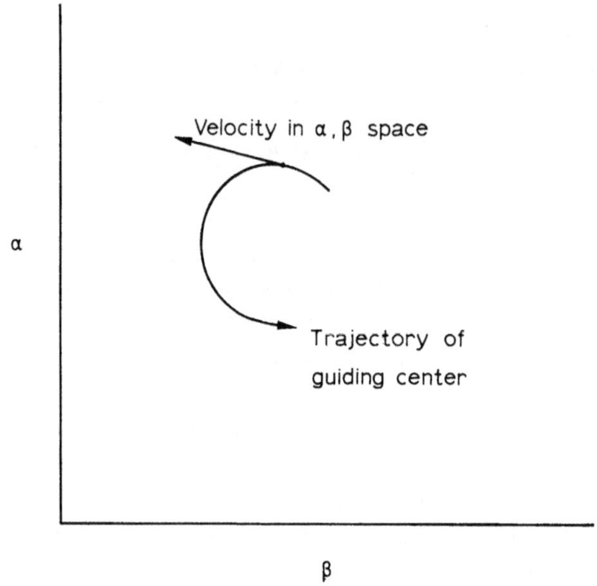

Fig. 4. The twice averaged velocity of the particle.

Before the third and last adiabatic invariant is discussed, an aside concerning the velocity of a field line will be worthwhile. NEWCOMB (1958) proved that for a vector $\mathbf{w}(\mathbf{r})$ to be satisfactory for a line velocity it must satisfy the condition that

$$\nabla \times \left(\mathbf{E} + \frac{\mathbf{w}}{c} \times \mathbf{B} \right)$$

vanish. This condition assures that two points on the same field line at one time will be on the same field line a short time later. In other words, field lines do not become fragmented. The condition also assures that field lines do not disappear (flux preservation). Both of these are reasonable conceptual requirements if the picture of moving field lines is to be of any use in visualizing a process. If fields are time dependent, α and β are functions of time as well as position. The velocity

$$\mathbf{w} = \left(\frac{\partial \beta}{\partial t} \nabla \alpha - \frac{\partial \alpha}{\partial t} \nabla \beta \right) \times \frac{\mathbf{B}}{cB^2}$$

satisfies the condition for a line velocity. The major difficulty is in general to find functions $\alpha(\mathbf{r}, t)$ and $\beta(\mathbf{r}, t)$ given $\mathbf{A}(\mathbf{r}, t)$ or $\mathbf{B}(\mathbf{r}, t)$, in other than very simple situations. It should be emphasized that $\mathbf{w}$ and $(c\mathbf{E} \times \mathbf{B})/B^2$ are not the same. The latter need not vanish even if the situation is static, since there may be a static **E**. However, **w** does vanish in a static situation, and it does seem more appealing to have a definition of line velocity that vanishes in a static situation. Moreover, $\mathbf{u}_E$ does not satisfy the line velocity condition unless $\nabla x \mathbf{E}_{\|}$ vanishes.

To return to the third invariant, suppose that the surface on which J is constant is topologically a cylinder, in which case the trajectory in Figure 4 closes on itself. Then at a later time the guiding center returns to a line traversed earlier, and there is a third periodicity and a third invariant – namely, the magnetic flux passing through the cylinder (NORTHROP and TELLER, 1960). Because the surface is made up of field lines, the flux Φ of **B** entering one end equals that emerging from the other (and equals that everywhere inside). If the fields are static, this flux is trivially invariant, but more importantly, it is invariant if the field varies slowly compared to the time for the guiding center to drift once around the surface. If the magnetic field is deformed slowly enough that Φ is conserved and then brought slowly back to its original configuration, it can be shown that the particle has the same energy finally as initially and has returned to the original drift surface. As another application of the invariance of Φ, consider the rotation of an asymmetric field such as the earth's. The field at any time has the same configuration as at an earlier time, but rotated, so that the energy K, which is a function of J, M, Φ, and t in general becomes independent of t. Thus there is no continuous acceleration of the particle by a field whose intrinsic form is constant in time, even though it slowly changes its position or orientation in space.

The dipole field will serve as an application of the new equations of motion (15). For an azimuthally symmetric field such as the dipole, α can be taken as the stream function and β as the longitude. For the dipole of strength μ

$$\begin{aligned} \alpha &= -\frac{\mu \sin^2\theta}{r} \\ \beta &= \varphi \end{aligned} \tag{17}$$

where r, θ, φ are the spherical coordinates. The average drift rate in longitude is

$$\langle\dot{\varphi}\rangle = -\frac{c}{eT}\frac{\partial J}{\partial \alpha} = -\frac{c}{eT}\frac{\partial}{\partial \alpha}\oint \mathrm{d}s(2m)^{\frac{1}{2}}\left[K - MB(\alpha,\beta,s)\right]^{\frac{1}{2}}. \tag{18}$$

If the expression for B in a dipole is inserted into (18) and the variable of integration changed to $\sigma = s/r_0(\alpha)$, a simple expression is obtained for the average drift rate; (r_0 is the equatorial distance of the field line from the origin).

$$\langle\dot{\varphi}\rangle = -\frac{mc}{eT}\frac{2r_0}{\mu}\int_{s_0}^{s_1} \mathrm{d}s\,\frac{v^2 + \frac{1}{2}v_\perp^2}{v_{\|}}. \tag{19}$$

The variable of integration has been changed back to s from σ in (19) after performing the differentiation; s_0 and s_1 are the values of s at the two mirror points. If (18) had been differentiated with respect to σ with s as the variable of integration, a different and more complicated integral would have been obtained. Finally, if the usual drift expression (5) is integrated over the longitudinal motion, as done by HAMLIN *et al.* (1961), a third and even more complicated integral is found. It is

$$\langle\dot{\varphi}\rangle = \frac{2mc}{eT}\int_{s_0}^{s_1}\frac{\mathrm{d}s}{v_\parallel}\,\frac{v^2 - \frac{1}{2}v_\perp^2}{BRr\sin\theta} \tag{20}$$

where R is the radius of curvature of the field line, a rather lengthy expression. All three integrals for $\langle\dot{\varphi}\rangle$ must of course be equal and related by integration by parts. The connection between (19) and (20) follows from the following relation, which is valid for a dipole field:

$$\frac{\dot{\varphi}}{v_\parallel} = -\frac{mc}{e}\,\frac{v^2 - \frac{1}{2}v_\perp^2}{v_\parallel BRr\sin\theta} = -\frac{mc}{e}\,\frac{r_0}{\mu}\,\frac{v^2 + \frac{1}{2}v_\perp^2}{v_\parallel} + \frac{mc}{e}\,\frac{r_0}{\mu}\,\frac{\partial}{\partial s}\left(\frac{v_\parallel\cos\theta}{r^2 B}\right) \tag{21}$$

The expression under the $\partial/\partial s$ is r times the radial component of the parallel guiding center velocity, and vanishes at the endpoints of the integration. Thus the average drift rate can be simplified to (19) from (20). The second or intermediate form of the integral is nearer to (20) than (19) in complexity; the considerable simplification occurring in (19) is a consequence of using the variable $\sigma = s/r_0$, and depends on the fact that for a dipole all field lines are of similar shape. The same simplification undoubtedly occurs for any field with azimuthal symmetry in which all field lines are of the same shape.

Equation (19) provides a simple algebraic relation connecting the average drift rate and the second adiabatic invariant J:

$$\begin{aligned}\langle\dot{\varphi}\rangle &= -\frac{mc}{eT}\,\frac{2r_0}{\mu}\int_{s_0}^{s_1}\frac{\mathrm{d}s}{v_\parallel}\left(v^2 + \frac{1}{2}v_\perp^2\right)\\ &= -\frac{mc}{eT}\,\frac{2r_0}{\mu}\int_{s_0}^{s_1}\frac{\mathrm{d}s}{v_\parallel}\left(\frac{3}{2}v^2 - \frac{1}{2}v_\parallel^2\right)\\ &= -\frac{mc}{eT}\,\frac{2r_0}{\mu}\,\frac{1}{T}\left(\frac{3}{2}v^2\frac{T}{2} - \frac{1}{2}\frac{J}{m}\right)\\ &= -\frac{mcr_0v^2}{e\mu}\left(\frac{3}{2} - \frac{J}{4KT}\right)\end{aligned} \tag{22}$$

As a second application of the expressions for $\dot{\alpha}$ and $\dot{\beta}$ derived previously, consider the effect of a curl-free electric field which is everywhere perpendicular to the magnetic field. The electric potential Ψ adds the term $-c(\partial\Psi/\partial\beta)$ to Equation (12) for $\dot{\alpha}$, and $c(\partial\Psi/\partial\alpha)$ to $\dot{\beta}$. If $\mathbf{E}\cdot\mathbf{B}$ is zero, then magnetic field lines are equipotentials and Ψ is a function of α and β, but not of s. The result is that the contribution of the electric field to $\dot{\alpha}$ and $\dot{\beta}$ is independent of the guiding center position along a field line; the electric field drift is thus towards the same adjacent field line at all s. The other two drifts are not always towards the same line except in special geometries, such as azimuthally symmetric ones, so that the net drift is not either. For a dipole, with $\mathbf{E}$ in the azimuthal direction SWIFT (to be published) has pointed out that $\dot{\alpha}$ is independent of s, since the other two drifts are orthogonal to $\mathbf{u}_E$.

3. A Formal Theory of Adiabatic Invariance

The theory of KRUSKAL (1962) will be reviewed in this section and the first two terms of the second invariant as obtained by this method given. The general theory supplies asymptotic solutions to coupled first order differential equations of a particular type, but adiabatic invariants exist only if the equations are canonical. Therefore the theory will be described for such equations. Suppose the canonical equations are

$$\begin{aligned} \dot{p}_i &= -\frac{\partial H}{\partial q_i}(\mathbf{p},\mathbf{q},\varepsilon) \\ \dot{q}_i &= \frac{\partial H}{\partial p_i}(\mathbf{p},\mathbf{q},\varepsilon) \end{aligned} \tag{23}$$

where $i=1\ldots n$, and suppose the solutions of (23) for $\varepsilon=0$ are closed loops in phase space. (The equations for a charged particle can be put in this form.) Then there exists a transformation $\mathbf{p}=\mathbf{p}(\mathbf{z},\varphi,\varepsilon)$ and $\mathbf{q}=\mathbf{q}(\mathbf{z},\varphi,e)$, where $\mathbf{z}$ has $2n-1$ components, such that the equations (23) assume the transformed form

$$\begin{aligned} \dot{\mathbf{z}} &= \varepsilon\mathbf{h}(\mathbf{z},\varepsilon) \\ \dot{\varphi} &= \omega(\mathbf{z},\varepsilon)\,. \end{aligned} \tag{24}$$

The essential point about (24) is that φ does not appear in $\mathbf{h}$ or ω; the theory shows how to obtain the transformations (as series in ε) from $(\mathbf{p},\mathbf{q})$ to $(\mathbf{z},\varphi)$ that put the equations of motion into this form. The transformation is actually made *via* intermediate variables $(\mathbf{y},\nu)$ where $\mathbf{y}$ is any vector constant of the unperturbed ($\varepsilon=0$) motion and ν is an angle-like variable about the unperturbed path. There is much freedom in the choice of $\mathbf{y}$ variables, but the $(\mathbf{z},\varphi)$ follow uniquely from the $(\mathbf{y}, \nu)$. Some choices of $\mathbf{y}$ result in a much simpler calculation than others, so that although the procedure for calculating the invariants is in principle straightforward for any $\mathbf{y}$, a shrewd choice for $\mathbf{y}$ greatly reduces the labor.

The adiabatic invariant is

$$I(\mathbf{z}, \varepsilon) = \oint \mathrm{d}\varphi \mathbf{p}(\mathbf{z}, \varphi, \varepsilon) \cdot \frac{\partial \mathbf{q}}{\partial \varphi}(\mathbf{z}, \varphi, \varepsilon) \tag{25}$$

Note that $\mathbf{z}$ is held constant in doing the integral and that I is a sum of action integrals over the degrees of freedom, and not any one alone. The path described in phase space as φ varies at constant $\mathbf{z}$ is closed but is not the same as for $\mathbf{y}$ constant, which is the unperturbed trajectory. Nor is $\mathbf{z}$ constant the perturbed trajectory, which usually is not even closed in phase space. The proof of the invariance of I consists of showing that

$$\frac{\mathrm{d}I}{\mathrm{d}t} \equiv \frac{\partial I}{\partial z_i} \frac{\mathrm{d}z_i}{\mathrm{d}t}$$

is zero, or more exactly that $\mathrm{d}/\mathrm{d}t(I_0 + \varepsilon I_1 + \ldots \varepsilon^n I_n)$ equals zero, plus terms of order ε^{n+1}, since everything is done by expansion in ε. Thus $I_0 + \varepsilon I_1 + \ldots \varepsilon^n I_n$ is constant, plus terms of order ε^{n+1}. The three invariants M, J and Φ of the preceding section are really just M_0, J_0 and Φ_0 of their respective series. The difficulty in going to higher order without a systematic theory like the one described here is to recognize what the order ε^{n+1} term is $\mathrm{d}/\mathrm{d}t$ of. The systematic theories assure us that it is always $\mathrm{d}/\mathrm{d}t$ of something (plus $0(\varepsilon^{n+2})$) and tells what that something is; hence this is an "all orders" theory.

To get a second invariant, the procedure of finding $\mathbf{z}$ variables must be repeated, but with what are called the reduced variables. It is possible that the equation $\dot{\mathbf{z}} = \varepsilon \mathbf{h}(\mathbf{z}, \varepsilon)$ can be cast into the required form (23) in terms of new canonical variables $\mathbf{P}$ and $\mathbf{Q}$. Since the Poisson bracket $[\varphi, I]$ can be demonstrated to equal 1, it is known that φ and I can be used as conjugate variables in a canonical transformation from $(\mathbf{p}, \mathbf{q})$ to $(\mathbf{P}, \mathbf{Q})$ with I and φ as parameters in the transformation; $\mathbf{P}$ and $\mathbf{Q}$ will have $n-1$ components each and are the reduced variables. The new Hamiltonian $H'(\mathbf{P}, \mathbf{Q}, I, \varepsilon)$ is independent of φ, since $\dot{I}$ equals $-\partial H'/\partial \varphi$, and $\dot{I}$ is zero. The reduced equations of motion are

$$\begin{aligned} \dot{Q}_i &= \frac{\partial H'}{\partial P_i}(\mathbf{P}, \mathbf{Q}, I, \varepsilon) \\ \dot{P}_i &= -\frac{\partial H'}{\partial Q_i}(\mathbf{P}, \mathbf{Q}, I, \varepsilon) \end{aligned} \tag{26}$$

If $\varepsilon=0$ gives closed curves in **P**, **Q** space for solutions of (26) (which means that $\dot{\mathbf{z}}=\varepsilon\mathbf{h}$ could also be cast into a form where the unperturbed solution is periodic), then a second invariant is

$$I'(\mathbf{z}') = \oint \mathbf{P}(\mathbf{z}', \varphi') \cdot \frac{\partial \mathbf{Q}}{\partial \varphi'} \, \mathrm{d}\varphi' \tag{27}$$

There are shortcuts which make it unnecessary to find **P** and **Q**. The variables and transformations needed to calculate the first and second invariants for a charged particle are shown in Figure 5.

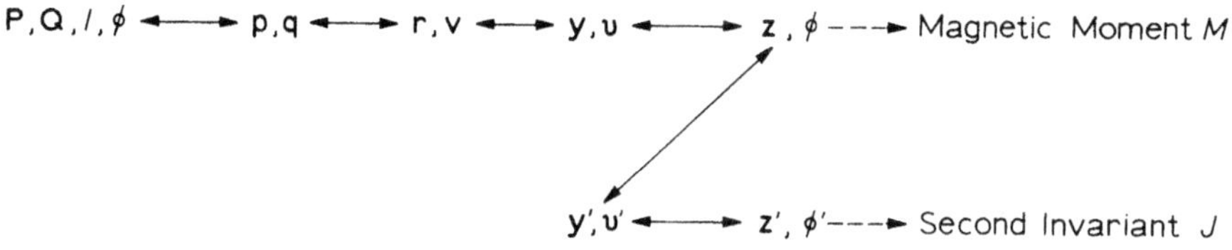

Fig. 5. Transformations used to obtain first and second invariants.

Invariants are obtained as functions of **z** or **z**′, but can be expressed in terms of the original dynamical variables (**p**, **q**) or (**r**, **v**) by reversing the above transformations.

The second term of the magnetic moment series εM_1 has been given by Kruskal (1958). Recently the εJ_1 term of the second invariant has been obtained (Liu and Northrop) for the special case of a static magnetic field and zero electric field. In the course of the calculation the guiding center equations of motion have been found consistent through order ε to be (**R** = guiding center position; **R**, Σ, H are the **z** variables):

$$\frac{\mathrm{d}\mathbf{R}}{\mathrm{d}t} = H\mathbf{e}_1(\mathbf{R}) + \frac{\varepsilon}{B(\mathbf{R})} H^2 \mathbf{e}_1 \times \frac{\partial \hat{\mathbf{e}}_1}{\partial s} + \frac{\varepsilon\Sigma^2}{2B^2} \mathbf{e}_1 \cdot \Delta B$$

$$+ \frac{\varepsilon\Sigma^2}{2B} \left[\tfrac{3}{2} \mathbf{e}_2 \cdot (\mathbf{e}_3 \cdot \nabla) \mathbf{e}_1 - \tfrac{1}{2} \mathbf{e}_3 \cdot (\mathbf{e}_2 \cdot \nabla) \mathbf{e}_1\right] \mathbf{e}_1 + \frac{\varepsilon}{B} \Sigma H \mathbf{e}_1 \mathbf{e}_2 \cdot (\mathbf{e}_1 \cdot \nabla) \mathbf{e}_1 + 0(\varepsilon^2)$$

$$\frac{\mathrm{d}\Sigma}{\mathrm{d}t} = -\frac{\Sigma H}{2} \nabla \cdot \mathbf{e}_1 + \text{a long term of order } \varepsilon$$

$$\frac{\mathrm{d}H}{\mathrm{d}t} = \frac{\Sigma^2}{2} \nabla \cdot \mathbf{e}_1 + \text{another long term of order } \varepsilon, \tag{28}$$

where $\mathbf{e}_1$ is the unit vector $\mathbf{B}/B$.

To lowest order in ε, Σ and H are the components of particle velocity perpendicular and parallel to the field line at the particle position. The components of $\mathrm{d}\mathbf{R}/\mathrm{d}t$ perpendicular to $\mathbf{e}_1$ are the usual drifts. From the second and third of equations (28) follow conservation of energy and of M_0, which is proportional to $\Sigma^2/2B$. The mirroring of the guiding center also is predicted by these last two equations.

A heuristic derivation of the εJ_1 term will be given now, prior to giving the results of the systematic method. In the proof of the invariance of J_0 earlier, the instantaneous rate of change of J_0 was

$$\frac{\mathrm{d}J_0}{\mathrm{d}t} = -\frac{eT}{c}[\langle\dot{\beta}\rangle\dot{\alpha} - \langle\dot{\alpha}\rangle\dot{\beta}]$$

$$= -\frac{eT}{c}\frac{\mathrm{d}}{\mathrm{d}t}\int_0^t [\langle\dot{\beta}\rangle\dot{\alpha} - \langle\dot{\alpha}\rangle\dot{\beta}]\,\mathrm{d}t,$$

or

$$\oint (2/m)^{\frac{1}{2}}(K - MB)^{\frac{1}{2}}\,\mathrm{d}s + \frac{eT}{mc}\int_0^t [\langle\dot{\beta}\rangle\dot{\alpha} - \langle\dot{\alpha}\rangle\dot{\beta}]\,\mathrm{d}t$$

is a constant, plus $0(\varepsilon^2)$.

This result, so easily obtained compared to the exact calculation, suffers from several deficiencies and uncertainties. In the first place, should M be M_0 or $M_0+\varepsilon M_1$? It makes a difference to the order we are interested in. Secondly, the integral $\int_0^t \ldots$ is not a local quantity – i.e., one must know the guiding center trajectory to evaluate the integral; it cannot be evaluated from the present position alone. But the usefulness of the invariants is precisely to give information about the trajectory.

These problems are resolved in the correct result, obtained by the systematic theory:

$$\text{constant} = \oint 2^{\frac{1}{2}}[K - (M_0 + \varepsilon M_1)B(\alpha,\beta,s)]^{\frac{1}{2}}\,\mathrm{d}s + T^2(\alpha,\beta,KM)/\varepsilon \\ \left[\langle\dot{\beta}\rangle \int_0^{U'} \dot{\alpha}(\alpha\beta KMU'')\,\mathrm{d}U'' - \langle\dot{\alpha}\rangle \int_0^{U'} \dot{\beta}\right] \tag{29}$$

where U' is $1/T \int \mathrm{d}s/v_{\|}$ and is an angle-like variable along the field line on which the guiding center is located at the instant (Figure 6). Likewise the $\int_0^{U'}$ is along the same line. The invariant is now a local quantity in the sense that the previous history of the guiding center is not needed. In addition the magnetic moment appears correct through order ε. In the $\oint$, the limits are the s at which $K-(M_0+\varepsilon M_1)\, B$ vanishes.

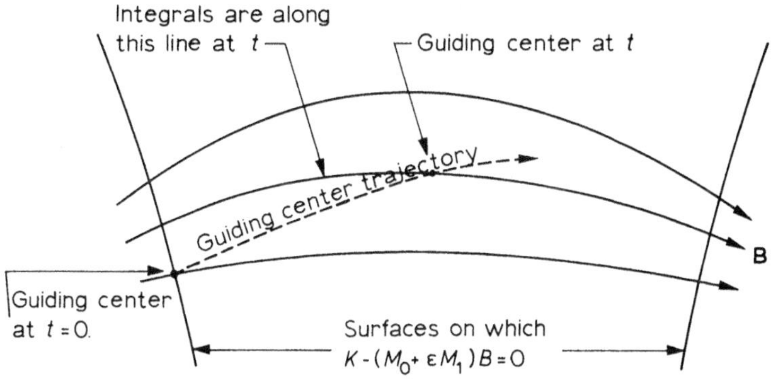

Fig. 6. The correction to J_0 is a local quantity.

At this point, the reader may ask why in the earlier proof (Equation 11) of the invariance of J_0 a term in $\dot{M}_0$ was not included, to give

$$\frac{\mathrm{d}J_0}{\mathrm{d}t}=\frac{\partial J_0}{\partial\alpha}\frac{\mathrm{d}\alpha}{\mathrm{d}t}+\frac{\partial J}{\partial\beta}\frac{\mathrm{d}\beta}{\mathrm{d}t}+\frac{\partial J_0}{\partial M_0}\frac{\mathrm{d}M_0}{\mathrm{d}t},$$

since $\dot{M}_0$ equals $-\varepsilon\dot{M}_1$ and is of the same order as $\dot{\alpha}$ and $\dot{\beta}$. The answer is that in finding the J_0 surface on which a particle moves on the average, one picks a number for M_0 and holds it constant in a numerical computation to find the surface. In effect one is using $M_0+\varepsilon M_1$, which *is* constant.

4. How Invariant are the Invariants?

The answer depends partly on what is meant by "the invariant". If only the lowest term of the series in ε, such as M_0 or J_0, is meant, the invariance will not be so good as if the εM_1 and εJ_1 terms are included. However, the invariant series are not convergent in general, hence taking more terms may not continue to produce better correspondence with reality for a given ε. As with any asymptotic series, the sum of a fixed number of terms becomes a better approximation as ε is reduced, and the more terms the larger the ε at which the approximation becomes of a given accuracy.

In addition to manifestations of higher terms, such as εM_1, another type behavior has been seen in numerical computations of particle trajectories (GARREN *et al.*, 1958; DRAGT, 1965) designed to study the magnetic moment in laboratory mirror geometry and a dipole respectively. Even the lowest order M_0 seems to have no validity, and the onset of the observed erratic changes in M_0 is sudden as ε is increased. For the mirror this occurs for ε such that the radius of gyration is more than about 0.05 of the system's dimensions. At the

Fig. 7. Intersections of J_0 surfaces and guiding center trajectories with midplane.

level of the second invariant J a similar study has been made by SIAMBIS (1964) for Ioffe-type mirror geometry. This is an ordinary laboratory mirror geometry with four or six conductors carrying currents parallel to the axes of the mirror coils. Figure 7 lies in a plane perpendicular to the axis of the system and midway between the mirror coils. Shown schematically is the intersection of a J_0 surface computed numerically by Siambis, and the dots show the guiding center position each time the midplane is traversed. The guiding center appears to oscillate about its J_0 surface, the oscillations being of greatest amplitude under the longitudinal currents. This oscillation may be in part a manifestation of εJ_1; it may also be an artifact arising from the use of the usual "guiding center equations", which are truncated versions of Equation (28) obtained by omitting the lengthy $O(\varepsilon)$ terms in $\mathrm{d}\Sigma/\mathrm{d}t$ and $\mathrm{d}H/\mathrm{d}t$, but retaining the $O(\varepsilon)$ drifts in $\mathrm{d}\mathbf{R}/\mathrm{d}t$ but not the $O(\varepsilon)$ term parallel to $\mathbf{B}$. If the complete equations through order ε were used, some of this oscillation might disappear, leaving the true εJ_1 effects visible.

No numerical study has been made of the invariance of the third invariant; it would be worthwhile doing because of the interest in diffusion by violation of Φ.

References

ALFVÉN, H: 1950, *Cosmical Electrodynamics,* Clarendon Press, Oxford, England.
DRAGT, A. J.: 1965, *Rev. Geophys.* **3**, 255.
FERMI, E.: 1949, *Phys. Rev.* **75**, 1169.
FERMI, E.,: 1954 *Astrophys. J.* **119**, 1.
GARDNER, C. S.: 1960 *Phys. Rev.* **117**, 215.
GARREN, A. *et al.*: 1958, *Proc. 2nd Intern. Conf. Peaceful Uses At. Energy,* **31**, 65.
HAMLIN, D. A., KARPLUS, R., VIK, R. C., and WATSON, K. M.: 1961, *J. Geophys. Res.* **66**, 1.
HELLWIG, G.: 1955, *Z. Naturforsch.* **10a**, 508.
KADOMTSEV, B.: 1958, *Plasma Physics and the Problem of Controlled Thermonuclear Reactions,* Vol. III, Akad. Nauk SSSR, Moscow.
KRUSKAL, M.: 1958, "The Gyration of a Charged Particle", Princeton University, Project Matterhorn Report PM-S-33 (NYO-7903), March (1958).
KRUSKAL, M.: 1962, *J. Math. Phys.* **3**, 806.
LIU, C., and NORTHROP, T. (to be published).
NEWCOMB, W.: 1958, *Ann. Phys.* **3**, 347.
NORTHROP, T. G. and TELLER, E.: 1960, *Phys. Rev.* **117**, 215.
NORTHROP, T. G.: 1963, *The Adiabatic Motion of Charged Particles,* Interscience Publishers, N. Y.
ROSENBLUTH, M. and LONGMIRE, C.: 1957, *Ann. Phys.* **1**, 120.
SIAMBIS, J.: 1964, *Guiding Center Motion of Particles in Combined Mirrormultipole Cusp Magnetic Fields,* University of California Ph. D. Thesis.
SWIFT, D. (to be published in *J. Geophys. Res.*)
VANDERVOORT, P.: 1960, *Ann. Phys.* **10**, 401.

Discussion

Several questions were raised concerning the degree with which the adiabatic invariants were conserved both in theory and in practice. Dr. NORTHROP described numerical work done by Garren, Riddell, Smith, Bing, Henrich, Northrop, and Roberts and similar work by Dragt in which the particle trajectory was studied during successive bounces in mirroring fields. It was found in these calculations that for a wide range of values of the magnetic moment the particle motion was regular and presumably continued trapped indefinitely. However, for a range of conditions in which the initial trajectory was near the loss cone, the behaviour was irregular and within a few reflections the particle would be lost through a mirror.

The boundary between the regular and irregular behaviour seemed rather sharp, and the size of the irregular region increased with increasing gyration radius. In an experiment done by Lauer, Gibson, and Jordan it was found that particles were trapped effectively an infinite time unless the pitch angle was too near the loss cone. The influence of axial symmetries on the conservation of adiabatic invariants was brought up by HESS who pointed out that Garmire in calculations similar to those of Dragt found that protons of 100 MeV were more easily lost from a Finch and Leaton magnetic field than from a dipole field. Dr. NEILSON added that in the DCX machine at Oak Ridge confinement for several minutes was observed when the field was axially symmetric. However, when a magnetic bump was added to the field, the confinement time was reduced substantially.

Drs. CAIN and MCILWAIN stressed the difficulty in applying the general theory developed for the scalar α, β representation to the actual geomagnetic field. Dr. McIlwain encouraged everyone to try to find such a representation for the earth as this would greatly simplify tracing field lines as required for the B, L calculations.

MAGNETIC COORDINATES

CARL E. McILWAIN
Department of Physics, University of California, San Diego, La Jolla, California

Abstract. The definitions and properties of the magnetic coordinates B, L, K, Φ_0, R, λ and Λ are reviewed. The roles these parameters play in the study of geomagnetically trapped particles are discussed in detail. The intrinsic accuracy of the magnetic shell parameter L is examined as are the accuracies which are desired, available, and possible. A plea is made for the adoption of standard methods of presenting results.

1. The Magnetic Shell Parameter

A. PURPOSE

Two possible incentives for introducing any new parameter are (a) to reduce the number of variables needed to describe a physical situation and (b) to organize data in a fashion which facilitates its physical interpretation. The need for a more accurate and succinct method of organizing, displaying, and interpreting the trapped radiation data being obtained from earth satellites led directly to the introduction of the magnetic shell parameter L (McIlwain, 1961).

B. THE ADIABATIC INVARIANTS

The existence, properties, applicability, and importance of the adiabatic invariants have been thoroughly discussed by Northrop and Teller (1960), Northrop (1963a, 1963b), and others.

The first invariant (often called the magnetic moment or Alfvén invariant) is given by

$$\mu = \frac{p_\perp^2}{2m_0 B} \tag{1}$$

where $p_\perp$ is the component of the relativistic momentum perpendicular to the magnetic field vector $\mathbf{B}$, m_0 is the particle rest mass, and B is the scalar magnitude of $\mathbf{B}$.

The second invariant, usually called the integral invariant, and sometimes called the "longitudinal" invariant because of its initial use in studying the properties of mirror machines, (for geomagnetically trapped particle studies, "latitudinal" would have been a more appropriate word) is given by

$$J = \oint p_\parallel \, \mathrm{d}s \tag{2}$$

where $p_\parallel$ is the component of the relativistic momentum parallel to $\mathbf{B}$ and the integral is taken along the motion of the guiding center over a complete oscillation from one mirror point to the other and back again.

Assuming μ is constant, equation (1) gives

$$p_{\|} = p\sqrt{1 - p_{\perp}^2/p^2} = p\sqrt{1 - B_l/B} \tag{3}$$

where B_l is the magnitude of the magnetic field at an arbitrary point along the particles path and B is the mirror point field.

If there is no electric field, and $\partial B/\partial t$ is small, then the total momentum p is constant. In this case, equation (2) can be written

$$J = 2p\int_{A}^{A'} (1 - B_l/B)^{\frac{1}{2}}\, \mathrm{d}s \tag{4}$$

where the integral is taken over a half oscillation, i.e., from one mirror point to its conjugate point.

The quantity

$$I \equiv J/2p \tag{5}$$

can be seen to be a function of the mirror point A and the magnetic field configuration and is independent of any properties of the particle. J is often thought of as being a property of the particle which is retained throughout its motion and is therefore a function of pitch angle as well as position. It is convenient to consider I to be a function of position only, that is, to be equal to the $J/2p$ of a particle which mirrors at that point in space. Stated differently, J=constant, μ=constant, specifies the surface on which a particle moves, while I=constant, B=constant, specifies the lines on which the particle mirrors. Connecting these lines (in the geomagnetic field case they form a ring in each hemisphere) with magnetic lines of force produces the surface specified by J=constant.

The third or flux invariant is equal to the flux of **B** enclosed by the surface J=constant and can be written

$$\Phi = \oint \mathbf{A} \cdot \mathbf{dl} \tag{6}$$

where **A** is the vector potential of the magnetic field and the integration is along a closed curve lying on the surface specified by J=constant. Two possible curves over which the integration may be taken are specified by I=constant, B=constant, (providing of course that p is constant). *Note*: The letters J and Φ are often also used to specify omnidirectional intensity and geographic longitude respectively.

C. PROPERTIES DESIRED FOR MAGNETIC COORDINATES

Over a wide range of conditions actually encountered in the geomagnetic field, the values of μ, J, and p for individual particles remain essentially constant over many complete cycles of motion. In these cases, $j_{\perp}$, the directional intensity of particles perpendicular to **B**, will be equal at all locations which have the same B and I values, providing the particles are uniformly distributed in longitude. Stated differently: to the extent that longitude can be considered a degenerate variable, all points in space which have the same B and I values can be considered equivalent. Stated still differently: provided certain conditions are met, B and I will organize $j_\perp$ measurements perfectly. When these conditions are not met, B and I will still organize measurements as well or better than any other parameters yet proposed and provide a framework within which deviations can be studied.

It can be seen that any two independent functions of B and I will also provide the desired property of having the same values at all of the possible mirror points of a given particle. Parameters which are not exact functions of B and I cannot be expected to provide this equivalence of mirror points.

Unfortunately, the quantity I has no simple physical significance or analog. Data plotted as a function of B and I are quite difficult to interpret or to mentally transform into their actual dependence in real space. It is therefore desirable to replace I with some function of B and I which has a simple physical analog in order to facilitate the interpretation of data and the understanding of theory.

In addition to organizing mirror points, it is often desirable to organize data along magnetic lines of force (or more exactly "magnetic shells"). Several reasons why this is desirable are:

(a) The motion of the guiding center of trapped particles tends to be along lines of force.

(b) It is often desirable to convert measurements of $j_\perp$ (B) along lines of force into $j(B, \alpha)$ and $J_0(B)$ and vice versa, where α is the pitch angle and J_0 the omnidirectional intensity.

(c) Diffusion by atmospheric scattering and by certain forms of electric and magnetic field perturbations occurs along lines of force.

(d) Many other phenomena such as whistlers tend to occur along magnetic field lines.

Unfortunately, there is no single function of B and I which is constant along all field lines in the earth's magnetic field (the significance of this will be discussed below). Fortunately, however, functions of B and I do exist which vary only slightly along lines of force.

D. DEFINITION OF L

In a pure dipole magnetic field, there is a class of functions of B and I which are exactly constant along all lines of force. A dipole line of force is described by the simple equation

$$R = L_{\mathrm{d}} \cos^2 \lambda \tag{7}$$

where R is the radial distance from the dipole, λ is the latitude (i.e. the angle from the direction of the north magnetic pole minus ninety degrees) and L_d is the maximum value of R, which is of course found at $\lambda=0$, the equator.

The magnetic field in the presence of a magnetic dipole is

$$B_d = \frac{M}{L_d^3 \cos^6 \lambda}(1 + 3 \sin^2 \lambda)^{\frac{1}{2}} \tag{8}$$

where M is the dipole magnetic moment.

Using equations (7) and (8) in conjunction with equations (4) and (5) and letting $Y=\sin \lambda$, it is found that for a dipole field

$$I_d = 2L_d \int_0^Y \left\{1 - \left(\frac{1 + 3Y_a^2}{1 + 3Y^2}\right)^{\frac{1}{2}} \left(\frac{1 - Y^2}{1 - Y_a^2}\right)^3\right\}^{\frac{1}{2}} (1 + 3Y_a^2)^{\frac{1}{2}} \mathrm{d}Y_a \tag{9}$$

or

$$I_d = L_d h_1 (\lambda) \tag{10}$$

Now by equation (8) we can write

$$\lambda = h_2 \left(\frac{L_d^3 B_d}{M}\right) \tag{11}$$

Substituting equation (11) into (10) we have

$$I_d = L_d h_3 (L_d^3 B_d / M) \tag{12}$$

or more symmetrically,

$$I_d^3 B_d / M = h_4 (L_d^3 B_d / M) \tag{13}$$

Inverting equation (13) gives

$$L_d^3 B_d / M = F(I_d^3 B_d / M) \tag{14}$$

or

$$L_d = f(B_d, I_d, M) \tag{15}$$

Thus in a static dipole field, there exist functions of B and I which depend only on L_d and are therefore precisely constant along lines of force.

Now take the case of the real geomagnetic field and represent by $I=g_i(B)$ the functional dependence of I versus B along a line of force designated by i. If two points in space a and b are found for which $B_a=B_b$ and $I_a=I_b$ and $I=g_a(B)$ and $I=g_b(B)$ are computed along the lines of force which pass through a and b respectively, it is found that

$$g_a(B) \neq g_b(B) \tag{16}$$

unless a and b are on the same line of force. Thus there is no function of B and I which is constant along the lines of force in the Earth's magnetic field. It is observed however that the differences between g_a and g_b are typically quite small.

One way to proceed would be to compute a large set of functions $g_i(B)$ corresponding to equivalent mirror points uniformly distributed in longitude and then to find their average $\bar{g}_i(B)$. When this is done, it is found that the result is almost identical with the dipole function, equation (12), providing of course that L_d and M are suitably chosen. This result implies that the dipole function f in equation (15) will vary along real lines of force as little as any function of B and I that is likely to be found.

The magnetic shell parameter L is therefore defined as

$$L^3B/M = F(I^3B/M) \tag{17}$$

where B and I are to be computed using the best available representation of the Earth's magnetic field, M is chosen such as to minimize the variation of L along real lines of force, and F is the dipole function given by equation (14). The form of equation (14) is preferable to that of equation (15) since it is a function of only one variable. The effect of using different values of M is relatively small and as yet there appears to be no reason to deviate from the value initially chosen: 8.06×10^{25} gauss cm^3, the approximate value of the Earth's dipole moment in 1960. The most convenient unit for L is earth radii. Taking one earth radius to be $R_e = 6371.2$ km, gives the value of M to be used in equation (17) to be $M = 8.06 \times 10^{25}/(6.3712 \times 10^8)^3 = 0.311653$ gauss R_e^3.

To compute L, given B and I, accurate values of the function F have been computed and used to generate sets of coefficients a_n for the equation

$$Y = \sum_{n=0}^{n=9} a_n X^n \tag{18}$$

where

$$Y = \ln\left(\frac{L^3B}{M} - 1\right) \quad \text{and} \quad X = \ln\left(\frac{I^3B}{M}\right).$$

It is sometimes necessary to obtain I from B and L. For this purpose, sets of coefficients b_n for the inverse of equation (18):

$$X = \sum_{n=0}^{n=9} b_n Y^n \tag{19}$$

have been computed. The sets of coefficients obtained are listed in Tables 1 and 2.

E. VARIATION IN L_0

The magnetic equator is best defined as that surface on which the value of B along every line of force is a minimum. In the earth's field, this surface is far from being a plane and at radial distances greater than 6 R_e it becomes time dependent and irregular to the extent of forming strange looking shapes and may even break into multiple sheets.

TABLE I

Coefficients a_n for Different Ranges of X

	$X < -22.$	$-22. < X < -3.$	$-3. < X < 3.$	$3. < X < 11.7$	$11.7 < X < 23.$	$X > 23.$
a_0	3.0062102×10^{-1}	6.2337691×10^{-1}	6.228644×10^{-1}	6.222355×10^{-1}	2.0007187	-3.0460681
a_1	3.33338×10^{-1}	4.3432642×10^{-1}	4.3352788×10^{-1}	4.3510529×10^{-1}	$-1.8461796 \times 10^{-1}$	$1.$
a_2	0	1.5017245×10^{-2}	1.4492441×10^{-2}	1.2817956×10^{-2}	1.2038224×10^{-1}	0
a_3	0	1.3714667×10^{-3}	1.1784234×10^{-3}	2.1680398×10^{-3}	$-6.7310339 \times 10^{-3}$	0
a_4	0	8.2711096×10^{-5}	3.8379917×10^{-5}	$-3.2077032 \times 10^{-4}$	2.170224×10^{-4}	0
a_5	0	3.2916354×10^{-6}	$-3.3408822 \times 10^{-6}$	7.9451313×10^{-5}	$-3.8049276 \times 10^{-6}$	0
a_6	0	8.1048663×10^{-8}	$-5.3977642 \times 10^{-7}$	$-1.2531932 \times 10^{-5}$	2.8212095×10^{-8}	0
a_7	0	1.0066362×10^{-9}	$-2.1997983 \times 10^{-8}$	9.9766148×10^{-7}	0	0
a_8	0	$8.3232531 \times 10^{-13}$	2.3028767×10^{-9}	-3.958306×10^{-8}	0	0
a_9	0	$-8.1537735 \times 10^{-14}$	$2.6047023 \times 10^{-10}$	$6.3271665 \times 10^{-10}$	0	0
E	2.0×10^{-4}	1.0×10^{-6}	7.0×10^{-8}	6.0×10^{-7}	1.2×10^{-5}	5.0×10^{-4}

E = Maximum Error in Y.

TABLE II

Coefficients b_n for Different Ranges of Y

	$Y < -7.$	$-7. < Y < -.7$	$-.7 < Y < 2.$	$2. < Y < 8.5$	$8.5 < Y < 20.$	$Y > 20.$
b_0	$-9.0176314 \times 10^{-1}$	-1.5035469	-1.5035665	-1.4921674	-1.3415203	3.0460681
b_1	2.999966	2.516886	2.5166432	2.4799708	2.5733781	1.
b_2	0	$-1.5473094 \times 10^{-1}$	$-1.5579012 \times 10^{-1}$	$-1.0499454 \times 10^{-1}$	-2.747596×10^{-1}	0
b_3	0	-1.482533×10^{-2}	$-1.6991733 \times 10^{-2}$	$-5.6314931 \times 10^{-2}$	2.9961512×10^{-2}	0
b_4	0	5.7665822×10^{-3}	3.3726437×10^{-3}	2.16412×10^{-2}	$-2.1911161 \times 10^{-3}$	0
b_5	0	2.5165553×10^{-3}	9.4374468×10^{-4}	$-3.9980282 \times 10^{-3}$	1.0870528×10^{-4}	0
b_6	0	4.6975588×10^{-4}	$-1.6048742 \times 10^{-4}$	4.5340102×10^{-4}	$-3.5717807 \times 10^{-6}$	0
b_7	0	4.9336212×10^{-5}	$-5.7606054 \times 10^{-5}$	$-3.2040569 \times 10^{-5}$	7.2957464×10^{-8}	0
b_8	0	2.8332675×10^{-6}	1.6653982×10^{-5}	1.3002586×10^{-6}	$-8.0754426 \times 10^{-10}$	0
b_9	0	6.9447496×10^{-8}	-1.114463×10^{-6}	$-2.3206968 \times 10^{-8}$	$3.3797513 \times 10^{-12}$	0
E	5.0×10^{-4}	5.4×10^{-6}	5.2×10^{-7}	1.5×10^{-6}	1.1×10^{-5}	5.0×10^{-4}

E = Maximum Error in X.

At the position of the minimum field value, B_{min}, along a line of force, the conjugate points A and A' in equation (4) become identical. Thus at the magnetic equator $I=0$. Now the value of F with a zero argument is exactly 1.0, thus by equation (14), the equatorial value of L is

$$L_0 = (M/B_{min})^{1/3} \tag{20}$$

Since L_0 is, by definition, constant along a line of force, it can be used to label lines of force.

The inequality (16) implies that the minimum values of B along a guiding center trajectory will depend upon the longitude, Φ_0 (See Section 2A for the definition of Φ_0). The values of L_0 reached by a particle mirroring at a particular B and L will therefore also depend upon Φ_0:

$$L_0 = L_0(B, L, \Phi_0) \tag{21}$$

Thus two particles a and b which mirror at the same values of L but at different values of B will drift in longitude through lines of force labelled by different values of L_0. Explicitly:

$$L_{0a}(B_a, L_a, \Phi_0) \neq L_{0b}(B_b, L_b = L_a, \Phi_0) \tag{22}$$

except for a few singular values of Φ_0. It should be noted that this is a fundamental characteristic of charged particle motion in the earth's magnetic field about which little can be done other than the cancellation of the earth's non-dipole moments.

The difference $L_{0a} - L_{0b}$ is almost always a monotonic function of the difference of B_a and B_b and is largest when B_a is equal to B_{min} and when B_b is the maximum value available for durable trapping (usually set by the earth's atmosphere to be the value of B at a radial distance of about 1.1 R_e). Figure 1 is a contour plot of

$$\frac{100}{L}[L_0(B_{min}) - L_0(B_{max})],$$

which is the percent difference between the maximum and minimum values of L_0 for constant L, but which can also be interpreted to be the percent variation of L along lines of force, i.e.

$$\frac{100}{L_0}[L(B_{max}) - L(B_{min})].$$

The above discussion and conclusions closely parallel that of STONE (1963). The points of view however, are somewhat different. Stone suggests that L be used to label particle trajectories instead of position in space. This would have the effect of making L a function of the angle to $\mathbf{B}$ as well as the spatial coordinates. Such a drastic complication cannot be justified on the basis of the

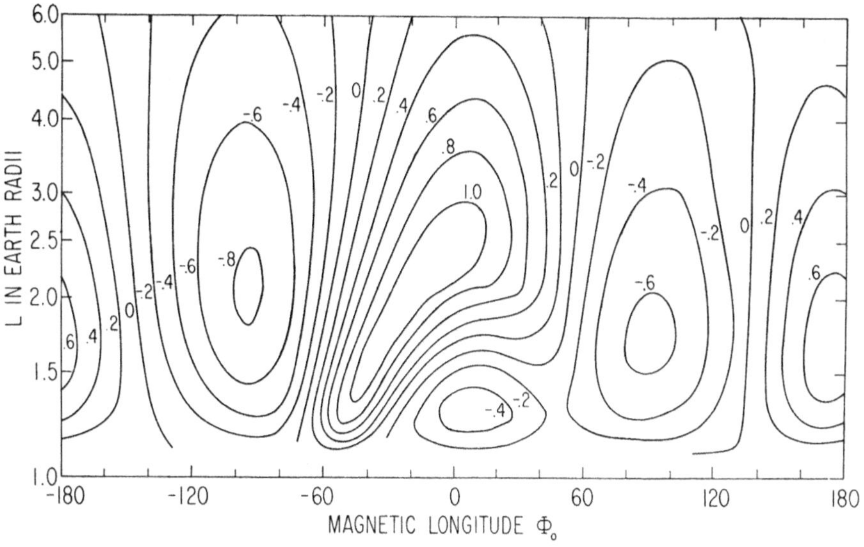

Fig. 1. The percent difference between L_0 ($R = L$) and L_0 ($R = 1.1$) as a function of L and magnetic longitude Φ_0. (The JENSEN and CAIN (1962) 48 term representation of the earth's field was used in the computations for Figures 1 and 2.)

improvement it might yield. It seems far preferable to label each point in space with a unique value of L and then to remember that it has exact invariant properties only for those particles which mirror at that point.

Stone also suggests that L_0 may be almost as accurate a parameter as L. It is shown in Section 3A that in almost all practical cases, the intrinsic error introduced by using L for all particles found at each point in space is far less than the errors which would be introduced by the use of L_0.

2. Other Magnetic Coordinates

A. THE MAGNETIC LONGITUDE, Φ_0

While in most cases, longitude can be treated as a degenerate variable, several problems have arisen in which it is necessary to consider variations with longitude explicitly. One class of such problems is the study of particles which exhibit important intensity changes during one period of longitudinal drift, for example, electrons mirroring at low altitudes.

Another such problem is computation of the average of the atmospheric density over an entire cycle of particle motion. In the treatment of this problem, HASSITT (1965) has defined the magnetic longitude coordinate, Φ_0, to be the geographic longitude at the minimum value of B on the line of force. By definition, Φ_0 is constant along lines of force.

It would be convenient for some types of problems if the rate of drift of particles with respect to the longitude coordinate were constant. The relative rates of drift with respect to Φ_0 for various lines of force have been computed by

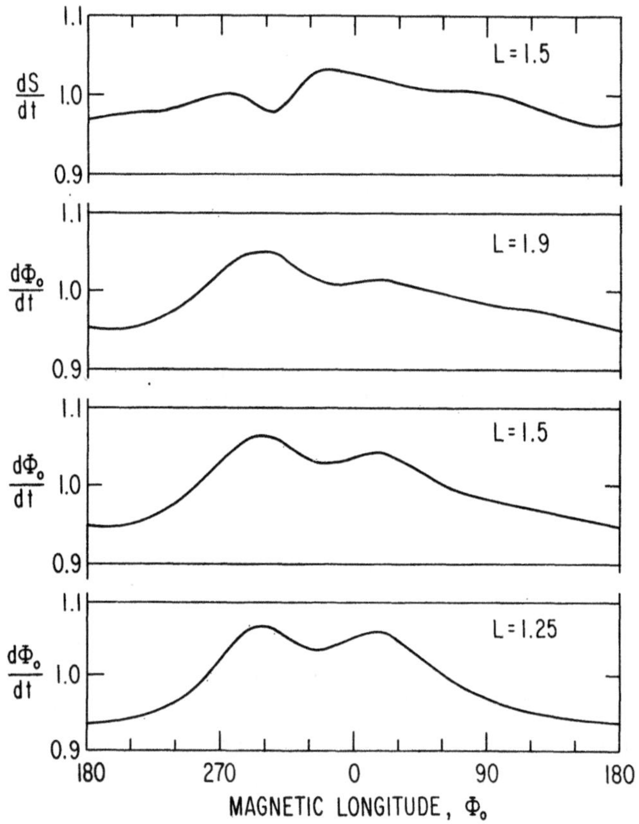

Fig 2. The lower three curves show the relative variation in the longitudinal drift ràte with respect to the magnetic longitude Φ_0. The uppermost curve shows the variation in the longitudinal drift rate with respect to the arc length along the magnetic equator.

HASSITT (private communication), three examples of which are shown in Figure 2. The maximum deviations can be seen to be about $\pm 7\%$. Part of these deviations are due to the geometrical differences in the geographic and magnetic equators. It is interesting therefore to note that the "true" drift rate – the drift rate with respect to arc length S along the magnetic equator – varies by $\pm 3\%$, as shown in the top curve in Figure 2. For cases in which the variation of the drift rate is considered important, $d\Phi_0/dt$ can be accurately determined by several methods (NEWKIRK and WALT, 1964; HASSITT, 1965).

B. THE INTEGRAL INVARIANT PARAMETER, K

As pointed out earlier, it is often desirable to organize data along lines of force. There is however, at least one type of study in which it is more desirable to organize data along traces of constant μ and J. This case is the one in which it is assumed that the first and second invariants μ and J are conserved while the third invariant Φ is being violated. While there is some question of whether this case ever actually occurs (G. HAERENDEL, private communication), it does seem plausible that the frequency spectrum of field fluctuations could be sufficiently large at low frequencies as to cause violations of Φ and yet be small at the higher frequencies corresponding to the gyrofrequency and the latitudinal bounce frequency associated with the first and second invariants and thus cause no important changes in μ and J.

If the flux invariant of a particle varies with time then the momentum, p, will vary proportionately, and $I = J/2p$ will also vary even when J is constant. It is possible however, to obtain a variable which is proportional to J, even if p varies, but which is independent of particle mass and momentum. Following the suggestion of KAUFMANN (1965), it is proposed that the parameter

$$K \equiv J/(8m_0\mu)^{\frac{1}{2}} \tag{23}$$

be used to uniquely identify the various traces which particle mirror points will follow when Φ, and therefore p, varies but J and μ are constant. By equations (1), (4), (5), and (23) it can be seen that

$$K = \frac{2pI}{(8m_0\mu)^{\frac{1}{2}}} = \frac{2(2m_0\mu B)^{\frac{1}{2}}I}{(8m_0\mu)^{\frac{1}{2}}} = B^{\frac{1}{2}}I \tag{24}$$

Thus K has a unique value for each point in space, and is completely independent of any characteristics of the mirroring particles being studied.

If B is in gauss and I is in earth radii, then K will lie between zero and about 100 (gauss)$^{\frac{1}{2}}$ R_e. Traces of constant K in B-L space are shown in Figure 3. It can be seen that these traces are not very different from the traces of constant magnetic latitude, and therefore constant B/B_0 (See Section 3B for the definitions of λ and B/B_0). Should it be the case (as Haerendel suggests it is) that the breakdown of the third invariant is almost always accompanied by a partial breakdown of the second invariant therefore making it necessary to consider diffusion in two dimensions, then it will probably be preferable to use B/B_0 rather than K since it is much easier to manipulate.

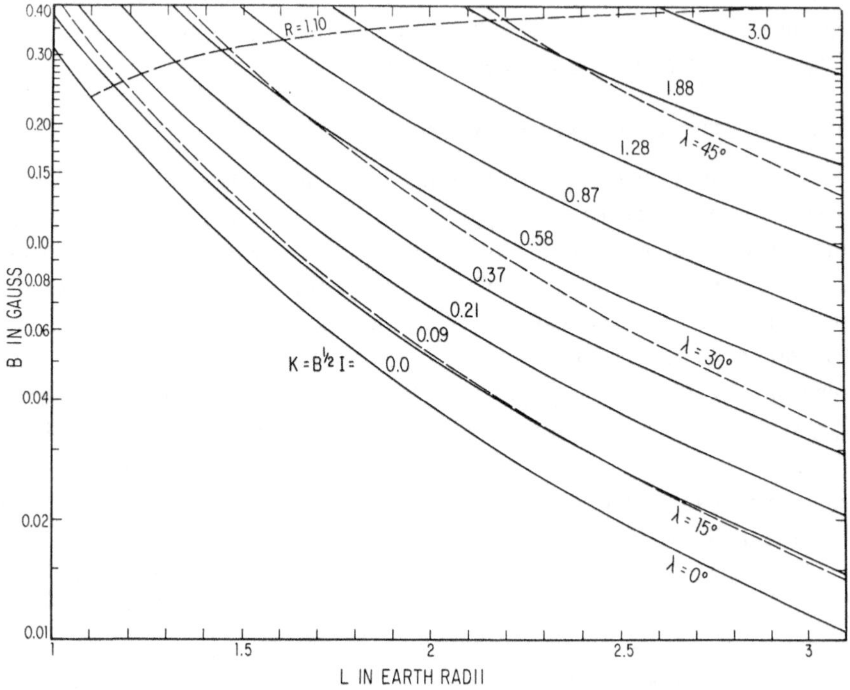

Fig 3. Curves of constant values of $K = J/\sqrt{8m_0\ \mu}$ in *B-L* space. Note that these curves are similar to the curves of constant magnetic latitude λ.

When B and L have been computed but the value of I has not been retained, it is much easier to reobtain the value of I by use of equation (19) than by direct integration.

3. Practical Considerations

A. Accuracy Desired and Attainable

Trapped particle fluxes are often found to vary faster than the 20th power of L. If it is desired to reduce the scatter of data in such cases to less than say 4%, then the errors in L must be less than 0.2%.

The errors in L, produced by errors in the magnetic field representation, are not easy to predict since they involve an integral over lines of force. It is probably a good guess however, that the RMS error in L will be between 1/2 and 1/4 of the RMS error in the magnetic field.

One source of errors in the magnetic field representations used in the computation of L arises from the fact that the field representations include at best only the secular changes in the field with time. Since field variations with time of over 0.001 gauss are observed during magnetic storms, errors in L of the order of

$$\frac{\Delta L}{L} \cong \frac{\Delta B}{3B} \cong \frac{0.001\,L^3}{3M} \cong 0.001\,L^3 \quad \text{or} \quad 0.1\,L^3\% \tag{25}$$

can be expected during these events. It can be reasonably expected however that the RMS error in L due to normal non-storm time variations will be less than 0.2% within $L=2.5$.

The major source of error in L at the present time is due to the errors in the quiet time magnetic field representations which result from inadequate distribution of the places proper measurements have been made. In a recent review, FOUGERE (1965) has shown that the RMS errors of the available representations with respect to the *measured* values are on the order of one percent. The maximum error in the regions with few or no measurements is presumably somewhat larger. When large quantities of satellite measurements of the earth's magnetic field become available, it can be reasonably expected that errors in L produced by field representation errors will be less than 0.2%.

As discussed previously, particles mirroring at the same L value but different B values in general do not travel on lines of force labelled by the same L_0 values. The values of L at the mirror points of the particles found traversing any given point will therefore depend upon the angle α, of the trajectory to the local field, and upon longitude. In principle, an experiment which exhibits the full $\pm$ 1% variation of L along lines of force could be performed by staying on the magnetic equator and measuring only those particles within the very narrow range of angles corresponding to mirror points at the highest B values available for durable particle trajectories. In practice however, many measurements are made over a moderate range of angles around $\alpha=90$ degrees. By the mirror equation

$$B = B_{\rm m} \sin^2 \alpha \tag{26}$$

it can be seen that $\alpha \cong 57°$ when $B_{\rm m}=1.4\,B$. Now L is rarely as much as 0.2% different at two points along a line of force whose B values are only a factor of 1.4 apart, thus any measurement made within $\pm$ 33 degrees of $\alpha=90$ degrees will have little intrinsic scatter due to the variation of L along lines of force.

Since the directional intensity perpendicular to $\mathbf{B}$ typically decreases as B^{-2} or faster, it can be shown that even omnidirectional measurements have little intrinsic scatter at constant B and L. Integrating the directional intensity $j_{\mathbf{0}}=K_1 B^{-2}$ from $\alpha=57$ to 90 degrees reveals that only 21% of the omnidirectional intensity comes from particles mirroring at B values more than 1.4 times the local B value. Should L_0 be used instead of the local L value, as suggested by STONE (1963), the ratio of the B values of the particles being measured to the equatorial B value will typically be much larger than 1.4. The limiting case is when measurements are made near the highest available B values. Since in this case all of the measured particles must mirror near the observation point, there will be much less than 0.1% intrinsic error in L. As shown in Figure 1 however, L_0 will have an intrinsic error of about $\pm$ 1%.

B. PRESENTATION OF DATA

A number of functions of B and L have been defined to facilitate the handling, presentation and understanding of data. Some of the commonly used variables of this kind are B/B_0, R, λ, and Λ, which are defined by the following equations:

$$B_0 = M/L^3 \qquad (\textit{not}\ M/L_0^3) \tag{27}$$

$$B = \frac{M}{R^3}\left(4 - \frac{3R}{L}\right)^{\frac{1}{2}} \tag{28}$$

$$\cos^2 \lambda = R/L \tag{29}$$

$$\cos^2 \Lambda = 1.0/L \qquad (\textit{not}\ R/L) \tag{30}$$

Since B/B_0, R, and λ are functions of both B and L, any of the twelve possible pairs of B, L or Λ, B/B_0, R, and λ may be used to uniquely identify a point in B-L space. Now there are three possible ways of making two dimensional plots to show the dependence of a quantity such as particle intensity upon two independent variables. There are also four ways of choosing linear or logarithmic scales for the two axes. Furthermore, simply changing the relative scales of the two axes can strongly affect the apparent significance of data plotted in any of the $12\times3\times4=144$ possible ways. It is clearly desirable to adopt a small number of standard ways of presenting data in order to facilitate comprehension and the comparison between different sets of data. The following three ways have been found to be particularly useful in exhibiting the information content of many different sets of data:

(a) Contours of constant values of the dependent variable in $R=\lambda$ space plotted in polar coordinates. Such plots can be used to exhibit the character of distributions in physical space.

(b) Contours of constant values of the dependent variable in B–L space using a linear scale for L and a logarithmic scale for B with one factor of ten in B corresponding to one earth radii in L. The use of a log scale for B prevents the collapse of the equatorial region at high L values into a small area on the graph.

(c) Particle intensities versus B for constant values of L with logarithmic scales for both intensity and B but with one factor of ten in B corresponding to four factors of ten in intensity. The use of a log scale B permits the direct comparison of the relative B dependences at different L values and is in general superior to a linear scale for B when data for covering wide range of L values are being plotted. Special graph paper for this method of plotting data is available from the Keuffel and Esser Co. of Los Angeles, California as Part LA-1231.

The difference in the appearance of contours plotted in B-L space as compared with the same contours plotted in R-λ space has been illustrated by KELLER and VETTE (1963), and amply demonstrate the need for minimizing the number of ways of presenting results.

It is not possible to express R and λ as functions of B and L in terms of well known functions. ROBERTS (1964) describes a method for obtaining R and λ when given B and L which involves a relatively small amount of computation.

The "invariant latitude" Λ, which was introduced by O'BRIEN (1962), has been found to be a particularly useful parameter for the study of phenomena at high magnetic latitudes. Since Λ is a function of L only, it uniquely labels magnetic shells, but at the same time remains a smooth function of position at high latitudes in contrast to the very non-linear behavior of L. It should be kept in mind that Λ has the properties of a latitude only when $R=1.0$ and that it is almost orthogonal to the magnetic latitude, λ, when λ is greater than 60 degrees.

Lines of force labelled by L greater than 6 R_e are known to be appreciably perturbed by external current systems. As more measurements of these current systems and the resulting magnetic fields become available, it may be possible to produce representations of the field in this region of space which have useful accuracy during magnetically quiet periods. To within the accuracy with which the magnetic field can be computed, B and L will be of course continue to organize measurements of particle fluxes perpendicular to the field providing there are no electric fields and the magnetic field variations are slow compared with the drift periods of the measured particles.

C. METHODS OF COMPUTING L

It has been found that various schemes of interpolation can produce L values with an accuracy of one or two percent but that these methods become unwieldy if it is desired that the maximum error is to be less than 0.2 percent. Accurate computations of L are therefore usually made by the more tedious procedure of direct integration along lines of force to obtain values for I. The computer program for computing L written in the Fortran language at the University of California at San Diego in 1962 has been widely distributed. This program attempts to minimize the number of computations of **B** required by using the first three points on the line of force to predict the location of the fourth point and then using the error in this prediction to establish the optimum arc length to the fifth point. A modified version of this program is now being prepared which incorporates the faster method of computing **B** devised by CAIN *et al.* (1964) and which simplifies the use of different spherical harmonic analyses of the earth's magnetic field.

The use of different approximations of the magnetic field for the computation of B and L could cause apparent differences between two sets of data which are in fact identical. It is desirable therefore that there be a minimum number of different representations in use at any one time. The re-computation of L for a large body of data can be quite expensive, therefore the "standard" magnetic field representation should be changed as infrequently as possible and only when a substantial gain in accuracy can be obtained.

Acknowledgements

The aid and advice of Dr. A. Hassitt, Mr. S. Rearwin, and Dr. W. Fillius in preparing this paper are gratefully acknowledged. This work was supported in part by the National Aeronautics and Space Administration Grant NsG-538.

References

CAIN, J. C., HENDRIKS, S., DANIELS, W. E., and JENSEN, D. C.: 1964, 'Computation of the Main Geomagnetic Field from Spherical Harmonics Expansion', Goddard Space Flight Center, X-611-64-316, Oct. 1964.
FOUGERE, P. F.: 1965, *J. Geophys. Res.* **70,** 2171.
HASSITT, A.: 1965, *J. Geophys. Res.* **70.**
HASSITT, A.: 1965, *J. Geophys. Res.,* **70,** 535–540.
JENSEN, D. C. and CAIN, J. C.: 1962, *J. Geophys. Res.* **67,** 3568.
KAUFMANN, L.: 1965, *J. Geophys. Res.* **70,** 2181.
KELLER, F. L. and VETTE, J. I.: 1963, *J. Geophys. Res.* **68,** 4866.
MCILWAIN, C. E.: 1961, *J. Geophys. Res.* **66,** 3681.
NEWKIRK, L. L. and WALT, M.: 1964, *J. Geophys. Res.* **69,** 1759.
NORTHROP, T. G. and TELLER, E.: 1960, *Phys. Rev.* **117,** 215.
NORTHROP, T. G.: 1963, *Review of Geophysics* **1,** 283–304.
NORTHROP, T. G.: 1963, *Interscience Tracts on Physics and Astronomy* 21.
O'BRIEN, B. J., LAUGHLIN, C. D., VAN ALLEN, J. A., and FRANK, L. A.: 1962, *J. Geophys. Res.* **67,** 1209–1225.
ROBERTS, C. S.: 1964, *J. Geophys. Res.* **69,** 5089–5090.
STONE, E. C.: 1963, *J. Geophys. Res.* **68,** 4157–4166.

Discussion

In response to questions on the applicability of the B-L coordinates at large radial distances, Dr. MCILWAIN stated that B and L will organize the data as far out as the earth's field is ordered. However, to compute B and L one needs to know B all along the line of force. Such computations based on surface fields are accurate for $L < 5$. At large distances in distorted fields one would not expect L to be constant along a line of force. However, B and L should still order the data.

Dr. Charles ROBERTS asked whether "I" might be a better ordering parameter than L for conditions where L was not constant along a field line.

McIlwain replied that "I" and L were equally good at great distance, but since L was preferable near the earth, L should be used throughout for consistancy. Also with L one has some feeling about the location. A large L denotes a position at great distance from the earth. While Dr. McIlwain was confident that the B-L system could eventually be extended to larger distances using a distorted magnetic field, he did not know whether an adequate knowledge of the field was available as yet.

For trapped particle considerations in which the equatorial pitch angle of a particle is of prime importance Dr. Roberts suggested the use of a new coordinate, to be called x, for specifying position on a line of force or on an L-shell. The properties of x, and the motivation behind its use are described in *J. Geophys. Res.* **70** (1965) 2517. Basically, x lables the point along a line of force by the cosine of the equatorial pitch angle that a particle would have if it mirrored at the point of interest. From the mirror equation, x is related to B by the formula $x=(1-B_0/B)^{\frac{1}{2}}$, where B_0 is the magnetic field intensity at the equator. The value $x=0$ always represents the magnetic equator, and surfaces of constant x correspond to surfaces of constant invariant dipole latitude, λ. The properties of x are such that it varies linearly with λ near the magnetic equator, whereas B varies quadratically with λ. For this reason x is a more sensitive coordinate in the vicinity of the magnetic equator.

FUTURE MAGNETIC FIELD FOR B-L SPACE

MARTIN WALT
Lockheed Missiles and Space Company Research Laboratories,
Palo Alto, California

Following this session a small group met to consider the advisability of using the 1965, 99 term expansion magnetic field model in place of the 1962 interim model now in current use. Dr. CAIN stated that based on the lower residuals calculated for the 1965 field, this new model was clearly an improvement over the earlier field. However, in about one year a new and substantially better model would be derived from POGO measurements.

MCILWAIN pointed out that the use of *B-L* values calculated with the 1965 field improved the ordering of his proton data. In analyses using the 1962 field a systematic longitudinal variation of counting rate at constant *B*, *L* was noted. This variation was reduced by a factor of 2 using the 1965 field. However, McIlwain stressed the fact that computation of *L* using the 99 term field would require approximately twice the computing time as for the 48 coefficient one, and unless the experimental data were of very high quality or especially sensitive to magnetic field values, this additional effort was not warrented. Furthermore, the *B-L* computer program now in wide use would have to be replaced to incorporate the 1965 magnetic field, and the improved accuracy resulting from such a change was probably not worth the general confusion and delays which would result, particularly in view of the pending change next year when a better field was available.

FREDEN and others agreed that most radiation belt data did not require the increased accuracy available with the new model, but that data taken at low altitude, where spatial gradients in trapped particle flux were large would need better field models.

In view of these discussions it was felt that formal adaptation of the 1965 field should not be made at this time although groups whose work was particularly sensitive to magnetic field accuracy should use the new or old models as they saw fit.

SESSION II

OBSERVED PARTICLE DISTRIBUTION IN THE NATURAL BELTS (I)

Chairman

DR. WILMOT N. HESS

Goddard Space Flight Center

SOME GENERAL ASPECTS OF GEOMAGNETICALLY TRAPPED RADIATION

J. A. VAN ALLEN

Department of Physics and Astronomy, University of Iowa, Iowa City, Iowa

Abstract. This paper provides a sketch of the physical principles of radiation belts and an abridged summary of the present state of knowledge of the geomagnetically trapped radiation. As such, it is intended as an introduction to detailed papers on specific topics which constitute the remainder of the Bergen conference.

Necessary and Sufficient Conditions for the Existence of a Planetary Radiation Belt

The possibility of the durable confinement (or trapping) of a moving, electrically-charged particle in the field of a static magnetic dipole was discovered by STÖRMER (1907). The rigorous theory of Störmer shows that permanently bounded motion is dynamically possible within, and only within, the bounded region $r_1 < r < r_2$ where r_1 and r_2 specify two surfaces of revolution about the axis of the magnetic dipole given by the following:

$$\frac{r_1}{b} = \frac{\cos^2 \lambda}{-(\gamma/b) + \{(\gamma/b)^2 + \cos^3 \lambda\}^{1/2}} \tag{1}$$

and

$$\frac{r_2}{b} = \frac{\cos^2 \lambda}{-(\gamma/b) + \{(\gamma/b)^2 - \cos^3 \lambda\}^{1/2}}. \tag{2}$$

In Equations (1) and (2) the position of a point is specified by spherical coordinates about the point dipole with the magnetic dipole **M** directed along the negative polar axis. The radius is denoted by r, the latitude by λ, and the longitude by ω. The quantity $b \equiv (ZeM/pc)^{1/2}$ where Ze is the electrical charge on the particle, p is the scalar momentum, and c is the velocity of light. γ is an arbitrary constant of integration of the equations of motion which specifies, essentially, the conditions of injection of the particle into the field. Both b and γ have dimensions of length.

For a dipole of moment equal to that of the earth ($M = 8.06 \times 10^{25}$ gauss cm^3), Table I (cf. VAN ALLEN, 1962) gives maximum possible values of r_2 and the corresponding values of r_1 (both at $\lambda = 0°$) for protons and electrons of the energies shown (1 earth radius $= 6.371 \times 10^8$ cm). The idealized model of Stör-

TABLE I

Störmer trapping boundaries

Maximum r_2 *(earth radii)*	*Corresponding* r_1 *(earth radii)*	*Kinetic energy E* 10^9 eV	
		Proton	*Electron*
1.00	0.41	58.6	59.5
2.00	0.83	14.0	14.9
4.00	1.66	2.90	3.72
6.00	2.49	0.96	1.65
10.0	4.14	0.173	0.60
15.0	6.21	0.036	0.264

mer has a basic intellectual appeal, places firm upper limits on the trapping capability of a dipole field, and guides the understanding of actual situations.

But the real geophysical system is found to have trapping capabilities which are so greatly inferior to those illustrated by Table I as to render the idealized model of little practical utility. The principal reasons for this fact are the following:

(a) Presence of the solid earth.

(b) Presence of the earth's atmosphere, extending as it does with non-negligible density to several earth radii.

(c) Interaction of the earth's magnetic field with hot ionized gas flowing radially outward from the sun through the solar system (the solar wind).

(d) Occurrence of electromagnetic perturbations of terrestrial origin.

Thus, the establishment of *sufficient* conditions for the durable trapping of particles having specified energy and injection conditions has become a dominantly observational undertaking.

The Störmer theory and other simple dynamical theories treat the motion of single, isolated particles, i.e., they neglect the interactions between particles and the magnetic perturbations which are caused by the electrical currents due to motion of the particles themselves. Thus, they yield no information whatever on the maximum intensities or number densities of particles which can be trapped, irrespective of their individual kinetic energies. Some crude guidance on this matter is provided by rudimentary theory of the magnetic confinement of plasma, which considers the diamagnetic effect of gyrating particles in a magnetic field and shows that the quantity

$$\beta = \frac{n\bar{E}}{(B^2/8\pi)} \tag{3}$$

must be always less than unity, and in particular cases, considerably less than

TABLE II

Number densities n and omnidirectional intensities J_0 corresponding to $\beta = 0.1$ in the geomagnetic field at $\lambda = 0°$

		Protons		*Electrons*			
		$E = 100$ keV		$E = 40$ keV		$E = 1.0$ MeV	
r (earth radii)	$B^2/8\pi$ (ergs cm^{-3})	n (cm^{-3})	J_0 (cm^2 sec)$^{-1}$	n (cm^{-3})	J_0 (cm^2 sec)$^{-1}$	n (cm^{-3})	J_0 (cm^2 sec)$^{-1}$
1.5	3.38×10^{-4}	211.	9.2×10^{10}	528.	5.93×10^{12}	21.1	5.96×10^{11}
2.0	6.01×10^{-5}	37.5	1.64×10^{10}	93.8	1.05×10^{12}	3.75	1.06×10^{11}
3.0	5.28×10^{-6}	3.30	1.44×10^{9}	8.25	9.26×10^{10}	0.33	9.32×10^{9}
4.0	9.40×10^{-7}	0.59	2.58×10^{8}	1.48	1.66×10^{10}	0.059	1.67×10^{9}
5.0	2.46×10^{-7}	0.154	6.75×10^{7}	0.385	4.32×10^{9}	0.0154	4.35×10^{8}
7.0	3.27×10^{-8}	0.0204	8.94×10^{6}	0.0510	5.61×10^{8}	0.00204	5.76×10^{7}

unity. In (3), n is the number density of charged particles having average kinetic energy $\bar{E}$ and B is the unperturbed magnetic field strength due to fixed currents. Table II gives some sample numerical values of n and of omnidirectional intensities J_0 in the equatorial plane in the geomagnetic field $(B=M/r^3)$ for three classes of particles, and for $\beta=0.1$.

The Störmer theory may be regarded as placing an absolute lower limit on the magnitude of M for a planet in order that particles of a specified energy can be durably trapped in a specified region. A considerably higher value of this lower limit is required when one considers the influence of the solar wind. The rough criterion for the value which M must have in order to permit trapping of particles of any energy out to a radial distance r is that the outward pressure of the planet's magnetic field $(B^2/8\pi)$ at r must equal or exceed the inward dynamic pressure of the solar wind $(n\,m\,v^2)$. In the foregoing, n is the number density of charged particles of mass m and directed velocity v in the solar wind (Van Allen, Frank, Krimigis, and Hills, 1965). Omitting factors of the order of unity, this criterion gives the maximum radial extent of a trapping region:

$$r_{max} \approx \left(\frac{M^2}{8\pi n m v^2}\right)^{1/6}. \tag{4}$$

An alternative formula in terms of the radius of the planet r_P and the magnetic field at its surface B_P is

$$\frac{r_{max}}{r_P} \approx \left(\frac{B_P^2}{8\pi n m v^2}\right)^{1/6}. \tag{5}$$

For the solar wind, $m=1.67\times10^{-24}$ g and at 1 AU $n\approx5$ cm^{-3} and $v\approx5\times10^7$ cm sec^{-1} (Snyder, Neugebauer, and Rao, 1963). Hence from (5), for the earth ($B_E\sim0.31$ gauss):

$$\frac{r_{max}}{r_E} \approx 7.5,$$

in rough agreement with observation.

In Equations (4) and (5), n is assumed to have an inverse square dependence on heliocentric radial distance and v, to be independent of this distance. A minimum, *necessary* condition for the existence of a planetary radiation belt is that M be sufficiently great that r_{max} exceeds the radial distance to the top of the appreciable atmosphere of the planet (to be defined more quantitatively in a later section).

It is generally believed that the foregoing criterion is also a *sufficient* condition for the presence of durably trapped, energetic particles in the magnetic field of a planet (or of a satellite of a planet). This belief has its foundation in the observational knowledge of the earth system and in the belief that all other planetary bodies of the solar system are subjected to qualitatively similar, external influences as is the earth, viz., the solar wind and solar and galactic cosmic rays (as well as the possibly irrelevant electromagnetic radiations from the sun).

There is no existing theory of a fundamental nature for predicting the detailed structure (particle composition, energy spectra, spatial distribution of intensities, etc.) in a planetary radiation belt, though Equation (3) with $\beta \sim 0.1$ is believed to provide an upper limit on the kinetic energy density of charged particles at any point therein.

To the author's knowledge, the theory of particle trapping in a magnetic field of higher multi-polarity than that of a dipole has not been worked out.

General Nature of the Earth's Radiation Belts

Knowledge of the particle populations in regions around the earth, or radiation belts, has developed almost entirely as an observational undertaking, with little detailed theoretical guidance. Specific observational work has been directed toward establishing:

(a) Sufficient conditions for durable trapping.

(b) Identification of trapped particles and spatial, temporal, and spectral characteristics of each identified species.

(c) Sources of observed particles.

(d) Effective trapping lifetimes.

(e) Losses of particles.

(f) Diffusion of particles.

(g) Acceleration processes.

There are, in addition, a large number of related matters such as the configuration of the geomagnetic field, heating and ionization of the atmosphere, and the relationship of trapped particles to the aurorae.

To a first crude approximation, all durably trapped particles in the geomagnetic field lie within a toroidal region bounded by two surfaces of revolution about its magnetic axis, the inner one a spherical surface at an altitude of ~ 700 km above the solid earth (VAN ALLEN, MCILWAIN, and LUDWIG, 1959; YOSHIDA, LUDWIG, and VAN ALLEN, 1960) and the outer one a magnetic shell of dipolar lines of force emanating from the earth at $\lambda \sim 70°$ and crossing the magnetic equator at $r/r_E \sim 8$ (FRANK, VAN ALLEN, and MACAGNO, 1963; FREEMAN, 1964; FRANK, 1965).

The nature of the inner boundary is reasonably well understood in terms of simple energy loss in the gaseous atmosphere of the earth (RAY, 1960; WALT, 1964). The general plausibility of this statement can be checked by simply determining the length of time which a specified particle can travel in a gas of appropriate density and comparing the result with a value of the order of 10^7 sec (VAN ALLEN, 1964). If the ratio is very much less than one, then the corresponding intensity in a quasi-stationary state will be very much less than the maximum value of that component anywhere in the radiation belt.

The general nature of the outer boundary of the durable trapping region is understood in terms of the criterion expressed by Equation (5).

Particle Populations

A complete observational knowledge of the particle populations in the radiation belt in a time-stationary state at a given moment may be shown (VAN ALLEN, 1962) to be equivalent to knowing

$$j_i(L, \alpha_0, E). \tag{6}$$

In Expression (6), j_i is the differential unidirectional intensity of the ith species of particle (i = proton, electron, alpha particle, etc.) at equatorial pitch angle α_0 in the kinetic energy range dE at E, and L specifies the magnetic shell (MCILWAIN, 1961).

There are now reasonably comprehensive bodies of observations of j_i (or the equivalent) for electrons of 40 keV$<E_e<$2 MeV (FRANK, VAN ALLEN, and HILLS, 1964; ARMSTRONG, 1965); for protons 100 keV$<E_p<$4.5 MeV and electrons 10 keV$<E_e<$100 keV (DAVIS and WILLIAMSON, 1963); for protons 10 MeV$<E_p<$200 MeV (FREDEN, BLAKE, and PAULIKAS, 1965); and for protons 1.1 MeV$<E_p<$63 MeV and electrons 150 keV$<E_e<$1.0 MeV (MCILWAIN, 1963). Recent work is described in papers by principal workers in the field at the present conference.

The identification of trapped alpha particles $E_\alpha \sim 2$ MeV has been reported (VAN ALLEN and KRIMIGIS, 1965), though only limited and preliminary information on their intensities and spatial distribution is available at this time.

There continues to be a glaring deficiency in the knowledge of protons $E_p<100$ keV and electrons $E_e<40$ keV.

General Energetic Considerations

The total magnetostatic energy in the field of a dipole of moment M external to a sphere of radius a centered on the dipole is

$$W_a = \iiint (B^2/8\pi)\,\mathrm{d}V = \frac{M^2}{3a^3} \tag{7}$$

(CHAPMAN and BARTELS, 1940).

Using (7) and taking a to be the radius of the solid earth, 6.371×10^8 cm, and $M = 8.06 \times 10^{25}$ gauss cm^3 the magnetostatic energy of the geomagnetic field external to the earth is found to be

$$W_a = 8.37 \times 10^{24}\,\text{ergs}. \tag{8}$$

It has been argued by DESSLER and VESTINE (1960) on the basis of the maximum admissible contribution of external currents to the geomagnetic field that an upper limit on the total kinetic energy of geomagnetically trapped particles is $\sim 6 \times 10^{22}$ ergs, or less than one percent of the external field energy. The observed distributions of particles do not exceed this limit (HOFFMAN and BRACKEN, 1965), which may be written as the volume average $\langle \beta \rangle = 0.01$.

More detailed examination of this matter is of interest. The magnetostatic energy between magnetic shells differing in L value by $\mathrm{d}L$ at L (measured in units of a) external to a sphere of radius a is found to be

$$\frac{\mathrm{d}W_a}{\mathrm{d}L} = \left(\frac{M^2}{2a^3}\right)\frac{(L-1)^{1/2}}{L^{5/2}}. \tag{9}$$

For the earth:

$$\frac{\mathrm{d}W_a}{\mathrm{d}L} = 1.26 \times 10^{25}\,\frac{(L-1)^{1/2}}{L^{5/2}}\,\frac{\text{ergs}}{\text{unit}\,L}. \tag{10}$$

The quantity $\mathrm{d}W_a/\mathrm{d}L$ has a maximum value at $L = 1.25$.

The area of an annular ring in the equatorial plane $\mathrm{d}A_0$ bounded by two circles of radius $a\,L$ and $a\,(L + \mathrm{d}L)$ is given by

$$\frac{\mathrm{d}A_0}{\mathrm{d}L} = 2\pi a^2 L \tag{11}$$

and at $r = a$ the area of a ring bounded by the same magnetic shells and perpendicular to B, $\mathrm{d}A_a$, is smaller by the factor:

$$\frac{\mathrm{d}A_a}{\mathrm{d}A_0} = \frac{B_0}{B_a} = \frac{1}{L^3(4 - 3L^{-1})^{1/2}}. \tag{12}$$

Thus, combining (9), (11), and (12),

$$\frac{\mathrm{d}W_a}{\mathrm{d}A_a} = \frac{M^2}{8\pi a^5 L}(4L^2 - 7L + 3)^{1/2} \tag{13}$$

(attributed equally to both hemispheres)

or in numerical form

$$\frac{\mathrm{d}W_a}{\mathrm{d}A_a} = 2.47 \times 10^6 \frac{(4L^2 - 7L + 3)^{1/2}}{L} \frac{\text{ergs}}{\text{cm}^2}. \tag{14}$$

The following is a numerical example of interest. At $L=5$ (or $\lambda_a=63.4°$), $\mathrm{d}W_a/\mathrm{d}A_a=4.07 \times 10^6$ ergs/cm^2. Thus if this magnetic shell were populated with charged particles to an average $\beta=0.01$ and the particles were subjected to a pure precipitation process involving only an appropriate change of their pitch angles, an average precipitation of 1 erg/cm^2 sec in both hemispheres could be maintained for 4×10^4 seconds or about 11 hours. This example provides a graphic illustration of the limited storage capability of the geomagnetic field.

Another formula of occasional use is the one for volume:

$$\frac{\mathrm{d}V_a}{\mathrm{d}L} = \frac{3}{35}\left(\frac{4\pi a^3}{3}\right)\frac{(L-1)^{1/2}}{L^{3/2}}(5 + 6L + 8L^2 + 16L^3). \tag{15}$$

At $L=5$,

$$\frac{\mathrm{d}V_a}{\mathrm{d}L} = 3.71 \times 10^{28} \frac{\text{cm}^3}{\text{unit } L}.$$

The greatest known "loading" of the geomagnetic field by energetic particles, as measured by β, at low L values was due to the Starfish nuclear burst of 9 July 1962. After the initial transients of the first few hours there was an omnidirectional intensity of 1.5×10^9 (cm^2 sec)$^{-1}$ of electrons (having a fissionlike spectrum) at $L=1.2$ and $\lambda=0°$. The corresponding value of β is $\sim 10^{-4}$. The natural loading of the field here is several orders of magnitude less.

By contrast, the natural loading of the field in the outer zone approaches values which suggest that the population there may be limited by plasma instabilities. The entire population of the radiation belts may be judged unstable against cooperative processes on the simple ground that the velocity distribution is anisotropic everywhere. The only question is the time rate of growth of instabilities and this question is not yet answered within a proper theoretical framework.

For example, at $L=4$ on the equator, values of J_0 for electrons $E_e > 40$ keV as great as 3×10^8 (cm^2 sec)$^{-1}$ have been observed (FRANK, VAN ALLEN, and HILLS, 1964) and values of J_0 for protons $E_p > 100$ keV of the order of 3×10^7 (cm^2 sec)$^{-1}$ are reported by Davis at this conference. The corresponding value of β is ~ 0.03, with the protons making the principal contribution. It is important to note that the true value, including particles below present experimental thresholds, is, of course, higher than that just estimated and may reasonably be as much as 0.10.

It is perhaps remarkable that the time variations of the population of outer zone electrons are much greater than those for protons despite the fact that the loading of the field is due principally to the latter. This observed fact may be taken to signify that the population of electrons is more closely coupled to the ambient medium, both electrostatically and electromagnetically, and hence exhibits a much more rapid growth of trapping instabilities than does the proton population.

Characteristic Time Scales

In principle, one would wish to remove all energetic particles from the earth's magnetic field (as by "de-gaussing" the earth for a brief period), then to inject known populations of known particles and to follow their subsequent time histories in observational detail. The closest practical approach to this ideal thus far has been by the injection of populations of fission-product decay-electrons from atomic bursts (see pertinent papers of this conference). This approach has yielded a large body of knowledge on the gross physical dynamics of populations of energetic electrons.

It is found that for $L<1.25$, the time histories of such particles can be well understood in terms of simple Coulomb scattering and energy loss in the residual atmosphere. Toward higher L values, other unknown loss processes dominate, and observed lifetimes beyond $L\sim 1.8$ become many orders of magnitude less than those attributable to atmospheric losses.

Extended studies of the natural population of trapped particles yield results of a generally comparable nature. The following is an abridged summary of characteristic lifetimes (VAN ALLEN and LIN, 1960; PIZZELLA, MCILWAIN, and VAN ALLEN, 1962; FORBUSH, PIZZELLA, and VENKATESAN, 1962; FRANK, VAN ALLEN, and HILLS, 1964; MCILWAIN, 1963; DAVIS, 1966):

(a) *For* $1.2<L<1.3$: Protons $E_p>20$ MeV, ~ 1 year;

(b) *For* $1.5<L<2.5$: Electrons $E_e>1$ MeV, months to days;

(c) *For* $L>2.5$: Electrons $E_e>1$ MeV, hours to weeks;

(d) *For* $L>2.5$: Electrons $E_e>40$ keV, hours to weeks;

(e) *For* $L>2.0$: Protons $E_p>100$ keV, markedly stable with small scale fluctuation in times of the order of days.

In the outer zone, $L\gtrsim 2.0$, fluctuations of particle populations are closely associated with geomagnetic disturbances, thus strongly suggesting a common cause – specifically fluctuations in one or more of the characteristic parameters of the solar wind.

Sources of Observed Particles

The evidence cited in the preceding section leaves little doubt that the dominant source of energy of the energetic particles in the radiation belts (as well as of aurorae and magnetic storms) is the solar wind. The typical power flux of the solar wind at the earth's orbit (SNYDER, NEUGEBAUER, and RAO, 1963) is 0.1 erg/cm^2 sec. The total geomagnetic system presents a circular frontal area of radius $\sim 12a$ and is, therefore, subjected to a typical power flow of $\sim 2 \times 10^{20}$ ergs/sec. There are fluctuations of approximately one order of magnitude, both positively and negatively, about this value.

The power flow in the solar wind is estimated to be adequate by a factor of the order of one hundred to power the sum of all known phenomena in the magnetospheric system.

The physical mechanisms for producing the known, natural populations of charged particles in the geomagnetic field are another matter. The current state of work on mechanisms of particle injection, diffusion, energy loss and energy gain, and precipitation is represented by other papers in this conference.

It now appears that only the protons having $E_p > 20$ MeV in the region $L \lesssim 1.8$ are not fundamentally attributable to the solar wind. These particles are probably due to the radioactive decay of neutrons which have been produced by galactic cosmic rays in the earth's atmosphere (LENCHEK and SINGER, 1962). Other weak sources of particles may be (a) solar cosmic rays either by direct injection and capture or by neutron production in the atmosphere; (b) energization of particles by the earth's rotation; (c) a neutral component in the solar wind (AKASOFU and MCILWAIN, 1963); and (d) electromagnetic coupling with whistlers and other ionospheric disturbances of terrestrial origin.

References

AKASOFU, S.-I., and MCILWAIN, C. E.: 1963, *Amer. Geophys. Union* **44,** 883.
ARMSTRONG, T.: 1965, *J. Geophys. Res.* **70,** 2077–2110.
DAVIS, L. R. and WILLIAMSON, J. M.: 1966, this volume, p. 215.
DAVIS, L. R. and WILLIAMSON, J. M.: 1963, Third International Space Science Symposium and COSPAR Plenary Meeting, Washington, D. C., April 30–May 9, 1962.
DESSLER, A. J., and VESTINE, E. H.: 1960, *J. Geophys. Res.* **65,** 1069–1071.
FORBUSH, S. E., PIZZELLA, G., and VENKATESAN, D.: 1962, *J. Geophys. Res.* **67,** 3651–3668.
FRANK, L. A.: 1965, *J. Geophys. Res.* **70,** 1593–1626.
FRANK, L. A., VAN ALLEN, J. A., and HILLS, H. K.: 1964, *J. Geophys. Res.* **69,** 2171–2192.
FRANK, L. A., VAN ALLEN, J. A., and MACAGNO, E.: 1963, *J. Geophys. Res.* **68,** 3543–3554.
FREDEN, S. C., BLAKE, J. B., and PAULIKAS, G. A.: 1965, *J. Geophys. Res.* **70,** 3113–3116.
FREEMAN, John W., Jr.: 1964, *J. Geophys. Res.* **69,** 1691–1723.
HOFFMAN, R. A. and BRACKEN, P. A.: 1965, *J. Geophys. Res.* **70,** 3541–3556.
LENCHEK, A. M. and SINGER, S. F.: 1962, *J. Geophys. Res.* **67,** 1263–1287.
MCILWAIN, C. E.: 1961, *J. Geophys. Res.* **66,** 3681–3691.
MCILWAIN, C. E.: 1963, *Science* **142,** 355–361.
PIZZELLA, G., MCILWAIN, C. E., and VAN ALLEN, J. A.: 1962, *J. Geophys. Res.* **67,** 1235–1253.
RAY, E. C.: 1960, *J. Geophys. Res.* **65,** 1125–1134.
SNYDER, Conway W., NEUGEBAUER, Marcia, and RAO, U. R.: 1963, *J. Geophys. Res.* **68,** 6361–6370.
STÖRMER, C.: 1907, *Arch. Sci. phys. nat.* **24,** 317–364.

VAN ALLEN, J. A.: 1962, *Trans. IAU* **IXB** 99–136.
VAN ALLEN, J. A.: 1964, *Nature* **203,** 1006–1007.
VAN ALLEN, J. A., FRANK, L. A., KRIMIGIS, S. M., and HILLS, H. K.: 1965, *Science* **149,** 1228–1233 and 1557.
VAN ALLEN, J. A. and KRIMIGIS, S. M.: 1965, *Trans. Am. Geophys. Union* **46,** 140.
VAN ALLEN, James A., and LIN, Wei Ching: 1960, *J. Geophys. Res.* **65**, 2998–3003.
VAN ALLEN, J. A., MCILWAIN, C. E., and LUDWIG, G. H.: *J. Geophys. Res.* **64,** 271–286.
WALT, Martin: 1964, *J. Geophys. Res.* **69,** 3947–3958.
YOSHIDA, S., LUDWIG, G. H., and VAN ALLEN, J. A.: *J. Geophys. Res.* **65,** 807–813.

SOME COSMIC RAY AND RADIATION BELT OBSERVATIONS BASED ON DATA FROM THE ANTON 302 G-M COUNTER IN ARIEL I

H. ELLIOT
Imperial College, London

Introduction

In this paper I shall discuss some of the data obtained by means of a shielded Anton 302 counter carried in Ariel I. In the direction in which the amount of absorbing material for this detector was a minimum the counter was effectively shielded by 0.73 g cm^{-2} Eccofoam, 1.13 g cm^{-2} lead, 1.01 g cm^{-2} aluminium, and 0.4 g cm^{-2} iron. Electrons with energies of 7.4 and 10.5 MeV have probabilities of direct penetration of 10 and 50% respectively. The corresponding threshold for protons is $\sim$ 50 MeV. The counter also responds, but with low efficiency of course, to bremsstrahlung from electrons stopping in the shielding material and to γ-rays. In the latter case the efficiency is about 0.4%. The product of counter efficiency for charged particles and the omnidirectional geometrical factor is $\sim$0.5 cm^2.

Ariel I was launched on April 26 1962 into an orbit with inclination 54°, apogee 1200 km and perigee 400 km. Included in the payload was a tape recorder capable of storing data over a complete orbit and playing it back on command. Tape recorded data from the counter are available from shortly after launch until three days after the Starfish event on July 9 1962, when operation became intermittent following solar cell damage.

When the spacecraft was outside the boundaries of the radiation belt, the counting rate of the G-M counter was expected to be due predominantly to cosmic ray protons and charged particle albedo. If an adequate correction can be made for the albedo contribution it is possible, in principle, to derive an integral rigidity spectrum for the cosmic ray protons from the measured latitude-dependence of the counting rate. It was of interest, therefore, to compare the spectrum deduced from the G-M counter data with that of the Čerenkov counter also carried in Ariel I – which was used to detect cosmic rays with $Z > 2$ and had negligible response to albedo.

Variation of Count Rate With Latitude and Longitude

Figure 1 (ENGEL, 1964) shows the flux values derived from the G-M count rate

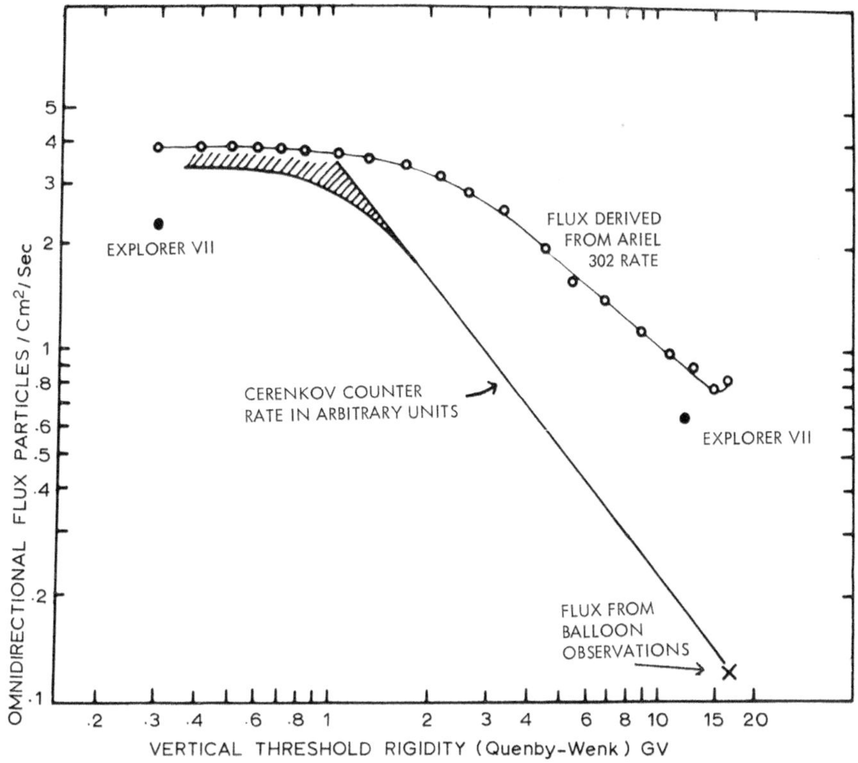

Fig. 1. Flux values derived from the Anton 302 counter in Ariel I plotted as a function of threshold rigidity. Two points from the Anton 302 counter in Explorer VII and the heavy particle spectrum derived from the Čerenkov counter in Ariel I are shown for comparison.

for altitudes less than 450 km plotted as a function of Quenby-Wenk threshold rigidity together with the rate for the cerenkov counter and two points from the Anton 112 counter in Explorer VII (Lin *et al.* 1963). A rudimentary correction has been applied to the Ariel data to take account of the contribution to the counting rate by electrons of the outer radiation belt. This correction was derived by making use of the altitude dependence of the radiation belt electrons and has been applied in the rigidity interval 0.5 to 2.0 GV. The contribution to the counting rate from this source is negligible in other rigidity intervals.

These data imply that the slope of the integral spectrum of the cosmic ray protons expressed as a power law would be about -0.74 in the region 5 to 15 GV compared with -1.2 to -1.5 for the heavy particles. Lin *et al.* (1963) obtained a value -0.83 from their data in fair agreement with that obtained here. Furthermore, the absolute values of the flux are not widely different bearing in mind the uncertainty in the geometrical factor of the detectors and the different epochs at which the measurements were made. The Explorer VII data were mostly obtained in 1960 whereas the Ariel data are more nearly representative of sunspot minimum conditions when the cosmic ray intensity is certainly higher.

Although the Explorer VII and Ariel results are broadly similar there does appear to be a real discrepancy between the flux values measured by G-M counters in both Explorer VII and Ariel, when near the equator, and those that would be expected on the basis of values deduced from balloon flights. The extent of this discrepancy is shown in Table I.

TABLE I

Collected Total Flux values at top of Atmosphere and near Equator

Approx. Threshold Rigidity (GV)	*Total Flux Particles/ m^2/sterad /sec*	*Date of Observation*	*Observer*	*Remarks*
16	170*	1953	RAO *et al.* (1953)	10 cm absorber, corrected for albedo
15	280 ± 40	1949	VAN ALLEN and GANGNES (1950)	3 gm/cm^2 absorber, rocket flight to 100 km, no albedo correction
16	200	1949	WINCKLER *et al.* (1950)	3 cm Pb absorber, Telescope with variable zenith
16	105 ± 12*	1957	MCDONALD (1958)	Lower Limit after albedo correction
17	126 ± 11*	1957	MCDONALD and WEBBER (1959)	No correction for reentrant albedo
16	135 ± 15*	1957	MCDONALD (1958)	α-particle flux × 7.5
16	127 ± 8*		WEBBER (1958)	α-particle flux × 7.5
12 (550–1100 km alt.)	630 ± 30	1959–61	LIN *et al.* (1963)	Explorer 7 geiger calculated from omnidirectional flux
15 (400 km alt.)	770 ± 30	1962	Ariel 302 geiger, calculated from omnidirectional flux	

* Results from vertical telescopes carried in balloons

In addition to the latitude dependence the cosmic ray flux is also, of course, longitude dependent, because of the departure of the geomagnetic field from strict axial symmetry. Accordingly, the longitude dependence of the Anton 302 count rate has also been investigated by Engel for the equatorial region. Figure 2 shows

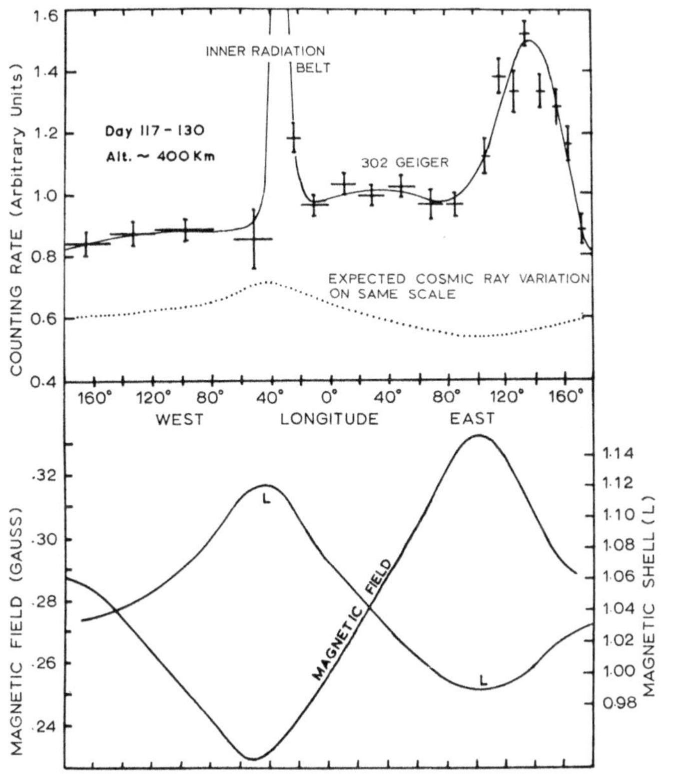

Fig. 2. Upper section shows the longitude dependence of the Anton 302 counting rate at constant altitude together with the variation in counting rate to be expected for cosmic rays alone. The lower section shows the corresponding magnetic field strength and L-value also at an altitude of 400 km. The data are for the period Day 117 to 130.

the rate plotted as a function of longitude at constant altitude of ~400 km. Also shown are the values of the B and L coordinates of the satellite. The high counting rate near 20°W is due to the radiation belt dippings below 400 km over the Atlantic. Instead of a minimum count rate in the neighbourhood of 100°E which would be expected, there is a second, and unexpected, maximum.

Although this maximum is much less marked than that due to the radiation belt it is nevertheless quite unmistakable. It occurs near the position where the lower boundary of the radiation reaches its maximum altitude.

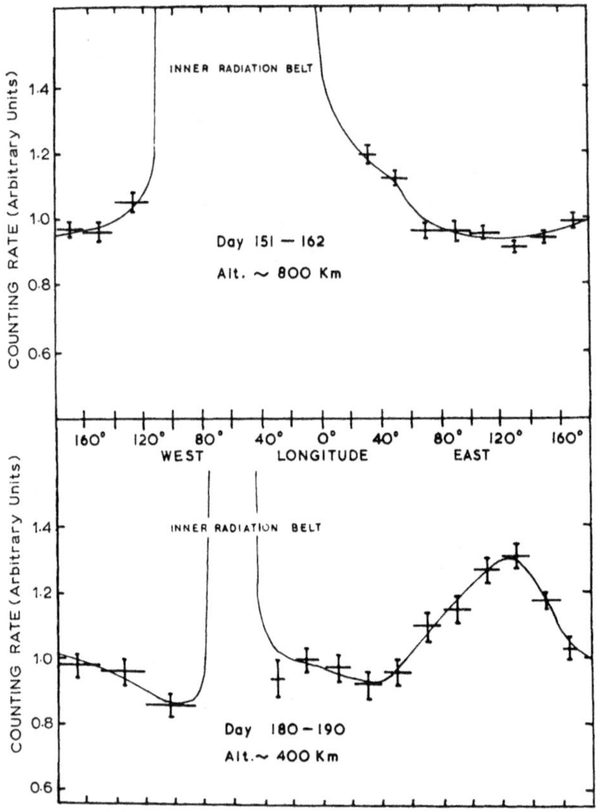

Fig. 3. Upper section shows the longitude dependence of the counting rate at 800 km altitude for the period Day 151 to 162. Data shown in the lower section are for 400 km altitude and the period Day 180 to 190.

Figure 3 shows two further data plots at different epochs, one for an altitude of 800 km and the other for an altitude of 400 km as before. At 800 km the peak due to the inner radiation belt is much broader than at 400 km as would be expected and at this height the count rate has a minimum in the neighbourhood of 100–120°E, there being no second maximum corresponding to that observed at 400 km. The data at 400 km are seen to be very similar during the two epochs corresponding to Days 117 to 130 and Days 180 to 190. This consistency indicates that the second maximum in the region 100° to 120°E is a permanent, or at any rate, a long lived feature of the data.

In Figure 4 the counting rate has been plotted as a function of altitude for the region of the second maximum. The counting rate at first decreases with increasing altitude and then increases sharply as the satellite enters the radiation belt in the region of 1100–1200 km. This variation with altitude is consistent

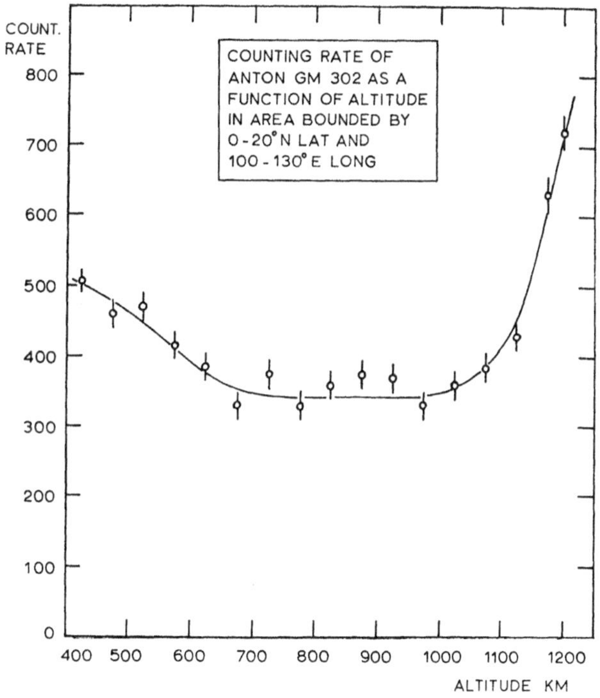

Fig. 4. Altitude dependence of the Anton 302 counting rate in the region of enhanced counting rate.

with the difference between the longitude plots for 400 km and 800 km shown in the two previous diagrams. The form of the altitude dependence is clearly inconsistent with an explanation of the second maximum either in terms of primary cosmic rays or of leakage downward of particles from the inner radiation belt. In either case the count rate would be expected to increase continuously with altitude.

Figure 5 shows the projection on a world map of four typical orbits which include segments passing across the geographical region, shown shaded, where the second maximum is observed. The enhanced counting rate in this region, it will be remembered, is observed at altitudes in the neighbourhood of 400 km which is close to perigee. Apogee for these orbits lies over the general area of the South Atlantic and Brazil and consequently it is on these orbits that the satellite is exposed to maximum inner belt proton fluxes. It therefore seemed that it might

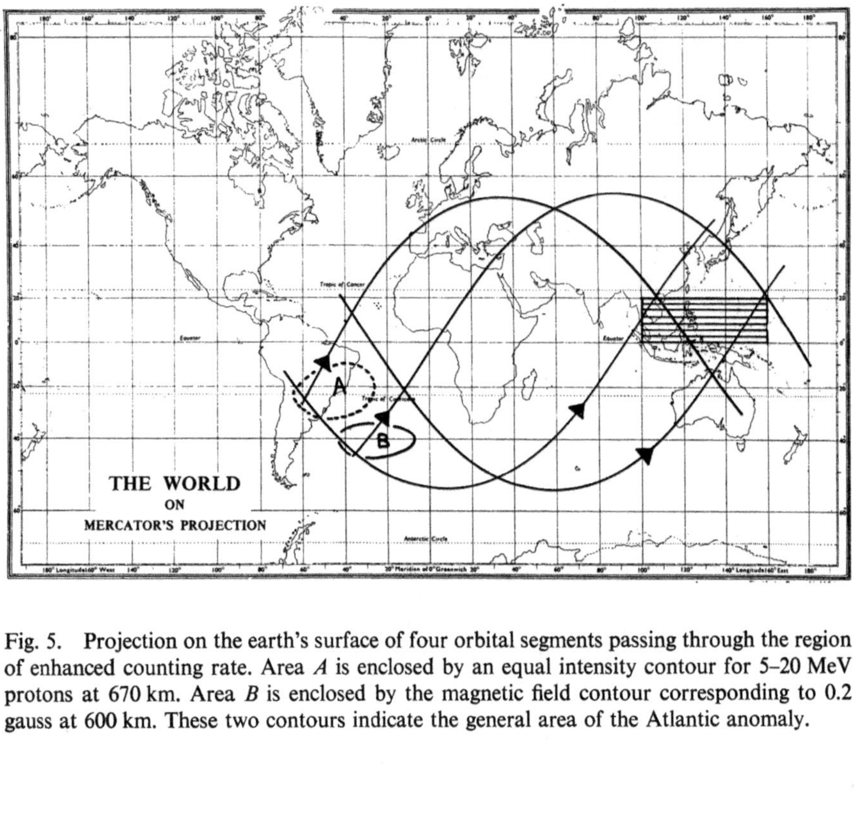

Fig. 5. Projection on the earth's surface of four orbital segments passing through the region of enhanced counting rate. Area *A* is enclosed by an equal intensity contour for 5–20 MeV protons at 670 km. Area *B* is enclosed by the magnetic field contour corresponding to 0.2 gauss at 600 km. These two contours indicate the general area of the Atlantic anomaly.

be possible to explain the enhanced counting rate in terms of the radioactive decay of spallation products in the satellite resulting from inner belt proton bombardment. The altitude dependence of the counting rate is readily accounted for on this explanation since increasing altitude of the satellite in the shaded region is associated with decreasing altitude, and therefore, lower flux values in the Atlantic anomaly region (i.e., when apogee is over the shaded area, perigee is in the Atlantic anomaly).

In an orbit of the kind shown in Figure 5 the total flux of radiation belt protons with energy greater than 50 MeV encountered by the satellite is $\sim 1.5 \times 10^6/\text{cm}^2$ for the complete orbit. Order of magnitude estimates indicate that this would be sufficient to account for the observed increase in equatorial counting rate.

In order to make a direct test of this explanation a spare detector was irradiated in the 150 MeV proton beam of the Harwell cyclotron by Dr Hynds and Dr Durney so as to produce an integrated flux through the counter of 4×10^6 protons/cm^2, which is about three times greater than the dose of 50 MeV protons in the radiation belt. Following this exposure the counting rate of the Anton 302 counter was observed to be far above the normal background rate and this enhanced rate diminished with time becoming roughly exponential after ~ 30 minutes. This exponential fall corresponded to a half-life of ~ 25 minutes. The enhanced rate was shown to arise from the lead shield surrounding the counter and was evidently due to the decay of spallation products in the lead. After a period of 50 minutes, corresponding roughly to the time the satellite takes to get from apogee to perigee, the enhancement in counting rate had diminished to ~ 1 count per second. This is to be compared with the rate of 0.1 to 0.2 counts per second observed in the second equatorial maximum.

We may conclude, therefore, that the equatorial counting rate of any lead-shielded G-M counter carried in a satellite is affected by exposure to the protons of the inner radiation belt and is likely to be appreciably enhanced over that due to cosmic rays alone. This would seem to account for the observed discrepancy between the flux values obtained from balloons and those measured in Explorer VII and Ariel. This enhanced rate is, of course, less evident at high latitudes where the cosmic ray flux is correspondingly higher. In the equatorial region it is most noticeable on those orbits where the fluxes encountered in the radiation belt are highest.

Intensity-Time dependence of the Proton Flux in the Inner Radiation Belt

Data from the Anton 302 counter have been used to study the dependence on time of the proton flux close to the lower boundary of the radiation belt. These observations relate to protons with energy greater than 50 MeV – the penetration threshold for the counter. Particle fluxes in the radiation belt are usually expressed in terms of the *B-L* co-ordinate system of McIlwain but here a preliminary analysis has been carried out by plotting the Anton 302 counting rate as a function of *B* only, using data from a short segment of the orbit in the neighbourhood of minimum *L* close to the geomagnetic equatorial plane. In the case of a dipole field *B* and *L* are directly related in the equatorial plane and consequently one co-ordinate would suffice to specify the flux in this plane for such a field. That this is very nearly true for the geomagnetic field also is borne out by Figure 6 where the counting rate data are seen to be well ordered in a plot of this kind which shows very clearly the lower profile of the radiation belt.

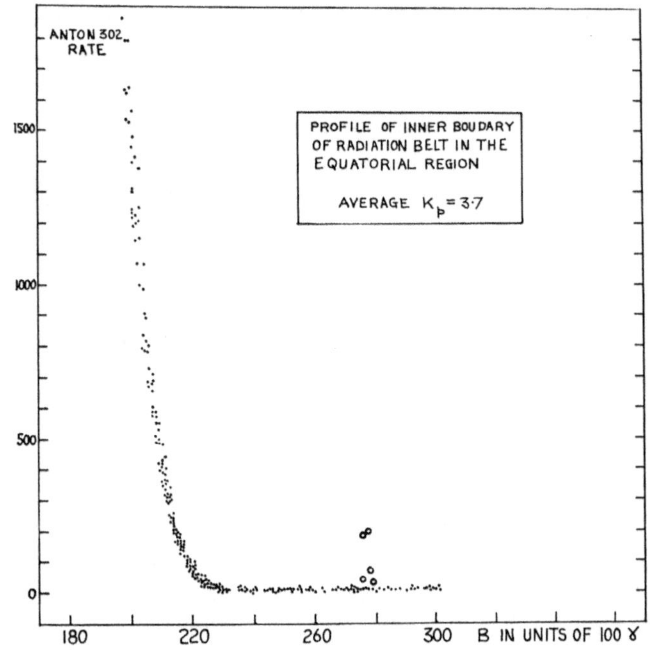

Fig. 6. Anton 302 counting rate plotted as a function of magnetic field strength B in the neighbourhood of the magnetic equator. The counting rate is expressed in counts per interval of 27 sec. Open circles represent anomalously high counting rates. The data are for a group of days with average daily total K_p of 3.7.

These data are for magnetically quiet days with an average daily total K_p of 3.7. Figure 7 shows comparable data for the most disturbed days during the period of observation with an average total daily K_p of 25.3. There is no discernible difference between Intensity vB profiles for these two groups of days. The points plotted as open circles are anomalous in that they represent counting rates occasionally observed that were far above the background rate due to cosmic rays and the induced radio activity discussed above. These anomalous points will be considered further below.

Figure 8 shows the counting rate plotted on a day to day basis for a narrow range of L after correction to a constant value of B. Shown on the same diagram are the corresponding magnetic character figures A_p and the atmospheric density, ρ_s at an altitude of 750 km. It will be seen that the counting rate remains

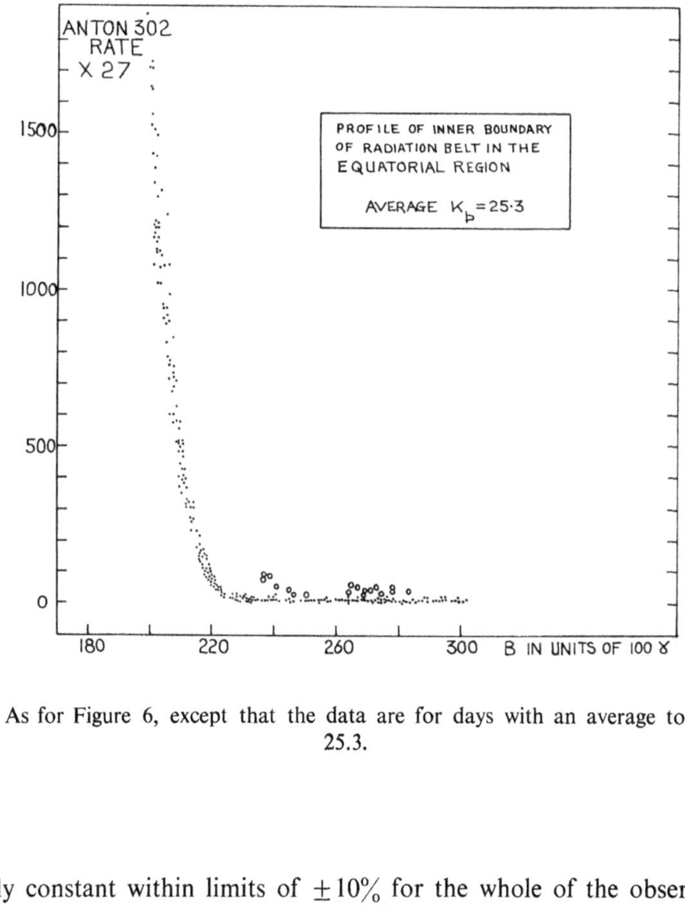

Fig. 7. As for Figure 6, except that the data are for days with an average total K_p of 25.3.

virtually constant within limits of $\pm 10\%$ for the whole of the observational period. The variations within these limits probably cannot be regarded as significant. There is certainly no correlation between the observed rate and either A_p or the atmospheric density ρ_s for the period under consideration. It must be emphasised, however, that there were no major magnetic disturbances or flares of importance greater than class 2 during this period. The high degree of constancy in the intensity of the energetic protons at low L-values apparent from these data is in line with the finding of other workers. (See for example PIZZELLA *et al.* 1962 and ZMUDA *et al.*, 1965.)

The anomalous values of counting rate represented by the open circles in Figures 6 and 7 occur well below the lower boundary of the radiation belt and Figure 9 shows a day to day plot of the counting rate for $1.0 \leqslant L \leqslant 1.1$. In this diagram the anomalous points stand out very clearly against the constant cos-

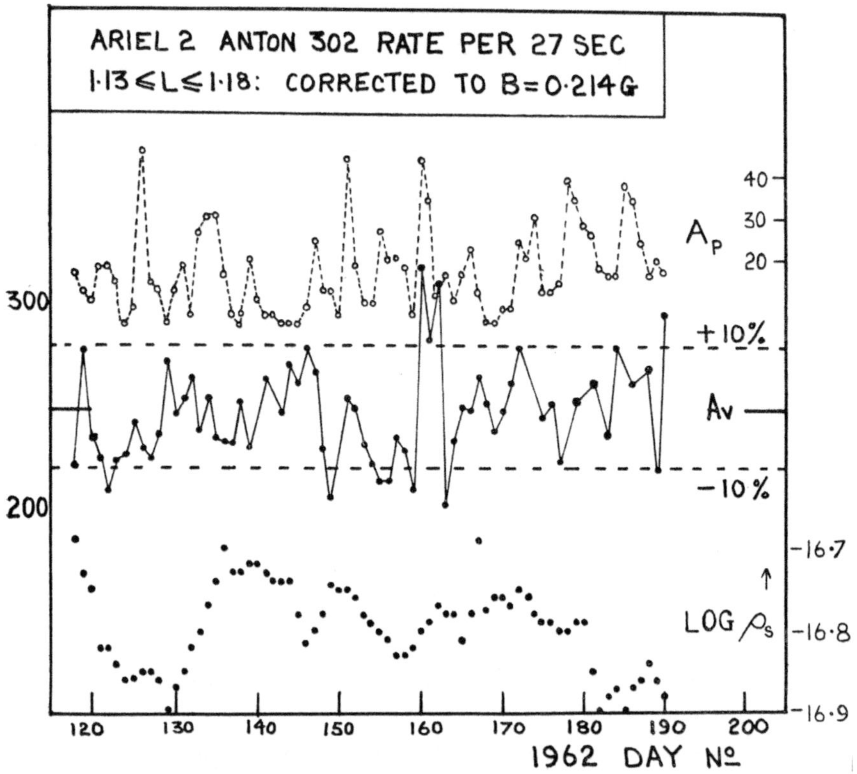

Fig. 8. (———) Anton 302 counting rate for inner belt protons.
(— — —) Magnetic character figure A_p.
(.........) Atmospheric density at an altitude of 750 km.
The counting rate is seen to remain constant within $\pm$ 10% or so during the whole period of observation and there is no significant correlation with either A_p or atmospheric density *ps*.

mic ray background and are seen to occur on six days only out of the total observation period. The dates are 4 May, 10, 12, 27, 28 and 30 June 1962. With the exception of 28 June, these dates correspond exactly to the dates of D.O.D. atmospheric nuclear explosions in the megaton range carried out in the neighbourhood of Christmas Island. This correspondence clearly indicates that the

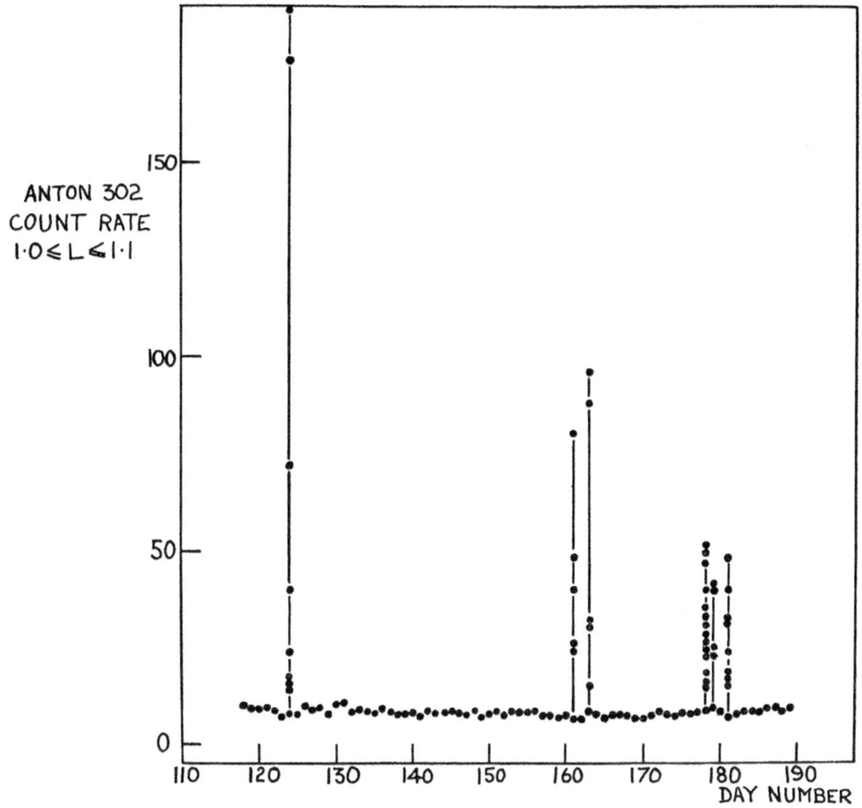

Fig. 9. Day to day plot of the Anton 302 counting rate close to the geomagnetic equator at altitudes below the radiation belt. The anomalous spikes correspond to the data points represented by open circles in Figure 6 and 7 and are due to D.O.D. atmospheric nuclear tests over Christmas Island.

enhanced counting rates are a direct consequence of these nuclear tests. The high values observed on 28 June were evidently due to the explosion on the 27th.

Long-Term Intensity Variation in the Inner Radiation Belt

It is now well established that the energetic proton intensity at low L-values is not subject to the rapid and violent fluctuations which characterise the outer belt electron population. Instead, we have a situation in which the proton intensity appears to remain constant, within 10% or so, at any rate, over periods up to a year or more.

This constancy is to be expected if we have a fixed source strength and long trapping times. The former is provided and the latter is required if cosmic ray neutron albedo is the dominant source of these particles. Consequently the stability of the inner belt proton fluxes has sometimes been advanced as evidence in support of the neutron albedo hypothesis. On this hypothesis the dominating loss process for the protons is atmospheric absorption giving trapping times of the order of years. The atmospheric density at the relevant altitudes shows a large variation over the solar cycle and it follows, therefore, that there should be an appreciable variation in the proton fluxes from sunspot minimum to sunspot maximum, maximum intensity occurring at sunspot minimum.

Detailed calculations of the magnitude of this variation have been made by BLANCHARD and HESS (1964) and these authors, and others, have compared various experimental data with the results of these calculations. Unfortunately these comparisons have hitherto been inconclusive. The difficulty has been that data from a particular detector have rarely been available for more than a year and in no case has a single detector been in operation for an appreciable part of the solar cycle. It has therefore been necessary to resort to comparison of flux values derived from totally different types of detector and this is not too satisfactory because of uncertainties in the geometrical factors. BLANCHARD and HESS (1964) and VERNOV *et al.* (1964), for instance, have concluded that the evidence supports the neutron albedo hypothesis, whereas WALT and MACDONALD (1964) and PIEPER *et al.* (1965) believe it to be inconclusive.

The satellite Explorer IV carried an Anton 302 counter with a not too dissimilar amount of shielding to that in Ariel I. The proton penetration threshold was $\sim$43 MeV compared with 50 MeV for Ariel. Explorer IV data on the inner belt protons were acquired during the period July to September 1958 close to sunspot maximum, whereas the Ariel I data are for the period April to July 1962, not far from sunspot minimum. Comparison of these two sets of data, obtained with closely similar detectors at widely separated points on the solar cycle, together with the calculations of Blanchard and Hess, might be expected to provide a conclusive test of the neutron albedo hypothesis.

Figure 10 shows a plot of the Anton 302 count rate for $L = 1.25$ as a function of B. The data points are shown as vertical bars, the lower end of the bar being the observed rate and the upper end the value obtained after making a correction for the different penetration threshold for the Ariel detector (50 MeV) relative to that for Explorer IV (43 MeV). This correction has been made using a differential energy spectrum of the form $E^{-1.7}$ and is seen to be quite small. Also shown are the corresponding data for Explorer IV which are displayed for clarity in the form of a smooth curve drawn through observational data taken from the paper by MCILWAIN (1961). The dashed line indicates the rates to be expected in May 1962 based on the Explorer IV curve and the calculations of

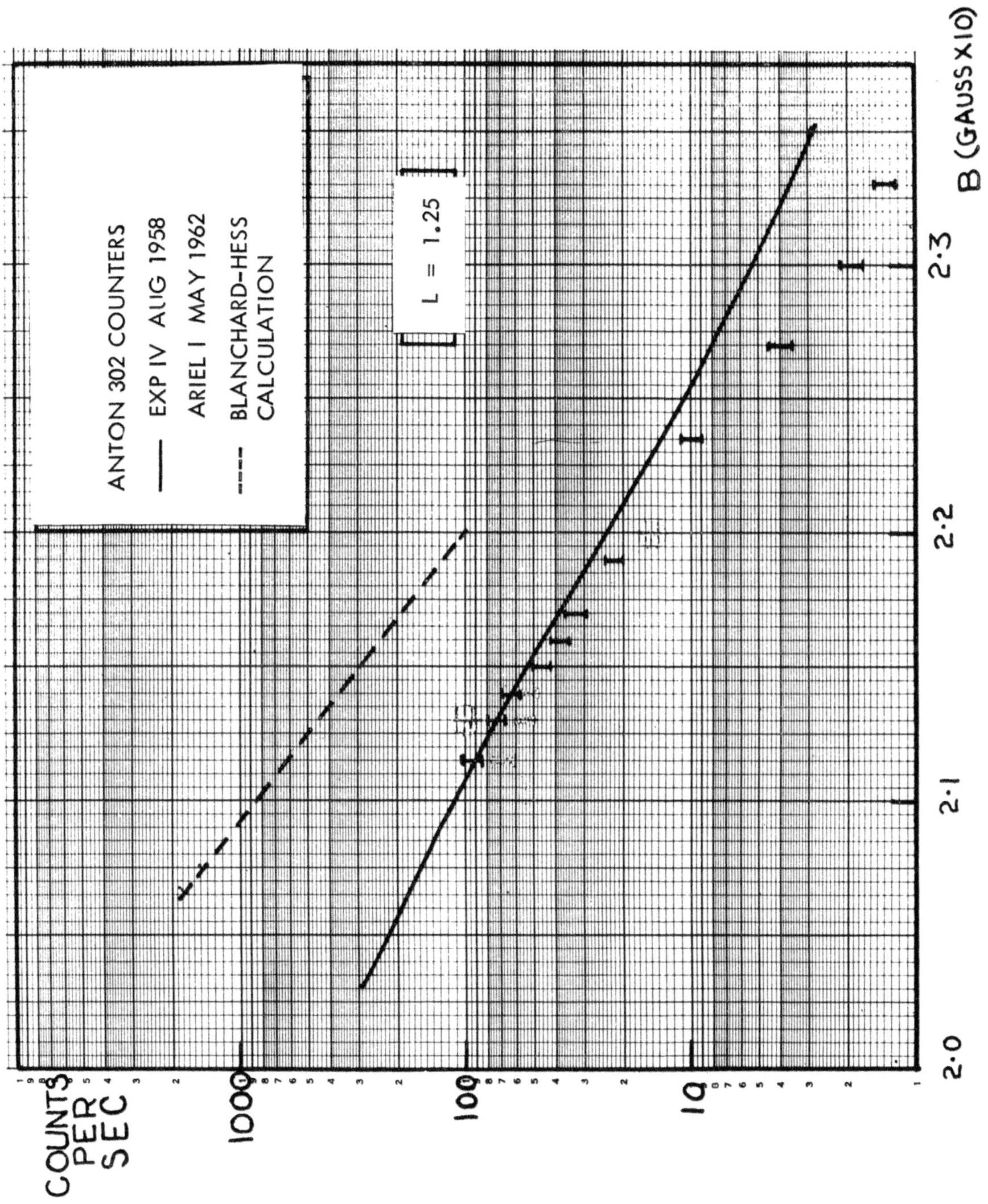

Fig. 10. Comparison between the Anton 302 rates in Explorer IV (————) and Ariel I (I I I I) for $L = 1.25$. The dashed line shows the expected counting rates for Ariel I based on the neutron albedo hypothesis and the calculations of Blanchard and Hess.

Blanchard and Hess. It will be seen that the Ariel points tend to lie below those from Explorer IV indicating that the flux values have diminished from 1958 to 1962 in qualitative contradiction to what we would expect if the neutron albedo

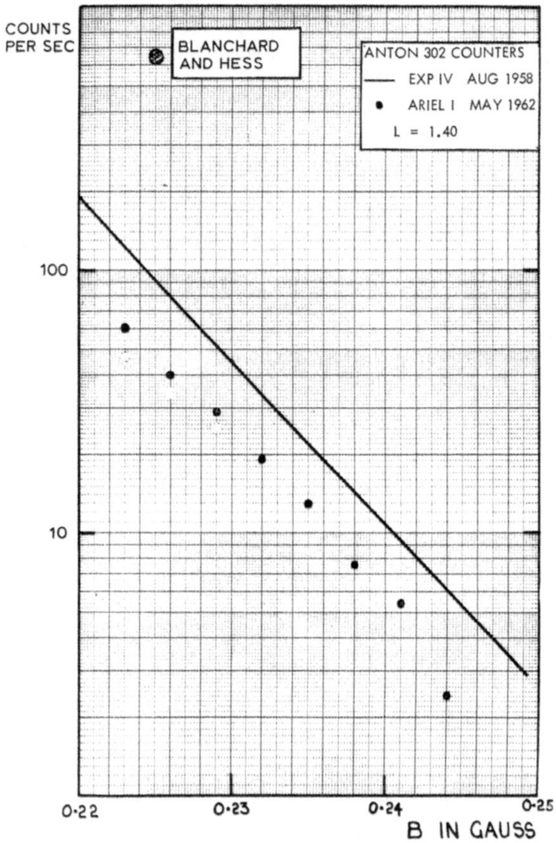

Fig. 11. Comparison between the Anton 302 rates in Explorer IV (———) and Ariel for $L=1.40$. The flux of protons with energy greater than 50 MeV on this L-shell has clearly decreased from mid 1958 to mid 1962.

hypothesis were correct. Figure 11 shows a similar set of data for $L=1.40$ where the decrease is even more marked. Here only a single point is available for comparison from the Blanchard-Hess calculations.

If these data are to be reconciled with the neutron albedo hypothesis and the Blanchard-Hess calculations it is necessary to suppose that the geometrical factor for the Anton 302 counter in Ariel I was a factor 8 or 9 less than that in Explorer IV. There is no conceivable reason for this and there is in any case direct evidence to the contrary which comes from the equatorial counting rate of the two detectors. McIlwain and Pizzella (1963), quoting Lin, give the background cosmic ray counting rate for the Explorer IV counter as Rate$=0.59L-0.41$ for $L<2.5$. Table II gives the rates for this counter at

TABLE II

L-value	*Explorer IV Background counts/sec*	*Ariel I Background counts/sec*
1.03	0.20	0.25
1.2	0.30	0.44
1.4	0.42	0.56
2.0	0.77	0.89

several L-values derived from this relation together with the observed rates from Ariel I. It will be seen that the Ariel I rates are higher than those for Explorer IV indicating that the geometrical factor for the Ariel detector is about 30% greater than that for Explorer IV. This would be a straight forward comparison if both detectors were counting cosmic rays only, which show a small change over the solar cycle at low latitudes, but as we have seen there is a complication because of induced radioactivity. The apogee of Explorer IV (2200 km) is higher than that for Ariel I (1200 km) and it is therefore to be expected, in view of the higher proton fluxes encountered, that the induced activity would be greater in Explorer IV. This would lead to a still higher geometrical factor of Ariel I relative to Explorer IV.

Summarizing, it seems that the Ariel I detector had a geometrical factor which was at least 30% higher than that of Explorer IV and we must conclude, therefore, that the intensity of inner belt protons with energy >50 MeV at $L=1.25$ and $L=1.40$ *decreased* from mid-1958 to mid-1962. This result is in qualitative disagreement with the increase to be expected if cosmic ray neutron albedo were the dominant source of these protons. Evidently they originate in some other way.

The Starfish High Altitude Nuclear Test

Following the high altitude nuclear test of 9 July 1962, a series of energetic particle phenomena were observed by Ariel. These have already been reported in detail in the literature (Durney *et al.*, 1962 and 1964), and a brief summary only will be given here. Relevant data are available from both the Anton 302 counter and the Čerenkov heavy particle detector.

The characteristics of the Anton 320 counter were as follows. In the direction where the amount of material is a minimum, electrons with energies of 7.4 and 10.5 MeV have, respectively, 10 and 50% probabilities of penetrating this amount of material. The counter will respond with a low efficiency to bremsstrahlung from electrons stopping in the shielding material. O'Brien *et al.* (1962a) have found this efficiency for an Anton 302, shielded by ca. 1 g/cm^2 magnesium and steel, to be ca. 10^{-5} at an energy of ca. 1 MeV. An approximate calculation yields an efficiency of ca. 10^{-6} at energies of ca. 0.5 MeV, and we regard the factor of 10^{-5} as an upper limit to the bremsstrahlung efficiency applicable here. The efficiency of γ-ray detection is about 0.4% and the product of the onmidirectional geometrical factor and the fast charged particle efficiency is about 0.5 cm^2. For technical reasons, the counter circuitry was designed so that the apparent counting rate fell to zero at a true rate of 88 counts/sec. All counting rates below 88/sec quoted in this paper have been corrected for dead time loss.

The Čerenkov detector consisted of a hollow Perspex sphere, 10 cm in diameter and 3 mm thick, viewed by a 5 cm photomultiplier through a hole cut in the base of the sphere. At the time of Starfish, the detector was counting pulses above two discriminator levels – one nominally equivalent to the light produced by relativisitic heavy nuclei of $Z^2 = 36$ and the other of $Z^2 = 65$, Z being the nuclear charge in units of the charge on the electron. The counting rates above the two levels were monitored in sequence. Inner belt protons arriving within a very small solid angle could also cause pulses above the discriminator levels. This was due to the scintillation light produced by protons traversing the width of the glass face of the photomultiplier backing the photocathode. Calibration with a cyclotron beam revealed that the effect started for protons of 70 MeV while maximum pulse height was reached at 105 MeV. The response of the Čerenkov detector to cosmic rays was reduced to zero if the anode current of the photomultiplier exceeded the amount which the e.h.t. generator could supply. For example, a strong γ-ray source could cause this saturation effect, presumably as a result of Compton electrons generated in the optical media which produced sufficient light at the photocathode. We estimated that 2.4 MeV electrons had a 50% probability of penetrating through the satellite skin and one thickness of the Čerenkov sphere, and that a flux of ca. 4×10^4 such electrons sterad^{-1}cm^{-2}sec^{-1} could produce sufficient Čerenkov light to cause saturation. The effects of electrons with lower energies are unknown, but it is likely that much higher fluxes would have been necessary to produce saturation.

Data on Starfish were obtained from 9 to 12 July, after which deterioration in the efficiency of the solar cells resulted in a serious loss of data. Up to 12 July, counting rates measured over $\frac{1}{2}$ min. intervals were telemetered for over 80% of the time in orbit.

At the moment of the explosion, Ariel I was at an altitude of 815 km at latitude 52°S and longitude 163°E – a distance of 7400 km from Johnston Island. Ariel lay roughly on the same geomagnetic longitude as the point of the explosion but at a higher L value, namely, $L=4.76$ as opposed to $L=1.12$.

TABLE III

Time at middle of interval (UT) 9 July	*Geiger-counter rate*	*Čerenkov-counter rate*	*Čerenkov-discriminator level*
08.58.55.9	64	4	high
08.59.26.7	80	4	high
08.59.57.4	72	8	high
09.00.28.1	40	4	high
09.00.58.8	0	8	high
09.01.29.5	0	8	high

Since the timing at the beginning of the event is important for some of our arguments later, we reproduce in Table III the counting rates of the two detectors, together with the time at the middle of the sampling interval. The sampling period is 30.72 sec but, because of the dead time associated with the encoder gate, the counts are in fact for a period of 26.9 sec in the case of the Geiger counter and 28.8 sec in the case of the Čerenkov detector.

These counting rates are plotted in Figure 12, the moment of the explosion (09.00.09 UT) being taken as the zero of the time scale. During the sampling interval centred on 09.00.28 UT and including the instant of the explosion, the Geiger-counter rate dropped to 40 from an average value of 75 just before the explosion, while in the next interval saturation had occurred. The simplest interpretation of these numbers is that the Geiger counter responded to natural radiation for 40/75 of the 30.72 sec interval and then saturated almost instantaneously due to the effects of Starfish. In this way a delay of 20 ± 5 sec from the explosion time is obtained before the counter records an increase. Two bursts of radiation are distinguishable in Figure 12: burst A, lasting about 6 min and showing on the Geiger counter only, and burst B, starting 6 min after the explosion, in which both detectors saturate.

An examination of the relative response of the two detectors during burst *A* yields some information on the type of particles causing the increase. Figure 13 shows the departure of the Čerenkov rate from the normal cosmic ray rate plotted against the Geiger rate for three different types of radiation. For

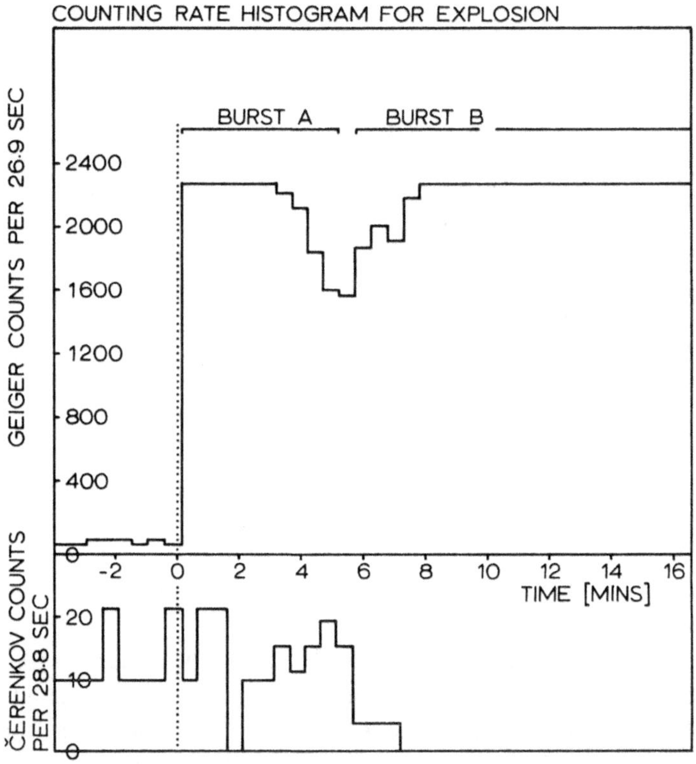

Fig. 12. The Anton 302 and Čerenkov counting rates plotted as a function of time with the moment of explosion, indicated by dotted line, taken as the zero of the time scale. During burst *A* only the G.M. counter saturates but in the case of burst *B* both detectors saturate.

protons in the inner belt, both rates increase until the Geiger saturates, while for the outer belt electrons only the Geiger counter shows any increase, the Čerenkov showing no response at all. The points for burst *B* are typical of the response to electrons from fission fragments released in Starfish. Here the Čerenkov rate is reduced towards zero as the Geiger rate increases because of the high current drain in the photomultiplier. The points for burst *A* lie close to the line applicable to outer belt particles and do not seem to be attributable to the energetic fission fragment β-particles.

Three possible explanations for the particles in burst *A* have to be considered:

(a) They are directly observed γ-rays from the nuclear explosion.

(b) They are electrons resulting from neutron decay.

(c) They are outer belt electrons whose mirror point distribution is perturbed by a hydromagnetic disturbance resulting from the explosion.

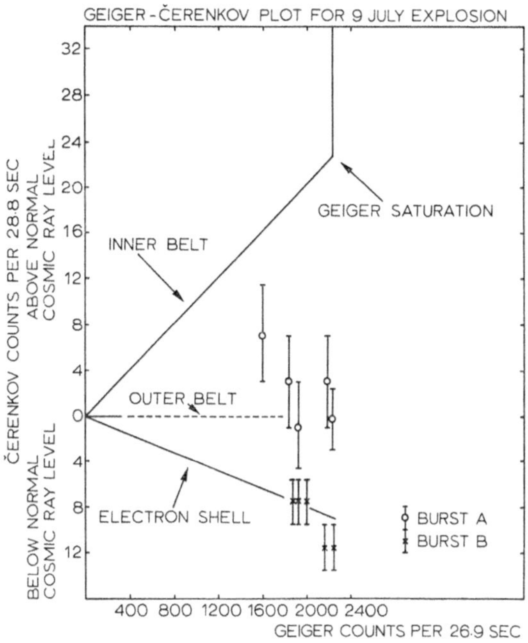

Fig. 13. Relation between the Anton 302 and Čerenkov counting rates for natural inner and outer belt particles. Points are also shown for burst *A* (○) and burst *B* (×), the latter being typical of the response of the detectors to fission fragment decay electrons.

Durney *et al.* (1964) favoured explanation (c) on the grounds that the electron flux from neutron decay could not account for the observed counting rate and explanation (a) seemed improbable since it required that a large fraction of the fission products should have appeared above the Ariel horizon in a matter of seconds following the explosion. However, in a recent paper D'Arcy and Colgate (1965) have presented evidence showing that *half* the debris from the explosion was guided by the magnetic field to the conjugate point with a transit time over the 4000 km path of between 2 and 20 seconds. The γ-ray flux from this amount of material would certainly be adequate to explain the observed counting rate and in view of this, it is not possible to say whether or not there was also some contribution from dumping of electrons from the natural radiation belt.

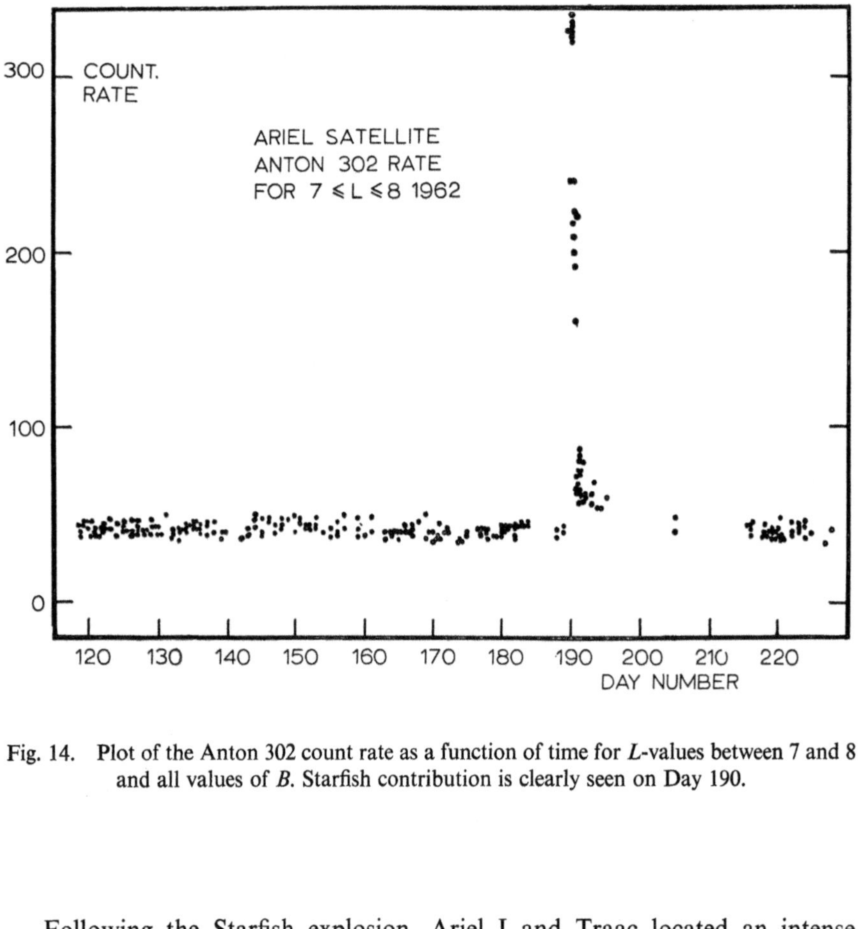

Fig. 14. Plot of the Anton 302 count rate as a function of time for L-values between 7 and 8 and all values of B. Starfish contribution is clearly seen on Day 190.

Following the Starfish explosion, Ariel I and Traac located an intense radiation flux on the $L = 1.12$ shell which passes 400 km above Johnston Island (DURNEY *et al.*, 1962; PIEPER, 1962). This belt caused the Ariel Čerenkov detector to saturate on the perigee segment of the orbit on 9 July. Clearly, the particles in this belt were β-decay electrons from fission fragments, and since 10% of these electrons have energies in excess of 2.7 MeV the Čerenkov detector probably saturated as a result of the light produced by these electrons penetrating to the Perspex sphere.

Because of the westward motion of the satellite relative to the Earth, we were unable to observe the eastern extent of the Johnston Island shell until 10 July, when the intensity was very much reduced. However, it is known that the $L=1.1$ shell intersects the atmosphere over South America and the lifetime of the electrons drifting eastward on the shell is only a few minutes. The shell must, therefore, have been continuously fed by the decay of fission fragments remaining over Johnston Island. The fission fragments could have been charged, in which case they could be trapped in the Earth's field, or they could have been neutral, in which case they would need to have a slow rate of diffusion in the upper atmosphere. The shell had a well defined western edge very close to the line joining Johnston Island and its magnetically conjugate point.

Peak intensity on this shell which lasted for several days, was located between $L=1.12$ and $L=1.18$. TRAAC observed the shell at about the same L-value and according to PIEPER (1963) maximum intensity occurred at $L=1.16$. It should not be confused with the main quasi-permanent belt of trapped β-particles with peak intensity at $L=1.2$ and which is generally referred to as the Starfish belt (O'BRIEN *et al.*, 1962b).

In addition to these two regions of high intensity, enhanced counting rates were observed out to the maximum L-value reached by Ariel. Figure 14, for example, shows the counting rate of the Anton 302 counter as a function of time for $7 \leqslant L \leqslant 8$ and the peak due to Starfish on Day 190 is clearly seen although these extra particles evidently disappeared in the course of the next few days. The counting rate in this region is seen to be normally rather constant since it is apparently almost entirely due to cosmic rays for a heavily shielded counter like that in Ariel. The process whereby these particles, presumably electrons, were injected at high L-values is not established but perhaps the most likely source is the decay of neutral fission fragments which travel outward from the explosion unimpeded by the geomagnetic field.

Acknowledgements

Drs Durney, Engel, Hynds and Quenby have all participated in the Ariel project and have made major contributions to the work described above. Ariel I was a joint US/UK undertaking and none of this work would have been possible without the generous co-operation provided by NASA. Miss W. Lloyd has been largely responsible for the computer programming and Mrs M. Jackson and her staff at UKAEA Risley have been particularly helpful in processing the data.

References

BLANCHARD, R. C. and HESS, W. N.: 1964, *J. Geophy. Res.* **69,** 3927.
D'ARCY, R. G. and COLGATE, S. A.: 1965, *J. Geophys. Res.* **70,** 3147.
DURNEY, A. C., ELLIOT, H., HYNDS, R. J., and QUENBY, J. J.: 1962, *Nature, Lond.* **195,** 1245
DURNEY, A. C., ELLIOT, H., HYNDS, R. J., and QUENBY, J. J. 1964, *Proc. Roy. Soc. A,* **281,** 565.
ENGEL, A.: 1964, London University, PhD, Thesis.
LIN, W. C., VENKATESAN, D., and VAN ALLEN, J. A.: 1963, *J. Geophys. Res.* **68,** 4885.
MCDONALD, F. B.: 1958, *Phys. Rev.* **109,** 1367.
MCDONALD, F. B. and WEBBER, W. R.: 1959, *Phys. Rev.* **115,** 194.
MCILWAIN, C. E.: 1961, *J. Geophys. Res.* **66,** 3681.
MCILWAIN, C. E. and PIZZELLA, G.: 1963, *J. Geophys. Res.* **68,** 1811.
O'BRIEN, B. J., VAN ALLEN, J. A., LAUGHLIN, C. D., and FRANK, L. A.: 1962a, *J. Geophys. Res.* **67,** 397.
O'BRIEN, B. J., LAUGHLIN, C. D., and VAN ALLEN, J. A.: 1962b, *Nature, Lond.* **195,** 939.
PIEPER, G. F.: 1963, *J. Geophys. Res.* **68,** 651.
PIEPER, G. F., BOSTROM, C. O., and ZMUDA, A. J.: 1965, *J. Geophys. Res.* **70,** 2021.
PIZZELLA, G., MCILWAIN, C. E., and VAN ALLEN, J. A.: 1962, *J. Geophys. Res.* **67,** 1235.
RAO, A. S., BALASUBRAHMANYAN, V. K., GOKHALE, G. S., and PEREIRA, A. W.: 1953, *Phys. Rev.* **91,** 764.
VAN ALLEN, J. A., and GANGNES, A. V.: 1950, *Phys. Rev.* **78,** 50.
VERNOV, S. N., SAVENKO, I. A., SHAVRIN, P. I., NESTEROV, N. E., PISARENKO, N. F., and SHARMA, K. N.: 1964, Cospar Symposium.
WALT, M. and MACDONALD, W. M.: 1964, *Reviews of Geophysics* **2,** 543.
WEBBER, W. R.: 1958, *Nuovo Cim.* **8,** Suppl. II, 532.
WINCKLER, J. R., STIX, T., DWIGHT, K., and SABIN, R.: 1950, *Phys. Rev.* **79,** 656.
ZMUDA, A. J., PIEPER, G. F., and BOSTROM, C. O.: 1965, *J. Geophys. Res.* **70,** 2045.

Discussion

FREDEN said he also does not find the time variations of proton fluxes predicted by Blanchard and Hess, but that he sees a small increase in disagreement with Elliot's decrease. Freden suggested that the most sensitive way to check on the time variation is to look for changes in the shape of the spectrum. Hess agreed and said that especially on the upswing to solar maximum, interesting spectral changes should take place. He outlined the calculations of Blanchard and Hess that show how large the expected effect is. The point was made that even though we don't know the magnetic field accurately enough to do the average atmosphere well, this would just shift the location of the calculated curves slightly in B-L space. But if the wrong sign effect or wrong size effect was really found over all space, then something else is going on.

MCILWAIN brought up the possibility that Explorer VII was counting electrons rather than protons at $L \sim 1.3$ as first suggested and therefore may not be applicable to the proton problem.

LENCHEK made the point that the secular variation of the main magnetic field shown by Cain was such that it would lower the particle mirror points by about one scale height in 4 years.

Questioning about the radioactivity detected from the series of low altitude bomb tests showed that the effect was not likely to be gamma rays because it was asymmetric. More activity was found east of the burst site indicating the possibility of the drift of electrons. WALT suggested the possibility that atmospheric winds might move the debris.

LOW ALTITUDE MEASUREMENTS OF TRAPPED ELECTRONS

W. L. IMHOF and R. V. SMITH
Lockheed Missiles and Space Company Research Laboratories, Palo Alto, California

1. Introduction

The region of space near the earth's atmosphere provides a unique opportunity for studying certain phenomena associated with particle trapping in the radiation belts. For example, by comparing theoretical calculations with the measured decay of electrons injected by the Starfish detonation of July 1962, it has been shown that the loss of electrons trapped on field lines below $L \approx 1.3$ is dominated by atmospheric scattering (WALT and MACDONALD, 1964). However, this comparison could be made only as an integral over a fission-like spectrum rather than as a function of energy, since there were no experimental data for monoenergetic electrons. More rigorous tests of atmospheric scattering calculations can be made by comparisons with detailed energy spectra and thorough study of the observed longitudinal variations in intensity and spectrum.

In addition to providing an opportunity for studying particle loss mechanisms in some detail, the presence of the atmosphere produces a unique position in space for studying natural source or redistribution mechanisms. Electron lifetimes vary extremely rapidly with L in this region, reaching values which are quite short below $L \approx 1.15$–1.20. Thus, the electron flux is a very steep function of L, and one is assured that electrons observed at a given location have only recently arrived. This paper will present some preliminary results obtained in this interesting region of space, in particular the observation of nearly monoenergetic electrons, in order to illustrate the studies which can be made here. Finally, there will be some speculation as to the wealth of potential information to be gained from an extensive program of measurements on very low L shells.

2. Description of Instrumentation

Measurements were performed during the period 30 October to 6 November 1963 from an earth-oriented satellite in polar orbit with 355 km and 288 km apogee and perigee, using the plastic scintillation spectrometer shown schematically in Figure 1. The scintillator (5.7 cm diameter by 3.2 cm) provided a minimum path length sufficient to permit spectral measurements based on total energy deposit to energies of at least 5 MeV. A large geometric factor and an acceptance solid angle of nearly 2π steradians were achieved with an aperture diameter of 2.54 cm.

The number and energy distribution of pulses in the plastic scintillator were recorded by means of a type 4439 photomultiplier and a 16-channel digital pulse height analyzer (ROWLAND *et al.*, 1963; REAGAN *et al.*, 1965). Individual

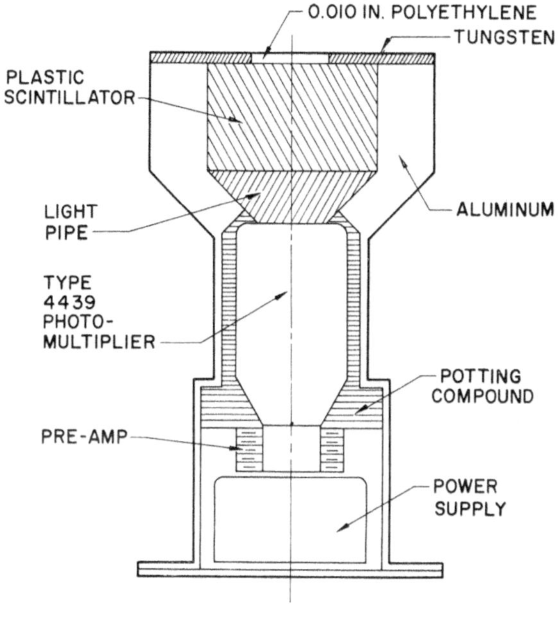

Fig. 1. Schematic diagram of the plastic scintillation spectrometer.

counts were stored as 4-bit words on a tape recorder, giving world-wide coverage for the data. The bandwidth limitation of this vehicle-provided storage and transmission system was only 32 such words per second at the tape recorder input. Therefore, the pulse height analyzer was deliberately arranged to have an accurately-known dead time of 31 milliseconds per analyzed pulse, and to sample a random fraction of the counts at higher rates. The analyzer address clock circuit was switched by a scaled factor of four on a continuous 27 sec–10 sec program, producing pulse height ranges of 0.28–2.61 MeV and 0.28–10.02 MeV for the 16 channels during the longer and shorter periods respectively.

This scale switching gave a broad range of energy response and allowed us to use the characteristic pulse height distribution from cosmic ray particles on the upper energy range as an in-flight gain calibration for both ranges. The overall gain remained within ± 3 percent for all of the data presented here, and the absolute energy calibration is estimated to have an accuracy of ± 15 percent. Laboratory measurements with monoenergetic electrons showed the energy resolution of the instrument to be 15 percent, 21 percent, and 30 percent full width at half maximum at 2.0, 1.0, and 0.5 MeV respectively. The bremsstrahlung contribution to the counting rate was found to be quite small, by means of laboratory measurements with high energy electrons from the beta decay of Li^8.

Identification of the types of particles producing the pulses was dependent upon the known responses of the detector and a knowledge of relative particle intensities and spectral shapes at the altitudes of concern. For electrons the threshold energy of the spectrometer was 0.28 MeV whereas for protons this energy was considerably higher, ~ 5 MeV, due primarily to a 0.025-cm thick polyethylene shield over the scintillator opening and partly to the lower light output efficiency of plastic scintillator to protons of such energies. In order to evaluate the proton contributions to the observed counting rates we have used the measurements of FREDEN and PAULIKAS (1964). Their reported fluxes of protons above 5 MeV at the altitudes of concern would produce counting rates in our spectrometer considerably smaller than the total rates observed. Therefore, on the basis of the known intensities and expected pulse height distributions for protons, the counting rates are assumed to represent electrons.

3. Energy Spectrum of Electrons

The energy spectrum at *B-L* values which reach minimum altitudes of 200–355 km is a strong function of *L*, as illustrated in Figure 2. The data are compiled over the period 2–3 November 1963 for $L > 2.5$ and over the longer period 30 October to 4 November 1963 for the lower *L* shells ($L \leqslant 2.5$) where no time variations were observed in the spectral shape (IMHOF and SMITH, 1965b).

In the region of the Starfish belt ($L \lesssim 1.7$) the spectrum is quite hard. Above ~ 2 MeV the shape is just slightly softer than (but within the experimental errors, not inconsistent with) that expected for trapped electrons soon after injection by decaying fission fragments (CARTER *et al.*, 1959). Atmospheric scattering calculations (WALT, 1964) predict a marked decrease in flux with decreasing energy. Preliminary comparisons between the present data and these theoretical spectral shapes have revealed differences at low energies (IMHOF and SMITH, 1964), which may be due to the presence of naturally-injected electrons with intensities which are not well known.

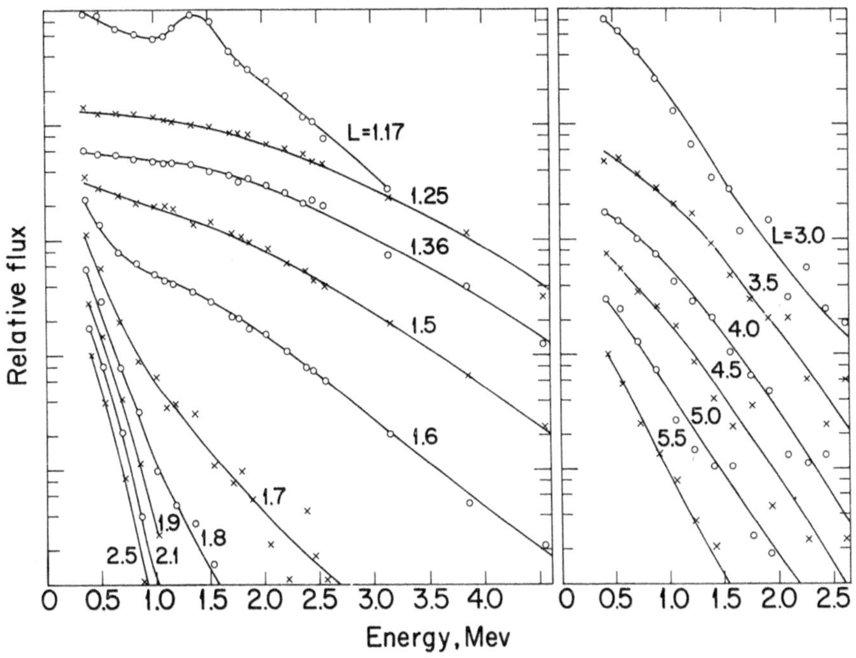

Fig. 2. The electron differential energy spectrum on several different L shells for the time periods 30 October to 4 November 1963 and 2–3 November 1963 for $L \leqslant 2.5$ and $L > 2.5$ respectively. The curves are arbitrarily displaced vertically and indicate shape only.

4. Longitudinal Variations of Electrons

Post-Starfish measurements made both in the fall of 1962 and in the fall of 1963 indicate a systematic dependence of flux upon B and L for points corresponding to minimum altitudes greater than 100 km in the region of the Brazilian anomaly. However, for large B values, where $h_{min} < 100$ km, the measured intensities were found to be dependent upon a third parameter, longitude. This dependence is illustrated by the isoflux contour maps shown in Figure 3, for electrons in the energy interval 0.28–0.71 MeV. The fluxes west of the anomaly are considerably larger than those at corresponding B-L points east of the anomaly. The sudden onset of a longitudinal dependence at values of h_{min} less than 100 km is presumably associated with the rapid increase in air density at such altitudes rather than with any sudden appearance of systematic errors in the values of B and L derived from the Jensen and Cain (1962) expansion of the earth's magnetic field.

The energy spectra observed at high B values, corresponding to electrons which dip quite low into the atmosphere in the Brazilian anomaly, are observed

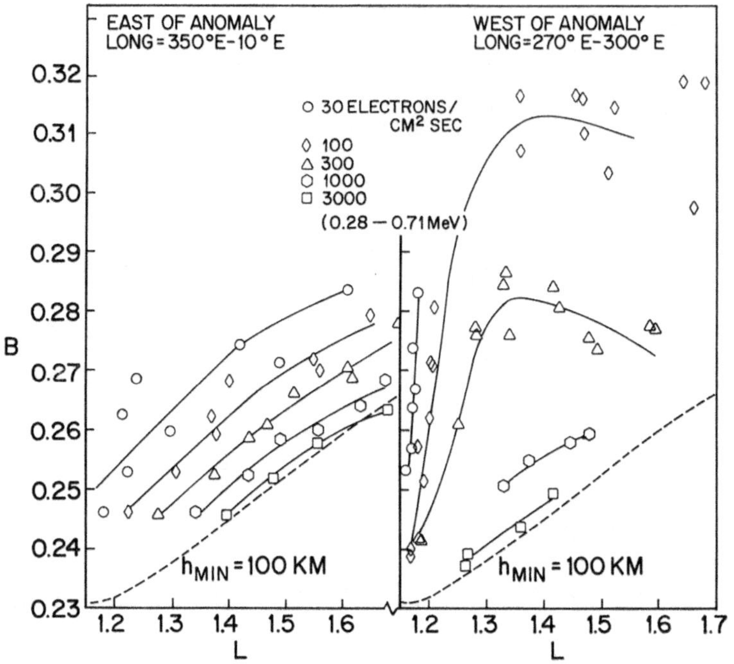

Fig. 3. Isoflux contour maps at two different longitudes for electrons in the energy interval 0.28–0.71 MeV for October–November 1963.

to be considerably softer than the spectra observed on the same L shells at points with $h_{min} > 200$ km. Details of the spectral softening for $L = 1.20$ to 1.40 are shown in Figure 4, where the ratio $\phi(E = 1.40\text{–}2.61\text{ MeV})/\phi(E = 0.28\text{–}0.71\text{ MeV})$ is plotted as a function of the local magnetic field strength minus the field strength at a minimum altitude of 100 km on the same L shell. The largest changes in spectral index are seen to occur for points with $h_{min} = 0\text{–}200$ km. In addition to the obvious spectral dependence upon B there is evidence for an east-west variation in the spectrum observed at a given B and L value. The "east" points display a harder spectrum, the larger statistical errors merely reflecting smaller fluxes and hence a lower counting rate.

Precipitating electrons have been observed before the Starfish detonation (Paulikas and Freden, 1964; Schrader *et al.*, 1963), but the available data are not sufficiently detailed to permit a critical evaluation of the non-Starfish contribution to the fluxes reported here. An indication that the electrons above 1.2 MeV at large B values represent a redistribution of the more stably trapped electrons rather than an injection of electrons from some external source is obtained from a comparison of measurements performed in September 1962 and in late 1963. Comparable intensity changes, associated with decay of the Starfish belt, had taken place for both the electrons observed at high B values and those trapped with mirror points at positions much closer to the equator.

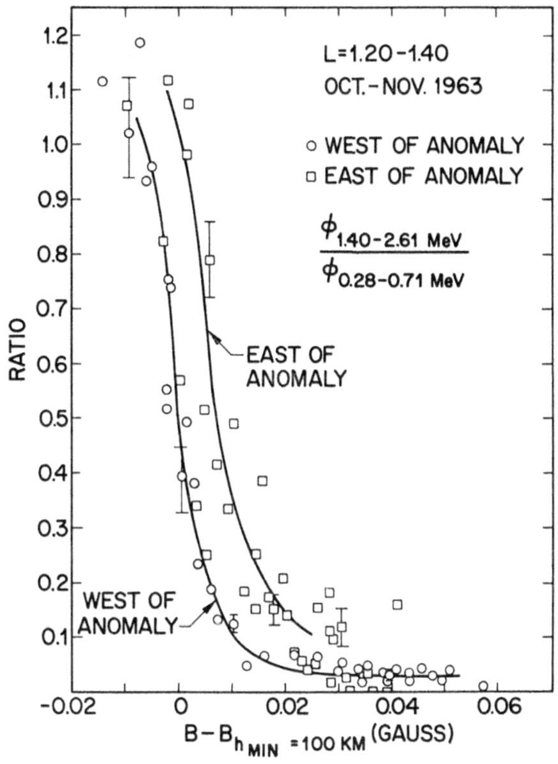

Fig. 4. The ratio of electron fluxes in the two energy intervals 0.28–0.71 MeV and 1.40–2.61 MeV plotted as a function of B–$B_{h_{\min}=100\ \mathrm{km}}$ for $L=1.20$–1.40.

Under the assumption that the lower energy electrons are also derived from those trapped on the same L shell, sizable low energy fluxes at large B values should arise from the natural electron belt.

The rate of redistribution or injection needed to maintain the fluxes observed at positions where $h_{\min} < 0$ km has been compared with the overall loss rate of trapped electrons by making use of the measured decay rate of the

Starfish belt. Assuming the electrons were trapped at the time of observation, it is found that the electrons observed at large values of B correspond to a small fraction of the total loss rate, ranging from ~ 4 percent at $L=1.18$ to ~ 0.04 percent at $L=1.60$. Therefore, the one or more effects of concern do not necessarily represent primary loss mechanisms, such as the diffusion of electrons due to atmospheric scattering which governs the loss of electrons trapped on magnetic shells with $L \lesssim 1.3$ (WALT, 1964).

There are several processes which might contribute to the fluxes of electrons observed at large B values where $h_{min} < 0$ km, such as large and small angle Coulomb scattering, the action of whistlers, and the possible presence of steady or slowly varying electric fields. A preliminary investigation has been made of part of the contribution from Coulomb scattering – that due to single large angle scatterings of electrons trapped with their mirror points near the magnetic equator (IMHOF and SMITH, 1965a). The observed buildup rate and the energy spectrum were found to be consistent with this process. However, for an overall evaluation of the relative contributions of the various processes, the longitudinal dependence of the injection rate to high B values must be considered in detail.

5. Observation of Nearly Monoenergetic Electrons on Very Low L Shells

During the period 30 October to 6 November 1963 we consistently observed rather sharply peaked spectra on very low L shells (IMHOF and SMITH, 1965). Processing the data in very short time intervals revealed a systematic and rapid variation of spectrum with L. Figure 5 shows the raw spectra observed during one of the many passes through this narrow B-L interval. At $L=1.150$ the peak is not much wider than our instrumental energy resolution of ~ 18 percent at 1.3 MeV.

Successive passes through a given L shell showed a systematic decrease in peak intensity, allowing comparison with the decay expected from atmospheric scattering theory as shown in Figure 6. Calculations performed at $L=1.138$ and $L=1.150$ have shown that an impulsively injected group of electrons at ~ 1.3 MeV will decrease in intensity much more rapidly than it spreads out in energy, consistent with the observations. For the period indicated, both the absolute value of the calculated decay constant and its variation with L are in agreement with the experimental results.

Although we were not able to observe the onset of this group of electrons, the decay curves can be extrapolated back in time to indicate a probable time of injection. This extrapolation to earlier times indicates a substantially constant intensity over a narrow L interval about one day before our first observation, suggesting an injection at the time of a major magnetic storm.

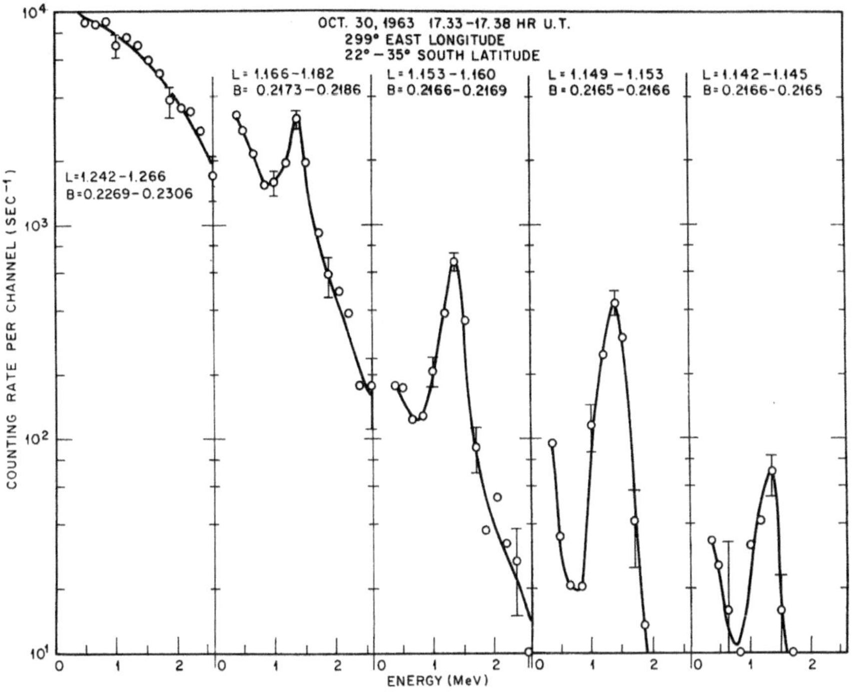

Fig. 5. Analyzer output per unit lifetime on different L shells for one pass through the Brazilian anomaly. Channels 1 and 16 are not shown.

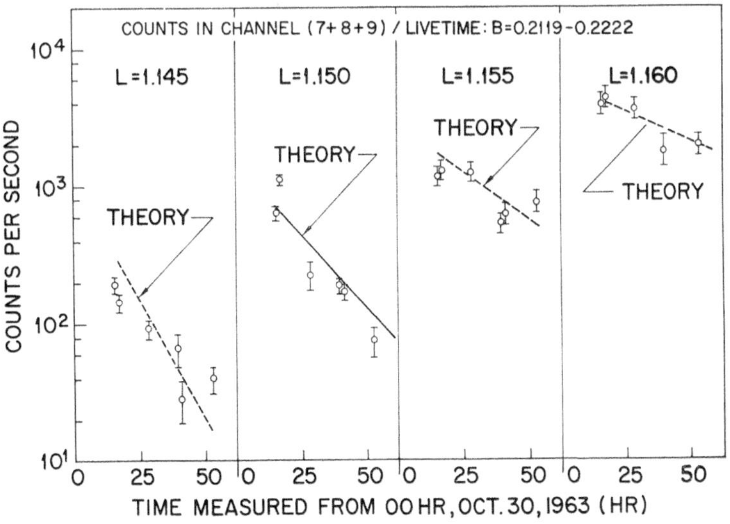

Fig. 6. The counting rate in channels $(7+8+9)$ as a function of time for various L shells. The data are restricted to a narrow B interval and within this interval corrections have been made for the variation of flux with B. The curves shown represent decay rates calculated from atmospheric scattering theory. The solid curve at $L=1.150$ is a calculation at that precise L shell whereas the dotted curves are based on calculations at $L=1.138$, 1.150, and 1.186.

The onset of another group of electrons at 0.75 MeV occurred between 1500 and 1630 UT on 2 November 1963, the times of successive passes of the satellite through the Brazilian anomaly, as seen in the raw spectra shown in Figure 7.

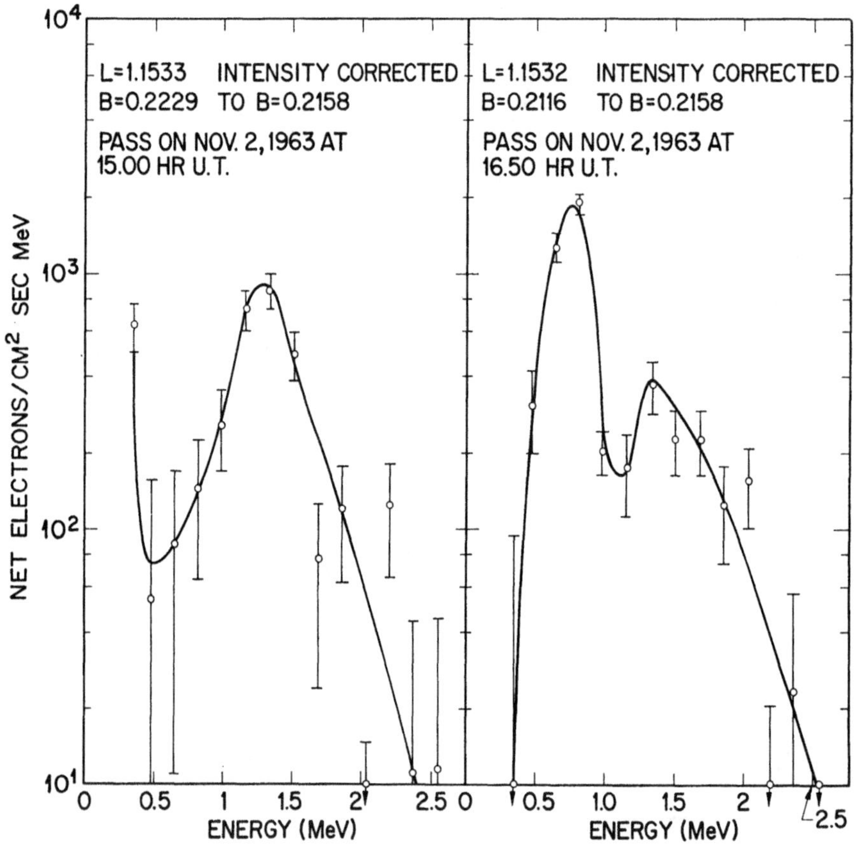

Fig. 7. The electron differential energy spectrum on two successive passes through the Brazilian anomaly.

For this comparison the intensities have been normalized, on the basis of the measured variation of flux with B, to a value of $B = 0.2158$.

Magnetograms from low-latitude magnetic observatories revealed that during both injection time intervals the horizontal component of the geomagnetic field increased periodically. Figure 8 shows the magnetograms recorded at Huancayo, Peru, where the amplitudes of the fluctuations were about a factor of six greater than those observed elsewhere. The phases of the oscillations were the same throughout the equatorial region of the earth, indicating that the magnetic-field variations were due to modulations of the solar wind intensity. In each case the period of the fluctuations was comparable to the longitudinal drift period for electrons of energy equal to that of the group.

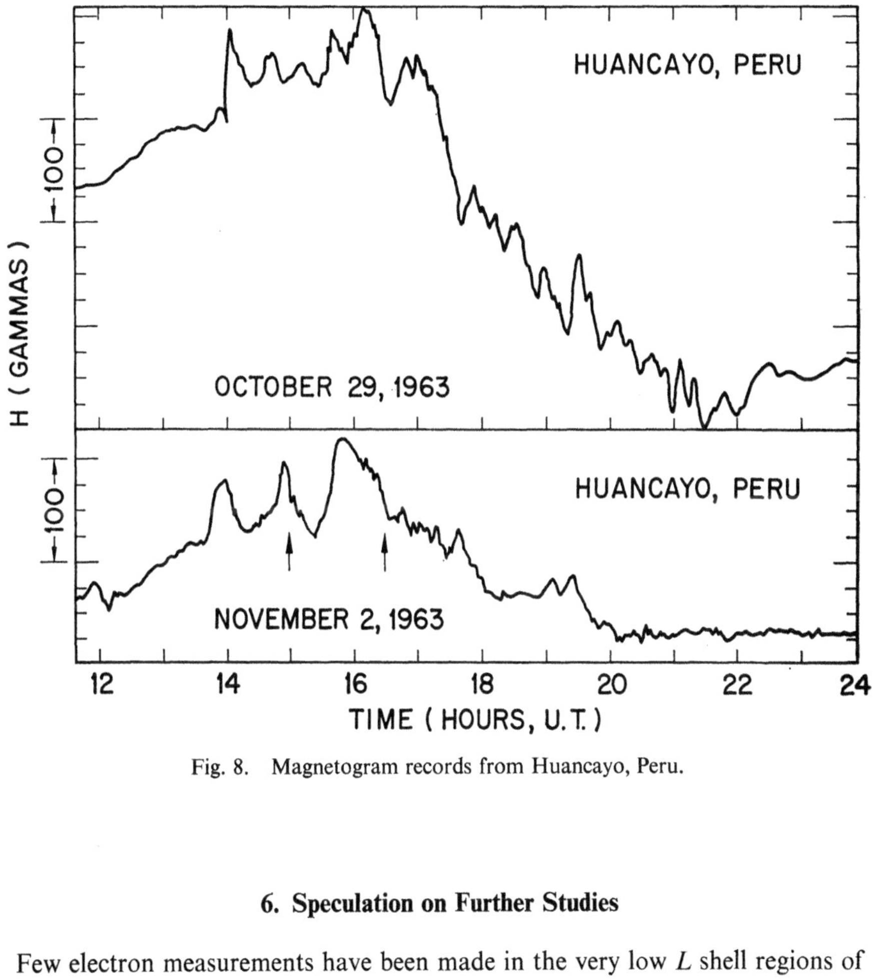

Fig. 8. Magnetogram records from Huancayo, Peru.

6. Speculation on Further Studies

Few electron measurements have been made in the very low L shell regions of space with counters of sufficiently high sensitivity to see small fluxes of injected electrons. In particular, with the exception of the data presented here, there are no published detailed energy spectrum measurements of electrons in very low flux regions. Since this area of endeavor has barely been touched, it is instructive to consider what information might be learned from an extensive series of such measurements.

The low L shell regions provide an extremely sensitive measure of particles which cross field lines in an inward direction, since the flux falls off very rapidly with decreasing L. By measuring detailed energy spectra one can study the energy selectivity of any redistribution processes. For example, one might see a resonance type phenomenon associated with time-varying fields. Although to date virtually all studies of particles crossing field lines have been made in the outer regions of the magnetosphere, there are many advantages to performing such studies at the lower edge of the inner belt.

(a) There is less overall activity – it is easier to separate individual cases.

(b) The inner edge is very sharp, hence one has an extremely high sensitivity to small changes in L, compared with the outer belt regions.

(c) In the lower edge of the inner belt the clearing mechanism (atmospheric scattering) is independent of the source mechanisms, rapidly provides a clean region for observing new injections, and is relatively well understood.

Because injections into very low L-shell regions apparently do occur naturally, this provides a convenient source of electrons with energy distributions which, in some cases, are nearly monoenergetic. Thus, one can perform studies of trapping lifetimes, due primarily to atmospheric scattering, as a function of L and energy. Furthermore, with monoenergetic electrons, a measure can be obtained of the relative losses due to energy degradation and scattering. The only previous electron lifetime measurements in the inner belt had come from the Starfish fission spectrum electrons, for which it was necessary to perform an integration over energy before comparing with theory.

From satellites in the altitude range $\sim 500 \rightarrow \sim 1000$ km it is possible to measure trapped fluxes on the low L shells as a function of longitude and thus perhaps obtain direct evidence relating to the existence of electric fields which can alter the energy and hence the L shell on which electrons are trapped. Such fields would probably be most effective in the region of the Brazilian Anomaly. Since it has not yet been possible to make direct measurements of electric fields in the radiation belts, this approach may perhaps yield important new information.

Acknowledgments

The authors wish to thank Drs. M. Walt and L. L. Newkirk for providing the atmospheric scattering calculations; Dr. J. B. Cladis for his ideas relating to the effects of electric fields; J. C. Bakke for his special role in electronic design; A. W. Crawford, E. E. Gaines, D. R. King, S. J. Norman, V. A. Olivier, J. B. Reagan, J. H. Rowland, and V. F. Waltz for assistance in varying degrees; V. P. Fenton, G. P. Minalga, D. Pefferle and associates for vehicle integration; and the Air Force Satellite Systems Division for providing vehicle space.

References

CARTER, R. E., REINES, F., WAGNER, J. J., and WYMAN, M. E.: 1959, *Phys. Rev.* **113,** 280–286.
FREDEN, S. C. and PAULIKAS, G. A.: 1964, *J. Geophys. Res.* **69,** 1259–1269.
IMHOF, W. L. and SMITH, R. V.: 1964, *Space Research* **5.**
IMHOF, W. L. and SMITH, R. V.: 1965a, *J. Geophys. Res.* **70,** 569–577.
IMHOF, W. L. and SMITH, R. V.: 1965b, *J. Geophys. Res.* **70,** 2129–2134.
IMHOF, W. L. and SMITH, R. V.: 1965c, *Phys. Rev. Letters* **14,** 885–887.
JENSEN, D. C. and CAIN, J. C.: 1962, *J. Geophys. Res.* **67,** 3568–3569.
PAULIKAS, G. A. and FREDEN, S. C.: 1964, *J. Geophys. Res.* **69,** 1239–1249.
REAGAN, J. B., BAKKE, J. C., IMHOF, W. L., and SMITH, R. V.: 1965, *IEEE Trans.*, NS-**12,** 83–88.
ROWLAND, J. H., BAKKE, J. C., IMHOF, W. L., and SMITH, R. V.: 1963, *IEEE Trans.*, NS-**10,** 178–182.
SCHRADER, C. D., KEIFER, R. C., WAGGONER, J. A., ZENGER, J. H. and BLOOM, S. D.: 1963, *Trans. Am. Geophys. Union* **44,** 83.
WALT, M.: 1964, *J. Geophys. Res.* **69,** 3947–3958.
WALT, M. and MACDONALD, W. M.: 1964, *Rev. of Geophys.* 543–577.

RESONANCE ACCELERATION OF PARTICLES IN THE INNER RADIATION BELT

J. B. CLADIS

Lockheed Missiles and Space Company Research Laboratories, Palo Alto, California

In this paper evidence is presented for the action of an acceleration mechanism which explains the sudden appearances of the two groups of rather mono-energetic electrons described by W. L. Imhof in the previous paper. Imhof shows that oscillations of the geomagnetic field occurred at the estimated onset time of the 1.35 MeV electrons on October 29, 1963 and during the observed onset time of the 0.75 MeV electrons on November 2, 1963. A detailed analysis of the particle data and the magnetograms indicates, (a) that the particles were accelerated by magnetic field variations over the western slope of the Brazilian geomagnetic anomaly which were in resonance with the longitudinal drift frequencies of the electrons, (b) that the accelerating magnetic fields were produced by a periodic modulation of local ionospheric currents, and (c) that the energy gained by the electrons by this process was sufficient to have caused the electrons to move down L shells to the region where they were observed ($L \approx 1.15$) from higher L shells where such electrons were abundant.

Magnetograms from low latitude observatories revealed that during both injection time intervals, oscillations of the horizontal component of the geomagnetic field occurred throughout the equatorial regions of the world and the phases of the oscillations were the same everywhere, indicating that the field variations were probably produced by a variation of the intensity of the solar wind. However, at Huancayo, Peru, which is located near the geomagnetic equator at a longitude of about 75° West, the field variations were found to be about a factor of six larger than those recorded at other equatorial regions. Data discussed below indicates that these large field variations were probably produced by local ionosphere currents.

The Huancayo magnetograms which were recorded on October 29, 1963, and November 2, 1963, are shown in Figure 7 of Imhof's paper. A sudden commencement of a magnetic storm at 1400 UT followed by a main phase can be seen in the October 29th magnetogram. The mean period of the magnetic field fluctuations following the sudden commencement is about the same as the longitudinal drift period of the 1.35 MeV electrons which were presumably accelerated during the time of these oscillations. The onset time of these electrons was estimated by extrapolating the decay curves as discussed by Imhof in the previous paper. Similarly, the mean period of the fluctuations seen in the November 2nd magnetogram is comparable to the drift period of the 0.75 MeV electrons which were first observed during the acceleration interval at about 1600 UT. In both cases therefore a resonance type acceleration process is indicated.

The measurement of the 0.75 MeV electrons during the acceleration interval was particularly fortuitous because it established the phase of the longitudinal motion of the electrons with respect to the phase of the magnetic field variations. The two arrows shown in the November 2nd magnetogram indicate successive times during which the polar-orbit satellite crossed the region of the Brazilian anomaly, which was the only region where trapped electrons could be observed because of the low altitude of the satellite. The 0.75 MeV electrons were not seen at $L \approx 1.15$ at the time given by first arrow, but they were seen there during the following pass over this region at a longitude of about 50° West. About 5 minutes earlier these electrons had been drifting over Huancayo. At that time the slope of the oscillation in the magnetogram was negative, indicating that the horizontal component of the magnetic field at Huancayo was decreasing with time. Similarly, at successive longitudinal drift periods earlier in time, the magnetograms show that the electrons had also been over Huancayo while the field was decreasing. If the phases of the field fluctuations above the ionosphere were the same as those on the ground, which would have been the case if the fluctuations were produced by currents in the magneto-pause, then the electrons would have been decelerated rather than accelerated by these field changes. It is therefore believed that these fluctuations were produced by ionospheric currents, which give the necessary phase reversal above the ionosphere to produce the acceleration observed.

The periodic variations of the ionospheric currents above Huancayo were probably produced by enhanced precipitations of radiation belt particles in the atmosphere above the Brazilian anomaly. Note that the fluctuations in both of the magnetograms began at about 1400 hours UT. This time corresponds to about 0900 local time, the time when the Sq ionospheric equatorial current normally flows toward the east. This current increases the horizontal component of the earth's field below the ionosphere in the manner depicted by the broad hump which appears in the November 2nd magnetogram between about 1200 to 1800 UT. (This hump in the October 29th magnetogram is distorted by the magnetic storm which occurred on that day.) Since the amplitude of this hump at Huancayo is generally larger than that observed at other equatorial stations, the increased current flow might be due to the increased height-integrated conductivity of the ionosphere over the Brazilian anomaly caused by the atmospheric absorption of radiation-belt particles which normally diffuse down

L shells. If this is true then the periodic variation of the solar wind intensity discussed above would be expected to produce an enhanced periodic dumping of particles over the anomaly, which would in turn modulate the flow of the ionospheric current.

The energy gain of an electron due to the magnetic field change can be estimated by taking the product of the magnetic moment of the electron and the magnetic field change as seen by the electron during its drift motion across the anomaly. This evaluation is strictly correct for particles in a uniform magnetic field with motions normal to the field, but it can be applied with sufficient accuracy to this case. By performing this calculation, using the peak to trough field changes given by the magnetograms from Huancayo energy gains per drift period of about 4 keV for the 1.35 MeV electrons and 3 keV for the 0.75 MeV electrons are obtained.

These numbers are in sufficiently good agreement with the energy gain per magnetic-field pulse required to enable the electrons to drift across L shells to $L \approx 1.15$ from above where appropriate fluxes of such electrons were present. The flux of energetic electrons increases sharply with L increasing along the low-altitude edge of the inner radiation belt as shown by Imhof in the previous paper. An examination of the flux distribution indicates that the observations would be consistent with L shell transfers of about 0.013 and 0.008 for the 1.35 MeV and the 0.75 MeV electrons respectively. Such L shell transfers require corresponding energy increases of 38 keV and 13 keV if the magnetic moment and integral invariants hold. By dividing these energies by the number of oscillations of the magnetic field intensity, the energy gain per drift period is about 6 keV for the 1.35 MeV electrons and about 4 keV for the 0.75 MeV electrons.

In this type of acceleration system, electrons which have longitudinal drift frequencies which are lower and higher harmonics of the magnetic field oscillations should also be accelerated. Such electrons were not observed. Electrons having the lower harmonic drift frequencies were absorbed too rapidly by the atmosphere at the lower L shell, and electrons having the higher harmonic drift frequencies probably were not observed because the flux of these particles was too small for comparable L shell transfers.

Since the magnetograms show that magnetic field oscillations occurred throughout the equatorial region of the earth, it is necessary to consider the influence of these world-wide oscillations on the energies of the particles. For this purpose the following type of resonance acceleration method was investigated. Assume that these oscillations throughout the world were due to a sinusoidal variation of the solar wind intensity, with the period of the oscillation equal to the longitudinal drift period of the electrons. An alternating electric field with an azimuthal component around the equator is established by the solar wind variation. This electric field accelerates particles when the longitudinal drift direction of the particle is in the same direction as the electric force acting on the particle. When the electric field reverses in sign, the energy of the particle is decreased. Clearly if the magnitude of the electric field is not dependent on local time and if the drift velocity is constant, the net energy change of the particle per drift period would be zero. Because of magnetic field irregularities, however, the drift period of a particle on a low L shell can vary with longitude by as much as 30%. Moreover, the electric field is somewhat stronger at local noon than it is at local midnight. Both of these effects can result in a net energy gain of a particle during a drift period. However, computation of this energy gain using the most favorable phase relationship and the longitudinal drift velocity variation given by NEWKIRK and WALT (1964) revealed that it was only about 0.5 keV per drift period for the case of the 0.75 MeV electrons. This acceleration mechanism is therefore not as important for this case as that produced by the ionospheric currents. However, this type of acceleration mechanism should become increasingly more important toward higher L shells in the magnetosphere.

The field fluctuations described in this paper are not too uncommon. PAI and SARABHAI (1964) show that they commonly occur during the sudden commencement and main phase of magnetic storms. The periods of the fluctuations range from about 30 to 60 minutes. They might therefore have an important effect on the acceleration and distribution of high energy protons and electrons in the magnetosphere.

References

NEWKIRK, L. L. and WALT, M.: 1964, *J. Geophys. Res.* **69,** 1759.
PAI, G. L. and SARABHAI, V. A.: 1964, *Planet. Space Sci.* **12,** 855.

Discussion

CAIN said Sugiura has shown the enhanced amplitude of the normal daily variations of Huancayo is probably due to normal ionospheric conductivity and winds and not due to enhanced precipitation.

ENERGY SPECTRUM OF INNER ZONE PROTONS

STANLEY C. FREDEN

Aerospace Corporation, El Segundo, California

Abstract. A summary of the early measurements of the proton spectra in the inner radiation zone is presented. A brief discussion is then given of the theoretical calculations which have been made to explain the measured proton spectra. Some recent data are presented in which proton spectra were measured over a wide range of *B-L* space from satellites and the experimentally measured spectra are compared with the theoretical calculations. Several conclusions are drawn about the source and loss mechanisms for protons in the inner radiation zone.

1. Introduction

A major objective of present day investigations on trapped radiation is to understand the injection and loss processes for each different species of particles. If a theory for the presence of a particular species of particles is to be considered correct and complete, it should be possible to explain both the absolute intensities of the particles as a function of position and time and the shape of the energy spectrum which may also be a function of position and time. In some cases, in fact, the shape of the energy spectrum may be a more sensitive test of a particular injection or loss mechanism than the absolute flux itself. A particular example of this is the Cosmic Ray Albedo Neutron Decay (CRAND) theory where the calculation of the absolute flux involves both the intensity of the neutron decay source and the absolute averaged air densities which the particles see in their trajectories around the earth. Both of these quantities are largely uncertain. The energy spectrum, however, in the steady-state condition, does not depend strongly on the density of the atmosphere or its composition. Thus, one may be able to obtain different information about the source and the loss mechanisms by looking at the energy spectrum than one obtains by measuring the absolute total flux of particles above a given energy threshold.

It is the purpose of this paper to present a summary of the experimental data about the proton spectrum in the inner radiation belt. A comparison of the measured spectral shapes with what might be expected from the Cosmic Ray Albedo Neutron Decay theory will also be presented. Little or no emphasis will be placed on whether or not the absolute flux agrees with that expected from the theory. For this discussion the inner radiation belt is considered to be at $L \lesssim 2.5$. This limit is somewhat arbitrary, but does represent the approximate position above which protons with energies greater than 40 MeV are no longer present in significant numbers.

2. Early Measurements

The energy spectrum of protons from 75 to 700 MeV was first measured in 1959 when a small stack of nuclear emulsions was flown to an altitude of about

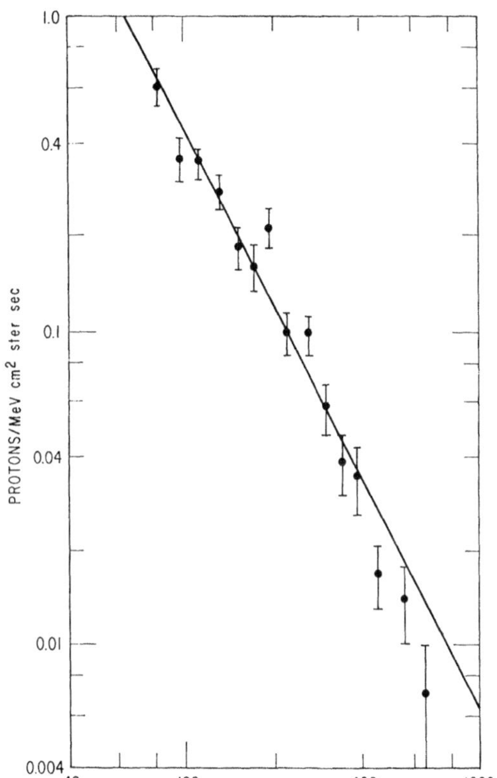

Fig. 1. Differential proton spectrum from 75 to 700 MeV measured in 1959 at an altitude of 1200 km. The solid line is a least squares fit to the data points (FREDEN and WHITE, 1959).

1200 kilometers on a Thor Able ballistic missile (FREDEN and WHITE, 1959). This was also the first direct evidence that energetic protons were the main penetrating component of the trapped radiation in the inner radiation belt. A plot of the measured proton spectrum is shown in Figure 1. In 1960, an Atlas vehicle was flown along a similar trajectory and carried two sets of nuclear emulsion stacks. One was supplied by HECKMAN and ARMSTRONG (1962), the other by FREDEN and WHITE (1962). On this flight, the energy spectrum was remeasured and was extended down to about 12 MeV. Although the nuclear emulsions integrated the flux over the trajectory of the vehicle the "mean" position where one can consider the measurement to have been made is at

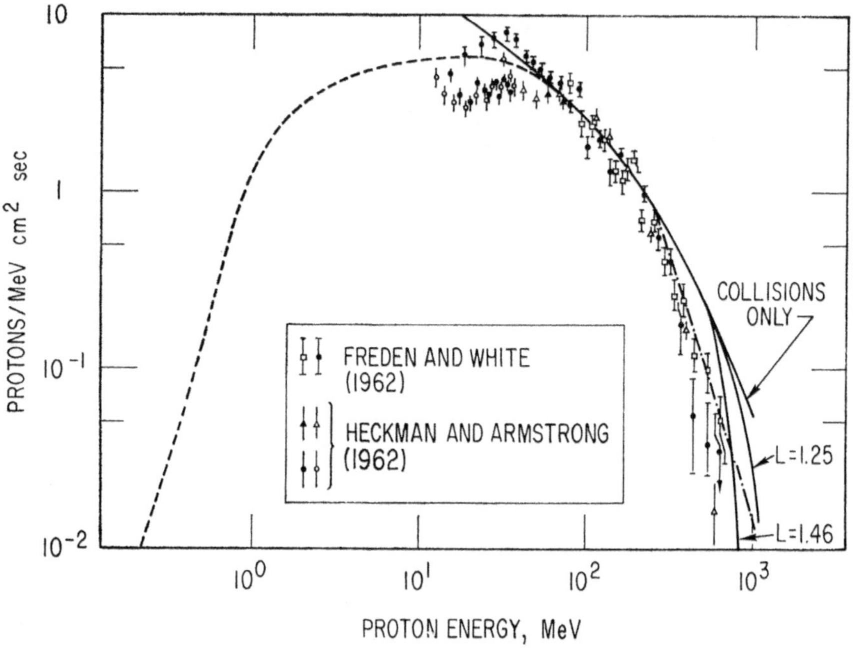

Fig. 2. Proton spectra from the nuclear emulsion experiments of FREDEN and WHITE (1962) and HECKMAN and ARMSTRONG (1962) at $L \simeq 1.30$, $B \simeq 0.200$, $B/B_0 \simeq 1.41$. The curves are those calculated by HAERENDEL (1964). The solid curves were calculated using losses due to atmospheric interactions and assuming the breakdown of the magnetic moment, μ, in a static dipole field ($L = 1.25$, $L = 1.46$). The dot-dash curve includes the effects of the breakdown of the third invariant Φ. The dashed curve includes losses due to charge exchange.

$L \simeq 1.30$, $B \simeq 0.200$. The data derived from these two emulsion stacks are presented in Figure 2. The solid lines in Figure 2 are theoretical curves and are discussed below. Although the geometry of the two exposures was similar, the two sets of data derived from them differ somewhat. This difference must therefore be attributed to different experimental techniques for the data analysis. Another rocket flight in 1960 carried a rotating nuclear emulsion stack to an altitude of approximately 1900 kilometers (NAUGLE and KNIFFEN, 1963). This experiment allowed a measurement of the proton spectrum as a function of position along the rocket trajectory. It showed very strong changes in the proton spectrum as a function of L, especially for protons below about 30 MeV. Figure 3 presents a sum-

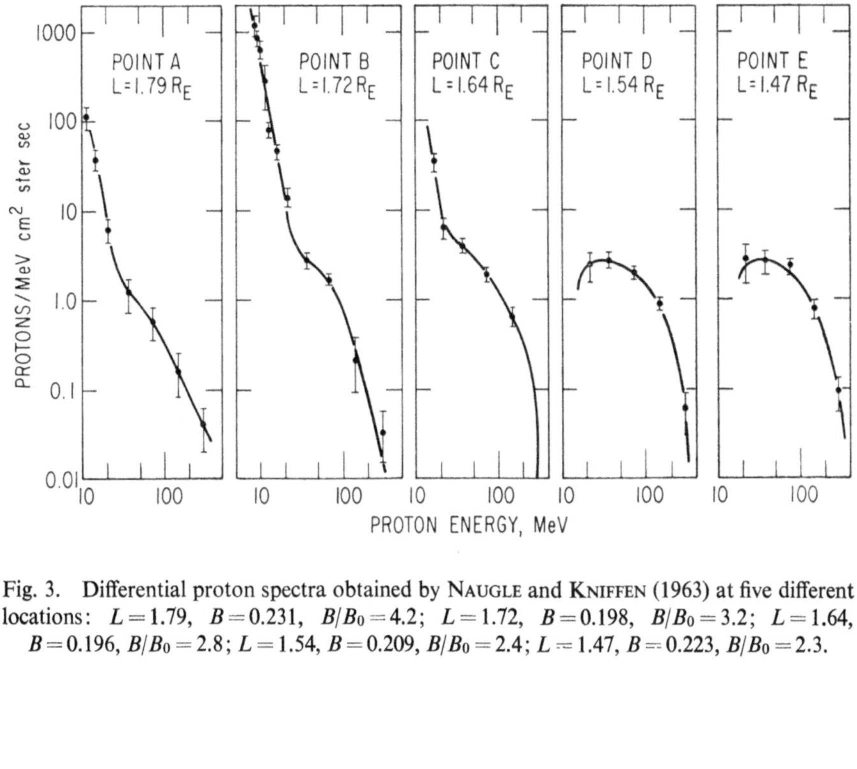

Fig. 3. Differential proton spectra obtained by NAUGLE and KNIFFEN (1963) at five different locations: $L=1.79$, $B=0.231$, $B/B_0=4.2$; $L=1.72$, $B=0.198$, $B/B_0=3.2$; $L=1.64$, $B=0.196$, $B/B_0=2.8$; $L=1.54$, $B=0.209$, $B/B_0=2.4$; $L=1.47$, $B=0.223$, $B/B_0=2.3$.

mary of the proton spectral measurements from this experiment. The variation of the proton spectral shape with L in the inner zone has further been demonstrated by MCILWAIN and PIZZELLA (1963) using Explorer IV data and by BOSTROM *et al.* (1965) using data from Injun I. The latter data also suggested a dependence of the spectral shape on B, the value of the local magnetic field, at a given L.

The spectrum of protons at $L=2.50$, $B=0.069$, $B/B_0=3.46$ was measured in the energy range from 1 to 7.3 MeV from a rocket by BAME *et al.* (1963). These measurements confirmed and extended those of NAUGLE and KNIFFEN at $L=1.79$ and showed that the proton spectrum continues to rise steeply at the higher L values at least down to energies of 1 MeV. The results of this experiment are illustrated in Figure 4.

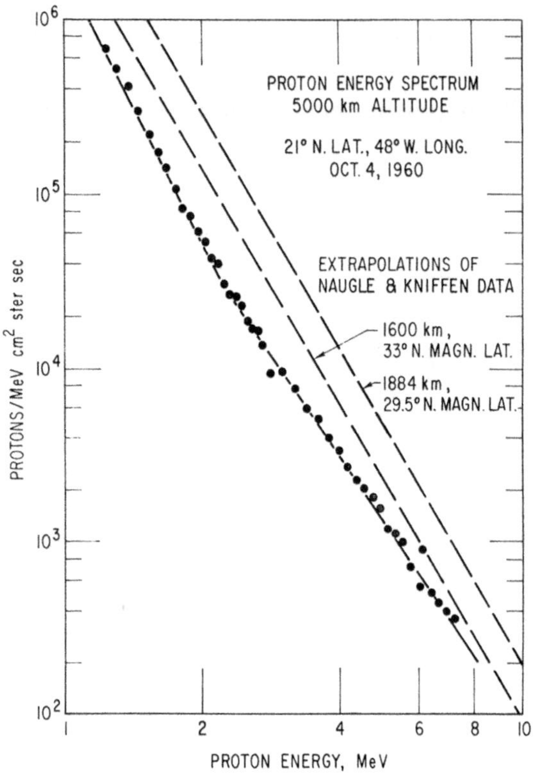

Fig. 4. Differential proton spectrum at 5000 km, $L = 2.50$, $B = 0.069$, $B/B_0 = 3.46$, measured by BAME *et al.* (1963) for protons with pitch angles between 60° and 90°. The dashed curves are extrapolations of the data of NAUGLE and KNIFFEN (1963) at $L = 1.79$ (1600 km) and at $L = 1.72$ (1884 km).

3. Theoretical Considerations

Before continuing with the presentation of the experimental data on the proton spectrum, it is worth pausing at this point to discuss briefly some of the theoretical work which has been carried on to explain the shape of the measured spectrum. The first calculations which gave good agreement between the experimental results and the theory were carried out by FREDEN and WHITE (1960). In these calculations it was assumed that the source of the protons was the decays of albedo neutrons produced in the earth's atmosphere by energetic cosmic rays. The loss mechanisms were assumed to be atmospheric energy loss and nuclear interactions; other loss mechanisms were neglected. The input albedo neutron spectrum was assumed to have the form $\Phi_n(E) = kE^{-2.0}$

neutrons/MeV/cm^2/sec (HESS, 1959). It was also assumed that the injection coefficient for protons from the decays of the neutrons was constant. One can then write down the continuity equation which allows for the flow of particles along the energy axis which, using the source and loss functions described above, in the steady state condition, becomes,

$$\frac{kE^{-2.0}}{\beta\gamma c\tau_n}\left[\frac{R_e}{R}\right]^2 - \frac{d}{dE}\left[N_p(E)\frac{dE}{dt}\right] - \eta\sigma N_p(E)\beta c = 0.$$

In this equation, β and γ are the usual velocity dependent terms for the proton, assumed to be the same as for the neutron, τ_n is the neutron lifetime, R is the radial distance of observation, N_p is the trapped proton *density*, η is the atmospheric density and σ is the nuclear interaction cross section for the trapped protons. Solution of this equation for $N_p(E)$ yields the desired proton spectrum, $\Phi_p(E)$, protons/MeV/cm^2/sec, which can be directly compared with the data, since $\Phi_p = N_p\beta c$. This theory gave rather good agreement with the early data (FREDEN and WHITE, 1960) in spite of the rather simplified model that was used. Several refinements and extensions to this theory have been made by LENCHEK and SINGER (1963) and by HAERENDEL (1964). These calculations take into account the effects of the anisotropy of the albedo neutron source and the effects of other loss mechanisms such as the breakdown of the various adiabatic invariants. Figure 2 shows the results of the calculations of Haerendel. The two sets of data shown are those of FREDEN and WHITE (1962) and HECKMAN and ARMSTRONG (1962) referred to earlier. The solid curves represent the expected spectra at various radial distances assuming a breakdown of the magnetic moment, μ, in a static dipole field. The dot-dash curve includes the effects of the breakdown of the third adiabatic invariant, Φ, the integral invariant. The dashed curve includes also the effects of charge exchange at the lower energies. One sees reasonable agreement between the data points and the theoretical calculations. It should be emphasized that the input neutron spectrum is not a well known function over the energy region considered and therefore the small discrepancies, both below about 50 MeV and above about 300 MeV, could be due to a neutron source function which is different than that assumed in the calculations (FREDEN and WHITE, 1962).

Although the present theory, as briefly described above, appears to be adequate to explain the shape of the proton spectrum at low L values, at least in the energy region above 10 MeV, there is a considerable deviation in the shape of the proton spectrum at higher L values especially at energies below 30 MeV. This can be seen by referring to the data of NAUGLE and KNIFFEN (1963) in Figure 3 at $L = 1.79$. It will also be apparent in data to be presented later in this paper. It has been proposed (ARMSTRONG *et al.*, 1961; NAUGLE and KNIFFEN, 1961; LENCHEK and SINGER, 1963; LINGENFELTER and FLAMM, 1964) that the excess of protons at lower energies and at the higher L values could be explained as due to the injection of protons from neutrons produced at the polar caps by solar protons arriving from solar flares. The absolute calculations on this point appear as yet inconclusive (LENCHEK and SINGER, 1963).

4. Recent Results

Although the previously described nuclear emulsion measurements of the trapped proton spectrum gave relatively fine spectral detail they were very limited in spatial coverage. With the relatively recent availability of large

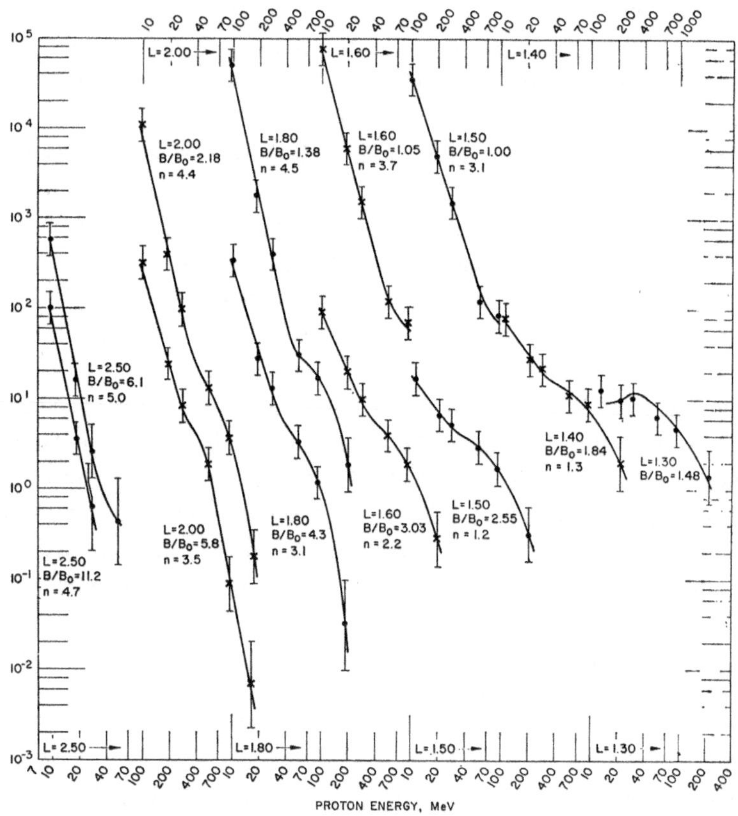

Fig. 5. Differential omnidirectional proton spectra as functions of the magnetic shell parameter L and the ratio B/B_0. The solid curve through the data points at $L = 1.30$ is the spectrum of protons at $L \simeq 1.30$, $B/B_0 \simeq 1.41$ measured using nuclear emulsions in 1960 (FREDEN and WHITE, 1962). The exponent n is obtained by fitting the spectrum in the 10 to 30 MeV region by a power law of the form E^{-n} (FREDEN *et al.*, 1965).

satellite payloads it has become possible to fly rather detailed counter experiments to measure the proton spectrum as a function of position in the radiation belt. Such experiments were carried out aboard satellite 1964-45A by FREDEN *et al.* (1965) and by FILLIUS and MCILWAIN (1964) aboard Relay I. The former experiment, which consisted of a series of five variously shielded solid state detectors, measured the omnidirectional differential proton energy spectrum in six energy intervals from 10 to 200 MeV. The satellite was in a near polar orbit with a perigee of 266 km and an apogee of 3765 km. This orbit permitted wide spatial coverage in the inner radiation zone. Some typical omnidirectional differential proton spectral measurements are shown in Figure 5 for L-values between 1.3 and 2.5. The quantity B/B_0 is the ratio of the value of the magnetic field at the point of observation to the value at the magnetic equator on the same L shell.

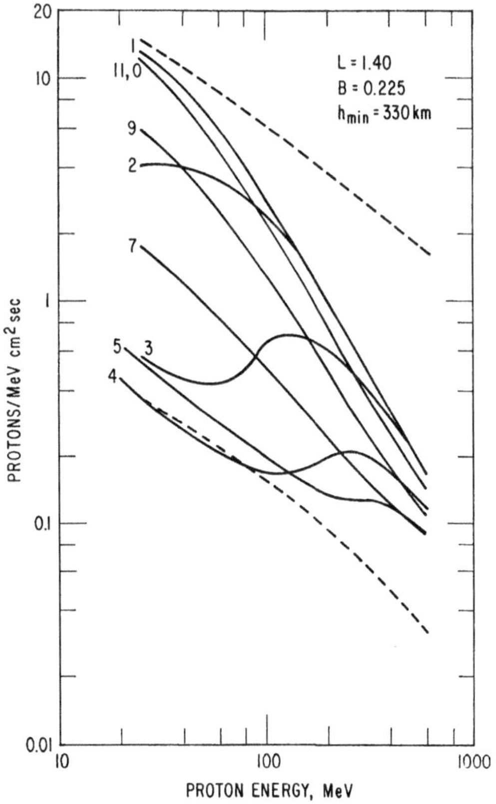

Fig. 6. Calculated proton energy spectra at different times in the solar cycle for $L = 1.40$, $B = 0.225$, $B/B_0 = 2.0$. The curves are labeled by the time in years from solar minimum (BLANCHARD and HESS, 1964).

The data at $L = 1.30$, $B/B_0 = 1.48$ can be directly compared with the spectrum at $L \simeq 1.30$, $B/B \simeq 1.41$ measured in 1960 using nuclear emulsions (FREDEN and WHITE, 1962). One sees no significant change in the spectrum above $\simeq 15$ MeV in this region of space during the four year interval, even though the earlier measurements were made near solar maximum and the present ones near solar minimum. This is in contrast to the calculations of BLANCHARD and HESS (1964) who predicted large changes in the proton spectrum during the course of the solar cycle because of the changes in the density of the upper atmosphere. These changes in spectral shape would be expected in those regions of space where the proton lifetimes are controlled by the atmospheric density and are short compared to the period of the solar cycle. An example of the changes which might be expected are shown in Figure 6 for $L = 1.40$, $B = 0.225$, $B/B_0 = 2.00$.

The spectra at $L \geq 1.4$, shown on Figure 5, can best be considered in two energy regions. Below 35 MeV there is a definite increase in the flux of protons compared to the spectrum at $L = 1.30$, as one goes to higher L-values. This increase also depends quite strongly on B/B_0. For $L \geq 1.4$ the spectra between 10 and 30 MeV can be well represented by expressions of the form $\Phi(E)\,\mathrm{d}E = KE^{-n}\,\mathrm{d}E$ at all values of B/B_0. The value of n has been derived as a function of B/B_0 for the various L values and is presented in Figure 7. The straight lines

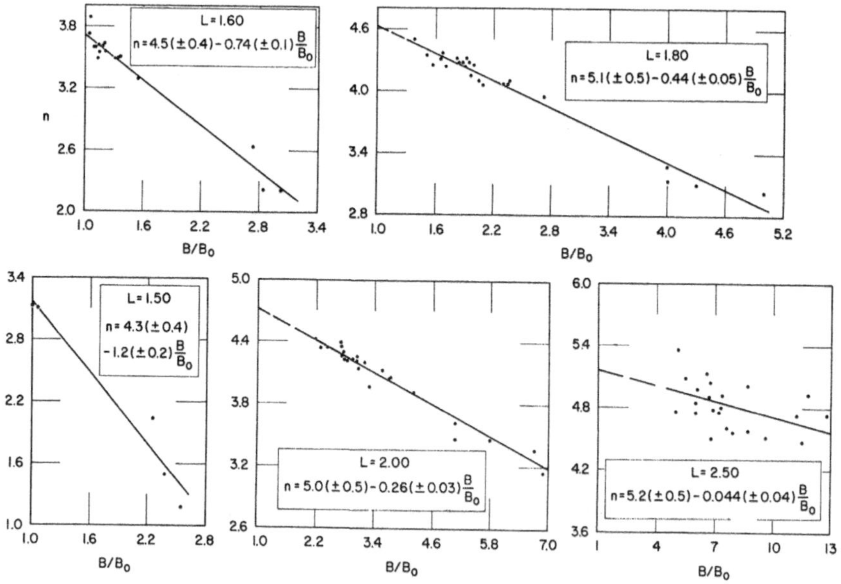

Fig. 7. Variation of the value of n with L and B/B_0 for protons from 10 to 30 MeV. The differential spectra in this energy region are represented by a power law of the form E^{-n}. The solid lines through the points are least squares fits to the data points (FREDEN *et al.*, 1965).

through the points are least squares fits to the data. It is seen that the dependence of n on B/B_0 decreases with increasing L.

The spectra above 55 MeV can be represented by expressions of the form $\Phi(E)\,dE = Ke^{-E/E_0}\,dE$. Figure 8 represents E_0 as a function of B/B_0 for $L=1.80$

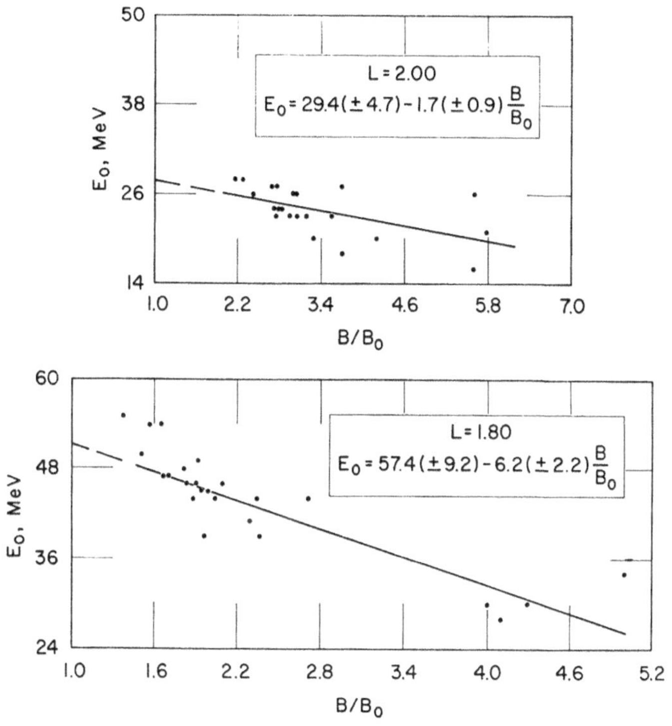

Fig. 8. Variation of the value of E_0 with L and B/B_0 for protons above 55 MeV. The differential spectra in this energy region are represented by the form $\exp(-E/E_0)$. The solid lines through the points are least squares fits to the data points (FREDEN *et al.*, 1965).

and $L=2.00$. It is seen that E_0 decreases both with increasing L and with increasing B/B_0.

The following qualitative interpretation of these data is proposed here. At $L \approx 1.3$ the proton spectrum between 10 and 200 MeV is reasonably well predicted from the Cosmic Ray Albedo Neutron Decay injection theory (CRAND) with losses due to atmospheric energy loss and nuclear interactions (FREDEN and WHITE, 1960; LENCHEK and SINGER, 1963; HAERENDEL, 1964).

Above 200 MeV the deviation of the data from that theory can be explained by a breakdown of the trapping conditions or by a steeper neutron source function. The deviation below $\simeq 35$ MeV may well be due to a neutron source function which is different than that used in the calculations (FREDEN and WHITE, 1962; LENCHEK and SINGER, 1963; HAERENDEL, 1964). For $L \gtrsim 1.4$ there is at least one other source of trapped protons. The energetic ones, above an energy which may be L (and B) dependent but which is of the order of

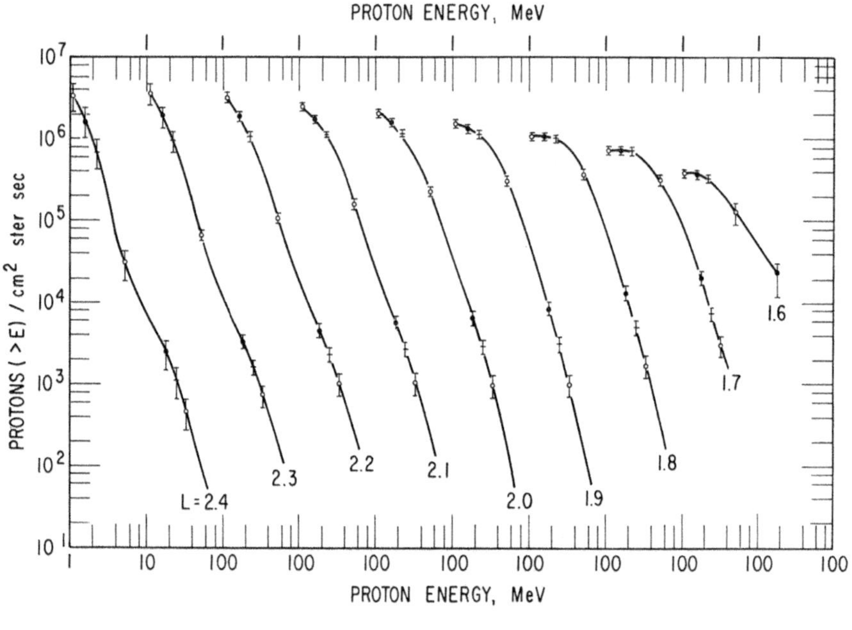

Fig 9. Integral proton energy spectra for pitch angles of 90° at the magnetic equator. Each spectrum extends from 1 to 60 MeV, but their starting points have been shifted to conserve space (FILLIUS and MCILWAIN, 1964).

40 to 50 MeV at L values below about 2, still come mainly from CRAND. The spectrum gets softer with increasing L in a manner which would be expected from losses due to non-adiabatic effects. The spectrum also gets softer, at least at L values of 1.8 and 2.0, with increasing B/B_0. This is consistent with the experimental results of GARMIRE (1963) and with theoretical calculations concerning the breakdown of the first adiabatic invariant in a static dipole field.

For protons below 35 MeV at $L \gtrsim 1.4$ another important source exists. This source, or sources, and/or additional loss mechanisms give strongly L and B/B_0 dependent spectra. This apparent source could be a redistribution or diffusion of protons whose origin was the decay of neutrons produced in the atmosphere by galactic or solar cosmic rays. It is worth noting that the variation of spectral shape as a function of B/B_0 for protons below 35 MeV is opposite to that for the higher energy protons.

The experiment of FILLIUS and MCILWAIN (1964) also shows the anomalous behavior of the proton spectrum below about 40 MeV. These authors have plotted the integral directional spectrum of protons measured perpendicular to the magnetic equator on the Relay I satellite in 1963. Their results, shown in Figure 9, again illustrate a rather large excess of protons below about 35 MeV compared to what would be expected from the Cosmic Ray Albedo Neutron Decay theory.

5. Conclusions

The following general conclusions can be drawn concerning the data pertaining to the proton spectrum in the inner radiation belt.

(1) The spectrum at low L values, about 1.3, is in reasonable agreement with what one would expect on the basis of the Cosmic Ray Albedo Neutron Decay source and losses due to atmospheric energy loss and nuclear collisions. The small deviations of the data from the theoretical predictions can be explained as due to additional loss mechanisms such as the breakdown of the magnetic moment in a static dipole field and/or, as due to an input neutron source spectrum which is different than that assumed in the calculations.

(2) As one goes to higher L values, a strong component of protons between 10 and 35 MeV appears. The flux of these particles is a strong function of B and L increasing with increasing L at the magnetic equator up to $L \approx 1.6$ but decreasing at a given L as one moves to higher B values. This component could be due to an additional source for the trapped protons or could possibly be the result of a redistribution or diffusion of particles across magnetic lines of force.

(3) The proton spectrum above about 55 MeV becomes steeper as one goes to higher L values. It also becomes steeper, at a given L value, as one goes to higher values of the magnetic field strength. This is expected if these particles are being lost because of the breakdown of their magnetic moments in the static field.

References

ARMSTRONG, A. H., HARRISON, F. B., HECKMAN, H. H., and ROSEN, L.: 1961, *J. Geophys. Res.* **66**, 351.
BLANCHARD, R. C. and HESS, W. N.: 1964, *J. Geophys. Res.* **69**, 3927.

Bame, S. J., Conner, J. P., Hill, H. H., and Holly, F. E.: 1963, *J. Geophys. Res.* **68**, 55.
Bostrom, C. O., Zmuda, A. J., and Pieper, G. F.: 1965, *J Geophys. Res.* **70**, 2035.
Fillius, R. W. and McIlwain, C. E.: 1964, *Phys. Rev. Letters* **12**, 609.
Freden, S. C., Blake, J. B., and Paulikas, G. A.: 1965, *J. Geophys. Res.* **70**, 3113.
Freden, S. C. and White, R. S.: 1959, *Phys. Rev. Letters* **3**, 9.
Freden, S. C. and White, R. S.: 1960, *J. Geophys. Res.* **65**, 1377.
Freden, S. C. and White, R. S.: 1962, *J. Geophys. Res.* **67**, 25.
Garmire, G.: 1963, *J. Geophys. Res.* **68**, 2627.
Haerendel, G.: 1964, *Fortschritte der Physik* **12**, 271.
Heckman, H. H. and Armstrong, A. L.: 1962, *J. Geophys. Res.* **67**, 1255.
Hess, W. N.: 1959, *Phys. Rev. Letters* **3**, 11.
Lenchek, A. M. and Singer, S. F.: 1963, *Planet. Space Sci.* **11**, 1151.
Lingenfelter, R. E. and Flamm, E. J.: 1964, *J. Geophys. Res.* **69**, 2199.
McIlwain, C. E. and Pizzella, G.: 1963, *J. Geophys. Res.* **68**, 1811.
Naugle, J. E. and Kniffen, D. A.: 1963, *J. Geophys. Res.* **68**, 4065.

Discussion

Hess agreed with Freden that the high energy neutron energy spectrum is not well known. His highest measurement was at about 200 MeV and theoretically the spectrum should get steeper beyond this. This might explain the apparent deficiency of high energy trapped protons. Walt said measurements by Hughes at Durham on neutron multiplicity disagrees with the Hess spectrum. Hess said Singer's neutron spectrum derived from studies of stars in emulsions agrees with his spectrum. There was general agreement that the high energy spectrum was not well known and should be measured.

Harris questioned the need for breakdown of the adiabatic invariants at high energies to explain the spectral shape. Freden answered that Haerendel had assumed this and he said that various calculations of orbits showed that breakdown did occur for high enough energies and L-values.

ELECTRON AND PROTON OBSERVATIONS

LUDWIG KATZ

Air Force Cambridge Research Laboratories, L. G. Hanscom Field, Bedford, Massachusetts

Introduction

The Air Force Cambridge Research Laboratories have over the past few years carried out a modest research program in trapped radiation. Taking into account the fact that at this meeting you are also being made aware of the results from the many other experimental groups in this field, I have chosen three specific experimental results for this lecture which I believe to contain important features somewhat unique to our experiments.

The three experiments I shall discuss in some detail are the following:

(A) Results on 1–4 MeV electrons as observed during the fall of 1963 on a polar satellite, with an emphasis on the changes this population is subject to even at a period of relative solar quiet.

(B) Results on 1 to 4 MeV protons obtained in a recent rocket flight near the equator up to L values of 3.7.

(C) Results on 55 MeV protons observed periodically over a period of 4 years mirroring at the South Atlantic anomaly.

1. Energetic Electron Measurements in July and August 1963

ORBIT-INSTRUMENTATION

The Air Force Sub-satellite Hitch-hiker I (1963-25B) was launched on 1 July 1963 into a near polar orbit with perigee at 326 km and apogee at 4150 km. The orbital period was 133 min and the satellite was spin stabilized at 68 rpm.

Intensities, energy spectra and pitch angle distributions of electrons between about 0.7 MeV and 4 MeV were obtained from a 2.5 g/cm^2 plastic scintillator spectrometer with a 25° aperture and a 0.12 cm^2 ster. geometrical factor. The dynamic range was 8×10^1 to 10^6 particles/cm^2 sec ster. An auxiliary channel, which measured the power deposited in the scintillator by electrons above 0.2 MeV, had a range of 8×10^2 to 8×10^8 MeV/cm^2 sec ster and could be used to derive electron intensities in regions where the count rate channel approached saturation. Spectral information was obtained by pulse height analysis using a comparator with a variable voltage sweep. The spectrometer output was sampled 6 times/sec, while the spin frequency was 1.1 rotations/sec, so that pitch angle distributions with good angular resolution could be obtained. Two sets of 3-axis flux gate magnetometers were used for aspect determination. Detailed information on the various instruments aboard Hitch-hiker I was given by PAOLINI *et al.* (1965).

RESULTS

Unidirectional Intensities

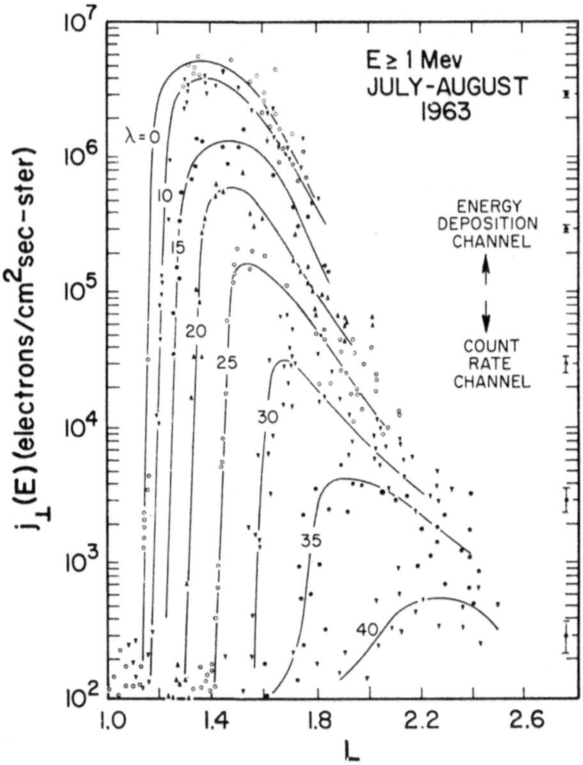

Fig. 1. Electron intensities in the inner belt at various λ.

Figure 1 presents perpendicular intensities of electrons above 1 MeV in the inner belt. Data obtained during the whole observation period are presented in curves of intensity vs. L at a constant B/B_0 or, equivalently, at a constant magnetic latitude λ, defined by

$$B/B_0 = \frac{(1 + 3\sin^2\lambda)^{1/2}}{\cos^6\lambda}$$

In regions where the count rate channel approached saturation electron intensities have been derived from the rate of energy deposition channel.

In the outer belt, time variations of one to two orders of magnitude occurred during the observation period. Figure 2 presents perpendicular intensities of electrons above 1 MeV as mapped on a day of high electron intensities (Figure 2a), and also during a period of low intensities (Figure 2b).

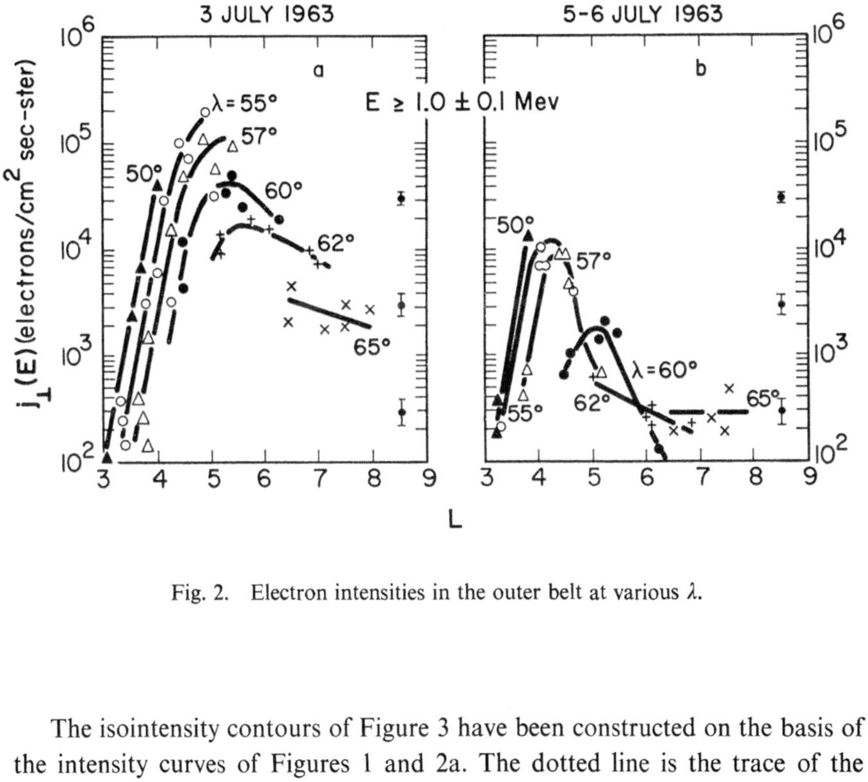

Fig. 2. Electron intensities in the outer belt at various λ.

The isointensity contours of Figure 3 have been constructed on the basis of the intensity curves of Figures 1 and 2a. The dotted line is the trace of the (R, λ) surface which has a minimum altitude of 100 km. In the horns of the outer belt at $\lambda \approx 55°$ to $62°$ significant electron fluxes reach beyond the 100 km minimum altitude surface being therefore scattered into the atmosphere at the Anomaly. Further information on electron intensities at altitudes below the levels reached by the satellite is provided by pitch angle distributions obtained during low satellite passes. On the basis of measured intensities and pitch angle distributions at specific times, loss rates up to 2×10^{17} electrons/sec or about 2×10^{11} ergs/sec have been computed for dumping of electrons above 1 MeV at the Anomaly.

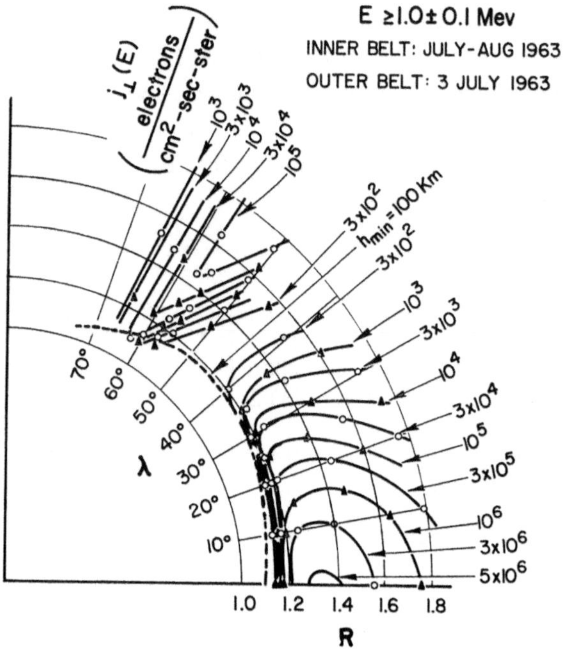

Fig. 3. Electron isointensity contours.

Pitch Angle Distributions and Integral Energy Spectra

Figure 4 shows a typical pitch angle distribution obtained in the inner belt. The dotted curve indicates the distribution expected on the basis of the perpendicular intensities of Figures 1 and 3. The background intensity measured within the loss cone is very likely due to electrons of energy above 10 MeV and/or protons above 65 MeV which penetrate the shielding around the scintillator. Figure 5 shows typical pitch angle distributions obtained in the horns of the outer belt. Measured distributions were, in general, consistent with those expected from the perpendicular intensity maps (dotted curves) within the accuracy limits determined by the instrument aperture and the uncertainty in aspect determination.

Energy spectra measured in the inner belt were usually distorted due to contamination by energetic particles which penetrated the shielding around the

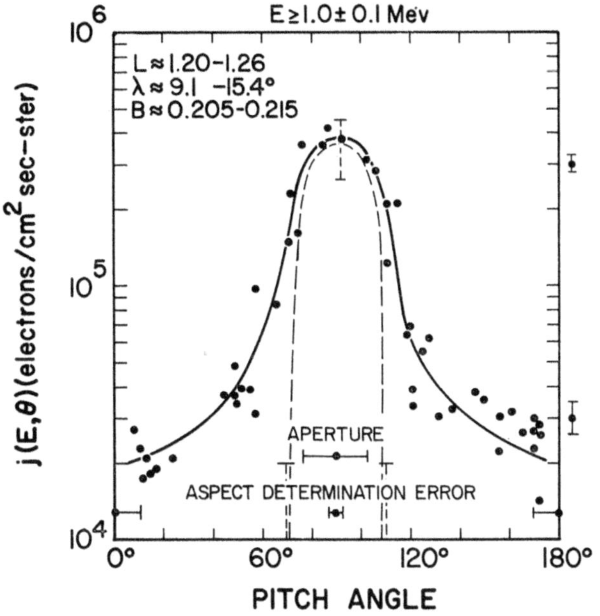

Fig. 4. Inner belt electron pitch angle distribution.

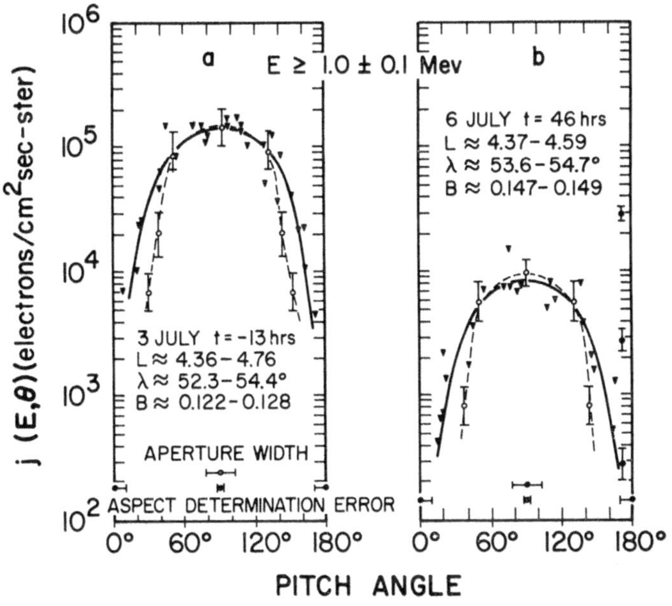

Fig. 5. Outer belt electron pitch angle distribution.

scintillator. At low altitudes where the loss cone is large (Figure 4) the omnidirectional background due to the penetrating particles could be corrected for by subtracting the intensity within the loss cone ($j_0(E)$) from the perpendicular intensity ($j_\perp(E)$). An integral energy spectrum obtained in this manner is shown in Figure 6. Measured inner belt electron spectra had approximately

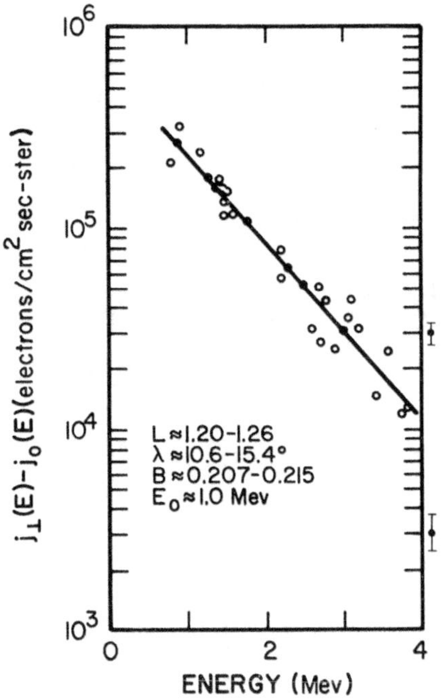

Fig. 6. Inner belt electron integral energy spectrum.

an exponential energy dependence exp $(-E/E_0)$ with E_0 between 0.7 and 1.2 MeV.

Figure 7 shows a typical outer belt electron integral energy spectrum between 0.7 and 4 MeV. Throughout the observation period and in outer belt regions with high enough intensities for an adequate spectrum determination, all spectra obtained could be fitted with exponential slopes $E_0 \approx 0.4 \pm 0.1$ MeV.

Outer Belt Time Variations

During the present observation period large time variations which correlated with solar and geomagnetic activity were observed in the outer belt. A summary plot of intensity levels at $L \approx 5$, $\lambda \approx 60°$ is given in Figure 8. The occurrence of solar flares and daily values of the magnetic variation A_p (LINCOLN, 1963a and b) are displayed in the same figure.

In the cases of the intensity decreases around 4 July, 17 July, 30 July and 19 August 1963, sudden increases in A_p had occurred at about the time of the

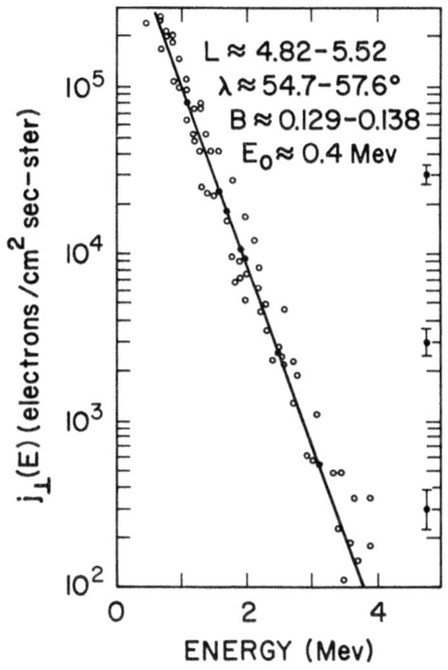

Fig. 7. Outer belt electron integral energy spectrum.

fall in intensity. Solar flares had been observed in three of these four instances. The intensity decrease during the magnetically quiet period between 10 August and 17 August occurred gradually with a simultaneous shift of the intensity peak toward lower L values. Following the depletions of 4 July, 17 July and 30 July the outer belt recovered to pre-storm intensity levels with recovery times which varied from a few to about 10 days.

For $L < \sim 4$ no significant time variations were observed during the observation period. The steep inner edge of the outer belt (Figures 2 and 3) remained relatively unchanged with time while the outer part of the belt underwent considerable variation. Due to the relative invariance of intensity levels at the low L edge of the outer belt, whenever the rest of the belt was depleted the intensity peak shifted toward low L shells.

The depletion and recovery cycle that occurred from 3 July to 15 July 1963 will now be examined in more detail. A class 1^+ solar flare occurred at 0734 UT, 4 July 1963 (4th quadrant, 8° latitude, 79° to meridian). Prior solar activity consisted of two flares on 28 June (class 1^+ in quadrant 4, class 3 in quadrant 1) and subsequent activity consisted of a flare on 25 July (class 1, quadrant 4). The

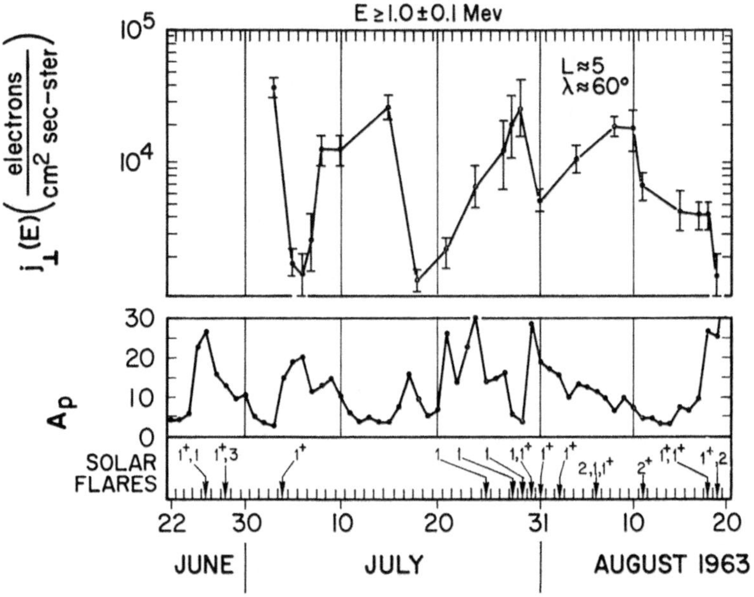

Fig. 8. Electron intensity fluctuations in the outer belt.

magnetic variation A_p was 3 on 2 and 3 July, went up to 14 on 4 July and stayed at 10–20 until 10 July; there was also Thule Riometer activity on 5 July. The occurrence of the flare will be taken as a time reference point in the subsequent discussion.

The change that occurred in the intensity profiles of the outer belt can be seen by comparing Figure 2a (data obtained 1 h to 5 h before the flare) with Figure 2b (25 to 54 h after the flare). Electrons above 1 MeV are depleted by one order of magnitude at $L \approx 4.5$ to about two orders of magnitude for $L > 5.5$. After the depletion, the electron intensity started to build up again until it essentially returned to the pre-storm level. Provided that the whole outer belt is depleted in the same way as the observed region and with the assumption of a constant logarithmic loss mechanism, it is found that a source of electrons of the order of $10^{13}-5\times10^{13}$ ergs/sec must be present in order to account for the observed recovery time.

During the observed time variations the relative position of various constant λ profiles changed considerably, particularly at higher L's, as can be seen, for instance, by comparing Figures 2a and 2b. This indicates that changes in mirror point distributions on certain magnetic shells occurred as well as variations in intensity alone. Such changes should also be evident from the pitch angle distributions on those magnetic shells. Unfortunately, for the regions where the intensity levels remained high enough to permit an adequate direct measurement of pitch angle distributions (i.e., around $L\sim4.5$) no significant changes were expected from the intensity profiles. Two such distributions obtained in about the same region 13 h before and 45 h after the flare have been given in Figure 5 together with the expected distributions on the basis of the intensity contours. Although the intensities are considerably different, the shape of the distribution remains essentially unchanged.

During the depletion and recovery cycle of 0.7 to 4 MeV electrons in the outer belt, the parameter E_0 of an exp $(-E/E_0)$ spectral shape, changes from about 0.4 MeV to 0.3 MeV after the flare returning later to 0.4 MeV. Since the very low intensity after the flare reduces the accuracy with which E_0 can be determined, the observed steepening may not be significant.

DISCUSSION

Electron intensities detected by Hitch-hiker I in the inner belt during July–August 1963 were at about the level reported by VAN ALLEN (1964) from Injun III data for June 1963. They are about one order of magnitude lower than electron fluxes measured by Explorer XV around January 1963 (MCILWAIN, 1963) and Relay I for May through September 1963 (MCILWAIN *et al.*, 1964). Electron data from a low altitude polar satellite reported by BOSTROM and WILLIAMS (1965) for October 1963 refer mainly to regions near the steep edge of the inner belt where comparison is rather difficult, although their intensity levels seem to be comparable to the ones reported here.

Measured inner belt spectra with exponential slope $0.7\text{ MeV}\leqslant E_0\leqslant1.2\text{ MeV}$ are not much different from the fission like spectra observed after the Starfish explosion (MOZER *et al.*, 1963; BROWN and GABBE, 1963).

The range of observed intensities for electrons above 1 MeV at the horns of the outer belt was roughly 10^3-10^5 electrons/cm^2 sec ster. Intensities measured with polar satellites near the horns of the outer belt by WILLIAMS and SMITH (1965) and PAULIKAS and FREDEN (1964) vary roughly around the same levels, while considerably lower hard electron intensities are reported by ARMSTRONG

(1965). Similar intensities for energetic electrons are reported in outer belt regions closer to the magnetic equator by FRANK (1965), FRANK *et al.* (1964), FRANK *et al.* (1963), and HOFFMAN *et al.* (1962).

The observed location of the intensity peak at the horns of the outer belt was between $L \approx 4.5$ and 6 with a very sharp and rather stable cutoff at $L \approx 4$. During Explorer VII measurements (FORBUSH *et al.*, 1962) the peak of the outer zone at the horns was considerably lower, at $L \approx 3$ to 5.

Spectral data for the outer belt electrons in the literature refer mostly to the energy range from about 40 keV to 1.6 MeV. The steep spectra observed (WILLIAMS and SMITH, 1965; FRANK *et al.*, 1964; PAULIKAS and FREDEN, 1964; MOZER *et al.*, 1963 and ROSSER *et al.*, 1962) are not inconsistent with the 0.7 MeV to 4 MeV spectra reported in this paper.

The directly measured pitch angle distributions in the inner and outer belt are seen to be consistent with the distributions derived from $j_\perp$ within uncertainties due to the angular resolution of the instrument and the accuracy of the aspect determination and the limitations imposed by the existent omnidirectional background. Some directly measured pitch angle distributions near the horns of the outer belt are given by CLADIS *et al.* (1961). Their measurements were taken at a lower altitude and consequently the distributions are considerably narrower than those presented here.

The observed presence of energetic electrons in high latitude portions of outer belt magnetic shells that plunge into the atmosphere at the Anomaly is consistent with observations by FORBUSH *et al.* (1962) and PAULIKAS and FREDEN (1964). Based on the data of this paper, the estimated loss rate of energetic outer belt electrons due to such dumping (about 2×10^{11} ergs/sec) is about two orders of magnitude smaller than the electron source that is estimated from the recovery time constants after outer belt depletions (about 10^{13} to 5×10^{13} ergs/sec). Therefore loss of hard outer belt electrons through this mechanism must be of minor importance in the electron balance of the outer belt compared to other loss mechanisms such as scattering into the loss cone and immediate dumping into the atmosphere or escaping from the trapping region through broken magnetic lines (DESSLER and KARPLUS, 1961).

The pattern of time variations observed in the outer belt is consistent with previous observations performed at high magnetic latitudes as well as close to the equator. Observations near the equator seem to agree on a pattern of a decrease in the intensity of energetic electrons around the time of magnetic disturbances by factors ranging from less than 10 (FRANK *et al.*, 1964; HOFFMAN *et al.*, 1962; ARNOLDY *et al.*, 1960) up to 1000 (ROSSER, 1963). Subsequently the intensities recover over a period of a few to several days, often going up to several times the pre-storm levels (FREEMAN, 1964; FRANK *et al.*, 1964; HOFFMAN *et al.*, 1962). In the region of the horns of the belt FORBUSH *et al.* (1962)

have observed about a 10-fold intensity decrease for $L > 3.5$ at the time of magnetic storms with a simultaneous increase at lower L's. WILLIAMS and SMITH (1965) observe a rise of hard electron intensities a few days after magnetic disturbances.

Although no major magnetic storm occurred during the present observation period a number of substantial depletions and subsequent recoveries of the hard electron outer belt component were observed, associated with solar and magnetic activity. Considering the general pattern indicated by the quoted references the observed intensity decreases in the horns seem to be part of a general depletion of the energetic electrons in the belt rather than mere redistributions of magnetic shells in space.

The recovery times observed after outer belt depletions are similar to times mentioned in the quoted papers. The estimated electron source required ($\sim 10^{13}$ to 5×10^{13} ergs/sec) would have to arise from the solar wind ($\sim 10^{20}$ ergs/sec) with an acceleration mechanism.

The integral electron spectra measured between 0.7 MeV and 4 MeV are relatively constant in shape during the observed time variations, with possibly a slight steepening at the time of the intensity drop. This is to be contrasted with the considerable spectral steepening that occurs at lower energies where the soft electron intensities actually increase at the time of the hard electron depletion (e.g., FRANK *et al.*, 1964).

2. Measurements of 1 to 4 MeV Trapped Protons in March 1965

INSTRUMENTATION

A Blue Scout Junior vertical probe launched on 30 March 1965 from Cape Kennedy reached a peak altitude of about 16500 km. It crossed magnetic shells with L from about 2 to 3.7, at magnetic latitudes λ from 40° to the magnetic equator. The probe carried instruments to measure the flux, energy spectra, and pitch angle distributions of trapped protons and electrons in various energy ranges.

The data presented in this paper concern protons between 1 and 4 MeV, and were obtained from a solid state spectrometer. This instrument consisted of two ORTEC solid state detectors each with a depletion depth of about 200 microns. The first detector had a sensitive area of approximately 25 mm^2 and its thickness of 50 mg/cm^2 corresponds to the range of a 5 MeV proton. Pulses recorded in the first detector alone, in anticoincidence with pulses from the second detector, correspond to protons of energy below 5 MeV which are stopped in the first detector and therefore produce a pulse proportional to their total energy.

Spectral information was derived by pulse height analysis. The pulses were fed to one input of a comparator, the other input of which was fed with a varying DC voltage sweep of about 6 sec period. The comparator output signal repetition rate thus corresponded to fluxes of protons of energy greater than some threshold energy, proportional to the sweep voltage.

The field of view of the detector was a cone with half angle of 3.3° and the geometrical factor was $A\Omega \approx 0.0027\ \text{cm}^2$ ster. The output circuit was a logarithmic count rate meter with a dynamic range of 4×10^3 to 4×10^7 protons/cm² sec ster.

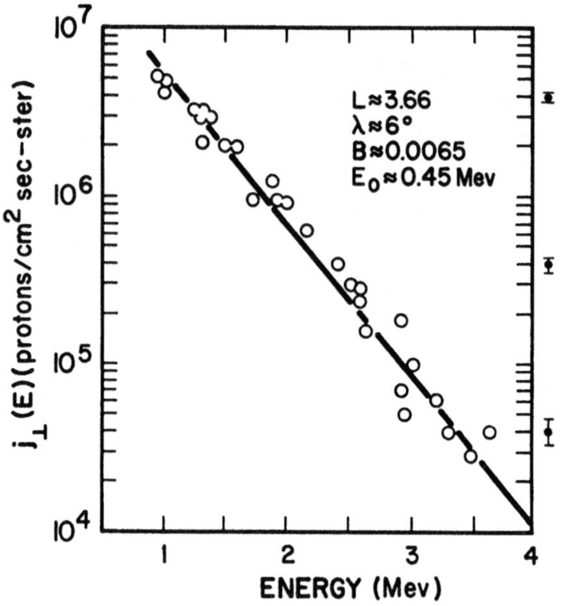

Fig. 9. Proton integral energy spectrum.

The spectrometer output was sampled 60 times/sec and the probe spin frequency was about 3 rotations/sec. Thus pitch angle distributions with good angular resolution could also be obtained.

RESULTS

Complete integral spectra over the energy range from about 1 to 4 MeV have been obtained for magnetic shells from $L \approx 2.5$ to $L \approx 3.7$. A typical spectrum is shown in Figure 9. Within the accuracy of the measurements, obtained spectra agreed with an exponential energy dependence exp $(-E/E_0)$. In the sampled region the slope parameter E_0 assumed values between approximately 0.4 and 0.6 MeV.

Figure 10 gives a pitch angle distribution of protons above 1 MeV. Due to the nearly perpendicular orientation of the probe spin axis with respect to the local magnetic field vector a complete scanning of pitch angles was possible during the largest part of the trajectory. On the basis of pitch angle distributions measured near the equator, information can be derived about perpendicular intensities at points situated at higher magnetic latitudes on the same magnetic

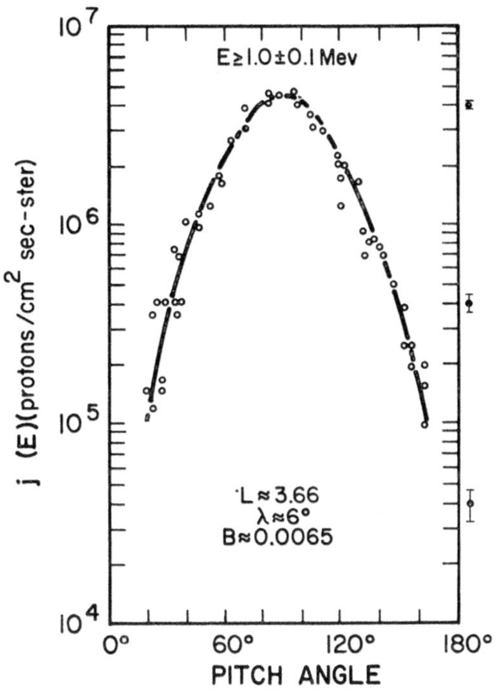

Fig. 10. Proton pitch angle distribution.

shell. Using perpendicular intensities derived in this manner from measured pitch angle distributions, and perpendicular intensities directly recorded along the rocket path, contours of intensity versus L at constant λ (or, equivalently, at constant B/B_0) can be constructed. Some of these are given in Figure 11.

DISCUSSION

The region sampled by the vertical equatorial probe in the present experiment was located in the "slot" between the inner and outer radiation belts

(2.3 ⩽ L ⩽ 3.7) at rather low magnetic latitude (0° ⩽ λ ⩽ 30°). Proton data obtained by DAVIS and WILLIAMSON (1963), with Explorer XII agree well with the intensity levels of protons above 1 MeV around $L \approx 3$ reported here. Spectral slopes in this region are in agreement as well. Low energy proton data from Relay I (FILLIUS and MCILWAIN, 1964 and MCILWAIN *et al.*, 1964) and Injun I (BOSTROM *et al.*, 1965) refer mainly to lower magnetic shells than those presently

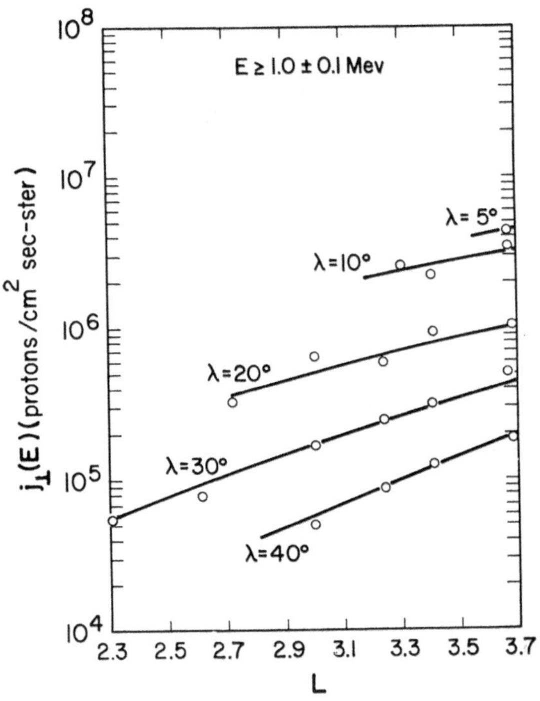

Fig. 11. Proton intensities at various λ.

sampled but intensity levels are roughly comparable. Similar proton fluxes were measured also by FRANK *et al.*, 1964 with Explorer XIV.

In summary, low energy proton perpendicular intensities up to about $5 \times 10^6/\text{cm}^2$ sec ster were observed at $L = 3$ to 3.5 near the magnetic equator. Measured pitch angle distributions were used to derive the perpendicular intensities at higher magnetic latitudes. Integral spectra between about 1 and 4 MeV exhibit an exponential energy dependence with a slope $E_0 \approx 0.4$ to 0.6 MeV.

3. 55 MeV Protons

In this section I wish to report on 55 MeV trapped protons detected in nuclear emulsions flown on oriented Air Force recoverable satellites since 1961 at a rate of almost one per month. Due to the fixed orientation, the ellipticity of the orbit and the fact that the geomagnetic field is offset by about 420 kilometers from the earth's center causing trapped protons to dip low in the atmosphere in the South Atlantic, specialized directional measurements are possible which allow the

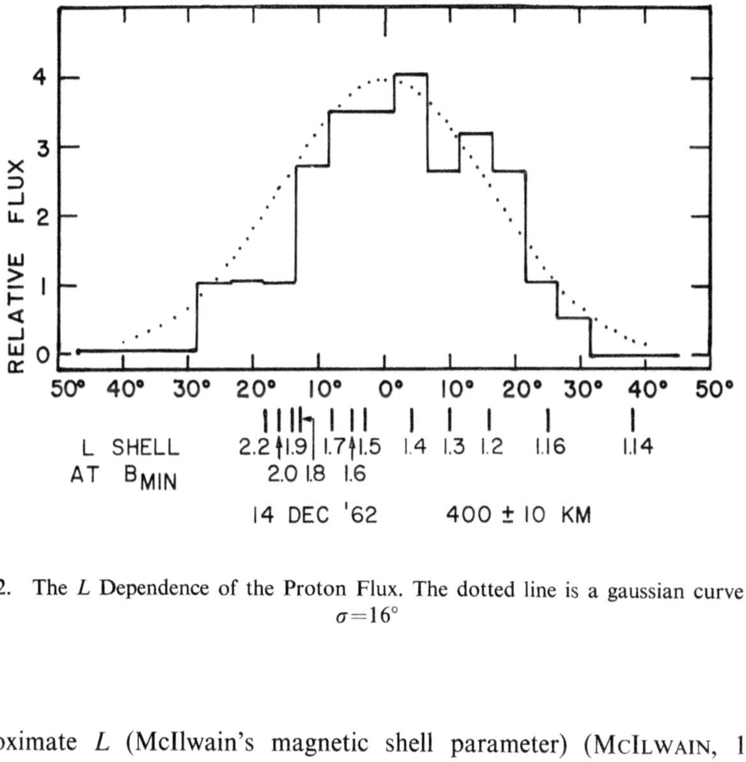

Fig. 12. The L Dependence of the Proton Flux. The dotted line is a gaussian curve with $\sigma = 16°$

approximate L (McIlwain's magnetic shell parameter) (McILWAIN, 1961) dependence and the altitude dependence of the flux to be determined by measurements on a single exposure.

Measurements on the angular distribution of the arrival directions of protons have been performed on emulsions flown on a number of different satellites. Extensive measurements (~ 1500 proton tracks) performed on a single vehicle (14 December '62) were carried out. The main feature of the angular distribution measurements is the marked anisotropy in the proton flux. The majority of the terminating protons make small angles with a mirror plane, indicating the detection of trapped protons mirroring in the vicinity of the South Atlantic anomaly as noted previously (Filz *et al.*, 1962, 1963 and 1964) and independently observed by Heckman and Nakano (1963).

In order to determine the L dependence of the proton flux, we assume for simplicity that the majority of the protons detected for a given L are incident when the satellite passes over the point of minimum B. From the intersection of the satellite trajectory with the locus of minimum B for fixed altitude but variable L we determine the great circle corresponding to the mirror plane. We then derive the proton flux as a function of the angle in the magnetic meridian plane as shown in Figure 12.

On a single flight, two independent flux measurements can be obtained by taking into account the fact that the slightly elliptical orbits cause the satellite to pass through the South Atlantic high flux region at two different altitudes in opposite directions. The protons detected by the upper cassette strike the vehicle during the high altitude pass while the protons detected by the lower cassette strike during the low altitude pass for most exposures. A "simultaneous" measurement is thus obtained at two different altitudes.

FLUX VS. ALTITUDE

The 55 MeV omnidirectional proton flux is plotted in Figure 13 as a function of altitude for the period August 1961 to June 1962. A least squares fit with a power law of the form $J=C(H_c)^K$ yields $C=4.99\times10^{-11}$ and $K=4.792$, where H_c is the guiding center altitude in the South Atlantic and J is the omnidirectional flux in protons/meter^2 second MeV, using the total orbiting time of the satellite. This relation with the parameters given represents the flux to about 11% accuracy. Since the satellites spend on the order of 1% of their orbiting time in the South Atlantic high flux region, a general average flux can be obtained by multiplying these numbers by 100. If cosmic ray neutron albedo is the primary source of protons of 55 MeV energy in the radiation belt and the primary loss mechanism is energy loss by ionization, then the directional proton flux should increase as the inverse of the atmospheric density (Singer, 1958). Cornwall *et al.* (1965), have integrated the atmospheric density over particle trajectories and obtained $\bar{\rho}$ the average atmospheric density in atomic electrons/cm^3 traversed as a function of B and L for 1959 and 1962. They used the Jensen and Whitaker 512 coefficient expansion of the magnetic field and "The Harris-Priester Model of the Atmosphere below 1000 km, smoothed into the Johnson model above 1000 km". The reciprocal of $\bar{\rho}$ for $L=1.3$ has been plotted on an arbitrary scale as a function of altitude along with our data in Figure 14. Even though the absolute magnitude of the densities differ by a factor of about ten, both the 1959 and 1962 atmospheres have been fitted at 420 km to our data. It is seen from Figure 14 that while no real choice can be made, the 1959 atmosphere appears to fit our data better than the 1962 atmosphere. Since our data were taken in

1961–1962, this might be expected since there will be a delay in the time required to approach the shape of the 1959 density versus altitude function. Thus, within the limits of the uncertainties of this experiment, it is established that the 55 MeV

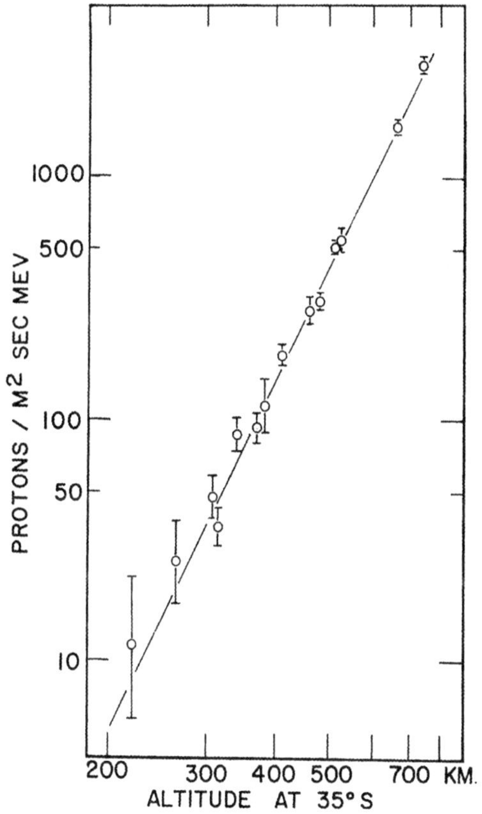

Fig. 13. The altitude dependence of the 55 MeV omnidirectional flux for the period August 1961 through June 1962. The time used is the total orbiting time and the altitude is the guiding center altitude. The solid line is the function $J = 4.99 \times 10^{-11} (H_c)^{4.792}$ and is weighted least squares fit to the data.

trapped proton flux increases as the reciprocal of the atmospheric density between 221 km and 738 km altitudes in the South Atlantic.

FLUX VS. TIME

It is also seen in Figure 13 that there is no significant time variation in the flux during the period August 1961 to June 1962. All of the data agree with this single curve as well as expected within the given statistical experimental errors which are less than about 15% in most cases. During this period for $L = 1.4$ and $B = 0.225$ gauss $H_{min} = 330$ km, BLANCHARD and HESS (1964) predict about a

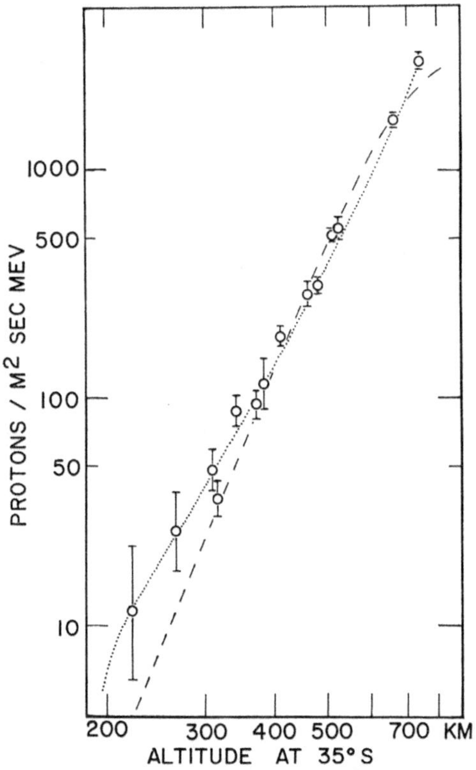

Fig. 14. Comparison with average atmospheric density. The same data are plotted as in Figure 13. The reciprocals of $\bar{\rho}$, the average atmospheric densities determined by Cornwall, are plotted on an arbitrary scale normalized to fit at 420 km. The dotted line is the 1959 atmosphere and the dashed line is the 1962 atmosphere.

60% increase in flux due to an anticipated decrease in atmospheric density during the period. Thus, either this increase is not valid throughout the lower edge of the proton belt, the atmospheric model used is in error or other sources than neutron albedo may contribute at low altitude.

Table I shows a comparison between our measured proton fluxes and the flux anticipated on the basis of Explorer IV data (MCILWAIN, 1961), Injun I

TABLE I

55 MeV omnidirectional proton flux integrated over the South Atlantic-Comparison of our results with those obtained by other experimenters. The fluxes have been uniformly reduced to units of protons/m^2 sec MeV using total vehicle orbiting time, and an energy of 55 MeV.

	Altitude kilometers	(1) *B at $L = 1.4$, 30°W*	(2) *Explorer IV*	(3) *Injun I*	(4) *This Experiment*	(5) *Injun III*	(6) *This Experiment*
Dates			Aug–Sep 1958	Jul–Dec 1961	Aug–1961 Jun–1962	Dec–1962 Sep–1963	Dec–1962 Sep–1963
Measurement Energy Range (7)			>43 MeV	>40 MeV	55 MeV	40–110 MeV	55 MeV
$\frac{J\text{ (Meas.)}}{J\text{ (55 MeV)}}$ (8)			138	141		77	
	220	0.238	16		8.4		
	340	0.227	61	52	68	190	180
	440	0.219		130	232	520	643
	500	0.213	340	240	427	910	1228
	700	0.197	1680	1060	2150	2440	2821

(1) Computed from Jensen and Cain 48 term expansion (Harrison *et al.*, 1963).
(2) Computed from results published by McIlwain *et al.* (1961).
(3) Computed from results published by Bostrom *et al.* (1965).
(4) Computed from $J = (4.99 \times 10^{-11})\, H^{(4.792)}$.
(5) Computed from results published by Valerio (1964).
(6) Computed for $T = 12$ months from decay curves derived in Section 4.3.
(7) Sensitive energy region for protons; note that Explorer IV and Injun I detectors are geiger tubes and are sensitive to electrons if present.
(8) Computed from a power law fit to Heckman and Nakano (1965) energy spectrum.

data (BOSTROM *et al.*, 1965) and Injun III data (VALERIO, 1964). The integral fluxes in protons/m^2 sec MeV are computed for five representative satellite orbits, using the omnidirectional proton intensities as a function of B and L as they were published. The first column gives the altitude at 35° south latitude. The second column gives our flux measurements calculated from $J = 4.99 \times 10^{-11} H^{4.792}$. The fourth column gives Injun I data obtained during the latter part of 1961.

The comparison between the Injun I data and the results of this experiment shown in column 5 is reasonably good considering the difficulties of interpolating published data. None of these calculations can be considered more accurate than 50% or so.

Comparison of our data with Explorer IV (Column 3) taken in 1958 indicates no substantial change in proton flux over a period of four years. During this period for $L = 1.4$, $B = 0.225$ gaus, $H_{min} = 330$ km, BLANCHARD and HESS (1964) predict an increase of a factor of about 12 in the 55 MeV proton flux due to atmospheric density changes. This increase apparently did not take place, or the Explorer IV counter is counting electrons instead of protons.

Columns 6 and 7 show a comparison between our data taken during the period December 24, 1962 to September 28, 1963 and Injun III data for the same period. The comparison is again good (within 30% at all altitudes). There is, however, a factor of three increase over 1961–62 levels. This increased proton flux at low altitudes has also been observed during late 1962 by HECKMAN and NAKANO (1965) and might be attributed to atmospheric density decreases (WALT and MACDONALD 1964). That this conclusion is not acceptable is borne out by our earlier discussion and our data for the July-August 1962 period which will now be discussed.

Table II shows a comparison between August 1961 to June 1962 data and proton fluxes determined by extrapolating back to July 9, 1962 using the decay curves derived below. None of the conclusions arrived at here depend on this extrapolation, it is simply done for consistency. It is seen that a significant increase has occurred in a period of less than one month, as has been previously noted (FILZ *et al.* 1962). The general flux level at 738 and 364 km was verified to be 2950 ± 240 and 18 ± 46 protons/m^2 sec MeV respectively by emulsions flown on satellite 1962 Alpha Gamma which was launched on 28 June 1962. Data taken on 29 May 1962 which are more accurately known at the low altitude show a flux of 92.5 ± 12.5 protons/m^2 sec MeV for an altitude of 368 km.

The fluxes given in Table II show a 1660% increase at 275 km, a 390% increase at 440 km and only a 67% increase at 665 km altitude. Looked at in this manner, there is no simple comparison between the increase at different altitudes. This also rules out any simple acceleration mechanism as the cause of the increase. If, however, we make the tentative assumption that the increase is due to a non-adiabatic pitch angle redistribution which moved one-half of the particles down the field line from higher altitudes (the other half would go up) and ask what decrease in altitude of inner belt protons would cause the observed increase, we arrive at Column 5, Table II, which shows altitude decreases required of from 304 km to 192 km for measurement altitudes from 275 to 665 km. If

TABLE II

The Differences Noted in Proton Flux Following July 1962

Altitude kilometers H_c	*Proton Flux**		*Percent Increase* $\Delta J/J(B)\times 100$	ΔH** *kilometers*	$\Delta\alpha_{eq}$† *Degrees*	ΔB_{eq}†† *Gamma* (10^{-5} *gauss*)
	Before $J(B)$	*After* $J(A)$				
275	24.5	432	1660	304	3.5°	697
350	77.6	612	689	272	3.3°	656
440	232	1136	390	268	3.6°	716
500	427	1708	300	270	3.4°	677
544	646	2000	210	252	3.4°	677
665	1685	2821	67	192	2.3°	457

* The units are protons/m² sec Mev where the time is the total orbiting time. $J(B)$ is taken from the relation $J(B) = 4.99 \times 10^{-11}\ (H_c)^{4.792}$, while $J(A) = J(E,0) + J(B)$ is calculated by extrapolating the decay curves back to July 9, 1962.

** ΔH is the decease in altitude required in the low altitude portion of the innerbelt to give twice the flux observed after the bomb.

† $\Delta\alpha_{eq}$ is the pitch angle change at the equator for $L = 1.4$ which is required to give the increase.

†† ΔB_{eq} is the change in the component of the magnetic field required perpendicular to the equatorial field to cause a pitch angle change of $\Delta\alpha_{eq}$ at $L = 1.4$.

one looks at the pitch angle change required at the equator, one obtains Column 6, Table II, which shows pitch angle changes required at $L = 1.4$ of 3.5°, 3.3°, 3.6°, 3.4°, 3.4°, and 2.3°. For lower L shells, the pitch angle change required would be greater while for higher L shells it would be less.

Thus, a pitch angle change of about 3.5° or more depending on the mechanism and its efficiency at the equator could cause the effect observed. A pitch angle change of 3.5° could be caused at the equator for $L = 1.4$ by a magnetic field perturbation with a component of about 0.007 gauss perpendicular to the magnetic field. Thus, our data strongly suggest that such a perturbation occurred during July 1962. No naturally occurring magnetic storms of this magnitude occurred during July 1962. The only event known to us of sufficient energy release in a reasonable scale size to cause the observed flux change was the U.S. high altitude thermonuclear test over Johnston Island on July 9th.

For a non-adiabatic effect to occur as proposed, the magnetic field change of about 0.007 gauss must be sustained in a wave about the order of the proton gyroradius (~ 100 km at the equator) in length and extending to a distance of the order of 3000 km from the bomb. The total energy would be about 2×10^{19} erg or three orders of magnitude less than the total energy release for a 1.4 megaton bomb which is about 6×10^{22} erg.

This calculation follows that made by DURNEY *et al.* (1962) who require a hydromagnetic shock wave with energy of the order of 10^{20} erg to explain a dumping of outer belt electrons which they observed following the bomb. VLF electromagnetic radiation has been observed following the bomb by CROOK *et al.* (1963) on the ground at a distance of 5400 km from Johnston Island with frequencies of 17.5 and 20 kc/sec ($\lambda \sim 5$ km) and peak amplitudes of the order of ~ 9 volts/m. This magnitude of the electric vector corresponds to a magnetic vector of about 0.0003 gauss so that if one allows a factor of about 20 or so for ionospheric losses, one could easily get a 0.007 gauss magnetic vector at $L = 1.4$. Also a significant electromagnetic signal is expected theoretically (KARZAS and LATTER, 1962). Thus, while we cannot really pinpoint the cause of the redistribution, we can say that the energy is available and was released in the right scale size in the correct order of magnitude by the July 9th high altitude burst.

DECAY

In order to study the decay of the 55 MeV protons, the data have been divided into four altitude intervals (275 ± 25 km, 350 ± 50 km, 440 ± 20 km and 500 ± 25 km). The data have been normalized to 275, 350, 440 and 500 km using a power law altitude dependence with an exponent of 4.792 as derived above and are plotted in Figure 15a and b. In this way no flux value has been altered more than about a factor of two from its measured value.

The data are compared to the theoretically expected decay of the additional flux of 55 MeV protons assuming ionization loss only (RAY, 1960). We use the form given by LENCHEK and SINGER (1962):

$$J(E,T) + J(B) = J(E,0)[1 + 7.7 \times 10^{-12} \bar{\rho} T E^{-3/2}]^{-2(n+1)/3} + J(B)$$

where $J(E, 0)$ is the initial flux, $J(E, T)$ is the flux at time T, $J(B)$ is the equilibrium background flux prior to 9 July 1962, $\bar{\rho}$ is the atmospheric density integrated of the particle trajectory in atomic electrons/cc, T is in seconds, $E = 55$ MeV.

n is taken to be 1/2 which is about the slope of the energy spectrum in the neighborhood of 55 MeV determined by HECKMAN and NAKANO (1965) in emulsions exposed in a similar way at this time. The parameters used in the

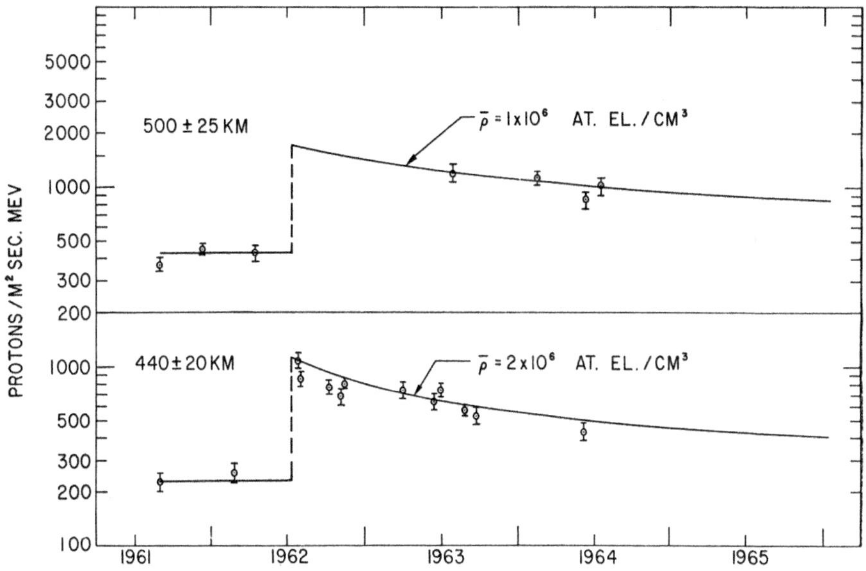

Fig. 15a.

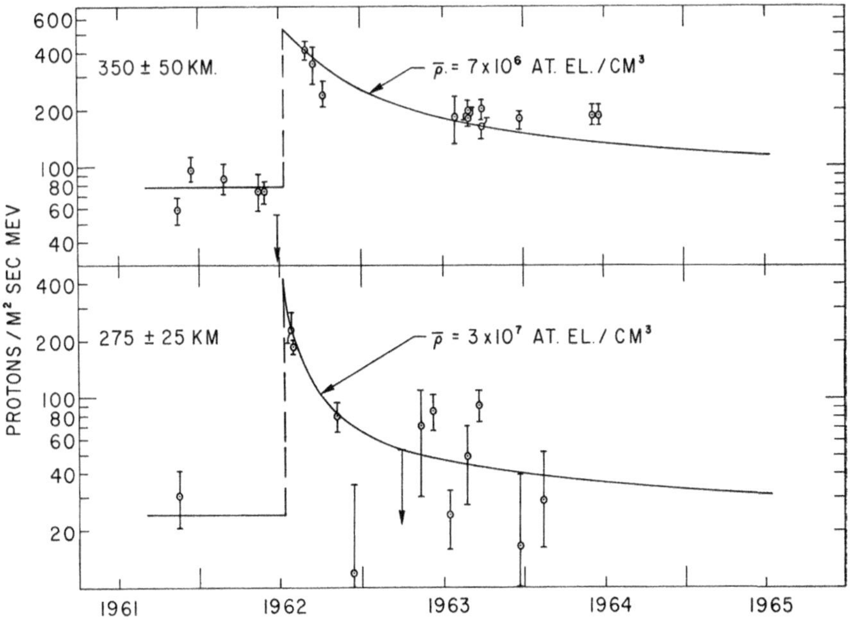

Fig. 15b.

Fig. 15. The time variation of the 55 MeV proton flux for altitudes of $H_c = 275$, 350, 440 and 500 km from August 1961 to July 1964. The solid curves drawn for the period after July 1962 are the theoretical decay curves based on the first data point following July 1962.

theoretical decay curves shown in Figure 9 are given in the first four columns of Table III. Column 5, Table III, gives $J(55\text{ MeV}, 0)$ assuming a July 9, 1962 starting time. The values of $\bar{\rho}$ were determined by Cornwall *et al.* (1965) at $L=1.3$ for 1962 by integrating over the total particle trajectory with the Jensen and Whitaker 512 coefficient expansion of the magnetic field and "the Harris-Priester model of the atmosphere below 1000 km, smoothed into the Johnson model above 1000 km."

TABLE III

Proton Decay Parameters

H_c *Altitude km*	$\bar{\rho}$ *Atomic Electrons per cm*^{3}	T *Time in Months past 9 July '62*	$J(B)$* *Equilibrium Background Flux*	$J(E,T)$* $+J(B)$ *Measured Flux*	$J(E,O)$* $+J(B)$ *Extrapolated to 9 July '62*
275	3×10^7	0.657	24.5	230 ± 55	432
350	7×10^6	1.708	77.6	412 ± 44	612
440	2×10^6	0.657	232	1080 ± 110	1136
500	1×10^6	12.7	427	1210 ± 134	1708
544	6×10^5	10.3	646	1679 ± 125	2000
665	2.4×10^5	2.3	1685	2790 ± 200	2821

* Flux in units of Protons/m^2 sec MeV based on total orbiting time.

The agreement between the theoretical and experimental decay is strikingly good and suggests that the model atmosphere, magnetic field model and theory are adequate to calculate trapped proton lifetimes at low altitudes.

Acknowledgments

The sections of this lecture referred to as 1 and 2 resulted from investigations carried out by Drs. G. Theodoridis and F. R. Paolini of American Science and Engineering, Cambridge, Mass. jointly with Mr. D. F. Smart of AFCRL and myself. The research under Section 3 was carried out by Mr. R. C. Filz of AFCRL and Mr. E. Holeman of Emmanuel College, Boston, Mass. A more detailed description of the work in Section 3 has been published in the *Journal of Geophysical Research* **70,** no. 23. Papers for eventual publications containing some of the aspects of Sections 1 and 2 are also in preparation.

The research reported has been funded in part by the United States Air Force In-House Laboratory Independent Research Fund.

References

ARMSTRONG, T.: 1965, *J. Geophys. Res.* **70**, 2077.
ARNOLDY, R. L., HOFFMAN, R. A., and WINCKLER, J. R.: 1960, *J. Geophys. Res.* **65**, 1361.
BLANCHARD, R. C. and HESS, W. C.: 1964, *J. Geophys. Res.* **69**, 3927.
BOSTROM, C. O. and WILLIAMS, D. J.: 1965, *J. Geophys. Res.* **70**, 240.
BOSTROM, C. O., ZMUDA, A. J., and PIEPER, G. F.: 1965, *J. Geophys. Res.* **70**, 2035.
BROWN, W. L. and GABBE, J. D.: 1963, *J. Geophys. Res.* **68**, 607.
CLADIS, J. B., CHASE, L. F., IMHOF, W. L., and KNECHT, D. J.: 1961, *J. Geophys. Res.* **66**, 2297.
CORNWALL, J. M., SIMS, A. R., and WHITE, R. S.: 1965, *J. Geophys. Res.* **70**, 3099.
CROOK, G. M., GREENSTADT, E. W., and INOUYE, G. I.: 1963, *J. Geophys. Res.* **68**, 1781.
DAVIS, L. R., WILLIAMSON, J. M.: 1963, *Space Research* **3**, 365.
DESSLER, A. J. and KARPLUS, R.: 1961, *J. Geophys. Res.* **66**, 2289.
DURNEY, A. C., ELLIOT, H., HYNDS, R. J., and QUENBY, J. J.: 1962, *Nature* **195**, 1245.
FILLIUS, R. W. and MCILWAIN, C. E.: 1964, *Phys. Rev. Letters* **12**, 609.
FORBUSH, S. E., PIZZELLA, G., and VENKATESAN, D.: 1962, *J. Geophys. Res.* **67**, 3651.
FILZ, R., HOLEMAN, E., and YAGODA, H.: 1962, *Bull. Amer. Phys. Soc.* II, **7**, 604.
FILZ, R., HOLEMAN, E., and YAGODA, H.: 1963, *Trans. Amer. Geophys. Union* **44**, 78.
FILZ, R., YAGODA, H., and HOLEMAN, E.: 1964, *Space Research* **4**, 642.
FRANK, L. A.: 1965, *J. Geophys. Res.* **70**, 1593.
FRANK, L. A., VAN ALLEN, J. A., and HILLS, H. K.: 1964, *J. Geophys. Res.* **69**, 2171.
FRANK, L. A., VAN ALLEN, J. A., WHELPLEY, W. A., and CRAVEN, J. D.: 1963, *J. Geophys. Res.* **68**, 1573.
FREEMAN, J. W.: 1964, *J. Geophys. Res.* **69**, 1691.
HECKMAN, H. H. and NAKANO, G. H.: 1963, *J. Geophys. Res.* **68**, 2117.
HECKMAN, H. H. and NAKANO, G. H.: 1965, *Space Research* **5**, 329.
HOFFMAN, R. A., ARNOLDY, R. L., and WINCKLER, J. R.: 1962, *J. Geophys. Res.* **67**, 4543.
KARZAS, W. J. and LATTER, R.: 1962, *J. Geophys. Res.* **67**, 4635.
LENCHEK, A. M. and SINGER, S. F.: 1962, *J. Geophys. Res.* **67**, 1263.
LINCOLN, J. V.: 1963a, *J. Geophys Res.* **68**, 6200.
LINCOLN, J. G.: 1963b, *J. Geophys. Res.* **68**, 6537.
MCILWAIN, C. E.: 1961, *J. Geophys. Res.* **66**, 3681.
MCILWAIN, C. E.: 1963, *Science* **142**, 355.
MCILWAIN, C. E., FILLIUS, R. W., VALERIO, J., and DAVE, A.: 1964, *NASA TND*-2516.
MOZER, F. S., ELLIOT, D. D., MIHALOV, J. D., PAULIKAS, G. A., VAMPOLA, A. L., and FREDEN, S. C.: 1963, *J. Geophys. Res.* **68**, 641.
PAOLINI, F. R., GIACCONI, R., WATERS, J. R., KATZ, L., and SMART, D.: 1965, *Space Research* **5**, 466.
PAULIKAS, G. A. and FREDEN, S. C.: 1964, *J. Geophys. Res.* **69**, 1239.
RAY, E. C.: 1960, *J. Geophys. Res.* **65**, 1125.
ROSSER, W. G. V.: 1963, *J. Geophys. Res.* **68**, 3131.
ROSSER, W. G. V., O'BRIEN, B. J., VAN ALLEN, J. A., FRANK, L. A., and LAUGHLIN, C. D.: 1962, *J. Geophys. Res.* **67**, 4533.
VALERIO, J.: 1964, *J. Geophys. Res.* **69**, 4949.
VAN ALLEN, J. A.: 1964, *Nature* **203**, 1006.
WALT, M. and MACDONALD, W. M.: 1964, *Revs. Geophys.* **2**, 543.
WILLIAMS, D. J. and SMITH, A. M.: 1965, *J. Geophys. Res.* **70**, 541.

Discussion

In reply to a question KATZ said he has no data on the September 22, 1963 magnetic storm since his satellite was inoperative by then. HESS asked if the recent emulsion data of Filz did show an increased proton flux and Katz said apparently yes but that the data analysis was not complete yet.

TEMPORAL STABILITY OF INNER ZONE PROTONS

GEORGE F. PIEPER
Goddard Space Flight Center, Greenbelt, Maryland

Introduction

The purpose of this paper is to present experimental evidence for the temporal stability of inner zone protons as determined from measurements made with instruments aboard the satellite Injun I in the period July–December, 1961. The analysis reported here was carried out at the Applied Physics Laboratory of The Johns Hopkins University and is a part of a more extensive study carried out with C. O. BOSTROM and A. J. ZMUDA and published elsewhere (PIEPER *et al.*, 1965; BOSTROM *et al.*, 1965; ZMUDA *et al.*, 1965).

The University of Iowa satellite Injun I (O'BRIEN, 1963) was launched June 29, 1961, into a nearly circular orbit at an altitude of approximately 1000 km and an inclination of 67°. The data used in this study were taken in the vicinity of the South Atlantic magnetic anomaly at two receiving stations, one in Brazil, at São José dos Campos, and the other in Salisbury, Southern Rhodesia.

The Injun detectors used in this analysis were the Applied Physics Laboratory package of four silicon *p-n* junctions, and the University of Iowa spectrometer background Geiger counter. These devices have been described in detail elsewhere (PIEPER *et al.*, 1962; O'BRIEN *et al.*, 1962). In this work we consider the *p-n* detectors to be sensitive to protons in the energy range of approximately 1 to 15 MeV and the SpB Geiger counter to be sensitive to protons of energy greater than 40 MeV.

Since we wish to deal here exclusively with the proton distribution, it is important to establish the lack of sensitivity of these detectors to the electrons present in the inner zone at the time of the measurements. Consideration of the magnitude of the electron flux at Injun altitude in 1961 and the measured response of the detectors to electrons in the laboratory allows this conclusion (PIEPER *et al.*, 1965).

Long Term Observations

Figure 1 is included to illustrate how the flux data look when plotted as a function of the magnetic shell parameter, L, with the magnetic field dependence indicated symbolically. For the protons with $E > 40$ MeV, the flux spread in similar B-L regions is generally small, showing the omnidirectional character

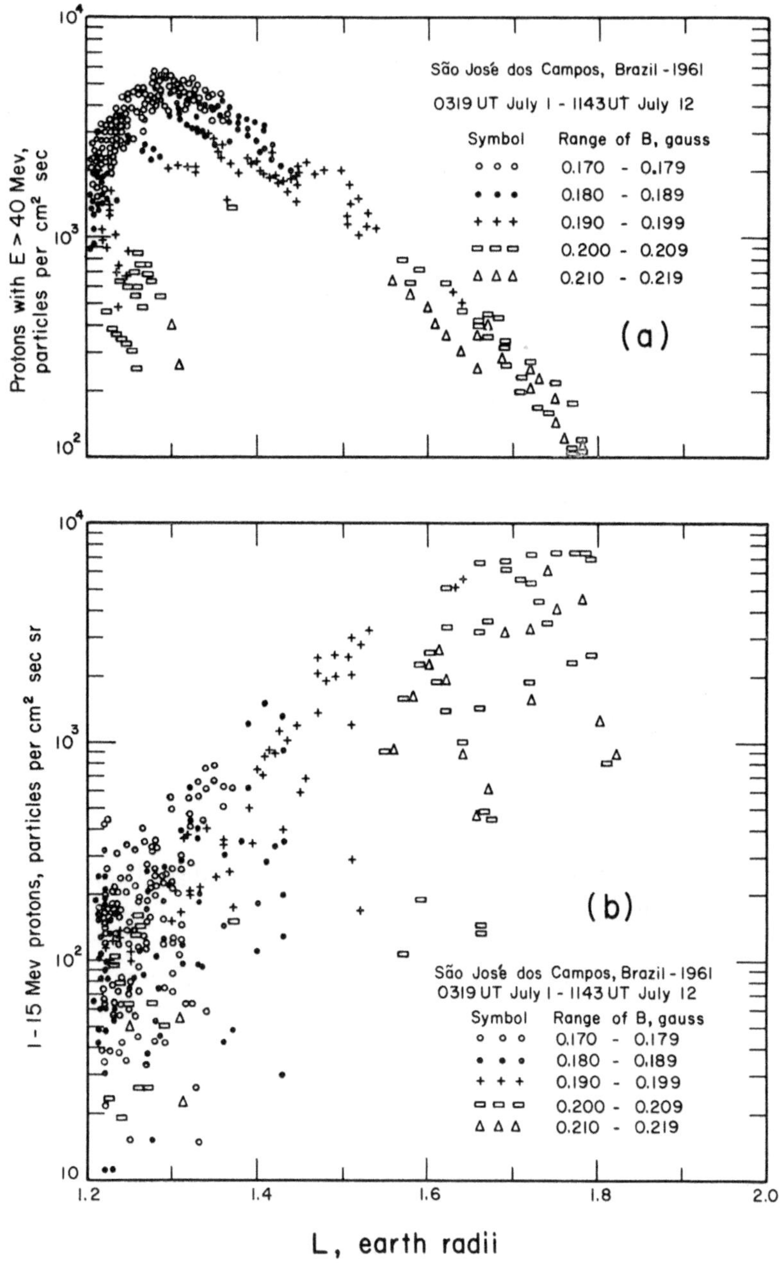

Fig. 1. Proton fluxes at 1000 km altitude near São José dos Campos, Brazil, for July 1–12, 1961.

of the detector. For the 1 to 15 MeV protons, the fluxes at fixed L generally extend over a wide range of values, although the B-dependence of the different values is similar. This spreading of flux values at the same point in B-L space results from the slow rolling-tumbling motion of the satellite which swings the

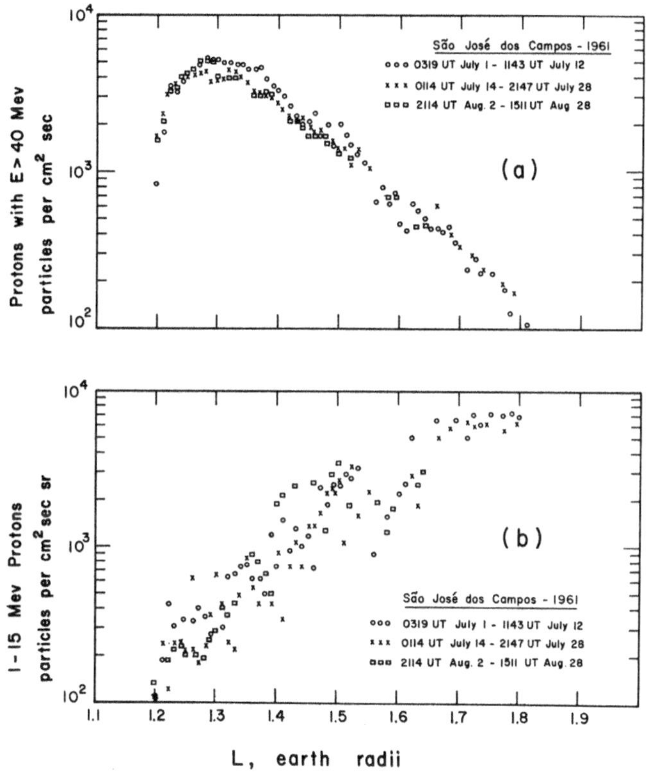

Fig. 2. Envelope fluxes near the Brazil station for three different time periods, July–August, 1961.

unidirectional p-n detectors in and out of the wedge of trapped particles. (Although the satellite contained a permanent magnet installed to align it along the local geomagnetic field, this stabilization was not achieved because Injun I failed to separate from another spacecraft in the multiple launch.) In this circumstance, the maximum flux seen at each L – the envelope flux – represents the directional flux of 1–15 MeV protons perpendicular to the field line.

Figure 1 shows the fluxes observed in the period July 1–12 from the Brazil station. Figure 2 shows the envelopes from Figure 1 and in addition similar envelopes for the periods July 14–28 and August 2–28. Figure 1 indicates explicitly the dependence of the fluxes on the magnetic induction, B. In Figure 2 this dependence is not shown explicitly, but at any fixed L, points for the three periods fall within a spread of 0.01 gauss for about 70% of the L's and within a spread of 0.02 gauss for the remaining L's.

It is of interest now to review the solar proton events and magnetic storms that occurred during these periods, that could possibly have affected these fluxes: (1) July 1–11 was a period of solar and magnetic quiet with no flares of importance 2 or 3, no recorded solar proton events and only one moderate magnetic storm. (2) The period July 12–28 was a violent one; there were twelve flares of importance 2 or 3, six magnetic storms and four major solar proton events, two of which were among the twelve largest solar cosmic ray outbursts during the 5-year period 1956–1961 and two of which had enough high energy particles to cause increases in the counting rates of high latitude neutron monitors. (3) The month of August 1961 was essentially quiet, with only one sudden commencement storm and no solar proton events. The point to be made here is based on the observation from Figure 2 that the envelope fluxes for the three periods are essentially completely interspersed. We conclude that the intense solar-terrestrial activity of July 12–28 produced no substantial concurrent or delayed (within a month) changes in the spatial dependence of the flux of trapped inner belt protons. Quantitatively, if we define the mean value at each L as the average of the local maximum and mimimum, then for the $E>40$ MeV protons, the rms deviation from the mean varies from 10% to 30% and averages 19%, while it varies from 9% to 46% and averages 29% for the 1–15 MeV protons.

A similar analysis can be made of data taken from the Salisbury receiving station between August 12 and December 28, 1961. The entire interval is divided into three periods, August 12 – September 28, October 4 – November 6, and November 10 – December 28. The envelope fluxes for these periods are shown in Figure 3, along with the fluxes for two particular passes that will be discussed below. As in the preceding figure, the magnetic field dependence of the fluxes is not shown explicitly in Figure 3. The magnetic induction lies in the range 0.200–0.219 gauss in $1.16<L<1.50$ and then increases nearly monotonically to a value in the range 0.250–0.259 gauss in $2.5<L<2.6$. Points for all three periods fall within a range of 0.01 gauss for about 15% of the L's and within a range of 0.02 gauss for the remaining L's.

These three periods are not so clear cut as the ones in July and August with regard to the vigor of the solar-terrestrial activity. The first period precedes and the other two follow the well known solar proton-magnetic storm event of September 28 – October 4, 1961, but all three periods contain magnetic disturbances and small solar proton events. In Figure 3b the interspersing of the envelope fluxes for 1–15 MeV protons is quite complete; the deviation from the mean varies from 11% to 69% and averages 31%. In Figure 3a, for the $E>40$ MeV protons, in the region $1.40<L<2.10$, the interspersing of fluxes from the three

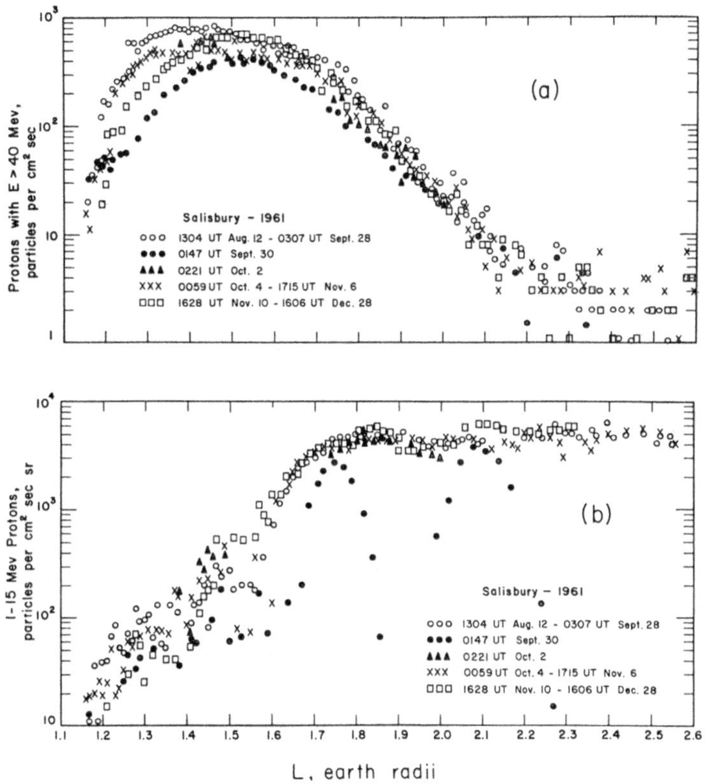

Fig. 3. Envelope fluxes for three different time periods, August–December, 1961, and two individual passes, September 30 and October 2, from the Salisbury station.

periods is very good, averaging 30%. For $L>2.10$, the fluxes are very small, less than 10 protons/cm^2 sec, and are intermixed very well. In the range $1.16<L<1.40$, the points are not well intermixed; those from the first period generally tend to be low. Except for this possibly anomalous situation, it seems safe to conclude that the gross characteristics of the trapped protons seen by Injun I from Salisbury did not change during the entire period considered, despite one major and four minor solar proton events and ten magnetic storms.

The Inner Van Allen Zone During the Disturbance of September 28 to October 4, 1961

So far the proton data have been treated in groups representing the accumulation of observations over periods of two weeks to about a month. It is desirable to narrow this time span, and especially important to look at inner zone behavior during a solar proton magnetic storm event. We do have two individual Salisbury passes that occurred during the September 28, 1961, event that can be examined. The first of these occurred at 0147 UT September 30, after the solar proton event had begun at high latitudes, but before the start of the sudden commencement storm of 2108 UT September 30 and the large influx of low energy particles. The second pass took place at 0221 UT October 2, in the recovery phase of the geomagnetic storm, while low energy particles were still being seen at high latitude. The proton fluxes for these passes are given by the solid points in Figure 3, and at a given L, each has a B value within the range covered by the envelope fluxes in the figure.

As seen in Figure 3b, the flux of 1–15 MeV protons on September 30 shows the effect of satellite rotation on the unidirectional detectors. The peak intensities at $L = 1.75$ and 2.1 agree quite well with the envelope fluxes for the preceding and following periods. The October 2 pass had a more favorable orientation and agrees at nearly all points.

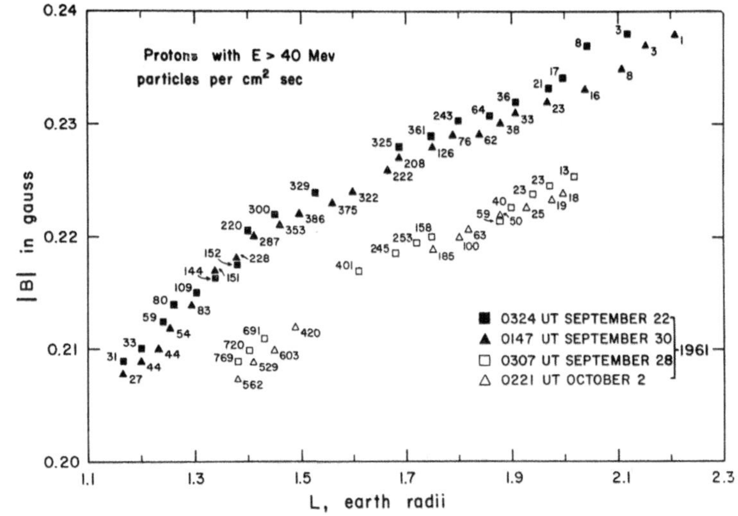

Fig. 4. A comparison of proton fluxes > 40 MeV obtained before and during the solar proton magnetic storm event of September 28–October 4, 1961.

In Figure 3a, the flux of $E>40$ MeV protons for the September 30 pass agrees well with the envelope fluxes only for $L>1.85$. Below $L=1.85$, the intensity is anomalously low, since one would expect the higher L shells to be affected most. The flux for October 2 agrees well at all available data points.

A further detailed comparison can be made of these two passes during the September 28 event with a pair of pre-event passes chosen because they cover nearly identical traces in B-L space. The comparison is shown in Figure 4, in which it is seen that the fluxes measured on September 30 are the same as those observed at nearly the same point in B-L space on September 22, with an average deviation of $\pm 10\%$, excepting a narrow region at $L=1.80$. Similarly, the September 28 and October 2 passes are the same within an average deviation of $\pm 11\%$.

The above analysis is certainly not conclusive, but it does clearly suggest that the inner zone was not affected by the disturbed conditions of the September 28 – October 4 event. The integrated flux of $E>40$ MeV solar protons during this interval was $\sim 3\times 10^6$ protons/cm^2, but this flux is too low by 2 or 3 orders of magnitude to produce a noticeable effect, according to the work of DRAGT and WHITE (1963).

Protons with E > 40 MeV at Identical B-L Points

In this section, we compare the omnidirectional intensity of protons with $E>40$ MeV at 17 specific points in B-L space, each point being sampled more than once by the satellite during the period July–December 1961. These points are shown in Figure 5. Each flux point within a set lies within $\Delta B=0.001$ gauss and $\Delta L=0.01$ R_e of the indicated value, and the time span within each set includes at least one and sometimes several solar proton magnetic storm events. Including the statistics of counting (generally less than 5%), the spread of fluxes around the local means of the individual sets varies from 0 to $\pm 33\%$ with an average of $\pm 7\%$, the constancy indicating the time stability of these inner belt protons. Unfortunately, a similar analysis is not possible for the low energy protons because of the satellite orientation problem.

Outer Boundary of the Trapping Region for E > 40 MeV Protons

Figure 6 shows the fluxes of trapped protons of $E>40$ MeV in the region $0.230<B<0.244$ gauss and $2.10<L<2.60R_e$ taken from thirteen passes between September 8 and November 15, a period containing eight magnetic storms and four solar proton events.

Comparison of the fluxes for different times shows the stable nature of this region, the outer boundary of the inner zone of high energy trapped particles. This stability occurs despite the region's sensitivity to the detection of changes: The fluxes are low and this volume of space is the part of the inner zone most

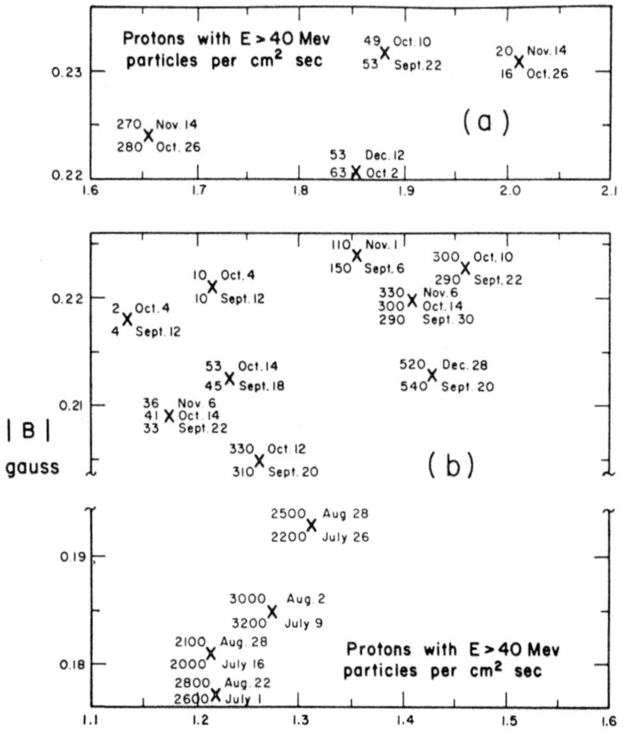

Fig. 5 L, earth radii

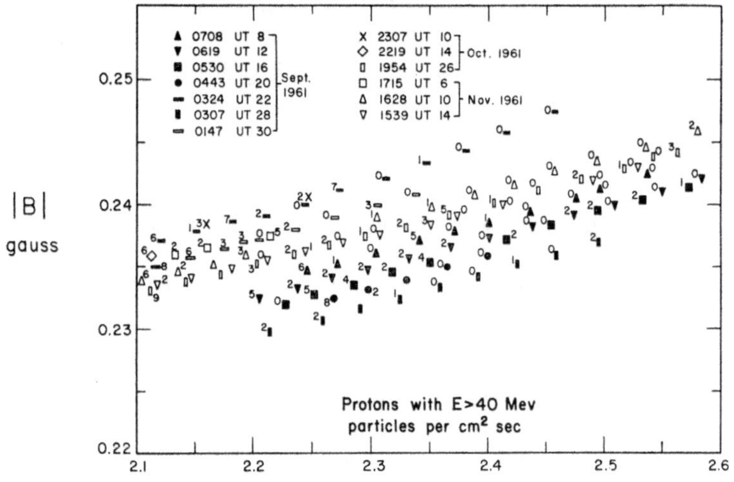

Fig. 6 L, earth radii

likely to be affected by the solar proton albedo neutron mechanism. As before, the satellite orientation problem prevents a similar analysis for the 1–15 MeV protons, and in addition the fluxes are at their maximum as seen by Injun I.

Discussion

Trapped protons in the inner Van Allen belt in the energy ranges 1–15 MeV and >40 MeV were not affected by the geophysical disturbances that occurred between July and December 1961, a period containing nine solar proton events and seventeen magnetic storms. While the activity in the diferent events varied, these were periods of intense disturbance by normal standards. To the several analyses considered here, it should be added that the comparison by Bostrom *et al.* (1965) of the Injun I data for July–December 1961 with the Relay I data from December 1962–January 1963 implies that 1–15 MeV protons were stable in the inner zone throughout the entire period from July 1961 to January 1963. It is also possible, by comparing the Injun I $E > 40$ MeV data with McIlwain's (1961) Explorer IV data taken in July–December 1958 to extend the time back to 1958. (The comparison is all inside $L = 1.80$ and there is a 2.5 year gap without coverage.) Since the Relay satellites, the coverage of the inner zone has been continuous (McIlwain, this volume). Thus there is a reasonable presumption that the inner Van Allen zone has been stable since July 1958 until the present (summer 1965), possibly inside $L = 2.1$, the depth of penetration of the huge event of September 23, 1963, and probably inside $L = 1.8 R_e$.

References

Bostrom, C. O., Zmuda, A. J., and Pieper, G. F.: 1965, *J. Geophys. Res.* **70**, 2035.
Dragt, A. J. and White, R. S.: 1963, *Trans. Am. Geophys. Union* **44**, 881 (abstract).
McIlwain, C. E.: 1961, *J. Geophys. Res.* **66**, 3681.
O'Brien, B. J.: 1963, *Advances in the Astronautical Sciences,* **12**, 99.
O'Brien, B. J., Van Allen, J. A., Laughlin, C. D., and Frank, L. A.: 1962, *J. Geophys. Res.* **67**, 3970.
Pieper, G. F., Bostrom, C. O. and Zmuda, A. J.: 1965, *J. Geophys. Res.* **70**, 2021.
Pieper, G. F., Zmuda, A. J., Bostrom, C. O. and O'Brien, B. J.: 1962, *J. Geophys. Res.* **67**, 4959.
Zmuda, A. J., Pieper, G. F. and Bostrom, C. O.: 1965, *J. Geophys. Res.* **70**, 2045.

Fig. 5. Protons with $E > 40$ MeV at specific points in *B-L* space sampled more than once by Injun I during the period July–December, 1961.

Fig. 6. The temporal stability of the outer boundary of the inner Van Allen zone as determined in 13 satellite passes between September 8 and November 15, 1961. Cosmic ray background has been subtracted.

Discussion

FREDEN asked if studies of low energy protons of 1–15 MeV at high latitudes had showed any evidence of trapping after solar events. PIEPER said there was no indication of trapping of low energy particles but that the sensitivity of this test was low. Discussion with MCILWAIN showed that the comparison with Relay data could not be used to argue about solar cycle flux changes because of background problems.

SOME OBSERVATIONS OF THE DISTRIBUTION OF ENERGETIC PROTONS IN THE EARTH'S RADIATION BELTS BETWEEN 1962 AND 1964

J. D. GABBE and W. L. BROWN

Bell Telephone Laboratories, Inc., Murray Hill, New Jersey

Abstract. Flux distributions for protons with energies between 4 and 13 MeV, 18 and 27 MeV, 26 and 33 MeV, 50 and 135 MeV, and 49 and 145 MeV measured from July 1962 to September 1963 at geocentric distances in the range 1.01 to 3.4 earth radii are presented. Using data extending to June 1964, secular temporal changes are discussed, some effects of the magnetic storm of September 22, 1963 are described, and some aspects of the energy spectrum are noted.

1. Introduction

The purpose of this paper is to present several bodies of data on the proton distribution. These data were gathered between 1962 and 1964 by means of instruments on Telstar 1, Explorer XV, and Telstar 2.

Distributions in the form of log flux vs log B for various specified values of L are used to summarize the measurements. B is the magnetic induction in Gauss, and L is the magnetic shell parameter (McIlwain, 1961). These coordinates and scales have been selected because they are well suited for making comparisons of flux values within and among the bodies of data in this paper and also between these results and those of other experimenters.

Some portions of these data have been published previously (Brown, Gabbe, and Rosenzweig, 1963; Brown and Gabbe, 1964; Brown, 1965). This article collects the various distributions in one place, supplements them with previously unpublished information, and presents them all in the same coordinate system and on the same scale.

Extensive analyses of the data from the various instruments have been undertaken. During the course of this work, simple mathematical functions which give convenient and statistically accurate descriptions of the proton distributions have been evolved. Some of the analyses have been completed, others are still in progress. Because they are computationally so convenient, the formulae describing the Telstar 1 data are included in this article. The details of the completed analyses are in the process of publication elsewhere.

2. The Satellites

Information about the satellites is given in Table I, and the regions of R-λ (McIlwain, 1961) space accessible to the spacecraft are illustrated in Figure 1. As may be seen in the figure, the larger region of space explored by Telstar 2 includes almost all the smaller portion of space covered by Telstar 1. Explorer

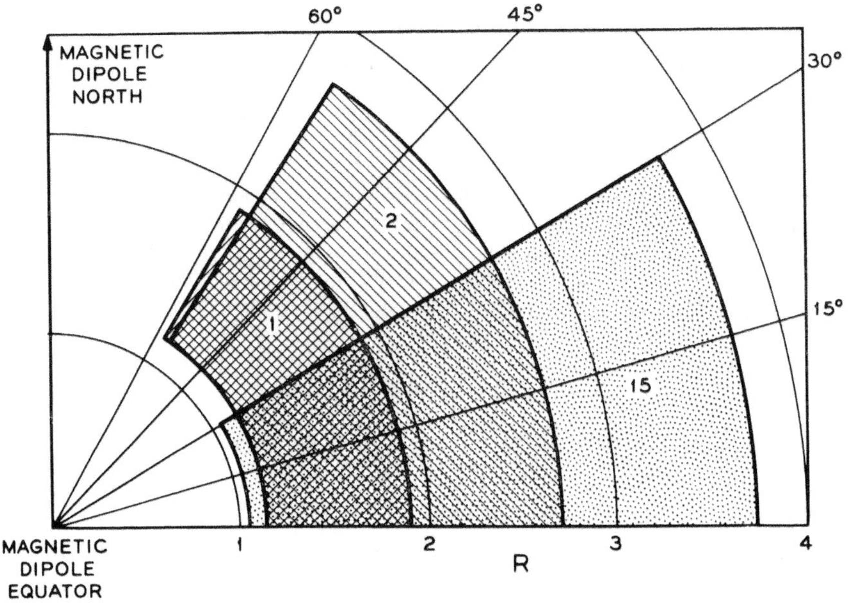

Fig. 1. The regions of R-λ space explored by Telstar 1 [1]; Telstar 2 [2]; and Explorer XV [15].

XV maps out higher altitude regions of space than the Telstar satellites, in addition to overlapping the portions covered by the Telstars at lower altitudes and latitudes. All the Explorer XV measurements were recorded during the active life of Telstar 1, but measurements from the two Telstar satellites do not overlap in time. The three vehicles were spin stabilized and all the detectors were mounted so that their symmetry axes lay in planes that were approximately perpendicular to the spin axes of their respective satellites.

3. The Instruments

The detecting elements in all of the instruments were semiconductor diodes made by diffusion of phosphorus into high resistivity p-type silicon. The sensitive

TABLE I

Satellites, Orbits, and Detectors

Satellite	Period covered by data		Orbit**			Spin Rate RPM		The detectors discussed here			
	from	*to*	*Perigee in R_e*	*Apogee in R_e*	*Incl. deg.*	*Max*	*Min*	*Energy range MeV*	*Directional Characteristics*	*f*	*Counting interval sec*
Telstar 1	7/10/62–	2/21/63	1.15	1.90	45	180	85	50–135	Omnidirectional	87	11
1962 $\alpha\varepsilon 1$								26–33	Omnidirectional	1700	15
Explorer XV	10/27/62–	1/27/63	1.05	3.72	18	72	72	4–13	cone, 10° half-angle	4000	1.45
1962 $\beta\lambda 1$											
Telstar 2	5/ 7/63–	5/7/65*	1.15	2.70	43	180	65	49–145	Omnidirectional	76	11
1963 13A								18–27	Omnidirectional	950	15

* Only data previous to 6/30/64 reported here.

** $R_e = 1$ earth radius $= 6371.2$ km.

regions of these diodes were disk shaped, about 2.9 mm in diameter and 0.4 mm in thickness at the nominal reverse bias voltage at which the diodes were operated. These devices have been described by BUCK *et al.* (1964) and the complete detector assemblies by BROWN *et al.* (1963). The energy range of the particles detected was determined by the shielding of the detector assemblies, the pulse height level set by the associated electronics, and the operating bias on the diodes. The conversion from the counting rate of the detectors to proton flux was made by multiplying the counting rate by the factors f in Table I. These constants take into account the geometric factors and the average efficiencies of the detectors. For a differential power-law spectrum of the form

$$J(E) \sim E^{-n} \tag{1}$$

where $J(E)$ is the flux of particles with energies between E and $E+\mathrm{d}E$, and the stated energy intervals, the quantities f deviate less than 5 per cent from those given in Table I for values of n between 0 and 7.5. Although the conversion factors are almost independent of the energy spectrum, variations in the conditions under which the detectors were made, assembled and operated lead to uncertainties of about 50 per cent in the absolute values of these constants for the several detectors.

4. The Data

The basic telemetered data are the number of counts recorded by a detector during a counting interval. The counting intervals for the various detectors are given in Table I. On the Telstar satellites the counting intervals occurred once a minute for both the detectors considered here. On Explorer XV the counting interval occurred approximately 5 times a minute. In all cases the date and universal time at which the data were recorded were inserted by the ground station. These times together with the ephemerides of the satellite allow one to associate the geographic position of the satellite with the data. The magnetic coordinates, B and L were calculated from the geographic position by means of McIlwain's FORTRAN computer program INVAR using the JENSEN and CAIN (1962) Epoch 1960 coefficients for the 48-term spherical-harmonic expansion of the earth's main magnetic field (MCILWAIN, 1962). In addition, bias voltages, satellite temperatures, and other housekeeping data appeared in the telemetry and were associated with the appropriate counting-rate data.

The Telstar detectors are intrinsically omnidirectional and the analyses performed have treated the data point by point. The Explorer XV detector is directional, and the data have been interpolated to specific L shells and corrected to equivalent omnidirectional flux using a method developed by ROBERTS (1965).

5. The flux vs B Contours

Figures 2 through 6 are presentations of the various proton distributions in the form of log J (where J is the omnidirectional flux) vs log B plots for selected L shells. For reasons connected with the equipment or analysis these presentations suffer from various small defects. Space does not allow a detailed discussion of these anomalies, but the most significant of them are mentioned in the text.

The results of the analysis of the data on protons with energies between 50 and 135 MeV taken with Telstar 1 are shown in Figure 2. The lines were produced from the equations and coefficients in Table II. The coefficients have

TABLE II

The Functional Fit to the 50–135 MeV Proton Data from Telstar 1

These expressions are valid only for $1.15 < L < 3.0$, $R < 1.95\ R_e$*, and $J > 5$ proton/cm² sec.

$$C(B, L)\begin{cases} = A^2\left[\dfrac{Z^2 - x^2}{Z^2(1-x^2)}\right]^{1/2}\left[1-\left(\dfrac{x}{Z}\right)^2\right]^{2S}, x \leqslant Z \\ = 0,\ x > Z \end{cases} \qquad \text{(a)}$$

where C = omnidirectional counting rate, counts/sec, of protons with energies above between 50 and 135 MeV; B = magnetic induction, Gauss; L = magnetic shell parameter; $x = \sqrt{1 - B_0/B}$; $B_0 = 0.311653/L^3$.

$$A\begin{cases} = \dfrac{A_1(L-L_0)}{A_2 + (L-A_3)^{A_4}},\ L \geqslant L_0 \\ = 0,\ L < L_0 \end{cases} \qquad Z = \sqrt{1-\left(\frac{Q}{L}\right)^3\left[4 - \frac{3Q}{L}\right]^{-1/2}}$$

and $Q = L_0 + R_1(L-L_0) + R_2(L-L_0)^2 + R_3(L-L_0)^3$.

$$J(B, L) \approx 87\, C(B, L),\ J > 5 \qquad \text{(b)}$$

where J = omnidirectional flux, protons/cm² sec, of protons with energies between 50 and 135 MeV.

$$j(u, L) \approx 87\,\frac{A^2\left[1-\left(\dfrac{u}{Z}\right)^2\right]^{2S}}{2\pi Z\beta(\frac{1}{2}, 1+2S)} \qquad \text{(c)}$$

where j is the equatorial unidirectional flux, protons/cm² sec ster, of protons with energies between 50 and 135 MeV

$u = \cos\alpha$

α = equatorial pitch angle

and $\beta(p, q) = \int_0^1 w^{p-1}(1-w)^{q-1}dw$ is the beta function.

$L_0 =$	1.132792	$R_1 =$	0.2636094
$A_1 =$	10.197	$R_2 =$	−0.6011559
$A_2 =$	0.24662	$R_3 =$	0.4746468
$A_3 =$	0.74397	$S =$	0.32986
$A_4 =$	5.3565		

* 1 R_e = 6371.2 km

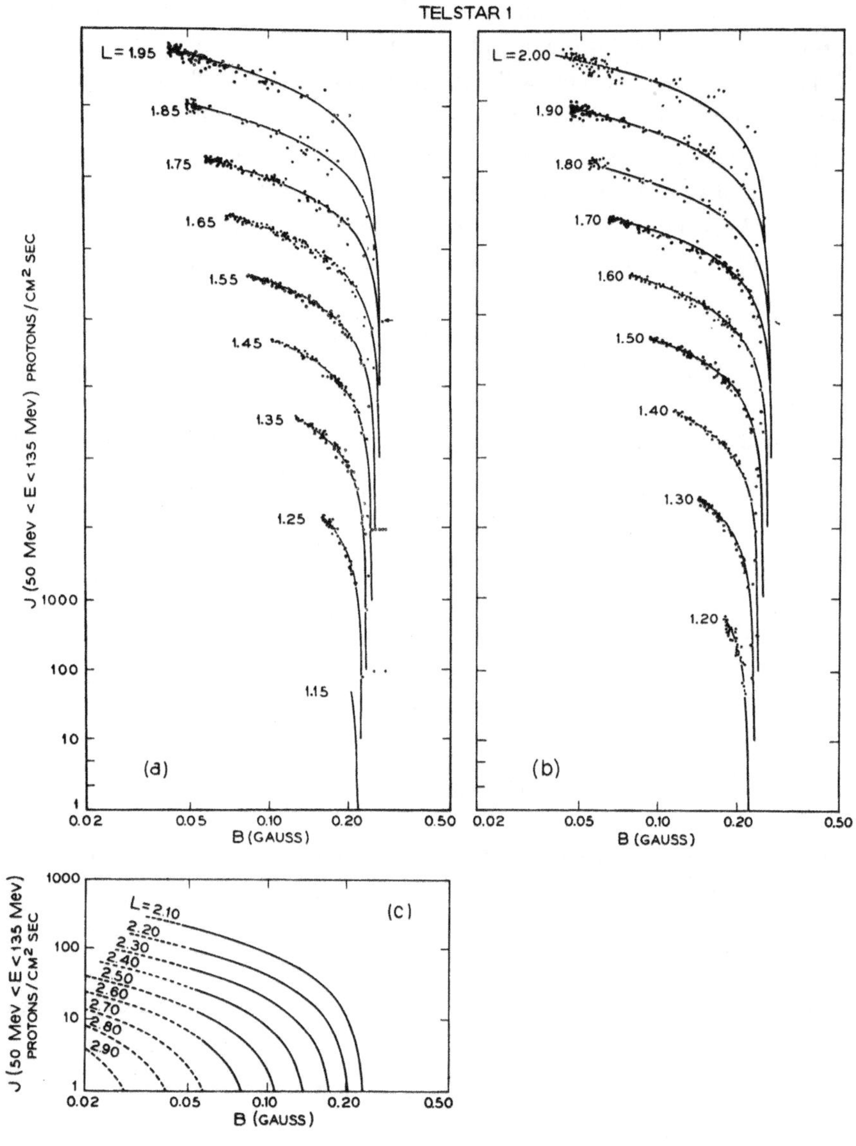

Fig. 2. Flux vs B contours for constant L for the 50–135 MeV protons measured on Telstar 1 (July 1962–February 1963). Adjacent contours are slipped one decade in J; and each contour rises from $J = 1$. The points are data points near the appropriate L. Data points representing $J = 0 (\log J = -\infty)$ are plotted at $J = 1$. The lines were generated from the model described in Table II. Dashes indicate extrapolations beyond the region in which data were acquired.

been determined by making a multidimensional least squares fit of the mathematical model in Table II to the data (Gabbe, Wilk, and Brown, 1966). The model with the tabulated coefficients summarizes the distribution of these protons to an extent reasonably approaching expected statistical fluctuations. The points in Figure 2 are individual data points having L-values close to the L-values of the lines. How close depends on the density of the data, but the differences between the L-values of the points and that of the line is usually less than 0.005. These results should not be extended to fluxes below 5 protons/cm^2 sec, a value that approximately represents the lower limit of the sensitivity of the detector. The actual data points associated with L-lines having values greater than $L=2.0$, shown in section (c) of the figure, have been omitted. This has been done because the overlap among observations belonging to different L-lines that results from the statistical scatter at low counting rates tends to confuse the presentation when the points are included. The quality of the fit remains satisfactory.

The distribution of protons with energies between 26 and 33 MeV was also measured on Telstar 1, and is shown in Figure 3. The model described in Table III has been fitted to the 26–33 MeV measurements and the resultant coefficients appear in the table. The geometric factor of the detector was quite small and the instrument is effectively insensitive to fluxes below 100 protons/cm^2 sec.

Figure 4 is a presentation of the data on the 4–13 MeV protons measured on Explorer XV, as reduced to omnidirectional flux. The lines are the result of a preliminary least squares fit to the reduced data. They cannot realistically be extrapolated to values of flux beyond the data points. The points plotted in Figure 4 are only a representative sample of the existing data, but they cover the full range of observations.

The data from Telstar 2 have been divided into two time periods. The dividing line is the magnetic storm of September 22 (day 265) 1963, which substantially affected the proton distribution for $L \gtrsim 2.2$. The data preceding this storm were acquired between days 127 and 265, 1963 and this time will be referred to as "Summer 1963". The period extending from day 265, 1963 to day 182, 1964 (day 547, 1963) is "Winter 1964". Flux distributions from the summer of 1963 are shown in Figure 5 for 49–145 MeV protons and in Figure 6 for the 18–27 MeV protons. The points in Figures 5 and 6 are data which fall within thin L slices, i.e. ranges of L defined by $L \pm \Delta L$, where L is the nominal value for the measurements. The value of ΔL is usually 0.005. The curves for the Telstar 2 data are not least squares fits, but are hand drawn through the points. This procedure is somewhat uncertain at the high B end of the higher L lines, where the points are scattered because the fluxes are too small to give good counting statistics; and the detailed shape at this end of the line contains this uncertainty.

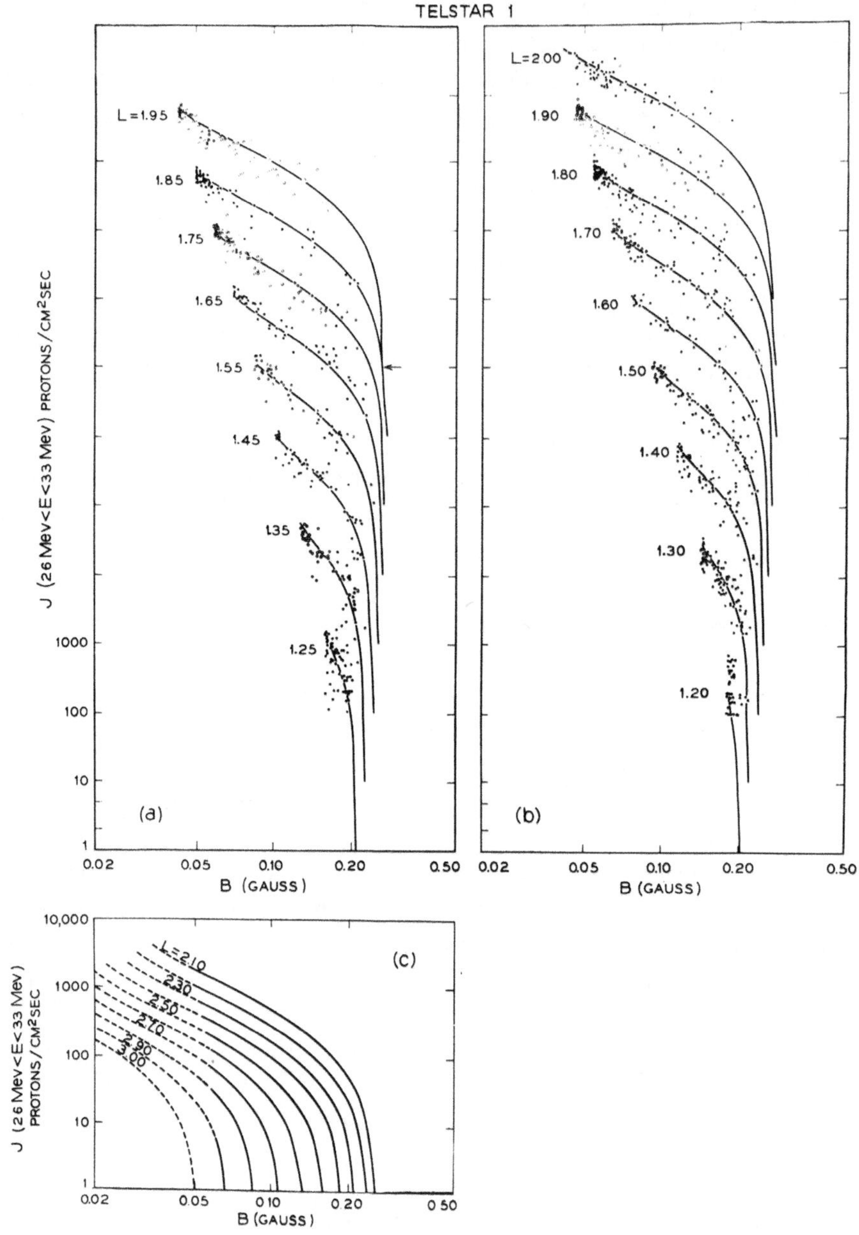

Fig. 3. Flux vs B contours for constant L for the 26–33 MeV protons measured on Telstar 1. The lines were generated with the model described in Table III. The remaining conditions in the caption of Figure 2 apply.

TABLE III

The functional fit to the 26–33 MeV proton data from Telstar 1

These expressions are valid only for $1.15 < L < 3.00$, $R < 1.95\, R_e$*, $J > 100$ protons/cm^2 sec.

$$C'(B, L) \begin{cases} = D^2 \left[1 - \left(\dfrac{x}{G}\right)^{5/4}\right]^{2N}, & x \leqslant G \\ = 0, & x > G \end{cases} \qquad \text{(a)}$$

where C' = omnidirectional counting rate, counts/sec, of protons with energies between 26 and 33 MeV

B = magnetic induction, Gauss

L = magnetic shell parameter

$x = \sqrt{1 - B_0/B}$

$B_0 = 0.311653/L^3$

$$D = \frac{D_1(L - L_1)}{D_2 + (L - L_1)^{D_3}}, \quad L \geqslant L_1 \qquad D = 0, \quad L < L_1$$

$$G = \sqrt{1 - \left(\frac{P}{L}\right)^3 \left[4 - 3\frac{P}{L}\right]^{-1/2}}$$

and $P = L_1 + P_1(L - L_1) + P_2(L - L_1)^2 + P_3(L - L_1)^3$.

$$J'(B, L) \approx 1700\, C'(B, L), \quad J' > 100 \qquad \text{(b)}$$

where J' = omnidirectional flux, protons/cm^2 sec, of protons with energies between 26 and 33 MeV.

$L_1 = 1.1667423$
$D_1 = 1.695$
$D_2 = 0.14926$
$D_3 = 2.5859$
$P_1 = 0.0923283$
$P_2 = -0.278932$
$P_3 = 0.2655505$
$N = 0.62703$

* 1 R_e = 6371.2 km

6. Equatorial Fluxes

Figure 7 summarizes the behavior of the omnidirectional flux J, as a function of L on the magnetic equator. Equatorial values of J have been taken from Figures 2 through 6. To these have been added the equatorial values of J for the later (Winter 1964) measurements of 49–145 MeV protons and the 18–27 MeV protons made with Telstar 2. The Telstar 1 50–135 MeV proton curve has been omitted from Figure 7 because it is virtually coincident with the Telstar 2 49–145 MeV proton curve for the summer of 1963 below $L = 1.95$. The dashes indicate extrapolation into regions where equatorial data are not available.

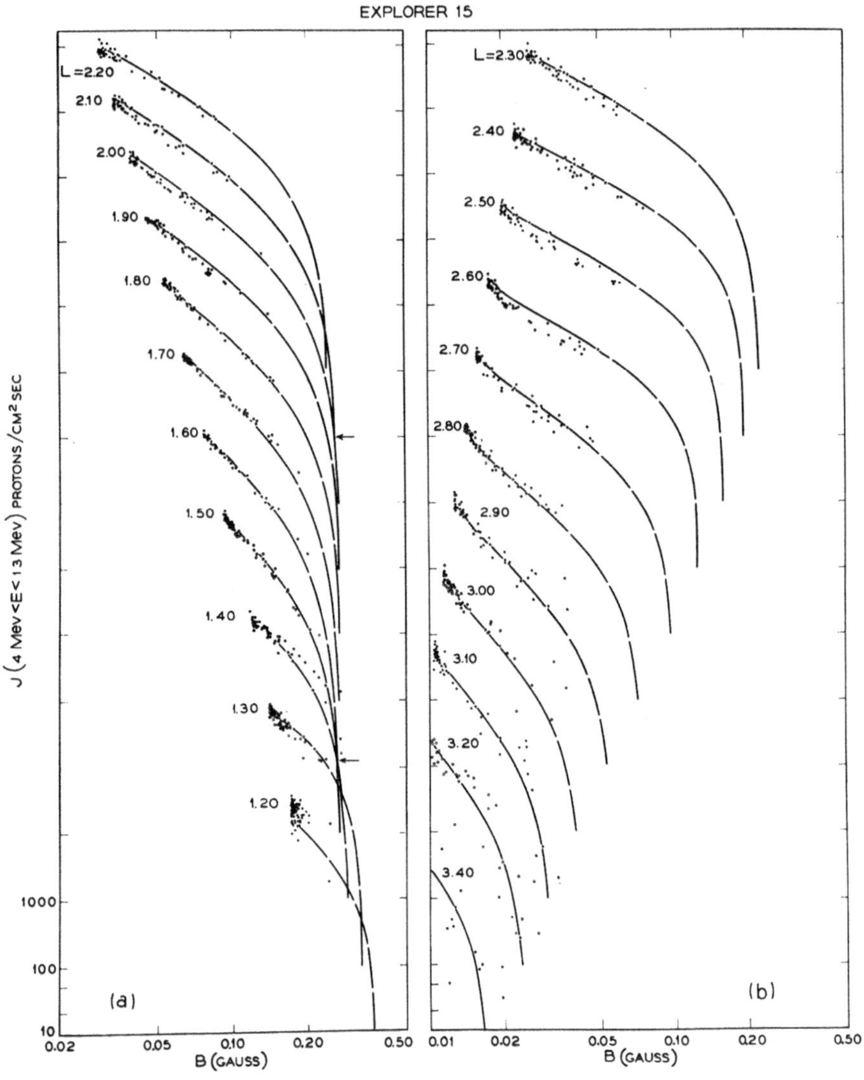

Fig. 4. Flux vs B contours for constant L for the 4–13 MeV protons measured on Explorer XV (November 1962 to January 1963). Each contour rises from $J = 10$. Only a representative sample of the interpolated data points are plotted. The lines come from a least squares fit to the data, and are extrapolated at the high-B end.

The substantial increase in equatorial flux as the energy of the protons decreases, and the movement of the maximum in the equatorial flux to lower values of L for the higher energy particles is observed in these results as in the results of previous experiments (DAVIS and WILLIAMSON, 1962; and FILLIUS and MCILWAIN, 1964). The effect of this behavior on the energy spectrum is treated later.

MCILWAIN (1963) observed a secondary maximum at the equator at $L \sim 2.2$ in the 40–110 MeV proton distribution measured by him with Explorer XV in late 1962. This secondary maximum was not observed by Telstar 1, whose *equatorial* coverage reached only to $L = 1.95$. The Telstar 2 measurements of 49–145 MeV protons (made subsequent to May 1963) show no secondary maximum, although the Summer 1963 and Winter 1964 curves in Figure 7 display a well defined plateau near $L = 2.2$. Apparently, the secondary maximum

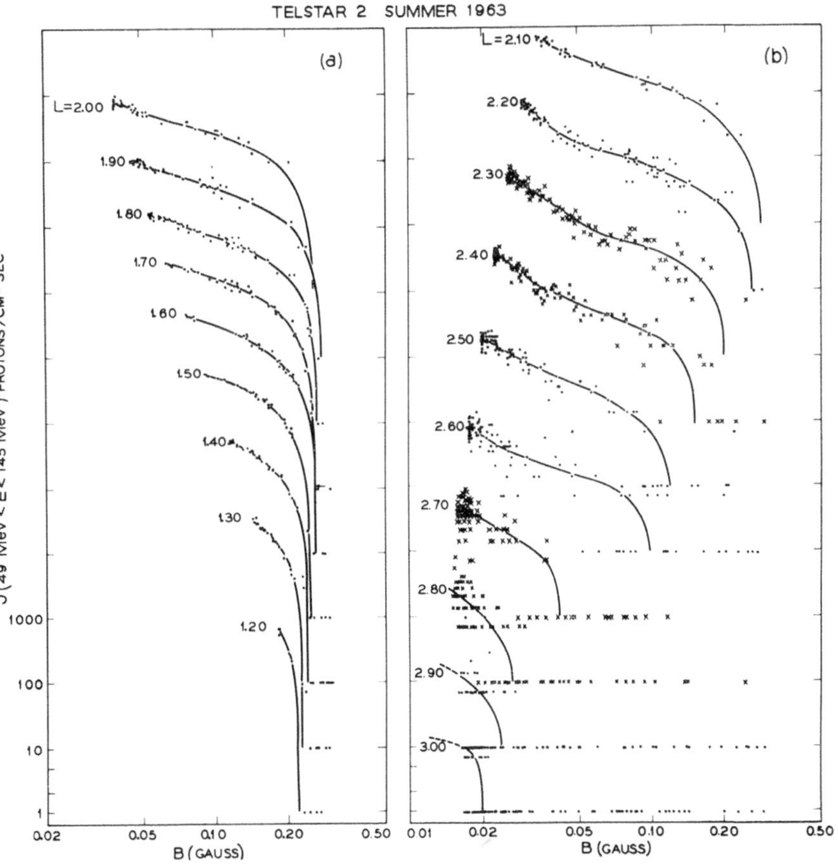

Fig. 5. Flux vs B contours for constant L for the 49–145 MeV protons measured on Telstar 2 from May to September 1963. The representation is as in Figure 2. The lines were hand drawn through the points.

has decreased with time relative to the main part of the proton distribution.

Temporal effects are also apparent in Figure 7. The equatorial flux of both the 49–145 MeV and 18–27 MeV protons is higher for the Winter 1964 curves than for the Summer 1963 curves for L less than about 2.2, while the inverse is true for higher values of L. These curves, however, are averages over several months of measurement, and the data will be considered in more detail below to see what underlies these changes.

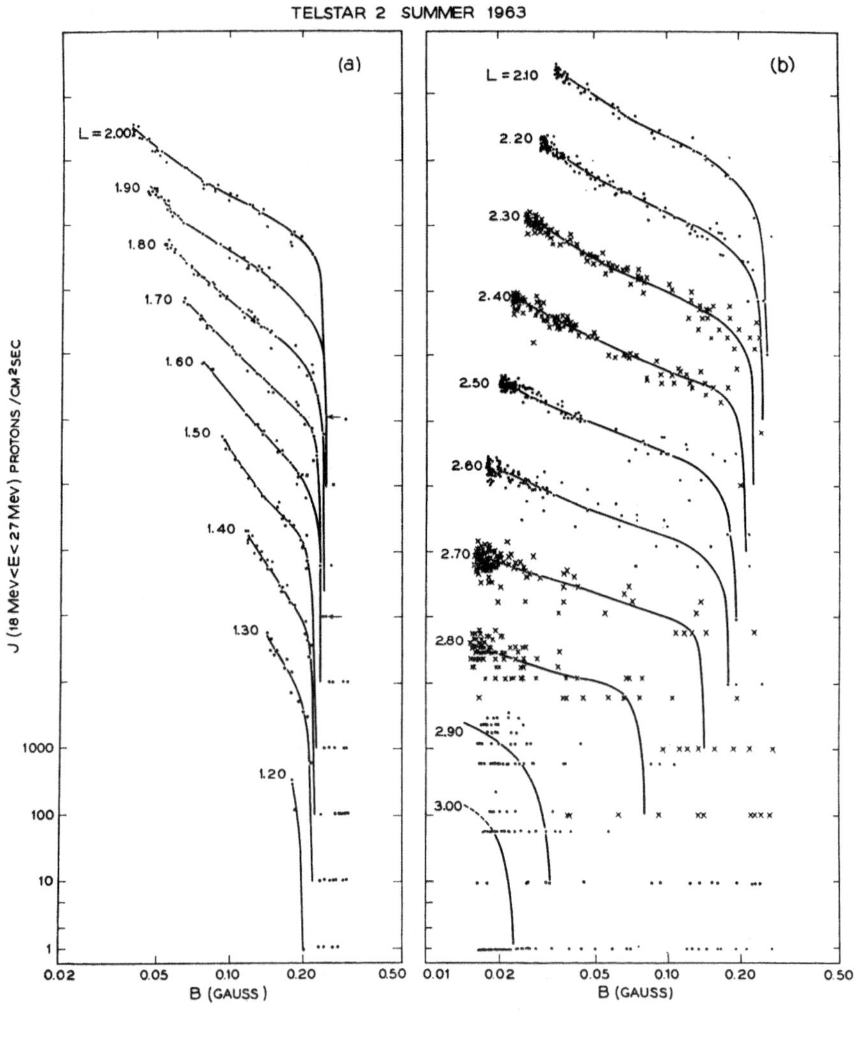

Fig. 6. Flux vs B contours for constant L for the 18–27 MeV protons measured on Telstar 2 from May to September 1963. The representation is as in Figure 2. The lines were hand drawn through the points.

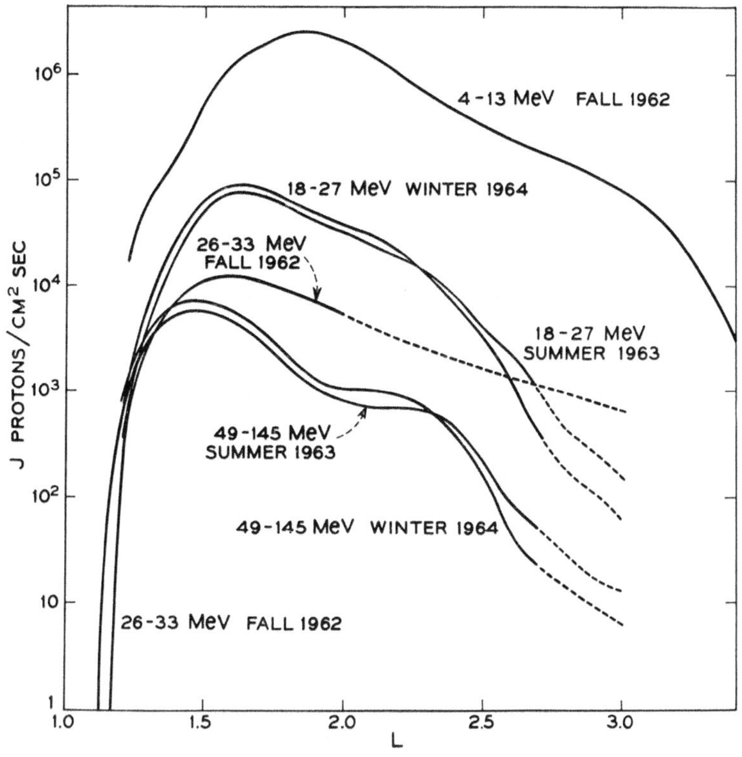

Fig. 7. Flux vs L contours at the magnetic equator. Seasons are the approximate midpoints of the time intervals during which the measurements were made. Dashes are extrapolations.

7. Temporal Effects

The character of the temporal effects in the 49–145 MeV protons on Telstar 2 is displayed in Figure 8, which is a plot of J vs time in days in the various B-L cells indicated in the legend.

Figure 8a, for $1.85 \leqslant L \leqslant 1.90$, is typical of the temporal behavior of the Telstar 2 protons in both (18–27 MeV and 49–145 MeV) energy ranges for $L \lesssim 2.2$. The flux increases steadily with time, and the rate of increase appears independent of B. A careful analysis of the Telstar 1 protons in the energy ranges 26–33 MeV and 50–135 MeV shows no corresponding secular increase between July 1962 and February 1963. The secular temporal effects observed by Telstar 2 are surprisingly large and include at least some long term changes in instrumental sensitivity. A more detailed examination of these effects, based

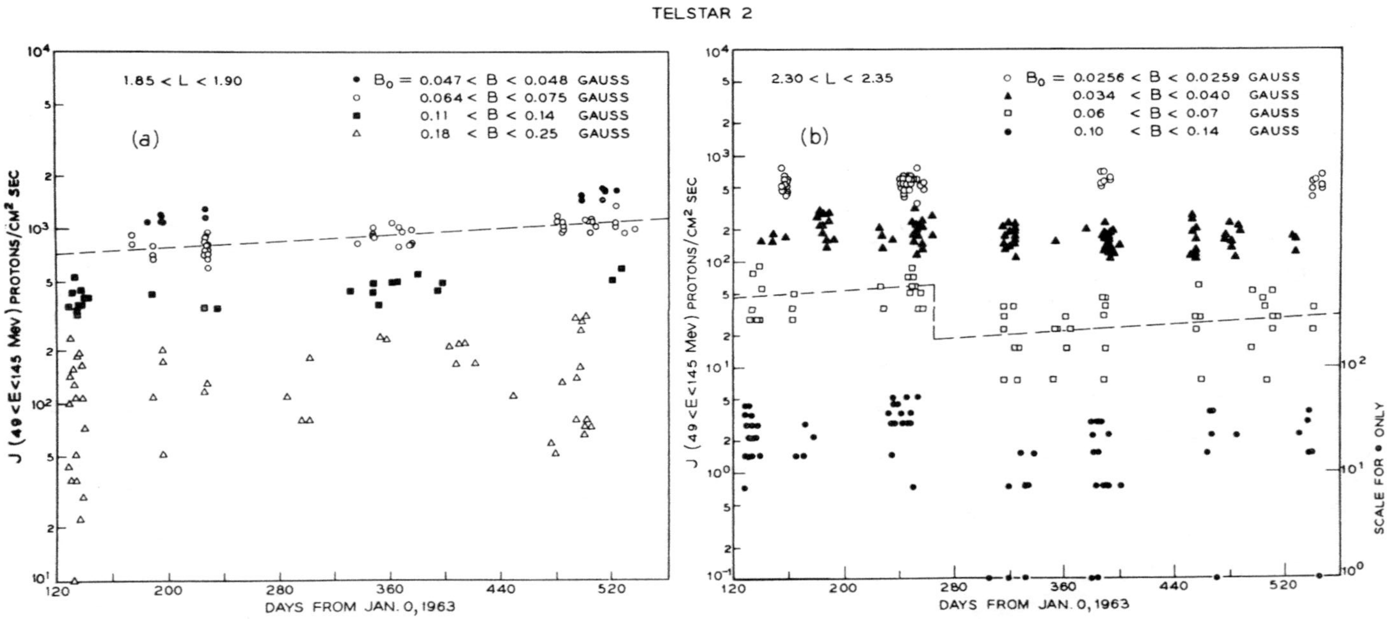

Fig. 8. Flux vs time in days for the 49–145 MeV protons measured on Telstar 2. The points are data taken when the satellite was in the *B-L* cells specified in the legend. In (a) note that the flux increases with time and the increase is approximately independent of *B*. In (b) note the decrease in the flux caused by the magnetic storm of day 265, 1963. The fractional decrease is greater in cells for which *B* is larger.

on residuals from computer fits is planned in an attempt to separate out the instrumental component.

In the more sophisticated analysis of the Telstar 2 data presently in progress, a parametric expression of the form

$$y(x) = A\left[1 + K(t - 265)\right]\left[\frac{x_e^2 - x^2}{x_e^2(1 - x^2)}\right]^{1/4}\left[1 - \left(\frac{x}{x_c}\right)^2\right]^S \tag{2}$$

where y=least-square prediction of $\sqrt{\text{counts/sec}}$, $x=\sqrt{1-B_0/B}$; $B_0=(0.311653/L^3)$; t=time in days in 1963; and A, K, x_c, and S are fitted parameters, is fitted to the square root of the counting rate of data in narrow

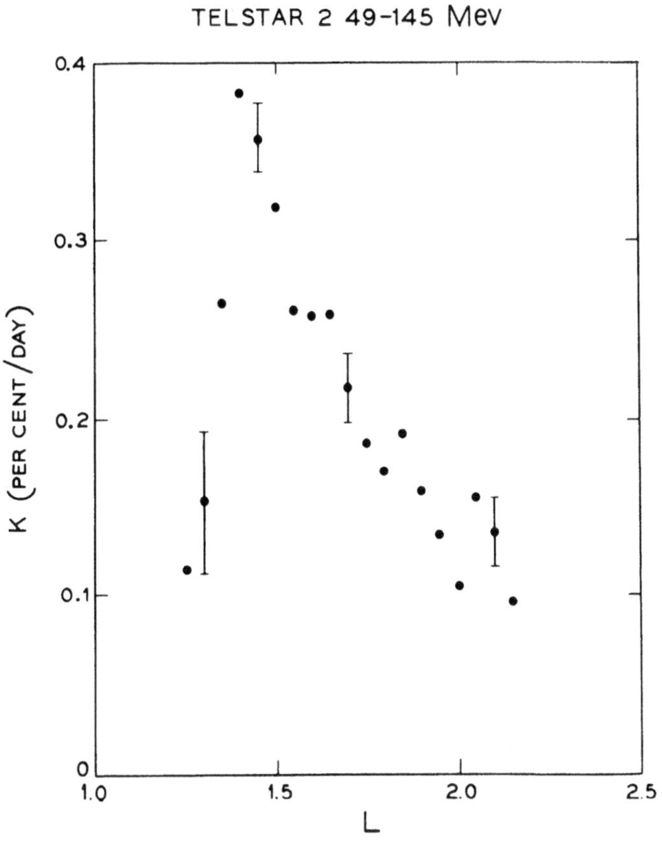

Fig. 9. The values of K, Equation (2), plotted against L. K is the time rate of change of the *square root* of the flux of 49–145 MeV protons from May 1963 to June 1964. The bars are two standard deviations. The per cent change in flux is $\sim 2K$.

L-slices. The parameter K gives the time dependence of the data in the L-slice. The time variation of the data shows no statistically significant dependence on x (or thus B) within L-slices. The values of K from a few preliminary fits to the 49–145 MeV data are plotted against L in Figure 9. The per cent increase in

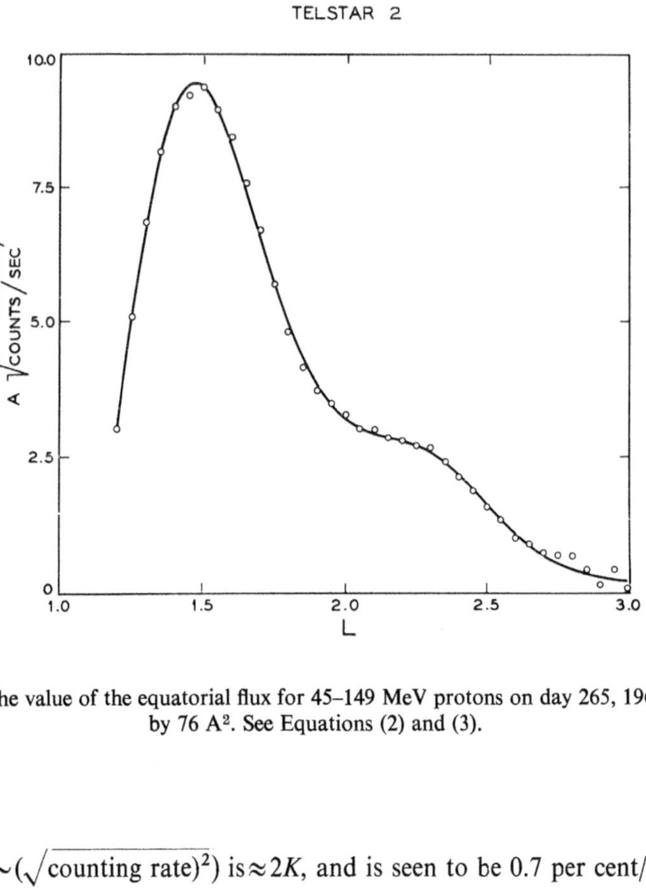

Fig. 10. The value of the equatorial flux for 45–149 MeV protons on day 265, 1963 is given by 76 A^2. See Equations (2) and (3).

flux (flux $\sim(\sqrt{\text{counting rate}})^2$) is $\approx 2K$, and is seen to be 0.7 per cent/day near $L=1.35$, but to drop to 0.2 per cent/day near $L=1.2$ and $L=2.2$. The secular behavior of the 18–27 MeV protons follows the same pattern. These changes are quite large and are presently unexplained.

The values of A in Equation (2) which represent the square roots of the prestorm equatorial counting rates corrected to day 265, 1963 are plotted against L in Figure 10. The line in Figure 10 is given by the expression

$$A(L) = \frac{7.07(L-1.124)}{0.168+(L-0.838)^{5.26}} + 1.49\exp\left[-\left(\frac{L-2.30}{0.313}\right)^2\right] \tag{3}$$

in which the values of the coefficients are the result of a least-squares fit to the points.

For values of $L \gtrsim 2.2$ the unusual effect of the magnetic storm of day 265, 1963 is superimposed on the secular effect. The Telstar observations of the storm effect extend to higher equatorial altitudes than those of Relay 1, but qualitative agreement between the Telstar 2 and Relay 1 measurements (McIlwain, 1964) is good where the data overlap in space. The storm decreased the number of trapped protons with energies between 49 and 145 MeV and between 18 and 27 MeV for $L \gtrsim 2.2$, and the depletion was completed in less than a day. The magnitude of the storm effect was heavily dependent on both B and L. Although the B-L cells in Figure 8b do not have data immediately subsequent to the storm, the B dependence of the depletion in the L range 2.30–2.35 for the 49–145 MeV protons is apparent. The equatorial fluxes were not noticeably affected, but there is a substantial decrease in flux at the higher B values. Some recovery of the fluxes is also evident. That the effect of the storm was more pronounced at higher values of L may be seen from Figure 7.

The reaction of the 18–27 MeV flux distribution to this storm was qualitatively very similar to that of the 49–145 MeV distribution, although somewhat less pronounced.

8. Energy Spectra

The equatorial flux values prior to the magnetic storm of day 265, 1963 have been corrected to integral flux above the minimum energy of the detector. In the higher energy detectors this has been done by making self-consistent determinations of M in an integral power law spectrum of the form

$$J(>E) \sim E^{-M} \tag{4}$$

in the 18 to 145 MeV region. The 4–13 MeV points have been corrected to integral flux above 4 MeV by extrapolating the power law spectrum for the higher energy detectors down to 13 MeV and adding this contribution to the 4–13 MeV flux. The equatorial results for various values of L are plotted as $J(>E)$ vs E on a log-log scale in Figure 11. To prevent overlap, each set of points is slipped one decade to the right. The values of $M(\pm \approx 0.3)$ are given on their respective curves. The equatorial spectrum in the energy range above 18 MeV is softest near $L=1.9$, hardens slightly with increasing L up to $L=2.8$, and hardens notably with decreasing L between $L=1.9$ and 1.3.

The character of the spectral dependence on B for fixed L may be seen in Figure 12. The points in Figure 12 were derived from the flux maps in the same manner as those in Figure 11. On a given L shell the spectrum hardens with increasing B.

It is clear from the plots that a simple power law is *not* in general sufficient to describe the spectrum down to 4 MeV, although it does reasonably represent the higher energy data. These trends as well as the values of the slopes defining

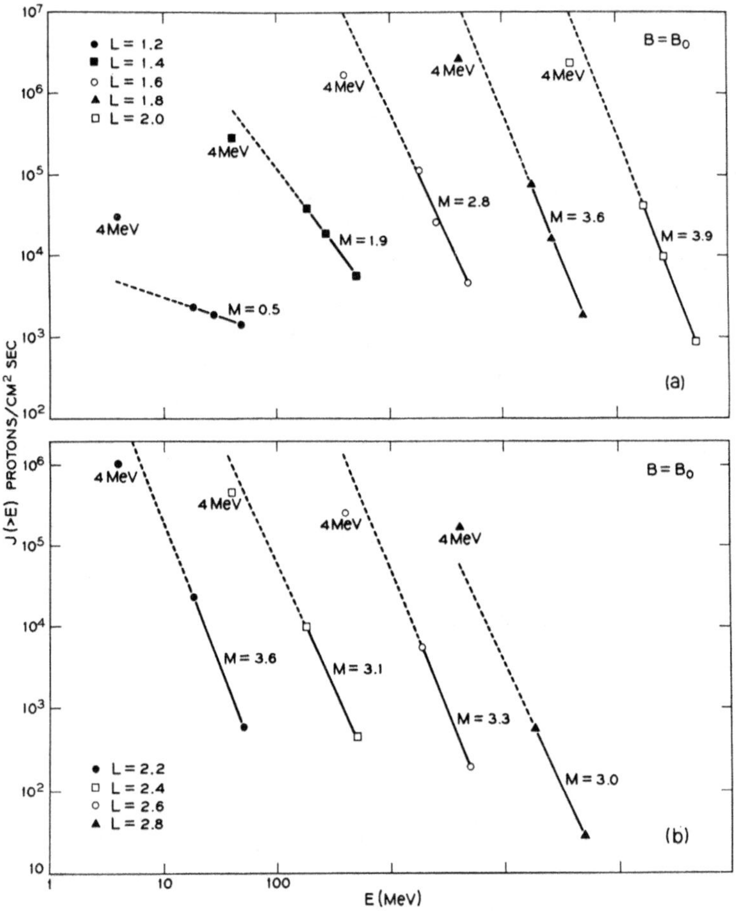

Fig. 11. Integral energy spectra on the magnetic equator.

the spectra are in very good agreement with the published results of other investigators (Fillius and McIlwain, 1964; and Freden *et al.*, 1965) where the data overlap. The present results, however, extend to higher altitudes than were accessible in the earlier experiments.

Various processes for introducing and removing protons from the radiation belt and for influencing the spacial and energy distributions of those present are

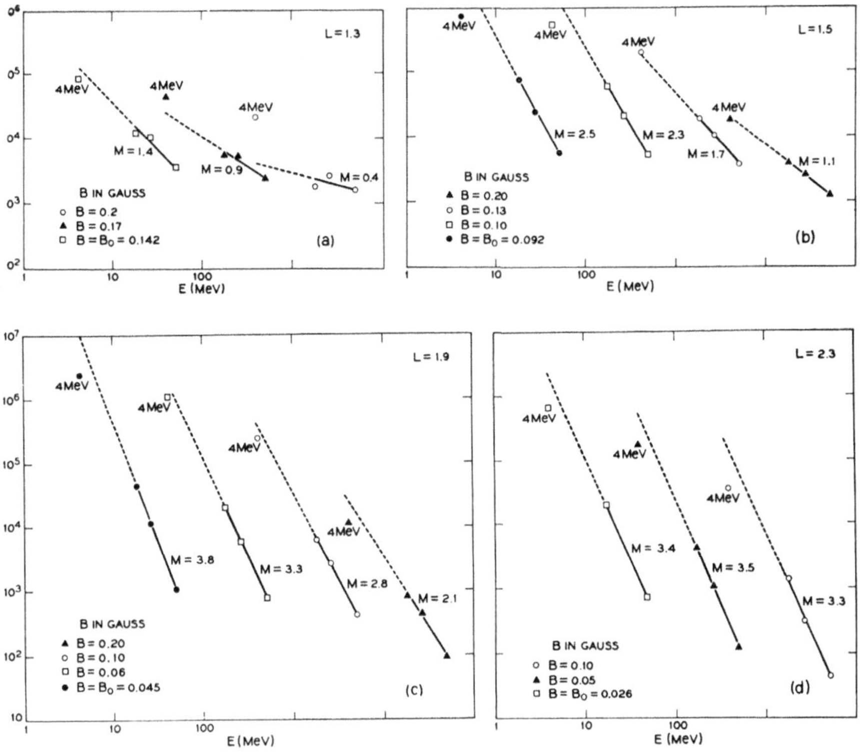

Fig. 12. Integral energy spectra on four *L* shells.

discussed at length in other papers in this symposium. However, the underlying mechanisms are not yet sufficiently well understood to serve as a basis for a quantitative explanation of the observed flux distributions and energy spectra presented here.

9. Acknowledgement

The authors are indebted to a large number of people for their contributions to this work. In particular, the nonlinear, multidimensional fitting procedures used were developed by Dr. M. B. Wilk. Major contributions to the data analysis were also made by Mrs. M. Becker, Miss E. A. Blake, Mrs. N. L. Graham, Mrs. W. Mammel, Mr. S. Polk, and Mrs. E. E. Yamin.

References

Brown, W. L.: 1965, *Symposium Proceedings of the Plasma Space Science* (ed. by C. C. Chang and S. S. Huang), Dordrecht, D. Reidel Publ. Comp., 1965, p. 189.

BROWN, W. L., BUCK, T. M., MEDFORD, L. V., THOMAS, E. W., GUMMEL, H. K., MILLER, G. L., and SMITH, F. M.: 1963, *Bell System Tech. J.* **42**, 899.

BROWN, W. L. and GABBE, J. D., *Project Telstar Report on Communications and Radiation Experiments with Telstar II* (Bell Telephone Laboratories, Inc., Murray Hill, N. J., 1964), Section 5.

BROWN, W. L., GABBE, J. D., and ROSENZWEIG, W. : 1963, *Bell System Tech. J.* **42**, 1505.

BUCK, T. M., WHEATLEY, H. G., and RODGERS, J. W.: 1964, *IEEE Transactions Nuclear Science* **11**, 294.

DAVIS, L. R. and WILLIAMSON, J. M.: 1963. *Space Res.* **3**, 365.

FILLIUS, R. W. and MCILWAIN, C. E.: 1964, *Phys. Rev. Letters* **12**, 609.

FREDEN, S. C., BLAKE, J. B., and PAULIKAS, G. A.: 1965, *J. Geophys. Res.* **70**, 3113.

GABBE, J. D., WILK, M. B., and BROWN, W. L.: 1966, to be published.

JENSEN, D. C. and CAIN, J. C.: 1963, *J. Geophys. Res.* **67**, 3568.

MCILWAIN, C. E.: 1961, *J. Geophys. Res.* **66**, 3681.

MCILWAIN, C. E.: 1962, private communications.

MCILWAIN, C. E.: 1963, *Science* **142**, 355

MCILWAIN, C. E.: 1964, 'Redistribution of Trapped Protons During a Magnetic Storm' (Department of Physics, University of California at San Diego, April, 1964).

ROBERTS, C. S.: 1965, *J. Geophys. Res.* **70**, 2517.

SUMMARY OF INNER ZONE PROTONS

STANLEY C. FREDEN
Aerospace Corporation, El Segundo, California

A supplementary evening session was conducted on inner zone protons. Many points were clarified for some of the Institute participants who had not directly performed research on the inner zone protons. From the wide variety of experimental data that had been presented by many lecturers, it seemed important to derive a brief summary of the current state of understanding of the inner zone protons.

Omnidirectional fluxes of protons with energy thresholds of 40 MeV, 5 MeV, 100 keV, and 1 keV were discussed as a function of L at the magnetic equator. It was pointed out that the isoflux contours are terminated at low L values because of the atmospheric loss mechanisms and are terminated at high L values by some other loss mechanisms, as yet unknown.

The fluxes of protons below about $L = 2$ appear to be very stable with time and insensitive to magnetic activity. Above $L = 2$ decreases in the proton flux were observed associated with the large magnetic storm in September 1963. On a given L shell, these decreases were energy dependent, decreasing with decreasing energy, and very dependent on B. The proton flux above 40 MeV tended to recover slowly to a pre-storm level in a manner which is consistent with cosmic ray albedo neutron decay injection.

For protons with energies above 40 MeV there are two peaks in intensity at $L \approx 1.5$ and $L \approx 2.2$. The peak at the higher L value is only observed near the equator and suggests a second source which could be due to a diffusion process. Secular variation of the higher L shell peak is suggested by comparing observations between 1963 and 1965 and indicates a shift to a lower L value during that time period.

At $L \approx 1.3$ the measured energy spectrum agrees reasonably well with that calculated using an albedo neutron decay injection source and atmospheric losses. The small discrepancies could be explained as due either to nonadiabatic losses at the high energies or to an albedo neutron spectrum which is different from that used in the calculations. This latter assumption is reasonable since the neutron spectrum is really not well known. At L values above 1.3, the spectrum above about 50 MeV becomes steeper with L as would be expected from non-adiabatic losses. Below about 30 MeV, however, a strong component appears which is not predicted by the albedo neutron decay theory. This flux is L dependent and strongly peaked at the equator, also suggestive of a diffusion process. The possibility of a solar proton albedo neutron decay source to explain the increased particle flux below 30 MeV seems to be ruled out because of the spatial distribution and the lack of a correlation between the flux and solar proton events.

The experimentally observed east-west asymmetry of mirroring protons at low altitude confirms the atmospheric loss process and provides a measurement of the atmospheric scale height. The temporal decay of the trapped proton flux at 55 MeV at low altitudes following the sharp increase presumably produced by the 1962 Starfish high altitude nuclear explosion also supports the atmospheric loss mechanism.

CONCLUSIONS ON NEUTRON ALBEDO DECAY SOURCE

WILMOT N. HESS
Goddard Space Flight Center, Greenbelt, Maryland

At the end of the Session, Dr. McCormac requested that the chairman try to summarize the experimental material presented dealing with neutron decay. Dr. Hess put the following chart (Table I) on the board as his estimate of the findings of the significant experiments. He stated that the definition of agreement that he was using was "agreement within a factor of three".

TABLE I
Neutron Source and Atmospheric Loss

Agree	*Inconclusive*	*Disagree*
(1) Freden and White proton energy spectrum	(5) Freden and White proton flux	All comparisons of experiments trying to find time variations at low altitudes
(2) Heckman east-west flux ratio*	(6) Pizzella Explorer VII time variations at low L	(8) Elliot
(3) Filz proton flux vs. altitude	(7) Filz time variations of 55 MeV protons on series of emulsions	(9) Freden
(4) Filz decay of transient proton flux*		(10) McIlwain
		(11) Gabbe

* These experiments bear only on whether the atmospheric loss is dominant (not on the source).

There was considerable discussion of the entries in the chart by Freden, Katz, McIlwain, Gabbe, Hess and others. There was not complete agreement that low altitude fluxes behaved exponentially with altitude as one would expect to be the case for atmospheric loss. Filz curves [(3)] show exponential behavior for $h^{\min}$ 200–700 km but McIlwain said his data don't seem to agree well with exponential atmospheric loss in most regions.

It was agreed by everybody that the time variations observed at low altitudes [(8) (9) (10) and (11)] do not seem to agree with the predictions of Blanchard and Hess. The time changes were either too small or even of the wrong sign. Hess said the calculated time variations were too large by about a factor of two due to the atmosphere used but that variations roughly as predicted should be observed somewhere in BL space even if the average atmosphere had been calculated wrong providing that the atmosphere dominated the process. It appears that something else is needed at low altitude.

An attempt was made to bracket the region in which there seemed to be agreement between experiments and the neutron decay process. At $L=1.3$ there seems to be agreement from 30–500 MeV. At $L=1.5$ there are disagreements. At low altitudes there are disagreements in the solar cycle variations.

SESSION III

OBSERVED PARTICLE DISTRIBUTION IN THE NATURAL BELTS (II)

Chairman

DR. BRIAN O'BRIEN

Rice University

THE ALOUETTE SATELLITE RESULTS

D. C. ROSE
Division of Pure Physics, National Research Council, Ottawa, Canada

Introduction

The Alouette satellite was originally planned as a topside sounder for measurement of the ionosphere electron density from above. The relation between the energetic particle flux striking the upper atmosphere and the ionosphere is so close that a proposal to include a package of particle detectors in the satellite was accepted as a primary experiment in the early stages of planning. In this paper some of the results obtained from this energetic particle package will be presented. A complete review of the results of measurements from Alouette, including both ionosphere and energetic particle information, would be out of the question in a short paper and even the data to be presented here are limited to only a fraction of that which could be made available. The satellite has been phenomenally successful. It was the first satellite to be designed and built in a country other than the U.S.A. or Russia and the policy of the U.S. National Aeronautics and Space Administration in giving such an opportunity to other countries is appreciated.

The energetic particle package contained six particle detectors whose properties are shown in tabular form in Table I. The Anton 302 Geiger counter was in a shielded part of the package and could be used only for omnidirectional particle counting with the properties shown in the table. The two Anton type 223 Geiger counters are thin window counters located at the end of cylindrical brass collimators. A magnetic field crossed the collimator tube of one counter. The axis of the collimator for the counter with the magnetic broom was parallel to the planned spin axis of the satellite. The other collimator with no field was set 10° off the spin axis. The opening in front of the silicon junction detector was also collimated and inclined 10° to the spin axis. The Geiger telescope and the scintillator which was within the telescope looked at right angles to the spin axis but were heavily shielded and included a wide angle.

The results used here will be those obtained only from the three Geiger counters, the other detectors are mentioned because some information from the silicon junction detector was necessary to show that the particles we are

TABLE I

Properties of particle detectors flown by the Alouette satellite

Detector	Directional properties			Omnidirectional properties		
	Geom. Factor	*Particle Response*	*Efficiency*	*Geom. Factor*	*Min. Shielding*	*Particle Response*
302 Geiger	–	–	~ 80% rel. electrons	0.55 cm^2	~ 1.4 g cm^{-2} over approx. 2.4 sterad	Electrons: 2.8 MeV 10% trans. 3.9 MeV 50% trans. Protons: > 33 MeV
223 Geiger	5.05×10^{-4} cm^2 sterad	electrons > 40 KeV protons > 500 KeV	~ 80% rel. electrons	0.22 cm^2	~ 1.4 g cm^{-4} over approx. 2.0 sterad	Electrons: 2.8 MeV 10% trans. 3.9 MeV 50% trans. Protons: > 33 MeV
223 Geiger with mag. field	2.33×10^{-3} cm^2 sterad	electrons > 250 KeV protons > 500 KeV	~ 80% rel. electrons	0.22 cm^2	~ 1.4 g cm^{-3} over approx. 0.5 sterad	Electrons: 2.8 MeV 10% trans. 3.9 MeV 50% trans. Protons: > 33 MeV
Silicon junction	(1) 1.57×10^{-3} cm^2 sterad (2) 1.57×10^{-3} cm^2 sterad	protons 1.3–7 MeV α-particles 4.3–28 MeV	100%	0.04 cm^2	~ 4.0 g cm^{-2} over approx. 3.5 sterad	Protons 55–60 MeV
Geiger telescope	3.5 cm^2 sterad	protons > 100 MeV α-particles > 400 MeV	9% rel. protons 65% rel. α-particles 100% heavier particles	–	–	–
Plastic Scintillator	3.5 cm^2 sterad	protons > 100 MeV α-particles > 400 MeV	9% rel. protons 65% rel. α-particles 100% heavier particles	–	–	–

examining were in fact electrons and the contribution from protons and Bremsstrahlung and the background of cosmic ray particles was either negligible or could be subtracted with no difficulty. You will note in this regard that the type 302 counter had over a large aperture the same shielding as the non-collimated part of the flux entering the two type 223 counters. Bremsstrahlung and cosmic ray background effects should be the same in all three when corrected for geometric factor. The silicon junction detector was insensitive to electrons and would record most of the protons accessible to the other two collimated counters. A comparison of counting rates in these four detectors shows that background corrections were necessary only when considering the data from the counter with the magnetic broom and even then were small compared to the data presented.

The satellite was launched at 0605 UT on September 29, 1962. It is still operating in August 1965 though the energetic particle data we have used were all taken within the first year of operation. The equipment included a magnetometer of the flux gate type at right angles to the spin axis. This magnetometer combined with the known spin at launch and a history of the spin decay made it possible to know the direction of the collimators with respect to the earth's field at any time. When spin decayed and a tumbling motion took over, the ambiguity in direction in relation to the field lines made the data too difficult to analyze, particularly in view of the fact that by that time more data were available than our analytical facilities could cope with in a reasonable time.

The orbit of the satellite is very nearly circular at a height of 1025 ± 20 km at an inclination of 80°. When injected, the spin period was 42 seconds and the spin axis was planned to be normal to the ecliptic but it was a little off this. It had a precession about a flattened cone whose maximum half angle was about 90°. The orbit plane precesses with respect to the sun-earth line about 2° per day, therefore, all local times can be covered in a period of about three months.

The results to be presented here were taken mainly from ground stations in Canada and at College, Alaska. The data will be presented in the well known McIlwain coordinates B and L or R and λ. Since the orbit was nearly circular the coordinate R can be considered as constant having a mean value of 1.14 R_e, the actual values lying between 1.12 and 1.16. We also have used in most instances for invariant latitude the quantity $\lambda = \cos^{-1}\sqrt{R/L}$ rather than that used by O'Brien (1962) $\Lambda = \cos^{-1}\sqrt{1/L}$. The data which will be shown are divided where applicable into trapped and precipitated components. Where the pitch angle measurements were $90 \pm 20°$ the particles would be mirroring at or near the satellite height and are classed as trapped even though evidence will be shown that their lifetime is short. Particles with pitch angles less than 45° will certainly be lost in the atmosphere. The data processing was largely automatic though the magnetometer records had to be read manually and put on punched cards.

Results

Although the energetic particle package of instruments carried in Alouette included equipment for detecting protons and heavier particles the review of results here will be limited to electron flux measurements in the outer belt at the satellite orbit altitude of 1000 $\pm$ 25 kilometers. Further, because of data handling limitations and the convenience of using records from our own and other northern ground stations we have up to the present limited our analyses to the northern "horns" of the outer belt.

Practically all the results to be discussed here have been or will very soon be published in a series of papers by McDiarmid, Burrows and others, as listed in the bibliography. Further, the results will be presented more or less in the order in which the analyses were carried out. Average of rather highly variable quantities will be examined first, then diurnal variation of intensity and of the northern boundary of the layer and the relation of these to the magnetic disturbance index, then the longer term or day by day variation of the peak intensity in the outer layer.

THE AVERAGE BEHAVIOUR OF THE OUTER LAYER INTENSITY

In the introduction, reference was made to the various detectors which made it possible to be sure that the measurements to be described here represented an electron flux with the minimum energies as shown in the data to be presented. The data will, therefore, be presented without further reference to the measuring techniques or the corrections used. These are described in more detail in the reference papers by McDiarmid *et al.*

Figure 1 shows the average omnidirectional intensity of electrons in the outer radiation belt with energies greater than 3.9 MeV. Curve (a) represents periods where the magnetic disturbance index K_p was less than 4 and curve (b) where it was greater than 4. Since these data were taken in late 1962 or early 1963 the high K_p index included no intense storm periods and never went above 6 or 7. The left part of curve (a) shows the effect of the nuclear explosion of October 1962, the lower broken curve represents the data taken from a few passes before the explosions. Similar data for the (b) curve is not shown because of inadequate data.

The numbers on these curves and subsequent figures represent the number of passes used for each point. The data were grouped in 2° intervals of invariant latitude and averaged, this being done by machine methods. The short vertical bars with crosses at each end represent the statistical or Poisson error while the larger broken bars are the root mean square deviation of the data from pass to pass. The position of the peak about 60° of invariant latitude should be noted and the average omnidirectional intensity of a few hundred electrons per cm^2

per second. The L values corresponding to the latitudes are shown at the top of the curve. The peak is centered at an L value of about 4.4.

Figure 2 shows the same kind of data for electrons with energies greater than 250 keV. At these energies the pitch angle could be measured since the opening to the counters was collimated and these curves represent particles

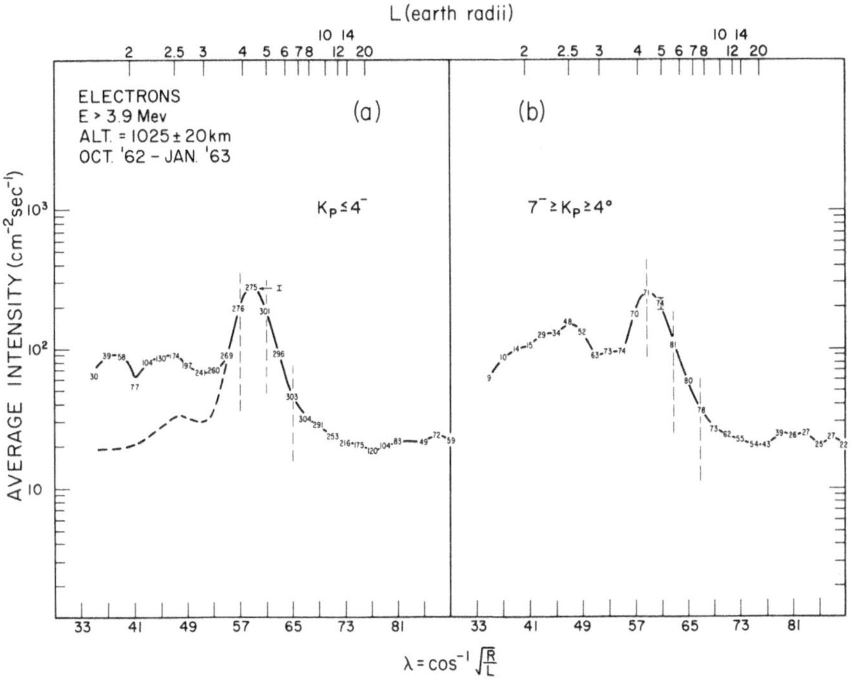

Fig. 1. Average omnidirectional intensities determined from the 302 Geiger counter. The dashed curve in (a) was obtained before the October, 1962, high-altitude nuclear tests. The dashed vertical lines are r.m.s. deviations of individual passes about the mean and the solid vertical lines are estimates of the statistical error in the mean. The numbers on the curves are the number of passes used to calculate the averages.

with pitch angles in the range 90 $\pm$ 20° or trapped particles mirroring at or close to the satellite altitude of about 1025 km. The curve tends to show two peaks but this should not necessarily be taken as a suggestion of a double layer since an examination of the variation bars shows large intensity changes particularly in the region of the northern boundary and the lower latitude peak. This peak does not bear any relation to the inner belt which at 1000 km would be observed only at appreciably lower latitudes. The intensity at the maximum is a few times 10^4 particles per cm^2 per second per steradian, two or more decades higher than the peaks in Figure 1, representing particles above 3.9 MeV. The position of maximum intensity in latitude or L value is not appreciably different and like the previous figure the mean value of the maximum intensity is not appreciably higher for curve (b) representing magnetically disturbed conditions than in curve (a).

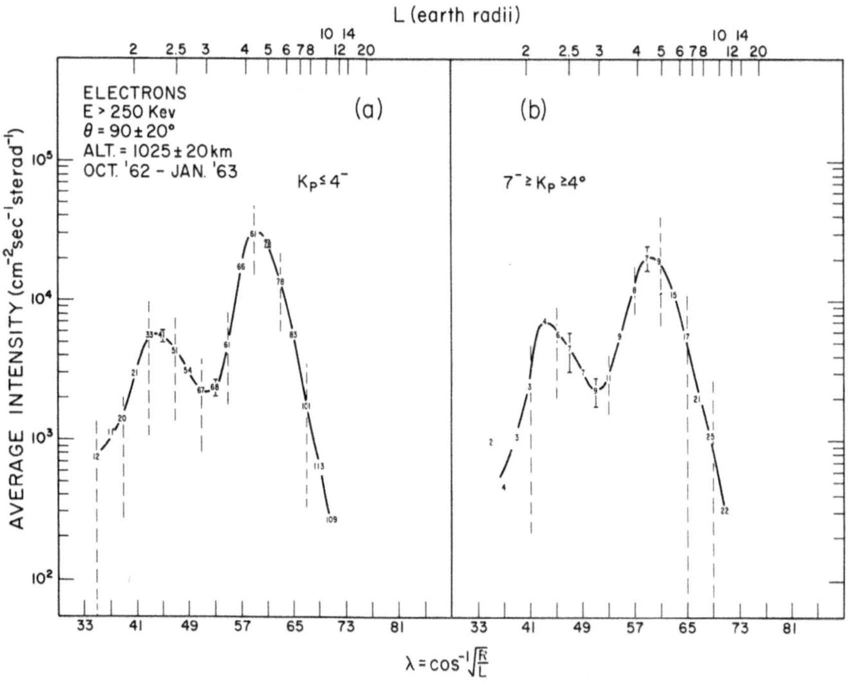

Fig. 2. Average directional intensity of trapped electrons determined from the 223 Geiger counter with the magnetic field.

Figure 3 shows similar data for trapped electrons with energies greater than 40 keV. The double peak in the average values is again apparent but because of the great variability, it is again not clear what significance should be given to it. The position of maximum intensity is shifted to higher latitudes. Unlike the previous figures the intensity under magnetically disturbed periods is appreciably higher than during quiet periods. The very large intensity fluctuation should be noted particularly at the northern boundary. Some of this is due to diurnal variations which will be discussed later.

Figure 4 shows similar data for precipitated electrons with energies greater than 40 keV. The variability shows that precipitated electrons are not always present. The averages show they are much more likely to occur during magnetically disturbed periods than during quiet periods. Precipitation occurs

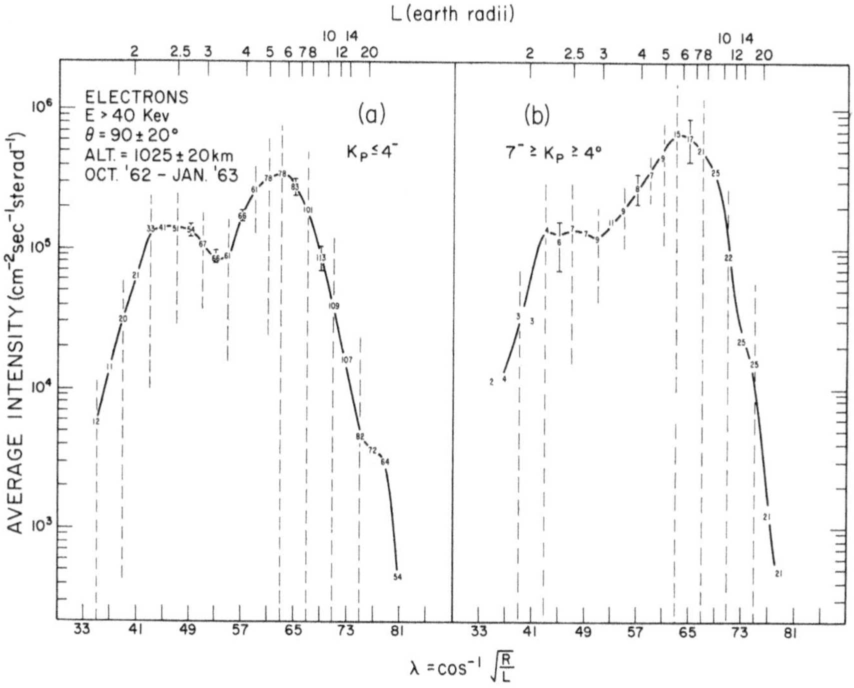

Fig. 3. Average directional intensity of trapped electrons determined from the 223 Geiger counter.

mainly on the high latitude side of the layer, the double peak appearance being missing. The position of the peak in the average curve is closely at the same latitude as that for trapped particles. The average intensity of precipitated particles is within an order of magnitude, even in quiet times, of the intensity of trapped particles with the same minimum energy. The amount of data available for precipitated particles at higher energies is somewhat less but again the single peak at an invariant latitude of 65° is shown. McDiarmid *et al.* (1963)

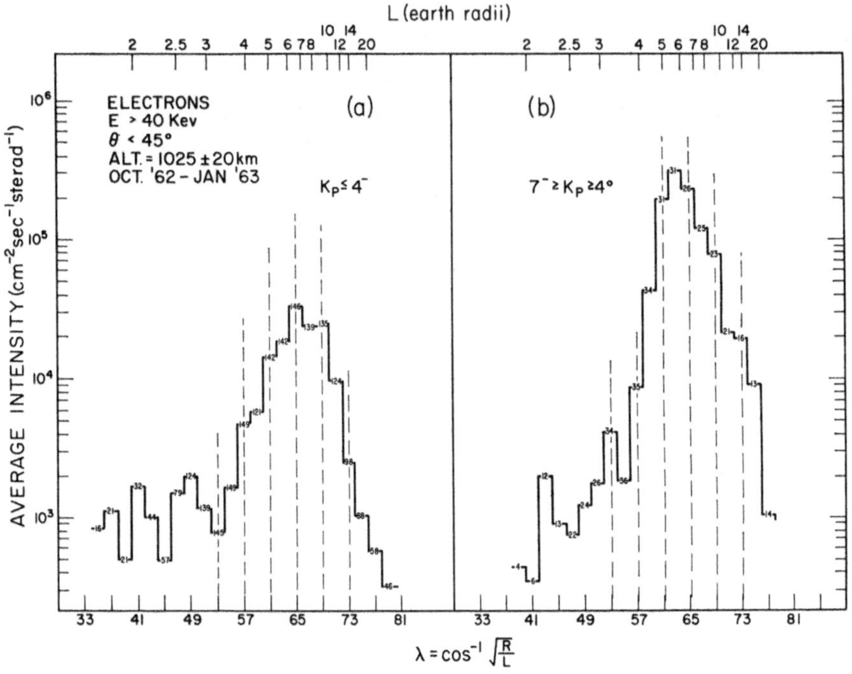

Fig. 4. Average intensity of precipitated electrons determined from the 223 Geiger counter plotted aginst invariant latitude.

show an appreciable softening of the spectrum at latitudes where precipitation is prominent.

THE DIURNAL VARIATION OF THE LAYER BOUNDARY

Figure 5 shows the percentage of passes in which the intensity of trapped particles is greater than 1.5×10^4 particles/cm^2/sec/sterad. This is plotted at different invariant latitudes and the data are divided into different hours of the day, the day being divided into six four-hour periods. The approximate numbers of passes used for each 2° interval are shown on the graph. Each of these lines represents the probable position of the boundary at the latitude and time concerned.

A horizontal cut across these curves plotted on a clock dial diagram shows the diurnal variation of the latitude of the boundary. This is shown in Figure 6.

The circles show the 10% probability level and the 50% probability level for trapped electrons above 40 keV and the 10% zone for precipitated electrons. The latter is less accurate than the data for trapped particles and it is uncertain

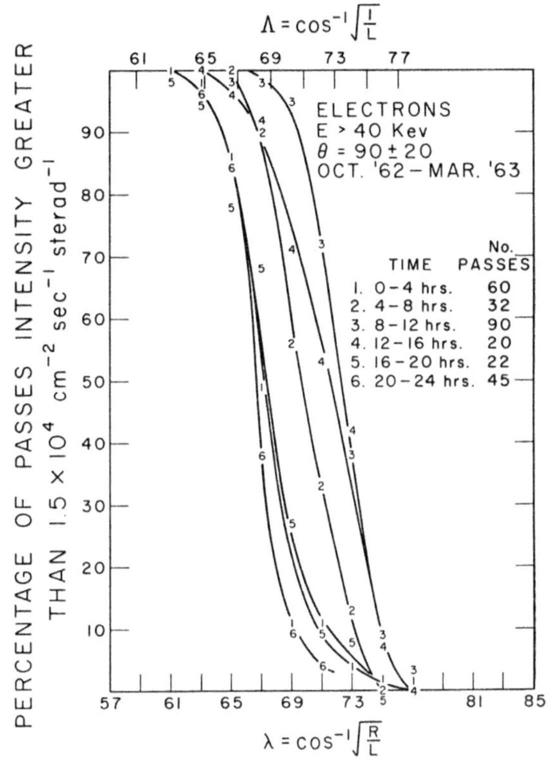

Fig. 5. Percentage of passes with trapped intensities greater than 1.5×10^4 cm^{-2} sec^{-1} sterad^{-1} vs. invariant latitude. The numbers representing the data points refer to the different local time intervals shown in the table. Also shown in the table are the approximate number of passes in each interval.

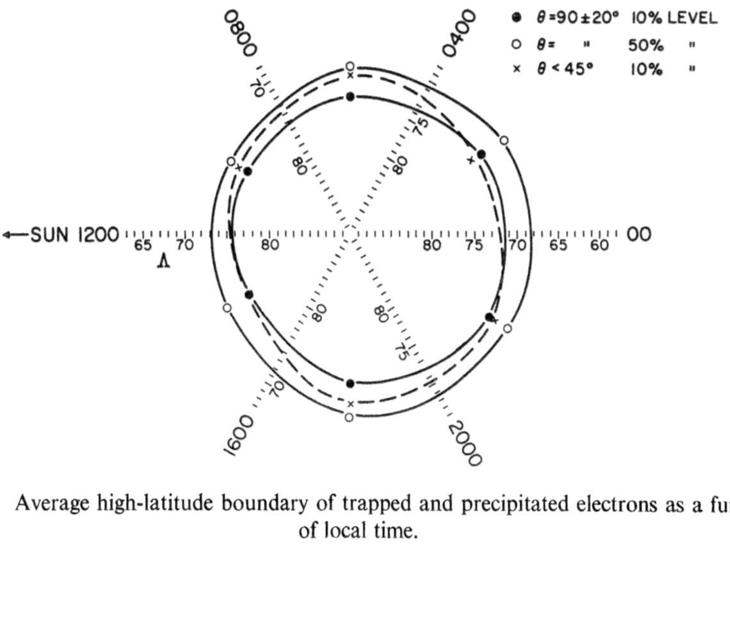

Fig. 6. Average high-latitude boundary of trapped and precipitated electrons as a function of local time.

if the different shape is significant. Both sets of data show an asymmetry with respect to local time, the boundary occurring at highest latitudes about noon and the difference between noon and midnight being close to 5° of latitude.

In examining diurnal variations, particularly averages of quantities with high variability, the results should be considered with some caution because of this high variability. Another point that should be mentioned is that because of the slow precession of the satellite orbit, the measurements at different times of day in local time are far from simultaneous in universal time. The precession of the orbit is such that it takes about three months to get observations that cover all local times. A long term variation therefore might account for the results. McDiarmid *et al.* (1964) have shown, however, that this is very unlikely to be so. By plotting the data using percentage of passes above the fixed level, versus orbit number or day of the year and separating the data occurring in local time for eight hours around midnight and eight hours around noon, over a long period there is a consistency in the indication of a diurnal variation and no suggestion of other long period variation that would eliminate the diurnal effect. For more detailed discussion, reference should be made to the original publication.

At this point one conclusion can be presented. That relates to the short lifetime of particles in the outer belt. The diurnal variation of the boundary, namely about 5°, is too great to be due to distortion of the field lines at 1000 km by the solar wind. It is in the right direction to suggest a diurnal distortion in the boundary of the adiabatic shell on which such a group of trapped electrons would move but calculations on the amount of distortion that should be expected indicate it should not be more than about 2°. This has been indicated by Malville (1960) and Hones (1963) both of whom show that the latitude shift of mirror points between noon and midnight would be appreciably less than 5°, namely about 2°, and the accuracy of the present measurements are considered to be about 1° in the case of the trapped radiation and 2° for precipitated particles. Connecting the 50% boundary with L values the change in the boundary from noon to midnight is equivalent to an L value of 15 to about 9. The actual L values have no particular meaning so close to the boundary of the magnetosphere but are convenient in comparing the variation in the latitude of the boundary at 1025 km as measured here with the extent of the trapped radiation belts at the equator as measured by others. O'Brien (1963) has also observed about a 6° variation in the latitude of the trapping layer at about 1000 km. Other measurements by the Iowa group (see Frank, 1965) show that the boundary near the equator is about 10 R_e on the day side and about 8 R_e on the night side.

Particles which mirror near the equator would be expected to drift around the earth at constant B which would be farther out on the day side than at night. On the other hand particles which mirror in the Alouette orbit at high latitude

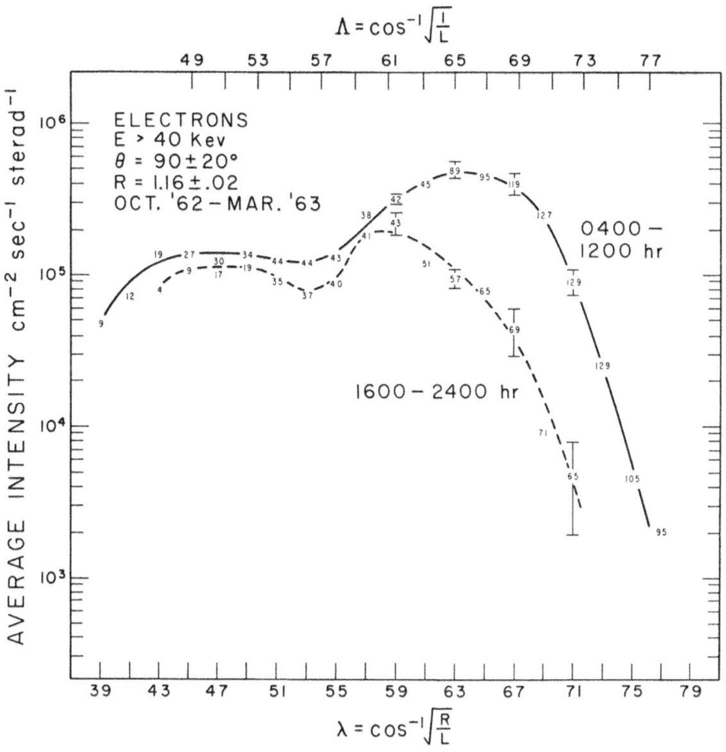

Fig. 7. Average intensity of trapped electrons versus invariant latitude for local times of 0400–1200 and 1600–2400 hours. The numbers on the graph representing the data points are the number of satellite passes used to calculate the averages and the vertical lines are estimates of the statistical uncertainties in the averages.

would be expected to stay on a shell that moves in and out from day to night as the distorted lines of force move. This would put them much farther out on the night side but if the calculations on the distorted field have any meaning the latitude shift should be only about 2° and the trapped boundary at night seems to be at lower R_e than by day, though there is a variable region outside that at night extends into the neutral zone as shown by FRANK (1965). The variation in the boundary of the belt is reasonable if one accepts a short lifetime for the trapped particles at 1025 km. In fact a lifetime that is short compared to the drift period of the particles is suggested. Of course, some other models of the distorted field than those of Malville and Hones might allow a greater distortion at 1025 km but there is other evidence that is consistent with a very short lifetime. This is the fact that the diurnal and boundary variation of precipitated particles behaves closely in the same way as for trapped particles, and the

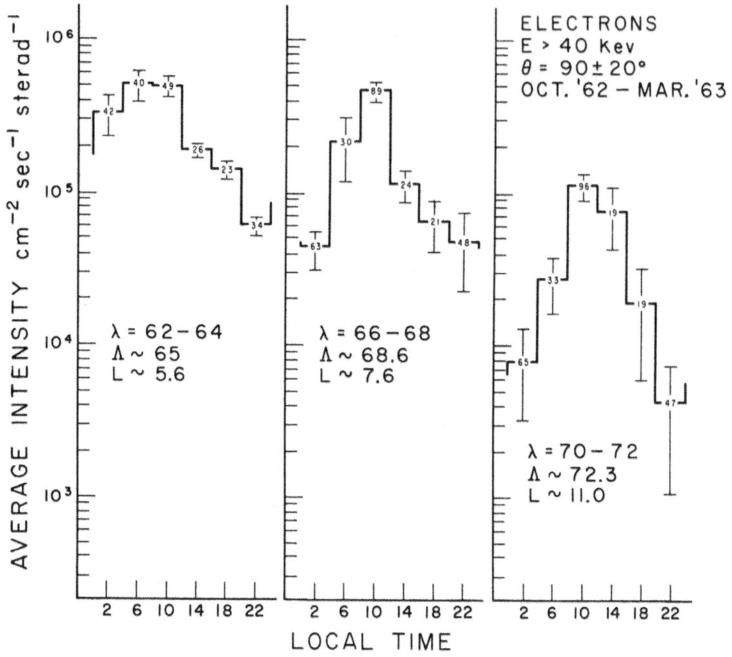

Fig. 8. Average intensity of trapped electrons versus local time for three different latitude intervals. The vertical error bars are estimates of the statistical uncertainties in the averages.

average intensity of precipitated particles is quite sufficient to deplete the layer very quickly. One pictures, therefore, a continued lowering of the mirror points by some form of scattering which is more rapid at low altitudes on any particular shell than a high, and at 1025 km results in an average lifetime of less than one drift period.

THE DIURNAL VARIATION AT OTHER LATITUDES

Having looked at the diurnal variation of the boundary of the outer belt, let us now look at the diurnal variation at different invariant latitudes. Figure 7 shows the average intensity of trapped particles with energies greater than 40 keV selected at two different times of day, one centered four hours before noon and the other just before midnight. The diurnal variation at different latitudes is shown. The remarks made in the previous section about the fact that it takes about three months to cover all local times apply here but again we are satisfied that no long term variation can account for these results.

Figure 8 shows the average intensity of trapped electrons versus local time at three different latitudes. The data were divided into six four-hour periods. It is noted that the diurnal variation is greatest at the highest latitude range. It is centered around noon in the latitude near the boundary but is earlier in the morning at latitudes near the intensity maximum. Figure 9 shows the same type of information for precipitated particles. The variation is not so apparent at the highest latitude but the morning maximum is even more clear at maximum average intensity.

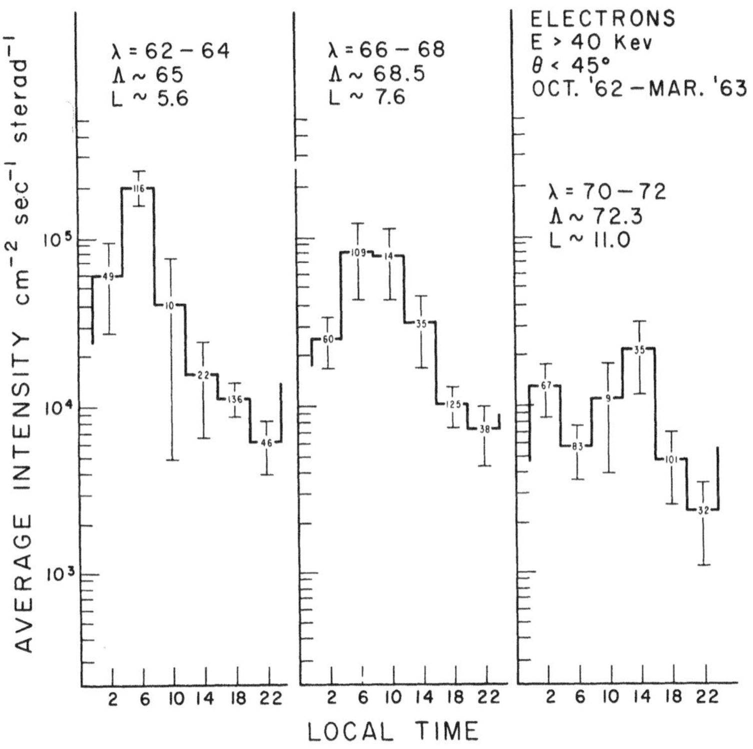

Fig. 9. Average intensity of precipitated electrons versus local time for three different latitude intervals.

Some comparisons can be made between these and other data. In Figure 10, curves (a) show the percentage of passes having intensities greater than 1.5×10^4 particles/cm^2/sec/sterad. plotted against invariant latitude for two four-hour periods about at maximum and minimum. The curves in (b) represent the percentage of time in which radio absorption occurred greater than 0.5 dB. The absorption curves shown are scaled off contour plots by HOLT (1963). BASLER (1963) and HARTZ *et al.* (1963) have shown plots of the daily variation at a number of stations and their results show a strong seasonal variation. But at the times shown here, they show a maximum in absorption occurrences a few hours before local noon.

An explanation of these diurnal variations based solely on distortion of the magnetic field by the solar wind hardly fits the results, as such an effect should show a symmetrical variation about local noon. A preferable explanation is one

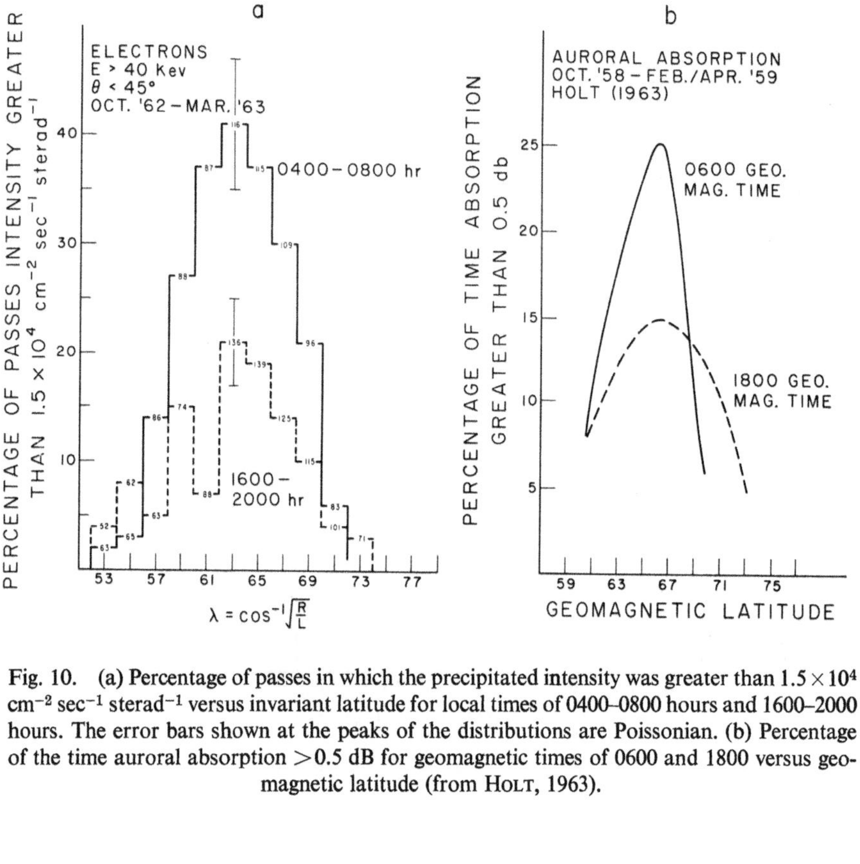

Fig. 10. (a) Percentage of passes in which the precipitated intensity was greater than 1.5×10^4 cm^{-2} sec^{-1} $sterad^{-1}$ versus invariant latitude for local times of 0400–0800 hours and 1600–2000 hours. The error bars shown at the peaks of the distributions are Poissonian. (b) Percentage of the time auroral absorption >0.5 dB for geomagnetic times of 0600 and 1800 versus geomagnetic latitude (from Holt, 1963).

that connects this diurnal variation with a diurnal variation in the source of particles supplying the outer layer and a short lifetime in the outer layer. An explanation based on distortion of the shell of guiding centers would result in a constant peak intensity though the peak should be at different latitudes on the day and night side. The results, however, show that the intensity on the night side at all latitudes above 63° is lower than on the day side and this is not consistent with merely a distorted drift path.

ELECTRON SPIKES OUTSIDE THE BELT

Before discussing intensity changes with a longer term variation there is one other phenomenon that should be presented. That is, intense fluxes of electrons

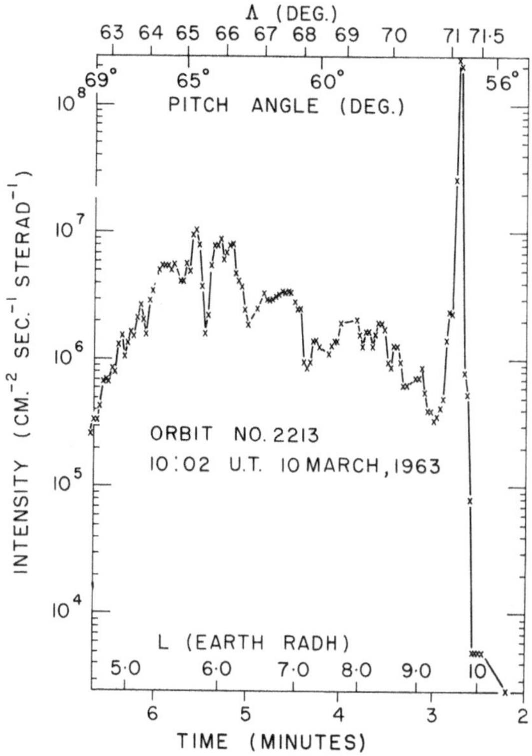

Fig. 11. Directional intensity of electrons with energy greater than 40 keV as a function of time for a satellite pass over a high latitude telemetry station. Also shown are the magnetic shell parameter L, the invariant latitude Λ and the pitch angles θ of the electrons being sampled. This is an example of a very intense electron spike at or near the high latitude boundary of the outer radiation zone.

that occur near the boundary of the outer layer and sometimes well outside the boundary. We speak of these as "high latitude electron spikes". One of these is shown in Figure 11. This figure shows an intense spike at the edge of the boundary of the outer layer. This is for electrons with energies greater than 40 keV.

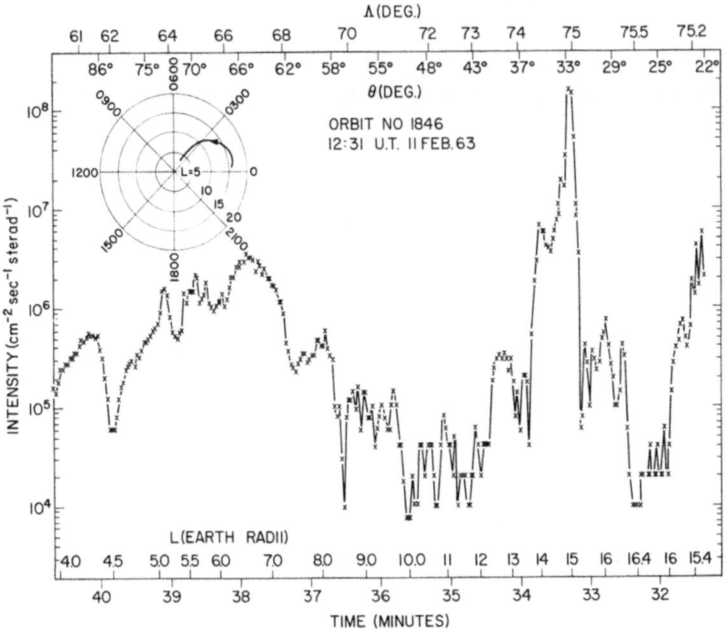

Fig. 12. Directional intensity as a function of time for a pass in which telemetry was received on the descending leg and on part of the ascending leg of the pass. The inset shows the satellite trajectory in local time and in L. This is an example where electron spikes are observed on both legs of the pass at approximately the same L value.

Figure 12 shows a more complicated electron spike structure. The same spike can be seen twice or partly seen a second time as the satellite goes up to high latitude and down again. This may be interpreted as showing that this spike had considerable spread in longitude though its width in latitude is only about 1°, or a more unlikely possibility of two spikes at the same latitude but at different longitudes and only a few minutes apart in time. Figure 13 shows a spike at a very high latitude. The features of these spikes may be summarized as follows:

(1) The peak intensity is sometimes very high, the highest is as much as two orders of magnitude above the highest intensity observed in the outer belt at 1000 km.

(2) They occur mainly on the night side of the earth.

(3) They only occur at relatively high K_p that is moderate or high magnetic activity.

(4) The flux is, as far as can be told, omnidirectional.

(5) They have a spectrum appreciably softer than the outer radiation zone electrons.

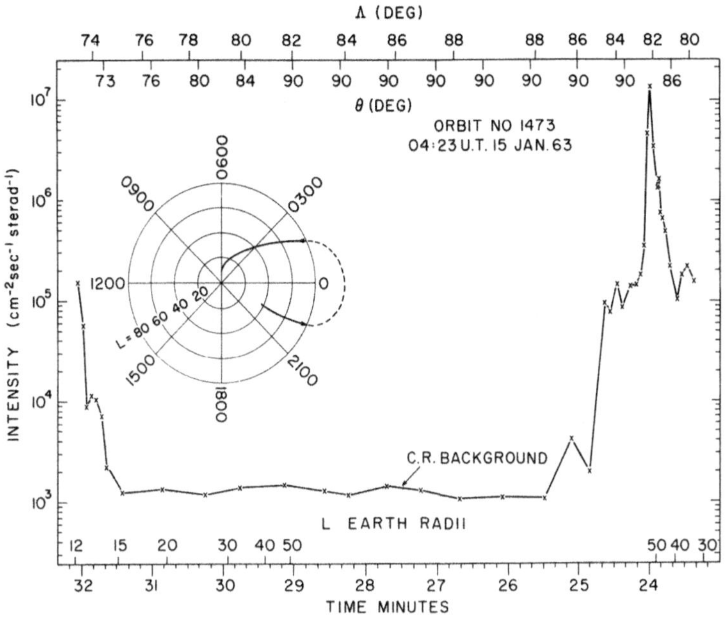

Fig. 13. Plot similar to Figure 12 showing intensity vs. time on the ascending and descending legs of a pass. This is an example where an electron spike is observed on one leg of the pass and not on the other. The intensity scale does not apply to the cosmic ray background shown.

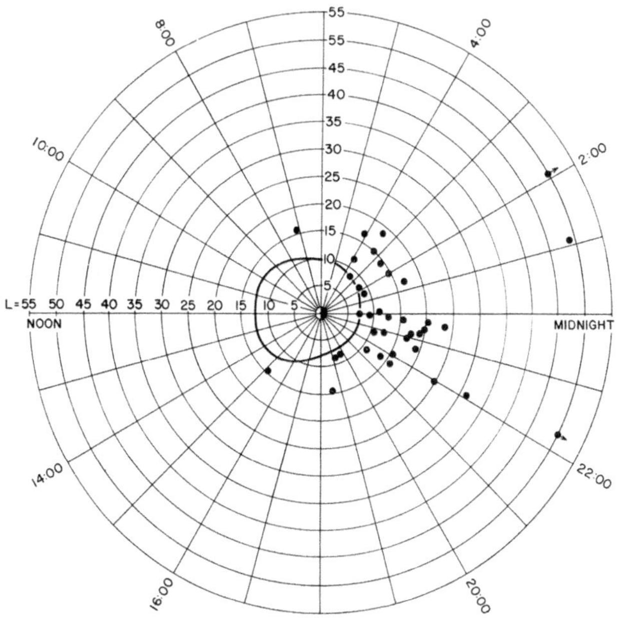

Fig. 14. Polar plot of the positions of the electron spikes in local time and L. The smooth curve is the average high latitude boundary of the outer radiation zone at 1000 km.

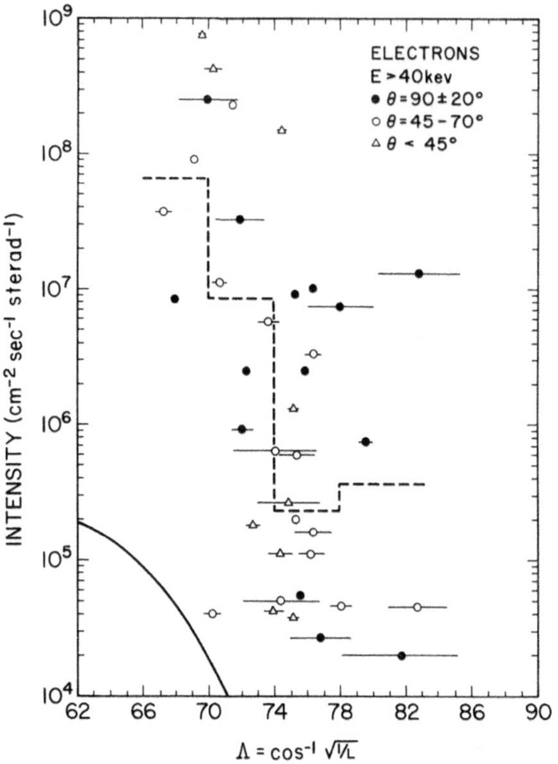

Fig. 15. Peak intensities in the electron spikes as a function of invariant latitude Λ. The different symbols refer to different ranges of pitch angles as shown. The dashed lines are median values and the length of the lines through the points give the latitude widths of the spikes. The smooth curve is the average night time intensity of the outer radiation zone.

It should be added that they are comparatively rare as they have only been seen on 7% of all passes, during which the satellite was in a position to see them: that is, at latitudes at or above the boundary and telemetry turned on and data being received.

Figure 14 shows the time and L-value distribution of the spikes observed. Note this is a sort of inside-out polar projection. The trapped radiation belt is inside the solid curve and spikes moving outward are at higher latitude. The local time distribution is shown. There seems to be reasonable symmetry about midnight and very few spikes occur on the sunlit side, those that were seen being near sunrise or sunset.

Figure 15 shows the distribution of the peak intensity of these spikes with latitude and pitch angle. The broken curve is the median. There is a tendency to lower intensity at the highest latitudes but with no obvious deviation from isotropy in pitch angle. McDiarmid and Burrows have made a rough calculation

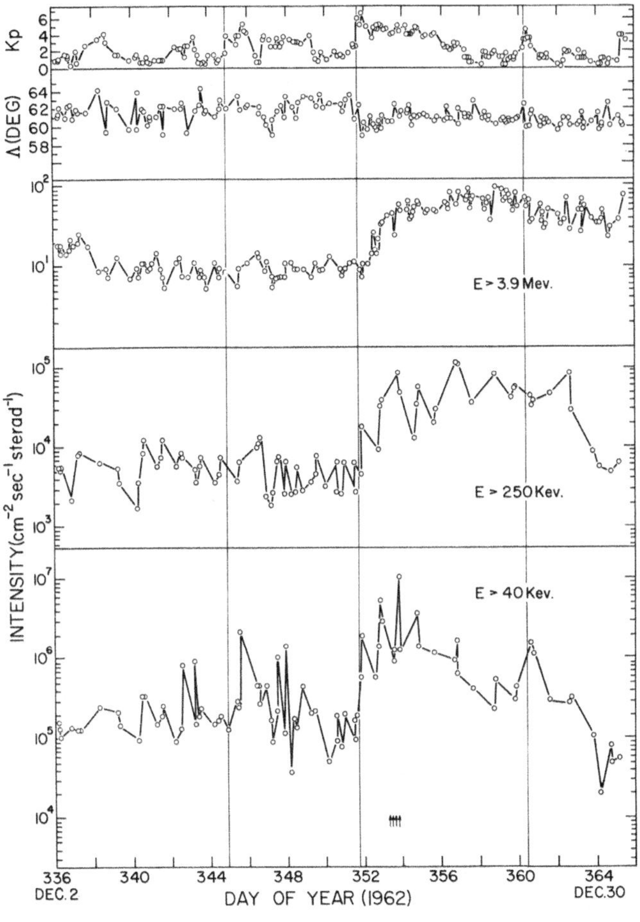

Fig. 16. Peak intensity of electrons with energies greater than 40 keV, 250 keV and 3.9 MeV in the outer radiation zone at 1000 km, versus time for the period 1 December to 31 December 1962. Shown at the top of the graph is the 3 hourly magnetic index K_p. Also shown is the latitude of maximum intensity for electrons greater than 3.9 MeV.

of the average intensity of electrons arriving in these spikes and it compares within a factor of three or four with the average loss of electrons by precipitation as measured by the Alouette satellite. We infer from this that spikes of this sort could represent a source of electrons for the outer belt. These particular spikes, of course, if they follow field lines come from the tail of the magnetosphere and would not be part of the trapped radiation. We have also suggested that these spikes may be connected with the islands of high intensity found mainly in the tail of the magnetosphere by ANDERSON *et al.* (1965).

LONGER IRREGULAR INTENSITY VARIATION

Having reviewed average conditions, we should look at the time-like variation of the peak intensity in the outer belt. The data on punched cards could easily be sorted on the computer to show the peak intensity of the outer belt for each pass and this plotted against time or pass number shows interesting features in the intensity variation of the peak.

There was some attraction in planning this paper so that the following data were presented before the average behaviour was studied, but when the satellite was launched and we started to look at data we had very little knowledge of the structure of the outer layer and its intense variability led us to look for more order in averages than in the time-like behaviour of the peak. The results therefore were presented here in the order in which analyses were carried out.

Figure 16 and subsequent figures show the intensity of the outer layer peak for three discrimination levels of energy plotted against the day of the year. At the top of the slide K_p is plotted and just under it the invariant latitude of the peak intensity of electrons with energies greater than 3.9 MeV. At the bottom arrowheads show the time of occurrence of the spikes which we have discussed in the previous section.

The first thing to note is the rapid increase in intensity occurring with an increase in K_p. In this case the increase occurs at all three energy levels and there is a lowering of the latitude of the peak of the greater than 3.9 MeV electrons after the increase in K_p. The behaviour of the lower energy level is consistent with all sudden increases in K_p and similarly, in the 250 keV level there is reasonable consistency. The intensity at the higher energy level, however, is not consistent at all. At the time of an increase in K_p on January 13th (see Figure 17) this level did not change appreciably and at a later increase on day 41 in February (see Figure 18) it dropped suddenly while the band greater than 40 keV rose sharply. We have no explanation of the behaviour of the highest energy band but from examining data for electrons greater than 40 keV a new source of electrons always appears in the belt when there is an increase in K_p, and a group of the so-called spikes occurs at the same time.

An examination of periods of quiet K_p such as the one between days 32 to 40 on Figure 18 shows that during such periods of very low K_p there is a consistent decay in intensity in the electron flux at the peak. If one considers this as an exponential decay there is sufficient data to work out a time constant using a least squares fit to the curve. An average of the time constant for five such periods of several days duration is 3.7 days.

McDiarmid has recently made a simple calculation based on the level of

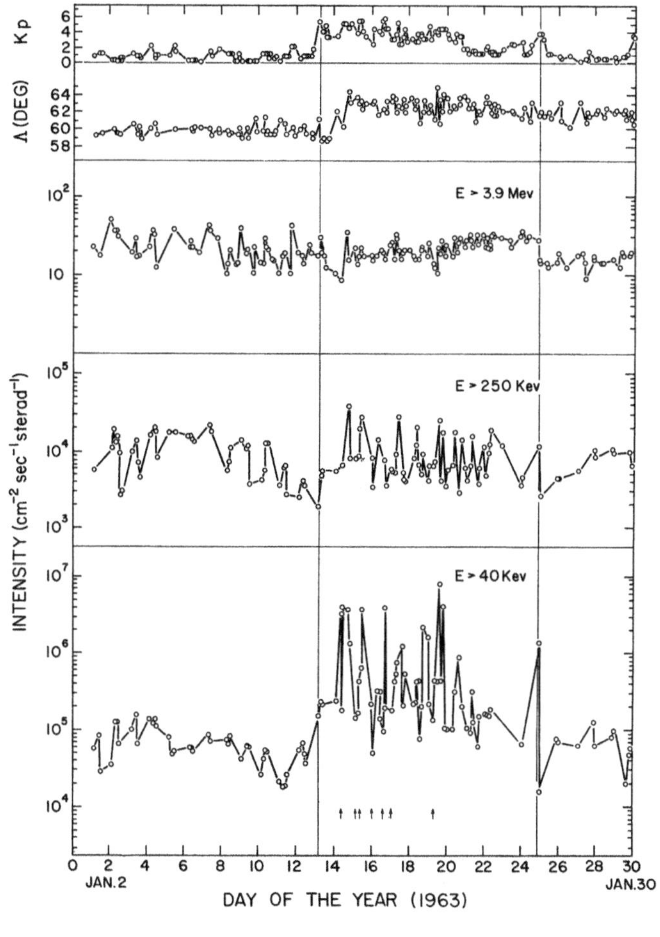

Fig. 17. Plot similar to Figure 16 for the period 1 January to 30 January 1963.

intensity published by FRANK (1965) using results from Explorer XIV. The intensities shown by Frank at $L=6.5$ at the equator can be compared with the average precipitation rate found at $L=6.5$ by Alouette. A time constant for the emptying of the equatorial intensity can be calculated. This gives a period of about one day rather than three but agreement even to this accuracy supports the suggestion of a strong scattering to lower mirror points and a very short lifetime for particles that are mirroring as low as 1025 km. The lowering

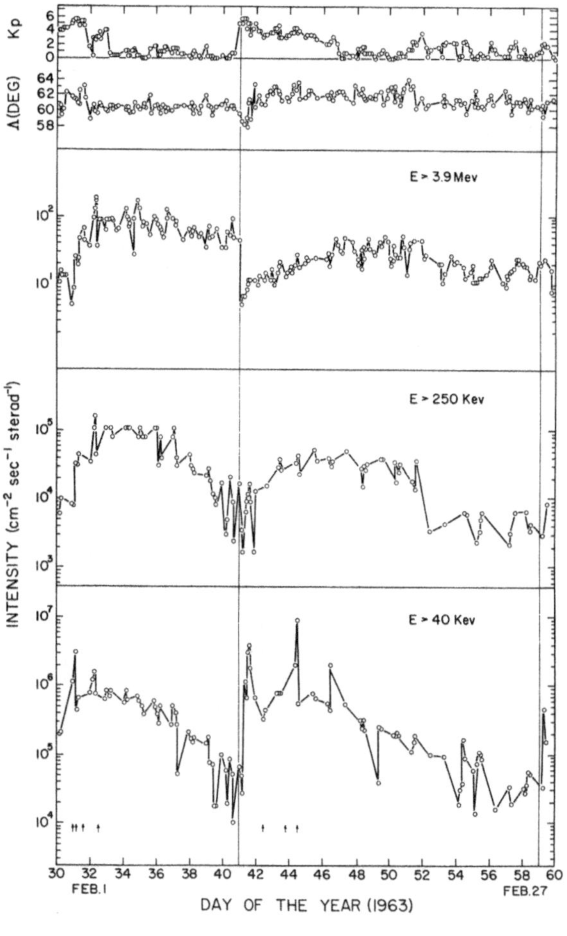

Fig. 18. Plot similar to Figure 16 for the period 30 January to 28 February 1963.

of the mirror point is, of course, enhanced with high K_p. A much shorter period than three days has been suggested for the particles observed at 1025 km. The decay rate will, of course, be controlled by the intensity where the lifetime and density are highest, namely, at the equator. Therefore, there is no inconsistency in the fact that the decay rate measured at 1025 km should be about the same as that at the equator.

References

ANDERSON, K. A., HARRIS, H. K., and PAOLI, R. J.: 1965, *J. Geophys. Res.* **70**, 1039.
BASLER, R. P.: 1963, *J. Geophys. Res.* **68**, 4665.
FRANK, L. A.: 1965, *J. Geophys. Res.* **70**, 1593.
HARTZ, T. R., MONTBRIAND, L. E. and VOGAN, E. L.: 1963, *Can. J. Phys.* **41**, 581.
HOLT, O.: 1963, *Norwegian Defence Research Establishment Report* **46**.
HONES, E. W.: 1963, *J. Geophys. Res.* **68**, 1209.
MALVILLE, J. M.: 1960, *J. Geophys. Res.* **65**, 3008.
MCDIARMID, I. B., BURROWS, J. R., BUDZINSKI, E. E. and ROSE, D. C.: 1963, *Can. J. Phys.* **41**, 1332.
MCDIARMID, I. B., BURROWS, J. R., BUDZINSKI, E. E. and WILSON, Margaret D.: 1963, *Can. J. Phys.* **41**, 2064.
MCDIARMID, I. B. and BURROWS, J. R.: 1964, *Can. J. Phys.* **42**, 616.
MCDIARMID, I. B. and BURROWS, J. R.: 1964, *Can. J. Phys.* **42**, 1135.
MCDIARMID, I. B. and BURROWS, J. R.: 1965, *J. Geophys. Res.* **70**, 3031.
O'BRIEN, B. J.: 1962, *J. Geophys. Res.* **67**, 3687.
O'BRIEN, B. J.: 1963, *J. Geophys. Res.* **68**, 989.

Discussion

Dr. MEAD commented that the magnetospheric model which he and Williams have developed (as reported by Williams in this session) is capable of quantitatively explaining the magnitude of the diurnal variation of latitude of trapped electrons observed by Alouette. However, this model cannot explain any dawn-dusk asymmetry such as observed with Alouette. FRANK reported that 1 MeV electrons observed with Explorer XIV also showed a two-hour dawn-dusk asymmetry towards local morning in the equatorial plane. DUNGEY mentioned that electric fields can give dawn-dusk asymmetries in the lower energy particles. MICHEL commented that Walters' work predicts a one-hour dawn-dusk shock-front asymmetry by virtue of the non-radial direction of the interplanetary magnetic field.

The point was made by Williams that, whereas Alouette data show a difference in peak intensity of trapped 40 keV electrons from day to night, the 280 keV electrons do not (see Williams' paper below). The "relativistic" (i.e. 280 keV) electrons thus display complete consistency with the Mead-Williams model, and perhaps there are other effects that modify the 40 keV fluxes. MCILWAIN then commented that this effect points out the need to distinguish between local-time effects in the coordinates and local-time effects in the particles. He considered it probable that the apparent boundary of the 40 keV electrons observed at such low-altitudes (e.g. 1000 km) may not be the real boundary of the magnetosphere but rather may be a consequence of a local-time dependent diffusion, which does not significantly affect the higher-energy (280 keV) electrons.

The question was raised as to whether the very intense spikes of radiation might be actually more intense than they appear because of saturation effects in the geiger counter, but Dr. ROSE said that at these counting rates of order 10^5 counts per second such effects were relatively small. VAN ALLEN commended the continued use of L as a coordinate system because the actual boundary of trapping does move back and forth over a given city (say) at high latitudes due to local time effects.

OUTER ZONE PROTONS

L. R. DAVIS and J. M. WILLIAMSON
Goddard Space Flight Center, Greenbelt, Maryland, U.S.A.

Abstract. Charged particle detectors aboard satellites Explorer XII, XIV and XV have provided measurements of the intensity and spectra of 100 keV to 1700 keV protons in the region above two earth radii during 1961 and 1962. These results as well as those of other measurements are summarized.

Additional measurements made on Explorer XXVI in February 1965 are presented and compared with the previous results. They show that: (1) between December 1962 and February 1965 the intensities of protons having energies greater than 500 keV increased by a factor of two or more over most and possibly all of the region from 2.5 to 5 earth radii, and (2) over the same region the 100 keV to 300 keV protons are greatly depleted in 1965.

During a magnetic storm on 18 April 1965 and while Explorer XXVI was near apogee (5 earth radii) with L almost constant, the low energy protons were observed to increase dramatically over a ten minute period. During the same period the 10 keV to 100 keV electrons decreased in intensity and then over the next 30 minutes built back up with higher average energy. Two such events were observed, both coincident with high latitude negative bays. The results indicate that the $>$ 100 keV protons are not directly related to the ring current particles.

Introduction

Energetic protons are present more or less continuously throughout the trapping region of the earth's magnetic field, i.e. from a few hundred kilometers altitude out to about ten earth radii in the sunward direction. There are no major discontinuities in the intensities or spectra of the protons to require or justify a division into innerbelt and outerbelt protons. The present use of such a division only reflects the difference in instrumentation and satellite orbits required for studies of protons in these two regions.

Summary of Previous Results

The first studies of protons in the region above $L=2$ earth radii were the 1960 rocket measurements of BAME *et al.* (1963), who measured the differential-directional intensity of protons in the energy range 1.02 to 7.3 MeV on the $L=2.5$ shell at low altitudes. Their results showed that the proton intensity increased rapidly with decreasing energy down to 1.02 MeV, and that while the intensity increased as one moved up the field line there was little or no change in the spectrum.

The measurements of 100 keV to 1.7 MeV protons by DAVIS and WILLIAMSON (1963) on the Explorer XII satellite showed: (1) that protons are trapped throughout the region from $L=2$ to the outer boundary of the trapping region, (2) that the maximum directional intensity of protons (>100 keV) was $6 \pm 2 \times 10^7$ protons per cm^2 sec sterad for protons mirroring near the equator at about $L=3.5$ earth radii, (3) that the proton energy density in the region 4 to 8 earth radii tended to follow the field energy density with a ratio of field energy density to proton energy density of approximately 6 in the equatorial plane, (5) that the average energy of the protons decreased with increasing radial distance, (6) over a month and a half period, during which there were no magnetic storms, the proton intensities in the region $L=2$ to 7 earth radii were constant to within $\pm 30\%$, and (7) during a magnetic storm the proton intensities, sampled only at high latitudes, were enhanced by a factor of three in the $L=3$ to 4.5 region.

The omnidirectional intensities of protons (>500 keV) have been measured near the equator for $2.8 < L < 3.6$ by FRANK *et al.* (1964). They reported typical intensities of 7×10^6 protons per cm^2 sec and that the intensities did not vary with time by more than a factor of 2 during the five month period studied (October 1962 through February 1962).

BOLYANOVA *et al.* (1965) have measured protons (>150 keV) in the region above $L=2$ on the Electron 1 and 3 satellites. These results on the spatial distribution and spectra of protons agree with the Explorer XII results.

VERNOV *et al.* (1965) have flown semiconductor counters on Cosmos 41 sensitive to protons in two ranges, $0.4 < E < 7$ MeV and 3MeV $< E < 8$ MeV. They reported: (1) finding temporal variations only with $L > 4.5$, (2) Proton intensities ($E > 0.4$ MeV) for several orbits for the region $2.5 < L < 10$, and (3) observations on two proton enhancement events at $L=8$ and 9.5 which occurred during small magnetic storms.

Additional measurements by Davis and co-workers on Explorers XIV and XV have been reported by DAVIS (1965) and Davis *et al.* (1964). The results are in good agreement with the Explorer XII data and thus show that the proton intensities during change to magnetically quiet times were the same in 1962 as in 1961 to within $\pm$ 25%.

A large number of measurements have been made on higher energy protons ($E > 1$ MeV) in the region of the inner belt. (BOSTROM *et al.*, 1965; BROWN *et al.*, 1963; FILLIUS, 1965; FILLIUS and MCILWAIN, 1964; FREDEN and PAULIKAS, 1964; FREDEN and WHITE, 1959; IMHOF and SMITH, 1964; MCILWAIN, 1963; and VALERIO, 1964). Since these results have been presented or reviewed in other papers at this meeting, they will not be discussed here.

Instrumentation

The scintillation detectors flown on Explorers XII, XIV, XV, and XXVI were nearly identical. Details of the detector have been given by DAVIS and WILLIAMson (1963) and Davis (1965). The detector measures the directional intensity of 100 keV to about 10 MeV protons in a pulse counting mode and the directional energy flux of 10 keV to 100 keV electrons in a current mode. Proton energy spectra are obtained by interposing, in time sequence, seven absorbers (eight absorbers on Explorer XXVI) in front of the detector to change the low-energy cut off. Figure 1 shows the proton count rate efficiencies as a function of proton energy for the seven absorbers used on Explorer XII. The curve labeled

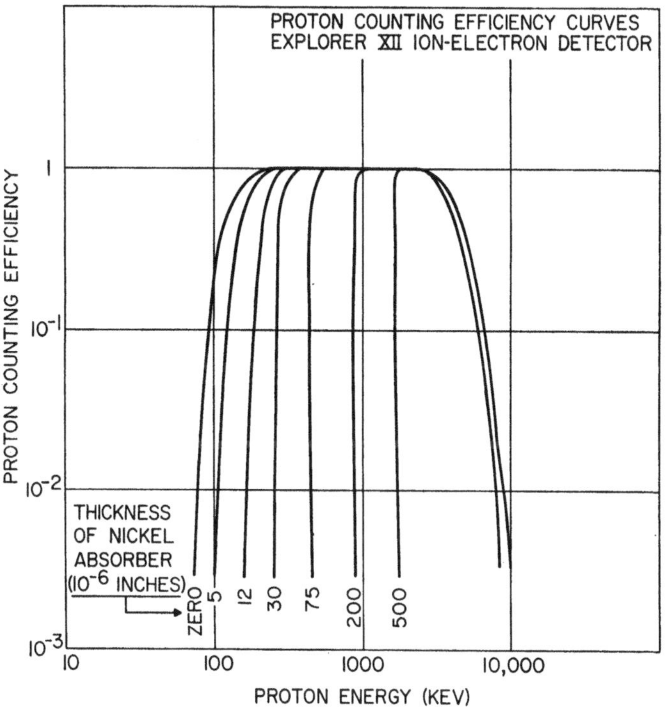

Fig. 1. Detector Count Rate Efficiency Curves.

zero thickness is the response of the open detector and the six remaining curves are those obtained with nickel foils ranging from 5 to 500 millionths of an inch thick. The energies assigned the integral intensities are the values where the efficiencies fall to 50 per cent on the low energy end. It can be shown that for exponential spectra this value is correct to within 5 keV. From analysis of flight data we have found that the spectra are steep enough that the measured intensities should closely approximate the true integral intensities.

The detector axis was oriented at 45 degrees to the satellite spin axis and particle counts were accumulated and telemetered in a fraction of a roll period to permit measuring intensity as a function of pitch angle. Because of this we have chosen to order our data in an *L*-equatorial pitch angle (EPA) coordinate

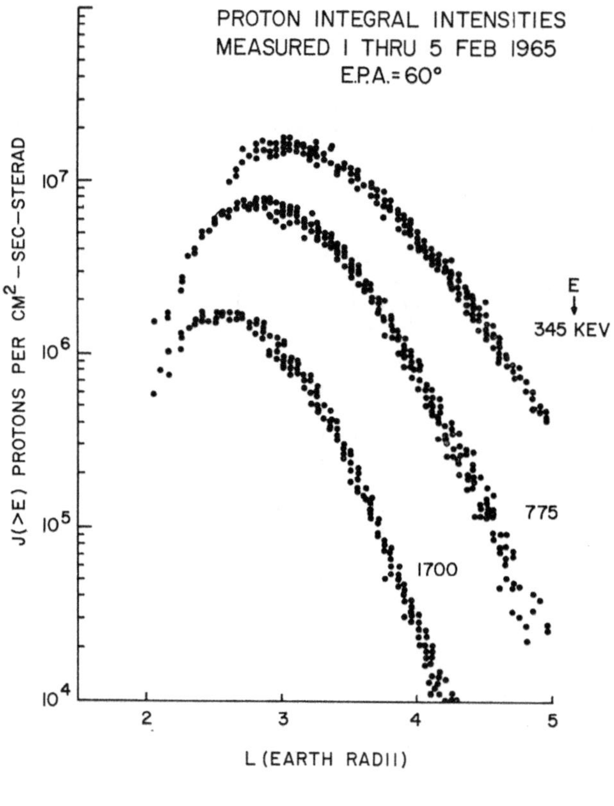

Fig. 2. Scatter Plot of proton intensity against *L*.

system (Davis, 1965). It should be noted that the *L*-values have been computed for the spacecraft location and not for the mirror point of the particles being sampled. However, the errors thus introduced are small and probably less than the errors that result from using the Jensen and Cain (1962) field model which does not include terms for external currents. In general, we find that this system adequately orders the proton data in the region below $L=5$ but not on the higher shells. The disordering on the higher shells is due both to the inadequate field model and to time variations of the proton intensities (DAVIS, 1965).

The degree of ordering achieved below five earth radii is illustrated in the scatter plot shown in Figure 2. Here proton intensities measured at EPA = 60

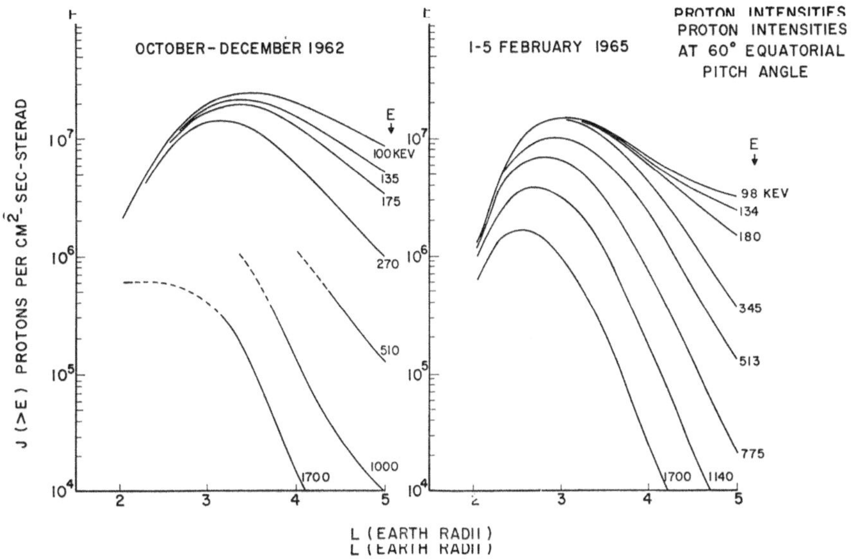

Fig. 3. Intensity curves for EPA = 60 against L for 1962 and 1965.

and for three selected energies are plotted against L from 2 to 5 earth radii. The data were obtained on twelve orbits of Explorer XXVI in the time period 1 through 5 February 1965. The scatter in the data has an RMS value of about 10% over most of the region, becoming somewhat larger at the higher L-values.

Observation of Long Term Variations

The Explorer XII, XIV, XV, and XXVI data provide samplings of the protons in 1961 and 1962, when solar activity was still relatively high, and in 1965, which is near solar minimum. As previously reported (DAVIS, HOFFMAN, and WILLIAMSON, 1964 and DAVIS, 1965) a comparison of the low energy proton intensities measured in August and September of 1961 with those measured in October, November, and December 1962 in the region $L = 2$ to 5 show that the protons were constant to within $\pm 25\%$. Data obtained in February 1965 will now be presented which show that significant changes in the proton intensities occurred between December 1962 and February 1965.

In Figure 3 the proton intensities measured at EPA = 60 degrees are plotted

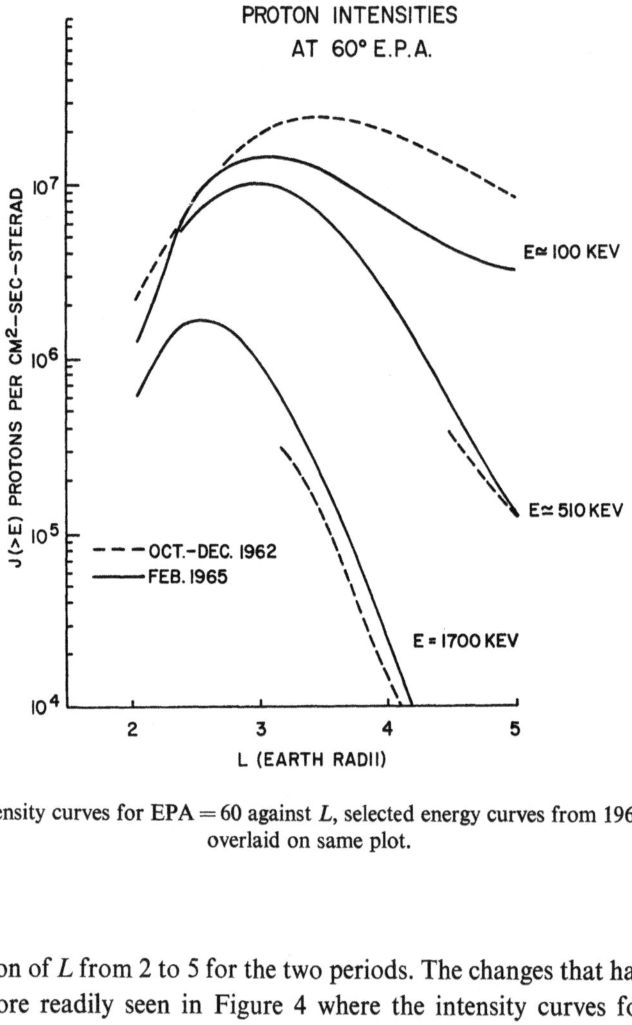

Fig. 4. Intensity curves for EPA = 60 against L, selected energy curves from 1962 and 1965 overlaid on same plot.

as a function of L from 2 to 5 for the two periods. The changes that have occurred are more readily seen in Figure 4 where the intensity curves for proton energies greater than 100 keV, 510 keV, and 1700 keV are overlaid. The changes to note are the decreased intensities of the protons > 100 keV and the increased intensities of the higher energy protons. The pitch angle dependence can be seen in Figures 5, 6, and 7 which show, for L-values of 2.5, 3.5, and 4.5 respectively, the proton intensities plotted against EPA for the two periods. They show that the changes have occurred at all pitch angles and that the shapes of the curves have not changed appreciably.

The proton integral spectra measured in 1965 and 1962 are shown in Figure 8 for EPA = 60 degrees and selected L-values. The spectra for some of the lower L shells have been multiplied by the factors shown in the figure in order to separate the data. The shape of the spectra above 500 keV are essentially the same. Above an L-value of about 3 and for energies below 300 keV the 1965

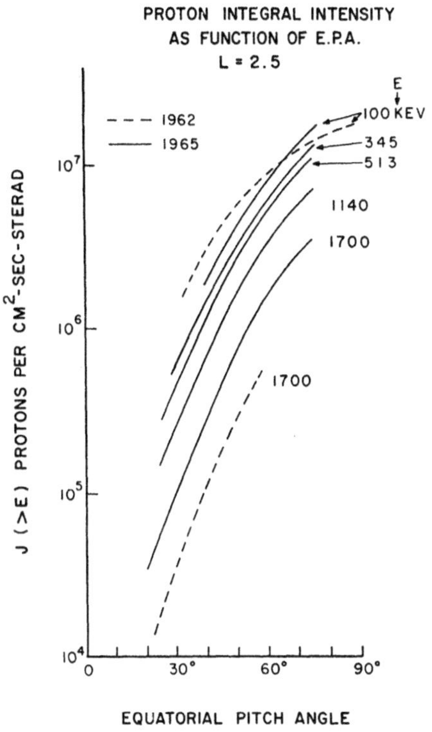

Fig. 5. Intensity against EPA for $L = 2.5$.

spectra are much flatter. Thus there has been a very large depletion of 100 to 300 keV protons.

In summary we find: (1) that the intensity of > 1700 keV protons throughout the region $L = 2.5$ to 5 are larger by factors of 2 to 3 in 1965 than in 1962. The enhancement extends more or less uniformly down to 500 keV, at least in the region above $L = 3.5$. No comparison can be made on lower shells because the Explorer XIV and XV detectors were saturated on the 500 keV channel in the region, (2) that throughout the region $L = 3$ to 5 the 100 keV to 300 keV protons are greatly depleted in 1965.

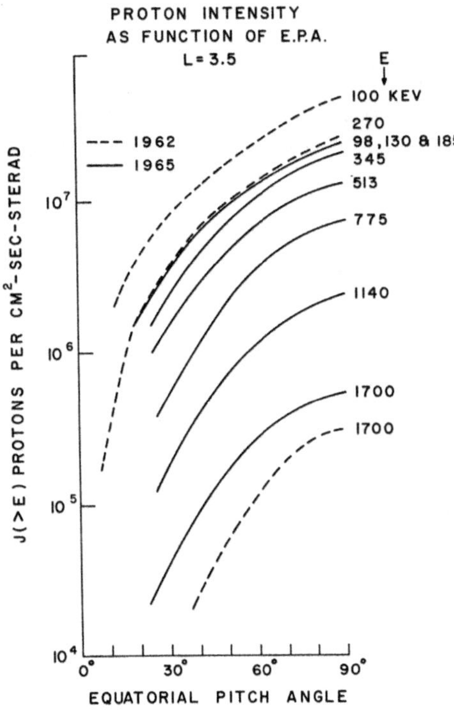

Fig. 6. Intensity against EPA for $L = 3.5$.

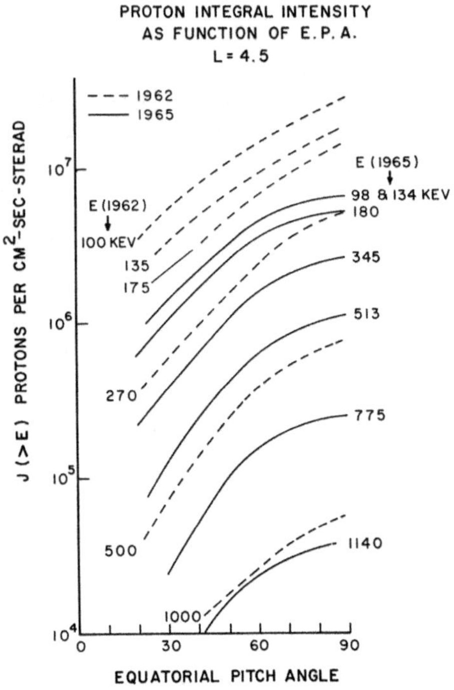

Fig. 7. Intensity against EPA for $L = 4.5$.

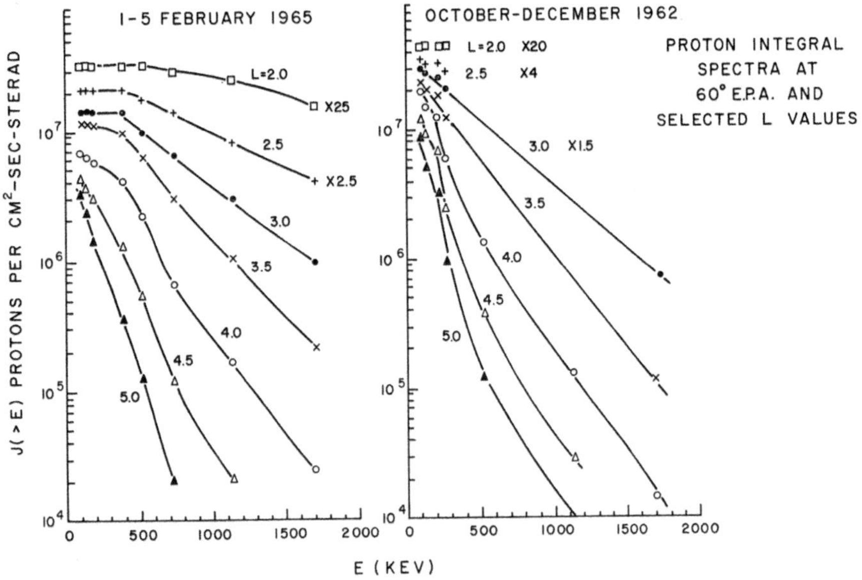

Fig. 8. Integral spectra for EPA = 60 and selected L-values for 1962 and 1965.

Observations of Short Term Variations

The first time variation in the intensity of outerbelt protons was observed on Explorer XII (DAVIS and WILLIAMSON, 1963). At 2108 UT on 30 September 1961

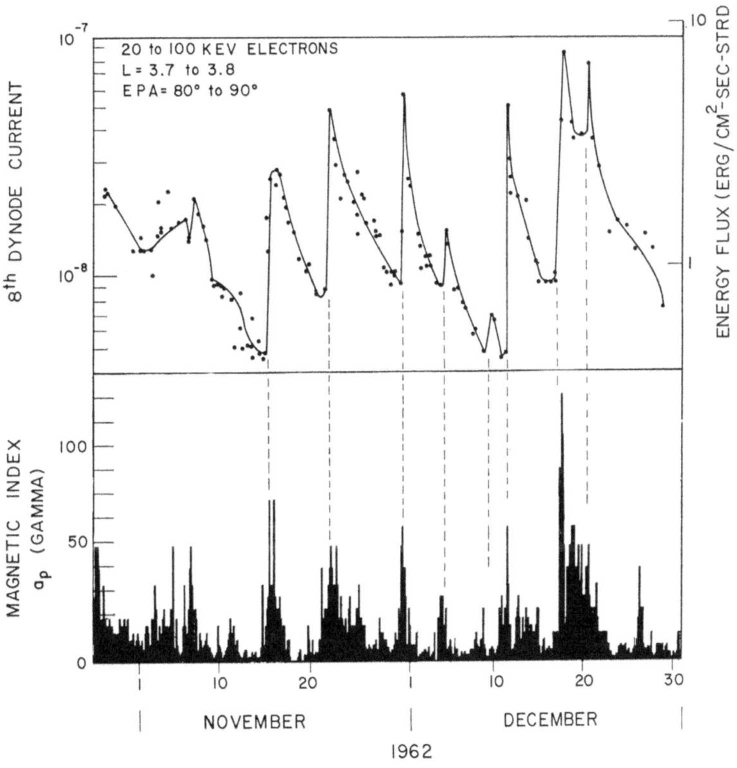

Fig. 9. Top, Time variation of 20 to 100 keV Electrons in November and December 1962. Bottom, Magnetic Index, a_p, for November and December 1962.

a sudden commencement occurred which was followed by a magnetic storm lasting several days. At the time of the sudden commencement Explorer XII was nearing apogee. The next pass through the proton belt occurred some 17 hours later. On this and three succeeding passes on 1 and 2 October, the proton intensities measured in the region 3 to 4.5 earth radii were greater by a factor of three than the pre-storm values. However, the satellite crossings of this region were at high geomagnetic latitudes so that only about 4% of the protons trapped on these field lines were measured. Thus with these data it could not be determined whether the total proton population on a field line was increased or whether there was simply a redistribution in pitch angles of the pre-storm population. A similar event was observed on one other occasion during the life of Explorer XII and again only high latitudes were sampled. No magnetic storms occurred during the lives of Explorer XIV and XV.

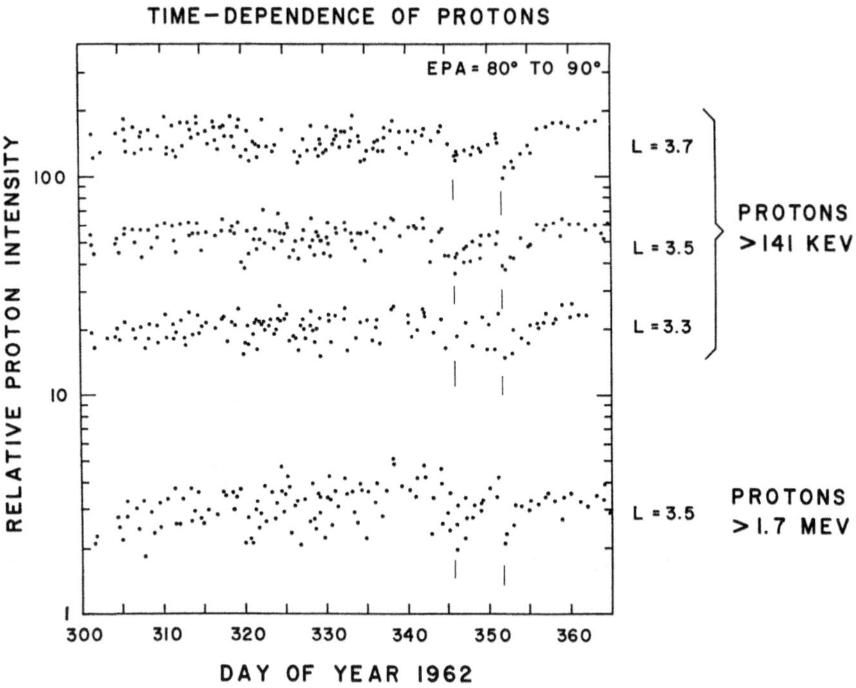

Fig. 10. Time-dependence of protons in last 65 days of 1962.

Explorer XV, whose five hour orbital period (apogee at 3.7 earth radii) provided relatively frequent sampling of the outer belt, has provided data which are well suited for studying the time variations in the low energy protons and electrons. Figure 9 shows a time plot of the intensity of 20 keV to 100 keV electrons measured in the region $3.7 < L < 3.8$ and $80° < \mathrm{EPA} < 90°$ for the months of November and December 1962. Also plotted in Figure 9 is the magnetic index. As can be seen the electrons undergo almost instantanious enhancements by factors of 3 to 10 at the beginning of each of the small *m*-region type magnetic disturbances. The time variations of the low energy protons are shown in Figure 10 for the same period. In contrast to the electrons the protons are almost constant to within the scatter of the data. However, the protons exhibit small decreases of about 30% at the beginning of the disturbances on day 345

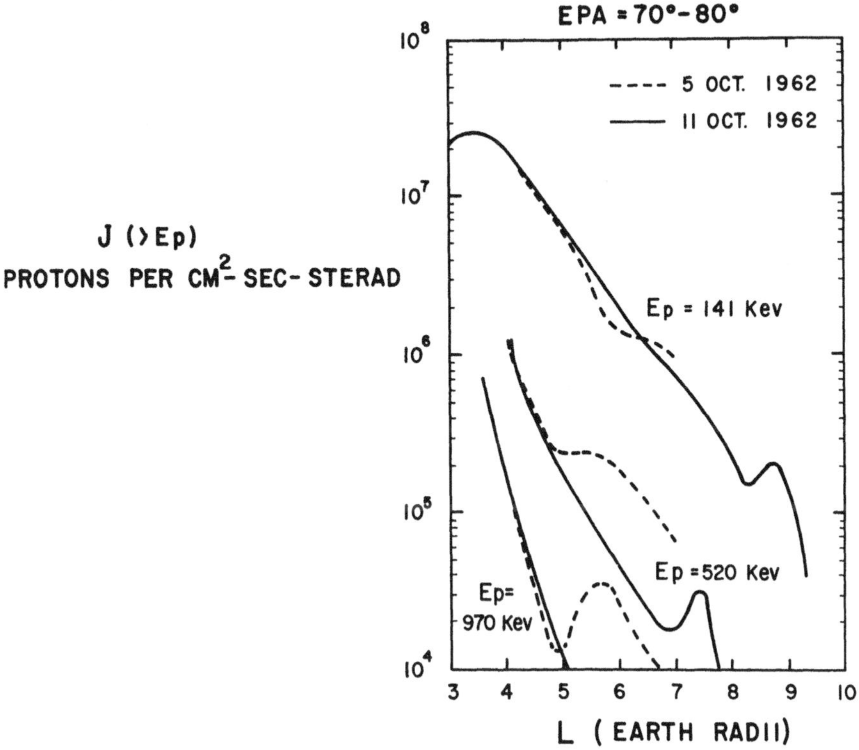

Fig. 11. Intensity curves against L, $70° < EPA < 80°$, showing enhanced high energy protons at $L = 5.8$ on 5 October 1962.

and 351 indicated by the vertical tic marks. The decreases are discernible at all energies in the region $L > 3$. We also find that the decreases occur at all pitch angles. It is possible that similar decreases occurred during the other disturbances and are hidden by the scatter in the data. In both events the protons return to their pre-disturbance values within about five days.

Davis *et al.* (1964) and Davis (1965) have reported that the protons intensities above $L = 5$ undergo frequent fluctuations. Explorer XII and XIV moved rapidly through this region and because of their long orbital periods, 26 and 36 hours respectively, did not sample the region often enough to follow the fluctuations. Figure 11 shows such an event. The proton intensities

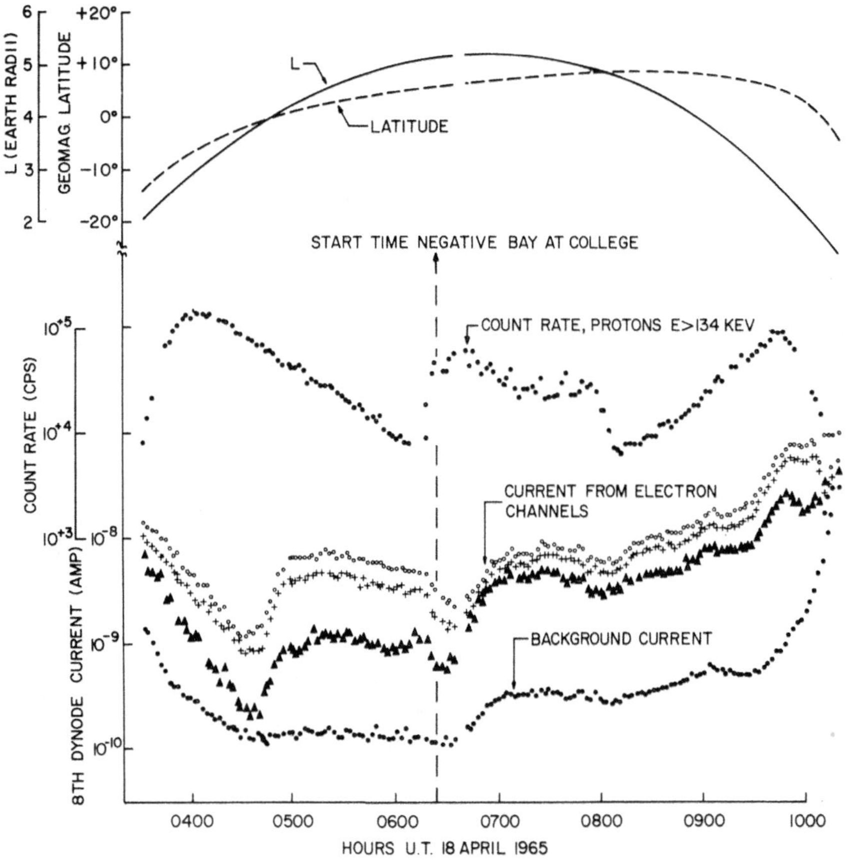

Fig. 12. Explorer 26 L-value, geomagnetic latitude, count rate > 134 keV protons, responses of three electron channels, and phototube background current for period 0330 UT to 1000 UT 18 April 1965.

$70° < \mathrm{EPA} < 80°$ for 5 and 11 October 1962 are plotted as a function of L for proton energies of 141 keV, 520 keV, and 970 keV. The 5 October data show the >970 keV protons enhanced at about $L = 5.8$, where a separate peak is visible. The >520 keV protons also are enhanced whereas the >141 keV protons are only slightly disturbed. By 11 October the enhancements have decayed away. Similar events have been reported by VERNOV *et al.* (1965).

An enhancement of both protons and electrons has been observed during a magnetic storm on Explorer XXVI. The orbit of Explorer XXVI is inclined about 20° to the equator, has an apogee of 5.1 earth radii, a period of about seven hours, and at the time of the storm, apogee was at 1400 local time. The storm began on 17 April 1965 with a sudden commencement at 1310 UT. The start of the main phase was delayed, occurring at about 0400 UT on 18 April. By 0600 UT the horizontal intensity at Guam was depressed 150 gamma and the maximum depression of about 250 gamma occurred at 0900 UT. Figure 12 shows Explorer XXVI data obtained during the development of the main phase. The satellite geomagnetic latitude, L-value, count rate of >134 keV protons, the three electron channel responses, and the background phototube current (obtained with a 1 g/cm^2 plug in the detector aperture) are plotted for the period 0330 UT to 1000 UT 18 April. As the satellite moved out the proton count rate increased in a normal fashion to a peak at $L=3$ and then declined as the satellite approached apogee. The proton intensities registered are perhaps a little lower than normal. At 0610 UT, when the satellite was on the $L=4.9$ shell, the proton intensity started to rise. Over the next ten minutes the four lowest energy proton channels (>98 keV, >134 keV, >180 keV, and >345 keV) showed increases of about a factor of 10. On the other hand, the electron channels started to show an abnormally high flux of soft electrons at 0440 UT (the spectrum of the electrons is indicated by the spacing between the electron channel responses, the wider the spacing the steeper the spectrum). While the protons were increasing, the electron flux was decreasing. This was followed by a buildup of much harder electrons, which even show up in the background current.

The first conclusion we draw from this event is that the >100 keV protons are not directly related to the ring current particles producing the main phase. At 0600 UT the >100 keV protons at $L=4.9$ were not enhanced whereas the main phase had started to develop more than one drift period earlier.

Examination of high latitude magnetometers has shown that the proton enhancement is nearly coincident in time with the start of a large negative bay. At College, Alaska, the bay started at 0620 UT as indicated in Figure 12.

A similar sequence was observed on the next orbit, as is shown in Figure 13. The start of this event was also nearly coincident with the start of a smaller negative bay. Two other bays occurred during the storm but in both cases the satellite was near perigee.

As yet, no detailed analysis and correlation with ground data have been made. However, the data obviously indicate that trapped electrons and protons are accelerated during negative bays.

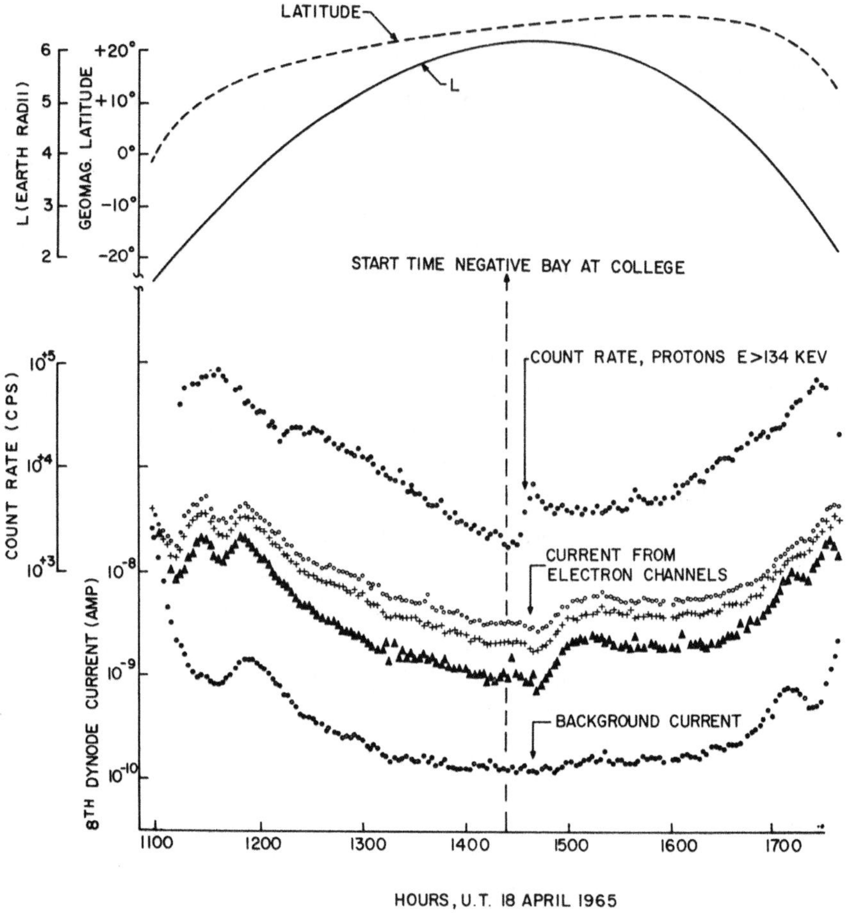

Fig. 13. Explorer XXVI L-value, geomagnetic latitude, count rate > 134 keV protons, responses of three electron channels, and phototube background current for period 1050 UT to 1730 UT 18 April 1965.

References

Bame, S. J., Conner, J. P., Hill, H. H., and Holly, F. E.: 1962, *J. Geophys. Res.* **67**, 1628.

Bolyunova, A. D., Vaisberg, O. L., Galperin, Yu. I., Potapov, B. P., Temny, V. V., and Shuiskaya, F. K.: 1965, COSPAR Symposium, Buenos Aires, 1965.

Bostrom, C. O., Zmuda, A. J., and Pieper, G. F.: 1965, *J. Geophys. Res.* **70**, 2035.

Brown, W. L., Gabbe, J. D., and Rosenzweig, W.: 1963, *The Bell System Technical Journal* **42**, No. 4, Part 2, 1505.

Davis, L. R. and Williamson, J. M.: 1963, *Space Res.* **3**, 365.

Davis, L. R., Hoffman, R. A., and Williamson, J. M.: 1964 (abstract), *Trans. Am. Geophys. Union* **45**, 84.

DAVIS, L. R.: 1965, *Proceedings of the Plasma Space Science Symposium* (held at Catholic Univ., Washington, D.C., June 1963), D. Reidel Publishing Co., Dordrecht, p. 212.
FILLIUS, R. W. and MCILWAIN, C. E.: 1964, *Phys. Rev. Letters* **12**, 609.
FILLIUS, R. W.: 1965, *State Univ. Iowa Res. Rept.* 65-16.
FRANK, L. A., VAN ALLEN, J. A., and HILLS, H. K.: 1964, *J. Geophys. Res.*, **69**, 2171.
FREDEN, S. C. and WHITE, R. S.: *Phys. Rev. Letters* **3**, 9.
FREDEN, S. C. and PAULIKAS, G. A.: 1964, *J. Geophys. Res.* **69**, 1259.
IMHOF, W. L. and SMITH, R. V.: 1964, *J. Geophys. Res.* **69**, 91.
JENSEN, D. C. and CAIN, J. C.: 1962, *J. Geophys. Res.* **67**, 3568.
MCILWAIN, C. E.: 1963, *Science* **142**, 355.
VALERIA, J.: 1964, *J. Geophys. Res.* **69**, 4949.
VERNOV, S. N., SAVENKO, I. A., TEL'TSOV, M. V., and SHAVRIN, P. I.: 1965, COSPAR Symposium, Buenos Aires.

Discussion

In response to BAGGERLY's question about the energy dependence of enhancements of the proton fluxes, DAVIS stated that the threshold of his detector's sensitivity prevented study of any energy above about 345 keV, so they may have been enhanced.

WESTCOTT asked whether the satellite was near to the longitude of College Alaska when the two magnetic bays were observed. DAVIS said that in the first bay they were about six hours apart in local time and for the second bay about twelve hours or so.

CAIN asked about the relative energy density of the trapped protons with respect to the local magnetic field. DAVIS replied that in 1961–1962 the fraction was about one part in five to ten, and currently (1965) it was about half that value.

DUNGEY drew attention to the question of whether magnetic bays or impulses are the cause of diffusion, and he asked whether Davis had studied the effects of impulses, as had Cahill with Explorer XII. DAVIS answered that he has not yet examined this.

MCILWAIN reiterated his earlier claim that the time variations of all particles of all energies have the same characteristic signature, wherein there is an initial decrease, then an increase (perhaps to a higher level) with then a subsequent return to "normal". He suggested that the time changes of the "signature" need to be adjustable for particle energy and time. DAVIS responded that he hadn't seen a decrease in the lower-energy electrons in one sector while he saw it in another local-time sector, but that if McIlwain said that the signature was then undiscernable but still present he (Davis) could not refute it.

HESS questioned whether the double peak in the February event had a time separation like that of the drift period, but DAVIS replied that the time separation was about 30 minutes for protons with a drift period of about 66 minutes. HESS joked that perhaps they were seeing a second harmonic.

O'BRIEN asked the audience to express an opinion as to whether the main phase of a magnetic storm is due to a ring current of trapped protons. DESSLER replied that there is no alternative theoretical suggestion and no experimental evidence to the contrary, although the principal ring current particles certainly have not been detected yet. Cain stated that he regarded the possibility that Davis' protons at $L \lesssim 3$ were the high-energy tail of a ring-current flux as implausible, and he suggested that maybe the magnetospheric tail was worthy of examination here. But DESSLER pointed out that, to produce a main phase decrease of (say) 250 gammas or $2 \cdot 5 \times 10^{-3}$ gauss, one needs to put the particles into a region where the field is at least 250 γ, so that the diamagnetic gas can even produce this much hydromagnetic stress. Any such effect in the magnetospheric tail, where the field is only 20 to 30 γ, could not produce a main phase decrease of more than 20 to 30 γ.

MICHEL asked whether Davis used the term "acceleration" to include possible transport effects from another region, because the rise and fall in intensity were so large that maybe the effect is just redistributing the particles in space. DAVIS replied that to get the intensity change required an energy increase by a factor of three and he thought this to be not likely due to a magnetic field effect. O'BRIEN pointed out that the physically important distinction between the introduction of a new particle into a region (i.e. a source) and the energization of a previous resident of the region (i.e. an acceleration) is not made in the experiments, which only see a local change in the fluxes of given energies. DESSLER said that simply radial movement of the protons (e.g. by a ring current) from say 3 to 3.5 R_e can change the flux of protons at a given point simply because of the existence of a strong radial gradient in the flux. Davis replied that the effect occurred after the main phase had developed, so it was not apparent to him that this could be the explanation.

There was general discussion as to where the peak intensity of 1 MeV protons occurred. In a rocket flight by Theodoridis and Katz the peak appeared to be at $L \sim 3.7$, whereas in Davis' work, and in that of McIlwain, Van Allen, and Brown the peak was about $L \sim 2.5$ to 3.0. Dr. FREDEN pointed out later that this is not really a discrepancy, because the rocket flight did not take an equatorial sample through L, and hence it could perceive an apparent peak at various L values depending on its trajectory through B-L space.

VELA SATELLITE MEASUREMENTS OF PARTICLES IN THE SOLAR WIND AND THE DISTANT GEOMAGNETOSPHERE*

JAMES H. COON

Los Alamos Scientific Laboratory, Los Alamos, New Mexico

1. Introduction

The satellite orbit chosen for the Vela Satellite Program was such as to avoid the primary regions of particle radiation trapped in the earth's magnetic field. The orbit was chosen to be circular at about 17 earth radii distance from the center of the earth in order to be outside the regions of durable trapping of particles. During each orbit the satellites traverse a region of interplanetary space on the sunward side of the earth beyond the influence of the earth's magnetic field. Then they pass into the transition region, through the magnetospheric boundary, across the tail of the magnetosphere, and back out through the transition region into interplanetary space again. It appears likely that particles in these regions of space constitute sources or sinks of particles for the trapped radiation zones, and are therefore valid subject matter for the present discussions of the trapped radiations.

Information about the characteristics of the low and medium energy particle radiations in the regions of space traversed by the Vela satellites has been and is being acquired by the satellites; the present report will summarize observations to the extent that the present state of analysis permits. As with many space programs, complete analysis has not kept up with the acquisition of data on magnetic tape. Some of the results to be presented here will be observations which, in a preliminary analysis, appear to be typical or representative of conditions which are varying. Because of the fluctuating nature of radiations in space, their characterization frequently requires statistical study over long periods of time, and detailed correlations with other related observations are required for purposes of interpretation. Also, special periods of extreme or unusual conditions should be selected for more detailed study and for comparisons with additional types of correlated signals. These kinds of more complete analysis have so far been carried out to a limited extent only, but are being continued with accelerated effort. Although some of this work has been published, (SINGER *et al.*, 1964; and MONTGOMERY *et al.*, 1965) and reported at meetings, (ARGO

* This work was performed as a part of the Vela Satellite research and development program for detection of nuclear explosions in space. This is a joint program of the Advanced Research Projects Agency of the U.S. Department of Defence, and the U.S. Atomic Energy Commission.

et al., 1964; BAME *et al.*, 1965; ASBRIDGE *et al.*, 1965; and STRONG *et al.*, 1965) it is anticipated that as analysis continues and more data are acquired, more complete reports will be published by members of the Space Science Group at Los Alamos Scientific Laboratory. This staff has provided the information on the space radiation measurements presented in this report.

In this report there will be no attempt to review the present state of knowledge relevant to the diverse observations reported. Instead, the terminology of recent literature will be employed, and a few brief remarks, references, and comparisons will be given to provide some orientation as to the significance of the observations reported.

2. Vela Satellite Program

To provide an understanding of some of the motivations of the work to be reported here, a few brief statements about the program, and in particular about the character of the instrumentation onboard the satellites may be useful. A primary objective of the Vela Satellite Program is to carry out research and development on methods of detecting nuclear explosions in space by means of satellites. A major goal is to develop capabilities for policing an international treaty banning the testing of nuclear weapons.

Selection of types of satellite-borne instruments, and determination of their design to detect nuclear explosions, depends critically on two factors. The first of these is the nature of the output radiations from a nuclear explosion in space. The second is the nature of the background radiations in the environment encountered by the instruments. Radiations from a nuclear explosion are comparatively well known, and are of sufficient intensity to be easily detected by state-of-the-art instruments over distances comparable to the earth-sun distance, unless background radiations produce interfering signals. Many features of the radiation background signals are still unknown, particularly over the full 11-year cycle of solar activity. Thus the Vela Satellite Program conducts basic investigations of those types of background radiations to which the nuclear weapon detectors are thought to be most susceptible.

From the above statements it may be correctly inferred that there are two general types of instruments on the Vela satellites. One type, to be referred to as type *A*, is designed specifically to detect and identify radiations characteristic of a nuclear explosion, in spite of background radiations. Type *B* instruments on the other hand, are designed to study selected natural radiations, which may interfere with the operation of the type *A* instruments, or with the interpretation of data recorded by the instruments. It is, in some cases, particularly advantageous to observe signals from the background radiations simultaneously with both types of instruments on the same satellite. It has been found, as might be expected, that type *A* instruments, with their associated signal processing and data handling circuits, are relatively poor or even useless for investigating the natural radiations. Thus the total complement of radiation sensors on the satellites is not effectively devoted to scientific study, but instead a rather small fraction of it.

To date there have been three pairs of Vela satellites put into orbit. The first pair carried only type *A* instrumentation, while the second and third carried additional instrumentation of type *B*. The first pair, referred to as the Vela 1 satellites, have obtained some information of scientific interest, but the additional instruments of type *B* onboard the Vela 2 and 3 satellites greatly increase the capability of obtaining basic information on the natural radiations.

3. Orbital Coverage and Coordinate Systems

Because of its importance in presenting results in a later section of this chapter, a description of orbital coverage and coordinate systems will now be presented. As mentioned above, there have been three Vela satellite launches. The launch dates were 17 October 1963, 17 July 1964, and 20 July 1965. Each launch has placed a pair of 300 pound satellites into approximately circular orbits, all of which are approximately coplanar, and the direction of motion is counter clockwise as viewed from above the northern hemisphere, i.e., from above the ecliptic plane. The orbit parameters are as follows: the period is 100 to 111 h; apogee is in the range 17 to 20 R_e (distance from center of earth in earth radii); perigee is in the range 16 to 18 R_e; inclination of the orbit plane to the ecliptic plane is 55° to 62°; and the inclination of the orbit plane to the earth's equatorial plane is 35° to 40°. The satellites are spin stabilized with a spin rate of 2.0 to 2.2 rps, and the spin axis makes an angle of $\sim 60°$ with the ecliptic plane. In addition, on launches 2 and 3, a smaller third satellite weighing approximately 10 pounds was lofted into a highly elliptical orbit. Experiments on these small satellites will not be reported on here.

Some features of the geometrical sampling of points in space by satellites in the orbit described above are illustrated in Figure 1, in which circular orbits are drawn as they would appear at different seasons of the year in the solar ecliptic coordinate system. The earth is taken to be at the origin which is the center of this sphere, the *X* axis extends from the earth toward the sun, the *Y* axis extends in the ecliptic plane toward the "dusk" meridian, and the *Z* axis extends perpendicular to the ecliptic plane into the northern hemisphere. The earth moves in the minus *Y* direction in its orbit around the sun. In this coordinate system the satellite orbit plane rotates seasonally, so that during two periods of the year the satellites cross the ecliptic plane near the dawn and dusk meridians, and three months later the crossings are near the subsolar and antisolar points. The regions of space sampled by the Vela satellites once every six months are points within a spherical shell of 17 to 18 R_e average radius, except

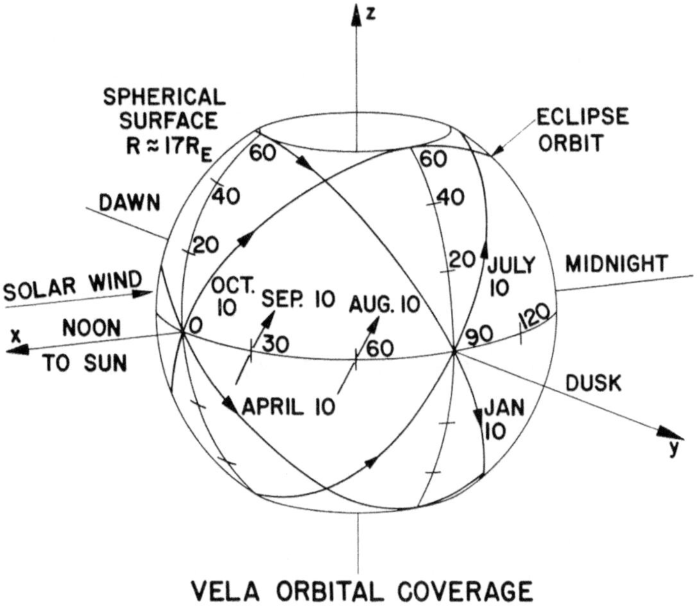

Fig. 1. Spatial coverage by the Vela satellite orbits. In this three dimensional view, selected circular orbits of the Vela satellites around the earth, whose position is at the center of the figure, are indicated, along with the *x*, *y*, and *z* axes of the solar ecliptic coordinate system. See text for further descriptive comments.

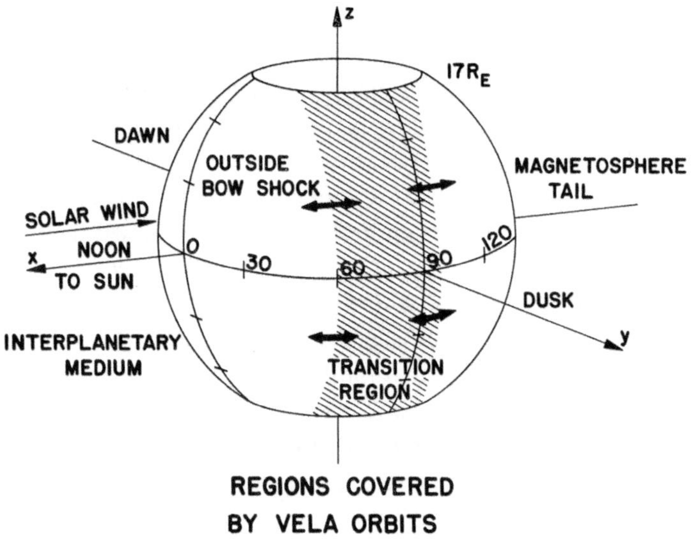

Fig. 2. Spatial coverage by the Vela satellites in terms of regions in the magnetosphere and beyond. The form of the diagram is similar to that of Figure 1.

points for which ecliptic latitude exceeds $+60°$ in the northern hemisphere or $-60°$ in the southern. Figure 2 is similar to Figure 1, except that areas of coverage are indicated in terms of regions in the magnetosphere and beyond.

The unusual features of the Vela orbits are: (1) points are sampled in the magnetopause and nearby transition region at distances up to 16 R_e from the ecliptic plane; compared with most space research spacecraft, this is relatively far from the ecliptic plane; and (2) because of the 4-day circular orbit, points on opposite sides of the magnetosphere are sampled in rapid sequence. Relatively little information can be acquired about dependence of radiations on radial distance from the earth; such information is a major objective of the Explorer series of satellites with very eccentric orbits.

As of August 1, 1965, the six Vela satellites continue to function and provide useful data.

4. Instrumentation

In this section will be presented a brief description of the characteristics of the instruments employed on the Vela satellites for making observations of low or moderate energy charged particles. These instruments detect electrons, protons, or alpha particles, and measure parameters such as flux, energy spectrum, and direction distribution over a selected range of values. Five types of instruments are employed as described below. The first type listed, the M4 detector, was on both Vela 1 and Vela 2 satellites. The remaining four types were on the Vela 2 satellites. The Vela 3 satellites carried improved instruments of these same types, and also some new types. These will not be described here, because no results from the Vela 3 satellites are presented in this report. Although all the Vela 3 instruments are functioning as planned, and the observations are significantly more sophisticated, there was not sufficient time since launch to prepare results for this report.

Data are obtained from the instruments by two very different methods. The first of these is during actual real time tracking of a satellite by a ground station. During such times the information rate is 256 bits per second. In the second method a relatively small fraction of information from the instruments, 256 times less than during real time tracking, is held in storage on the satellite for subsequent transmission. Most of the useful data of scientific interest is acquired during real time tracking, which is typically one hour out of every four. There is, however, a considerable amount of additional real time tracking for special purposes of obtaining data during particular times or regions of the orbit.

A. M4 SCINTILLATION DETECTORS

There are eight to ten of these detectors on each of the Vela 1 or Vela 2 satellites. They employ thin screens of cesium iodide scintillator viewed by photomultiplier tubes, and covered with thin films of beryllium. They respond to electrons of energy greater than ~85 keV (~50% transmission of electrons thru 3.4 mgm cm^{-2} Be) and to solar X-rays. Differentiation of these two types of radiation is based on (1) the character of signal modulation, or on (2) comparisons with signals from other instruments onboard the same satellite, or on (3) comparisons with signals from instruments on one of the other Vela satellites.

The solid angle viewed by these detectors is $>2\pi$ steradian and no attempt is made to investigate directional effects for charged particles. The flux threshold is $\sim 2 \times 10^4$ electrons cm^{-2} sec^{-1} omnidirectional. There is a capability for measuring flux level in seven steps on the Vela 1 satellites or fourteen steps on the Vela 2 satellites up to $\sim 10^6$ electrons cm^{-2} sec^{-1}.

B. M5 SCINTILLATION DETECTORS

These detectors on each of the Vela 2 satellites are similar to the M4 detectors described above, but have elements with greater sensitivity, wider dynamic range of measurement, and lower electron energy threshold. These detectors measure electrons of energy greater than ~50 keV (~50% transmission of electrons thru 2.3 mg cm^{-2} Be) with a flux threshold of $\sim 10^3$ electrons cm^{-2} sec^{-1} omnidirectional.

C. ELECTROSTATIC ANALYZER

The electrostatic analyzer employs a pair of concentric hemispherical electrodes between which a potential is applied for deflection of charged particles into an electron multiplier particle counter. Polarity of the applied potential may be reversed, so that either positive or negative particles may be detected. Particle energy per unit charge over the range 0.3 to 20 keV is measured in fourteen steps by successively stepping the magnitude of the applied potential. Qualitative information is also obtained about electrons of ~100–200 eV energy.

Information about directional distribution of particle motions is obtained by employing the spinning motion of the satellite for direction scanning. Details of the directional information obtained depend in a complicated way on several factors including the geometry of the acceptance aperture of the instrument, satellite spin rate, satellite spin axis orientation, and on the scanning time sequence whereby particles are observed at successive azimuthal angles by successive time delays after a signal is received by a sun sensor. In the simplified case for which particle motion is assumed to be parallel to the ecliptic plane, eight directions in the ecliptic plane are viewed, five of these being equally spaced within $\pm 12°$ of the sunward direction, and one each toward the dawn meridian, toward the dusk meridian, and in the antisolar direction. In this case angular resolution is $\sim \pm 3°$, and angular error is less than one degree. For particles moving in directions not parallel to the ecliptic plane the angular acceptance is broad, the total angle being 95°. This broad angle is related to the fact that particles may enter the instrument from a fan-shaped region over a total angle range of ~5° by 95°.

The time required to acquire information for a complete energy spectrum and directional distribution is 128 seconds during real time tracking. Because successive portions of the data are acquired at 1 second intervals, a change in the character of the radiation can sometimes be sensed with 2 second time resolution.

D. G.M. COUNTER

A G.M. counter is employed for measuring electrons of energy greater than 40 keV. Particles are accepted from a cone of 35° half angle. The axis of this cone is perpendicular to the spin axis of the satellite and parallel to the axis of the entrance aperture of the electrostatic analyzer. Information about direction of motion of particles is obtained in the same manner as for the electrostatic analyzer, but the capability is modified by the difference in geometry of the acceptance aperture. In particular, the angular resolution for observing directional distributions in the ecliptic plane is much broader.

E. SOLID STATE DETECTOR

A single element gold surface-barrier silicon particle detector was carried on the Vela 2 satellites. It was designed to measure electron energy spectrum in the range 50 to 475 keV for electrons, or in the range 180 to 570 keV for protons. Electrons and protons can be distinguished by comparison of solid state and G.M. count rates. Energy spectrum is measured in seven steps, by successive stepping of a pulse height analyzer. The area-solid angle product is 0.04 cm^2 sterad, and although the entrance aperture cone is about 20° half angle, no directionality is attempted; counts are accumulated and averaged over more than one satellite revolution period. Minimum detectable flux of electrons is $\sim 100\ cm^{-2}\ sec^{-1}$ and flux levels up to $\sim 10^7\ cm^{-2}\ sec^{-1}$ may be measured.

5. Observations and Results

A summary of some of the more significant results obtained by the Vela satellites to date will be presented primarily from an observational point of view. In order to introduce a degree of coherence in presenting observations which are in some cases not closely associated in terms of basic relationships, the results will be presented in a sequence corresponding to the gross flow of particles through the regions of space traversed by the satellites. These regions are illustrated in Figure 3 which shows the general topography of the radiation zones in and beyond the distant magnetosphere as developed by space research during the past few years. The observations to be reported here begin with the flow of particles in interplanetary space, where the solar wind is not disturbed by the earth's magnetic field. From this region particles flow through the bow shock into the

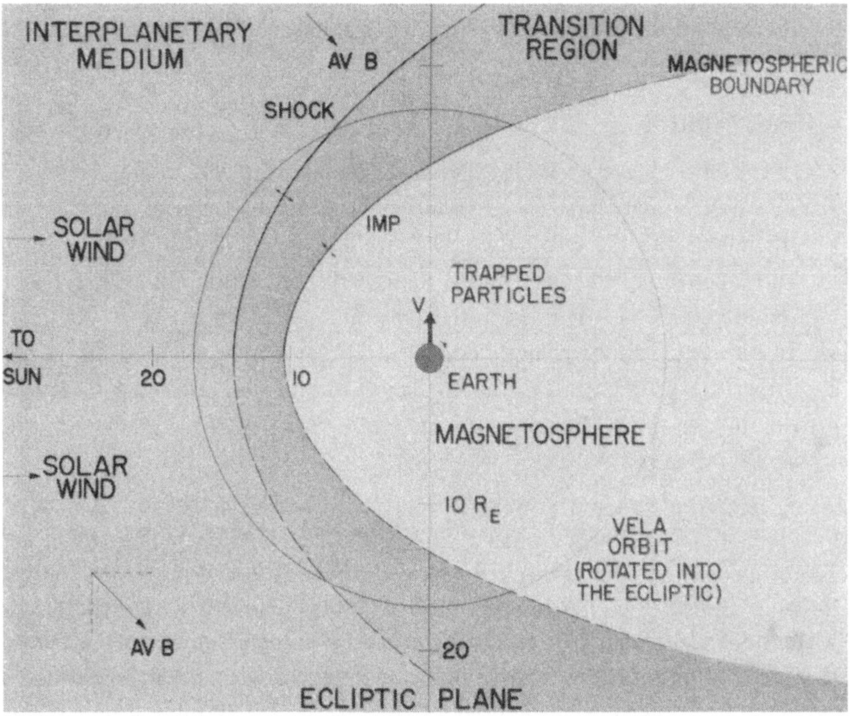

Fig. 3. Regions in the ecliptic plane sampled by the Vela satellites. The circle in this figure is not a satellite orbit, but is rather the approximate locus of points sampled in the ecliptic plane during 6 months time.

transition region, and around the magnetospheric boundary (magnetopause); some of the particles may then pass through the magnetopause into the tail of the magnetospheric cavity, although most of the plasma apparently flows on parallel to the magnetopause for much greater distances. Thus the results will be presented in five sections entitled: (a) Interplanetary Medium; (b) Bow Shock; (c) Transition Region; (d) Magnetospheric Boundary; (e) Magnetospheric Cavity.

A. INTERPLANETARY MEDIUM

On the sunward side of the orbit the Vela satellites traverse a region of interplanetary space where the solar wind is undisturbed by the earth's magnetic field, except possibly for rare precursors on the sunward side of the bow shock. It is not clear whether there are precursors extending out as far as the sunward apogee points reached by the satellites. During a part of the time when the Vela 2 satellites are in this region, the electrostatic analyzers measure energy spectra, flux density, and directional distributions of the positive ions in the solar wind. From these observations preliminary analysis allows some statements about bulk velocity, flow directions, effective or apparent temperature, ion density, and the proportion of alpha particles in the solar wind. In addition, limited observations on electrons can be reported, though most of the solar wind electrons have energies too low to be detected by the instruments.

Data for the period from launch on 17 July 1964 through February 1965 has been analyzed in some detail. About 25% of the time during this period at least one of the Vela satellites was in the interplanetary solar wind region, with electrostatic analyzers operating to observe the solar wind positive ions. Examples of three observed positive ion spectra are shown in Figure 4, which illustrates the variable character of the solar wind. These spectra are similar to those observed by others, (SNYDER and NEUGEBAUER, 1963; WOLFE and SILVA,

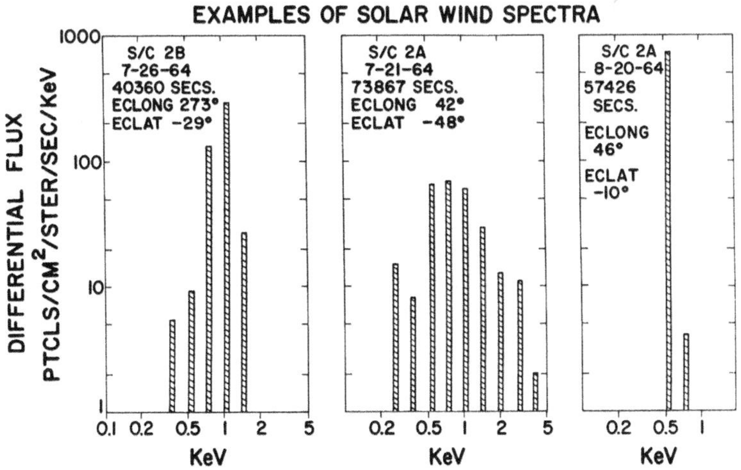

Fig. 4. Examples of three positive ion spectra observed by the Vela 2 satellites in the interplanetary medium. Note successive energy channels are equally spaced logarithmically. The width of the vertical bars indicates instrumental channels width, and $\Delta E/E$ is constant. At intermediate energies between bars the instrument is not sensitive.

1965). The data for one spectrum is acquired in about 2 minutes, during which time directional information is also acquired. More than 10^5 positive ion energy and directional distributions in the solar wind were recorded during the 7-month period mentioned above. The important features of the spectra illustrated in

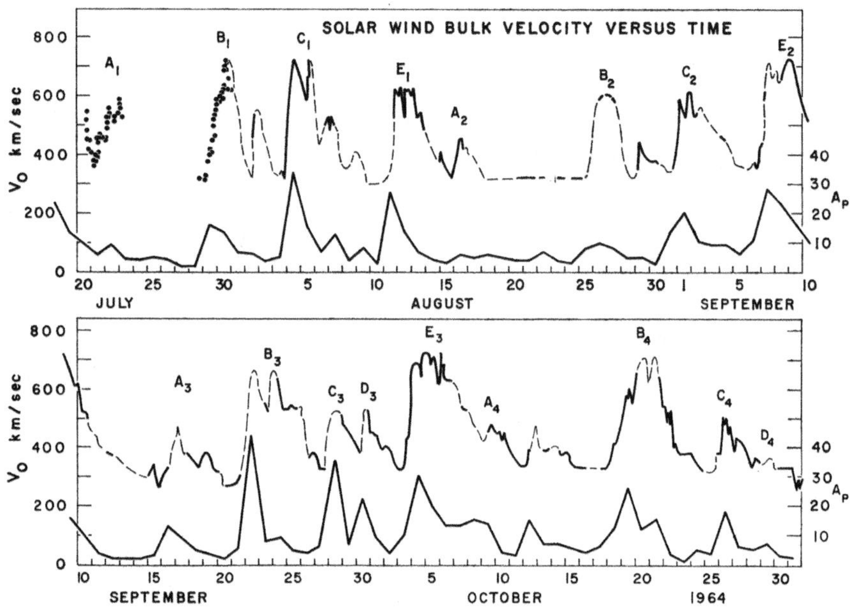

Fig. 5. Bulk velocity of solar wind plasma in interplanetary space versus time, derived from observed proton energy spectra. The points plotted near the symbol A_1 represent the nature of individual determinations which are made every two minutes, though in this figure only a few points per hour can be plotted. The heavy solid lines are smoothed curves through such points. The faint dashed lines are usually estimates based on observations in the transition regions, during times when there was no data being acquired in interplanetary space. Also plotted are reported values of A_p, the daily average magnetic disturbance. In this figure the A_p curve sometimes appears to start rising before a rise in V_0. This is not actually observed, and appears on this plot only because of the one-day resolution of the A_p values.

Figure 4 are: (1) the width of the spectrum which is influenced by the temperature; however it is found that usually the main group in the velocity distributions can be fitted to Maxwellian distributions only qualitatively, and sometimes not even qualitatively; in addition, a high velocity tail of $\sim 10\%$ is usually present; (2) the absolute proton energy values, from which can be derived an average proton velocity, which is a measure of bulk plasma velocity; and (3) the integral under the curve which is a measure of flux density. Of the three spectra shown in Figure 4, the left one shows intermediate spectrum width, while the others are extreme, but not uncommon, examples of broad and narrow spectra. Spectra of this and intermediate character are observed to be present whenever the instrument is operating in the positive ion mode. Analysis is underway to look

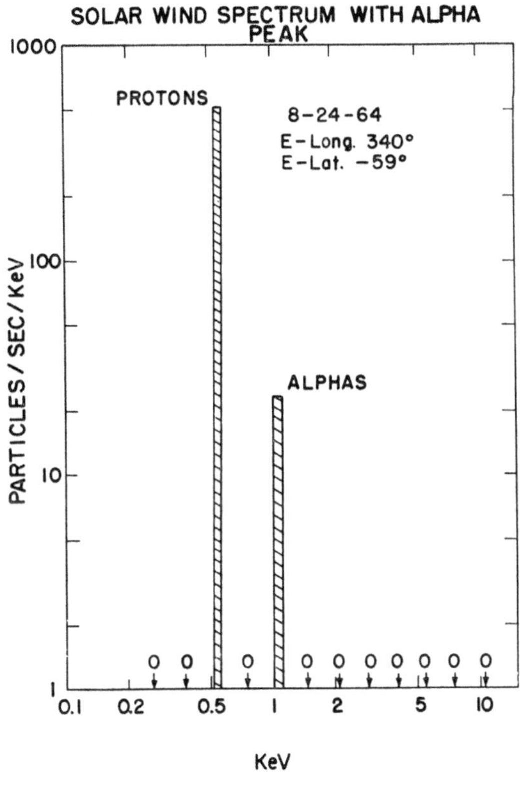

Fig. 6. A non-typical positive ion energy distribution in the interplanetary medium, showing a distinct alpha particle group. The zeros indicate locations of an instrumental energy channel, where the observed count was zero.

for possible correlations between a broad spectrum (high temperature), high average proton energy (high bulk velocity), and high flux density.

The observed bulk velocity versus time derived from data of this kind for a 3-month period is plotted in Figure 5, along with reported values of A_p representing the daily average magnetic disturbance observations at ground level.

Several interesting features are apparent in this figure. There is an indication of a base line bulk velocity of 320–330 km per sec which may be characteristic of the solar wind escaping from the quiescent sun during this epoch. Usually the bulk velocity is enhanced and variable, with a variable period of the order of hours or days, and, although detailed correlation studies have not yet been made,

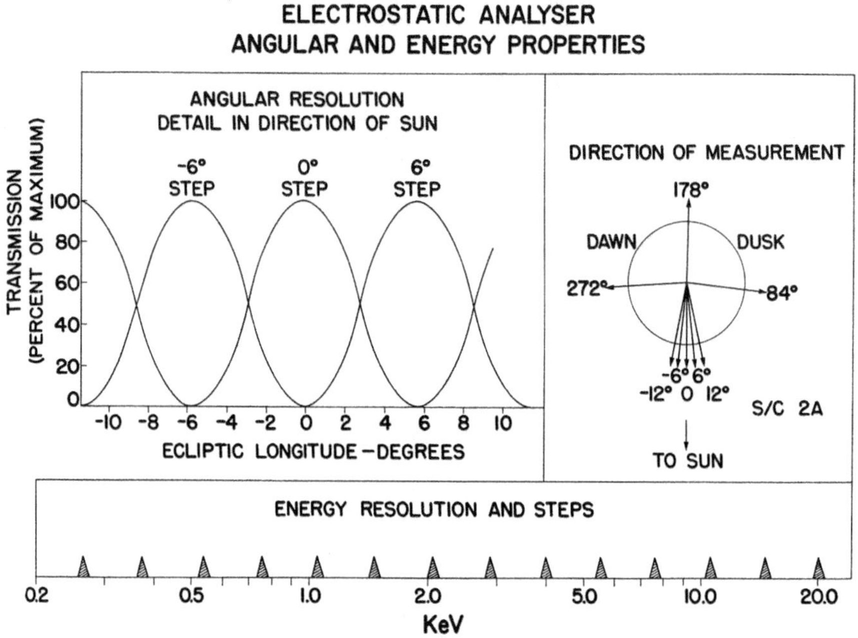

Fig. 7. Instrumental characteristics of the electrostatic analyzer on the Vela 2 satellites.

it is apparent that these enhancements are correlated with solar activity near the central meridian as seen from the earth. The greatest observed velocity is 720 km per sec. The peaks in the bulk velocity plot generally recur at intervals of ~27 days, persisting through several solar rotation periods. The correlation with A_p is good, though it is interesting to note that A_p tends to be a maximum at the time of a sharp rise, rather than at the peak of the bulk velocity.

Figure 6 is another type of observed positive ion energy distribution, where a distinct alpha particle group is observed at the position where the energy per unit charge is twice that of the proton. The zeros along the absissae line in this figure indicate the positions of other instrumental energy channels, where the observed count was zero. Because the instrumental channel widths are narrow compared with their spacing, there is a deficiency such that there are gaps where

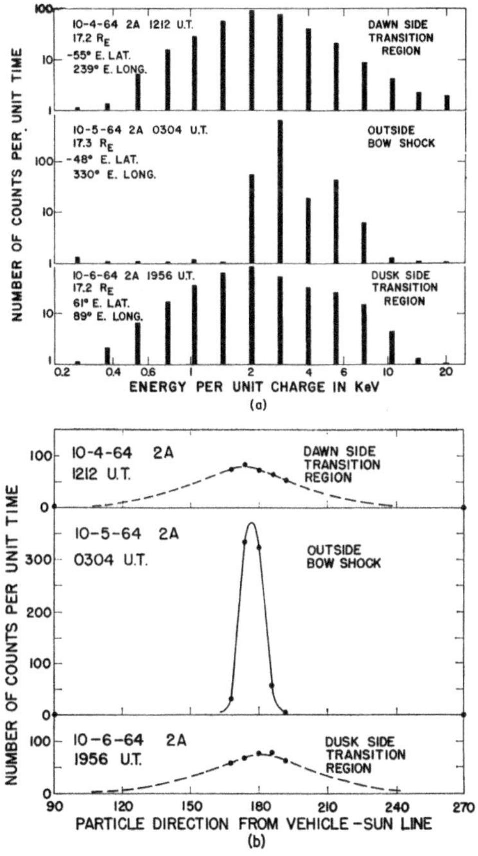

Fig. 8. Examples of three observed positive ion spectra are shown in the upper half of the figure. These observations were made during a part of one orbit on October 4, 5, and 6, 1964. The three directional distributions, one corresponding to each energy distribution and measured at the same time, are shown in the lower half of the figure. In these directional plots 180° signifies a component of motion in the ecliptic plane in the direction from the sun toward the satellite, or earth.

particles cannot be observed. At the time of this observation the proton and alpha particle energies were narrow and happened to fall into two channels. Usually the alpha particles are not so well resolved and sometimes they escape detection. However, the analysis of many such observations shows the alpha particle proportion is variable, from less than 1% up to ~20%, and the estimated average is 3 to 6%.

In the lower part of Figure 7 is illustrated the instrumental deficiency mentioned above, i.e., the narrow energy channels with wide gaps between them. A similar instrument carried on the Vela 3 satellites fills in these gaps with four times as many channels.

Figure 7 also shows the directional capabilities of the electrostatic analyzer, and the lower half of Figure 8 shows three typical observed proton directional distributions. The center one of these three distributions is typical of the solar wind in interplanetary space, and it can be noted that the component of motion in the ecliptic plane was nearly at 180°, i.e., nearly along the sun-earth line. These directional distributions are typically ~10° full width at half maximum, but are sometimes as narrow as ~2°. When corrected for the aberration due to the earth's motion around the sun, examination of many distributions of this kind indicates the average solar wind direction (its component in the ecliptic plane) is radially outward from the sun within ±2°. Unambiguous fluctuations of ±6° or more, i.e., non-radial flow direction, are commonly observed.

Systematic analysis of temperatures and flux densities in the solar wind have not been made. Effective longitudinal temperatures, derived from the width of the energy spectra, are typically ~10^5 °K or less, and typical positive ion densities are 3 to 5 per cm^3. These values are in general agreement with other observations (SNYDER and NEUGEBAUER, 1963; and WOLFE and SILVA, 1965).

Other observations by the Vela satellites on particles in the interplanetary medium include occasional energetic electrons of energy >40 keV, and protons >180 keV, with energy spectrum information measured by the solid state detector. Further analysis is required to attempt to determine whether these are associated with precursors on the sunward side of the bow shock, or associated with disturbances inherent in the solar wind as it propagates away from the sun.

In addition, fluxes of electrons of energy ~150 to 500 eV are not infrequently observed with the electrostatic analyzer, and the measured flux density of electrons of $E > 350$ eV is in the neighborhood of 10^7 cm^{-2} sec^{-1}. These electrons appear to be solar wind electrons, which are more readily observed with this instrument when the temperature of the solar wind makes excursions to relatively high values.

B. BOW SHOCK

Twice during each orbit the Vela satellites pass through the region of the so-called bow shock, or hydromagnetic shock wave (AXFORD, 1962; KELLOGG, 1962; NESS *et al.*, 1964) where the solar wind plasma, carrying the interplanetary magnetic field, is subjected to violent interactions as a result of collision with the earth's magnetic field. Reference to Figures 1, 2, and 3 again will provide orientation as to sampling of points near the bow shock by the Vela satellites. It may be noted that points near the subsolar points are not sampled.

As the Vela satellites traverse the expected bow shock regions, a few R_e beyond the magnetopause, sudden and drastic changes in the character of the solar wind particles are observed. Figure 8 illustrates how the observed proton energy and directional distributions change form at the bow shock. These data were obtained by the electrostatic analyzer during a part of one orbit on October 4, 5, and 6, 1964. The trend shown here repeats itself consistently on other orbital passes. Within the transition region on the dawn side of the orbit the energy spectra are relatively broad; as the satellite moves in its orbit the energy spectra suddenly become much narrower, the average proton energy increases, and the spectrum is characteristic of the interplanetary solar wind; then as the satellite continues around to the dusk side of the orbit, the spectrum suddenly becomes broad again, with a decrease in proton energy. As one passes from interplanetary space through the bow shock into the transition region, the positive ion density is observed to increase, typically by a factor of 1.5 to 2. Referring to the directional distributions in the lower part of Figure 8, we find a similar sequence of sudden changes, where the distribution is relatively broad ($\sim 50°$ full width at half maximum) in the transition region, and narrow ($\sim 10°$ full width at half maximum) in interplanetary space. These changes are sometimes observed to occur in times no greater than 2 seconds. On some passes the changes occur several times within a few tens of minutes, as if the bow shock is shifting out and in rapidly because of fluctuations in the solar wind. If this interpretation is correct, the position of the bow shock is observed to shift by distances up to several R_e.

Electron data for this same orbital pass of October 5, 1964 are shown in Figure 9. These observations will not be described in detail, but the character of the electron flux also goes through sudden and violent changes at the bow shock.

An overall view of the changes in proton and electron fluxes is consistent with the concept that as the solar wind encounters the bow shock, a turbulence is established in which energy is removed from the solar wind protons, solar wind electrons acquire energy, and the direction of motion of the particles is made more random. The omnidirectional flux of electrons of energy sufficient to be observed by the instruments increases from $\sim 10^7$ cm^{-2} sec^{-1} or less for $E > 350$ eV outside the bow shock, up to $\sim 5 \times 10^8$ cm^{-2} sec^{-1} inside the transition region. The temperature of the protons rises from typical values of $\sim 10^5$ °K or less in interplanetary space to 10^6 °K or more. The observed temperature of the electrons inside the bow shock is typically $\sim 10^6$ °K, and $\sim 5 \times 10^6$ °K is not uncommon.

C. TRANSITION REGION

Many of the characteristics of the charged particles observed by the Vela satellites in the transition region have been described in the above section on the bow shock, and will not be repeated. Two additional results will be given here. One of these concerns the plasma streaming direction as derived from the proton directional distributions observed by the electrostatic analyzer. These directions apply, of course, only to points in the transition region traversed by the Vela satellites. Although a statistical analysis has not yet been made, inspection of

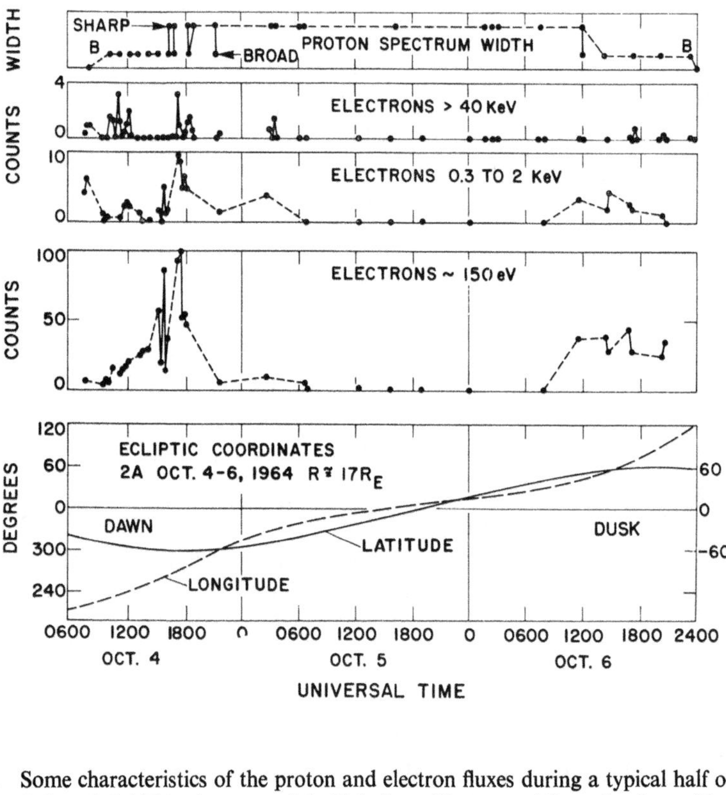

Fig. 9. Some characteristics of the proton and electron fluxes during a typical half orbit on October 4, 5, and 6, 1964, when one of the Vela 2 satellites passed from the dawn side proton boundary, out through the transition region into the interplanetary medium, and back into the dusk side transition region. The time scale at the bottom applies to all five portions of the figure. In the top portion of the figure, the points labeled "*B*" at the beginning and at the end of the plot represent positions of the "proton boundary". The region where the proton spectrum is indicated to be broad is interpreted as the transition region, while the long central region where proton spectrum is sharp is interpreted as interplanetary space. Oscillations between sharp and broad near 1800 hour on October 4 is interpreted as possibly a shifting of the bow shock position.

scores of distributions indicates the plasma streaming direction usually deviates from the sun-earth line by about 10° or more in a sense so as to indicate that the plasma is streaming parallel to the magnetopause, as observed by others (Bridge *et al.*, 1965). It should be remembered that the electrostatic analyzer senses the component of motion parallel to the ecliptic accurately, but not the perpendicular component of motion.

The second result to be reported here is the observation of a population asymmetry in fluxes of electrons on the dawn and dusk sides of the transition region. This effect is observed with all five types of detectors described in Section 4 above on instrumentation. Especially the electrons of $E > 40$ keV are observed relatively rarely in the dusk side transition region, at least an order of magnitude less frequently than on the dawn side. The question of the persistence of this asymmetry is an important one. The asymmetry observations over rather long periods of time will be reported on more fully in Section E below, because the asymmetry extends into the magnetospheric cavity.

D. MAGNETOSPHERIC BOUNDARY

It appears that the location of the magnetospheric boundary, or magnetopause, can be identified with the electrostatic analyzers on the Vela satellites by sudden appearance or disappearance of ~0.5 to 5 keV streaming protons which persist in the transition region, but which are usually absent in the magnetospheric cavity. The topmost portion of Figure 9 illustrates this observation and shows characteristics of the proton spectrum observed by the electrostatic analyzer for the pass of October 4, 5, and 6, 1964. The interpretation of this chart is as follows. In the center of the chart the proton spectra are observed to be sharp, and the satellite is in interplanetary space. Toward the edges of the chart the proton spectra become broad, and the satellite is in the transition region. On the left hand side, which is the dawn side, the spectrum fluctuates several times between sharp and broad as if the bow shock is shifting in and out. At the positions labeled B the protons disappear completely, and the interpretation is that the satellite is inside the magnetospheric cavity. This "proton boundary" appears to coincide with the magnetopause locations as determined by magnetic field and particle observations of others, (Ness *et al.*, 1964; Frank and Van Allen, 1964). By observations on successive satellite orbits, it is found that the position of the proton boundary varies by distances of up to several R_e, indicating location instability to about the same degree as is observed for the bow shock.

For the one year period following the Vela 2 launch in July 1964, a dawn-to-dusk asymmetry in the location of the proton boundary has been studied. Many determinations of the proton boundary, during seasons when the satellites cut through the ecliptic plane near the magnetopause, indicate the average boundary locations were at ~95° ecliptic longitude on the dusk side, and at ~255° (instead of the symmetrical angle of 265°) on the dawn side. This implies a tilt of ~5° in the ecliptic plane. The magnitude and direction of this tilt provide evidence for symmetry of the magnetospheric boundary with respect to the solar wind direction, which suffers an aberration of 3–5° because of the earth's orbital motion around the sun.

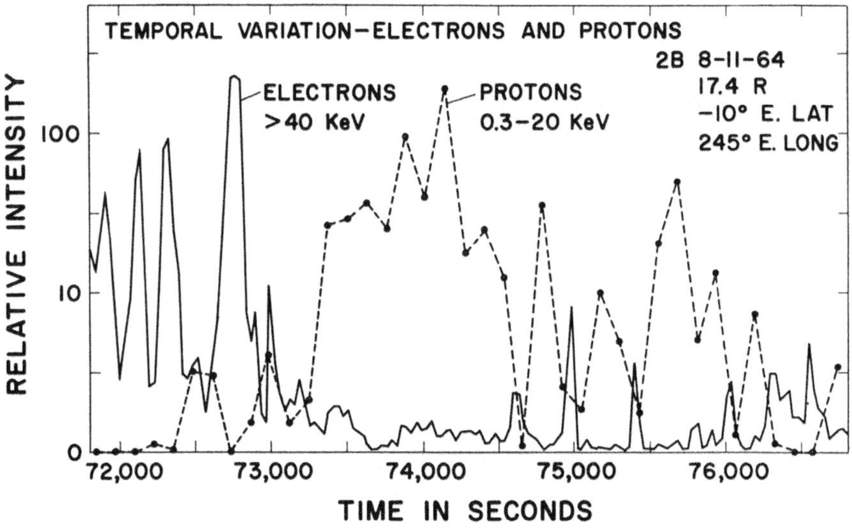

Fig. 10. Example of observed inverse correlation of energetic (>40 keV) electron flux and proton flux near the dawn side magnetopause.

At the time of a dawn-side proton boundary traversal by one of the satellites, particularly when the orbit crosses the proton boundary near the magnetic equatorial plane, there sometimes appear striking relationships between the proton fluxes and energetic (>40 keV) electron fluxes. An example is illustrated in Figure 10, at a time when the satellite is emerging from the magnetospheric cavity into the dawn-side transition region. Energetic electrons were being observed by the G.M. counter. As time progressed, simultaneous observations of 0.3–20 keV protons were made. There is a striking correlation between a high flux of electrons and a low flux or absence of observable protons, and vice versa. Observations of this phenomenon have also been made in which the flux of lower energy electrons of 0.3 to 20 keV correlate directly with the flux of >40 keV electrons. This phenomenon of inverse correlation between proton and electron fluxes may well be related to the location instability of the magnetopause described above. The time interval covered in Figure 10 is about 4000 seconds, during which time the satellite has moved about 1.5 R_e.

E. MAGNETOSPHERIC CAVITY

A little less than half the Vela satellite orbit is in the tail of the magnetosphere. Points near 17 R_e are sampled at all ecliptic latitudes up to 60° and magnetic latitudes up to about 50°, and at distances up to ~16 R_e from the ecliptic plane. The "null plane" or "neutral sheet" (NESS, 1965) is traversed at points fairly near the antisolar point once each orbit during seasons of one to two months before and after the time when the earth can eclipse the satellite. Several results on the particle observations by the Vela satellites in these regions of the magnetosphere will now be presented.

Interesting aspects of the topography of energetic electrons have been found (MONTGOMERY *et al.*, 1965). Energetic electrons are frequently observed in the tail of the magnetosphere by the M4 and M5 scintillation detectors which are sensitive to electrons of energy greater than ~85 keV and ~50 keV respectively (see Section 4). The spatial distribution of fluxes $>2\times10^4$ cm^{-2} sec^{-1}, as observed by the M4 detectors on the Vela 1 satellites is shown in Figures 11 and 12. These observations were made on about 70 successive orbits for each of the two satellites, from October 1963 through July 1964. It appears that these events are related to the events reported by ANDERSON *et al.* (1965). The presence of the electron flux persisted at the moving satellite for periods from less than one to more than 20 hours, with the satellite moving at a speed of about one R_e per hour. As one satellite followed the other into nearly the same region, about two days later, it sometimes encountered a similar electron flux environment, and sometimes not. Although detailed temporal persistence is lacking, it is evident from Figures 11 and 12 that there is a persistent clustering of the electron flux within ±6 R_e ($\pm20°$ latitude) of the plane of the magnetic equator, and no strong ordering about the ecliptic plane. Thus it appears the earth's magnetic dipole field plays an important role in the population distribution of these energetic electrons. A plot of the data in solar magnetospheric coordinates shows poorer clustering about magnetospheric latitude than about simple geomagnetic latitude. Although the spatial distribution plots of Figures 11 and 12 are not weighted to account for the time during which the satellites occupy the different regions, an investigation of this point leads to the conclusion that the general statements made above are unaltered.

Another striking result apparent from Figure 12 is a dawn-to-dusk asymmetry in the population of the energetic electrons, there being a relative void on the dusk side for ecliptic longitudes less than ~130°. A similar asymmetry has also been demonstrated by analysis of data from the G.M. counters on

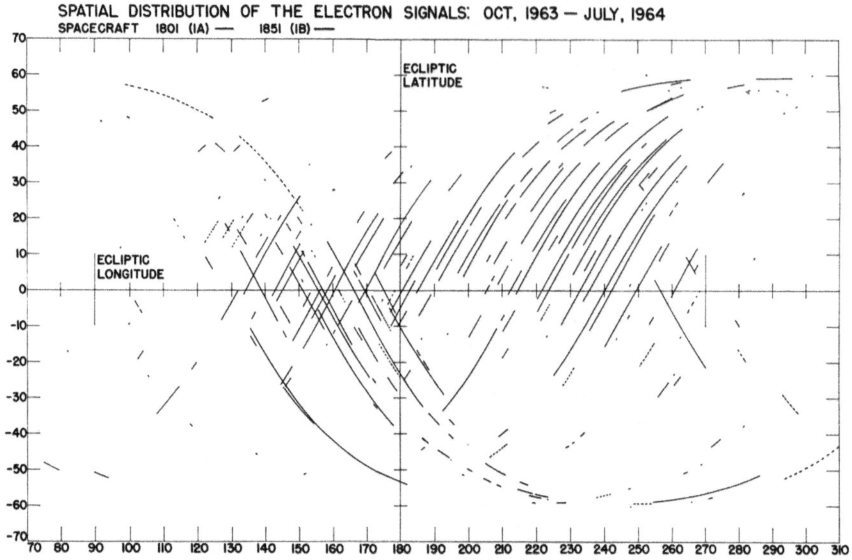

Fig. 11. Spatial distribution of energetic (> 85 keV) electron fluxes observed by M4 scintillation detectors, in solar ecliptic coordinates. Solid lines represent regions where the electron flux was $\geqslant \sim 10^5$ cm^{-2} sec^{-1} as the satellite moved along the indicated coordinate positions. Dashed lines represent regions where flux was $\geqslant \sim 2 \times 10^4$ cm^{-2}sec^{-1}, which is the instrumental threshold. Limit of ecliptic latitude reached by the satellite is 60°.

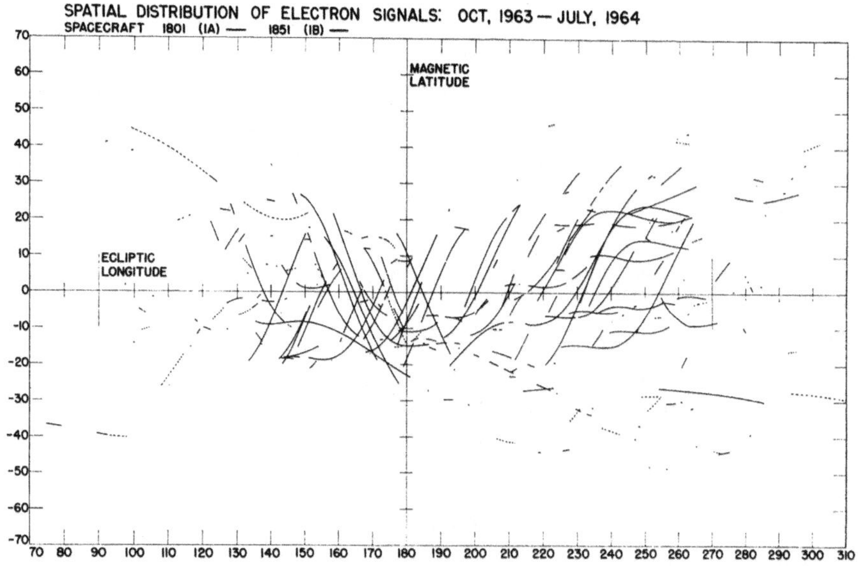

Fig. 12. The data of Figure 11 is replotted with satellite positions located in terms of geomagnetic latitude instead of ecliptic latitude. Limit of geomagnetic latitude reached by the satellite is 47°.

Vela 2 satellites during a 3 month period from August through October 1964, and is illustrated in Figure 13. This figure also illustrates the asymmetry in the location of the observed proton boundaries as described earlier. In these energetic electron observations the presence of energetic electrons beyond the magnetopause in the transition region on the dawn side was commonly observed.

More detailed observations on the quantitative character of the electron fluxes near the region of magnetic field reversal or "neutral sheet" (NESS, 1965; AXFORD *et al.*, 1965) have been made by the electrostatic analyzer and by

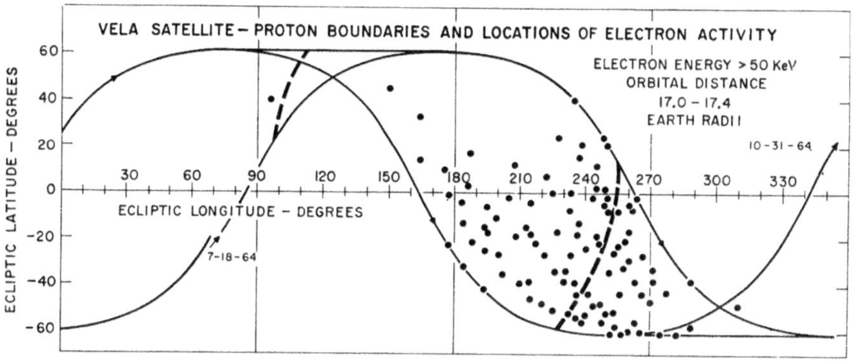

Fig. 13. Spatial distribution of energetic (> 40 keV) electrons observed by the G.M. counters during more than a 3 month period from August through October 1964. On this plot a satellite moves along paths having the form of the two solid lines shown, these two lines being paths on 7–18–64 and 10–31–64 as indicated. The satellites traversed successive paths between these two lines, and points are plotted where the energetic electron flux was observed to exceed $\sim 10^4$ cm^{-2} sec^{-1}.

simultaneous observations with the G.M. counter. These observations include spatial mapping, energy spectrum, flux, and time variations. The data provide information about an observed physical feature which may be termed the "plasma sheet", and which is located in the general region of the magnetic field reversal. In this plasma sheet the data furnish information about the plasma kinetic energy density which may be compared with the known magnetic field energy density, about effective electron temperatures, and about suggested impulsive energization of the entire electron population which appears to behave as a coherent group. Some of these data and results will now be presented.

As noted earlier in this chapter, the electrostatic analyzers on the Vela 2 satellites measure the energy spectrum and flux of low energy electrons in the energy region 0.35 to 20 keV. At the same time the G.M. counters measure the more energetic electrons of energy greater than 45 keV. Observations of this kind have been analyzed in some detail for a 7-week period in September and October 1964, when each of the Vela 2 satellites passed through the magnetospheric tail near the antisolar point 11 times. A sample of data for one of these passes on October 4 and 5, 1964, is shown in Figures 14 and 15. Scanning such

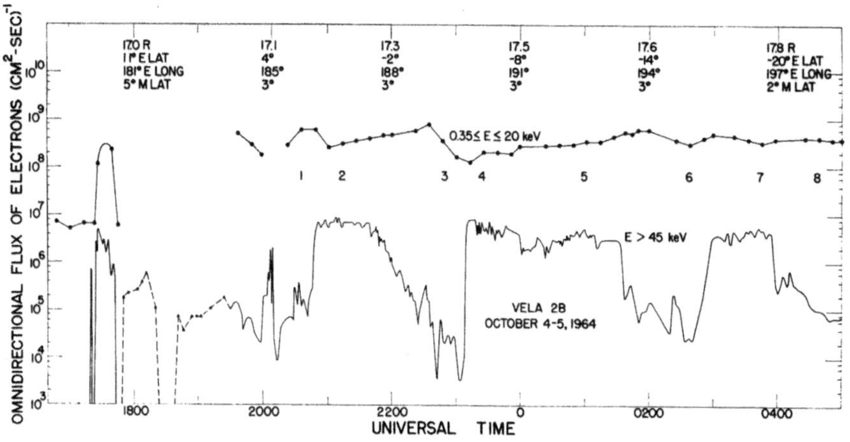

Fig. 14. Time history of omnidirectional fluxes of electrons for a portion of a pass through the tail of the magnetosphere. Electron flux with $0.35 \leq E \leq 20$ keV is measured with an electrostatic analyzer; flux with $E > 45$ keV is measured with a thin-window G.M. counter. Radial distance, ecliptic latitude and longitude, and geomagnetic latitude of the spacecraft are given at the top of the graph. Analyzer data are not available for those times without points. Integral energy spectra for the numbered times are given in Figure 15.

data for the 22 passes of the satellites through this region of space indicates that there is a "plasma sheet" in which there is always present a flux of electrons in the energy range 0.35 to 20 keV with flux generally in the range 10^8 to 10^9 cm^{-2} sec^{-1}; outside the region the flux drops by one or two factors of ten or more. This plasma sheet is centered about the expected region of magnetic field reversal, and has a thickness of about 6 R_e or more. The spatial location of this region is generally similar to the region populated by the more energetic electrons as observed by the M4 detectors, and shown in Figure 13. The flux of 0.35 to 20 keV electrons is much more stable than the flux of >45 keV electrons, the latter being subject to very sharp onsets when flux values increase by several orders of magnitude within a few minutes.

It may be noted that the spectra presented in Figure 15 include the >45 keV point measured with the G.M. counter. The smoothness of the spectrum at a given time, and the coherent energy changes over the whole spectrum is a consistent characteristic observation, suggesting all the electrons belong to one family which is impulsively accelerated by some mechanism, which is as yet not understood. Sudden drops in the energy of this family of electrons are also sometimes observed, though characteristically the drops are more gradual.

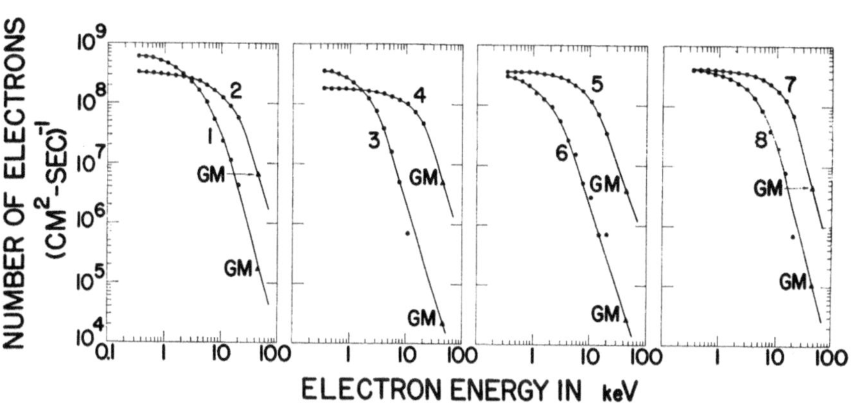

Fig. 15. Integral energy spectra of electrons determined at the numbered times of Figure 14. Points below 20 keV are determined with an electrostatic analyzer, the points at 45 keV with a G.M. counter. Large increases in electron flux for $E>45$ keV seem to be caused by an impulsive change in the energy distribution of the total electron population.

These fluctuations are of a type similar to those reported from IMP 1 observations (ANDERSON *et al.*, 1965), although it should be noted that the present measurements refer to changes in the entire electron energy spectrum, not simply the related change in the flux of >45 keV electrons.

The observed electron energy distributions are quasithermal in character, being usually somewhat broader than a pure thermal distribution. Effective electron temperatures derived from the observed energy distributions range between $\sim 5\times10^6$ and 50×10^6 °K.

The shape of the integral energy spectra, being flat at low energy, suggests that an integration may in fact provide a reliable total electron energy density, though of course a significant rise in electron flux at energies below $\sim$100 eV is not precluded. On the assumption that such a rise does not exist, the eight spectra presented in Figure 15 yield the following numbers: the average electron energy within a given spectrum ranges from 1.4 to 12 keV; the average electron density ranges from 0.08 to 0.23 cm^{-3}; and the integrated plasma energy density, evaluated on the arbitrary assumption that the photon and electron energy densities are equal, ranges from 0.26 to 1.0×10^{-9} erg cm^{-3}. An equal magnetic field energy density would be provided by B values in the range 11 to 22×10^{-5} gauss, which is in the range of values measured by the IMP 1 satellites (NESS, 1965).

References

ANDERSON, K. A., HARRIS, H. K., and PAOLI, R. J.: 1965, *J. Geophys. Res.* **70**, 1039.

ARGO, H. V., CONNER, J. P., BONNER, T. I., EVANS, W. D., MONTGOMERY, M. D., SINGER, S., STOGSDILL, E. E., JEAN, G., ASBRIDGE, J. R., BAME, S. J., FELTHAUSER, H. E., OLSON, R. A., MICHAEL, D. E., and STRONG, I. B.: 1964, Eight papers on the scientific results from the Vela Satellite Program, *Trans. Am. Geophys. Union* **45**, 623–625.

ASBRIDGE, J. R., BAME, S. J., FELTHAUSER, H. E., OLSON, R. A., and STRONG, I. B.: 1965, *Trans. Am. Geophys. Union* **46**, 142.

AXFORD, W. I.: 1962, *J. Geophys. Res.* **67**, 3791.

AXFORD, W. I., PETSCHEK, H. E., and SISCOE, G. L.: 1965, *J. Geophys. Res.* **70**, 1231.

BAME, S. J., ASBRIDGE, J. R., FELTHAUSER, H. E., OLSON, R. A., and STRONG, I. B.: 1965, *Trans. Am. Geophys. Union* **46**, 142.

BRIDGE, H., EGIDI, A., LYON, E., MORENO, G., OLBERT, S., and ROSSI, B.: 1965, *Trans. Am. Geophys. Union* **46**, 119.

FRANK, L. A. and VAN ALLEN, J. A.: 1964, *J. Geophys. Res.* **69**, 4923.

KELLOGG, P. J.: 1962, *J. Geophys. Res.* **67**, 3805.

MONTGOMERY, M. D., SINGER, S., CONNER, J. P., and STOGSDILL, E. E.: 1965, *Phys. Rev. Letters* **14**, 209.

NESS, N. F., SCEARCE, C. S., and SEEK, J. B.: 1964, *J. Geophys. Res.* **69**, 3531.

NESS, N. F.: 1965, *J. Geophys. Res.* **70**, 2989.

SINGER, S., CONNER, J. P., EVANS, W. D., MONTGOMERY, M. D., and STOGSDILL, E. E.: 1964, *Proc. Intern. Space Sci. Symp.* **5**, Florence.

SNYDER, C. W. and NEUGEBAUER, M.: 1963, *Proc. Intern. Space Sci. Symp.*, **4**, Warsaw.

STRONG, I. B., ASBRIDGE, J. R., BAME, S. J., FELTHAUSER, H. E., and OLSON, R. A.: 1965, *Trans. Am. Geophys. Union* **46**, 134.

WOLFE, J. H. and SILVA, R. W.: 1965, Private communication.

Discussion

BEARD asked whether the solar wind temperature or alpha-proton ratio changed with bulk velocity, but COON replied that this study was not yet completed. Coon replied to another question by stating that the solar wind spectrum was similar as seen by satellites some 20 R_e apart.

DESSLER showed a sketch that illustrated how one might transfer energy into the magnetosphere through the processes that occur when a fast solar wind overtakes a slow one. At the interface in the model, there is non-radial irregular flow and at the tangential velocity discontinuity there should be a Helmholtz instability which would give turbulence and irregularities in direction. Then the solar wind would beat on the magnetosphere "like raindrops on a tin roof".

This would explain how it is a variation in solar wind velocity or the magnetic field direction that leads to geomagnetic activity rather than simply high bulk velocities per se.

There was considerable discussion about the dawn-dusk asymmetry in the magnetopause presented by COON. MEAD stated that merely a readjustment of the axis of symmetry by 5° would eliminate the effect. Furthermore, he drew attention to the different degrees of organization or ordering of the data by different coordinate systems. Coon's data indicate that magnetic latitude appears best out to some 16 or 17 earth radii, and Ness then finds that beyond this distance a solar-magnetospheric coordinate system is best for organizing the neutral sheet, so that at large radial distances the tail direction should be solar controlled rather than geomagnetically oriented.

In response to a question by WALT, COON mentioned that the neutral sheet as observed by particle fluxes at 17 R_e is about 6 to 7 R_e thick, and so thicker than implied by Ness' measurements.

TRAPPED RADIATION ESTIMATES FOR THE OTHER PLANETS

L. REIFFEL*
IIT Research Institute, Chicago, Ill.

If one were to construe the title of this paper narrowly, one could restrict our subject almost entirely to Jupiter, for this planet is the only one other than the earth about which we can make reasonably reliable estimates of trapped radiation intensities. It will be more useful, however, to take a broader view and discuss a number of topics relevant to planetary radiation belts and/or planetary magnetic fields and thus convey the flavor of the entire field at this time. We shall cover the solar system planet by planet, reserving somewhat more of the commentary for Jupiter, since we know the most about it.

Obviously there are only two or three ways to get information on planetary magnetic fields and trapped radiation at the great distances involved. One can, at least in principle, make various types of optical observations, but these have not provided any quantitative clues for planets other than the earth. Optical searches for auroral emissions from the planets, which could be related to trapped radiation phenomena, have thus far yielded only negative results and data have not been reported on searches for Zeeman effect or other spectral signatures which might yield the desired data. A second classical technique is, of course, to study radio emissions and such work has indeed been quite fruitful both in the decimeter and decameter wavelength regions. To these methods, we may now add the planetary rocket probe experiment as represented by the successful Venus fly-by of the U.S. Mariner II spacecraft and the successful Mars fly-by of the U.S Mariner IV. It is almost certain that before the end of another decade, we will have more data about these planets from planetary probes, orbiters, and soft landers, and we shall, furthermore, probably have direct spacecraft observations of both Jupiter and Mercury. It is also likely that our capabilities for gathering data on the planets from the neighborhood of the earth will be vastly improved as a result of the application of space technology to the establishment of orbiting radio and optical astronomical equipment. It is thus quite clear that this field is very much in its infancy and sweeping changes in our understanding of planetary radiation belts, the nature and sources of planetary magnetic fields, and related matters lie just ahead.

* Now at NASA Headquarters, Washington, D.C.

Mercury

Evidence for a trapped radiation belt around Mercury might take the form of an increasing brightness temperature as a function of wavelength, thereby allowing one to argue for non-thermal sources contributing at the longer wavelengths. In fact, in the case of Mercury, a bolometric temperature of 610 °K has been obtained for the sub-polar point on the planet for a wavelength of ten microns, while centimeter radio emission studies have yielded 1100° plus or minus 300 °K at 3.45 and 3.75 centimeters (McElheny, 1964). The presently accepted explanation of this change in apparent temperature does not rely upon emissions from trapped radiation, but rather on the presumption of a very tenuous atmosphere, possibly radiogenic Argon, which would resolve the discrepancy by warming the dark side of Mercury by convection. The radio temperature of 1100° was obtained using an assumption that the dark side of the planet resided at zero degrees Kelvin. If convection raised the dark side temperature, the bolometric and radio data could be consistent. The required atmosphere is not in conflict with light polarization data and new eleven centimeter radio data for the whole disk (Kellerman, 1965).

It would appear, from the evidence available at present, that Mercury has no appreciable trapped radiation belt and, hence, presumably no appreciable magnetic field. It may be noted that Mercury rotates about four times faster (Peale and Gold, 1965) than Venus which we know to have a very small magnetic field, if any at all. Mercury's rotation rate is almost two orders of magnitude slower than that of the earth. However, Mercury's density is equal to that of the earth, while Mars, which we also know has no substantial field or trapped radiation belt in spite of its rapid rotation, is much less dense and this may imply quite different conditions in the cores of Mars and Mercury. Mars may in fact have a radically different history than the earth and Mercury, a point to which we shall return later. It is also possible that before tidal friction reduced the rotational angular momentum of Mercury its core composition and configuration were such as to permit it to have a considerable magnetic field and trapped radiation belt. Evidence of this ancient radiation belt could conceivably be sought on the surface of the planet in the form of residual radiation effects of various kinds.

Venus

The Mariner II fly-by of Venus which passed at a distance of 41000 kilometers or 6.6 radii from its center gave no indication whatsoever of a planetary magnetic field or trapped radiation. The original deductions (Goldstein and Carpenter, 1963) from this data led to an upper limit for the magnetic moment of Venus which was one-tenth that of the earth. Subsequent considerations, based especially on the shape and symmetry of the boundary regions between a planetary magnetosphere and solar wind have allowed a reduction of this upper limit by an order of magnitude (VAN ALLEN, 1965). In addition to this strong direct evidence against an appreciable magnetic field on Venus, decametric radio bursts have been searched for between 18 and 20 megacycles, but have not been found. In this frequency regime, where Jupiter is very active, no emissions from Venus above a flux of 2×10^{-22} watts/meter^2/cycle/second have been found over several hundred hours of high quality listening time (CARR *et al.*, 1961; SMITH *et al.*, 1965). All available evidence thus points to a negligible magnetic field and the non-existence of a radiation belt for Venus.

Mars

Thanks to Mariner IV, which passed Mars at a distance of 6118 miles, we now know that the magnetic moment of that planet is much less than that of the earth. The particle detectors and magnetometer equipment aboard Mariner IV all yielded null indications which the several experimenters have interpreted to mean the Mars magnetic moment is not more than 10^{-3} to 10^{-4} that of the earth. This result is again consistent with the negative results recently reported in a decametric emission search (SMITH *et al.*, 1965). There has been one report of non-thermal 21 centimeter decimetric radiation from Mars (DAVIES, 1964), but more recent observations show no such effect (KELLERMAN, 1965).

The mounting evidence against a Martian magnetic field is contrary to the expectations of most workers and may be connected with the relatively low density of the planet. ALFVÉN (1963) has suggested that the moon and Mars belong to one family while Mercury, Venus and the earth belong to another and possibly younger group of planets. The latter group has a density ranging from five to six while the former bodies have densities between three and four. A possible common origin of the moon and Mars may be related in some way to the high crater density on both objects and may account, in part at least, for the absence of a magnetic field associated with either.

Saturn

Saturn is similar to Jupiter in several respects and has a short rotational period of somewhat less than eleven hours making it a promising candidate for possession of a radiation belt. On the other hand, Drake and others have felt quite reasonably that the ring structure around the planet will make it difficult for an extensive radiation belt to develop. Some increase in brightness temperature with wavelength has been observed as one proceeds from the infrared to wavelengths of the order of ten centimeters, but this could simply be the result of thermal emission from different levels in the planetary atmosphere. If the entire decimetric effect is ascribed to non-thermal phenomena, the absolute radio luminosity of Saturn must be less than one-sixth that of Jupiter.

Hints of decametric radiation from Saturn have been obtained from time to time (SMITH *et al.*, 1965), but the evidence is not definite enough to make positive statements as yet. The situation with Saturn, therefore, is that there is just the suggestion of both non-thermal decimetric and decametric radiation which may imply a weakly developed radiation belt. We clearly do not have enough information to make any meaningful estimates of either radiation fluxes or magnetic moment.

The Outer Planets

About the outer planets, we can really say nothing definitive in the context of this paper except to note that recent work (PEEBLES, 1964) suggests that Uranus and Neptune have heavy cores containing about ten earth masses and it is, therefore, possible that conditions exist which would be favorable to the development of planetary magnetic fields.

Jupiter

It is well-established that Jupiter has an extensive and well-developed planetary magnetic field and an accompanying radiation belt system. While no doubt remains to the existence of a Jovian belt, our knowledge of the particle intensities and other detailed properties is in a very unsatisfactory state. In the decimeter region from 10 to 70 centimeters, Jupiter radiates strongly (especially in the region from 20 to 30 centimeters), while in the decametric regime the planet displays a very rich and complex array of phenomena which is only partially unravelled at present. (There have been many recent reviews of observations and theories for Jupiter, see for example, WARWICK, 1964; ROBERTS, 1963. While these reviews remain valuable, the recent discovery of the influence of the Jovian moon Io on decametric emissions and the greatly refined decimetric observations of recent date make them somewhat obsolescent.)

The decimetric radiation from the planet is now presumed to arise from synchrotron radiation originating from relativistic electrons mirroring in a Jovian dipole field. CHANG and DAVIS (1962), working with a thin shell synchrotron model, have made estimates of the relativistic particle density in the belt, assuming the belt volume is ten times the volume of Jupiter and that the spectrum goes as $1/E$ between two cut-off "critical" energies. Their results can be summarized as follows:

TABLE I
Magnetic Fields and Particle Densities for Jupiter

Assumed Jovian Field =	0.1 *gauss*	1	10
Critical energy* in MeV			
corresponding to $\lambda =$ 3 cm	75 MeV	25	7.5
10 kmc			
$\lambda = 300$ cm	8	2.5	0.8
0.1 kmc			
For loss of 1/2 Energy	30 years	1 year	1/30 year
Electrons/cm³ in belt	2×10^{-2}	2×10^{-3}	2×10^{-4}

* Critical $E = 0.55$ the energy such that maximum emission power per cycle per second occurs at λ.

Recent data presented by BERGE and MORRIS (1965), however, indicates that a thin shell approximation is of dubious applicability and, furthermore, the observed distribution of electron pitch angles may be different than that considered by Chang and Davis. It seems probable that the beautiful decimetric interferometer results (ROBERTS and KOMESAROFF, 1964) now becoming available will greatly clarify the morphology of the Jovian belts, but substantial work remains to be done. The decametric radiation from Jupiter is very complicated in both the time and frequency domain. Extensive work by WARWICK (1963) and by Ellis (see for instance, ELLIS and MCCULLOUGH, 1963) and co-workers attempts to account for the decametric radiation by means of Čerenkov and cyclotron radiation respectively. Indeed many features of the decametric radiation can be accounted for by these theories and excellent correlations had been obtained between the time dependent behavior of certain aspects of a decimetric cyclotron radiation and the decametric bursts. This correlation has been so precise as to lead to the definition of a new system of coordinates (λ_{III}) (MORRIS and BERGE, 1962) with a precise rotational period of nine hours, 55 minutes, 29.37 seconds on which both decimetric and decametric phenomena are anchored. The rocking of the plane of polarization which has been found at decimetric wavelengths is very satisfactorily explained by such anchoring combined with a tilt in the dipole orientation relative to the rotational axis of about nine degrees. In addition, Warwick's theories have defined the position of the asymmetric dipole apparently required to account for the time dependence and other properties of the decametric emission. The more recent work of Berge and Morris cited earlier, however, seems to require that the dipole be in a different octant of the planet than prescribed by Warwick's theory. This may imply that the decametric radiation propagates directly from its point of origin rather than reflecting off the planet's surface or ionosphere as assumed by Warwick. It is probably impossible as yet to make a clear decision as to whether Čerenkov or cyclotron processes are the seat of the decametric radiations and, in fact, other mechanisms such as the amplified whistlers considered by Chang may be important (CHANG, 1963).

One must proceed somewhat cautiously in accepting a magnetic moment for Jupiter of 4.2×10^{30} gauss cm^3, as arrived at by Warwick's theory. The basis of the calculation is essentially geometric with the moment set by using the approximate upper limit of observed decametric burst frequencies. Nevertheless, the analysis does lead to a field strength of 0.43 gauss at the shell $L=3$. This decametric result, of course, falls nicely into the middle of the range of magnetic fields considered by Chang and Davis, who worked with decimetric radiation as shown in Table I.

Any theory of the Jovian radiation belt and any inferences concerning its properties must account for the recent and somewhat startling discovery that the Jovian moon Io exerts a strong influence on Jupiter emission (BIGG, 1964).

It is to be observed that an important element in both the Warwick and Ellis theories of the decametric radiation has been some sort of geometric effect to account for the strong dependence on Jovian longitude of the properties of the radiation as received at the earth. In Warwick's theory, a specular reflection of the Čerenkov radiation is invoked along with relatively narrow cones of emission. In Ellis's theories a "generalized Doppler" equation is applied to electrons presumed to be emitting cyclotron radiation and, when the effect of a suitable Jovian ionosphere is included, a geometric beaming of the emission develops which again is argued to account for the results as seen from earth. Neither theory, of course, has considered that emission could be somehow triggered by Io and that this could be a predominant factor in controlling the time dependent character of the radiation. The precise consequences of the Io effect for our purposes – that is, its effect on estimates of the magnetic field of Jupiter and the properties of any associated radiation belt – are far from clear. Its effect, however, on the magnetic moment quoted above may be quite significant since the calculation was based on processes originating at $L=3$, whereas Io orbits at some 6.6 radii from the center of the planet.

The current status of this rather dynamic situation is well summarized in a recent paper by DULK (1965). Following up on Bigg's original work, he finds that Io can induce emission into the ecliptic plane when it is over the hemisphere of Jupiter bounded by $110° < \lambda_{III} < 290°$. The implication is that over the other hemisphere the radiation does not travel along the ecliptic and hence does not reach the earth. This is consistent with the picture of a tilted and displaced Jovian dipole suggested by other evidence as described earlier. Previous decimetric and decametric results have indicated that $\lambda_{III} \cong 200°$ is the plane containing both the asymmetric dipole and the rotational axis of the planet. Indeed, Dulk derives an emission pattern for Io-related events on Jupiter which exhibits two 40-degree cones of emission which may even originate from a common source. The emission cones are activated once every 12.9 hours near the time when Jupiter's magnetic axis sweeps passed Io. Both broad band and narrow band emission are observed and it is reasonable to suppose that the former corresponds to an effect over a fairly large longitude range, a fairly thick L shell, or both, while the latter would involve a small group of field lines quite near the satellite.

Presumably, Io disturbs the orbits of particles trapped on field lines and this, in turn, results in decametric emissions. The character of the various emissions produced suggests that Io has either an internally-generated or an induced magnetic field and magnetospheric tail. Using Io as a probe for study of the Jovian magnetic field and its associated trapped particles, we shall no doubt learn considerably more about the planet and also acquire further insight into important processes occurring in the earth's own radiation belt as well.

References

Alfvén, H.: 1963, *Icarus* **1**, 357.

Berge, G. L. and Morris, D.: 1965, Private communication – to be published.

Bigg, E. K.: 1964, *Nature* **203**, 1008.

Carr, T. D. *et al.*: 1961, *Astrophys. J.* **134**, 105.

Chang, D. B. and Davis, L.: 1962, *Astrophys. J.* **136**, 567.

Chang, D. B.: 1963, *Astrophys. J.* **138**, 1231.

Davis, R. D.: 1964, *Thirteenth Intern. Astron. Union General Assembly,* Hamburg.

Dulk, G. A.: 1965, *Science* **148**, 1585.

Ellis, G. R. A. and McCullough, P. M.: 1963, *Australian J. Phys.* **16**, 380.

Goldstein, R. M. and Carpenter, R. L.: 1963, *Science* **139**, 905.

Kellerman, K. I.: 1965, *Nature* **205**, 1091.

Kellerman, K. I.: 1965, *Nature* **206**, 1033.

McElheny, V. K.: 1964, *Science* **146**, 627.

Morris, D. and Berge, G. L.: 1962, *Astrophys. J.* **136**, 276.

Peale, S. J. and Gold, T.: 1965, *Nature* **206**, 1240.

Peebles, P. J. E.: 1964, *Astrophys. J.* **140**, 328.

Roberts, J. A.: 1963, *Planet. Space Sci.* **11**, 221.

Roberts, J. A. and Komesaroff, M. M.: 1964, *Nature* **203**, 827.

Smith, A. G. *et al.*: 1965, *Astrophys. J.* **141**, 475.

Van Allen, J.: 1965, Private communication and paper "University of Iowa Measurements, 1958–1965."

Warwick, J. W.: 1963, *Astrophys. J.* **137**, 41.

Warwick, J. W.: 1964, "Radio Emission from Jupiter," *Ann. Rev. Astron. Astrophys.* **2**, 1.

OUTER ZONE ELECTRONS

DONALD J. WILLIAMS*
The Johns Hopkins University, Applied Physics Laboratory, Silver Spring, Maryland

Introduction

Satellites in low altitude, high inclination orbits have been used to study near-earth effects of perturbations occurring well out in the magnetosphere by observing the spatial and temporal behavior of energetic electrons. Studies concerned with $\geqslant 40$ keV electrons include: the initial observation of electrons precipitated into the atmosphere from altitudes >1000 km (O'BRIEN, 1962); the initial observation of a diurnal variation in the trapped electron population (O'BRIEN, 1963); further, more detailed studies of trapped and precipitated electrons and their diurnal variations (O'BRIEN, 1964; O'BRIEN and TAYLOR, 1964; MCDIARMID and BURROWS, 1964 a and b; FRANK *et al.*, 1964; ARMSTRONG, 1965); and the recent observation of the existence of "spikes" of 40 keV electrons at latitudes above the boundary of the outer zone and preferentially during local night (MCDIARMID and BURROWS, 1965). A recent study at 10 keV has shown that intense fluxes of these lower energy electrons occur predominantly at local night and high latitudes (FRITZ and GURNETT, 1965).

We present and review in this paper the results of an initial series of studies concerned with the observation and interpretation of the spatial and temporal behavior of high energy ($E_e \geqslant 280$ keV) trapped electrons in the outer zone. These observations were obtained at low altitude and high latitudes from the polar orbiting satellite 1963 38C.

Differences in the behavior of the lower energy ($\gtrsim 40$ keV) and higher energy ($\gtrsim 280$ keV) electrons have been noted previously (O'BRIEN, 1964; WILLIAMS and MEAD, 1965). It is possible that an improved understanding of the behavior of the low energy electron population may be achieved by a fuller understanding of the high energy electron population. We will present evidence, using the $\gtrsim 280$ keV electron data obtained from instruments aboard satellite 1963 38C, which leads us to believe that a few characteristic features of the behavior of the high energy electron population in the outer zone can be understood if the distorted geomagnetic field is taken into consideration.

* Now at the NASA Goddard Space Flight Center, Greenbelt, Maryland.

The data concerning $\geqslant 280$ keV electrons will be presented in three sections: Dayside variations which show a striking 27 day correlation in trapped electron intensities and appear to be closely related to the passage of the sector boundaries of the interplanetary field as observed by NESS and WILCOX (1965); Diurnal variaions which are shown to be consistent with charged particle motion, under conservation of the adiabatic invariants, in a distorted field similar to that measured by IMP-1 (NESS *et al.*, 1964); and Storm-time variations which show that the behavior of these trapped electrons during a main phase magnetic storm is consistent with the simultaneous behavior of the magnetic field in the earth's magnetic tail (NESS and WILLIAMS, 1966).

Satellite and Detector

The satellite 1963 38C and the detector of interest have been described in detail previously (WILLIAMS and SMITH, 1965).

Briefly, the satellite 1963 38C was launched on September 28, 1963 into a nearly circular polar orbit having a 1140 km apogee, a 1067 km perigee, an 89.9° inclination and a 107.5 min. period. Just after launch the satellite orbital plane made an angle of $\sim 6°$ with the noonmidnight meridian and was moving toward the noon-midnight meridian at the approximate rate of 1° per day due to the earth's motion about the sun. The satellite was magnetically aligned and displayed an oscillation of $\lesssim 6°$ about the local line of force some three days after launch.

The experiment of interest is an integral electron spectrometer comprised of five, one millimeter thick, surface barrier solid state detectors. A combination of discriminator levels and absorbing foils yields the spectrometer characteristics shown in Table I. Monitoring an onboard proton spectrometer has shown that proton contamination in these outer zone electron data is negligible.

The spectrometer is oriented to observe the satellite alignment axis. Therefore, after alignment has been achieved the instrument measures the intensity of trapped electrons mirroring at (or very near) the point of observation.

Results from the lower energy channels, $E_e \geqslant 280$ keV and $\geqslant 1.2$ MeV, will be presented in this report. Directional flux levels ($\#/\text{cm}^2$ sec ster), accurate to $\sim \pm 50\%$, can be obtained by multiplying the $\geqslant 280$ keV and $\geqslant 1.2$ MeV count rates by 500 and 1000 respectively.

As of July, 1965 satellite 1963 38C was still in operation and transmitting data of high quality.

Dayside Variations

In this section we shall present some data and correlate its behavior with other phenomena. We do not know the mechanisms responsible for the observed particle behavior in this case, even though we can qualitatively predict this behavior given the behavior of an additional parameter, e.g., the K_p values.

TABLE I

Electron Spectrometer Characteristics: Oriented 90° to Satellite Alignment Axis

Detector	*Half Angle (deg)*	*Geometric Factor (cm² ster)*	*Foil Thickness (mg/cm²)*	*Particle Energies yielding pulses between 250 keV and 1 MeV discriminator levels*		
				Electrons	*Protons (energies in MeV)*	
1	6.4	$2.8\,(10)^{-3}$	10.3 AL	$\geqslant 0.28$ MeV	$2.0 \leqslant E_p \leqslant 2.3$	$E_p \geqslant 178$ MeV
2	6.4	$2.8\,(10)^{-3}$	412 Cu	$\geqslant 1.2$ MeV	$14.4 \leqslant E_p \leqslant 14.5$	$E_p \geqslant 179$ MeV
3	6.4	$2.8\,(10)^{-3}$	946 Cu	$\geqslant 2.4$ MeV	$23.27 \leqslant E_p \leqslant 23.34$	$E_p \geqslant 181$ MeV
4	6.4	$2.8\,(10)^{-3}$	1470 Cu	$\geqslant 3.6$ MeV	$30.01 \leqslant E_p \leqslant 30.70$	$E_p \geqslant 183$ MeV
5	3.2	$4.9\,(10)^{-4}$	10.3 AL	$\geqslant 0.28$ MeV	$2.0 \leqslant E_p \leqslant 2.3$	$E_p \geqslant 178$ MeV

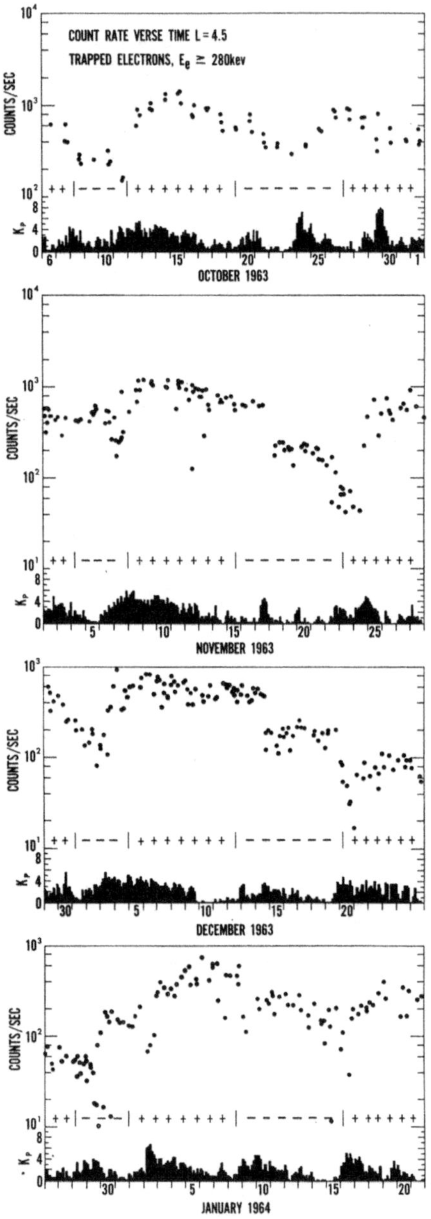

Fig. 1. Trapped electron intensities for the four consecutive solar rotations beginning October 6, 1963. The data are for $L = 4.5$ and 1100 km. The occurrence of the interplanetary magnetic field sector boundaries is also shown, $+$ = field away from sun, $-$ = field toward sun (NESS and WILCOX, 1965). The trapped electron intensities are seen to display a clear 27-day variation with increasing intensities occurring as a sector boundary sweeps past the earth. There is also evidence for a second increase occurring $\sim 180°$ (13 1/2 days) after the initial increase in a given rotation period.

It appears that even to eliminate certain mechanisms from consideration, simultaneous observations are required, not only of magnetic and particle phenomena, but also at various positions on lines of force throughout the magnetosphere. Hopefully, in the future more use will be made of simultaneous observations to track down the mechanisms responsible for the charged particle behavior throughout the trapping regions.

An initial study of trapped, ⩾280 keV electron intensities, observed on the sunlit hemisphere by satellite 1963 38C, was presented by WILLIAMS and SMITH (1965). A correlation of trapped electron intensities with magnetic activity (enhanced electron intensities correspond to high K_p values) was presented along with the characteristics of the electron responses to magnetic activity as a function of the magnetic shell parameter, L. Time delays in the electron response to magnetic activity were obtained and found to increase with both increasing energy and L. A rough correlation of the energy spectrum with magnetic activity was obtained, showing that the spectrum becomes softer as the magnetic activity increases.

It was also observed that during times of magnetic quiet, the trapped electron intensities exhibit a steady decay, implying that a steady solar wind does not supply fresh particles to these high latitude ($L \geqslant 3$), low altitude trapping regions. Perturbations in the trapped particle population seem to occur mainly when the solar wind and/or the interplanetary field direction changes.

It was further observed that electron intensity increases occurred during the time of the 27-day recurring increase in magnetic activity. However, only one solar rotation period was included in the analysis and no definite correlation with a 27-day period could be made.

We have now obtained the trapped, ⩾280 keV electron data at 1100 km on the sunlit hemisphere for the first four solar rotation periods after the launch of 1963 38C. The solar rotations analyzed are the four beginning on October 6, 1963, November 2, 1963, November 29, 1963, and December 26, 1963. The data show a definite 27-day periodicity in that the high energy ($E_e \geqslant 280$ keV) trapped electrons throughout much of the outer zone ($L \geqslant 3.5$ at 1100 km) undergo an intensity increase every 27 days coincident with the recurring magnetic activity.

This is shown in Figures 1 and 2 where the trapped electron intensities at $L = 4.5$ are displayed for $E_e \geqslant 280$ keV and $E_e \geqslant 1.2$ MeV respectively, for the four consecutive solar rotations beginning October 6, 1963. The data have been reduced as intensity (counts per sec) versus time plots at half integral values of L from $L = 3.0$ to $L = 8.0$. The $L = 4.5$ shell is shown as being typical of the center of the outer zone. Shown along with the electron data are the 3 hour averages of K_p. The correspondence between the electron intensities and magnetic activity is readily discernible.

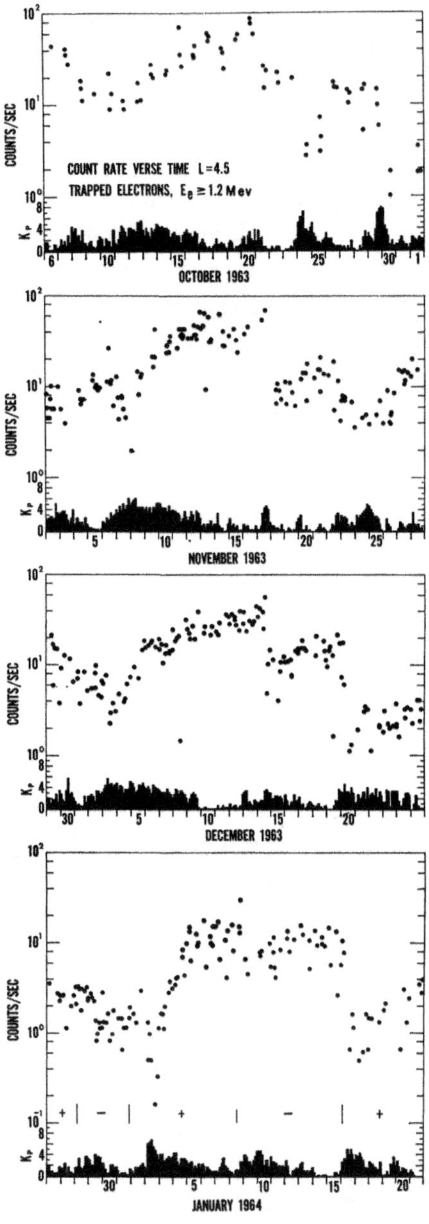

Fig. 2. Same as Figure 1: $L = 4.5$ $E_e \geqslant 1.2$ MeV. Note time lag between appearance of these higher energy electrons and the lower energy electrons in Figure 1.

During the latter two solar rotation periods shown in Figures 1 and 2, November 29 and December 26, 1963, the magnetometers aboard the NASA IMP-1 satellite were able to monitor the interplanetary magnetic field. For these two solar rotations plus the succeeding rotation beginning January 22, 1964, NESS and WILCOX (1965) have reported the existence of a regular longitudinal sector structure in the interplanetary field. The total longitudinal structure corotates with the sun and is separated into four sectors. In two of these sectors, each occupying about 2/7 of the total longitude, the field is directed away (+) from the sun. In the two remaining sectors, 2/7 and 1/7 of the total longitude, the field is directed toward (−) the sun. The sector boundaries are neutral sheets separating regions of oppositely directed field and produce, as they sweep past the earth, a rapidly changing transitional field configuration which may interact with the magnetosphere and its environs.

The sector boundaries as measured by NESS and WILCOX (1965), are included in the plots of Figures 1 and 2, and it can be seen that the increase in trapped electron intensities observed near the beginning of a solar rotation, occur at or just after the arrival of a sector boundary. It is interesting to note that for the rotations of November 2, November 29, and December 26, 1963, there is an additional particle increase at the arrival of the sector boundary located $\sim 180°$ from the initial particle increases. The passage of both of these boundaries produces a similar transitional field, i.e., the field changes from − to +. The latter portion of the October 6 rotation was obscured by two large storms and thus does not exhibit this effect.

These results show clearly the fact that the trapping region is very sensitive to conditions in the interplanetary medium. Whether the interplanetary field configuration or the solar plasma intensity is the primary trigger for these magnetospheric and particle intensity responses is not yet known. It is somewhat surprising, though, that these mild perturbations (compared to the storms near solar-maximum) are able to significantly alter the relativistic electron population deep in the magnetosphere ($L = 3.5$).

Finally, we show in Figure 3 a plot of the trapped, $\geqslant 280$ keV electron intensity at $L = 3.0$ for these solar rotations. No discernable response can be seen at the sector boundaries. Only intense magnetic activity yields particle increases at $L = 3$. This occurs for two storms at the end of October, 1963 and not again until January, 1964. During the relatively quiet periods, the intensities show a steady decay right to the detector count rate threshold.

Diurnal Variations

Measurements of the diurnal shift of trapped, $\geqslant 280$ keV electrons in the outer zone (WILLIAMS and PALMER, 1965) showed that these higher energy electrons

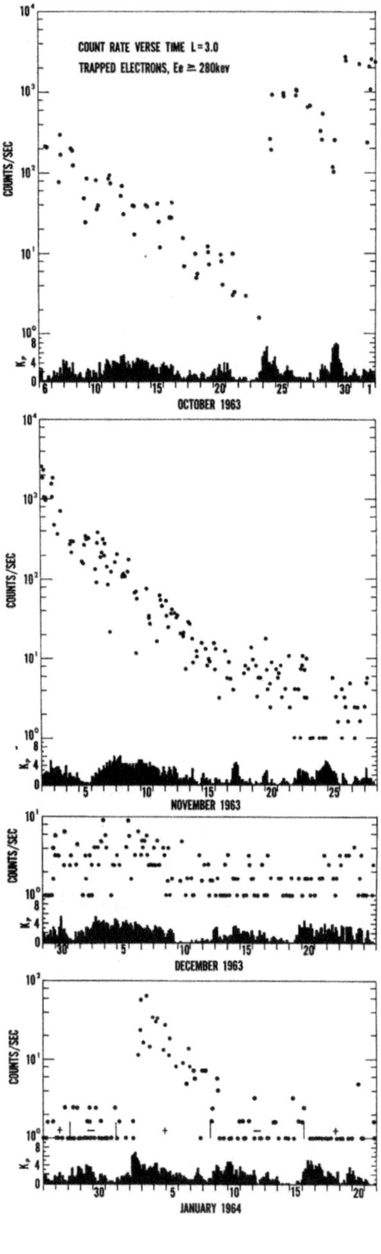

Fig. 3. Same as Figure 1: $L = 3.0$ $E_e \geqslant 280$ keV. Note steady decay except for occurrences of sudden intense magnetic activity.

display a significantly smaller diurnal latitude shift during periods of magnetic quiet than do trapped, $\geqslant 40$ keV electrons in the outer zone. An initial qualitative analysis by WILLIAMS and PALMER (1965) suggested that the diurnal shift of the trapped, $\geqslant 280$ keV electrons might be explained by particle motion, under conservation of the adiabatic invariants, in a distorted magnetosphere such as described by MEAD (1964).

A recent, more detailed quantitative study of these diurnal shifts (WILLIAMS and MEAD, 1965) has shown that the addition of a current sheet in the anti-solar hemisphere of Mead's model is found to fit the observed latitude shifts, assuming charged particle motion under the conservation of the adiabatic invariants, μ and J, and the energy E. The addition of this region of enhanced plasma density in the tail field leads to a quiescent field configuration similar to that recently suggested by DESSLER and JUDAY (1965) and AXFORD *et al.* (1965) and recently measured by the magnetometer aboard IMP-1 (NESS, 1965).

We shall briefly review the investigations of WILLIAMS and MEAD (1965) leading to the result that the behavior, during periods of magnetic quiet, of energetic ($E_e \geqslant 280$ keV) electrons trapped in the outer zone is consistent with their motion in a distorted magnetosphere similar to that suggested and observed, under the conservation of the adiabatic invariants, μ and J, and the energy, E. A recent study of trapped high energy ($E_e \geqslant 1.6$ MeV) electrons near the geomagnetic equator has led to a qualitatively similar result (FRANK, 1965). In the following section we shall discuss some characteristic features of outer zone electron behavior during magnetically disturbed periods.

Figure 4 shows the noon and midnight latitude profiles, displayed as count rate vs. invariant latitude plots, obtained for the magnetically quiet period October 2 through October 12, 1963. The curves of Figure 4 represent the average values obtained for 47 dayside passes and 28 nightside passes received during this time period (WILLIAMS and PALMER, 1965). In this time interval, the satellite orbital plane remained within 8° of noon-midnight meridian.

It can be seen from Figure 4 that both energies, $E_e \geqslant 280$ keV and $\geqslant 1.2$ MeV, behave in a similar manner and show a distinct separation of the noon and midnight latitude profiles. To test the hypothesis that this separation is simply a latitude shift of the trapped electron population due to their drift in a distorted magnetosphere under conservation of μ, J and E, the amount of latitude shift was obtained as a function of noon time latitude by obtaining the noon, Λ_D, and midnight, Λ_N, latitudes which yield the same trapped electron intensity.

The results are shown in Figure 5 where the amount of shift, $\Delta\Lambda = \Lambda_D - \Lambda_N$, is plotted versus the noontime latitude of observation, Λ_D. Since the $\geqslant 1.2$ MeV electron data displayed the same behavior, only the $\geqslant 280$ keV electron data are shown. Two methods of analyzing the data, for reasons of accuracy

(WILLIAMS and MEAD, 1965), are shown in Figure 5. The matched pass data are from pairs of passes, each pair of which yields data continuous in time that traces out both noon and midnight latitude profiles within a period of ~30 minutes. This is as close as one can come with one satellite to simultaneous

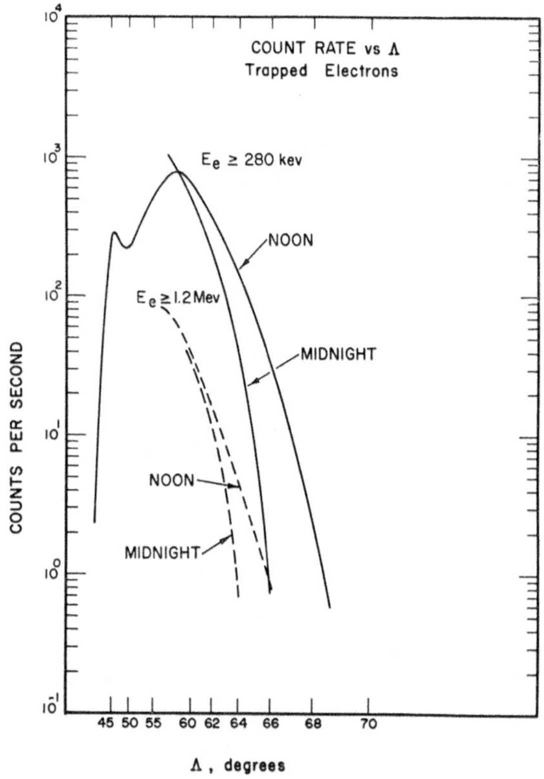

Fig. 4. Day and night count rate versus Λ plots for the magnetically quiet period October 2–12, 1963. Curves are shown for both $E_e \geqslant 280$ keV and $\geqslant 1.2$ MeV. Λ is defined at satellite altitude: $\cos \Lambda = \sqrt{1.17/L}$. (From WILLIAMS and MEAD, 1965.)

observation of the noon and midnight profiles and eliminates a great deal of scatter due to time variations associated with magnetic activity.

It is now possible to fit this observed diurnal variation by adjusting the magnetic field configuration to yield the appropriate drift paths while conserving μ, J and E. This was done by using a dayside configuration (MEAD,

1964) which agrees with experimental observations (NESS *et al.*, 1964) and adjusting the nightside configuration to fit the results of the charged particle observations.

The solid line through the data in Figure 5 shows the predicted amount of shift from the model magnetosphere considered to yield the most reasonable

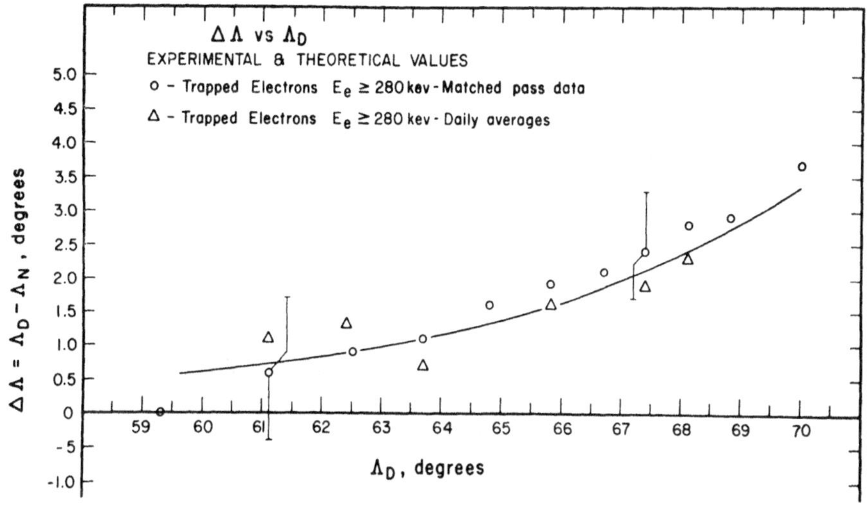

Fig. 5. Plot of latitude shift, $\Delta\Lambda = \Lambda_D - \Lambda_N$, versus noontime latitude, Λ_D. Bars on data points represent entire spread of data seen during the period October 2–12, 1963. Solid curve is the predicted latitude shift obtained from the magnetosphere model of Figure 6. (From WILLIAMS and MEAD, 1965.)

results of the configurations tested. The resultant magnetosphere is shown in Figure 6. It is seen to contain a current sheet coincident with the nightside magnetic equator which separates solar directed fields in the northern hemisphere from anti-solar directed fields in the southern hemisphere. The dashed lines show the field due to the current sheet alone, B_{CS}, while the solid lines give the resultant field consisting of B_{CS}, the earth's dipole field (B_d), and the field due to magnetospheric boundary currents (B_S). The midnight meridian high latitude trapping boundary, defined by field line closure, occurs at 67° at an altitude of 1100 km and agrees with the observed nightside boundaries for both $\geqslant 40$ keV and $\geqslant 280$ keV electrons.

The current sheet shown in Figure 6 extends from 10 R_e (earth radii) to 40 R_e and has a strength of 40 γ immediately adjacent to the sheet. These parameters cannot be uniquely determined by this type of analysis. We therefore

TABLE II

Current Sheet Characteristics

Inner Edge (R_e)	*Outer Edge (R_e)*	*Strength (gamma)*
10	40	40
8	40	33
8	100	23

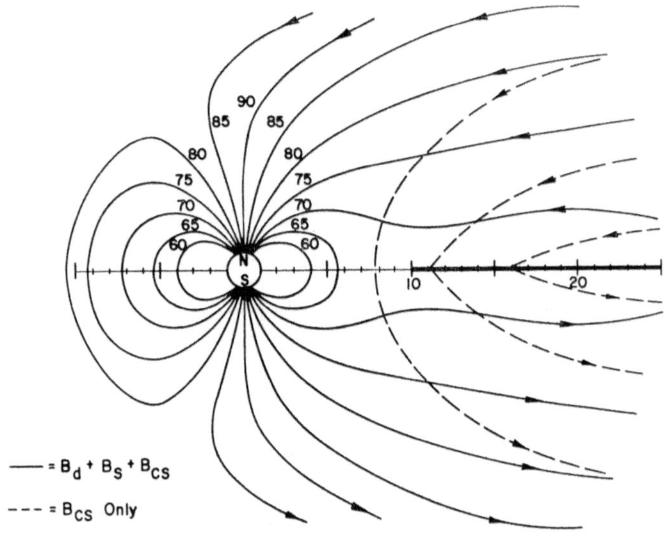

Fig. 6. Magnetospheric configuration fitting the trapped electron diurnal variation observations. Solid lines show field lines resulting from addition of a current sheet in the tail of the field model originally discussed by MEAD (1964). Dashed lines show field lines due to the current sheet alone. (From WILLIAMS and MEAD, 1965.)

show in Table II additional current sheet configurations which will fit the observed diurnal variations. Note that these field strengths are somewhat larger, on the average, than those observed from $\sim 10\,R_e$ to $30\,R_e$ in the tail (NESS, 1965).

This field configuration, its characteristics and shortcomings have been discussed in some detail by WILLIAMS and MEAD (1965). Improvements include the use of a more appropriate current sheet in the model, the inclusion of a possible ring current, and a readjustment of the coefficients defining the field due to the boundary currents.

However, the similarity of the field shown in Figure 6 to recent observations (NESS, 1965) and recent theoretical suggestions (DESSLER and JUDAY, 1965; AXFORD *et al.*, 1965) leads us to conclude that during periods of magnetic quiet, energetic ($E_e \geqslant 280$ keV) trapped electrons in the outer zone behave in a manner consistent with their drift in the distorted geomagnetic field, under conservation of the adiabatic invariants, μ and J, and the energy, E.

We next present observations of the storm time behavior of energetic outer zone electrons which demonstrate the close interplay existing between the trapping regions and the extended geomagnetic tail.

Storm Time Variations

Early measurements of the high latitude boundary of trapped, $\geqslant 40$ keV electrons at low altitude showed that this boundary moves to lower latitudes during periods of intense magnetic activity (MAEHLUM and O'BRIEN, 1963). This effect seems to occur for 40 keV electrons, on both the dayside and night-side hemispheres. A movement to lower latitudes during high magnetic activity, of the midnight meridian high latitude boundary of trapped, $\geqslant 280$ keV electrons has also been reported (WILLIAMS and PALMER, 1965). However, for these higher energy electrons, the noontime high latitude boundary does not seem much disturbed throughout the period of magnetic activity. Sometime after the peak of activity (a few days), the noontime trapping boundary appears to move slightly to higher latitudes, an effect which may be simply a manifestation of an overall increase in trapped particle intensities (WILLIAMS and PALMER, 1965).

The initial results of a study now underway have shown that during a magnetic storm the midnight latitude profile of trapped electrons at 1100 km behaves in a manner consistent with the behavior of the magnetic tail field at 30 R_e (NESS and WILLIAMS, 1966). These results are obtained by using simultaneous observations of trapped electrons by the APL satellite 1963 38C and of the magnetic field by the NASA satellite IMP-1. We shall review these results and present evidence that this storm time behavior appears to be a predictable pattern.

Figure 7 shows a sequence of midnight meridian passes, again displayed as count rate versus latitude plots, obtained prior to, during, and after a period of magnetic activity occurring on March 30, 1964. A plot of the planetary K_p index is given at the top of the Figure. We see that passes *a* and *b*, obtained prior to the disturbance, display latitude profiles similar to those observed during periods of magnetic quiet (WILLIAMS and PALMER, 1965; WILLIAMS and MEAD, 1965). Pass *c*, obtained during the disturbance, shows a distinct collapse of the outer boundary to lower latitudes. Pass *d*, obtained after the disturbance, shows that the boundary has moved back toward the pre-storm value.

A sequence of passes analyzed earlier for the magnetic disturbance occurring on April 1, 1964, shows the same characteristic behavior as above (NESS and WILLIAMS, 1966). In both cases (the storms of March 30, 1964 and April 1, 1964) no new particles had appeared at the time of the collapse of the outer

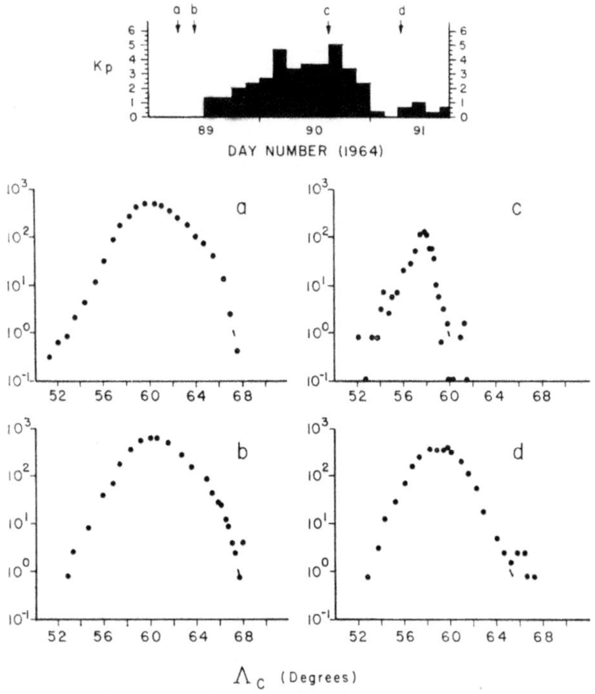

Fig. 7. Four passes displayed as count rate vs. Λ plots obtained near midnight meridian showing collapse of outer boundary during magnetic storm. Pass times indicated on the K_p plot are *a*: 0633 h March 29 (Day 89), 1964; *b*: 1001 h March 29 (Day 89), 1964; *c*: 1508 h March 30 (Day 90), 1964; *d*: 0651 h March 31 (Day 91), 1964.

boundary to lower latitudes. However, after the disturbance (passed in Figure 7), fresh particles appear throughout these trapping regions.

The behavior of the midnight high latitude boundary for trapped $\geqslant$280 keV electrons (defined at 1 count per sec and shown as small dashes in Figure 7) is shown in Figure 8 for the period March 29 through April 2, 1964. A plot of the K_p index is included. The dramatic lowering of the midnight trapping boundary during magnetically active periods is clearly observed. During this period, the satellite was within 4° of the midnight meridian.

The collapse of the trapping boundary to lower latitudes during a magnetic disturbance may be caused by formerly closed lines of force being extended into the tail region during the disturbance. Particles trapped on these field lines will be thus injected into the distant magnetic field, possibly forming some of the

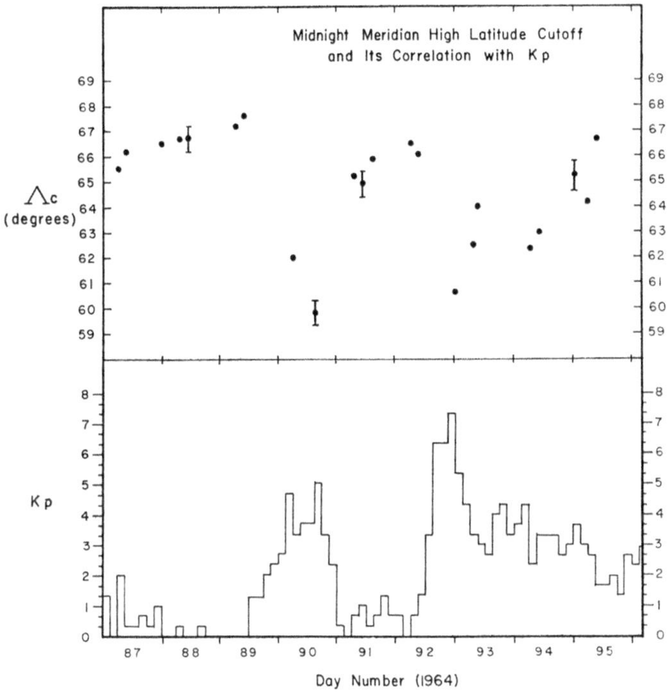

Fig. 8. Behavior of the outer trapping boundary (defined at 1 count per sec) during the period March 29 through April 2, 1964 and its correlation with K_p.

electron "patches" observed recently in these regions (ANDERSON *et al.*, 1965).

Analyzing the data for the April 1, 1964 disturbance, NESS and WILLIAMS (1966) found that coincident with the surface magnetic activity and the trapped electron boundary collapse, the field strength at 30 R_e in the tail increased. Using a time history of the field magnitude in the tail, it was possible to obtain a midnight trapping boundary for electrons by analyzing field line closure in the field model of Figure 6 (WILLIAMS and MEAD, 1965). These results are reproduced here in Figure 9.

The tail field magnitude is given by three hour averages and is seen to strongly correlate with K_p. The disturbance on April 1, 1964 is clearly seen in the tail field magnitudes and the K_p values. The predicted midnight trapping boundary is shown and was obtained using the following current sheet para-

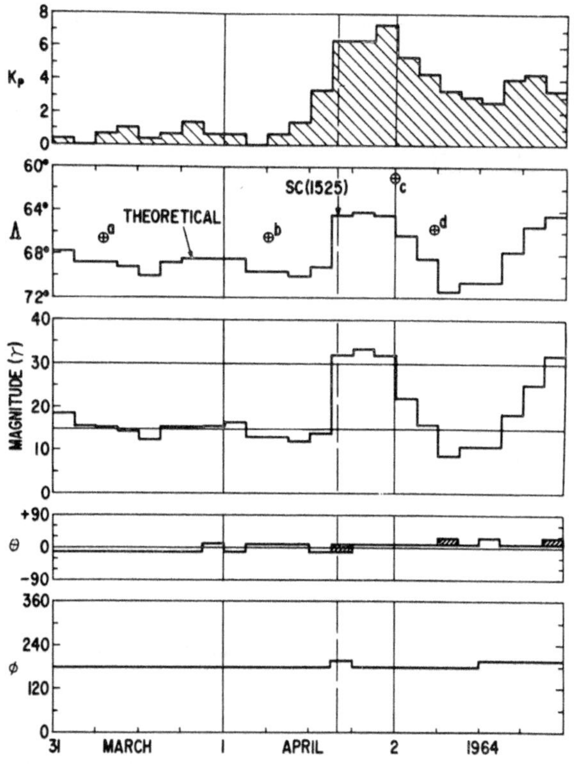

Fig. 9. Plot showing K_p variation, tail field magnitude and direction at 30 R_e and the predicted and observed outer trapping boundary throughout the disturbance of April 1, 1964. (From Ness and Williams, 1966.)

meters: inner edge, 8 R_e; outer edge, 200 R_e and field strengths as shown in Figure 9. The measured trapping boundaries *a*, *b*, *c* and *d*, correspond to passes obtained at *a*, 0651 h 31 March 1964; *b*, 0608 h 1 April 1964; *c*, 0000 h 2 April 1964; *d*, 0526 h 2 April 1964.

While the absolute position of the predicted and measured boundaries differ by an average of about five degrees, it is seen that the magnitude of the boundary collapse during the disturbance can be explained by the behavior of the tail field. The discrepancy in absolute position has been noted and discussed previously (WILLIAMS and MEAD, 1965; see model improvements listed in previous section).

We have thus seen that, using a tail field model of the magnetosphere and assuming motion of charged particles conserving μ, J and E, the behavior of the midnight latitude profile for energetic electrons as observed at 1100 km is consistent with the behavior of the magnetic field some 30 R_e in the tail. Moreover, the magnitude of the latitude shift is in fair quantitative agreement with the results of a field model using a tail like structure.

The data of Figures 7 and 8 imply that this is a predictable effect and is not a special event unique to the April 1, 1964 disturbance. A general study is now underway using all available simultaneous data from the satellites 1963 38C and IMP-1.

Concluding Remarks

Evidence of a 27 day cycle in the intensities of energetic ($E_e \geqslant 280$ keV) trapped electrons has been presented. This effect occurs throughout much of the outer zone, $L \geqslant 3.5$. Occurring near the minimum of solar activity, this effect shows clearly the sensitive interplay between the radiation cavity and the properties of the solar wind and of the interplanetary magnetic field. The correlation of these intensity increases with the passage of the recently observed interplanetary magnetic field sectors (NESS and WILCOX, 1965) was also demonstrated.

It was further shown that the behavior of trapped, energetic electrons during periods of magnetic quiet is consistent with their motion in the distorted geomagnetic field, under conservation of the adiabatic invariants. Using this result, it was then shown that, during magnetic storms, the observed behavior of the midnight meridian high latitude trapping boundary of $\geqslant 280$ keV electrons is consistent with the observed behavior of the tail field.

Many problems still exist in connection with these trapped energetic electrons. Where do they come from? Are the observed intensity increases due to injection at low energy from the solar wind into the magnetosphere and subsequent acceleration within the magnetosphere, or are the intensity increases due to local acceleration mechanisms? By observing the spatial and energy dependence of the low altitude trapped electron intensity increases, one "feels" that no single mechanism is reponsible for these changes throughout the entire trapping region (WILLIAMS and SMITH, 1965). Rather, different mechanisms may dominate in various regions of the magnetosphere. What are some of these "acceleration mechanisms" capable of accelerating electrons to many MeV in energy? What are the loss processes causing the observed lifetimes in the outer zone (WILLIAMS and SMITH, 1965)? Why isn't the dayside trapping boundary for ⩾280 keV electrons much perturbed during magnetic disturbances? Can electric fields, capable of accelerating electrons some tens of keV but not hundreds of keV (TAYLOR and HONES, 1965), explain the many differences between the behavior of ⩾40 keV and ⩾280 keV trapped electrons (O'BRIEN, 1964; WILLIAMS and MEAD, 1965)?

With improved measurements and the availability of quantitative models for comparison, we hope to see many of the questions answered during the oncoming solar maximum. One of the most important tools in these analyses will be the use of simultaneous observations at various observation points throughout the magnetosphere.

Acknowledgments

This work was supported in part by the Bureau of Naval Weapons, Department of the Navy, under contract NOw 62-0604-c and in part by a National Aeronautics and Space Administration Grant.

References

ANDERSON, K. A., HARRIS, H. K., and PAOLI, R. J.: 1965, *J. Geophys. Res.* **70,** 1039.
ARMSTRONG, T.: 1965, *J. Geophys. Res.* **70,** 2077.
AXFORD, W. I., PETSCHEK, H. E., and SISCOE, G. L.: 1965, *J. Geophys. Res.* **70,** 1231.
DESSLER, A. J. and JUDAY, R. D.: 1965, *Planet. Space Sci.* **13,** 63.
FRANK, L. A., VAN ALLEN, J. A., and CRAVEN, J. D.: 1964, *J. Geophys. Res.* **69,** 3155.
FRANK, L. A.: 1965, University of Iowa Report 65-14, May 1965.
FRITZ, T. A. and GURNETT, D. A.: 1965, *J. Geophys. Res.* **70,** 2485.
MAEHLUM, B. and O'BRIEN, B. J.: 1965, *J. Geophys. Res.* **68,** 997.
MCDIARMID, I. B. and BURROWS, J. R.: 1964a, *Can. J. Phys.* **42,** 616.
MCDIARMID, I. B. and BURROWS, J. R.: 1964b, *Can. J. Phys.* **42,** 1135.
MCDIARMID, I. B. and BURROWS, J. R.: 1965, *J. Geophys. Res.* **70,** 3031.
MEAD, G. D.: 1964, *J. Geophys. Res.* **69,** 1181.
NESS, N. F., SCEARCE, C. S., and SEEK, J. B.: 1964, *J. Geophys. Res.* **69,** 3531.
NESS, N. F. and WILCOX, J. M.: 1965, Goddard Space Flight Center Report No. X-612-65-157, April 1965.
NESS, N. F.: 1965, *J. Geophys. Res.* **70,** 2989.
NESS, N. F. and WILLIAMS, D. J.: 1966, *J. Geophys. Res.* **71,** 1.
O'BRIEN, B. J.: 1962, *J. Geophys. Res.* **67,** 3687.
O'BRIEN, B. J.: 1963, *J. Geophys. Res.* **68,** 989.
O'BRIEN, B. J.: 1964, *J. Geophys. Res.* **69,** 13.
O'BRIEN, B. J. and TAYLOR, H.: 1964, *J. Geophys. Res.* **69,** 45.
TAYLOR, H. E. and HONES, E. W.: 1965, University of Iowa Report 65-10, April 1965.
WILLIAMS, D. J. and SMITH, A. M.: 1965, *J. Geophys. Res.* **70,** 541.
WILLIAMS, D. J. and PALMER, W. F.: 1965, *J. Geophys. Res.* **70,** 557.
WILLIAMS, D. J. and MEAD, G. D.: 1965, *J. Geophys. Res.* **70,** 3017.

Discussion

A spirited discussion took place for more than an hour, and it was principally concerned with the relationship between boundaries of trapping and magnetic lines of force. Several participants vigorously defended the identity of the two, using topological and other arguments. However, when the discussions were continued on later informal occasions, these protagonists generally conceded their errors. Therefore, while the occasion was historically interesting for conference participants, there seems little point in documenting such erroneous views any further in the permanent literature. The following summary is therefore an edited version of about ninety minutes discussion by several dozen scientists and it also includes a short note by McIlwain which he presented twice during the discussion but which did not immediately convince some of those present, but which is the basis of their later conversion. It certainly should be reproduced in the permanent literature.

O'Brien pointed out that the Injun I discovery of the diurnal variation of the boundary of trapping was one of the first indications of a local-time distortion of the magnetosphere and thence of the existence of a continuous solar wind. It is a fascinating concept to use Van Allen radiation at an altitude of 1000 km as a probe of conditions out at 100000 km.

Dungey endeavoured to find someone to answer the following simple question: if one artificially injected a particle just beyond the boundary of trapping on the night side, where would it go? No-one cared to answer! However Williams stated that during storm times, particles could be trapped on the day side which would then drift into the tail to be untrapped or lost on the night side, rather than vice versa.

An extensive discussion took place in which many participants argued about the behaviour of energetic particles near the trapping boundary as they drift in longitude around the earth. A simple question was posed as to whether, on the day and night side, the equatorial boundary of trapping is connected by a magnetic field line to the low-altitude boundary of trapping. Part of the problem in answering this experimentally appeared to be that, whereas the equatorial boundary on the dayside is coincident for all electron energies (coincident with the magnetopause), there is experimental indication that the low-altitude boundary is apparently energy dependent, in that the 40 keV boundary is several degrees above the 280 keV boundary. Dessler suggested that there is no real difference at low altitudes, but simply that the reported differences are due to observations of different energies at different seasons. But there was no quantitative proof of this. Several experimenters expressed the view that this apparent energy dependence might be largely an effect of lower sensitivity of the 280 keV detector, but Williams claimed that the effect was too large for such an explanation.

Mead stated that if one started two particles on the dayside, one with large pitch angle and one with small pitch angle, and let them drift to the dark side while conserving p and I, then the particle with large pitch angle could drift closer to the earth (it follows constant B) while the other could drift into the tail. He therefore concluded that equatorial and low-altitude boundaries of trapping need not coincide. The invariant surface may then cut through field lines by several R_e rather than follow them.

NORTHROP made the point that even if magnetic field lines are nominally open, particles can still be trapped on them if the intensity changes appropriately at very high altitudes.

PARKER emphasized that one should not define a magnetic field line by such energetic particles, but instead by magnetic fields and plasmas frozen on them.

MCILWAIN presented a pictorial representation of the diurnal effect on two lines of force in his usual L-B coordinate system in order to explain how pitch angle or mirror altitude must be considered in any discussion of longitudinal drift and day - night effects (see Note and sketch below).

Comment on the Local-Time Distortion of the Geomagnetic Field Viewed in B-L Space

C. E. MCILWAIN

In a static magnetic field with no electric field, B and L will organize directional intensities perpendicular to the field perfectly. In the distant magnetosphere, the shape of lines of force appear to be grossly distorted from that of a dipole field. *This may cause L to vary by relatively large amounts along lines of force.*

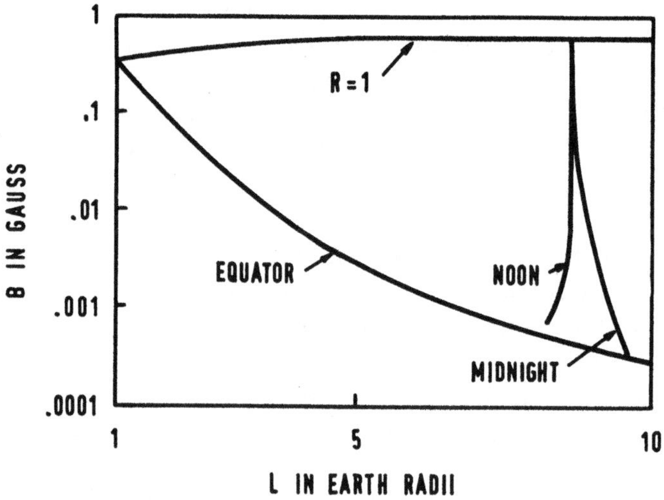

Fig. 1. The possible characteristic variations of L along lines of force in the noon and midnight meridians.

In particular, the lines of force in the noon and midnight meridians have very different shapes and it is to be expected that the variations of L along these lines of force will be different. If two lines of force are taken which have the same L value at the earth's surface, the variations may appear as shown in Figure 1.

Lines of force can be labelled by the L value at the magnetic equator: $L_0=(M/B_{min})^{1/3}$. It can be seen in the sketch that particles mirroring at high B values may drift through lines of force which are labelled by *quite different* L_0 values. Now particles initially mirroring at the magnetic equator will continue to do so on lines with the *same* L_0 values. It is thus concluded that the boundary of the magnetosphere defined by the lines of force with the same L_0 values may be quite different from the boundary defined by the lines of force through which particles with low altitude mirror points must drift in order to keep the L values of their mirror points constant.

The non-equivalence of the magnetic shells defined by particles with different mirror points travelling in magnetic fields without axial symmetry is well known (NORTHROP and TELLER). Within $L=5$ the differences are small and can usually be ignored. As pointed out above, the differences at higher L values may be large, in which case they cannot be ignored.

SESSION IV

SOURCES AND SINKS OF NATURAL TRAPPED RADIATION

Chairman

JAMES A. VAN ALLEN

University of Iowa

ORIGIN AND LOSS OF INNER ZONE PROTONS

A. M. LENCHEK

Department of Physics and Astronomy, University of Maryland, College Park, Maryland

1. Injection From Fast-Neutron Decay

A. INJECTION GEOMETRY

A trapped orbit may be thought of as a closed path $\oint ds$. Let $n\,dE\,dA\,d\Omega\,ds$ be the number of particles within the spatial volume $dA\,ds$ (where dA is an element of cross section drawn perpendicular to the velocity), with velocity vectors with the solid angle $d\Omega$ and with energies within dE. It can be shown that for adiabatic motion $dA\,d\Omega$ remains constant as a particle moves along the orbit (LENCHEK and SINGER, 1963). Since energy loss is negligible during a single bounce period, $dE\,ds$ remains constant. Since absorption and injection is negligible on a single orbit traversal, n is constant everywhere along the orbit. This quantity increases due to injection at the rate

$$\bar{q} \equiv \left(\frac{\partial n}{\partial t}\right)_{\mathrm{inj}} = \oint \mathbf{q}(s)\cdot d\mathbf{s} / \oint ds \tag{1.1}$$

where $\mathbf{q}(s)$ is the number of particles created per unit time within $dA\,ds\,d\Omega\,dE$. Since the neutrons generated by the decay of fast protons preserve the direction of the neutron,

$$\mathbf{q} = \mathbf{j}_n / t_n v \tag{1.2}$$

where $\mathbf{j}_n$ is the unidirectional, differential intensity of neutrons (n/cm^2 ster sec MeV), t_n is the decay lifetime (corrected for time dilation) and v is the neutron velocity. Note that $\mathbf{j}_n$ must be tangent to the orbit and directed in the same sense of motion as the trapped particle. The averages are over phase angle at fixed latitude and an average over longitude.

B. ALBEDO NEUTRONS

While much work, both theoretical and experimental, has been done on slow albedo neutrons, comparatively little has been done on fast neutrons ($E \gtrsim 10$ MeV).

WENTWORTH and SINGER (1955) estimated the energy spectrum of fast albedo neutrons by studying (a rather small amount of) data (CAMERINI, 1952)

McCormac (ed.), Radiation Trapped.

on the distribution of secondary particles in cosmic ray interactions in photographic emulsions. The analysis was limited to stars of the C, N, O group and yielded the spectrum of secondary *protons* (Figure 1). The neutron spectrum was assumed to be the same as the proton spectrum. Over the interval 50-400 MeV a power law was fitted to the curve. The result was a yield of $8E^{-1.8}$

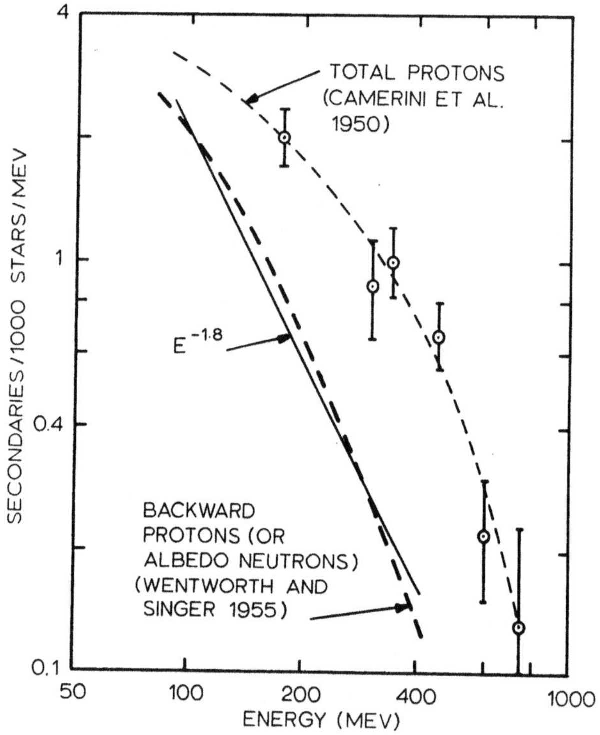

Fig. 1. Energy spectrum of fast cosmic ray neutrons calculated by WENTWORTH and SINGER (1955).

neutrons/MeV-star for the yield at $E>50$ MeV. (E is in MeV). Based on this result SINGER (1958) deduced a leakage flux $2E^{-1.8}$ neutrons MeV^{-1} cm^{-2} sec^{-1} above 50 MeV.

HESS (1959, 1960) and HESS *et al.* (1961) made very detailed diffusion calculations of *slow* neutrons in the atmosphere. For the energy spectrum of *fast* neutrons they used

$$\text{constant } E^{-2}e^{-160/E^2}$$

as an approximation to the data of Camerini and to several other proton spectra observed in high-energy experiments.

The energy spectrum of fast neutrons *above the atmosphere* should be steeper than the spectra within the atmosphere, i.e., at the source. Neutrons above ~ 10 MeV are generated predominantly by spallation, or knock-on, interaction. The products of such interactions are projected forward in the lab frame. The angular distribution of the secondaries becomes increasingly narrower with increasing primary energy. Therefore the higher energy neutrons are generated in a narrow angular distribution near the horizon. At a point outside the atmosphere the omnidirectional flux receives a proportionately larger contribution of lower energy neutrons, which arrive within a larger solid angle than higher-energy neutrons. The effect should produce an omnidirectional, differential energy spectrum approximately E^{-3} n/cm^2 sec MeV at $E > 50$ MeV (LENCHEK and SINGER, 1962).

Some support for neutron spectra *at the source* of either E^{-2} or $E^{-1.8}$ was provided by airplane measurements made with a bismuth fission counter (HESS *et al.*, 1959). The detector was sensitive to neutrons above 60 MeV, the fission taking place within a proportional ionization chamber. The pulse height spectrum depends (weakly) on the neutron spectrum, and was compatible with a spectrum *approximately* E^{-2}.

GAUGER (1964) flew the same detector in a series of flights at depths of 160 to 300 g/cm^2 and at geomagnetic latitudes from 0 to 55 °N. The flights were made in 1962. The neutron flux >60 MeV was found to be about six times smaller than that found by HESS *et al.* (1959) in 1957. (Since solar activity was much lower in 1962 than in 1957, the albedo intensity should have been higher during Gauger's flights than during those of Hess *et al.*) The energy spectrum was roughly $E^{-2.5}$.

TRAINOR and LOCKWOOD (1964) flew a moderated BF_3 counter on a polar orbiting satellite in 1962. At 300 km above the equator an albedo neutron flux 0.085 ± 0.03 n/cm^2 sec was measured. At the pole the flux was larger by a factor ~ 3. This latitude variation is much smaller than either of the latitude variations predicted by theory (HESS *et al.*, 1961; LINGENFELTER, 1963). The latter are ~ 4.5 to 1 and ~ 10 to 1 respectively.

NEWKIRK (1963) calculated the neutron flux in the atmosphere in the energy interval: thermal to about 20 MeV. He finds a slow neutron flux about three times smaller than that computed by HESS *et al.* (1961). The calculation gives no information on the fast albedo in the energy range of interest here. However, it is worth noting that Newkirk finds that the leakage flux due to these *slow* neutrons is 0.8 n/cm^2 sec at latitude 57 °N. This figure is about four times greater than the flux measured by TRAINOR and LOCKWOOD (1964).

LINGENFELTER (1963) has calculated the leakage flux by the multigroup diffusion theory. The results substantially agree with those of NEWKIRK (1963) for low energies. For energies greater than 10 MeV no calculations were made and the spectrum was assumed to be that of HESS *et al* (1961) (Figure 2). The neutron flux below 10 MeV was computed to be about three times smaller than that computed by HESS *et al.* (1961). The latitude variation assumed is about twice that used by HESS *et al.* (1961). The spectrum from 0.5 to 10 MeV

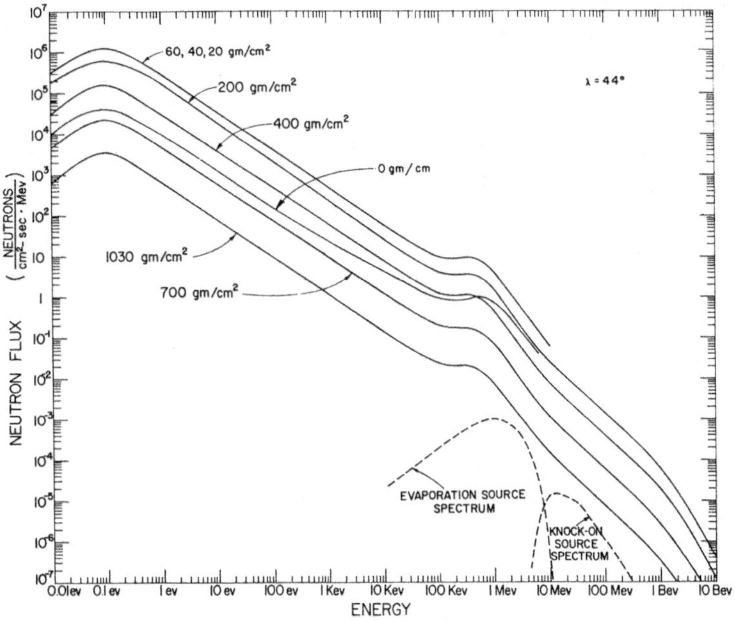

Fig. 2. Energy spectrum of cosmic ray neutrons calculated by HESS *et al.* (1961).

also falls off more rapidly than does that calculated by HESS *et al.* (1961). Therefore in fitting a high-energy tail to the diffusion result, Lingenfelter's flux above 10 MeV is less than one third that of HESS *et al.* (1961) at the equator.

It should be pointed out that in fitting a high-energy spectrum to the diffusion spectrum, a very small variation in the *form* of the high-energy spectrum will produce a large variation in differential flux at, say 100 MeV, even if the diffusion result is very accurate. This is because of the steep slope of the diffusion-calculated spectrum in the region 1 to 10 MeV where the fit must be made.

LINGENFELTER and FLAMM (1964) calculated the leakage flux *below* 10 MeV resulting from interactions of solar protons in the atmosphere.

LENCHEK (1962) calculated the same quantity and obtained a unidirectional leakage intensity in the vertical direction at the pole 6.4×10^{-4} n/cm^2 ster sec per solar proton/cm^2 sec.

The unidirectional intensity increases with increasing zenith angle ϕ about as $1-\cos\phi$. The leakage flux is $\sim 6 \times 10^{-3}$ n/cm^2 sec per solar proton/cm^2 sec. This figure is computed assuming a solar cosmic ray spectrum $E^{-2.7}\,dE$, for which the mean energy of the protons above the neutron excitation threshold is $\sim$25 MeV. In the calculation of LINGENFELTER and FLAMM (1964) a flatter proton spectrum is employed, for which the mean energy of the protons is much higher. This, plus the use of a much larger excitation cross section, leads LINGENFELTER and FLAMM (1964) to a leakage flux some ten times greater than that obtained by LENCHEK (1962).

The fast albedo neutrons generated by solar cosmic rays cannot populate L shells within $L \simeq 1.5$ and they cannot populate equatorial orbits anywhere.

BAME *et al.* (1963), using BF_3 proportional counters and Li^6I scintillation counters, measured the neutron flux above the atmosphere. Because the detector efficiencies are very low above 10 MeV and because the bulk of the response is due to neutrons below this energy, the experiment sheds no light on the flux density of high energy neutrons.

WILLIAMS and BOSTROM (1964) studied neutrons at $\sim$1000 km with a moderated B^{10} detector. The neutron flux observed was $\sim$2 times greater than that predicted by HESS *et al.* (1961) and $\sim$10 times greater than that predicted by LINGENFELTER (1963). The observations were made 1961–1962 and refer to latitudes about 40 °N geomagnetic.

HAYMES (1964) investigated neutrons in the range 1 to 14 MeV at balloon altitudes. Over a range of 2.5° the latitude variation was in agreement with the calculations of LINGENFELTER (1963).

In summary, three rather detailed calculations of the transport of slow neutrons generated in the atmosphere by cosmic rays are available. The results differ by a factor three in respect of leakage flux and by a factor two in respect of the latitude variation. From a peak at thermal energy, the energy spectrum falls off about as $E^{-1.3}\,dE$. Near 100 keV the spectrum flattens and tends to develop a peak near 1 MeV. At higher energies, where the spectrum is determined by the leakage of knock-on neutrons, the spectrum is poorly known. It is believed to be approximately $E^{-2}\,dE$ for the unidirectional intensity. The strong anisotropy of the intensity leads to an omnidirectional spectrum more like $E^{-3}\,dE$.

Solar cosmic rays, being relatively low in energy, generate few knock-on neutrons. The evaporation neutrons have been the subject of two theoretical studies. The yield of leakage neutrons per solar proton is strongly dependent on the form of the proton spectrum, ranging from $\sim 6 \times 10^{-3}$ neutrons/proton to $\sim 10^{-1}$ neutrons/proton.

C. INJECTION FROM SOLAR COSMIC RAYS OR SOLAR WIND

Solar cosmic rays and the solar wind offer more or less attractive sources of protons which might diffuse into the trapping region, possibly being locally accelerated in the process. These sources may, in fact, play a role in populating the outer belt where protons up to about 1 MeV are found. However, the absence of trapped *alpha particles* in the inner region, when both these sources contain large fractions of alphas, seems to rule them out as important sources at $L \lesssim 2$.

2. Loss Mechanisms

A. COULOMB INTERACTIONS

A fast particle (mass m', charge $Z'e$, velocity v) loses energy to stationary field particles (mass m'', charge $Z''e$) at the rate

$$\frac{\mathrm{d}E}{\mathrm{d}t} = -\,4\pi r_0^2 (Z'Z'')^2\,\rho c (m/m'')\,mc^2\,\Lambda/\beta \tag{2.1}$$

where $r_0 = e^2/mc^2$, m = electron mass, ρ = density of field particles, $\beta = v/c$. Note the inverse dependence on the mass of the target particle. Nuclei are therefore inefficient sources of energy dissipation compared to electrons. In the presence of free electrons

$$\Lambda = \ln (\lambda_{\mathrm{D}} p/\hbar) \tag{2.2}$$

where λ_{D} = Debye shielding distance and $\hbar/p$ represents the theoretical minimum impact parameter. Under magnetospheric conditions this term is quite slowly varying and has the value ~ 20. In neutral material

$$\Lambda = \tfrac{1}{2}\left[\ln\left(\frac{\pi^2 m^2 c^4}{(1-\beta^2)^{3/2} I^2}\right) - 2.9\right] \tag{2.3}$$

for a singly-charged fast particle. I is the atomic ionization potential averaged over all electrons.

The scattering rate is

$$\langle(\Delta\theta)^2\rangle = 8\pi r_0^2 (Z'Z'')^2\,\rho c \left(\frac{m}{m'}\right)^2 \frac{(1-\beta^2)}{\beta^3}\Lambda' \tag{2.4}$$

where $\Lambda' \simeq \Lambda$. Note here the dependence on the mass of the *fast* particle rather than the *target* particle.

The relative rates of deflection and fractional energy loss for a single-charged fast particle moving through material of atomic number Z is

$$\frac{\text{fractional deflection rate}}{\text{fractional energy-loss rate}} = \frac{2(Z+1)(\gamma-1)}{\beta^2\gamma^2}\,\frac{m}{m'} \tag{2.5}$$

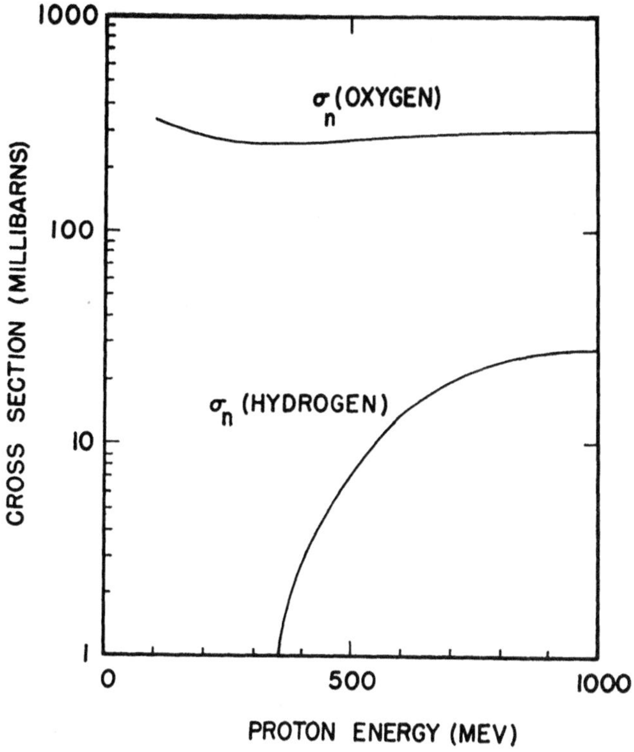

Fig. 3. Nonelastic cross sections for protons ((LENCHEK and SINGER, 1963).

where $\gamma = (1 - \beta^2)^{-1/2}$. We see that for protons deflection is negligible compared to energy-loss at all energies.

B. NUCLEAR INTERACTIONS

Absorption in nonelastic interactions give a mean-free-path $l_n = M/\sigma_n$ g/cm^2, where M is the mass of the atmospheric atom and σ_n is the cross section (Figure 3). Energy-loss and scattering in elastic interactions can be shown to be negligible. Nuclear interactions in air became more important than ionization at $E \gtrsim 300$ MeV.

C. CHARGE EXCHANGE

Charge exchange with neutral atoms is an important loss process at energies below 1 MeV. In hydrogen the mean-free-path is

$$l_x = 5 \times 10^{-3} E^{5.5}$$

g/cm^2 for E in MeV ($0.3 < E < 2$) (FITE *et al.*, 1958; BATES and DALGARNO, 1953). This path length is shorter than the ionization range for $E < 600$ keV. The cross section for oxygen is comparable.

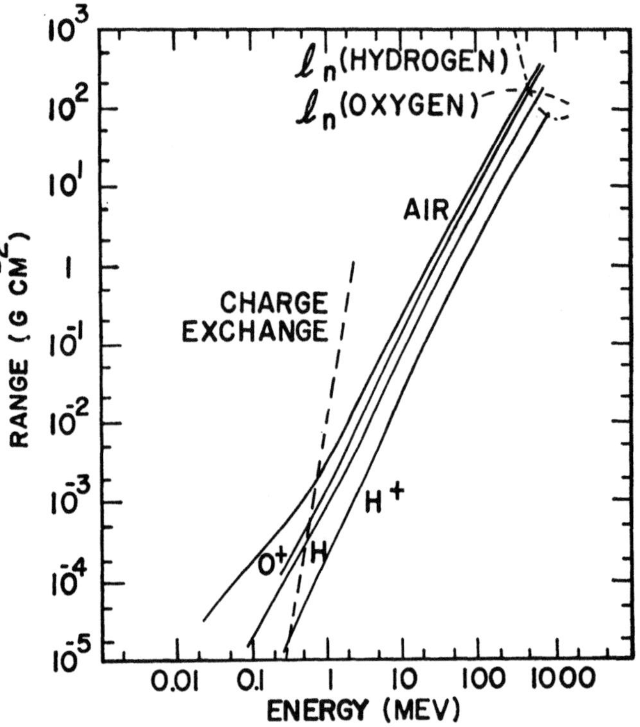

Fig. 4. Range and mean free path of protons. The solid curves are ionization ranges. The curve labeled "charge exchange" refers to a hydrogen atmosphere and is based on data of FITE *et al.* (1958) and BATES and DALGARNO (1953). The mean free paths (l_n) are based on the cross sections of Figure 3.

Ionization loss, nuclear interaction and charge exchange are summarized in Figure 4.

D. LIFETIMES

The concept of a lifetime for a trapped particle can be very useful but some caution must be exercised in its use. For example, the time needed to slow down from an initial energy to say 10 per cent of that energy need not be related to the time needed for the particle density at the initial energy to decay to 10 percent of its initial value. The latter time depends in part on the energy spectrum of the particle population.

The least ambiguous lifetime one can define is the lifetime for catastrophic removal. This time is $1/\sigma\rho v$ where σ is the relevant cross section, ρ is the density of absorbing particles and v is the velocity of the trapped particle. Charge

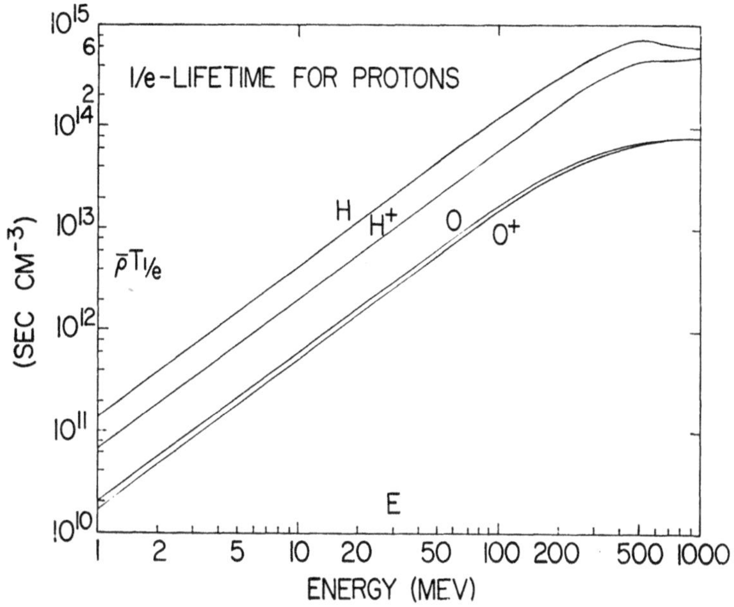

Fig. 5. Lifetime for reducing E to E/e by ionization or for reducing the number of protons to $1/e$ by nonelastic nuclear interactions.

exchange and nonelastic nuclear interactions can be treated by this simple lifetime. (It should, however, be remembered that nonelastic nuclear interactions generate secondary particles at lower energy, for which a correction may be needed).

A lifetime for energy loss can be defined as the time required to slow down from initial energy E to final energy E/e. For nonrelativistic protons this time is

$$T_{1/e}(\text{energy-loss}) = 6.7 \times 10^{10} E^{3/2}/\rho \tag{2.6}$$

seconds in fully-ionized hydrogen with density ρ ions cm^{-3} and E in MeV.

These two lifetimes may be combined to give the $1/e$-lifetime (the harmonic mean of the two) shown in Figure 5. In this figure, $\bar{\rho}$ is the atmospheric density averaged over the orbit of the trapped particle. For an orbit confined to the heart of the inner belt $\bar{\rho} \simeq 10^4 - 10^5$ cm^{-3}, implying a lifetime of from decades to hundreds of years for a 100-MeV proton.

E. NON-ADIABATIC PROCESSES

Hydromagnetic waves propagating through the magnetosphere are capable of perturbing the orbits of trapped particles. In an arbitrary, i.e., nonresonant, interaction between a wave and gyrating particle, the variation of relative phase of the particle and the perturbation field causes the net effect on the magnetic moment to be very small. However, under conditions of resonance the effect may become large (WENTZEL, 1961; DRAGT, 1961).

Consider a particle whose gyrofrequency is ν_c and whose velocity parallel to **B** is $v_{\|}$. Upon encountering a hm wave of frequency ν_{hm} propagating along the field line with velocity V_{A}, the frequency of the wave as seen in the rest frame of the guiding center, will be Doppler shifted to the frequency

$$\nu' = \nu_{\text{hm}} v_{\|} / V_{\text{A}}$$

If $v_{\|}$ is such that $\nu' = \nu_c$ then the particle and wave are in resonance. DRAGT (1961) has computed the amount by which the magnetic moment is changed in such an interaction. It is evident that given a sufficient number of waves of sufficient amplitude, particles can be rapidly removed.

Hydromagnetic waves are believed to be generated in the outer magnetosphere by the interaction between the magnetosphere and the solar wind. Since the waves do not propagate readily across the lines of force unless ν_{hm} is less than the local gyrofrequency, the spectrum of waves near the earth should be cut off at frequencies higher than some ν_{max}, equal to the gyrofrequency at the point of generation. DRAGT (1961) estimated $\nu_{\text{max}} = 3$ cps. Then protons with energies greater than $E_{\text{max}}(L)$, a function of L, will be able to satisfy the resonance condition and can be affected. Based on a particular model of V_{A} versus L, Dragt concluded that $E_{\text{max}} = 160$ MeV at $L = 2$ and that

$$E_{\text{max}} \propto L^{-11}$$

Little is known about the strength, frequency spectrum and frequency of occurrence of hydromagnetic waves in the magnetosphere. Nevertheless, it cannot be denied that if such waves are present often enough and with a frequency spectrum and amplitude such as postulated by Dragt, they must very rapidly remove high energy protons from at least the outer edge of the proton belt.

A more controversial question is that of the long-term stability in the *static* geomagnetic field. The adiabatic approximation must break down at some point as the ratio of gyroradius to the scale of inhomogeneity of the field increases with increasing L (SINGER, 1959). But the details of the breakdown of adiabaticity are not clear. Analytic studies employing higher order approximations than the Alfvén approximation (HAYAKAWA and OBAYASHI, 1963; DRAGT, 1965) conclude that the mirror point oscillates about a mean value with excursions that increase in amplitude as the particle's momentum is increased. However, the effects computed by these authors are not large enough to account for the absence of 100-MeV protons beyond $L=2$. Trajectory analyses (GARMIRE, 1963; DRAGT, 1965) seem to confirm the oscillation of the mirror point. Moreover, such analyses reveal the existence of "unstable" orbits in the following sense. One evaluates the magnetic moment on successive equatorial crossings. For one class of initial conditions, a plot of magnetic moment *versus* phase angle (or distance from the dipole) will produce a smooth curve. However, a class of orbits is also found for which the plot becomes scattered. The transition from "smooth" to "scattered" plot is quite sharp as the initial conditions are varied. Analyses of this sort applied to the stability of orbits in magnetic mirror machines (GARREN *et al.*, 1958) conclude that orbits of the "smooth" type enjoy long-term stability while those of the "scattered" type are unstable.

Both GARMIRE (1963) and DRAGT (1965) find that, for a given dipole line of force, protons of a given energy are stable if their mirror points lie above a critical altitude, and that the maximum-energy proton that can be stable on a given line decreases with increasing L. For example, Dragt finds that 100-MeV protons are stable on the field line $L=3.8$ up to a latitude of 40°. GARMIRE (1963) finds that for mirror points held at constant altitude, E_{max} varies approximately as L^{-4} while for mirror points in the equatorial plane E_{max} varies approximately as L^{-2}, so that the maximum-trappable energy increases along a field line toward higher altitude.

However, both of these trajectory analyses predict stable trapping to higher L than is observed. The maximum energy of protons trapped at $L\sim 2$ is about 100 MeV (MCILWAIN, 1963) but the orbit analyses referred to above predict that protons of up to $\sim$500 MeV should be trappable at this distance. The discrepancy may be due to the action of transient perturbations such as the scattering by hydromagnetic waves.

3. The Trapped Protons

A. TRANSPORT EQUATION

The distribution function $n(E, \mu, L, t)$ is defined such that $n v\, \mathrm{d}E 2\pi \,\mathrm{d}\mu$ is the number of particles crossing a square centimeter (oriented perpendicular to $\mathbf{v}$) per second with energy in $\mathrm{d}E$ and with $\mathbf{v}$ within the solid angle $2\pi \sin \alpha_e \,\mathrm{d}\, \alpha_e$ at the equatorial crossing. We write μ for $\cos \alpha_e$. It can be shown (LENCHEK and

SINGER, 1963) that

$$\frac{\partial n}{\partial t} = 4\pi r_0^2 mc^3 \bar{\rho}(L,\mu)\frac{\partial}{\partial E}[nA/\beta] - nv\sigma\bar{\rho} + \bar{q}$$
$$+ \oint \mathrm{d}s \left\{ \frac{B(s)}{B_e}\rho(s) \int_E^\infty \mathrm{d}E' v\sigma_n P \int_{\mu_1(s)}^1 \mathrm{d}\mu n\mu \left[1 - \frac{B(s)}{B_e}(1-\mu^2)\right]^{-1/2} \right\} \Big/ \oint \mathrm{d}s \tag{3.1}$$

where B_e is the equatorial field strength, $\mu_1(s)=(1-B_e/B(s))^{1/2}$ and $P(E', E)\mathrm{d}E$ is the probability of generating a secondary proton in $\mathrm{d}E$ at E following a collision at E'. The last term in (3.1), representing secondary production, makes a negligible contribution. The maximum contribution occurs for orbits lying entirely within the oxygen-helium region of the atmosphere ($L<1.2$). For such orbits the secondary injection is less than 5 per cent of direct albedo neutron injection.

B. ENERGY SPECTRUM

The proton energy spectrum under equilibrium conditions and neglecting secondary injection is given by

$$j(E,\mu,L) = \frac{v}{\langle -\mathrm{d}E/\mathrm{d}t\rangle} \int_E^{E_{max}} \bar{q}(E')e^{-f}\mathrm{d}E' \tag{3.2}$$

where

$$f = \int_{E'}^{E_{max}} \frac{\bar{\rho}\sigma v}{\langle -\mathrm{d}E/\mathrm{d}t\rangle}\mathrm{d}E'' \tag{3.3}$$

The brackets indicate averages over the orbit. Note that E_{max} is the energy above which trapping is not stable.

Near 100 MeV, where ionization is dominant, the proton spectrum is flatter than the neutron spectrum. If the neutron spectrum, averaged over the orbit is $E^{-\alpha}\,\mathrm{d}E$ then the proton spectrum will be approximately $E^{1.5-\alpha}\,\mathrm{d}E$. At several-hundred MeV, where nuclear interactions dominate, the proton spectrum may be slightly *steeper* than the average neutron spectrum. In all cases the proton intensity is inversely proportional to the atmospheric density averaged over the orbit.

Injection by neutron albedo from solar cosmic rays produces an equilibrium proton spectrum with a peak near Θ, the so-called nuclear evaporation temperature (LENCHEK, 1962). Theoretical estimates (LENCHEK and SINGER, 1963) of Θ yield 4 MeV although the true value may be a few MeV lower (HESS *et al.*, 1961).

Proton spectra and spatial distributions computed on a variety of slightly different models (SINGER, 1958; HESS, 1959; LENCHEK and SINGER, 1962, 1963; FREDEN and WHITE, 1960; WHITE *et al.*, 1965) all predict a differential proton spectrum that varies from about $E^{-1.5}\,\mathrm{d}E$ near 50 MeV to about $E^{-3}\,\mathrm{d}E$ near

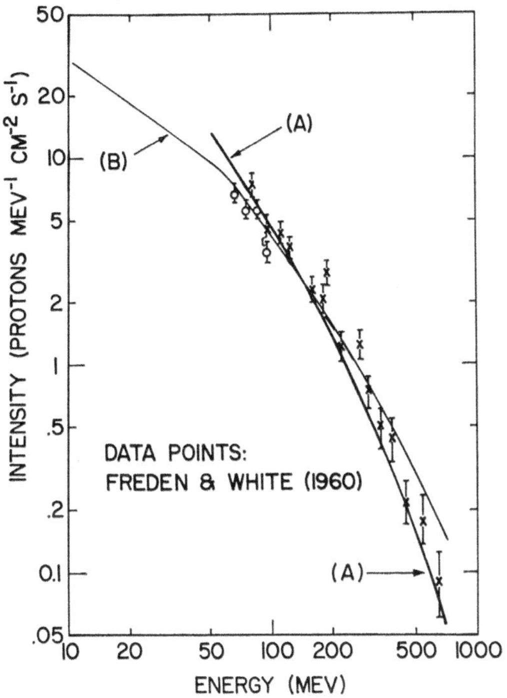

Fig. 6. Comparison with observation. The data are from two nuclear emulsion flights to $L \sim 1.4$ (FREDEN and WHITE, 1960). Curve *A* is the spectrum computed by LENCHEK and SINGER, (1962). Curve *B* is computed by FREDEN and WHITE (1960).

300 MeV. The theoretical spectra are in fair agreement with observed spectra (FREDEN and WHITE, 1960; FREDEN and PAULIKAS, 1964; BROWN *et al.*, 1964; FILLIUS and MCILWAIN, 1964; FREDEN *et al.*, 1965) (Figure 6). The spatial distribution has been examined in greatest detail by WHITE *et al.* (1965) who computed the spatial distribution of cosmic ray albedo neutron decay (CRAND) injection. CORNWALL *et al.* (1965) computed the atmospheric density experienced by trapped protons and compared the predicted fluxes with observations.

For $L<1.9$ the CRAND data and theory agree. At higher L an additional (non-adiabatic) loss mechanism is needed.

Observations of protons above 10 MeV (NAUGLE and KNIFFEN, 1963) at high latitudes near $L=1.6$ were initially interpreted as being of solar proton albedo neutron decay (SPAND) origin (LENCHEK, 1962). More recent observations (FILLLIUS and MCILWAIN, 1964; FREDEN *et al.*, 1965) show several features that cannot be explained by the SPAND mechanism: the proton intensity in the interval 1 to 10 MeV at *equatorial latitudes* is at least two orders of magnitude too large to attribute to SPAND, equatorial orbits are heavily populated where SPAND would leave them empty, and the energy spectrum shows no sign of developing a peak as it is explored to lower and lower energies down to 1 MeV. Thus, it is now clear that the so-called anomalous component that appears below about 20 MeV and at $L \gtrsim 1.5$ cannot be of SPAND origin.

C. CONCLUSION

The first particles detected in the radiation belt were energetic protons near $L=1.2$ (VAN ALLEN *et al.*, 1958). The suggestion made by SINGER (1958) that the radiation consisted of protons in the 100-MeV range generated by the decay of cosmic ray albedo neutrons seems still to be valid. The theory is able to account for observations to better than an order of magnitude. However, the region of space in which the theory works is very small, less then one percent of the magnetosphere. Scattering by hydromagnetic waves seems to limit stable trapping to this small region above the magnetic equator. Low energy protons of unknown origin are superimposed on the proton belt of cosmic ray albedo origin.

References

BAME, S. J., CONNER, J. P., BRUMLEY, F. D., HOSTETTER, R. L., and GREEN, A. C.: 1963, *J. Geophys. Res.* **68,** 1221.

BATES, D. R. and DALGARNO, A.: 1953, *Proc. Phys. Soc. (Lond.)* **A66,** 972.

BROWN, W. L., DAVIDSON, L. W., and MEDFORD, L. V.: 1964, *Bell Lab. Rec.*

CAMERINI, V.: 1952, *Progress in Cosmic Ray Physics* (Ed. by J. G. WILSON), North-Holland Publ. Co., Amsterdam, Vol. 1, p. 3.

CORNWALL, J. M., SIMS, A. R., and WHITE, R. S.: 1965, *J. Geophys. Res.* **70,** 3099.

DRAGT, A. J.: 1964, *J. Geophys. Res.* **66,** 1641.

DRAGT, A. J.: 1965, *Reviews Geophys.* **3,** 255.

FILLIUS, R. W. and MCILWAIN, C. E.: 1964, *Phys. Rev. Letters* **12,** 609.

FITE, W. L., BRACKMAN, R. T., and SNOW, W. R.: 1958, *Phys. Rev.* **112,** 1161.

FREDEN, S. C. and PAULIKAS, G. A.: 1964, *J. Geophys. Res.* **69,** 1259.

FREDEN, S. C. and WHITE, R. S.: 1960, *J. Geophys. Res.* **65,** 1377.

FREDEN, S. C., BLAKE, J. B., and PAULIKAS, G. A.: 1965, *J. Geophys. Res.* **70,** 3111.

GARMIRE, G.: 1963, *J. Geophys. Res.* **68,** 2627.

GARREN, A., RIDDELL, R. J., SMITH, L., BING, G., HENRICH, L. R. NORTHROP, T. G., and ROBERTS, J. E.: 1958, *Proceedings Second International Conf. Peaceful Uses Atomic Energy, Geneva,* United Nations, N. Y., p. 65.

GAUGER, J.: 1964, *J. Geophys. Res.* **69,** 2209.
HAYAKAWA, S. and OBAYASHI, H.: 1963, *J. Geophys. Res.* **68,** 3311.
HAYMES, R. C.: 1964, *J. Geophys. Res.* **69,** 841, 853.
HESS, W. N.: 1959, *Phys. Rev. Letters* **3,** 11.
HESS, W. N.: 1960, *J. Geophys. Res.* **65,** 3107.
HESS, W. N., PATTERSON, H. W., WALLACE, R., and CHUPP, E. L.: 1959, *Phys. Rev.* **116,** 445.
HESS, W. N., CANFIELD, E. H., and LINGENFELTER, R. E.: 1961, *J. Geophys. Res.* **66,** 665.
LENCHEK, A. M.: 1962, *J. Geophys. Res.* **67,** 2145.
LENCHEK, A. M. and SINGER, S. F.: 1962, *J. Geophys. Res.* **67,** 1263.
LENCHEK, A. M. and SINGER, S. F.: 1963, *Planet. Space Sci.* **11,** 1151.
LINGENFELTER, R. E.: 1963, *J. Geophys. Res.* **68,** 5633.
LINGENFELTER, R. E. and FLAMM, E. J.: 1964, *J. Geophys. Res.* **69,** 2199.
NAUGLE, J. E. and KNIFFEN, D. A.: 1963 *J. Geophys. Res.* **68,** 4065.
NEWKIRK L. L.: 1963 *J. Geophys. Res.* **68,** 1825.
SINGER S. F.: 1958, *Phys. Rev. Letters* **1,** 171.
SINGER S. F.: 1958, *Phys. Rev. Letters* **1,** 181.
SINGER S. F.: 1959, *Phys. Rev. Letters* **3,** 188.
TRAINOR J. H. and LOCKWOOD J. A.: 1964, *J. Geophys. Res.* **69,** 3115.
VAN ALLEN, J. A. LUDWIG, G. H., RAY, E. C., and MCILWAIN, C. E.: 1958, *Jet Propulsion* **28,** 588.
WENTWORTH, R. C. and SINGER, S. F.: 1955, *Phys. Rev.* **98,** 1546. (A).
WENTZEL, D. G.: 1961, *J. Geophys. Res.* **66,** 359, 363.
WHITE, R. S., AUSTIN, M. M., and DRAGT, A. J.: 1965, *Space Res.* **5,** 498.
WILLIAMS, D. J. and BOSTROM, C. O.: 1964, *J. Geophys. Res.* **69,** 377.

PARTICLE EFFECTS IN THE GEOMAGNETIC FIELD

E. N. PARKER

Enrico Fermi Institute for Nuclear Studies and Department of Physics University of Chicago Chicago, Illinois, U.S.A.

Introduction

The purpose of these two lectures is to summarize present theoretical understanding of the behavior of charged particles and plasma in the geomagnetic field. The summary is necessarily brief, and is at best only a guide to the published literature. It must be understood that the present understanding of particle behavior in the geomagnetic field is in an elementary form. The understanding does not yet provide a theory for particle acceleration and for the aurora and the associated Van Allen radiation belts, nor does it give a unique picture of what gas inflates the geomagnetic field during the main phase of a storm, nor does it give a clear understanding of the geomagnetic tail structure, etc., etc. But the understanding does give a picture of many simpler effects.

In putting these lectures together, it became evident that the behavior of the particles cannot be discussed properly outside the context of their reaction on their environment. Therefore, some space is devoted to the formulation of the geomagnetic effects of particles and plasma in the field of Earth.

1. Deformation of the Geomagnetic Field

A. GENERAL CONSIDERATIONS

I begin with the point (PARKER, 1956, 1958a; DESSLER and PARKER, 1959) that the geomagnetic field is filled with a thermal plasma. Hence it is a hydromagnetic medium. It is an elastic body and the strains in it are caused by the stresses exerted on it by the energetic particles and gases in space. PIDDINGTON (1959, 1960a) has re-emphasized this point.

Consider a magnetic dipole located at the origin with its axis pointing in the positive z-direction. The dipole simulates the geomagnetic field, insofar as the field is generated within the solid body of the Earth. The xy-plane is the equatorial plane of the dipole. Denote by B_0 the magnetic field density across the equatorial plane at a radial distance a from the dipole. Take a as equal to the radius of Earth (6.4×10^3 km) and B_0 about 0.35 gauss. The magnetic moment of the dipole is $B_0 a^3$.

The dipole field is permeated by the terrestrial atmosphere and is, of course, firmly gripped by the solid body of Earth so far as short period fluctuations are concerned, but for lack of space we will not go into these effects (see review in ELSÄSSER, 1956). The dipole field is "assaulted" from space by the solar wind. It is this "assault" with which we shall be concerned. The solar wind is fully ionized, with the result that, to a first approximation, it does not mix with the geomagnetic field. The result is the well known confinement of the geomagnetic field by the wind to a comet shaped region (CHAPMAN and FERRARO, 1931; DUNGEY, 1958; PARKER, 1958a) whose boundary, the magnetopause, lies at about $10a$ in the direction of the sun. The length of the comet tail stretching out behind Earth is not known, the estimates running from $10^2 - 10^5 a$. The wind is responsible, in some way, for the generation of the aurorae and for the associated energetic particle belt in the geomagnetic field. At times of high activity the wind leads to superheated plasma pressure within the geomagnetic field, seriously distorting the field to produce the geomagnetic storm.

The basic principles describing the distortion of a magnetic field by a gas pressure p_{ij} are relatively simple, but have been widely misstated in the literature. The correct equations are the hydromagnetic equations (PARKER, 1956; DESSLER and PARKER, 1959). The coupling of the gas and field is through the electric currents automatically induced in the gas by the field. Equilibrium is achieved when there is a balance between the magnetic stresses and the plasma or particle pressure. The magnetic stresses are described by the well known Maxwell stress tensor $M_{ij} = -\delta_{ij}B^2/8\pi + B_iB_j/4\pi$, representing an isotropic pressure $B^2/8\pi$ and a tension $B^2/4\pi$ along the magnetic lines of force. In the absence of stresses exerted on the field by the plasma, the magnetic field assumes the form $B_i = -\partial\psi/\partial x_i$, $\nabla^2\psi = 0$ for which $\nabla \times \mathbf{B} = 0$. Stresses exerted on the field introduce an elastic strain, indicated by the nonvanishing of $\nabla \times \mathbf{B}$. Equilibrium is described by

$$\frac{\partial}{\partial x_j}(-p_{ij} + M_{ij}) = 0 \quad \text{or} \quad -\nabla p + (\nabla \times \mathbf{B}) \times \mathbf{B} = 0\,. \tag{1}$$

A stress newly introduced into the field produces a strain which propagates with the characteristic Alfvén speed $B/(4\pi\rho)^{1/2}$ where ρ is the density of the medium tied to the field.*

Pressure gradients exerted on the field produce a $\nabla \times \mathbf{B}$ which is perpendicular to $\mathbf{B}$, as is evident from (1). Torsion exerted on a tube of magnetic flux produces a $\nabla \times \mathbf{B}$ which is parallel to $\mathbf{B}$, and is transmitted along the tube to wherever an

* Distortion of the field can also be computed by applying the Biot-Savart Law to the currents produced by the particle motions. The formulae for this are given by ALFVÉN (1950) and are discussed by DESSLER and PARKER (1959).

opposite torque relieves the stress. The geomagnetic field can support no permanent torsion because of the nonconducting atmosphere which has no static resistance to shear. Thus, in the geomagnetic field there is the simplification that $\nabla\times\mathbf{B}\cdot\mathbf{B}\cong 0$. This permits (1) to be solved immediately for the strain $\nabla\times\mathbf{B}$ by taking the vector product with $\mathbf{B}$. The result is $\nabla\times\mathbf{B}=4\pi\mathbf{B}^{-2}\mathbf{B}\times\nabla p$ for the simple case of a scalar pressure p, with $\varepsilon_{ijk}\partial B_k/\partial x_j=4\pi B^{-2}\varepsilon_{ijk}\partial p_{kl}/\partial x_l$ for the tensor pressure, where ε_{ijk} is the usual permutation tensor used to write vector products, equal to ± 1 according as ijk is an even or odd permutation on 123, and zero otherwise. It should be noted (PARKER, 1957b) that in a collisionless plasma the equilibrium pressure force $\partial p_{ij}/\partial x_j$ is perpendicular to B_i with the special form $+\nabla_\perp p_\perp+(p_\parallel-p_\perp)\,[(\mathbf{B}\cdot\nabla)\mathbf{B}]_\perp/B^2$ where the subscripts $\parallel$ and $\perp$ represent the components parallel and perpendicular to the magnetic field, respectively. Some may find it easier to work with the currents $(c/B^2)\mathbf{B}\times[\nabla p_\perp+(p_\parallel-p_\perp)B^{-2}(\mathbf{B}\cdot\nabla)\mathbf{B}]$ produced by the particles. Great care must be used in employing this expression to be sure that $p_\parallel$ and $p_\perp$ are correctly computed. The static equilibrium distribution of $p_\parallel$ and $p_\perp$ along a magnetic line of force depends upon the pitch angle θ distribution $\Psi(\theta)$. If $\Psi(s,\theta)\sin\theta\,\mathrm{d}\theta$ is taken to be the number of particles per unit volume with pitch angles in $(\theta,\theta+\mathrm{d}\theta)$ at a distance s along a magnetic line of force, then it can be shown (PARKER, 1957b) from the Liouville equation that $\Psi(s,\theta)\propto\sin^\beta\theta$ yields $\Psi(s,\theta)\propto B^{-\beta/2}$ for static equilibrium along that same line of force. In general

$$\Psi(s,\theta)=\sum_\beta C_\beta\sin^\beta\theta\left[\frac{B(0)}{B(s)}\right]^{\beta/2}.$$

This representation of Ψ is probably as convenient as any. Fortunately, as will be shown, it is usually not necessary to go into such detail in the general discussion of the effects of plasma and particles in the geomagnetic field.

B. DISTORTION BY INTERNAL FORCES

In principle the distortion of the field $\mathbf{B}(\mathbf{r})$ by a gas pressure $p(\mathbf{r})$ can be computed from (1) and the appropriate boundary conditions. One may treat (1) directly as a nonlinear differential equation for $\mathbf{B}(\mathbf{r})$ (CHANDRASEKHAR, 1956). But the general solution is mathematically intractable. To avoid vast numerical computations, one limits the effects to small plasma pressures and treats the resulting strain $\Delta\mathbf{B}(\mathbf{r})$ as a small perturbation on the dipole field. The easiest way to carry this out is to note that there is a vector identity that the curl of a field has associated with it the field

$$\Delta\mathbf{B}(\mathbf{r})=\frac{1}{4\pi}\iiint\mathrm{d}^3\mathbf{r}'\,\frac{[\nabla'\times\mathbf{B}(\mathbf{r}')]\times\mathbf{R}}{R^3}$$

in addition to the irrotational portion $-\nabla\psi$ of the field (PARKER, 1962). In this expression $\mathbf{R}\equiv\mathbf{r}-\mathbf{r}'$ and ∇' represents differentiation with respect to $\mathbf{r}'$. Replacing $\nabla\times\mathbf{B}$ in the integrand by the pressure stresses $4\pi B^{-2}\nabla p$ gives the strain $\Delta\mathbf{B}(\mathbf{r})$ in terms of the pressure stresses ∇p exerted on the field. In tensor notation the result can be written

$$\Delta B_i(\mathbf{r}) = \iiint \frac{\mathrm{d}^3\mathbf{r}'}{R^3 B^2}\left(B_j R_j \frac{\partial p_{ik}}{\partial x_k'} - B_i R_j \frac{\partial p_{jk}}{\partial x_k'}\right). \tag{2}$$

This equation tells us that the net compression or expansion $\Delta\mathbf{B}(\mathbf{r})$ is determined by the algebraic sum of the inward and outward forces exerted on the field. The integral representation is particularly useful in calculating $\Delta\mathbf{B}(\mathbf{r})$ when $p \ll B^2/8\pi$. For then $\Delta\mathbf{B}\ll\mathbf{B}$, and, as a first approximation, B_i under the integral sign may be approximated by the undistorted dipole $-\partial\psi/\partial x_i$. A number of general theorems can be established formally from (2).

Consider a gas pressure p which is an even function of x, y, and z, i.e. which has symmetry about the three coordinate planes. It follows from (2) that the distortion is symmetrical about the coordinate planes and therefore $\Delta\mathbf{B}(\mathbf{r})$ is stationary in the vicinity of the origin and is in the z-direction. In physical terms this means that the distortion $\Delta\mathbf{B}(\mathbf{r})$ of the dipole $-\partial\psi/\partial x_i$ by a belt of trapped particles or plasma extending around the dipole field is a field which is uniform in the vicinity of the origin. As a first approximation, then, the distortion $\Delta\mathbf{B}(0)$ computed at the origin characterizes $\Delta\mathbf{B}(\mathbf{r})$ over the entire region occupied by Earth. The presence of the solid body of the conducting Earth distorts the $\Delta\mathbf{B}$ computed from (2), of course, increasing the magnitude of $\Delta\mathbf{B}$ at the surface of Earth. But this is a refinement we will not enter into here.

A second theorem has to do with the location of the gas or particles producing the distortion. Suppose that the pressure p, which vanishes at $r=0$, ∞, is of the general form $L^{-3}f(r/L)$ where L is a scale factor. It is evident that the total energy $\frac{3}{2}\iiint\mathrm{d}^3\mathbf{r}\,p$ is independent of L. Substituting into (2) it is readily shown that $\Delta\mathbf{B}(0)$ is independent of L in a dipole field, in which B, $B_i \propto r^{-3}$. That is to say, the distortion of the field is independent of the scale of the particles or plasma and depends only upon the total energy and the shape of the particle distribution.*

The third theorem states that the distortion of the field, in fact, depends very little upon the shape of the distribution, so that altogether the distortion $\Delta\mathbf{B}(0)$ depends principally upon the total energy contained in the particles. Independence of the spatial distribution is only approximate, but good enough for most present purposes. Approximate independence is readily proved by the following examples.

* In n-pole fields $B \propto r^{-(n+2)}$ and $\Delta B \propto L^{n-1}$.

A belt of particles confined to the equatorial plane gives a pressure which can be written $p_{zz}=0$, $p_{xx}=p_{yy}=p(z, \varpi)$, where $p(z, \varpi)$ is the kinetic energy density and ϖ is distance $(x^2+y^2)^{1/2}$ from the dipole axis. The function $p(z, \varpi)$ is essentially a delta function of z. Then for the field strain $\Delta\mathbf{B}$ in the vicinity of the origin (see first theorem above), we have $\mathbf{r}=0$ so that $\mathbf{R}=-\mathbf{e}_\omega\varpi$ where the pressure is nonvanishing. Here $\mathbf{e}_\omega$ is a unit vector pointing away from the dipole axis. The unperturbed dipole field is $B_z=B_0(a/\varpi)^3$. Integration by parts with the boundary condition that p vanish at $\varpi=0$, ∞ gives

$$\frac{\Delta\mathbf{B}(0)}{B_0}=-\mathbf{e}_z\frac{2\mathscr{E}}{3\mathscr{E}_{\mathrm{n}}} \tag{3}$$

where $\mathscr{E}$ is the total energy of the particles (equal to the volume integral of p) and $\mathscr{E}_M$ is the total energy of the dipole field outside the solid body of Earth, equal to $\frac{1}{3}a^3B_0^2\cong 1\times 10^{25}$ ergs. The net outward distension $\Delta\mathbf{B}(0)$ is the sum of a compression of the field by the diamagnetic pressure of the particles and the expansion produced by the outward pressure of the particles. The details of this are in the literature.

As another example consider particles streaming steadily in both directions between latitudes $\pm\alpha$ along the magnetic lines of force (zero pitch angle and zero net current along the lines of force). $\Delta\mathbf{B}(0)$ can be computed as a general function of α (DESSLER and PARKER, 1959). When $\alpha\ll 1$,

$$\frac{\Delta\mathbf{B}(0)}{B_0}=\mathbf{e}_z\frac{2\mathscr{E}}{\mathscr{E}_M}. \tag{4}$$

When $\alpha=\pi/2$,

$$\frac{\Delta\mathbf{B}(0)}{B_0}=\mathbf{e}_z\frac{2\mathscr{E}}{3\mathscr{E}_M}. \tag{5}$$

The outward distension of the field is caused by the centrifugal force of the particle motion along the curved lines of force.

These extreme examples illustrate the small range of the proportionality constant in the general condition $\Delta\mathbf{B}(0)/B_0\propto\mathbf{e}_z\mathscr{E}/\mathscr{E}_{\mathrm{M}}$. For most particle configurations, the coefficient lies between about 2/3 and 4/3.

Consider application of these calculations to particles in the geomagnetic field. The energy content of the Van Allen belts has been worked out rather carefully by HOFFMAN and BRACKEN (1964) from present observations. They calculate from their results that the outward distension of the geomagnetic field by the trapped particles leads to the strain $|\Delta\mathbf{B}(0)|\cong 9\times 10^{-5}$ gauss (corresponding to about 3×10^{21} erg). As another example consider a typical main phase decrease of 10^{-3} gauss (100γ). This expansion is the result of an increase of 3×10^{22} ergs in trapped particles and plasma. It is clear from the observations that the increase is not in the Van Allen particles. Perhaps it is only a general heating of the ambient plasma to, say, 1 keV per ion (DESSLER and PARKER, 1959; PARKER, 1962).

C. DISTORTION BY EXTERNAL FORCES

The geomagnetic field is subject to external as well as internal stresses. The general rule is that pressures on the boundary compress the field and produce a strain $\Delta\mathbf{B}(0)$ which is approximately uniform in the vicinity of the origin and is oriented within about 30° of the dipole axis (PARKER, 1958a). Deformation by external stresses on the outer boundary of the field produces no $\nabla \times \mathbf{B}$ inside the boundary, where $\mathbf{B} = -\nabla\psi$ so far as the external stresses are concerned. It is instructive to consider first the distortion of the field by boundaries of fixed geometry. For instance the classical example of a plane boundary (CHAPMAN and FERRARO, 1931) perpendicular to the equatorial plane at a distance h yields a compression $\Delta\mathbf{B}(0) = +\mathbf{e}_z B_0(a/h)^3/8$. It can be shown (PARKER, 1958a) that the compression field is nearly in the z-direction no matter what the orientation of the confining plane. Enclosure by a sphere of radius h confines the field on all sides, giving a compression 16 times larger, $2B_0(a/h)^3$. The actual case lies somewhere in between, with enclosure on the sunward side at $10a$, and little or no enclosure on the night side. It is an easy matter to compute the field of a dipole confined inside a cylinder with open or closed ends (PARKER, 1962).

The real problem, however, is to compute the shape of the boundary given the nature of the stresses on the boundary. In two dimensions the problem can be solved rigorously for the case of a uniform beam of non-interacting particles reflecting from the boundary (ZHIGULEV and ROMISHEVSKII, 1959; DUNGEY, 1961; HURLEY, 1961). In three dimensions the problem of a dipole subjected to a uniform external pressure was solved by SLUTZ (1962) and subject to a uniform beam on the sunward side, with hydrostatic pressure on the night side, by SLUTZ and WINKELMAN (1964). Another method, involving moments, was developed by MIDGELEY and DAVIS (1962) and applied to the dipole surrounded by a uniform pressure and then to the dipole field subject to a beam of particles (MIDGELEY and DAVIS, 1963). Another method is a scheme of successive approximations worked out by BEARD (1960). We use it as an illustration. The approximation begins by noting that the tangential component of the field at a plane boundary is precisely double that component in the absence of the boundary. A concave boundary of the geomagnetic field gives a smaller increase of the tangential component, and a convex boundary gives a larger increase. In Beard's first approximation he uses the factor of two. Then if the field is subject to an incident beam of N particles/cm^3 with velocity v at an angle of incidence ϑ, the pressure on the boundary is $2NMv^2 \cos^2 \vartheta$. The normal component of the field vanishes at the boundary, since in the infinite conductivity approximation the wind and field do not mix. So the magnetic pressure of the tangential component must produce a magnetic pressure equal to the above wind pressure. Taking the tangetial component to be exactly twice the component B_t in the same direction in the absence of wind pressure, we have $(2B_t)^2/8\pi = 2NMv^2 \cos^2 \vartheta$, i.e. $B_t \propto \cos \vartheta$. Now if the dipole field is described in terms of the usual spherical coordinates (r, θ, ϕ), the undisturbed field is $B_r \propto 2 \cos \theta / r^3$ and $B_\theta \propto \sin \theta / r^3$. If ψ is the angle between the radius vector r and the boundary, then on the noon-midnight meridional plane $\tan \psi = r\,d\theta/dr$ where $r = r(\theta)$ is the equation for the boundary. If the wind is perpendicular to the dipole, it is readily shown that $\vartheta = \psi - \theta$. Then $B_t \propto \pm(2 \cos \theta \cos \psi - \sin \theta \sin \psi)(a/r)^3$ and $\cos \vartheta = \cos \psi \cos \theta + \sin \psi \sin \theta$. Since $B_t \propto \cos \vartheta$ it follows that

$$\frac{dr}{r}\frac{1 \mp 2/r^3}{1 \mp 1/r^3} = \tan \theta \, d\theta$$

when r is expressed in suitable units. Note that $r = 1$ is a solution of this differential equation. When $r \neq 1$, the solution is

$$\frac{r^3}{r^3 \pm 1} = C \cos \theta$$

where C is the integration constant. It is evident that $r = 1$ corresponds to the sunlit boundary. The solution for the night side boundary, for which $r > 1$ must pass continuously to $r = 1$. Hence $r^2/(r^3 + 1) = C \cos \theta$. The "suitable units" for r are readily deduced, without going through the algebra from the beginning, from the fact that at the subsolar point $\vartheta = 0$ and $B_t = B_\theta = B_0 (a/r)^3$. Then equating $(2B_t)^2/8\pi$ to $2NMv^2$ gives $r = a(B_0^2/4\pi NMv^2)^{1/6}$ there, which corresponds to $r = 1$ in "suitable" units. The units are obviously a $(B_0^2/4\pi NMv^2)^{1/6}$ or some $10a$. The reader is referred to Beard's papers for calculations of the boundary elsewhere around the Earth, and for the higher orders of approximation in which the factor of two in the first approximation is corrected for the non-planar boundary deduced from the first approximation (MEAD and BEARD, 1964; BEARD, 1964; BAKER *et al.*, 1964). Beard's method has been applied to deformations of the geomagnetic field under a variety of wind conditions by SPREITER and BRIGGS (1962) and SPREITER and HYETT (1963). A method for treating the boundary in the vicinity of the neutral points has been given by BLUM (1964).

2. Problems at the Magnetopause

There are a number of important and still unsolved theoretical problems connected with boundary of the geomagnetic field. The field is an obstacle in the supersonic wind, with the result that a bow shock is produced upstream from the magnetopause (GOLD, 1959; PARKER and TIDMAN, 1960; AXFORD, 1962; KELLOGG, 1962; SPREITER and JONES, 1963). The shock is collisionless and the structure of collisionless shocks is not clearly understood yet. There appear to be a number of effects which contribute to the shock transition (ADLAM and ALLEN, 1958; DAVIS *et al.*, 1958; GARDNER *et al.*, 1958; FISHMAN *et al.*, 1960; COLGATE, 1959; AUER *et al.*, 1961; AUER *et al.*, 1962; KELLOGG, 1964; WALTERS, 1964; NOERDLINGER, 1964; SCARF *et al.*, 1965). The shock transition is evidently responsible for the generation of fast particles (FAN *et al.*, 1964, see also FRANK *et al.*, 1963) and it has been suggested that the generation mechanism is Fermi acceleration (JOKIPII and DAVIS, 1964) and/or resonance with acoustic waves (FREDRICKS *et al.*, 1965).

The shock leads to disorder on various scales with the result that the plasma which is actually in contact with the outer boundary of the geomagnetic field is "turbulent" (BERNSTEIN *et al.*, 1964).

The importance of "turbulence" in the solar wind as it approaches Earth, and the importance of the "turbulence" generated by the shock transition, for agitating the geomagnetic field (geomagnetic activity) are still a subject of debate. It is evident that pounding on the magnetopause produces hydromagnetic waves within the magnetosphere. The propagation of hydromagnetic waves in the magnetosphere is a subject in itself (DESSLER *et al.*, 1961; MACDONALD, 1961). The question is to what extent the boundary of the geomagnetic field is itself unstable. If the boundary is stable, then geomagnetic fluctuations are presumably a direct consequence of turbulence in the wind. If the boundary is unstable, geomagnetic fluctuations may arise when the wind is steady. Considering the solar wind as a stream of free particles leads to a prediction of instability (DUNGEY, 1958; PARKER, 1958a), but treating the wind as a fluid leads to stability unless the fields in the wind are nearly parallel (or anti-parallel) to the geomagnetic field at the boundary (FEJER, 1964). Unfortunately the solar wind is neither a stream of free particles nor a fluid on the small scales associated with collision with the magnetosphere, so better theoretical treatment is needed. Observations show a direct correlation between wind velocity and geomagnetic activity (SNYDER *et al.*, 1963) which does not settle the question.

The question of stability of the magnetopause is not only fundamental to the understanding of geomagnetic activity, but in addition is fundamental to the rate of convection of the geomagnetic field, to the rate of convection and diffusion of particles in the geomagnetic field, and to the structure of the long geomagnetic tail extending out behind Earth. It was pointed out some years ago (PARKER, 1958a), from the calculations which led to the conclusion of instability, that the wind exerts a "viscous" drag around the sides of the magnetosphere and carries away the lines of force of the outer geomagnetic field just as a high wind blows smoke away from a chimney. How far the "smoke" extends out behind the Earth is not known. It was pointed out that the lines of force which are carried away in the wind exert an outward tension on the field as a whole, which counteracts a little of the inward pressure exerted on the field by the wind. We were so enamored of the effects of tension at that time as to suggest that so many lines of force might be blown away by a high wind as to cause the main phase of the geomagnetic storm. But the infatuation cooled rapidly and it was soon evident that the wind pressure must always dominate the outward tension. Hence we were surprised to find PIDDINGTON (1960b, 1962, 1963) asserting that the tension sometimes dominates and is responsible for the general expansion of the field causing the main phase of the geomagnetic storm. So far as we can see there is no basis for this opinion.

The two really interesting problems concerning the "viscous" drag of the wind on the field are the geomagnetic tail and the deep convection of the field. Consider first the geomagnetic tail. JOHNSON (1960) pointed out that, if the solar wind slides smoothly along the flanks of the geomagnetic field, then the pressure of the wind must close off the geomagnetic field at some distance behind Earth. The wind would close in at the Mach angle, equal to the reciprocal of the Mach number of the wind. In round numbers the Mach number is about 10, so the tail would extend back for a distance of about ten times the radius of the magnetosphere, for a total of $10^2 a$. Observations show that the tail is broader and longer than the smooth sliding model, indicating clearly a stronger interaction between the wind and the tail. The most striking observation showing the grip in which the wind holds the flanks of the magnetosphere is the null surface found by NESS (1965). Ness finds from the IMP-1 magnetometer data that the geomagnetic lines of force at a distance of $30a$ back in the tail are extended approximately parallel to the Earth-Sun line, with a neutral sheet between the regions of opposite field in each magnetic hemisphere. This is quite unlike the field calculated (PARKER, 1962) for a passive boundary. The wind must exert a tension of the order of 5×10^{10} dynes (equal to about one tenth of the total force of the solar wind against the geomagnetic field) in order to stretch out the observed tail.

The "viscous" grip in which the wind appears to hold the tail leads to the conclusion that the tail must be very long (see for instance DESSLER, 1964). It has been suggested that the neutral sheet, which is an intimate theoretical part of a long tail, plays an important part in the generation of auroral particles and the associated outer radiation belt (DUNGEY, 1958; AXFORD *et al.*, 1964; DESSLER and JUDAY, 1965).

Convection of the geomagnetic field is possible because of the layer of non-conducting atmosphere at the surface of Earth. AXFORD and HINES (1961, AXFORD, 1964) have pointed out that the "viscous" drag of the wind may drive a quasi-stationary convection of the field and that such convection may be important in transporting hot plasma and energetic particles in and out through the magnetosphere. The reader is referred to their paper for an extended discussion. A trapped particle with a velocity w perpendicular to the magnetic field B has $w^2 \propto B$ as a consequence of adiabatic compression while being convected by the geomagnetic field. The energy gain may be quite large because B is typically 4×10^{-4} gauss at the magnetopause, 40×10^{-4} gauss at $r=4a$, and 400×10^{-4} gauss at $r=2a$. The growth and decay of the hot plasma which inflates the geomagnetic field to produce the main phase of the magnetic storm may be the result of the inward convection. The redistribution of charged particles along a line of force during convection may be found in the literature (PARKER, 1961b). The main impediment to convection is the viscosity of the ionosphere, the upper part of which is dragged with the convection as a consequence of the high electrical conductivity.

A recent calculation by CHANG *et al.* (1965) shows that the small inflation of the geomagnetic field by the Van Allen radiation belts gives marginal stability against interchange where the ionosphere drag is low on the night side. They suggest therefore that a simple flute instability on the night side may maintain a limit on the density of the radiation belts. An over supply of particles inflates a tube of force so much that the tube is squeezed outward by its neighbors. Anything which tends to reduce the ionospheric conductivity, enhances the instability.

3. Diffusion and Acceleration of Trapped Particles

Energetic particles appear in nature wherever there is violent agitation of plasmas in the presence of a magnetic field. The solar wind and geomagnetic field are no exception to this general rule. Fast particles are generated in and around the bow shock and transition region, and appear as the aurorae and Van Allen belts within the geomagnetic field. The mechanism by which the particles are accelerated is not known.

A. PARTICLE DIFFUSION

The guiding center approximation permits a description of the individual trapped particle trajectories in terms of the magnetic moment invariant, the longitudinal invariant, and the flux invariant (NORTHROP, 1963). Violation of the invariants by geomagnetic fluctuations has been considered at some length (KELLOGG, 1959; PARKER, 1960, 1961a; DRAGT, 1961; WENTZEL, 1961, 1962; DAVIS and CHANG, 1962; TVERSKOY, 1964; FÄLTHAMMAR, 1965). It now appears (DUNGEY, 1965; NAKADA and MEAD, 1965) that violation of the flux invariant may play an important role in the transport of particles across magnetic lines of force.

The flux invariant is the simple consequence of the gradient drift of a particle in an inhomogeneous magnetic field. Consider a particle moving in the plane of the geomagnetic equator with velocity w. The field is in the z-direction with magnitude. $B = B_0(a/r)^3$. To a first approximation the particle moves in a circle of radius $R = Mwc/qB$. But the field is stronger on the inner side of the circle and weaker on the outer side, with the result that the radius of curvature of the particle trajectory is smaller on the inner side. The trajectory is thus a cycloid. The instantaneous center of the circle drifts perpendicular to the field and perpendicular to ∇B with a velocity $\frac{1}{2}wR/l$ where l is the characteristic scale of B, $1/l = |\nabla B|/B$ (ALFVÉN, 1950). For a dipole field, $l = \frac{1}{3}r$. The particle drifts along a line of constant field density and consequently circles the Earth in a time $(4\pi/3)(qB_0a^3/Mw^2c)/r$. It can be shown (NORTHROP, 1963) that a deformation of the geomagnetic field in a time long compared to the time in which the particle circles the Earth will alter the path of the particle in such a way as to leave the total magnetic flux through the path unchanged. If the deformation is removed slowly, the particle returns to its original path. It can also be shown (KELLOGG, 1959) that a rapid deformation will generally not conserve the flux, with the result that a net displacement of the particle will remain when the deformation is removed. Physically the process is simple. Suppose that the side of the magnetosphere is compressed where the particle happens to be. The particle is displaced inward with the lines of force of the field and, as a consequence of the general compression, finds itself in a stronger magnetic field. Suppose that the compression is maintained for some time. The particle drifts along the contour of constant B (since the drift is perpendicular to ∇B). When the particle has drifted around to the opposite and undisturbed side of the geomagnetic field, it is still at the same value of B as on the compressed side, and therefore nearer Earth than previously. If the compression is removed before the particle returns, the net effect is an inward displacement of the particle, representing a violation of the flux variant. An outward displacement is accomplished if a rarefaction is introduced instead of a compression. This is Kellogg's diffusion mechanism. Detailed calculations may be found in PARKER (1960), DAVIS and CHANG (1962), FÄLTHAMMAR (1965), NAKADA and MEAD (1965), DUNGEY (1965), KAUFMANN (1965).

The magnetic moment remains invariant while the particles diffuse across the field, so that the particle energy is proportional to B. It follows, therefore, that in the absence of collisions with atmospheric particles and other energy losses, the particle energy increases inward as $1/r^3$. A 1 keV solar wind particle folded into the field at $10a$ has an energy of 125 keV at $2a$. Particles convected in by the general turning over of the field proposed by AXFORD and HINES (1961) undergo the same adiabatic compressional heating. So that inward particle transport which violates the flux invariant leads to particle acceleration.

Comparison of theoretical and observational density and energy distribution of trapped particles suggests that such transport is important beyond about $2.5a$ (NAKADA *et al.*, 1965; NAKADA and MEAD, 1965; DUNGEY, 1965).

B. PARTICLE ACCELERATION

Besides adiabatic compression there are two known mechanisms which may accelerate particles in the geomagnetic field. They are wave acceleration and the Fermi mechanism. Wave acceleration is based on the idea that the frequency of the electromagnetic disturbances seen by a rapidly moving particle may occasionally be in resonance with the cyclotron period of the particle, thereby imparting large amounts of energy to the particle (PARKER, 1961c; STIX, 1964; SCARF *et al.*, 1965; FREDRICKS *et al.*, 1965). Fermi acceleration is based on the general idea that a few of the faster particles in a plasma are reflected from the moving hydromagnetic waves in the plasma (FERMI, 1949, 1954) leading to a statistical gain of energy.

1. *Fermi Mechanism*

Consider the Fermi mechanism. It is basically a hydromagnetic process, occurring in violently agitated plasmas as a consequence of the gas and field motions with characteristic frequencies well below the ion cyclotron frequency. Hence the electric field is the usual $\mathbf{E} = -\mathbf{v} \times \mathbf{B}/c$ where $\mathbf{v}$ is the plasma velocity and $\mathbf{B}$ the magnetic field. The nonrelativistic equation of motion of a particle with charge q, mass M, and velocity $\mathbf{w}$ is

$$M \frac{\mathrm{d}\mathbf{w}}{\mathrm{d}t} = q\left[\mathbf{E} + \frac{\mathbf{w} \times \mathbf{B}}{c}\right]$$

Eliminating $\mathbf{E}$ and taking the scalar product with $\mathbf{w}$ gives the energy equation, which upon permuting the triple product can be written

$$\frac{\mathrm{d}}{\mathrm{d}t} \tfrac{1}{2} M w^2 = \mathbf{v} \cdot \mathbf{F}$$

where $\mathbf{F}$ is the Lorentz force $q\mathbf{w} \times \mathbf{B}/c$ exerted on the particle by the magnetic field. This equation shows that in hydromagnetic fields the energy of a fast particle changes *only* when the motion of the gas does work against the Lorentz force. Thus a particle may be accelerated as it "bumps" against moving fields, or as it is adiabatically compressed or expanded in a changing field. We call this acceleration the Fermi mechanism.

In its simplest and original form the Fermi mechanism was analogous to the trend toward equipartition of energy of a light particle, of mass M and speed w, placed in among a number of randomly moving massive elastic objects of very large mass $\mathcal{M}$ and velocity v. The trend toward equipartition exists because head on collisions between the particle and the massive objects are slightly more probable, and the energy transfer is slightly larger, than overtaking collisions, as a consequence of the slightly higher relative velocity in the head on collision. The average fractional energy gain per collision is of the order of $(v/w)^2$. The ratio v/w is generally small compared to one, and $(v/w)^2$ much smaller.

Davis (1956) showed that if one does not restrict his attention to the average energy gain but instead considers the particle distribution, there are some particles which gain energy much faster than the average. Fan (1956) considered special wave forms in which the field grows exponentially with time between two approaching hydromagnetic waves. He found that the average gain is then first order, rather than second order, in v/w. Later it was shown that in the presence of sharp hydromagnetic waves, head on collisions are much more probable than overtaking collisions, so that the average fractional energy gain is much larger, $O(v/w)$ (Parker, 1958b).

The Fermi mechanism under a variety of circumstances has been worked out in the literature (Parker, 1955, 1957a, 1958a; Parker and Tidman, 1958; Davis, 1958; Schatzman, 1963; Wentzel, 1963, 1964) in connection with its many possible applications, to the solar flare, and cosmic rays. Jokipii and Davis (1964) have considered the Fermi mechanism as a possible origin of the fast particles observed in the transition region and bow shock outside the geomagnetic field. Their calculations suggest that the mechanism may play an important role there.

2. *Wave Acceleration*

Consider the idea that particles are accelerated when their velocity is such that they fall into resonance with plasma waves of one kind or another. This was proposed recently by Stix (1964) in connection with the production of fast electrons in laboratory plasmas. Stix suggested a resonance between the electron plasma wave, as seen by a fast electron, and the cyclotron frequency of the fast electron. Bernstein *et al.* (1964), Scarf *et al.* (1965), Fredricks *et al.* (1965) have taken up this general idea and applied it to the generation of the fast particles observed in the region of impact of the solar wind against the geomagnetic field (Freeman *et al.*, 1963; Frank *et al.*, 1963; Fan *et al.*, 1964). They suggest there a resonance of fast electrons with ion acoustic waves.

The principle of the acceleration is easily demonstrated with a simple example taken from an illustration in the problem of acceleration of electrons in the geomagnetic field by resonance with electromagnetic waves in the whistler mode (Parker, 1961c). Consider an electron with charge $-e$ and mass m, at $z=0$ moving perpendicular to a uniform magnetic field B in the z-direction.

Suppose that transverse to the magnetic field there is an electric field $E(z, t)$ in the x-direction. It follows at once that the equations of motion are

$$\frac{d^2x}{dt^2} = -\left[\frac{e}{m}E(z,t) + \Omega\frac{dy}{dt}\right], \frac{d^2y}{dt^2} = +\Omega\frac{dx}{dt} \tag{6}$$

where $\Omega = eB/mc$. The terms involving the magnetic field of the wave are small to second order and so they have been dropped. Integrate the equation for d^2y/dt^2 and choose the origin of the coordinates so that dy/dt vanishes where $x=0$. Then eliminate dy/dt from the equation for d^2x/dt^2, obtaining

$$\frac{d^2x}{dt^2} + \Omega^2 x = -\frac{e}{m}E(0,t).$$

The particle motion produced by $E(0, t)$ is

$$\frac{dx}{dt} = -\frac{e}{m}\int_{-\infty}^{t} d\tau \cos\Omega(t-\tau)E(0,\tau).$$

Now suppose that $E(0, t)$ is of the form

$$E(0,t) = f(t)\cos\int_{-\infty}^{t} du\,\omega(u).$$

It is readily shown that so long as $f(t)$, $\omega(t)$ and $\omega(t) - \Omega$ change by only a small fraction in each cyclotron period $2\pi/\Omega$, the velocity follows the amplitude of the wave, and subsides to zero with the final passage of the exciting wave packet. If, however, the wave should pass across resonance, $\omega = \Omega$, a residual velocity remains after the passage of the wave. Suppose that $\omega(t) = \Omega(1 \pm 2v^2\Omega t)$ during the passage across resonance, where v is a constant small compared to one. Then with the passage of the exciting wave there is left the residual motion

$$\frac{dx}{dt} = -\frac{ef(0)}{m}\int_{-\infty}^{+\infty} d\tau \cos\Omega(t-\tau)\cos\Omega(\tau \pm v^2\Omega\tau^2)$$
$$\cong c\frac{\pi^{\frac{1}{2}}f(0)}{vB}\cos(\Omega t \pm \pi/4).$$

In this expression $f(0)$ is the amplitude of the wave at the time of resonance.

The slower the passage through resonance, the smaller is ν and the larger is the residual velocity.

In the calculations of FREDRICKS *et al.* (1965) it is estimated that ion waves with a potential variation of a few hundred volts in the transition region outside the magnetosphere are sufficient to produce 50 keV electrons. The amplitude and frequency of the ion waves is subject to direct measurement from space vehicles.

Various other kinds of particle acceleration in plasma fields have been considered in the collisionless shock transition, but so little is known of the phenomenon that nothing definite can be said at the present time. In this connection the collisionless shock transition upstream in the solar wind from the magnetosphere, and at the head of interplanetary blast waves from the sun, may afford the one situation directly accessible to exploratory measurements.

Finally, I should remark that the neutral sheet, between two oppositely directed magnetic fields has been a favorite candidate for particle acceleration (DUNGEY, 1958). The neutral sheet is a region of enhanced field annihilation, perhaps unstable, with the added charm of eluding thorough analysis even to the present day. DUNGEY (1962) has pointed out that there is a neutral region at the subsolar point on the magnetopause whenever the interplanetary field points southward. He argues that the existence of the neutral point plays an important role in particle acceleration. I have been very pessimistic about the idea, partly because particle acceleration in a neutral sheet has not been demonstrated from first principles (Newton and Maxwell). There has been a recent development which suggests that there may be more to the neutral sheet than has been proved so far. A hydrodynamic Ohm's law calculation by PETSCHEK (1963) shows a field configuration giving much more rapid annihilation at a neutral sheet than previously calculated. The calculated fields flow into each other at about 10^{-1} the Alfvén velocity. Dessler has pointed out that the calculated annihilation is unfortunately so rapid that it takes place in only a small fraction of the electron collision time. So Ohm's law is inapplicable and the calculation needs to be done over again in some way. Nonetheless, the calculation indicates that there is perhaps a more rapid field annihilation than previously believed. So altogether, the neutral sheet is an interesting phenomenon and deserves both theoretical and observational attention.

The basic idea behind the rapid annihilation of field at the neutral sheet is the fact that the fluid squeezes out from between the two regions of oppositely directed fields, as shown in Figure 1, and permits the fields to come closer and closer together. The cross-hatched area represents the region of rapid Joule dissipation. The velocity V with which the fluid is ejected along the neutral sheet is readily shown from energy considerations to be of the order of the Alfvén speed $V_A = B/(4\pi\rho)^{1/2}$. Conservation of mass requires that $Vl \cong vL$ for steady conditions, where v is the velocity with which the fields of breadth L are moving together. For steady conditions v is equal to the velocity with which the fields dissipate across the thickness l of the transition, so that $v \cong c^2/4\pi\sigma l$ where σ is the electrical conductivity in esu ($10^7\ T^{3/2}$/sec). It follows from these

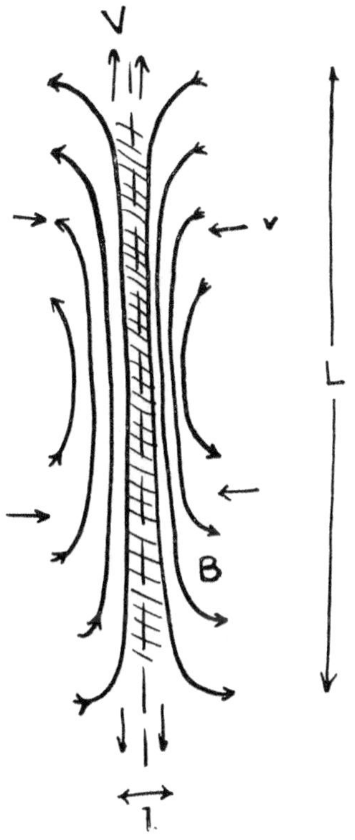

Fig. 1. A sketch of the magnetic lines of force of two oppositely directed fields pressed firmly together. The short arrows represent the fluid velocity. The cross hatching indicates the region of rapid resistive dissipation of the magnetic field.

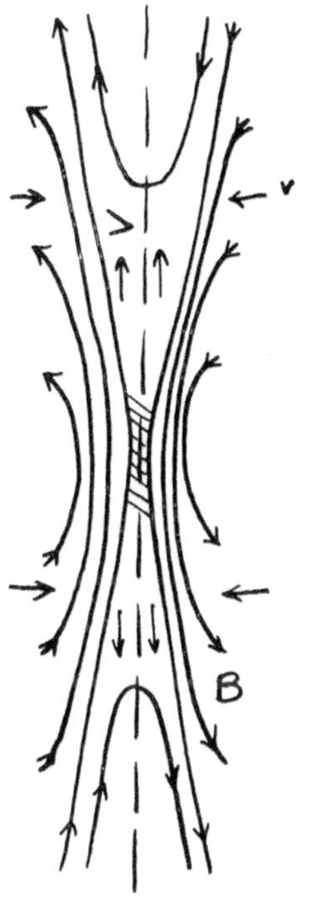

Fig. 2. A sketch of the magnetic lines of force of two oppositely directed fields as suggested by Petschek. The short arrows represent the fluid velocity. The cross hatching indicates the restricted region of rapid resistive dissipation of the magnetic field.

facts that the annihilation velocity may be written $v = V_A/\mathscr{R}^{1/2}$ where $\mathscr{R}$ is the characteristic magnetic Reynolds number $4\pi\sigma L V_A/c^2$, which is generally very large compared to one (PARKER, 1957c, 1963). Were it not for the steepening of the gradient of the field from L to l by the squeezing of fluid out from beneath the opposite fields, v would be equal only to $V_A/\mathscr{R}$.

Petschek points out that there seems to be a regime, as illustrated in Figure 2, where the merging takes place over much smaller dimensions. The cross-hatched area represents the region of rapid Joule dissipation. His calculations give $v = V_A/\ln\mathcal{R}$, which is so rapid as to require a more sophisticated approach than the one which he used.

Finally, I should remark that there has always been with us the idea that there may be strong electric fields parallel to the magnetic field or to neutral sheets in the field. It is evident that such fields produce enormous particle acceleration. For precisely this reason, such electric fields are forbidden whenever a significant thermal plasma is present. Hence, discussion of such electrostatic acceleration requires, in most cases, the assumption that a thermal plasma is absent. Unfortunately, not every author has realized this. It is not obvious to me that a thermal plasma can be absent from the geomagnetic field with the terrestrial atmosphere nearby. The absence of thermal plasma is the key question here, and the proponents of parallel electric fields should devote their attention to it rather than to the endless obvious acceleration possibilities if parallel electric fields should exist. Presumably the place to expect parallel electric fields is in a region where the required current density $c\nabla \times \mathbf{B}/4\pi$ is so large as to exceed the ability of the ambient free electrons to transport it.

This concludes my brief summary and appraisal of particle acceleration and diffusion. The references contain the foundation on which the summary was based. Other lecturers will treat some of the individual topics in considerable depth. As is evident from my remarks, there has been a lot of progress, both theoretical and experimental, but fundamental problems involving collisionless shocks, particle acceleration, and plasma blowing against a field, are still not solved.

References

Adlam, J. H. and Allen, J. E.: 1958, *Phil. Mag.* **3,** 448.

Alfvén, H.: 1950, *Cosmical Electrodynamics* (Oxford University Press, Oxford).

Auer, P. L., Hurwitz, H., and Kilb, R. W.: 1961, *Phys. Fluids* **4,** 1105.

Auer, P. L., Hurwitz, H., and Kilb, R. W.: 1962, *Phys. Fluids* **5,** 298.

Axford, W. I.: 1962, *J. Geophys. Res.* **67,** 3791.

Axford, W. I.: 1964, *Planet. Space Sci.* **12,** 45.

Axford, W. I. and Hines, C. O.: 1961, *Canadian J. Phys.* **39,** 1433.

Axford, W. I., Petschek, H. E., and Siscoe, G. L.: 1964, *Avco Everett Research Lab. Report* No. **190.**

Baker, J. C., Beard, D. B., and Young, J. C.: 1964, *Phys. Fluids* **7,** 504.

Beard, D. B.: 1960, *J. Geophys. Res.* **65,** 3559.

Bernstein, W., Fredricks, R. W., and Scarf, F. L.: 1964, *J. Geophys. Res.* **69,** 1201.

Blum, R.: 1964, *J. Geophys. Res.* **69,** 1765.

Chandrasekhar, S.: 1956, *Astrophys. J.* **124,** 232.

Chang, D. B., Pearlstein, L. D., and Rosenbluth, M. N.: 1965, *J. Geophys. Res.* **70,** 3085.

Chapman, S. and Ferraro, V. C. A.: 1931, *Terrest. Magnetism Atmospheric Elec.* **36,** 77, 171.

Colgate, S. A.: 1959, *Phys. Fluids* **2,** 485.

DAVIS, L.: 1956, *Phys. Rev.* **101**, 351.
DAVIS, L.: 1958, *Nuovo Cim. Suppl.* **8**, 126.
DAVIS, L., LÜST, R., and SCHLÜTER, A.: 1958, *Z. Naturforsch.* **13a**, 916.
DAVIS, L. and CHANG, D. C.: 1962, *J. Geophys. Res.* **67**, 2169.
DESSLER, J. A.: 1964, *J. Geophys. Res.* **69**, 3913.
DESSLER, A. J. and PARKER, E. N.: 1959, *J. Geophys. Res.* **64**, 223.
DESSLER, A. J., HANSON, W. B., and PARKER, E. N.: 1961, *J. Geophys. Res.* **66**, 3631.
DESSLER, A. J. and JUDAY, R. D.: 1965, *Planet. Space Sci.* **13**, 63.
DRAGT, A. J.: 1961, *J. Geophys. Res.* **66**, 1641.
DUNGEY, J. W.: 1958, *Cosmic Electrodynamics* (Cambridge University Press, New York).
DUNGEY, J. W.: 1961, *J. Geophys. Res.* **66**, 1043.
DUNGEY, J. W.: 1962, *J. Phys. Soc. Japan* **17**, *Supp.* A-II, 15.
DUNGEY, J. W.: 1965, *Space Sci. Rev.* **4**, 199.
ELSÄSSER, W. M.: 1956, *Rev. Mod. Phys.* **28**, 135.
FÄLTHAMMAR, C.-G.: 1965, *J. Geophys. Res.* **70**, 2503.
FAN, C. Y.: 1956, *Phys. Rev.* **101**, 314.
FAN, C. Y., GLOECKLER, G., and SIMPSON, J. A.: 1964, *Phys. Rev. Letters* **13**, 149.
FEJER, J. A.: 1964, *Phys. Fluids* **7**, 499.
FERMI, E.: 1949, *Phys. Rev.* **75**, 1169.
FERMI, E.: 1954, *Astrophys. J.* **119**, 1.
FISHMAN, F. J., KANTROWITZ, A. R., and PETSCHEK, H. E.: 1960, *Rev. Mod. Phys.* **32**, 959.
FRANK, L. A., VAN ALLEN, J. A., and MACAGNO, E.: 1964, *J. Geophys. Res.* **68**, 3542.
FREDRICKS, R. W., SCARF, F. L., and BERNSTEIN, W.: 1965, *J. Geophys. Res.* **70**, 21.
FREEMAN, J. W., VAN ALLEN, J. A., and CAHILL, L. J.: 1963, *J. Geophys. Res.* **68**, 2121.
GARDNER, C. S., GOERTZEL, H., GRAD, H., MORAWETZ, C. S., ROSE, M. H., and RUBIN, H.: 1958, *Proc. Second UN Intern. Conf. Peaceful Uses of Atomic Energy* **31**, 230.
GOLD, T.: 1959, *Nuovo Cimento Suppl.* X, **13**, 318.
HOFFMAN, R. A. and BRACKEN, P. A.: 1964, NASA Goddard Energetic Particles Preprint Series, X-611-64-186.
HURLEY, J.: 1961, *Phys. Fluids* **4**, 854.
JOHNSON, F. S.: 1960, *J. Geophys. Res.* **65**, 3049.
JOKIPII, J. R. and DAVIS, L.: 1964, *Phys. Rev. Letters* **13**, 739.
KAUFMANN, R. L.: 1965, *J. Geophys. Res.* **70**, 2181.
KELLOGG, P. J.: 1959, *Nature* **183**, 1295.
KELLOGG, P. J.: 1962, *J. Geophys. Res.* **67**, 3805.
KELLOGG, P. J.: 1964, *Phys. Fluids* **7**, 1555.
MACDONALD, G. J. F.: 1961, *J. Geophys. Res.* **66**, 3639.
MEAD, G. D.: 1964, *J. Geophys. Res.* **69**, 1181.
MEAD, G. D. and BEARD, D. B.: 1964, *J. Geophys. Res.* **69**, 1169.
MIDGLEY, J. E. and DAVIS, L.: 1962, *J. Geophys. Res.* **67**, 499.
MIDGLEY, J. E. and DAVIS, L.: 1963, *J. Geophys. Res.* **68**, 5111.
NAKADA, M. P., DUNGEY, J. W., and HESS, W. N.: 1965, *J. Geophys.* **70**, 3529.
NAKADA, M. P. and MEAD, G. D.: 1965, *J. Geophys. Res.* (submitted for publication).
NESS, N. F.: 1965, *J. Geophys. Res.* **70**, 2989.
NOERDLINGER, P. D.: 1964, *J. Geophys. Res.* **69**, 369.
NORTHROP, T. G.: 1963, *The Adiabatic Motion of Charged Particles* (Interscience Publishers, New York).
PARKER, E. N.: 1955, *Phys. Rev.* **99**, 241.
PARKER, E. N.: 1956, *J. Geophys. Res.* **61**, 625.
PARKER, E. N.: 1957a, *Phys. Rev.* **107**, 830.
PARKER, E. N.: 1957b, *Phys. Rev.* 924.
PARKER, E. N.: 1957c, *J. Geophys. Res.* **62**, 509.
PARKER, E. N.: 1958a, *Phys. Fluids* **1**, 171.
PARKER, E. N.: 1958b, *Phys. Rev.* **109**, 1328.
PARKER, E. N.: 1960, *J. Geophys. Res.* **65**, 3117.

PARKER, E. N.: 1961a, *J. Geophys. Res.* **66,** 693.
PARKER, E. N.: 1961b, *J. Geophys. Res.* **66,** 2641.
PARKER, E. N.: 1961c, *J. Geophys. Res.* **66,** 2673.
PARKER, E. N.: 1962, *Space Sci. Rev.* **1,** 62.
PARKER, E. N.: 1963, *Astrophys. J. Suppl.* No. 77, **8,** 177.
PARKER, E. N. and TIDMAN, D. A.: 1958, *Phys. Rev.* **111,** 1206; **112,** 1048.
PARKER, E. N. and TIDMAN, D. A.: 1960, *Phys. Fluids* **3,** 369.
PETSCHEK, H. E.: 1963, Avco-Everett Research Lab. Rep. AMP 123.
PIDDINGTON, J. H.: 1959, *Geophys. J.* **2,** 173.
PIDDINGTON, J. H.: 1960a, *J. Geophys. Res.* **65,** 93.
PIDDINGTON, J. H.: 1960b, *Geophys. J. Roy. Astron. Soc.* **3,** 314.
PIDDINGTON, J. H.: 1962, *Geophys. J. Roy. Astron. Soc.* **7,** 183.
PIDDINGTON, J. H.: 1963, *Planet. Space Sci.* **11,** 451.
SCARF, F. L., BERNSTEIN, W., and FREDRICKS, R. W.: 1965, *J. Geophys. Res.* **70,** 9.
SCHATZMAN, E.: 1963, *Ann. Astrofys.* **26,** 234.
SLUTZ, R. J.: 1962, *J. Geophys. Res.* **67,** 505.
SLUTZ, R. J. and WINKELMAN, J. R.: 1964, *J. Geophys. Res.* **69,** 4933.
SNYDER, C. W., NEUGEBAUER, M., and RAO, U. R.: 1963, *J. Geophys. Res.* **68,** 6361.
SPREITER, J. R. and BRIGGS, B. R.: 1962, *J. Geophys. Res.* **67,** 37.
SPREITER, J. R. and HYETT, B. J.: 1963, *J. Geophys. Res.* **68,** 1631.
SPREITER, J. R. and JONES, W. P.: 1963, *J. Geophys. Res.* **68,** 3555.
STIX, T. H.: 1964, *Phys. Fluids* **7,** 1960.
TVERSKOY, B. A.: 1964, *Geomagnetizm and Aeronomija* **3,** 351.
WALTERS, G. K.: 1964, *J. Geophys. Res.* **69,** 1769.
WENTZEL, D. G.: 1961, *J. Geophys. Res.* **66,** 359, 363.
WENTZEL, D. G.: 1962, *J. Geophys. Res.* **67,** 485.
WENTZEL, D. G.: 1963, *Astrophys. J.* **137,** 135.
WENTZEL, D. G.: 1964, *Astrophys. J.* **130,** 1013.
ZHIGULEV, V. N. and ROMISHEVSKII, E. A.: 1959, *Dokl. Akad. Nauk. SSSR* **127,** 1001; *Soviet Physics, Doklady* **4,** 859.

PRECIPITATION OF ELECTRONS AND PROTONS INTO THE ATMOSPHERE*

BRIAN J. O'BRIEN
Department of Space Science, Rice University, Houston, Texas

Introduction

A precipitated particle is defined here as one whose trajectory at the point of observation is such that it will plunge from a high altitude into the dense atmosphere, where it most likely will lose all or most of its kinetic energy. A precipitated particle may have once been trapped, but some may also never have been trapped.

We define at a given location a loss cone or cone of precipitation, whose axis is the local geomagnetic field vector **B** and which is bounded at a pitch angle α_D. A pitch angle (α) is the angle between **B** and the velocity vector of a charged particle. If the particle motion is such that its magnetic moment (μ) is conserved, then

$$\mu = \frac{\frac{1}{2}mv_\perp^2}{B} = \text{constant}$$
$$= \frac{\frac{1}{2}mv^2 \sin^2 \alpha}{B}$$

and hence $\sin^2 \alpha / B = \text{constant}$, if the kinetic energy of the particle does not change.

Suppose for simplicity we treat the dense atmosphere as ending at an altitude of 100 km, where $|\mathbf{B}| = B_{100}$. Then the loss cone at any point is bounded by α_D which is the pitch angle of a charged particle that would mirror (i.e. have $\alpha = 90°$) at 100 km. Consequently

$$\frac{\sin^2 \alpha_D}{B} = \frac{1}{B_{100}}$$

At an altitude of 100 km, $\alpha_D = 90°$. At an altitude of ~ 1000 km at high latitudes, $\alpha_D \sim 55°$. In the equatorial plane, the equatorial pitch angle α_{OD} of the loss cone is only a few degrees at $L \sim 6$ (see Figure 1). Note the artificiality of

* Research supported in part by the National Aeronautics and Space Administration under Contract NAS/r209.

Figure 1 in that it assumes that the very act of precipitation does not perturb the magnetic moment of the particle.

For this note, we distinguish four categories of precipitated particles, *viz.*

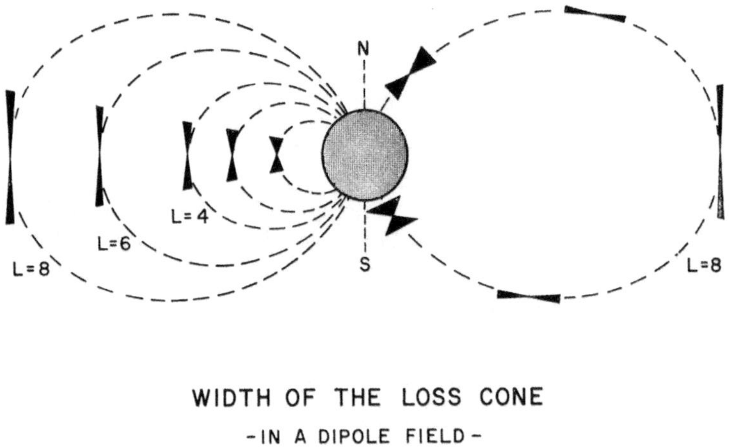

Fig. 1. Variation of the size of the nominal loss cone with altitude and with L.

(1) Those subject to attrition due to longitudinal drift into a region of low-altitude mirroring e.g. over the South Atlantic where $|\mathbf{B}|$ at a given altitude is a minimum or, conversely, where the altitude of a given $|B|$ is a minimum (Walt, 1964);

(2) Those precipitated from inside the magnetosphere, i.e. at latitudes below the auroral zone (O'Brien, 1964);

(3) Auroral-zone particles, which cause auroras (O'Brien and Taylor, 1964); and

(4) Polar-cap precipitation, which may be of three subcategories: (a) solar wind (O'Brien, unpublished; Michel 1965), (b) solar cosmic rays (Pieper *et al.*, 1962), (c) high-latitude "islands" or "spikes" (McDiarmid and Burrows, 1965).

LONGITUDINAL DRIFT ATTRITION (CATEGORY 1)

Category 1 is of interest both because of its causes and its effects. Its causes are known at L values of $L \lesssim 1.25$ (McIlwain, 1961) as being due simply to interaction with the residual atmosphere (Walt 1964). But at $L \gtrsim 1.25$ the loss rates appear faster than explicable by such a simple mechanism. Other papers in this meeting will cover this topic and we do not examine it further here. However, we draw attention to the fact that enhanced loss of trapped electrons and protons in such regions as the Capetown anomaly and the Rio minimum may produce localized quasi-permanent but weak and diffuse auroras. An accurate quantitative treatment of such effects remains to be made.

SOLAR WIND OVER THE POLAR CAP (CATEGORY 4a)

It appears to the writer that a number of models of the magnetosphere have as unstated implications the fascinating possibility that the polar cap atmosphere might be bombarded by the solar wind, probably thermalized with densities of tens of electrons and protons per cc and kinetic energies of order 100 eV. There is no direct experimental evidence known to us that refutes this possibility, and indeed there are several "anomalous" polar-cap characteristics (e.g. enhanced 6300 Å airglow) that might be regarded as indirect evidence of such an effect. Experimental resolution of this problem requires a suitably-instrumented polar satellite at an altitude of above some 500 km. Experimental resolution from ground-based observations is presently impractical e.g. we calculate that only some 1 to 0.1 rayleighs of H_β would be produced as a Balmer emission from the solar-wind-proton bombardment, and this is much below the geocoronal H_β intensity. This topic will be examined further at the 1966 Alpbach Polar Cap Studies Conference, and was treated by MICHEL (1965).

SOLAR COSMIC RAYS (CATEGORY 4b)

From M-region and solar-flare regions occasionally are ejected streams of solar cosmic rays of energies MeV's to GeV's. These bombard the polar cap and produce various phenomena beyond the scope of this note.

But recent treatments by MICHEL and DESSLER (1965) do demonstrate a particular point of relevance to the other categories of precipitation. Briefly, if one adopts an open tail model of the magnetosphere, then access of solar cosmic rays into the polar cap may be by diffusion back along the field lines. Specifically, the model predicts that above a certain magnetic latitude Λ_s (say), the cut-off rigidity should be zero, so that solar cosmic rays of *all* energies could enter the polar cap. Furthermore, this very same latitude (Λ_s) should be simultaneously both the boundary of trapping of Van Allen radiation and also should be the auroral zone. Experimental resolution of this important point also awaits an adequate satellite measurement during a solar cosmic ray event.

Historical Development

The effects of high-latitude precipitation are sometimes manifested visibly as auroras. The first detection of proton bombardment was the measurement of Doppler-shifted Balmer emissions in auroras (see CHAMBERLAIN, 1961). The first direct measurement of electron bombardment was in rockoon measurements in the auroral zone (VAN ALLEN, 1957). During the I.G.Y. two rocket-borne studies of auroras showed that a dominant cause of the bright visible auroral forms is bombardment of the atmosphere by electrons with energies of

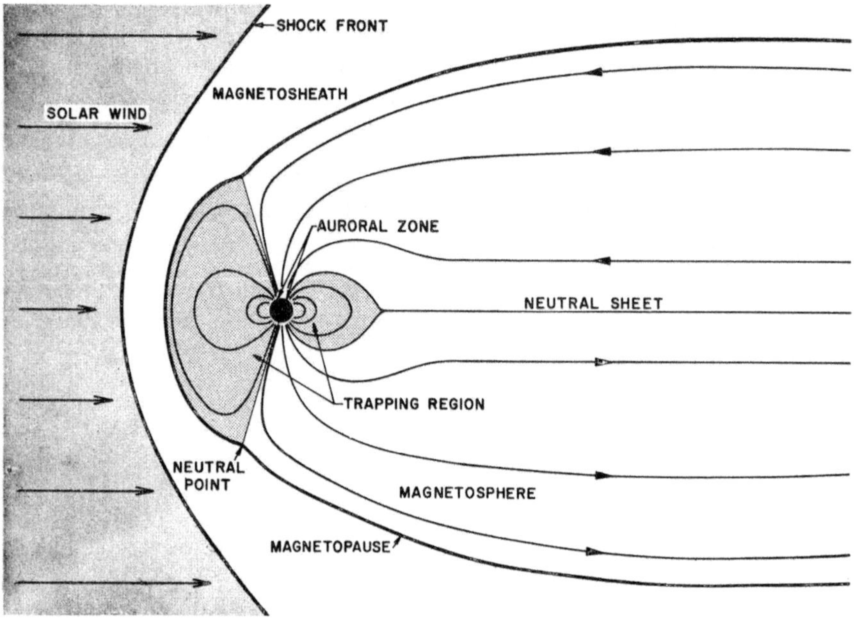

Fig. 2. Sketch of the magnetospheric configuration (after DESSLER and O'BRIEN, 1965). Note that the auroral zone is along the boundary of trapping of Van Allen radiation.

a few keV (MCILWAIN, 1960; DAVIS *et al.*, 1960). Balloon-borne studies of bremsstrahlung produced by electron bombardment of the atmosphere have increased knowledge of the spatial and temporal characteristics of this bombardment (see review by ANDERSON, 1965), at least for those electrons with energies above some ten or twenty keV.

Satellite-borne measurements of precipitated electrons (O'BRIEN, 1962) showed that they come from altitudes above ~1000 km, and hence they require a magnetospheric source. Characteristics of the precipitation, such as its tendency to approach isotropy over the upper hemisphere, have been summarized by O'BRIEN (1964). The fact that the electrons trapped as Van Allen particles are an inadequate reservoir to sustain precipitation was demonstrated with satellite-borne instruments (O'BRIEN, 1962; O'BRIEN, 1964) and with balloon-borne apparatus (ANDERSON, 1965). We review below the observed characteristics of precipitation, and then illustrate our fundamental ignorance of the causes of these characteristics.

The important fact that the high-latitude boundary of trapping delineates the auroral zone was found by a coordinated study with the satellite Injun 3 (O'BRIEN and TAYLOR, 1964). But it is not known whether the boundary is inside or outside or even straddling the auroral zone, and yet this information is vital to determination of such important points as whether the source of auroral radiation lies on so-called "closed" or "open" magnetic field lines, as sketched in Figure 2.

Parameters of Precipitation

The parameters involved in an adequate experimental description of a precipitation event are such that a complete experimental measurement would yield

$$j(i, E, \alpha, t, \text{L.T.}, K_p, \text{etc.})$$

where j is the directional flux in units of particles $\text{cm}^{-2}\ \text{sec}^{-1}\ \text{sterad}^{-1}$, of particles of type i, e.g. electrons, protons or alpha particles, of kinetic energy E, at a pitch angle α, at a (real) time t, and a local time L.T., with a planetary magnetic disturbance index K_p, and so on.

There are associated measurements that need to be made for a full experimental understanding of a precipitation event, e.g. the light, ionization, heating, etc. produced by it, the associated VLF radiation (e.g. GURNETT and O'BRIEN, 1964), the associated variations of the local magnetic field, of the solar wind, of the backscattered particle flux, of conjugate phenomena, and so on. Most of these are also needed to test a theoretical model or interpretation of precipitation, but as yet there are few existing models that provide a definite prediction that can be tested experimentally.

The division into the high-latitude categories 2 and 3 defined above is based on the *possibility* that they may have different causes, e.g. category 2 being on "closed" field lines and category 3 on "open" field lines, but it is not yet certain that they are actually different or even whether they differ from category 4c as we discuss below.

In the remainder of this note, we treat precipitation of particles of categories 2 and 3 and 4c and some of their effects and related phenomena.

Precipitation of Electrons

Several studies have been made of the precipitation of electrons with energy $E_e \geqslant 40$ keV, since this is the energy of penetration of the 1.2 mg cm^{-2} mica window of type 213 geiger tubes. A few studies of lower-energy electrons have been made with different apparatus carried in rockets and satellites. We summarize the experimental results according to the dependence of the flux on the several parameters above.

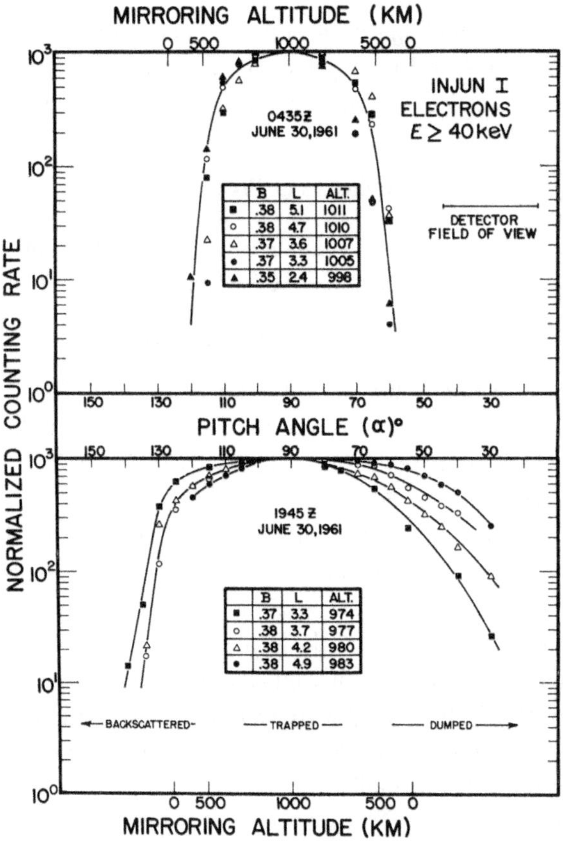

Fig. 3. Comparison of pitch-angle distributions with and without precipitation (after O'BRIEN, 1962).

KRASSOVSKII *et al.*, (1962) were the first to observe particle precipitation with satellite instrumentation. O'BRIEN (1962) then studied with Injun 1 the change in pitch-angle distribution associated with precipitation events, as shown in Figure 3. Then with Injun 3 the tendency towards isotropy and the finite backscatter were examined in further detail (O'BRIEN, 1964).

Figure 4 shows the variation with time in precipitation of electrons with $E_e \geqslant 40$ keV at $L=6.0$ and 6.5 in January 1963 observed by Injun 3 (O'BRIEN, 1964). The planetary magnetic disturbance index K_p is also shown. In general

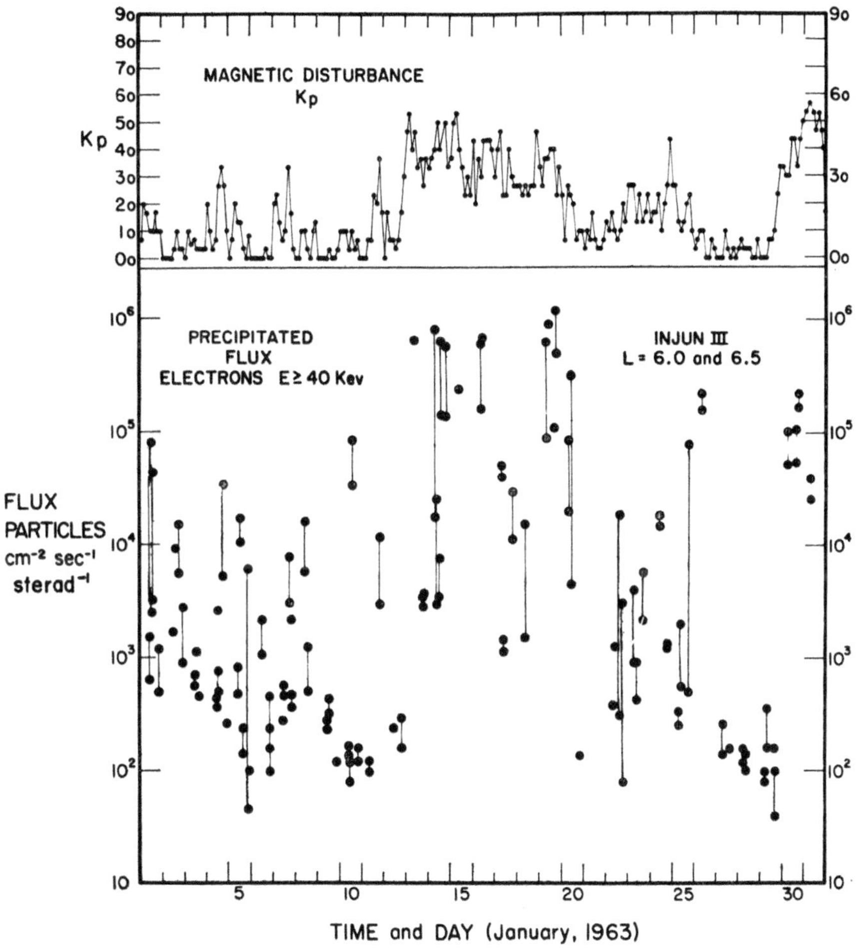

Fig. 4. Variation of precipitated flux of electrons with $E_e \geqslant 40$ keV (after O'BRIEN, 1964).

there is greater precipitation at times of larger K_p, and the precipitated flux at a given location increases on the average by a factor of about five for every unit increase of K_p between 0 and 5.

Figure 5 shows the variation with space or L of similar precipitation. Relevant conclusions are: (a) the precipitation increases on the average from $\sim 10^2$ to 10^3 particles/cm^2 sec at mid-latitudes to $\sim 10^5$ in the auroral regions; (b) the precipitation at any location is extremely variable with time (as Figure 4 shows); and

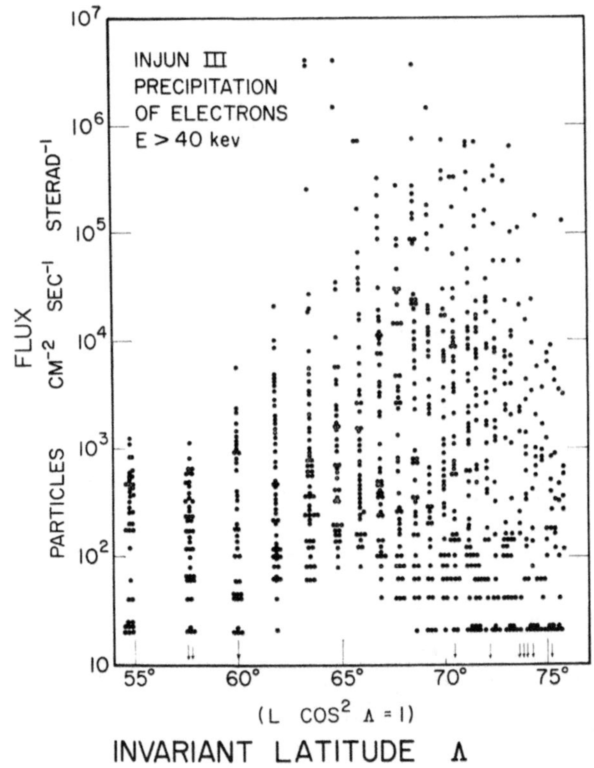

Fig. 5. Variation of precipitation with McIlwain's L parameter. Note that there is always precipitation at $L \sim 6$.

(c) in the region $5 \leqslant L \leqslant 7$ there is always finite precipitation, and therefore the precipitation or acceleration mechanism is always operating in this region.

Other features of particle precipitation discussed by O'Brien (1964) are:

(1) In intense precipitation the pitch-angle distribution tends to approach isotropy over the upper hemisphere at altitudes between ~ 250 km and ~ 1000 km, i.e., the precipitated flux tends to become equal to the locally-"trapped" flux.

(2) The energy spectrum varies in a complex way but, in general, tends to become softer towards higher latitudes (GURNETT and FRITZ, 1965; SHARP *et al.*, 1965). Experimental information on electron and proton spectra is generally poor, not only because of meagre experimental studies but also because of the enormous and rapid spectral variations that defy glib generalities.

(3) Of the electrons with $E_e \geqslant 40$ keV that are precipitated, at least about 10% are reflected or back-scattered and travel along magnetic field lines towards the conjugate region (O'BRIEN and others, 1965). Whether they reach the conjugate area is discussed below.

In Figure 6 we show simultaneous measurement (O'BRIEN and others, 1964) of an aurora, of the precipitated particles which were partly responsible for causing it, and of nominally trapped or Van Allen particles. Note that the particle flux tends to isotropy over the aurora, and that the aurora occurs near the high-latitude boundary of trapping of Van Allen particles. Many of the twenty-three detectors on the satellite Injun 3 have been used in studies of such auroral events e.g. the VLF detectors measured auroral hiss and so on (see O'BRIEN and others, 1964).

All of the above discussion treated electrons with energy $E \geqslant 40$ keV, and it is well known that such electrons contribute relatively little of the energy that excites aurora. With the Injun 3 photometers to detect auroras underneath the satellite, we were able to relate these high-energy electrons with auroral phenomena. The relationship was reasonably close when one considers that the photometers measured auroral light emitted from an area about 100-km diameter, while the particle detectors measured radiation characteristic of an area only approximately a cyclotron radius ($\sim$ tens of meters) about the satellite.

Probably the most important aspect of this Injun 3 study was the demonstration that the high-latitude boundary of trapping of 40 keV electrons is more or less along the immediate auroral zone. But it remains uncertain whether the auroral zone is outside or inside or straddling the boundary. Yet this is an extremely important point to resolve, since it immediately will determine whether the unknown auroral source or acceleration mechanism acts on "open" or closed magnetic field lines. I describe below a few experiments that may resolve this matter.

It is clear, however, from the above studies, that precipitation of 40 keV electrons certainly can take place within the magnetosphere (O'BRIEN, 1964). If one postulates a separate source for softer auroral electrons on open field lines, then one needs also a precipitation mechanism on closed field lines. This is the reason we specified categories 2 and 3 as separate categories above.

Recently GURNETT and FRITZ (1965) have given attention to the problem of variation of electron energy with latitude, and they present several examples where the low-energy ($\sim$10 keV) electron flux was a maximum at a higher-latitude than the boundary of apparent trapping of 40 keV electrons. If this was invariably the case, then the auroral particles might be ascribed to sources outside the stable magnetosphere. But it is not always the case, e.g. see Figure 10 of O'BRIEN and TAYLOR (1964), so the matter still remains unsettled.

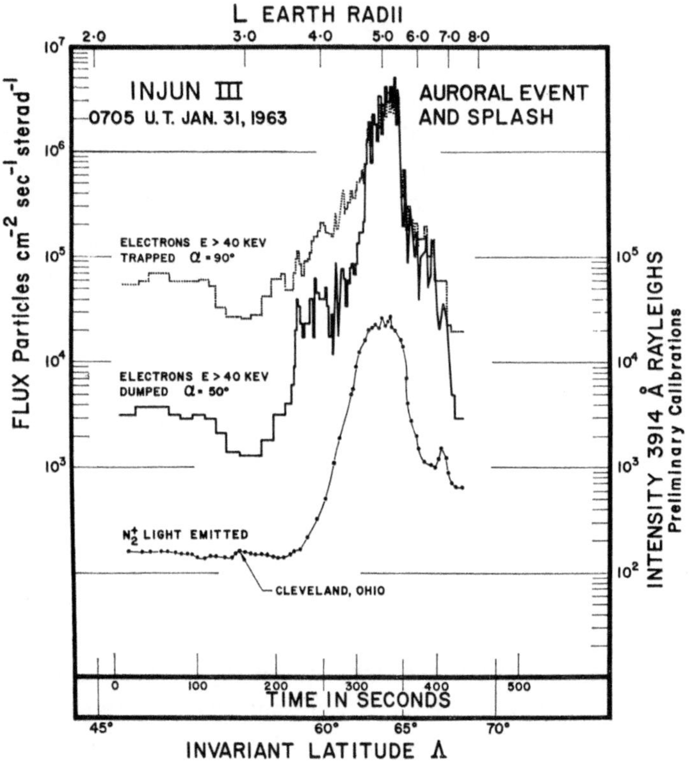

Fig. 6. Sample of Injun 3 simultaneous observation of an aurora and of precipitated and "trapped" electrons (after O'BRIEN and TAYLOR, 1964).

What gives some added plausibility to the above tentative assertion is the observation by FREEMAN *et al.* (1963) of a large flux (probably of electrons with several keV) outside the boundary of trapping. Such a flux, like that observed earlier by GRINGAUZ *et al.*, (1961) might conceivably be the source region from which auroral electrons are derived. Theoretical models of the magnetosphere, such as those discussed by Dessler and Michel at this conference, imply that indeed auroras occur at higher latitudes than the boundary of trapping of 40 keV electrons.

Another piece of experimental evidence that tends to support such a model is that the energy spectra of auroral electrons are typically much softer than those of Van Allen particles (McIlwain, 1960). However, such an effect could be ascribed to a dependence of the energy spectra on pitch angle on a given magnetic field line, rather than to the above dependence on latitude.

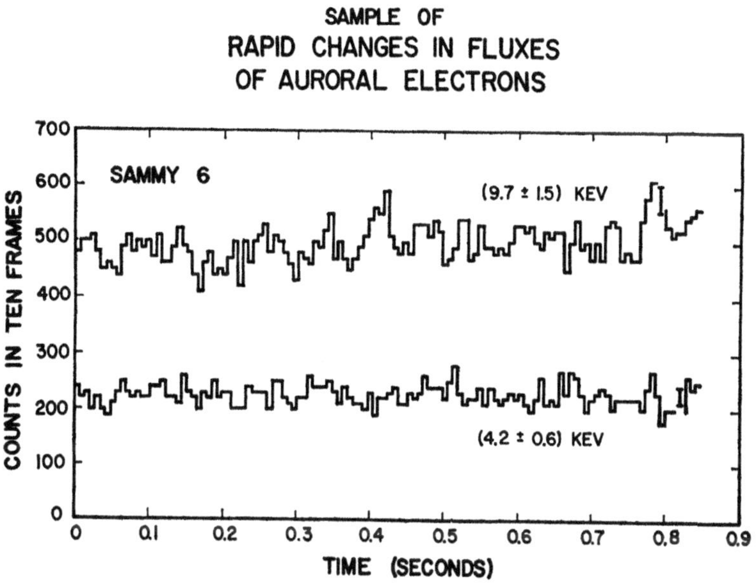

Fig. 7. Example of rapid spectral changes in auroral electrons.

One definitive experiment to resolve this problem would be a complete study of trapping, precipitation and auroras at nominally-conjugate points. The extensive balloon-borne studies of electron precipitation (e.g. Brown *et al.*, 1965) have not resolved this problem, and in part these cannot ever resolve the auroral problem because of their experimental bias towards detection of higher-energy electrons.

The Rice University group is working on the use of two simultaneously-operative "Owl" satellites to study this problem.

The energy spectra of auroral electrons have been studied by several groups since the original rocket flights of McIlwain (1960) and Davis *et al.* (1960). The two most recent definitive measurements have been made by the Lockheed group (Sharp *et al.*, 1965) with low-altitude satellites and by the Rice group with a Nike-Apache rocket Sammy 6.

The Lockheed measurements show that the spectra do not "blow up" at very low energies ~100 eV. This implies either (a) that the source of auroral electrons is filled with particles with initial energies of ~100 eV (e.g. a thermalized solar wind in the magnetosheath) or (b) that the acceleration and propagation conditions are such that all electrons are given a few hundred eV in the process.

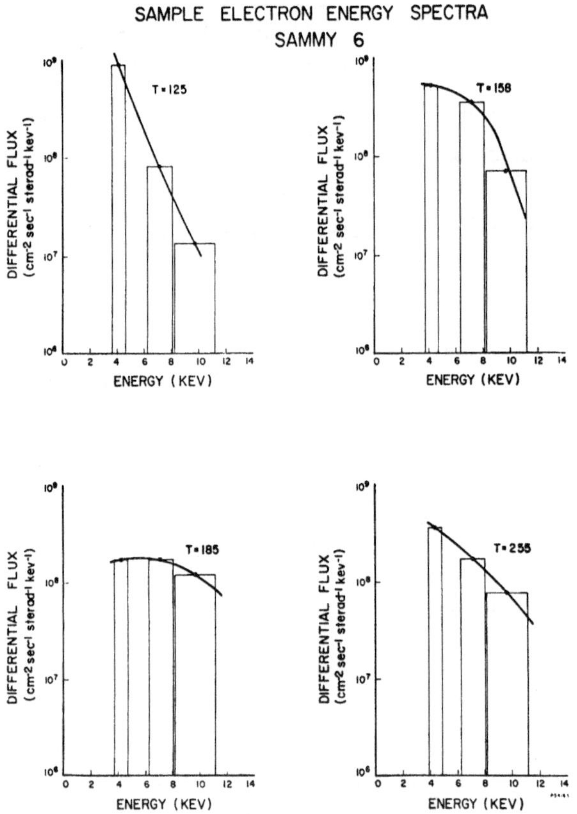

Fig. 8. Samples of energy spectra of auroral electrons.

The Rice group is currently analyzing many spectral samples made in a rocket flight into an aurora (WESTERLUND *et al.*, 1965; LAQUEY *et al.*, 1965). Samples are shown in Figures 7 and 8. This experiment was designed to study the time dispersion of arrival at the rocket of electrons with different energies. From such measurements and appropriate models one can seek to estimate the altitude of the auroral-electron source.

Precipitation of Protons

Early knowledge of proton precipitation came generally from ground-based studies of Doppler-shifted Balmer emissions (see CHAMBERLAIN, 1961). These indicated that fluxes of order 10^7 to 10^8 protons cm^{-2} sec^{-1} with energies of order 1 keV occasionally bombard the auroral atmosphere.

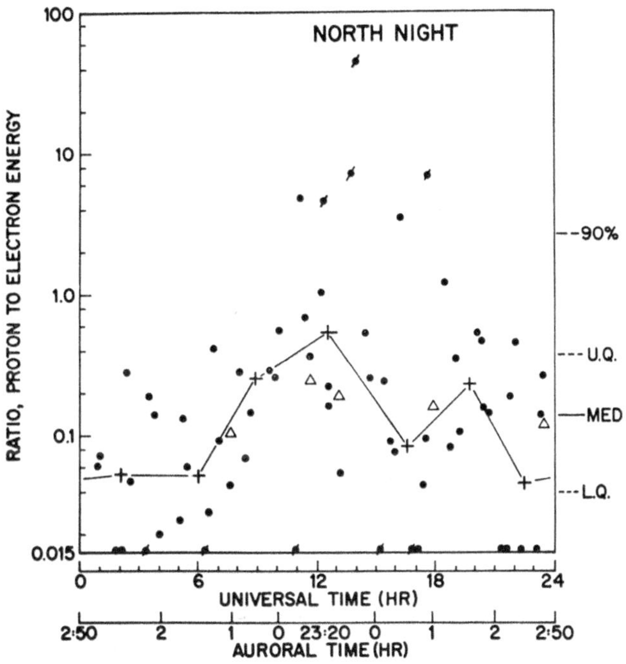

Fig. 9. Comparison of fluxes of precipitated protons and electrons (after EVANS *et al.*, 1965).

Direct measurement of the higher-energy ($E_p \gtrsim 100$ keV) component of this radiation was accomplished with rockets (MCILWAIN, 1960 and DAVIS *et al.*, 1960) and recently with polar orbiting satellites (MIHALOV and MOZER, 1963 and EVANS *et al.*, 1966). The very important finding made by EVANS *et al.* (1966) is that the total auroral energy input from protons is an appreciable proportion (some ten to twenty percent) of the electron energy input (Figure 9), a possibility discussed earlier (e.g. OMHOLT, 1964). One may therefore envisage bright well-defined auroral forms as being excited by electron bombardment where the cyclotron radius is only of order tens of meters, while more diffuse forms may be excited by protons with cyclotron radii some kilometers wide.

Now, there are appreciable fluxes of trapped protons e.g. 10^8 cm^{-2} sec^{-1} with energies above 100 keV (DAVIS and WILLIAMSON, 1963) just as there are appreciable fluxes of trapped electrons. So it is valid to examine whether the trapped protons might be a reservoir to supply the precipitated protons, just as the same speculation has been examined and found inadequate for electrons (e.g. see O'BRIEN, 1964). One can quickly deduce that the proton reservoir is not adequate either, unless it has a source sufficient to replenish it in a few hours.

The source and acceleration problem is therefore twofold, and it must concern itself with both electrons and protons.

Backscattering of Precipitated Particles

When particles are precipitated into the atmosphere, they may undergo Coulomb backscattering and then move back up out of the atmosphere. This effect is particularly significant only for electrons, but other "anomalous" backscattering processes such as mentioned below may affect protons also.

We define the reflection coefficient (R_α) at a given pitch angle (α) as being the ratio between the upcoming flux at (180°-α) and the downgoing flux at α. From simple Coulomb scattering alone one expects $R_\alpha \sim 0.1$ to 0.2 for electrons with a few keV to a few tens of keV. Such values have been found in the only satellite-borne study to date (O'BRIEN, 1964). But two rocket flights into auroras (MCDIARMID *et al.*, 1961 and CUMMINGS *et al.*, 1965) have found surprisingly high values of R_α approaching unity. MCDIARMID *et al.* (1961) attempted to explain their findings as being due to an upwards-directed electrostatic field.

CUMMINGS *et al.* (1965) found that R_α depended on the incident flux of electrons. When the precipitated flux of electrons with $E_e \geqslant 40$ keV was above $\sim 10^5$ particles cm^{-2} sec^{-1} sterad^{-1}, R_α approached the theoretical value of ~ 0.2. When the flux decreased, R_α increased to approach unity at fluxes of $\sim 10^2$ to 10^3. No explanation of this effect has been given, and numerous possible explanations were refuted by CUMMINGS *et al.* (1965). But one interesting and hitherto-untreated effect was discussed, wherein the magnetic field created by an auroral current system in the ionosphere can significantly perturb the size of the loss cone and the form of the pitch-angle distribution.

This type of effect can be very significant in particle precipitation and it provides a form of "atmospheric feedback". The current flows essentially where the ionization is greatest, and the ionization is greatest in turn where particle precipitation is greatest. So the magnetic field perturbation of the particle precipitation is greatest where the particle precipitation is greatest. The magnetic field perturbation even with a moderate current system, can enhance the magnetic field at higher altitudes and make it double-valued with altitude, so that particles are forbidden from mirroring over an altitude range of ~100 to 130 km (say) because they have mirrored already at higher altitudes.

High-Latitude Spikes

Recently McDiarmid and Burrows (1965) reported as a unique phenomenon the occurrence of spikes of electrons with $E_e > 40$ keV at latitudes somewhat higher than the boundary of trapping. They suggest that such spikes may represent the production of electrons in the tail of the magnetosphere, which they suggest may also be a source of Van Allen electrons and of electrons precipitated over the auroral regions. They suggest further that the "spikes" may be related to the "islands in time" observed by Anderson *et al.* (1964) with Imp 1 in the tail of the magnetosphere.

The particular relevance of this point here is that the suggestion by McDiarmid and Burrows (1965) explicitly is that most of the precipitated electrons in the auroral regions may have been energized in the neutral sheet of the magnetospheric tail and then injected into the trapping region of the earth's field. Hence, rather than having an "open-field" source and another "closed-field" source, one might envisage simply an open-field or a neutral-sheet source.

Concluding Comments

Although a great deal of experimental data about particle precipitation has been obtained, there are still numerous gaps in our experimental knowledge of precipitation, and there are even more gaps in theoretical understanding of the causes of precipitation.

For example, experimentally we need to know the answers to the following points:

(1) To what extent does conjugate particle precipitation occur in each of the above categories? E.g. is there conjugacy only in category 2?

(2) What is the dependence of the energy spectrum on the pitch-angle, e.g. is the spectrum softer as $\alpha \to 0$?

(3) What is the ratio of protons to alpha particles in auroral precipitation? From this knowledge one may hope to identify the source (an experiment by David Reasoner and the writer is underway to resolve this matter).

(4) What are the altitude and injection characteristics of the auroral-particle source? E.g. does it lie on "open" or "closed" field lines or in the neutral sheet?

(5) What is the energy balance between precipitated particles and their effects? E.g. only one percent or so of the particle energy deposited in an aurora is manifested as visible light, so how much is in the UV, in heat, etc.?

(6) What are the complete descriptions of the backscattering of auroral particles from the atmosphere and do any of these particles reach the conjugate area?

Experimental answers to these questions are required both for guidelines and for subsequent tests of auroral and precipitation theories.

References

Anderson, K. A.: 1965, *Auroral Phenomena* (ed. M. Walt), Stanford University Press.

Anderson, K. A., Harris, H. K., and Paoli, R. J.: 1964, Physics Dept. and Space Science Lab., University of California Report, Series **5**, Issue 50 (1964).

Brown, R. R., Barcus, J. R., and Parsons, N. R.: 1965, *J. Geophys. Res.* **70**, 2579.

Chamberlain, J. W.: 1961, *Physics of the Aurora and Airglow,* Academic Press, New York.

Cummings, W. D., LaQuey, R. E., O'Brien, B. J., and Walt, M.: 1965, *J. Geophys. Res.*, in course of publication.

Davis, L. R. and Williamson, J. M.: 1963, *Space Res.* **3**, 365.

Davis, L. R., Berg, O. E., and Meredith, L. H.: 1960, *Space Res.* **1** (ed. H. Kallman-Bijl), 721.

Dessler, A. J. and O'Brien, B. J.: 1965, *Satellite Environment Handbook,* 2nd Edition (ed. F. Johnson) Stanford University Press.

Evans, J. E., Joki, E. G., Johnson, R. G., and Sharp, R. D.: 1966, *Space Res.* **6.**

Freeman, J. W., Van Allen, J. A., and Cahill, L. J.: 1963, *J. Geophys. Res.* **68,** 2121.

Fritz, T. A. and Gurnett, D.: 1965, *J. Geophys. Res.* **70,** 2485.

Gringauz, K. I., Kurt, V. G., Moroz, V. I., and Shklovskii, I. S.: 1961, *Soviet Astronomy* **4,** 680.

Gurnett, D. and O'Brien, B. J.: 1964, *J. Geophys. Res.* **69,** 64.

Krassovskii, V. I., Shklovskii, I. S., Galperin, Y. I., Svetlitskii, E. M., Kushnir, Yu. M., and Bordovskii, A. G.: 1962, *Planet. Space Sci.* **9,** 37.

LaQuey, R. E., O'Brien, B. J., and Westerlund, L.: 1965, American Geophysical Union Meeting, Dallas (Sept. 1965).

McDiarmid, I. B. and Burrows, J. R.: 1965, *J. Geophys. Res.* **70,** 3031.

McDiarmid, I. B., Rose, D. C., and Budzinski, E. E.: 1961, *Canadian. J. Phys.* **39,** 1888.

McIlwain, C. E.: 1961, *J. Geophys. Res.* **66,** 3681.

McIlwain, C. E.: 1960, *J. Geophys. Res.* **65,** 2727.

Michel, F. C.: 1965, *Planet. Space Sci.*, in course of publication.

Michel, F. C. and Dessler, A. J.: 1965, *J. Geophys. Res.*, in course of publication.

Mihalov, J. D. and Mozer, F. S.: 1963, *Trans. Am. Geophys. Union* **44,** 881.

O'Brien, B. J.: 1964, *J. Geophys. Res.* **69,** 13.

O'Brien, B. J. and Taylor, H.: 1964, *J. Geophys. Res.* **69,** 45.

O'Brien, B. J.: 1962, *J. Geophys. Res.* **67,** 1227.

Omholt, A.: 1965, *Auroral Phenomena* (ed. M. Walt), Stanford University Press.

Pieper, G. F., Zmuda, A. J., Boström, C. O., and O'Brien, B. J.: 1962, *J. Geophys. Res.* **67,** 4959.

Sharp, R. D., Evans, J. E., Johnson, R. G., and Reagan, J. B.: 1965, *Space Res.* **5.**

Van Allen, J. A.: 1957, *Proc. Acad. Sci. U. S.* **43,** 57.

Walt, M.: 1964, *J. Geophys. Res.* **69,** 3947.

Westerlund, L., O'Brien, B. J., and LaQuey, R. E.: 1965, American Geophysical Union Meeting, Dallas (Sept. 1965).

LOSS RATES OF TRAPPED ELECTRONS BY ATMOSPHERIC COLLISIONS

MARTIN WALT

Lockheed Missiles and Space Company Research Laboratory, Palo Alto, California

1. Introduction

The collisions of energetic trapped electrons with constituents of the atmosphere has long been recognized as a possible loss mechanism. There is no question that collisions occur, and the influence of such collisions on trapped electrons is readily calculable, within the limitations of our knowledge of the atmosphere and geomagnetic field. The basic question of geophysical interest is the role of these collisions in the overall history of the trapped electrons. One wishes to know whether collisions are an important loss mechanism, or whether other processes such as plasma instabilities or interactions with electric and magnetic fields dominate the behavior.

From evidence such as that presented by Van Allen, Williams, and O'Brien on the time variations of trapped electrons in the outer belt and the precipitation of electrons into the atmosphere, it is apparent that atmospheric collisions play a negligible role in the behavior of electrons in the outer belt, except possibly at low altitude. However, in the inner belt where the magnetic perturbations are less intense and the atmosphere is denser, one would expect the atmosphere to be relatively more important. Furthermore, the rather good agreement between the high energy proton measurements and the theoretical predictions based on a neutron decay source and atmospheric losses (LENCHEK and SINGER, 1963; VALARIO, 1964) strongly suggests that the atmosphere is a major loss mechanism for protons. Since the high energy protons are more subject to some kinds of magnetic perturbations than are the trapped electrons, it is reasonable to expect that the atmosphere has an important influence on electrons also.

One has two goals in mind in calculating the effects of the atmosphere on trapped electrons and comparing the results of theory and experiment. If there are discrepancies between theory and experiment, these differences will indicate the existence and magnitude of the other loss processes and perhaps lead to an understanding of their nature. Secondly, if it is found that in some region of the magnetosphere the atmosphere is the only important loss mechanism, then from the equilibrium flux and energy spectrum one can deduce the magnitude and spectrum of the source. Such knowledge may hopefully lead to insight into the source of the electrons in the Van Allen radiation belt.

2. Effect of Atmospheric Collisions on Electron Distributions

A. ELECTRONS TRAPPED AT HIGH ALTITUDES

When an energetic trapped electron collides with an atmospheric atom, it will in general lose energy and be deflected. In the energy range of interest for the Van Allen belt (several tens of kilovolts to several MeV), the collisions can be treated as coulomb collisions in which the differential cross sections for deflections are strongly peaked in the forward direction. In a single collision the electron will lose some tens of volts energy. Thus, in an individual collision the electron will have its trajectory altered only slightly, and the effect of the atmosphere will be felt only by the cumulative effect of many such collisions. Since these collisions are stochastic, it is not possible to predict the behavior of an individual electron, it is only possible to assign probabilities to the future behavior, or equivalently, it is possible to describe the time dependence of a distribution function describing the behavior of an electron population. This stochastic nature of the motion of a trapped electron suffering collisions greatly complicates the mathematical treatment and forces one to use tedious numerical techniques to obtain quantitative results.

In deriving the diffusion equation for the electron distribution function, it will be assumed that (a) the scattering is described by two-body coulomb collisions between fast electrons and atmospheric constituents at rest, (b) between collisions, the adiabatic invariants of motion are constant, (c) collisional diffusion across magnetic L shells is neglected, and (d) straggling resulting from the statistical fluctuations in the amount of energy lost in individual collisions is neglected. While it is possible that straggling could be important for some situations, the broad energy spectra usually encountered in radiation belt problems makes this feature appear negligible. In the present discussion attention will be focused on electrons whose trajectories are altered only slightly during a longitudinal drift period. Thus, atmospheric densities can be averaged over a complete drift period, and the resulting electron fluxes will be independent of longitude. In the low altitude regions where the observed fluxes on a given B & L line vary with longitude, this longitude independent treatment is not applicable. Qualitative considerations of the low altitude case will be discussed below.

With these assumptions, the diffusion equation describing the time evolution of a distribution of electrons on a single magnetic shell may be written as follows. The derivation has been published elsewhere (MacDonald and Walt, 1961) and will not be repeated here.

$$\frac{\partial f}{\partial t}=\frac{\partial}{\partial T}\left[\frac{T+1}{T^{1/2}(T+2)^{1/2}}\mathscr{M}f\right]+\frac{T+1}{2T^{3/2}(T+2)^{3/2}}\frac{\partial}{\partial\mu}\left[S\mathscr{L}\mu(1-u^2)\frac{\partial}{\partial\mu}\left(\frac{f}{\mu S}\right)\right] \tag{1}$$

where

μ = cosine of pitch angle in the equatorial plane.

$T=E/mc^2$ = electron energy in rest mass units.

S = distance between mirror points as measured along the spiral path of the particle

In this equation $f(\mu, T, t)$ is the number of electrons at time t per unit ΔT and unit $\Delta\mu$ that have orbits passing through a unit area perpendicular to the magnetic field when crossing the equatorial plane. In terms of $f(\mu, T, t)$ the flux is given by

$$\phi(\mu,T,t)=\frac{f(\mu,T,t)}{4\pi\mu t_b}(\text{unit area}\cdot\text{unit }\Delta T\cdot\text{steradian})^{-1}$$

where t_b is the time required for an electron to move between mirroring points. Note that $f(\mu, T, t)$ is a distribution of orbits rather than a flux or a particle density. The functions $\mathscr{L}(\mu, T)$ and $\mathscr{M}(\mu, T)$ are time averages over the particle trajectory and give the effect of the atmosphere in scattering and slowing down the electrons. $\mathscr{L}$ & $\mathscr{M}$ thus contain all the information concerning the atmosphere and geomagnetic field. Explicitly,

$$\mathscr{L}(\mu,T)=4\pi(e^2/mc^2)^2c\left\langle\frac{\alpha^2(1-\mu^2)}{\mu^2(1-\alpha^2)}\sum_i n_iZ_i^2\ln\eta_i^{-1}\right\rangle$$

$$\mathscr{M}(\mu,T)=4\pi(e^2/mc^2)^2c\left\langle\sum_i n_iZ_i\ln\sigma_i\right\rangle$$

where m = mass of electron, e = charge on the electron, α = cosine of local pitch angle at any point in the orbit, c = velocity of light. The quantities n_i and Z_i are the number density and atomic number of the scattering constituent i, respectively, and the brackets denote time averages.

The logarithmic factors in the $\mathscr{L}(\mu, T)$ and $\mathscr{M}(\mu, T)$ functions result from integration over all possible impact parameters for the collisions, and thus depend on the scattering constituent. The values of these factors are of order 10 and vary slightly with electron energy. This weak energy dependence of $\mathscr{L}(\mu, T)$ and $\mathscr{M}(\mu, T)$ can usually be neglected. The calculation of $\mathscr{L}(\mu, T)$ and $\mathscr{M}(\mu, T)$ is a straightforward but tedious process, requiring the evaluation of the bracketed quantities in equation (1) at all locations of the particle bounce and longitudinal drift motion. The averaging of these quantities is carried out over the trajectory weighting each value proportionally to the time the electron spends in that interval. Thus, during the bounce motion, the longer residence time near the mirror point is important and during the drift motion, the variation in drift rate with longitude should be included. In general the largest contribution to $\mathscr{L}$ and $\mathscr{M}$ comes from that part of the trajectory near the South American magnetic anomaly. The dependence of $\mathscr{L}(\mu, T)$ and $\mathscr{M}(\mu, T)$ on μ is illustrated in Figure 1, for $L=1.185$ and $L=1.9$.

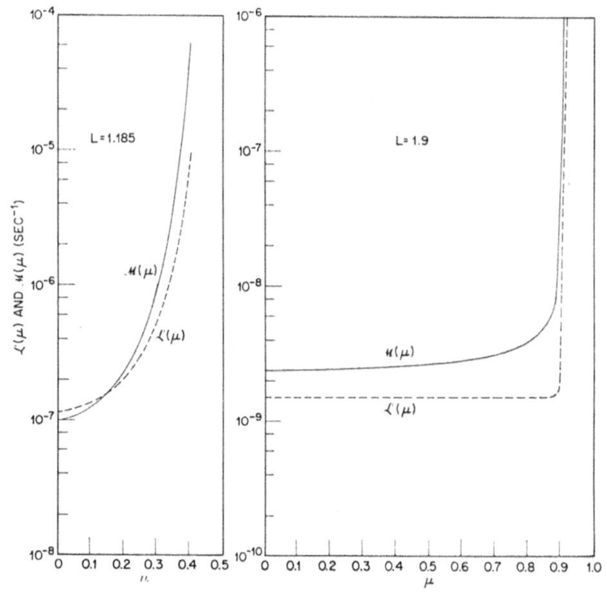

Fig. 1. The effective atmospheric density functions $\mathscr{L}(\mu)$ and $\mathscr{M}(\mu)$ for the magnetic shells $L=1.185$ and 1.9.

Since the variables of equation (1) cannot be separated, an analytical solution is not possible without serious approximations. With the approximation that $\mathscr{M}(\mu, T)$ is independent of μ, a solution can be obtained by expanding f in a series of eigenfunctions of the diffusion operator. Physically this procedure is equivalent to expanding the distribution function into spatial normal modes and the neglect of the μ variation in $\mathscr{M}(\mu, T)$ results in the various modes remaining independent. At any arbitrary time the electron distribution can be found by summing the contributions of the various modes. Since this approximation for $\mathscr{M}(\mu, T)$ is not valid for $L<2$ (see Figure 1), the analytic solution is of limited geophysical interest. However, the general concept of normal modes is a useful one as will be seen below.

For computations in the inner belt region, where the application of the electron scattering theory is of greatest promise, the evaluation of equation (1) must be carried out by numerical means. This has been accomplished using standard finite differences techniques (VARGA, 1962). A three-dimensional mesh was constructed with coordinate axes identified with the variables t, T, and μ.

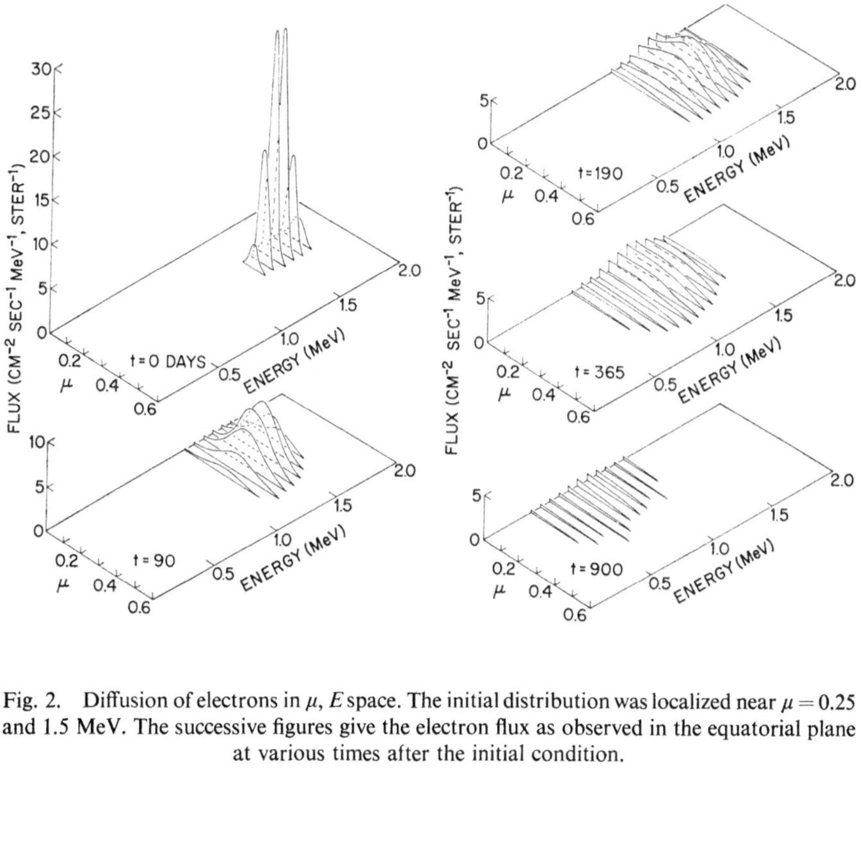

Fig. 2. Diffusion of electrons in μ, E space. The initial distribution was localized near $\mu = 0.25$ and 1.5 MeV. The successive figures give the electron flux as observed in the equatorial plane at various times after the initial condition.

By integrating (1) over a volume element in μ, T, t space, a finite difference approximation to the equation was obtained. This difference equation connects the values of $f(\mu, T, t)$ at five points in μ, T, t space. Points at $\mu = 0$ and μ_c, where μ_c is the maximum value of μ for which trapped orbits can occur, were related by slightly modified difference formulas owing to the boundary conditions applicable to these points. Given an initial distribution $f(\mu, T, 0)$ on the plane $t = 0$, it was then possible to generate the distribution at all planes $t > 0$ by application of the finite difference equations over each successive $t = \text{constant}$ plane.

The diffusive nature of the electron behavior is illustrated in Figure 2 which shows the time history of a distribution of 1.5 MeV electrons initially injected in a localized range of μ near 0.25. The distribution of flux as a function of both pitch angle cosine and energy is indicated by the curves along T=constant planes, the flux distribution being the envelope of the curves shown. The different figures give the distribution at the indicated times after the injection of the initial, peaked distribution. The most significant feature of Figure 2 is the dispersion in pitch angle resulting from scattering. This rapid spread in pitch angle causes the electron distribution to take on a configuration characterized by the shape of the magnetic field and the atmospheric density profile. Such behavior makes it extremely difficult to infer the initial pitch angle distribution from measurements made at some later time. Similarly, in the case of the natural belts, it appears unlikely that one can derive the angular distribution of the electron source function from measurements of the equilibrium distribution. From Figure 2, it is apparent that changes in pitch angle and in energy occur at comparable rates, so that approximations in which either process is neglected will give only qualitative information.

B. DIFFUSION AT LOW ALTITUDE

For electron orbits which reach altitudes below about 150 kilometers, the parameters describing the orbit are altered significantly during one longitudinal drift period. This rapid change causes the electron flux at constant B and L to be a function of longitude, being greater to the west of low field intensity magnetic anomalies. Data of this type have been presented by IMHOF and SMITH (1965b) and others. Although atmospheric collisions are very likely responsible for this longitude effect, a quantitative analysis has not been made thus far. A solution by diffusion equation as in the high altitude case would require a distribution function containing four independent variables, time, energy, pitch angle, and longitude. In the equilibrium case, three independent variables are needed. The density functions $\mathscr{L}$ and $\mathscr{M}$ would also be functions of longitude. Direct numerical integration of the time independent low altitude case would be as tedious as solving the time dependent high altitude case. However, because a considerable amount of low altitude data is available, such an analysis should be undertaken.

A simplified approach to the low altitude case was presented earlier (MACDONALD and WALT, 1962), and with the assumption of an exponential atmosphere and no energy loss, an analytic solution to the diffusion equation was found. The result showed that even for particles mirroring in the dense lower atmosphere, a rapid dispersion in pitch angle distribution is the characteristic behavior. However, the neglect of energy loss collisions was not justified and the longitude effect produced by magnetic anomalies was discussed only qualitatively.

A more detailed treatment of the longitude dependence has recently been undertaken by ROEDERER and WELCH (1966). Since their distribution function is fundamentally different from the one used here, the equations cannot be easily transformed into equivalent expressions although the mathematical treatment is similar.

3. Comparison of Theory and Experiment

A. TIME VARIATION

As described earlier, the goal of this work is to evaluate the role of atmospheric collisions on the long term behavior of electrons in the radiation belts. In the event that atmospheric collisions were found to be the dominant loss mechanism, one would use this knowledge to determine characteristics of the electron source. Unfortunately, in studying the natural radiation belts, the lack of information on the source function has made it impossible to reach conclusions about the role of the atmosphere in the electron history. It was possible to show that *if* atmospheric collisions were the major loss mechanism, neutron decay could not be an important electron source (KELLOGG, 1960; WALT and MACDONALD, 1961). It was also shown that the observed time changes in the electron fluxes in the outer belt could not be produced by atmospheric effects.

This impass was broken dramatically in 1962 with the occurrence of the Starfish nuclear detonation. The electrons injected by this high altitude nuclear explosion provided an "initial condition" from which the effects of the atmosphere on the trapped particles could be calculated. The calculated time evolution of the electron fluxes could then be compared with experimental values to give an unambiguous test of the atmospheric scattering theory.

The initial spatial distribution of the electrons from this explosion was measured for $L<1.5$ by the satellite Injun 1, and these measurements, together with the assumption that the energy spectrum of the electrons was that of fission fragments, specified the initial conditions for equation (1). The subsequent behavior of the electrons has been calculated, giving the spatial and energy distribution to be expected at any later time. Examples of the energy spectrum and intensity as a function of time are shown in Figure 3, which gives the relative omnidirectional flux as a function of energy for L shells of 1.185, 1.25, and 1.5 at different times after the detonation. On each shell data are given for two values of B. The monotonic decay of the flux is apparent, and the principal spectral change results from the preferential removal of the low-energy electrons. This spectral hardening is more pronounced at low L values, where the magnetic mirrors are less efficient and the ratio of leakage to energy-loss processes is greater than for higher L shells.

The detailed comparison of the initial time variation of the Starfish electrons with the behavior expected on the basis of the atmospheric scattering theory has been studied, and the comparison of observed and calculated electron fluxes for the first 40 days after the detonation are shown in Figure 4 for $L=1.185$ and in Figure 5 for $L=1.25$ (WALT, 1964). The solid lines are the theoretical curves, and the points are experimental values obtained with Injun 1 (VAN ALLEN and

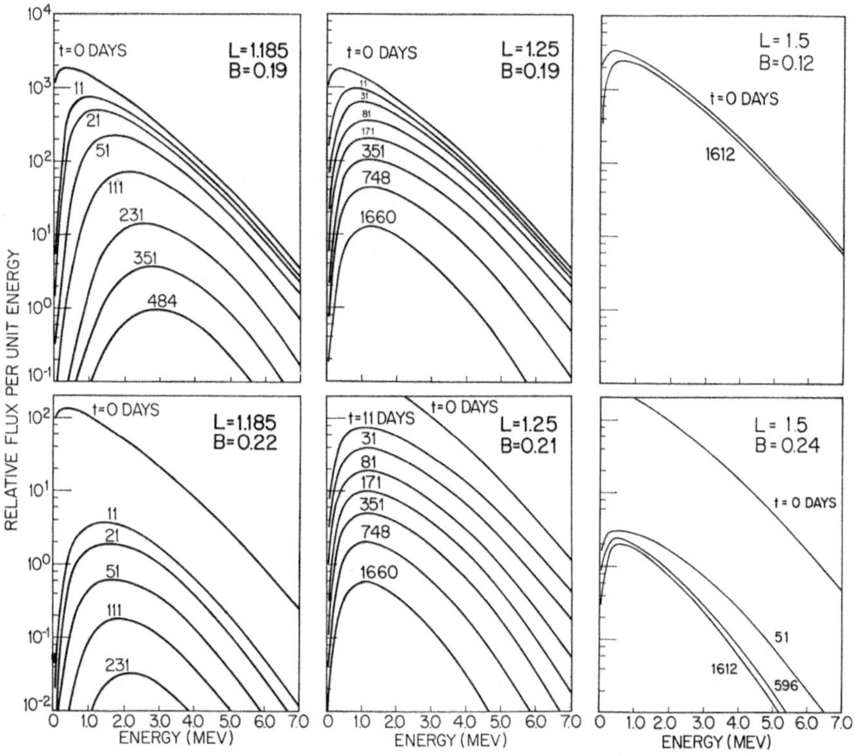

Fig. 3. Theoretical values of the electron flux produced by the high altitude nuclear detonation on July 9, 1962. The curves give relative values of the omnidirectional flux as a function of energy at various positions in the radiation belt and at various times after the detonation.

LIN, 1964). The experiment and theory coincide at $t=0$ since the experimental distributions were used as the initial conditions for the calculation. For $L=1.185$, the overall agreement between theory and experiment is excellent. Even such detailed behavior as the observed cross-over between the curves at $B=0.19$ and $B=0.195$ is reproduced by the theory.

At $L=1.25$ the agreement is also gratifying, although the deviation between experiment and theory at large values of B (low altitude) appears larger than can be accounted for by the uncertainties involved. However, the time dependence at high altitude appears to be accurately given by the atmospheric scattering theory. Recent work by Filz and Holeman on the increase in proton flux at low altitude following Starfish suggests that the background correction made to the Injun I data was not adequate. An increased background would explain the discrepancies observed at high B values.

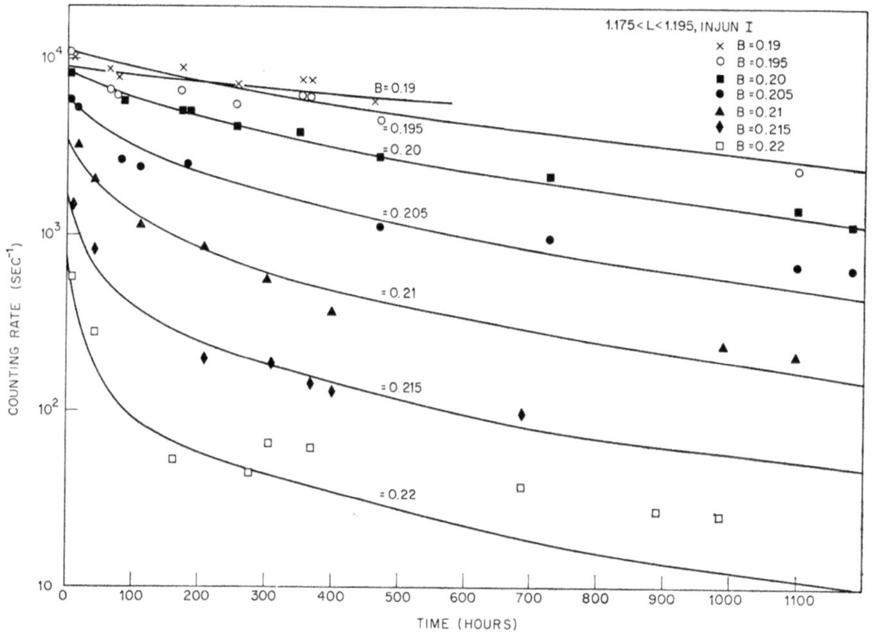

Fig. 4. Comparison of theoretical and experimental values for the time variation of the trapped electron flux resulting from the Starfish high-altitude nuclear explosion. Experimental points shown are from VAN ALLEN and LIN (1964); the solid lines are theoretical values calculated for the magnetic shell $L = 1.185$, assuming that the trapped electrons undergo collisions with the atmosphere.

In Figures 4 and 5 it is seen that the decay rate is high initially, particularly at the large B values. Subsequently, the decay becomes less rapid and asymptotically approaches an exponential function of time. In this later phase the decay rates at all B values on the same L shell are nearly equal. This behavior can be understood on the basis of the approximate or eigenfunction solution to the diffusion equation, where the eigenfunctions correspond to normal modes of the electron distribution. In the approximation of no coupling between modes, each mode will decay independently with an exponential time response, the decay rate being more rapid for the higher modes. Any initial distribution can be represented as a sum of normal modes which then decay concurrently, the higher modes becoming less important with increasing time. When only one mode remains, the flux at all values of B will decay exponentially with the same

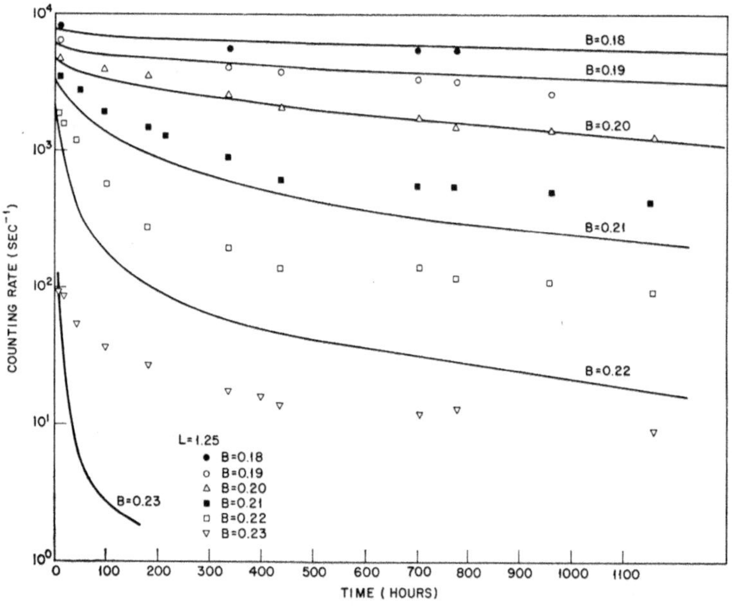

Fig. 5. Same as Figure 4, but $L = 1.25$.

time constant. Because the modes are coupled through the μ dependence of $M(\mu, T)$, this theoretical interpretation is not rigorously true, and the decay never becomes strictly exponential. However, over a limited period of time, the exponential fit to the decay is a convenient approximation.

In particular the time required for the electron flux to decrease by a factor of $1/e$, the characteristic decay time constant, is a useful parameter characterizing the loss rate of trapped electrons. A comparison of experimental values of the characteristic decay time with theory is shown in Figures 6 and 7, Figure 7 giving a more detailed view of the low L data of Figure 6. The experimental points are from the references indicated. While these measurements are not strictly comparable since they were made at different times, with different energy responses, and sometimes on different electron populations, such inherent differences are usually smaller than other uncertainties of the measurements. The dashed lines give the theoretical values obtained by numerical integration of equation (1). The two sections of the theoretical curve in Figure 7

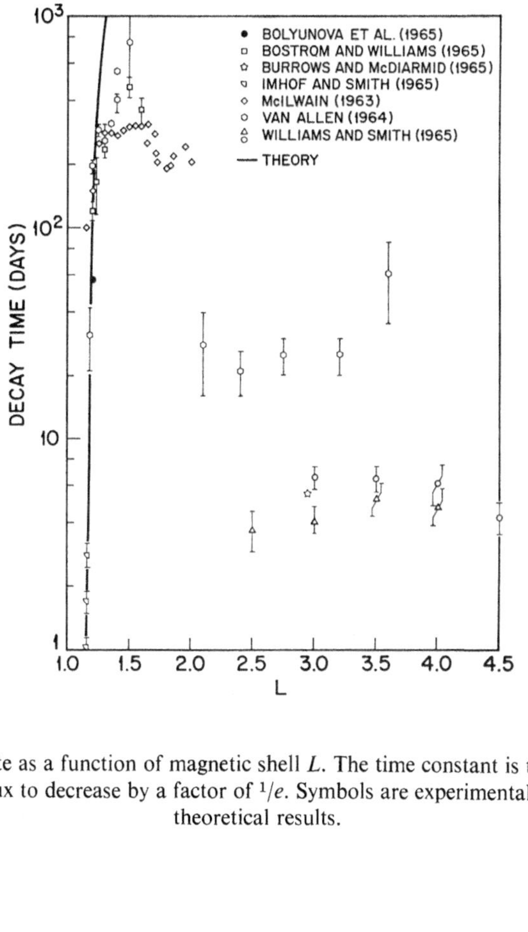

Fig. 6. Decay rate as a function of magnetic shell L. The time constant is the time required for the electron flux to decrease by a factor of $^1/e$. Symbols are experimental values, lines are theoretical results.

do not meet smoothly, since they were computed for different initial electron distributions. For $L > 1.185$ the initial electron distribution was a fission energy spectrum and the spatial distribution was selected to fit the Injun 1 measurements of the artificial radiation belt. This theoretical distribution function is the appropriate one for comparison since most of the experimental time constants were derived from measurements with Starfish. The theoretical curve near $L = 1.15$ was computed for a nearly monoenergetic group of electrons centered at 1.3 MeV, since the four experimental points in that region were derived from measurements by IMHOF and SMITH (1965a) of such a distribution.

From the comparison of theory and experiment as in Figures 4, 5, 6, and 7, it is apparent that the atmosphere has little effect on electrons trapped at $L \gtrsim 1.25$, except possible at low altitudes. However, in the lower part of the

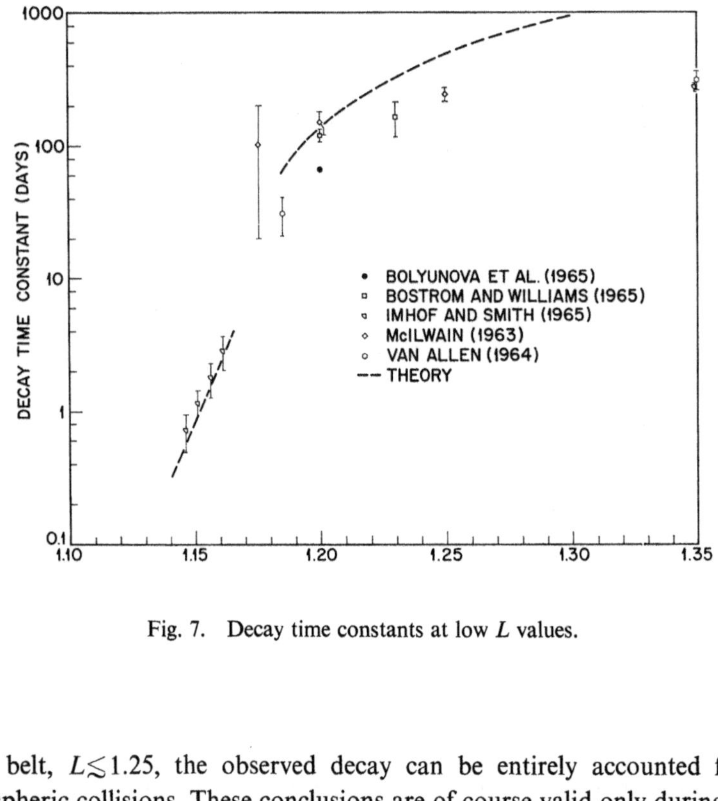

Fig. 7. Decay time constants at low L values.

inner belt, $L \lesssim 1.25$, the observed decay can be entirely accounted for by atmospheric collisions. These conclusions are of course valid only during years of minimum solar activity, a condition which has prevailed since the Starfish explosion. It is quite possible that the region of atmospheric dominance may be altered by increased solar activity since the atmospheric density and magnetic fluctuations will be significantly changed during that time.

4. Discussion and Source Characteristics

Although the evidence is strong that atmospheric collisions dominate the loss processes for electrons in the lower part of the radiation belt, one cannot conclude that atmospheric collisions are the only forces perturbing the motion. It is quite probable that the unknown mechanisms which control the loss at $L \cong 1.5$ are also present at $L < 1.25$; their effect is, however, overwhelmed by the more rapid loss through collisions. The most direct evidence for the presence of other forces at $L < 1.25$ comes from time changes in the electron radiation belts, for

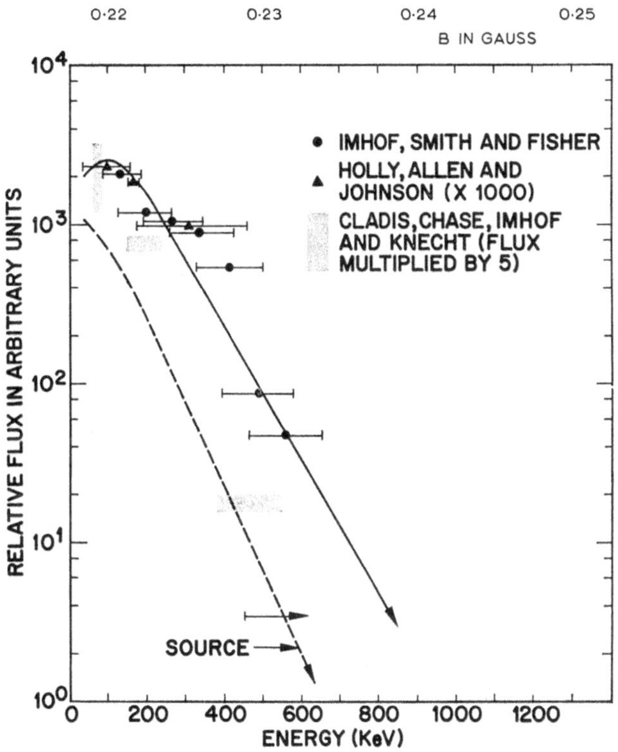

Fig. 8. The spectrum of the electron source in the lower part of the inner radiation belt. The dashed line gives the calculated source function while the solid line represents the normalized experiment spectra as presented by Imhof *et al.* (1963).

example, the spectrum measurements of Imhof and Smith (1965a). The nearly monoenergetic populations observed by them at $L \approx 1.15$ are very likely produced by some process associated with time variations in the geomagnetic field.

From indirect reasoning regarding the source of the natural electron belts, it also seems likely that processes which alter the adiabatic invariants must occur. The only known source mechanism for the inner belt electrons which does not require a change in the adiabatic invariants of the electrons is the decay of neutrons produced in the earth's atmosphere by cosmic ray collisions. Since this source is inadequate from the standpoint of both intensity and energy spectrum, some other source must exist, probably one which either accelerates electrons and/or transfers them across magnetic shells. Hence, by this indirect reasoning mechanisms other than atmospheric collisions probably alter the adiabatic invariants throughout the magnetosphere. Below $L=1.25$, however, the losses resulting from the collisions appear to be the dominant mechanism, at least during minimum solar activity.

In the region of space $L \lesssim 1.25$ the major loss mechanism for electrons results from collisions with the atmosphere. Therefore, experimental values of the natural flux of trapped electrons can be used in the diffusion equation to infer the conditions which the electron source must fulfill in order to maintain the observed belt. Although the spatial distribution diffuses to such an extent that the source distribution cannot be deduced from the electron flux distribution, both the energy spectrum and intensity of the source can be estimated. The source energy spectrum was calculated for the shell $L=1.185$ by an iteration procedure in which an assumed source spectrum in the diffusion equation was successively modified until the calculated equilibrium flux spectrum agreed with the experimental data. The resulting source flux spectrum and the calculated equilibrium flux spectrum, are both shown in Figure 8, together with a number of experimental values. It can be seen that the shapes of the source spectrum and the calculated equilibrium flux spectrum do not differ greatly. There are large uncertainties in the pre-Starfish experimental data which were used in the analysis. Thus, the peak displayed in the calculated equilibrium flux spectrum is not significant.

Direct measurements by FRANK and VAN ALLEN (1963) of the flux of natural electrons with energies greater than 40 keV for this region give a value of $\sim 2 \times 10^7$ electrons $\text{cm}^{-2}\ \text{sec}^{-1}$. In order to sustain an electron flux of this intensity, the source spectrum described above must have an intensity of at least 5×10^{-9} electrons $\text{cm}^{-3}\ \text{sec}^{-1}$ ($E > 40$ keV). Although this value of source intensity is a lower limit, it is approximately three orders of magnitude greater than can be supplied by decaying cosmic-ray albedo neutrons. The source spectrum required by the present analysis is also in severe disagreement with that produced by the beta decay of neutrons.

References

BOLYUNOVA, A. D., VAISBERG, O. L., GALPERIN, YU. I., POTAPOV, B. P., TEMNY, V. V., and SHUYSKAYA, F. K.: 1966, *Space Res.* **6.**

BOSTROM, C. O. and WILLIAMS, D. J., 1965, *J. Geophys. Res.* **70,** 240.

BURROWS J. R. and MCDIARMID, I. B.: 1965, Private communication.

FRANK, L. A. and VAN ALLEN, J. A.: 1963, *J. Geophys. Res.* **68**, 1203.
IMHOF, W. L. and SMITH, R. V.: 1965a, *Phys. Rev. Letters,* **14**, 885.
IMHOF, W. L. and SMITH, R. V.: 1965b, *J. Geophys. Res.* **70**, 569.
IMHOF, W. L., SMITH, R. V., and FISHER, P. C.: 1963, *Space Res.* **3** (ed. W. PRIESTER), 438.
KELLOGG, P. J.: 1960, *J. Geophys. Res.* **65**, 2705.
LENCHEK, A. M. and SINGER, S. F.: 1963, *Planet. Space Sci.* **11**, 1151.
MCILWAIN, C. E.: 1963, *Science* **142**, 333 and private communication.
MACDONALD, W. M. and WALT, M.: 1961, *Ann. Phys.* **15**, 44.
MACDONALD, W. M. and WALT, M.: 1962, *J. Geophys. Res.* **67**, 5025.
ROEDERER, J. G. and WELCH, J. A.: 1966, this volume, p. 373.
VALARIO, JOHN: 1964, *J. Geophys. Res.* **69**, 4949.
VAN ALLEN, J. A. and LIN, W. C.: 1964, *Trans. Am. Geophys. Union* **45**, 85.
VAN ALLEN, J. A.: 1964, *Nature* **203**, 1006.
VARGA, R. S.: 1962, *Matrix Iterative Analysis,* Prentice-Hall, Englewood Cliffs, New Jersey, 1962.
WALT, M.: 1964, *J. Geophys. Res.* **69**, 3947.
WALT, M. and MACDONALD, W. M.: 1961, *J. Geophys. Res.* **66**, 2047.

SOURCE OF OUTER ZONE PROTONS *

WILMOT N. HESS

Theoretical Division, Goddard Space Flight Center, Greenbelt, Md.

Introduction

In the $L=2$ to 5 range in the outer radiation belt, DAVIS *et al.* (1964) find that the spectra of the relatively stable 0.1 to 5 MeV protons show smooth but large variations with L and equatorial pitch angle, α_0, as in Figure 1. Protons near

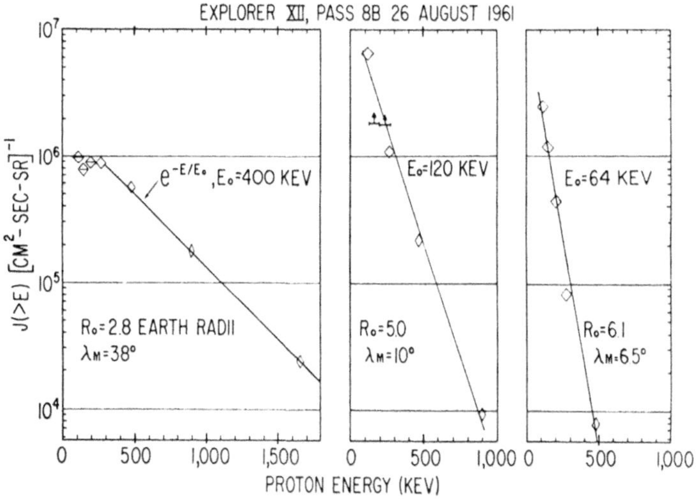

Fig. 1. Low energy proton spectra at various distances from the earth, from DAVIS and WILLIAMSON (1963).

the earth and at α_0 near 90° are more energetic than those at larger L and at smaller α_0. The spectra are well represented by e^{-E/E_0}.

Inspection of the experimental data showed that $E_0 \propto L^{-3} \propto B$. This was very suggestive since to have a characteristic energy of the system vary with the

* The first part of this paper follows closely the material in NAKADA, DUNGEY and HESS: 1965, *J. Geophys. Res.* **70**, 3529.

magnetic field suggested betatron acceleration and suggested that the particles were moving radially across the field. Such a process must violate at least the third adiabatic invariant of motion since the particle moving radially does not conserve the magnetic flux inside its orbit.

KELLOGG (1959) first suggested that the radiation belt might be formed through magnetic disturbances in which the third adiabatic invariant of trapped particles is violated without violating the first and second invariants. Violation of the third invariant allows motion in L-space. As particles move closer to the earth they tend to gain energy with the maintenance of the first invariant since, for example, E/B is a constant for 90° pitch angles. So, this process can introduce acceleration of protons. Kellogg's suggestion has been adopted for this study although the mechanism for motion in L space is unspecified. It has further been assumed that motion in L-space is rapid compared with atmospheric loss and scattering processes except very near the earth and that the geomagnetic field is sufficiently well represented by a dipole.

Energy and Angle Variations

If the first and second adiabatic invariants of trapped particles are maintained during motion in L space, changes in both the energy and equatorial pitch angle can be calculated. The first invariant is

$$\mu = \frac{E \sin^2 \alpha_0}{B_0} = \frac{EL^3 \sin^2 \alpha_0}{0.312} \tag{1}$$

where B_0 is the equatorial magnetic field. The second invariant is

$$J = m \oint v \cos \alpha \, \mathrm{d}l = mv \, \overline{\cos \alpha} l = \sqrt{E} LF(\alpha_0) \tag{2}$$

where m is the mass, v the velocity, α the local pitch angle and l is along the guiding center. The integration is over a complete north-south oscillation.

Since μ and J are constants, Equation (1) may be divided by the square of Equation (2) to give

$$L \left[\frac{\sin \alpha_0}{F(\alpha_0)} \right]^2 = \text{constant}. \tag{3}$$

From this, the changes in α_0 with L have been found and are shown in Figure 2. As DAVIS and CHANG (1962) have indicated, particles diffusing inwards assume flatter helices and move closer to the equator.

These changes in α_0 with L and Equation (1) may be used to find the variation in energy with L and α_0. Results are shown in Figure 3 for protons

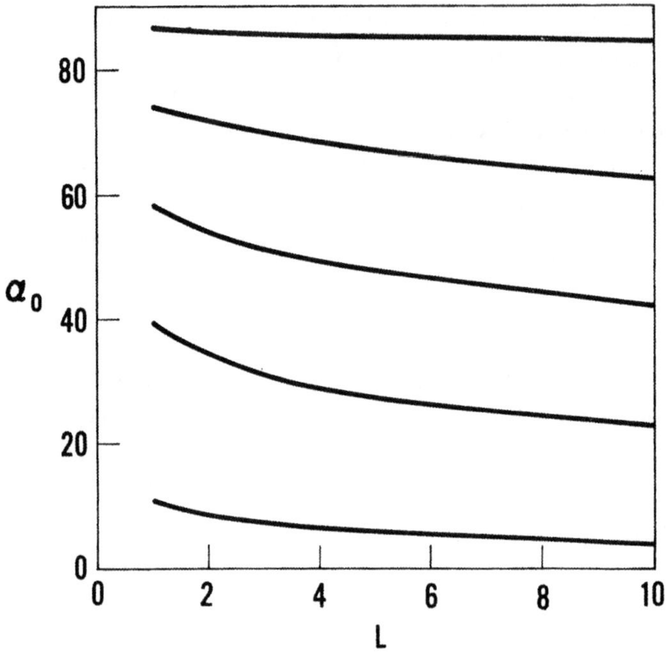

Fig. 2. Variation of equatorial pitch angle with L for motion conserving μ and J.

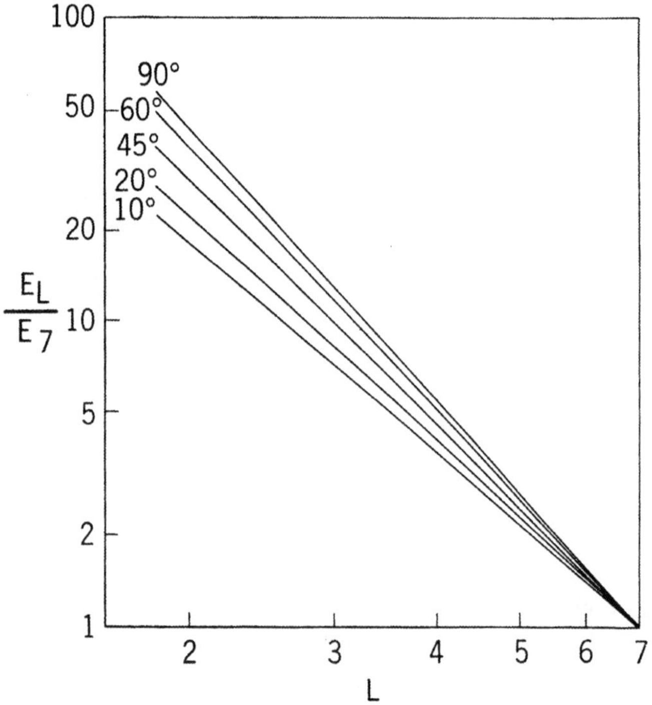

Fig. 3. The ratio of proton energies at L to that at $L=7$, assuming μ and J constant, for various equatorial pitch angles at $L=7$.

having α_0 values at $L=7$ as indicated on the curves. Energies are relative to energies at $L=7$.

The energy transformation is given by

$$EL^3 \sin^2 \alpha_{0L} = E_s L_s^3 \sin^2 \alpha_{0s} \tag{4}$$

where the subscript s refers to the source. If the source spectrum has an exponential form, the transformation is given by

$$\exp - E_s/E_{0s} \rightarrow \exp \frac{- EL^3 \sin^2 \alpha_0 L}{E_{0s} L_s^3 \sin^2 \alpha_{0s}} \equiv \exp - E/E_0 \tag{5}$$

which shows

$$E_0 L^3 \sin^2 \alpha_{0L} = E_{0s} L_s^3 \sin^2 \alpha_{0s} \tag{6}$$

From this it can be seen that an exponential source remains exponential after L space motion and that E_0 changes in the same way with L and α_0 as has been calculated for a single particle in the previous section.

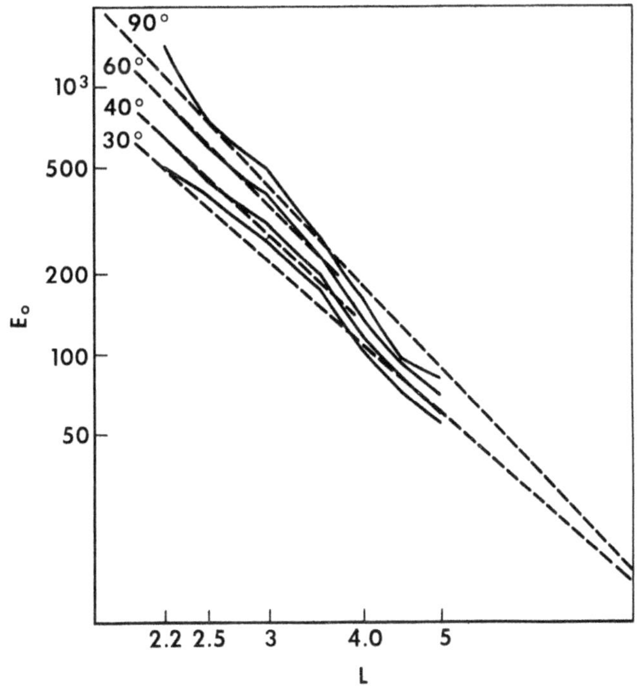

Fig. 4. Variation of E_0 with L for various equatorial pitch angles.

These two predictions of the model may be compared with experiment. The first prediction, that the spectra retains its exponential form, is in agreement with experiment. To test the second prediction, measured E_0 (DAVIS *et al.*, 1964) have been plotted in Figure 4 as a function of L with appropriate changes in α_0 with L. The labels on the curves refer to α_0 values at $L=7$. The dashed curves in Figure 4 are taken from Figure 3 for corresponding changes in E with L and α_0. The changes in E_0 with L show good agreement between the model and experimental results. The experimental results also give the same trend as the model in the change in the slopes of the curves with α_0.

If the dashed curves in Figure 4 are extended, they intersect near $L=10$. This intersection is where the spectrum is independent of α_0 and thus gives a source location with the simplest assumptions about the source.

Flux Variation with L

Now let us consider the way the flux of particles will vary with motion in L. The density of particles in phase space, f, is given by

$$f(v,\alpha) = \frac{\mathrm{d}N}{\mathrm{d}A\,\mathrm{d}t\,p^2\mathrm{d}p\,\mathrm{d}\Omega} \tag{7}$$

But, the pitch angle distribution in protons/cm^2-sec-ster-MeV measured by DAVIS *et al.* (1964) is

$$j(E,\alpha) = \frac{\mathrm{d}N}{\mathrm{d}A\,\mathrm{d}t\,\mathrm{d}E\,\mathrm{d}\Omega}, \tag{8}$$

This gives

$$f(v,\alpha) = \frac{j(E,\alpha)\,\mathrm{d}E}{p^2 v\,\mathrm{d}p} \tag{9}$$

or

$$f(E,\alpha) = \frac{1}{2mE} j(E,\alpha). \tag{10}$$

This shows the simple relationship between the measured fluxes and the velocity distribution function. Our model of the drift process has μ and J constants of the motion. We can study the way the particle flux varies with position by studying

$$f\left[E(\mu_1, J_1), \alpha(\mu_1, J_1), L\right]$$

The values of f were computed as functions of L for the data of DAVIS and WILLIAMSON (1963) for many pairs of the values of μ and J, and Figure 5 shows

f plotted against L for fixed μ and J. The curves for all μ and J were normalized in the region $L=4$ to compare their shapes and the following interesting empirical fact shows up

$$f(\mu, J, L) = g(L)\, h(\mu, J) \tag{11}$$

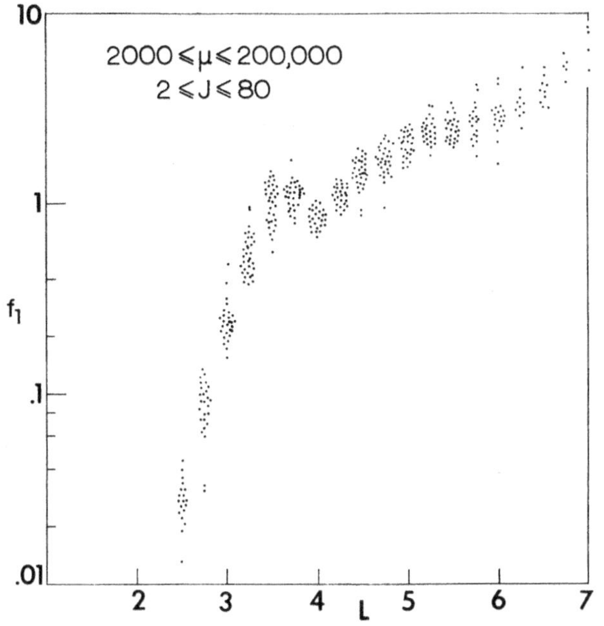

Fig. 5. The velocity distribution function f vs. L for various values of μ and J determined from Davis' data. The curves for different μ and J have been normalized in the Region $L \sim 4$.

That is to say, the function f is separable and the L dependence is essentially the same for all μ and I. This being the case we can write

$$\frac{f(\mu_1, J_1, L)}{f(\mu_1, J_1, 5)} = \frac{g(L)}{g(5)} = C(L) \tag{12}$$

where $C(L)$ can be read directly from Figure 5 and it can be used to transform fluxes from one L to another. Figure 5 shows f varying considerably with L. It is seen that $(\partial f/\partial L)_{\mu J}$ is always positive, suggesting that the particle source is at large L, the particles diffusing inwards and loss processes reducing f further in. The small slopes of Figure 5 at the larger L values implies that loss processes are probably relatively unimportant there, the slope probably being due to diffusion of particles away from the source at the outer boundary. The much larger slopes at the lower L values imply that loss processes are relatively important in this region.

We now have a scheme for transforming proton fluxes from one position in B, L space to another position. Using Equations (1) and (2) we know how the

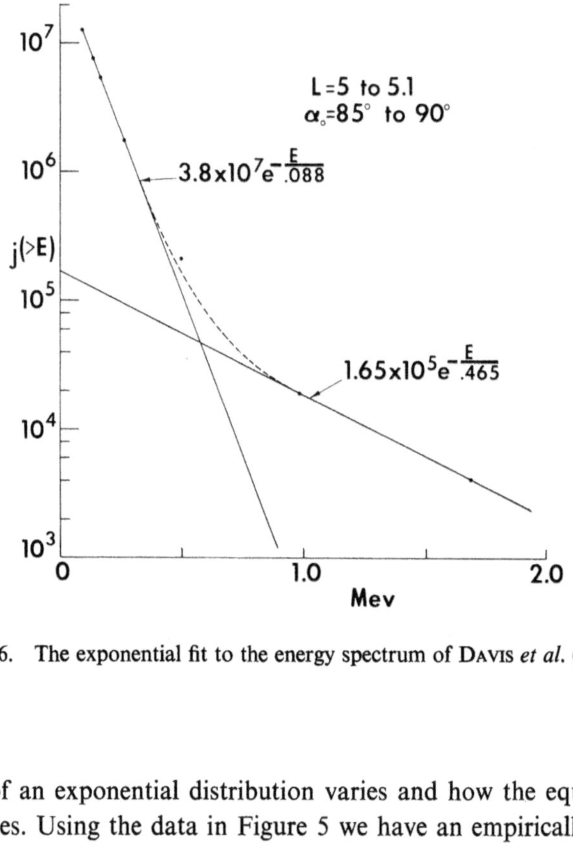

Fig. 6. The exponential fit to the energy spectrum of Davis *et al.* (1964).

energy E_0 of an exponential distribution varies and how the equatorial pitch angle changes. Using the data in Figure 5 we have an empirically determined scheme for transforming particle fluxes as given by Equation (12). Starting with the data of Davis *et al.* (1963) shown in Figure 6 for equatorial particles at $L=5$, we have transformed this to $L=2.4$. This equatorial data has been fit by an expression

$$j_i(E) = \sum_i a_i e^{-E/E_{0i}} = 3.8 \times 10^7 e^{-E/0.088} + 1.65 \times 10^5 e^{-E/0.465} \qquad (13)$$

Obviously the second term in Equation (13) is not very well known since from Figure 6 there are only two points that can be used to fit this. This transformed data is compared with the Relay 1 data of FILLIUS and MCILWAIN in Figure 7. Over the range where comparison is possible the agreement is very good. The transformed flux and the experimental flux at $L=2.4$ agree to within a factor of 2 which is as accurately as $C(L)$ is known from Figure 5. Beyond $E=18$ MeV the comparison is impossible. Davis' data runs out here. The experimentally

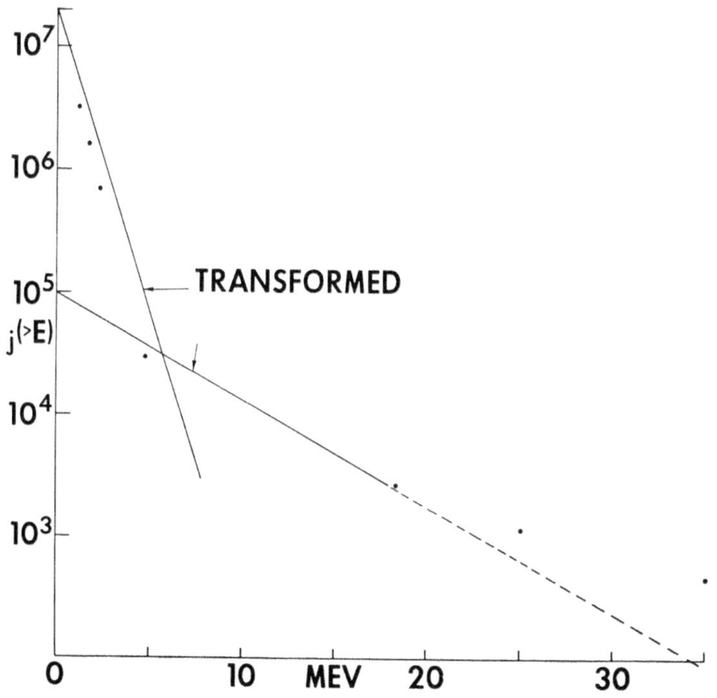

Fig. 7. A comparison of the transformed Davis, Hoffman and Williamson data with the spectrum measured by Fillius and McIlwain at $L=2.4$.

measured fluxes at $L=2.4$ fall significantly above the extrapolation of Davis' data shown dotted in Figure 7 suggesting that a third term should be added to the flux description in Equation (13). For part of the following discussion we have added a third exponential

$$j_3(E) = 1.0 \times 10^5 \, e^{-E/4.95} \tag{14}$$

Off-Equator Protons

Having found reasonable agreement between measured fluxes and transformed fluxes for equatorial particles, let us now examine off-equator particles having $\alpha_0 \neq 90°$. We start with a fit to the Davis *et al.* (1964) data at $L=5$ similar to that given in Equation (13) but with values of the constants $a_i(\alpha_0)$ and $E_{0i}(\alpha_0)$ varying with the pitch angle. Davis has made such a fit to his data so this initial data is well known. Using the same transformation as before involving equations (1), (2) and (12) the flux measured at $L=5$ has been transformed to

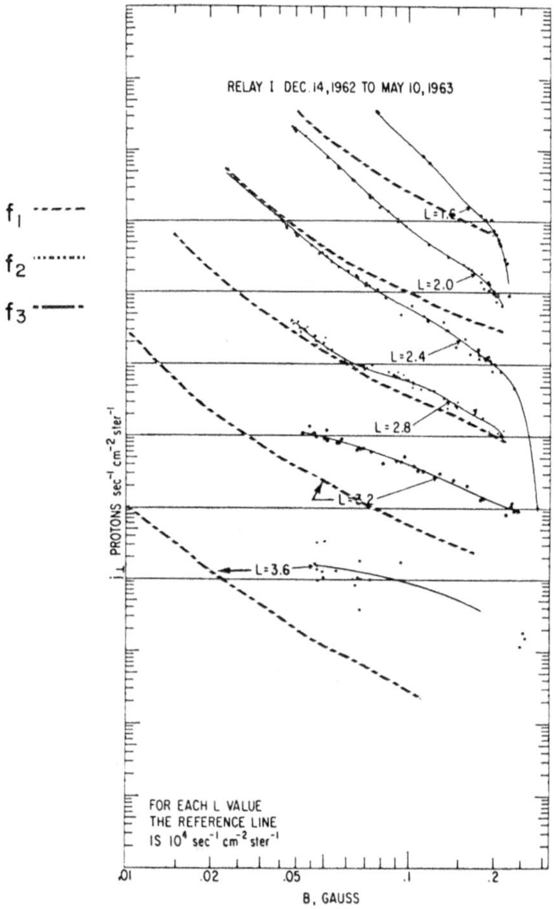

Fig. 8. A comparison of transformed Davis *et al.* (1964) data with measurements by Fillius and McIlwain of protons of 1.1 to 14 MeV. The curves are identified by which exponential term from Equations 13 or 14 dominates in the transformed data.

other L values and is compared with the Relay 1 directional intensity data of Fillius and McIlwain for off-equator locations in Figures 8 to 12. In these figures the transformed flux curves are shown labeled by which exponential 1,

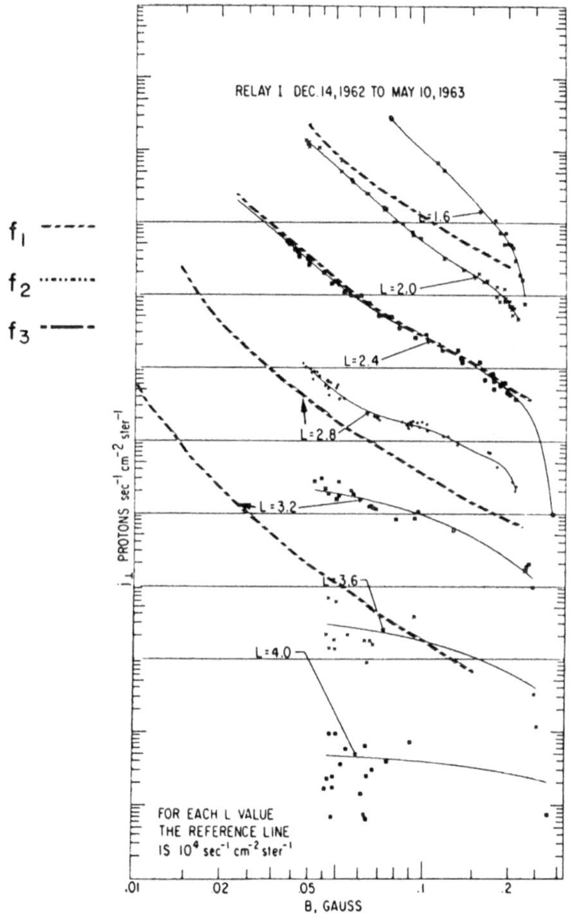

Fig. 9. A comparison of transformed Davis *et al.* (1964) data with measurements by Fillius and McIlwain of protons of 1.6 to 7.1 MeV. The curves are identified by which exponential term from Equations 13 or 14 dominates in the transformed data.

2 or 3 from equations (13) or (14) dominates in determining the flux above a certain energy at a certain location. From inspecting Figures 8 to 12 we can arrive at the following general statements:

(1) Near the equator the agreement of the measured fluxes and transformed fluxes is quite good where direct comparison can be made. This agreement remains quite good going off-equator up to a location dependent on energy and

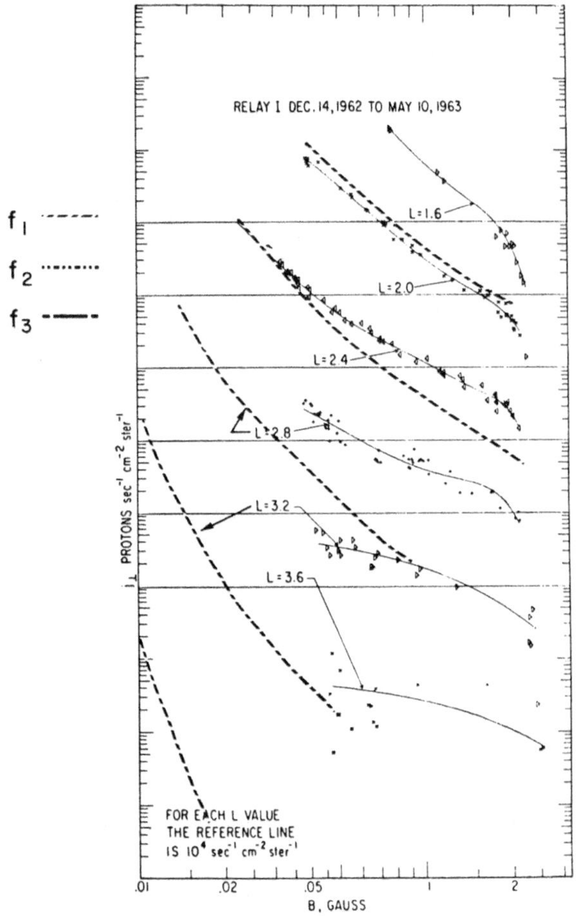

Fig. 10. A comparison of transformed Davis *et al.* (1964) data with measurements by Fillius and McIlwain of protons of 2.25 to 4.7 MeV. The curves are identified by which exponential term from Equations 13 or 14 dominates in the transformed data.

location but given very roughly by $B/B_0 \sim 3$ for proton energies of a few MeV. For higher energies (Figure 11 and Figure 12) the region of agreement is quite small.

(2) For locations well off-equator there is a decided discrepancy between

the measured fluxes and transformed fluxes. The measured fluxes are consistently higher.

There are at least two ways in which the off-equator fluxes might be made

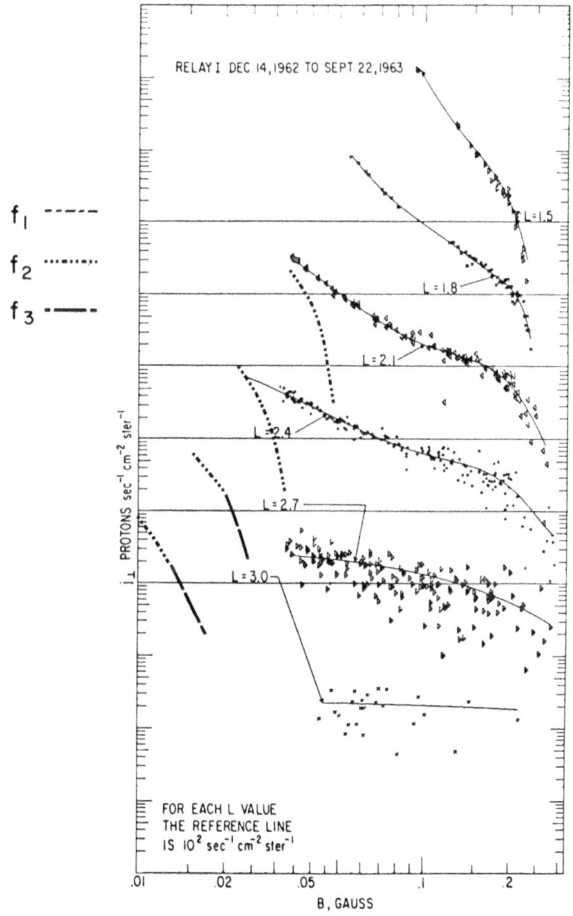

Fig. 11. A comparison of transformed Davis *et al.* (1964) data with measurements by Fillius and McIlwain of protons of 18.2 to 25 MeV. The curves are identified by which exponential term from Equation 13 or 14 dominates in the transformed data.

larger than those given by the transformation here. First there may be processes which violate either the first or second adiabatic invariant that usually will tend to move particles down field lines. In this way we could populate off-equator locations by moving particles from the equatorial region down field lines. We must be careful not to do this so efficiently that the equatorial fluxes are changed significantly, otherwise we would lose the agreement that we do have at the

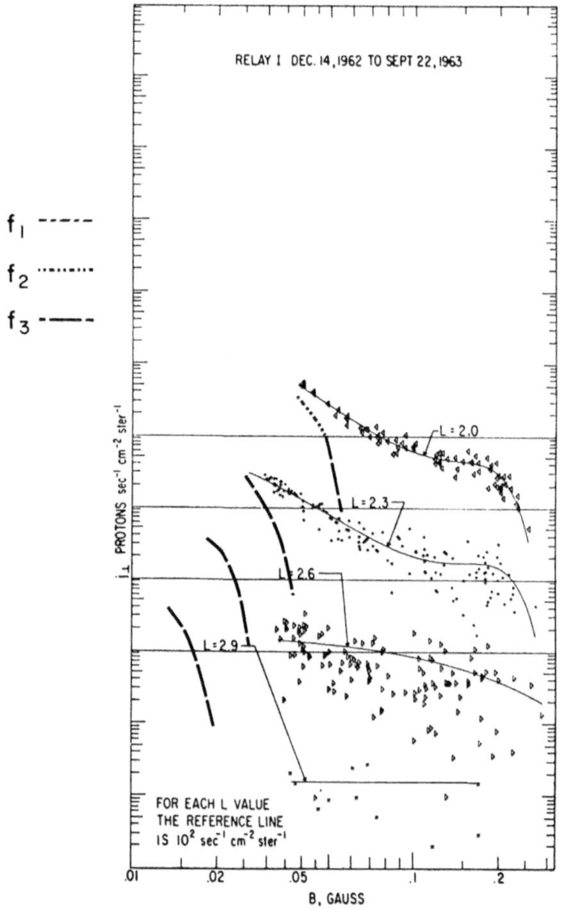

Fig. 12. A comparison of transformed DAVIS, *et al.* (1964) data with measurements by Fillius and McIlwain of protons of 35 to 63 MeV. The curves are identified by which exponential term from Equations 13 or 14 dominates in the transformed data.

equator. But most of the Davis protons in a tube of force are confined quite close to the equator so maybe supplying the lower altitude population can be accomplished without damaging the equatorial agreement.

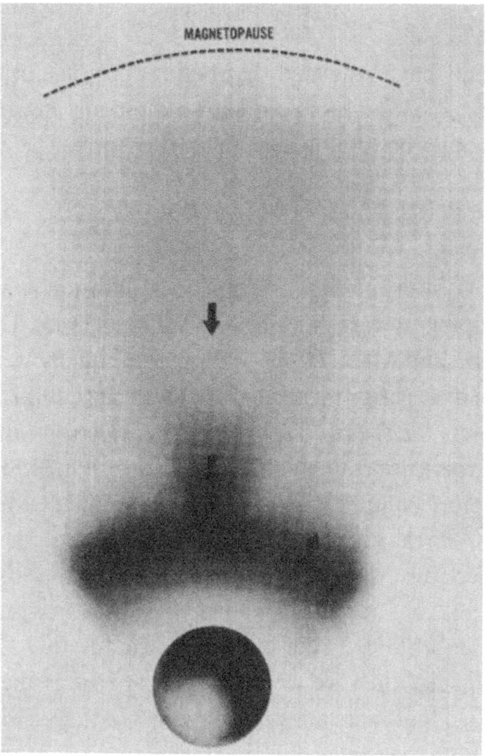

Fig. 13. A possible model to explain off-equatorial low energy protons. Protons travel in from the magnetopause near the equator and travel outward from the inner radiation zone off-equator.

A second way of supplying the off-equator proton flux, which seems more likely to me, would involve a second source of protons. It is well established that neutron decay protons produce most of the observed $E>30$ MeV protons for $L<1.5$. The neutrons decay source extends outwards into the outer belt, falling off about a R^{-2}. These protons will also be acted on by the process that violates the third invariant and these protons will also move in L. These will tend to diffuse outwards away from the source. Tverskoy (1964) suggested that this process might be responsible for the sharp outer edge of the inner belt proton distribution. We have the interesting possibility that outer zone protons diffuse inwards to fill the region near the equator at low L values and simultaneously inner zone protons are diffusing outwards to fill the higher B parts of field lines as shown in Figure 13. This is what is expected by analogy to gradient diffusion where particles tend to move away from regions of high flux. This idea has not been tested quantitatively.

Other Particles

Recently experiments have indicated that besides protons, other particles also undergo L diffusion. FRANK (1965) has shown a radially inwards moving wave of electrons in the outer belt following a magnetic storm. The wave had a radial velocity

$$v \cong kL^8 \tag{15}$$

This wave has the right properties to be due to L diffusion concerning μ and J. The magnetic pumping process described by PARKER (1960), DAVIS and CHANG (1962), and NAKADA and MEAD (1965) will operate equally well on all particles that have the same drift period τ_D around the earth since this is the only particle parameter that enters the theory. This means that electrons and protons of the same energy will experience roughly the same L diffusion. However the electron problem appears more complicated than the proton one. Other processes such as precipitation or pitch angle scattering and in general short lifetime seem characteristic of electrons, while for protons L diffusion appears to be dominant.

VAN ALLEN (1965) has recently identified α-particles in the outer radiation belt. If some of the protons in the outer belt are due to L diffusion inwards from the magnetopause, then we would expect α-particles there too. It is now quite well established that there are several percent α-particles in the solar wind. If these get reasonably thermalized in the transition zone, as seems to occur for protons and electrons, then we should have for the flux of α-particles in the transition zone

$$j_\alpha(E) \approx kj_p(E) \quad \text{where} \quad k \sim .05 \tag{16}$$

Assuming that these α-particles can get through the magnetopause, as seems to be the case for the protons, then by analogy to the known proton energy spectrum inside

$$j_p(E) = k_p e^{-E_p/E_{0p}} \tag{17}$$

we would expect the α-particle energy spectrum to be

$$j_\alpha(E) = k_\alpha e^{-E_0/E_{0\alpha}} \tag{18}$$

Also as a result of the thermalization we should have

$$E_{0\alpha} = E_{0p} \tag{19}$$

for all locations in the magnetosphere where loss processes are not too important. Loss processes will be different for protons and α-particles and probably will change the two spectra in different ways. Loss processes seem important for protons for $L<3$.

This situation where the proton and alpha particle energy spectra are the same should be true as long as we have steady state, without loss for both particles, that is, where the Fokker-Planck equation can be written (DAVIS and CHANG, 1962)

$$\frac{\partial f}{\partial t}=\frac{\partial}{\partial r}[\langle\Delta r\rangle f]+\tfrac{1}{2}\frac{\partial^2}{\partial r^2}[\langle(\Delta r)^2\rangle f]=0 \tag{20}$$

The spectra should be the same for steady state because we would expect the diffusion process to operate on the two particles in similar ways so that

$$\frac{\langle\Delta r_\alpha\rangle}{\langle(\Delta r_\alpha)^2\rangle}=\frac{\langle\Delta r_\mathrm{p}\rangle}{\langle(\Delta r_\mathrm{p})^2\rangle} \tag{21}$$

which from (20) would give $j_\alpha(E)=kj_\mathrm{p}(E)$. The two sets of particles don't have to diffuse with the same velocity. They can move through each other and probably will, because protons and α-particles of the same energy will have different drift periods

$$\tau\propto\frac{1}{v_\mathrm{D}}=\left[\frac{mc}{zeBR_\mathrm{e}}\left(\frac{v_\perp^2}{2}+v_\parallel^2\right)\right]^{-1} \tag{22}$$

From (22) for protons and α-particles of the same kinetic energy

$$\frac{\tau_\alpha}{\tau_\mathrm{p}}=2 \tag{23}$$

This means that a different set of magnetic disturbances will cause violation of the Φ invariant for protons and α-particles of the same energy.

A prediction of this model of L diffusion is that inside the magnetosphere the proton and α-particle energy spectra will be related by

$$\frac{j_\alpha(E)}{j_\mathrm{p}(E)}=\frac{k_\alpha e^{-E/E_0}}{k_\mathrm{p}e^{-E/E_0}}=k\sim .05 \tag{24}$$

in the region of L space where loss is unimportant (roughly $L>3$). Also if the suggestion put forward here is correct, that outer belt protons at large B/B_0, i.e. well off the equator, are drifting outwards from the inner belt, then we would not expect α-particles to exist in this region of space. If α-particles are found at large B/B_0, then maybe violation of the μ or J invariants is important for outer belt protons.

References

DAVIS, L. R., HOFFMAN, R. A., WILLIAMSON, J. M.: 1964, *Trans. Amer. Geophys. Union* **45**, 84.
DAVIS, L. R. and WILLIAMSON, J. M.: 1963, *Space Res.* **3**, 365.
DAVIS, L. and CHANG, D. B.: 1962, *J. Geophys. Res.* **67**, 2169–2179.
FILLIUS, R. W. and MCILWAIN, C. E.: 1965, 'A survey of inner zone protons', University of California (at San Diego) paper, April 13, 1965.
FILLIUS, R. W. and MCILWAIN, C. E.: 1964, *Phys. Rev. Letters* **12**, 609–612.
KELLOGG, P. J.: 1959, *Nature* **183**, 1295.
PARKER, E. N.: 1960, *J. Geophys. Res.* **65**, 3117–3130.
NAKADA, M. P., DUNGEY, J. W., and HESS, W. N.: 1965, *J. Geophys. Res.* **70**, 3529–3533.
TVERSKOY, B. A.: 1964, *Geomagnetism and Aeronomy* **3**, 351–366.
FRANK, L. A.: 1965, *J. Geophys. Res.* **70**, 3533–3541.
NAKADA, M. P. and MEAD, G. D.: *J. Geophys. Res.*, to be published.
VAN ALLEN, J. A., and KRIMIGIS, S. M.: 1965, *Trans. Amer. Geophys. Union* **46**, 140.

DIFFUSION OF PROTONS IN THE OUTER RADIATION BELT

M. P. NAKADA and G. D. MEAD

Goddard Space Flight Center, Greenbelt, Maryland

Introduction

Recent observations by DAVIS and WILLIAMSON (1963) and DAVIS *et al.* (1964) in the $L=2$ to 8 earth radii region of the outer radiation belt confirm the existence of large fluxes of 0.1 to 10 MeV protons. Except for the more energetic protons beyond $L=5$, the intensities appear to be very stable over times of the order of years. An outstanding feature of these protons is the large but smooth variation in their spectra with L and equatorial pitch angle.

Sudden impulses and sudden commencements have rise times of a few minutes and decay times of a few hours. Typical protons seen by DAVIS *et al.* (1964) have drift times of the order of a half hour or so. (The drift time of equatorially-trapped, non-relativistic particles is $44/LE$ minutes (LEW, 1961), with L in earth radii and E in MeV.) Thus the protons considered here should be diffused by this mechanism.

Review of Previous Analyses

The recent theoretical study of the spectra of these protons by DUNGEY *et al.* (1965) strongly suggests that the source is near the edge of the magnetosphere and that some process that violates the third adiabatic invariant for trapped particles (i.e., permits motion between L-shells) plays an important role in populating the radiation belt. PARKER (1960), DAVIS and CHANG (1962), and TVERSKOY (1964) have studied one such process which must surely operate according to present interpretations of geomagnetic fluctuations due to solar plasma.

PARKER (1960) evaluated the mean square broadening per magnetic disturbance and obtained the large-L portion of the steady state distribution of electrons by using a diffusion equation and considering many disturbances. DAVIS and CHANG (1962) found that for this mechanism the Fokker-Planck equation was appropriate (see also PARKER (1963), TVERSKOY (1964), and DUNGEY (1965)). They concluded that the mean displacement of a ring of particles, $\langle \Delta r \rangle$, had a large effect on the equatorial distribution. DAVIS and CHANG (1962) also obtained steady-state radial distributions by assuming a source near the magnetopause and assuming that the particle density vanished at a particular value of L near the earth. This latter assumption is quite reasonable, since the diffusion rate becomes small very rapidly as the earth is approached. However, unless specific loss rates are considered, it is not possible to determine where the density vanishes and, thus, the effectiveness of the mechanism.

TVERSKOY (1964) added Coulomb energy loss to the Fokker-Planck diffusion equation, obtained solutions for relativistic electrons, and concluded that this mechanism was important for these electrons. He also compared diffusion times with Coulomb energy-loss lifetimes of protons and concluded that if sources existed, MeV protons and α particles should exist in the outer belt. We also made a similar comparison (MEAD and NAKADA, 1964). Although this procedure does give an indication of the effectiveness of the mechanism, it is rather unsatisfactory, since the interpretation of "diffusion times" and Coulomb energy-loss lifetimes is vague and dependent on the proton spectrum.

Results of Present Analysis

In the present paper, therefore, we have obtained solutions to the Fokker-Planck equation containing terms describing energy losses due to Coulomb interaction with the ambient electrons as well as charge exchange losses with the neutral atmosphere. Both radial and energy distributions are obtained.

Instead of an image dipole model of the disturbed field, we used the more accurate magnetosphere model of MEAD (1964), based on the solution to the Chapman-Ferraro problem of a perfectly-conducting plasma incident upon a three-dimensional dipole field. Also, sudden commencements and sudden impulses were interpreted in terms of the compression of the magnetosphere from an initial quiet-time configuration, caused by an *increase* in the solar wind intensity, rather than assuming that the solar wind was absent between magnetic disturbances, as did Parker and Davis and Chang. To obtain a numerical value for the diffusion coefficient in the Fokker-Planck equation, we used an estimate of the actual number of observed sudden impulses and commencements of various amplitudes during the period 1958 to 1961.

The solutions to the equation were then converted to integral fluxes $j(>E)$ corresponding to the seven energy thresholds of Davis and Williamson's detector. The resulting theoretical curves were compared with their experimental fluxes, as shown in Figure 1. The curves are remarkably similar, indicating that a diffusion process of this general nature seems to be responsible for the observed outer belt proton fluxes.

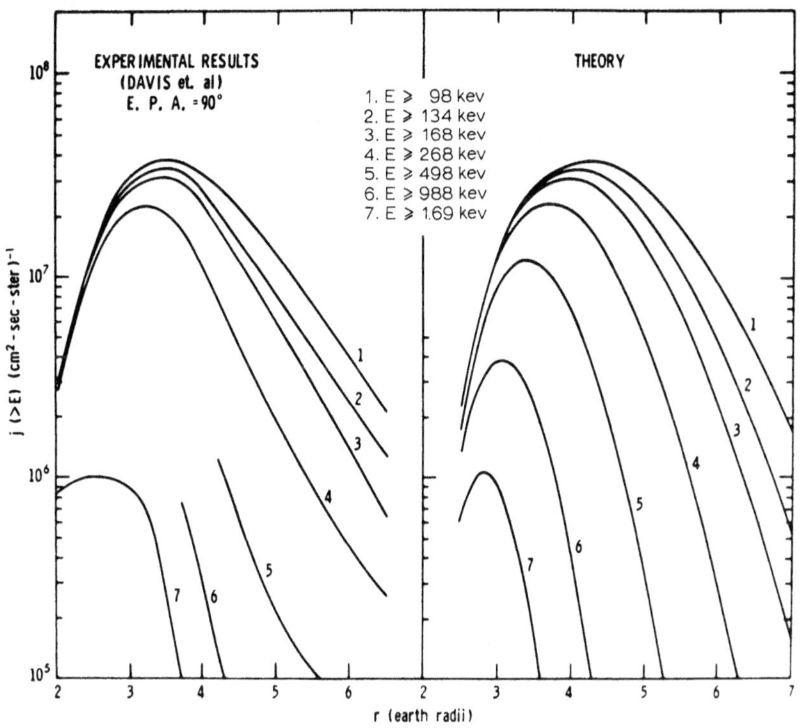

Fig. 1. Integral flux.

Discussion and Conclusions

These studies strengthen the suggestion by DUNGEY *et al.* (1965) and TVERSKOY (1964) that the source of the outer belt protons is near the magnetopause and that some process that violates the third adiabatic invariant is important in moving protons into the region of the outer radiation belt. Although uniqueness cannot be claimed, the similarity of the calculated fluxes to measurements gives support to the idea that some diffusion process is important.

Conclusions on whether or not this diffusion is caused exclusively or in major part by magnetic disturbances due to solar wind fluctuations must await further experimental and theoretical results. Similar conclusions may be drawn about the source and injection mechanism.

FRANK *et al* (1964), FRANK and VAN ALLEN (1964) and FRANK (1965) have reached similar conclusions on the importance of some diffusion process in their studies of energetic electrons ($E > 1.6$ MeV) in the outer radiation belt. They further find that large variations in the source of electrons are likely. In contrast, the source for protons is likely to exhibit considerably smaller time variations, since proton fluxes at $r<5$ earth radii are quite stable.

Although proton fluxes have been quite stable over the last few years, changes are likely to occur during other phases of the solar cycle if the processes considered here are important. Changes in parameters such as source strength, source spectrum, magnetic activity, neutral hydrogen density, and the electron density can cause predictable changes in proton fluxes and spectra. It should be interesting to follow changes in measurable parameters and proton fluxes as solar maximum is approached.

A detailed article will be published in *J. Geophys. Res.*, October 1, 1965.

Acknowledgements

We gratefully acknowledge the many discussions with Mr. Leo Davis and the use of his data prior to publication. Dr. M. Sugiura kindly compiled the data on sudden commencements and impulses for us. We also acknowledge the guidance of Mr. William Cahill, Dr. Arnold Stokes, and Mr. Robert Baxter in the solution of the partial differential equation.

References

Davis, Leverett, Jr., and Chang, D. C.: 1962, *J. Geophys. Res.* **67**, 2169.
Davis, Leo R., Hoffman, R. A., and Williamson, J. M.: 1964, *Trans. Am. Geophys. Union* **45**, 84.
Davis, Leo R., and Williamson, J. M.: 1963, *Space Res.* **3**, 365.
Dungey, J. W.: 1965, *Space Sci. Rev.* **4**, 199.
Dungey, J. W., Hess, W. N., and Nakada, M. P.: 1965, *Space Res.* **5**, 399. See also *J. Geophys. Res.* **70**, 3529.
Frank, L. A.: 1965, *Univ. Iowa Rept.* **65**-13.
Frank, L. A., and Van Allen, J. A.: 1964, *J. Geophys. Res.* **69**, 4923.
Frank, L. A., Van Allen, J. A., and Hills, H. K.: 1964, *J. Geophys. Res.* **69**, 2171.
Lew, John S., 1961, *J. Geophys. Res.* **66**, 2681.
Mead, G. D.: 1964, *J. Geophys. Res.* **69**, 1181.
Mead, G. D., and Nakada, M. P.: 1964, *Trans. Am. Geophys. Union* **45**, 85.
Parker, E. N.: 1960, *J. Geophys. Res.* **65**, 3117.
Parker, E. N.: 1963, *Interplanetary Dynamical Processes,* Appendix B, pp. 268–270, Interscience Publishers, New York, 1963.
Tverskoy, B. A., 1964, *Geomagnetizm and Aeronomija* **3**, 351.

LONGITUDE DEPENDENCE OF GEOMAGNETICALLY TRAPPED ELECTRONS

JUAN G. ROEDERER
Goddard Space Flight Center, Greenbelt, Md.
and
JASPER A. WELCH
U.S. Air Force Systems Command, Andrews Air Force Base, Md.

Abstract. This paper will describe a method for calculating the longitude dependence of geomagnetically trapped electrons at low altitude on basis of scattering and energy loss due to atmospheric interactions.

1. Geometrical Description of Longitude Dependent Fluxes

We shall describe trapped electrons by the distribution function used by WELCH *et al.* (1963). For a given L-shell we denote by $\delta n = U(B, T, \Phi, t)\delta\Phi\delta B\delta T$ the number of electrons contained at the time t in a tube of field lines of magnetic flux $\delta\Phi$ situated at a longitude Φ (geographic longitude of the equatorial, minimum B, point), with mirror points between B and $B+\delta B$, and with kinetic energies between T and $T+\delta T$. Relations of U with experimentally accessible quantities are given in WELCH *et al.* (1963). In particular U is connected to the differential, directional equatorial flux $j_0(\cos\alpha(B))$ in the form $U=2\pi j_0\tau/B^2$ where τ is the halfperiod of bouncing. In all which follows, we shall assume particle shells to be degenerate, i.e., independent of the particles' pitch angle.

We shall denote with δx the distance between equatorial points of two neighboring field lines on the same shell; δy is the distance between the equatorial points of two field lines belonging to neighboring shells ($\delta y \perp \delta x$). The quantity $x = \int \mathrm{d}x = \int (\partial x/\partial\Phi)$ can be used to label a field line on a given shell. With this notation, the flux element of a magnetic tube becomes $\delta\Phi = B\delta x\delta y$.

Our problem is to follow the history of trapped electrons as a function of time, at different longitudes, using a Fokker-Planck equation. However, we cannot take x or Φ as a suitable longitudinal variable, because U is a distribution in *flux*, and not in x alone. That is, in the flux expression $B\delta x\delta y$, not only δx, but also δy, is a function of longitude in the real geomagnetic field. In order to find a useful longitudinal variable, we have to transform the flux into an expression $B\delta\xi\delta\eta$, in which $\delta\eta$ – still related to the radial increment

* National Academy of Sciences – National Research Council, Senior Post Doctoral Research Associate, on leave from Centro Nacional de Radiación Cósmica, Buenos Aires, Argentina.

δy – is now longitude-independent, so that it may be ignored when following the particles during their longitudinal drift. In that case, U is a true distribution in the new longitudinal variable ξ. Notice further that ξ and η *must be purely geometrical* variables, i.e. they must not involve the particle's energy.

To obtain ξ and η, we use McILWAIN's (1961) relation between L and I: $R^3L^3B/M = F(I^3B/M)$. Differentiating at constant B, we obtain

$$\frac{\partial I}{\partial y} = \frac{(1/B_0)(\partial B_0/\partial y)}{3/RL}\frac{R^2L^2}{I^2F'} = \kappa_0\lambda(B),$$

where R is the radius of the Earth and B_0 the field strength at the equator. κ_0 is the field line's curvature at the equatorial point, in units of the corresponding dipole curvature, and λ is a function of B/B_0; $E = 1/6\lambda$ is tabulated by HAMLIN *et al.* (1961). If we now set $\delta\eta = \kappa_0$, $\delta y = \delta I/\lambda$, and $\delta\xi = \delta x/\kappa_0$ we obtain the two wanted invariant variables $\eta = I/\lambda$ and $\xi = \int(\mathrm{d}x/\kappa_0) = \int(1/\kappa_0)(\partial x/\partial\Phi)\mathrm{d}\Phi$. Notice that η is now longitude independent, $\eta = \eta(I, B)$ and can be dropped in our considerations. The variable ξ contains the complete information on the longitude dependence of a "tube" of monoenergetic particles drifting in the real field (in absence of interactions), and will be adopted as the longitudinal variable for our Fokker-Planck description.

Notice that with our notation, the mean equatorial drift velocity (NORTHROP and TELLER, 1960) has a very simple form:

$$\dot{x} = u_0 = \frac{p}{eB_0}\frac{\kappa_0\lambda}{\tau},$$

where p is the particle momentum and the only longitude dependent factor is the curvature κ_0. The half period of bouncing τ is independent of longitude, to the same degree as L is constant on a given shell. To show this, we notice that

$$v\tau = \int\frac{\mathrm{d}s}{\sqrt{(1 - B/B_m)}} = \int\sqrt{(1 - B/B_m)}\mathrm{d}s + \int\frac{B/B_m}{\sqrt{(1 - B/B_m)}} = I + 2B_m\frac{\partial I}{\partial B_m}.$$

It can be shown that $\partial I/\partial B_m$, taken along a field line, depends on longitude only as the quantity $\partial L/\partial B$ does, i.e., much less than κ_0. Numerical bouncing time calculations (e.g. HASSITT, 1965) *did* show a longitude dependence of τ for particles mirroring far from the equator; this, however, can be explained by the fact that these authors picked up their shell field lines at the equator (constant B_0 ring). Therefore, they do not correspond *exactly* to the field lines of a high B_m/B_0 shell (see shell splitting in STONE (1963)).

The expression for the drift velocity in ξ is now:

$$\dot{\xi} = \frac{u_0}{\kappa_0} = \frac{1}{\kappa_0}\frac{\partial x}{\partial\Phi}\dot{\Phi} = \frac{p}{eB_0}\frac{\lambda}{\tau}$$

2. The Longitude Dependent Fokker-Planck Equation

Our Fokker-Planck equation (WALT and MACDONALD, 1961; WELCH *et al.*, 1963) reads:

$$\frac{\partial U}{\partial t}+\frac{\partial}{\partial \xi}(U,\xi)=\frac{\partial}{\partial T}(U,\dot{\varepsilon})-\frac{\partial}{\partial B}(U\langle\beta\rangle)+\tfrac{1}{2}\frac{\partial^2}{\partial B^2}(U\langle\beta^2\rangle)+Q \qquad (1)$$

where

$$\langle\beta\rangle=\tau^{-1}\{\beta\},\quad \langle\beta^2\rangle=\tau^{-1}\{\beta^2\}\quad\text{and}\quad \dot{\varepsilon}=\tau^{-1}\{\varepsilon\}.$$

The curly brackets are average changes of mirror point field B, of B^2 and of energy (absolute value), per half-bounce, respectively. Q is the source. Notice that the convection term in longitude is $\partial/\partial\dot{\xi}(U\xi)=\dot{\Phi}(\partial U/\partial\Phi)$. This latter expression differs from what one would have obtained when using Φ in the original Boltzmann equation (i.e. $\partial/\partial\Phi(\dot{\Phi}U)$).

If one integrates Equation (1) over ξ, one obtains a longitude-average equation. However, if, and only if, we neglect the variation of U with longitude, does one obtain the equation used previously by several authors (WELCH *et al.*, 1963; WALT and MACDONALD, 1962). The longitude-average Fokker-Planck treatment of trapped particle diffusion is therefore valid only for particles mirroring at high altitude at *all* longitudes.

Notice now that $N=\int_{B_0}^{B}U\dot{\xi}\,\mathrm{d}B$ is the number of particles per unit time, flowing past a given line situated at ξ, and which mirror between the equator and B. The function $B=B(\xi)$ which keeps the value of the integral N constant throughout all longitudes, can be defined as "characteristic trajectory of a mirror point". The "velocity" of mirror points *along* a field line is

$$V_m=\frac{\mathrm{d}B(\xi)}{\mathrm{d}t}=-\frac{1}{U}\int\frac{\partial(U\dot{\xi})}{\partial\xi}\mathrm{d}B=\langle\beta\rangle-\tfrac{1}{2}\frac{\partial\langle\beta^2\rangle}{\partial B}-\tfrac{1}{2}\langle\beta^2\rangle\frac{\partial\ln U}{\partial\beta} \qquad (2)$$

Notice that UV_m gives the net number of particles flowing per second and per unit magnetic flux through a given B-L ring.

There are three physically quite distinct contributions to the mirror point flow along a field line. The *first* term in Equation (2) represents a steady increase in B, i.e. a steady lowering of mirror point altitude. The second and third terms arise in the dispersion mechanism. Their presence in (2) clearly shows that dispersion *also* contributes to a steady, average flow of mirror points along field lines. The *second* term in (2) is always negative and therefore represents an *upward motion* of mirror points (backscattering from the denser atmosphere). The *third* term represents mirror point lowering only if $(\partial U/\partial B)<0$. The average flow of mirror points therefore not only depends on the atmosphere, but also on the distribution U itself, as pointed out earlier by WALT and MACDONALD (1962).

The coefficients intervening in Equation (1) must be calculated numerically, integrating the proper atomic collision cross section functions along field lines of a given L-shell. The expressions for these coefficients are given by WELCH *et al.* (1963). It can be shown that for the high density region (high B, i.e., low altitude mirror points) the following approximations hold:

$$\langle \beta^2 \rangle = 2\Delta B \langle \beta \rangle \qquad \langle \beta \rangle \sim N_{es} \sqrt{\frac{\Delta B}{B}}$$

$$\langle \varepsilon \rangle \sim N_{ei} \sqrt{\frac{\Delta B}{B}}$$

$\Delta B = (\partial \ln N_{es}/\partial B)^{-1}$ is the scale "height" in B, N_{es} and N_{ei} are the effective atmospheric densities for Coulomb scattering and energy loss, respectively. In these approximation the mirror point velocity is $V_m \cong -\Delta B \langle \beta \rangle (\partial \ln U/\partial B)$. With this result, it is easy to predict qualitatively what will happen to a given distribution of electrons mirroring at high B values when they drift past the anomaly. Let us suppose a stationary distribution roughly inversely proportional to the atmospheric density. In that case, the mirror point velocity reduces to $V \cong \langle \beta \rangle$. Starting west of the anomaly, mirror points flow downwards; first very slowly, then faster, as the particle approaches the "center" of the anomaly (where B-L rings attain their lowest altitude). For high initial B values, electrons will precipitate down west of the anomaly. A certain portion of electrons, lying in a narrow window in B, will manage to get around the anomaly, but then precipitate at the east side. Finally, another portion again contained in a narrow B window will lower their mirror points into the "shadow" region west of the anomaly, populating B-L rings which originally dipped below sea level in the anomaly (process of replenishment). All lower B electrons will go through with only very small changes in mirror points. Only a relatively small region of the atmosphere controlls this mechanism; therefore, we must expect a *diurnal variation* for the process of replenishment and precipitation, provided atmospheric scattering is the only interaction present.

3. Numerical Integration and Results

The numerical calculations were divided in three parts: (1) determination of the shell geometry; (2) determination of the coefficients in Equation (1), and (3) numerical integration of Equation (1). Parts (2) and (3) are particularly difficult because almost all quantities vary over many orders of magnitude in the region of interest. The magnetic field model was based on the Jensen and Cain 48-coefficient expansion, Epoch 1960 (JENSEN and CAIN, 1962). The atmospheric model was taken from HARRIS and PRIESTER (1964). Two models were used, corresponding to high and low solar activity, respectively. For each model, the local time at the Anomaly was taken at 0400 and 1400 h, corresponding to lowest and highest density at about 300 km altitude, in the diurnal cycle.

The main scope of this work was to determine the flow of electrons through the anomaly. No attempt was made to determine lifetimes and other "long lasting" properties, involving great numbers of drifts around the earth. For this reason, the unknown source term and the time derivative in Equation (1) were neglected. The input at a line well West of the anomaly was either a delta function in B and T, or a given distribution in B and T. The delta function input was used for test runs only, for it does not lead to a steady state solution as required in our hypothesis.

4. Conclusions

The results can be summarized as follows:

(1) The time independent distribution U along a given line of force is roughly inversely proportional to the atmospheric density. The energy spectrum softens appreciably for electrons mirroring at high B values.

(2) For this "equilibrium" distribution, replenishment due to atmospheric scattering only occurs in a region of limited longitudinal extent, east of the anomaly. This replenishment does not explain the observed tail of the distribution at high B values.

(3) Precipitation of electrons occurs mainly 5–15° west of the anomaly, this displacement being greater for lower energies.

(4) For a given energy, there is a sharp transition in the lifetime of electrons at a critical B value, such that electrons mirroring at higher values will survive only a few longitudinal drifts.

References

HAMLIN, D. A., KARPLUS, R., VIK, R. C., and WATSON, K. M.: 1961, *J. Geophys. Res.* **66**, 1.
HARRIS, I. and PRIESTER, W.: 1965, *Cospar International Reference Atmosphere* (1965).
HASSITT, A.: 1965, *J. Geophys. Res.* **70**, 535.
JENSEN, D. C. and CAIN, J. C.: 1962, *J. Geophys. Res.* **67**, 3568.
MACDONALD, W. M. and WALT, M.: 1962, *J. Geophys. Res.* **67**, 5025.
MCILWAIN, C. E.: 1961, *J. Geophys. Res.* **66**, 3681.
NORTHROP, T. C. and TELLER, E.: 1960, *Phys. Rev.* **117**, 215.
STONE, E. C.: 1963, *J. Geophys. Res.* **68**, 4157.
WALT, M. and MACDONALD, W. M.: 1962, *J. Geophys. Res.* **67**, 5013.
WELCH, J. A., KAUFMANN, R. L., and HESS, W. N.: 1963, *J. Geophys. Res.* **68**, 685.
WALT, M. and MACDONALD, W. M.: 1961, *Ann. Phys.* **15**, 44.

ON THE OBSERVATION OF IONOSPHERIC EFFECTS DUE TO DUMPING OF TRAPPED PARTICLES

G. J. GASSMANN

Air Force Cambridge Research Laboratories, Office of Aerospace Research, Bedford, Mass.

C. P. PIKE, Jr.

Lowell Technological Institute Research Foundation, Lowell, Mass.

Abstract. Energy considerations lead to the conclusion that the Van Allen belts feed sizable amounts of ionizing energy into the ionosphere by precipitation of particles. Most suspicious, although not clearly unique in importance, are the western slopes of the magnetic anomalies and their conjugate areas. A map of prospective dumping areas of electrons due to longitudinal drift is presented.

The results from thirteen airborne measurements in an area conjugate to the suspected South Atlantic dumping area are presented. Vertical ionospheric recordings during a number of nights of magnetically quiet periods have been analyzed in search for local effects. Up to nine observations per location, obtained at various different local times, were averaged. The echo amplitudes show no local effect recognizable on top of the expected temporal fluctuations. A distinct local effect, however, in a large area centering 1000 km north of Puerto Rico and probably extending to the east consisted of low values of $f_{\max}\,E_{\text{s-f}}$ relative to those at other areas covered.

1. Introduction

The list of the energy sources which must be considered in order to understand the ionosphere is indeed not complete without the radiation belt. The energy demand for maintaining the whole ionosphere might be estimated as to about $2\cdot10^{19}$ erg/sec; that is an energy supply required per column of 4 erg/cm$^2\cdot$ sec (thickness of ionosphere 300 km, total volume of day-side ionosphere $v=10^{26}$ cm^3, electron density $n_e=2\cdot10^6$ cm^{-3}, recombination coefficient $\alpha=10^{-9}$ cm$^3\cdot$ sec^{-1}; ion-pair production rate $q=n_e^2\cdot\alpha=4\cdot10^3$ cm$^{-3}\cdot$ sec^{-1}; ionization energy required 30 eV/electron-ion-pair; total dose rate required $=30\cdot q\cdot v=1.2\cdot10^{31}$ eV/sec $=2\cdot10^{19}$ erg/sec). Most of this energy is provided from the solar electromagnetic radiation striking the earth (which amounts to about 10^{24} erg/sec) and by cosmic rays. On the other hand it is known that the solar wind provides a potential energy source of approximately 10^{20} erg/sec in the process of its interaction with the magnetosphere. Due to this interaction approximately 10^{18} erg/sec of the above energy are readily dumped into the auroral zones by electrons and protons (Van Allen, 1965, this volume). Electron precipitation takes place in addition to the above also at moderate latitudes almost continually and around the globe, as evident from observed pitch angle distributions of trapped electrons (O'BRIEN, 1965).

At the geomagnetic anomalies of lowest field intensity precipitation of charged particles is likely to be enhanced for two reasons. First, the loss cone is widest. Second, the increase $\Delta\alpha$ of the loss cone with every degree of longitude φ_0 (on the western slope of the anomaly) will eliminate and dump all trapped electrons of pitch angles within $\Delta\alpha$, if any have arrived during their eastward drift (O'BRIEN, 1965, this volume). The enhancement of dumping near the magnetic anomalies as compared to other areas has not been estimated at this time. The energy, however, which is being dumped during a quiet day at the South Atlantic anomaly may be estimated following GLEDHILL and VAN ROOYEN (1965). Those workers found from the pitch angle distribution observed by O'BRIEN (1962) over Iowa that more than 20% of the flux will be dumped at the conjugate point near the South Atlantic anomaly. With an assumed total omnidirectional flux of 10^7 electrons/cm$^2\cdot$sec, 100 keV each, the dumped energy per column is 0.35 erg/cm$^2\cdot$sec. This number, however, might be as large as 2 erg/cm$^2\cdot$sec if all electrons of energies above 1 keV were included. According to Gledhill and Van Rooyen such flux, if absorbed in an altitude range of 110 to 130 km, would give rise to an energy deposition of 10^{-6} erg/cm$^3\cdot$sec. This would produce ionization comparable to that by solar UV. Assuming arbitrarily an area of (3000 km)^2 for the anomaly the total rate of energy dumped in that area during a quiet day was $2\cdot10^{17}$ erg/sec.

Any attempts to estimate from satellite measurements further details of dumping, for example as to height and geographical location, would require the knowledge of angular flux rates and energy distribution for all shells and longitudes.

Although it is evident from the above that electron dumping plays its role in the formation of the ionosphere, not much evidence of this has been accumulated from observations on the ground. The conventional methods of observations have been utilizing ionospheric sounders, riometers, propagation effects or airglow phenomena. Since the predominant global variations of ionospheric features are fixed to the sun, the dumping effects could be separated from other ionization effects if they were fixed to the rotating earth. This, however, is clearly the case only for those dumping effects which are connected with the geomagnetic anomalies.

2. Electron Dumping Due to Longitudinal Drift

The location of areas of expected dumping due to longitudinal drift of trapped electrons may be predicted from magnetic maps. In absence of detailed knowledge of the electron injection mechanism, we may substitute a schematic assumption: in every volume element of the B, L, φ_0 space of the magnetosphere with $\Delta B=0.01$; $\Delta L=0.01$; $\Delta\varphi_0=1°$, one additional electron per second is assumed to mirror. All electrons whose mirror points during the eastward drift descend below 100 km altitude at a particular location may then be considered being dumped there. The resulting areas and dumping rate densities were obtained from a B and L projection at the 100 km altitude level, taken from DUDZIAK *et al.* (1963). This projection allows to follow from each geographical point in 100 km altitude, with a particular B and L, the longitudinal westward extent of the particular mirror point belt $\Delta B \cdot \Delta L$, for constant B and L, which lies above the 100 km level. It also allows to determine for each geographical point, the horizontal cross section of the $\Delta B \cdot \Delta L$ mirror point belt intersected by the 100 km horizontal level. The original magnetic map is reproduced in Figure 1 with superimposed lines of the derived dumping areas. The heavy lines give the number of electrons dumped per cm^2 column per second. The heavy dashed lines give the longitudinal extent, in degrees, of the mirror point belt whose content is emptied out at that location. The South American dumping area contour lines were found being open toward the south (where no B and L projections were available); their closing suggested in the map is merely intuitive and for flight planning purposes.

Although this map is constructed under an arbitrary assumption on the injection, it may be modified and adjusted to an actual injection mechanism which might become known. In any case, the dumping of electrons due to longitudinal drift is expected to occur within the areas marked in Figure 1.

The dotted lines on the map mark the geomagnetic conjugate areas corresponding to the dumping areas. Those areas may be called secondary dumping areas and arise from a redistribution of pitch angles due to Coulomb scattering at the primary dumping areas. From there a sizeable percentage, say 20%, of the electrons about to be dumped escape and are dumped at the conjugate point. Experimental evidence of this has been observed over Iowa by O'BRIEN (1962).

The map of Figure 1 may serve as a guide in planning of measurements. It should be kept in mind that the map deals only with dumping due to longitudinal drift. Other types of dumping may be present at the same time.

Fig. 1. Electron dumping due to longitudinal drift with an assumed injection scheme. Map with B and L contour lines in 100 km altitude adapted from DUDZIAK *et al.* (1963) with superimposed primary (heavy solid lines) and secondary (dotted lines) dumping areas. The heavy solid lines indicate the number of electrons dumped per cm^2 column per second. The dashed lines indicate the longitudinal extent, in degrees toward the west, of the mirror point belt whose content is emptied out at that location.

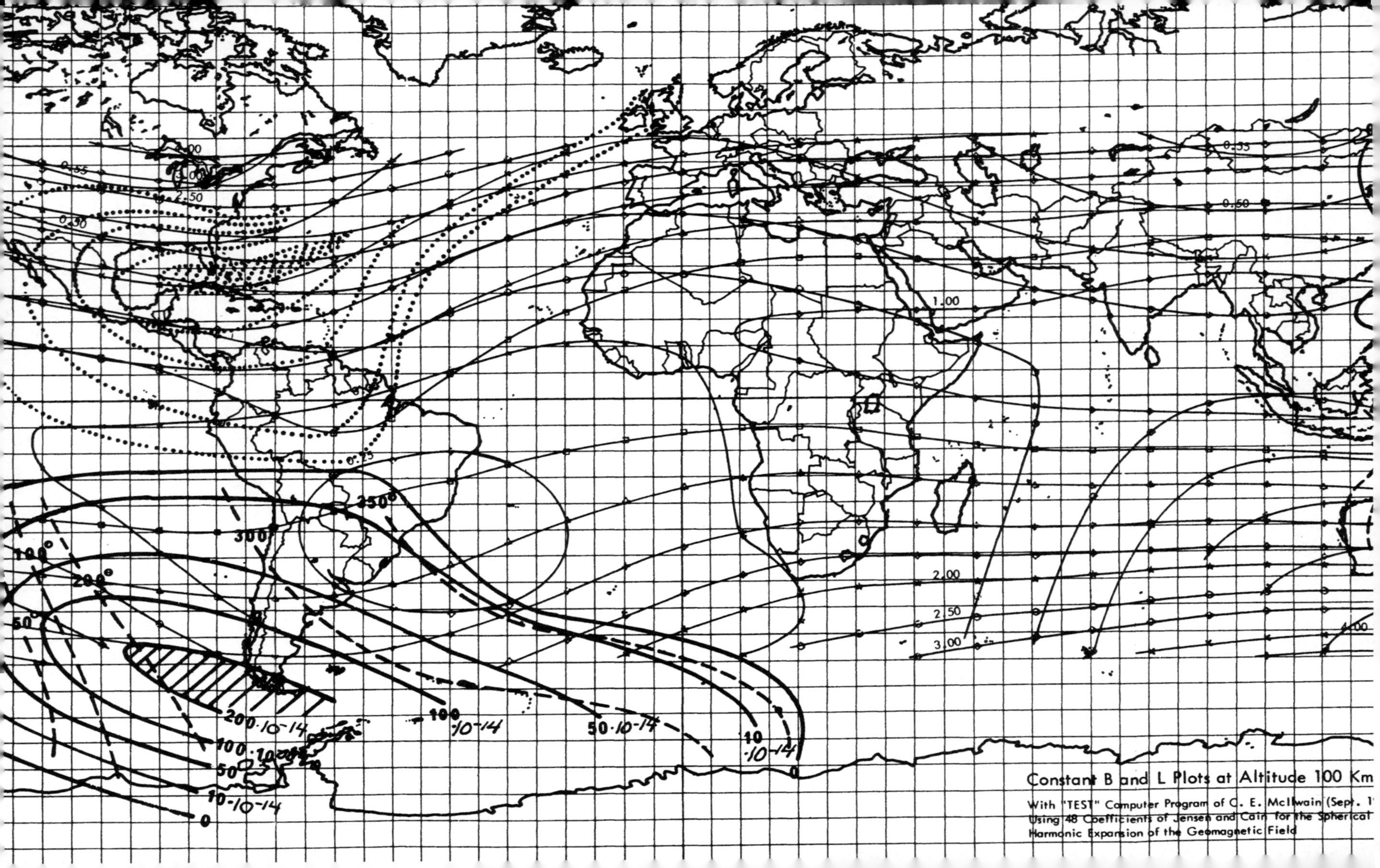

Constant B and L Plots at Altitude 100 Km

With "TEST" Computer Program of C. E. McIlwain (Sept. 1
Using 48 Coefficients of Jensen and Cain for the Spherical
Harmonic Expansion of the Geomagnetic Field

3. Some Experimental Results

The experimental approach used for the measurements reported here utilized a mobile ionosounder carried by an airplane fast enough to cover the area of interest within a time period shorter than the periods of ionospheric variations (mostly diurnal). The airborne equipment produced 50 μ second pulses while continually sweeping from 1 to 24 MHz within $\frac{1}{2}$ minute and recorded range and amplitude of all echoes. The PRF was 30 pulses/second. The gain of the system was controlled by frequent calibrations and was kept constant for most of the time.

The geographical area covered so far by a total of 13 flights comprised the Central Atlantic Ocean near 60 °W longitude. All measurements reported on here were taken during various hours of the night. The geomagnetic activity during those nights was quiet.

In order to arrive at a statistical review, the ionospheric data from each flight were treated as in the example plotted in Figure 2. After plotting the f_0F2 versus time, three frequencies were chosen for which the echo amplitudes (in a relative db*-scale) were read, averaged over about four pulses and plotted versus time. The quantities h' F2 and h' E_s were also plotted as well as f_{max} E_s. The particular type of E_s encountered was transparent and flat near 100 km with a blanketing frequency f_b E_s most of the time too low to be observable. For this reason the f_{max} E_s was plotted. Three of the plotted quantities were converted into block diagrams as indicated by the hatched areas: *A* for the average maximum frequency f_{max} E_s of sporadic *E* echoes; *B* for the occurrence of short-period lowerings of h' F2 lasting not more than 20 minutes; *C* for the average amplitude of F2-echoes. Each of the block diagrams was then replotted versus location, the latter in terms of *L*-shell numbers.

Then the arithmetic means of the quantities *A*, *B*, and *C* at the various *L*-shell numbers were obtained from the data of nine flights west of Bermuda and four flights east of Bermuda. (The procedure to average db values in order to obtain the mean amplitudes is a minor inaccuracy in this treatment). The result is plotted on the map of Figure 3. The numbers along the plots indicate the number of counts from which the plotted value at that location is averaged.

There is a distinct absence of sporadic E (curve *A*) in the area north of Puerto Rico. Judging from the flights east of Bermuda this area of absence or low occurrence of E_s may extend or may be even more pronounced toward the east. In good correlation with the absence of E_s is the lowering of h' F2 (curve *B*) This could be a true F layer effect or just the consequence of the varying group delay in the E_s layer. This will be clarified after more data have been obtained. The echo amplitude (curve *C*) shows no local effect of statistical significance. In order to find a possible local effect more data would be required because of the large short-time fluctuation of that quantity.

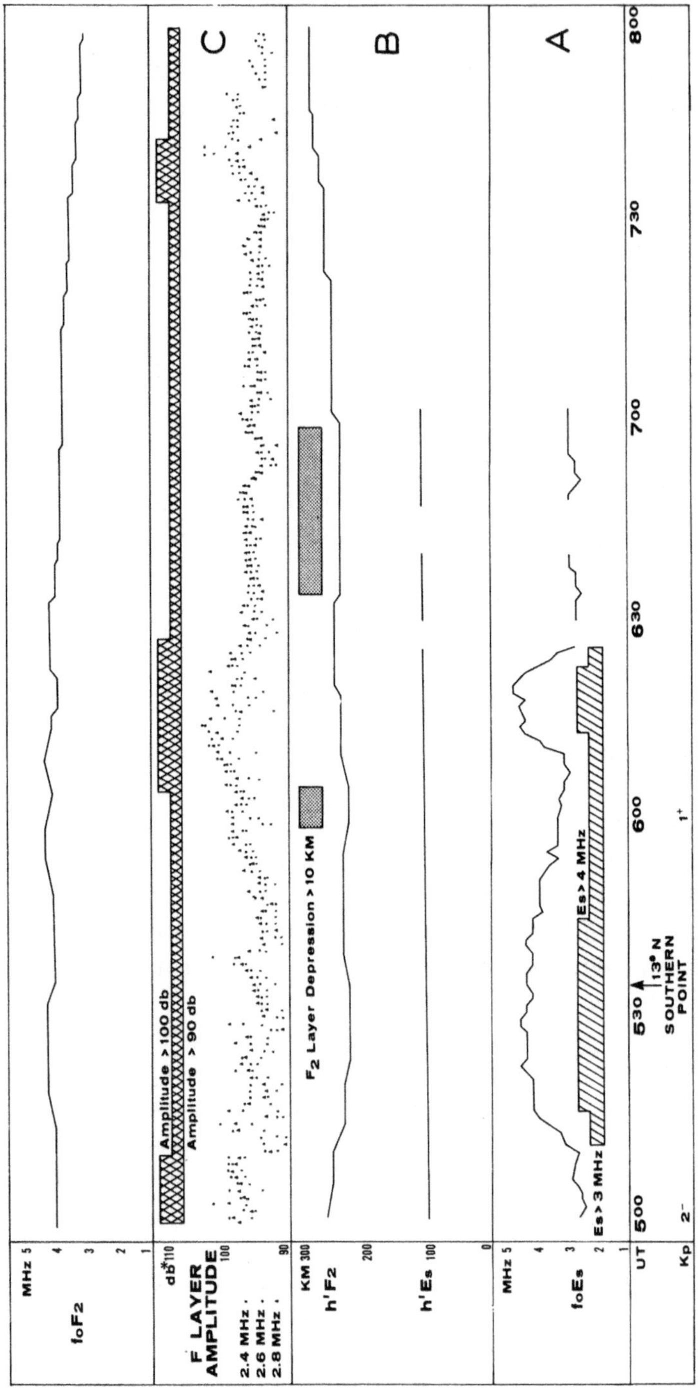

Fig. 2. Example of the ionosperic parameters plotted and how three parameters were converted into block diagrams *A*, *B*, and *C* for convenient averaging.

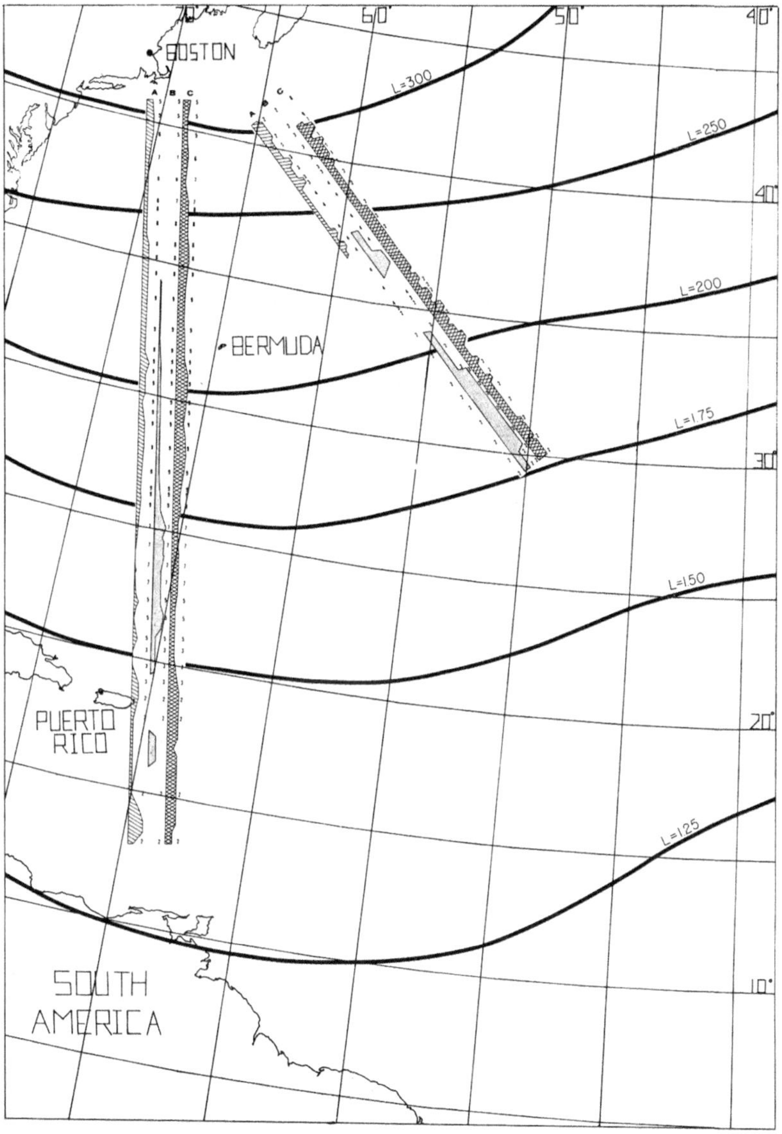

Fig. 3. Map of suspected secondary dumping area with results from thirteen flights. The figures beside the plots along the flight routes indicate the number of independent measurements from which the quantities *A*, *B*, and *C* (definition in Figure 2) were averaged.

4. Interpretation and Conclusions

The verification by ionospheric soundings of dumping of trapped particles is in its beginnings, except for the Arctic zones. Of the several kinds of dumping processes, the one due to longitudinal drift of trapped electrons, thus causing ionospheric effects near the magnetic anomalies and their conjugate areas, may be easiest to observe. Some experimental evidence has been collected from airborne vertical soundings in an area conjugate to the South Atlantic dumping area. The echo amplitudes show no local effect recognizable on top of the normal fluctuations, the latter being due to both irregularities at the reflection level and to variations in absorption due to fluctuating cosmic ray energy deposition in 60 to 80 km. This must mean that the energy deposition from prospective dumping in the same altitude range is small compared to that by cosmic rays (about 10^{-12} erg/cm$^3\cdot$ sec). The dumped electrons, if any, would have to have an energy of >50 keV each.

The regular E-layer behavior as to prospective dumping was not determined due to equipment limitations below 1.5 MHz. However, the occurrence of E_s showed a low at a location approximately 1000 km north of Puerto Rico and extending by ± 500 km north-south and possibly toward the east. The magnetic maps suggest that this might be an effect of dumping. If so, electrons of energies between 1 and 50 keV each would likely be responsible since those >50 keV have been ruled out according to above and those <1 keV do not penetrate down to altitudes below 100 km. Since the E_s effect shows negative correlation to prospective dumping the explanation is not obvious.

The above results pertain to an area only which may be considered a secondary dumping area. Figure 1 may be useful in planning measurements in one of the primary dumping areas. The equipment should include ionospheric sounders, riometers and airglow photometers.

Acknowledgements

The airborne equipment was designed, installed and operated under the supervision of R. W. Gowell of AFCRL. The data was taken by J. E. MacNeil of Boston College. The flights were coordinated by R. H. Dowd, Lt. Col., USAF and R. W. Whidden of AFCRL and flown under the aircraft Commanders R. E. Creager, Major, USAF and K. E. Crooks, Capt., USAF of Electronic Systems Division. Figure 1 was constructed by R. A. Wagner of Lowell Technological Institute Research Foundation.

References

DUDZIAK, W. F., KLEINECKE, D. D., and KOSTIGEN, T. J.: 1963, *Graphic Displays of Geomagnetic Geometry,* General Electric Company, Santa Barbara, California.
GLEDHILL, J. A. and VAN ROOYEN, H. O., Department of Physics, Rhodes University, Grahamstown, South Africa, 1965, private communication.
O'BRIEN, B. J.: 1962, *J. Geophys. Res.* **67**, 1227–1233.
O'BRIEN, B. J.: 1966, This volume p. 321.
VAN ALLEN, J. A.: 1966, This volume p. 65.

SECTION V

PHYSICS OF THE OUTER MAGNETOSPHERE

Chairman:

DR. J. W. DUNGEY

Imperial College, London

SURVEY OF ACCELERATION AND DIFFUSION

J. W. DUNGEY
Imperial College, London

1. Acceleration under static conditions. Adiabatic

There are several reasons for paying attention to the static problem. Observational information about the static field is better than information about temporal variations of the field. It is desirable to understand the static problem before tackling the temporal variations, just as the undisturbed motion in a dipole field was the first topic to be mastered in this Institute. No apology is necessary, however, as the static field does provide interesting acceleration. One important feature of static fields is the existence of an electrostatic potential, so that acceleration requires that a particle moves to a place where the potential is different from that where the particle was before its acceleration. The potential will be taken to be constant on a line of force and generally to be positive on the morning side and negative on the evening side as has been described many times before. Thus in the current sheets both in the dayside boundary and in the tail the electric field has the same direction as the current.

The motion when the adiabatic approximations are valid has been extensively discussed before and only a few points will be noted. The "partial trapping" on the night side may be well known, but it is not often mentioned in the literature. The point is that energetic particles well out on the night side are likely to drift out through the boundary. They are trapped "up and down" but not "round and round". This is essentially discussed by O'Brien (1963) in connection with the night side trapping boundary. On the night side also particles gain energy as they drift, and this is the acceleration mechanism used by Axford and Hines (1961). Less energetic particles may also be pulled in by the electric field sufficiently to reach the dayside magnetosphere but in a static model go out through the boundary on the day side. This Axford and Hines acceleration mechanism is limited by the adiabatic conditions. The case of 90° pitch angle is familiar, the energy being just proportional to B. More interesting for the aurora is small pitch angle and this is also amenable to a simple approximation, I being nearly $v\times$ the length l of the line of force between mirror points. Then $W \propto l^{-2}$. Now, if the adiabatic theory were valid for electrons on lines which close at a greater distance (say 30 R_e) in the tail, substantial acceleration would be possible by this mechanism. However, the validity requires B to be not much less than 1γ, and NESS' (1965) observations suggest that the minimum field on such a line is somewhat too small for the adiabatic condition.

2. Non-adiabatic

Non-adiabatic motion is important in the current sheets, the tail sheet being most interesting. The work of SPEISER (1965) will be outlined here and Hoh will speak on this subject. Speiser takes an x-axis normal to the sheet, a y-axis towards the sun, the axes are orthogonal and the current is in the z-direction. The basic field is B_y varying rapidly with x and changing sign at $x=0$ and a uniform E_z. To these must be added a small B_x varying slowly with y and possibly other small components. An important feature of the motion is expressed by

$$\ddot{x} = (e/m) B_y \dot{z}$$

remembering that B_y changes sign at $x=0$. This shows that the particle oscillates across the sheet if $\dot{z}$ maintains one sign, but is accelerated away from the sheet if $\dot{z}$ maintains the other sign. The change in energy also depends on $\dot{z}$ and is a gain in the former case.

The motion of protons in the current sheet is clearly nonadiabatic and indeed the thickness of the sheet may be controlled by the gyroradii of protons. SPEISER (1965) has computed trajectories of protons and finds that thin beams should bombard the auroral zones. These could not be observed at the altitudes so far examined by appropriate satellites because the beams would be spread by charge exchange (DAVIDSON, 1965). Such beams could of course also be spread by time-varying fields and may themselves be unstable. Since performing these computations Speiser has found an approximate formulation which clarifies the proton motion, but is more valid for electrons.

In his approximate treatment Speiser discusses the section of the trajectory in which the particle oscillates across the sheet and points out that the motion parallel to the sheet averaged over the oscillations is almost independent of the oscillation across the sheet, and so just depends on B_x. He neglects the dependence of B_x on y and concludes that particles which come into the sheet with small energy will be ejected with small pitch angles and speeds of about $2cE_z/B_x$. Numerically this might be $2.10^8 B_x^{-1}$ cm/sec with B_x in γ and the corresponding energy for electrons $10B_x^{-2}$ eV. If B_x becomes too small, the approximation loses validity, but it is easily seen that the particle energy then approaches the limit set by the potential across the tail of tens of keV. Speiser finds that electrons of more than 1 keV should have come from points with $B<0.1\gamma$ on this static model, and when mapped on to the ionosphere the north-south thickness is only about one kilometer. There are of course many mechanisms which spread the area of precipitation, and some of the particles will bounce on their first arrival.

It should be noted that similar acceleration may occur in the dayside current sheet, but fewer of the accelerated particles are likely to reach the ionosphere.

Concluding this section it should be emphasised that acceleration to 10 keV does not necessarily require time varying fields.

3. Auroral Substorms

Considerable evidence supports the belief that auroral substorms are associated with a sudden intensification of reconnection in the tail, this being directly related to the morning to evening electric field in the tail region. The most obvious consequence of sudden reconnection is a *pt* pulsation because reconnected tail lines catapult towards the earth and their stored energy goes into hydromagnetic waves. Now JACOBS and SINNO (1960) and KATO and SAITO (1962) find that any *pt* pulsation is invariably associated with a bay somewhere in the auroral zones. The correlation of bays with southward interplanetary field has been found by CAHILL and FAIRFIELD (1965). They found a southward component in the field in the transition region to be a necessary condition for bays. The sudden onset appeared to be associated with a sudden change in the direction of the transition region field, the initial field having a southward component, and a tentative explanation is offered here. It is assumed that the distant part of the tail adjusts its orientation to that of the interplanetary field, assuming this to be steady for a sufficient time. Then if field lines of some other orientation suddenly arrive, they will tangle round the tail and exert a pull on it away from the sun. This increases the tension in the tail and intuition suggests that reconnection is likely to occur, because it relieves the tension.

Figure 1 shows a guess at the view from the sun of the magnetosphere and tail, when the tail lines connect to an interplanetary field lying in the geomagnetic equatorial plane. The observed part of the tail is barely recognizable in this view, but lies in the central region of Figure 1 and the reader is assumed to be familiar with its geometry. The point here is that when the field lines curve out to join the interplanetary field there is an asymmetry, the geometry being something like that of an aircraft propellor. This asymmetry offers a possible explanation for the preponderance of southward field just outside the boundary on the day side (CAHILL, private communication). Figure 1 shows interplanetary field lines outside the tail region and suggests that they are distorted by the flow past the obstacle formed by the tail. Such distortion can occur only behind the shock, but it was not possible to show the three dimensional shock surface. The distortion suggested adds always a southward component to the field, and it is suggested that this causes a growing predominance of southward field as one moves from the shock to the boundary on the day side.

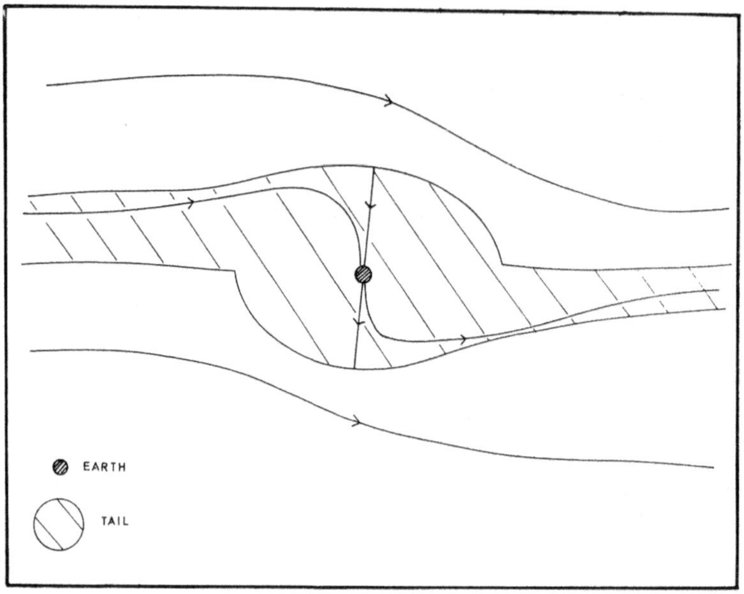

Fig. 1. View from the sun of the tail in a case with the interplanetary field in the ecliptic, showing the distortion of neighbouring interplanetary field lines.

4. Diffusion in General

The diffusion processes have been surveyed recently (DUNGEY, 1965, which will be referred to as (S)) and this section makes a few comments for emphasis and attempts to relate other papers at this symposium. For dependent variable the distribution function f (density in phase space) has particular virtues. It is the quantity used in any basic plasma theory and is the quantity conserved by Liouville's theorem. As shown in (S) the diffusion due to large scale disturbances has the nature of stirring and is expressed by a single classical diffusion term, provided the dependent variable is f. The independent variables can be any three of L, energy W, pitch angle α, and the adiabatic invariants. If the diffusion terms involve derivatives with respect to the adiabatic invariants the elements of phase space in terms of these variables are needed. These turn out to have quite simple expressions, the underlying reason being that they are derivable from a Hamiltonian. It may also be noted that the particle flux is roughly proportional to fW^2, so that f decreases with energy considerably faster than the flux. Measurable values of f differ by many orders of magnitude and it is useful to express observations in terms of f, if the spectrum is well enough known.

As seen in (S) the different kinds of diffusion all invoke resonance with one of the three periodicities in the motion of a trapped particle. Gyroresonance will be surveyed by ROBERTS (this proceedings) and some general aspects of bounce resonances are discussed in the next section. This leaves the case when μ and I are preserved and this will be discussed by Mead and was extensively discussed at the Institute see p. 481. The importance of this mechanism is strongly suggested by the thorough observations of Leo Davis on the outer belt protons. HESS (private communication) suggests that the protons observed by McIlwain in the slot fit on to Davis' protons and ELLIOT's (private communication) 40 MeV protons at very low L could still be a continuation.

FÄLTHAMMAR (1965) has discussed the drift resonance in a general way and emphasised that there are contributions both from the induction field associated with variations in B and the potential electric field, though it is difficult to make accurate estimates of either. The induction field is believed to come mainly from impulses as discussed originally by Parker. The potential field probably comes mainly from bays and the frequency of occurrence may be a few per day for both impulses and bays. Mead and Nakada use a detailed study of impulses by Sugiura which shows that the most important contribution may be regarded as arising from $\sim 10\gamma$ impulses occurring once every few days. The change in energy from impulses can be described as a betatron effect the order of magnitude being given by $\delta W/W \sim \delta B/B$ or $3.10^{-5}\mu L^3 \delta B$ with B in gammas. It is believed that δB increases with L, some evidence being provided by CAHILL and NISHIDA (1964), and this variation is computed from the Beard model by Mead and Nakada. No such detailed study of bays has been made, nor is any accurate assessment possible because the factors relating the electric field to observed quantities involve considerable uncertainties. It is curious, however, that such estimates as can be made show results remarkably similar to the effects of impulses, the most notable difference being that δW does not increase proportionally to W above a certain critical energy. The electrostatic potential associated with a bay is supposed at any L to have a maximum on the morning side and the magnitude of the potential difference varies on a time scale of $\sim \frac{1}{2}$ hour. For low energies δW results from the drift of the particle in the electric field just as in the mechanism of Axford and Hines referred to in Section 1. At high energies the particles drift several times round the earth during the course of a bay, and roughly δW is limited by the maximum potential difference on the particle's L shell, though a more accurate treatment following Fälthammar is desirable when detailed information about the electric field is available. The critical condition may be taken as $\tau_d = 1$ hour giving $W = 10^5 L^{-1}$ eV and roughly $\delta W/W$ is constant for lower energies and δW is constant for higher energies. The potential difference in bays increases with L, indeed bays are commonly thought of as high latitude phenomena, but the form of this variation is not well known.

Estimates of the potential at high L are usually tens of kV. Thus near the boundary either an individual bay or an individual impulse may give a value of δW which is a considerable fraction of W. At $L=5$ a 10γ impulse gives $\delta W/W = 4\cdot 10^{-2}$ and a 10 kV bay gives $\delta W/W = 5\cdot 10^{-2}$ for $W < 2\cdot 10^5$ eV. Bays at middle latitudes have received less attention, but are observed. At $L=2$ a 10γ impulse gives $\delta W/W = 2.4\cdot 10^{-3}$ and a 1 kV bay gives $\delta W/W = 2\cdot 10^{-3}$ for $W < 5\cdot 10^5$. These very uncertain numbers for bays suggest that the effect is worth pursuing and it is instructive to estimate the F region velocity corresponding to the potential at $L=2$. Taking the electric field as 0.25 volts/km, the velocity is only 6 metres/sec., and drift speeds for the ionosphere are observed to exceed this considerably, but unfortunately the east-west component of velocity dominates, and we are directly concerned with the east-west component of E. The required value is reasonable, however, and again it seems that the diffusion due to bays is comparable to that due to impulses.

Some mention should be made of disturbances seen by particles at frequencies much higher than their gyrofrequencies. If the disturbances to be described by SCARF *et al.* (1965) are indeed ion waves with wavelengths of metres, the frequency seen by an energetic particle must be megacycles. It seems unlikely that resonance is important when it must be of such high order, and as far as their effect on trapped particles go, these disturbances just contribute to the random noise, possibly being the major contribution. It may be noted that the stirring picture used in (S) is good when neighbouring particles see almost the same disturbance field and becomes less good when the wavelength gets smaller. For very short wavelengths there is very little correlation between neighbouring particles and so much correlation can be neglected in the statistics. The short wavelengths then contribute to the "normal Fokker-Planck" coefficients of (S).

In concluding this section let the need for measurements of electromagnetic disturbances be reemphasised.

5. General Properties of Bounce Resonances

The bounce resonance was discussed previously (S) on the basis of Fourier decomposition in time and longitude. This would be valid if the field were axially symmetrical and seems adequate for comprehension of the general properties. For a resonance the disturbance must last for many bounces, but this may still be only a fraction of τ_d and the departure from symmetry may be relatively unimportant even at quite high values of L. One then considers disturbances varying like $\exp i\,(\omega t + m\phi)$ and the resonance condition

$$\omega + 2\pi m/\tau_d = 2\pi N/\tau_b \tag{1}$$

is analogous to that for gyroresonance. Also analogously with gyroresonance, the longitudinal invariant is not changed if $N=0$, and disturbances with low values of N tend to be the most effective for a given amplitude. Again, assuming axial symmetry the disturbance is static in a certain frame, but in this case a rotating frame. The frequency seen by the particle, $\omega + 2\pi m/\tau_d$, is the same in any frame.

The use of the rotating frame has been investigated by R. Etherington and some of his conclusions follow here. SCHIFF (1939) discussed electromagnetism in general in rotating frames and showed that the Maxwell equation

$$\partial B/\partial t = -c \operatorname{curl} \mathbf{E}$$

is still valid. Consequently in the frame rotating with the disturbance the electric field is a potential field, the change in particle energy is determined by the potential, and the energy derived directly from the potential of the disturbance itself cannot be enhanced by resonance and so is not important. The important potential is that due to the rotation of the frame, which is constant on a magnetic shell (the field being axially symmetrical).

The angular velocity of the frame in which the disturbance is static is ω/m and the resonance condition (1) gives

$$\frac{\omega}{m} = \frac{2\pi N}{m\tau_b} - \frac{2\pi}{\tau_d}. \qquad (2)$$

Equation (2) shows that the angular velocity of the frame is much less than the gyrofrequency Ω unless N is very large, which case we except. On the other hand the period of rotation of the frame may be less than the bounce period, in which case the particle energy corresponding to the rotation of the frame may be quite large. Nevertheless the use of this frame does not involve any difficulty. The effect of the Coriolis force is equivalent to a uniform magnetic field corresponding to a gyrofrequency of ω/m and this is negligible compared to the actual magnetic field since $\omega/m \ll \Omega$. Similarly, although the centrifugal acceleration $r(\omega/m)^2$ may be quite large, the acceleration due the electric field is $\Omega r\omega/m$ and is again much larger. Also the centrifugal force has a potential which is just the energy corresponding to the rotation of the frame. Thus the change of energy in the original frame can be taken to be the same as that in the rotating frame.

The value of $\mathrm{d}W/\mathrm{d}L$ will now be discussed assuming that there is no steady electric field in the original frame. There may in fact be appreciable potential differences between shells in the magnetosphere, and certainly the rotation of the earth makes a moderate potential. Such potentials are easily added, but it must be remembered that they also change τ_d. Apart from this the electric field in the frame of the disturbance is simply $r\omega B/m$ giving

$$\frac{dW}{dL} = e\,R_e^2 L\omega B_{eq}/m \tag{3}$$

or using the resonance condition (1)

$$\frac{dW}{dL} = 2\pi e\,R_e^2 L B_{eq}(N/m\tau_b - \tau_d^{-1}). \tag{4}$$

Thus the value of dW/dL depends on N/m, but not on any other aspects of the geometry of the disturbance (see S). It is known (e.g. NORTHROP, 1963) that I is preserved when $N=0$ and (4) with $N=0$ relates $(\partial W/\partial L)_{\mu,I}$ to τ_d. This relation takes the Hamiltonian form if the flux invariant is used instead of L. It is noteworthy that such simple relations exist and the corresponding relation

$$(\partial I/\partial W)_{\mu,L} = \tau_b/2 \tag{5}$$

leads to a simple expression for $(\partial I/\partial L)_{\mu,W}$. Also for a dipole a further relation is available

$$(\partial I/\partial L)_{\mu,W} = (3W\tau_b - I/2L).$$

For a dipole $-L/W(\partial W/\partial L)_{\mu,I}$ is between 2 and 3. Equation (4) shows how dW/dL varies with N/m and since $N\tau_d/m\tau_b$ may well be large, large values of dW/dL are typical and in the extreme case when $m=0$, L does not change at all. In this latter case Störmer's integral exists and confirms this conclusion. When dW/dL is large the important effect of the disturbance is diffusion in energy at constant μ, leading to both acceleration and dumping.

For orders of magnitude for the effects of bounce resonances we may take, say for the betatron effect

$$\frac{\tau_b}{W}\frac{dW}{dt} \sim b/B$$

For the change in L due to resonant changes in the direction of the magnetic field (see Figure 1 in (S))

$$\frac{\tau_b}{L}\frac{dL}{dt} \sim b/B$$

These are probably the largest effects and of course dW/dL is fixed by (4).

In connection with bounce resonances it should be remarked that observations of VLF emissions (HELLIWELL, 1965) and perhaps ELF pulsations (TEPLEY, 1964) suggest that the particle flux may vary appreciably with bounce phase. This would provide a good opportunity to observe phase variations. It is difficult to observe gyro phase variations because the frequencies are high, and drift variations because of uncertainties in I and the need to have two satellites simultaneously on the same shell. Bounce phase variations by contrast could be observed by two identical detectors looking in opposite directions from the same satellite.

References

AXFORD, W. I., and HINES, C. O.: 1961, *Canadian J. Phys.* **39,** 1433.
CAHILL, L. J. and FAIRFIELD, D. H.: 1965, *J. Geophys. Res.* **70.**
DAVIDSON, G. T.: 1965, *J. Geophys. Res.* **70,** 1061.
DUNGEY, J. W.: 1965, *Space Sci. Rev.* **4,** 199.
FÄLTHAMMER, C.-G.: 1965, *J. Geophys. Res.* **70,** 2503.
HELLIWELL, R. A.: 1965, *Whistlers and Related Ionospheric Phenomena,* Stanford University Press.
JACOBS, J. and SINNO, K.: 1960, *Geophys. J.*, **3,** 333.
KATO, Y. and SAITO, T.: 1962, *J. Phys. Soc. Japan,* **17,** Supp A-II, 34.
NESS, N. F.: 1965, *J. Geophys. Res.* **70,** 2989.
NORTHROP, T. G.: 1963, *Rev. Geophys.* **1,** 283.
SCARF, F. L., CROOK, G. M., and FREDERICKS, R. W.: 1965, *J. Geophys. Res.* **70,** 3045.
SCHIFF, L. I.: 1939, *Proc. Nat. Acad. Sci. U.S.* **25,** 391.
SPEISER, T. W.: 1965, *J. Geophys. Res.* **70,** 4219.
TEPLEY, L.: 1964, *J. Geophys. Res.* **69,** 2273.

Discussion

HOH presented some possible constraints on temperature and density values in the tail. VAN ALLEN felt that this measurements demonstrate an order of magnitude higher temperature in the tail than Hoh's 600 eV. Hoh said that the density would then be lowered accordingly.

DUNGEY mentioned that the discussion involves whether we have neutral sheets, lines or points. In the neutral sheet the field is just very weak (and presumably in a direction perpendicular to the plane of the sheet), but unless there is some special condition such as symmetry, the field is strictly a zero vector only at discrete points. DESSLER emphasized that in the theory that does not allow magnetic field merging, the neutral sheath was predicted, whereas the Dungey model is modified to fit the experiments after the fact.

FÄLTHAMMAR has looked at the L dependence of the diffusion coefficients in the outer belt. A brief paper on this is presented below.

COEFFICIENTS OF DIFFUSION IN THE OUTER RADIATION BELT

CARL-GUNNE FÄLTHAMMAR
Royal Institute of Technology, Stockholm, Sweden

The importance of diffusion processes in the outer radiation belt seems now to be a matter of general agreement. As collisions between particles are infrequent in the outer magnetosphere, one has to consider diffusion processes due to interactions with fluctuating electric and magnetic fields. In the case of the soft outer-belt protons (DAVIS and WILLIAMSSON, 1963; DAVIS *et al.*, 1964) there are indications that the diffusion processes are such as to conserve the first two adiabatic invariants (DUNGEY *et al.*, 1965; HESS, 1966; NAKADA and MEAD, 1965, 1966). Diffusion of this kind was suggested by KELLOGG (1959) in connexion with inner-belt electrons. Steady-state particle distributions have been computed on the assumption that the diffusion is produced by particular kinds of disturbances, namely magnetic impulses and initial phases of magnetic storms (PARKER, 1960; DAVIS and CHANG, 1962; TVERSKOY, 1964; NAKADA and MEAD, 1965, 1966). The work of Nakada and Mead shows that, for better quantitative agreement between theoretical results and observational outer-belt proton data, a larger diffusion coefficient would be required than that deriving only from the special disturbances considered. DUNGEY (1965, 1966) has pointed out that electric potential fields, particularly those associated with magnetic bays, may provide another and perhaps equally important transport mechanism.

It is the purpose of the present contribution to present more general expressions for the diffusion coefficients due to electric and magnetic disturbances that can contribute to the transport of particles. The disturbances to be considered are on one hand time-dependent electric potential fields (associated e.g. with phenomena such as large-scale magnetospheric plasma motions and magnetic bays), and on the other hand magnetic time-variations and their associated electric induction fields. It is found that, unlike what is true for the special cases considered before, the diffusion coefficient is in general energy dependent and can have a different spatial variation than the r^{10} variation characterizing those special cases.

Only a brief summary will be given here. Details have been published elsewhere (FÄLTHAMMAR, 1965, 1966). Cf. also ALFVÉN and FÄLTHAMMAR (1963), where basic principles of the particle motion and acceleration are treated.

For simplicity we shall take the unperturbed magnetic field to be a dipole field and consider particles near the equatorial plane. Using the guiding-centre approximation and neglecting the inertia term, we have the following expression for the radial displacement of a particle

$$r(t) - r(0) = \int_0^t \left\{ -\frac{E_\varphi}{B} + \frac{\Omega r}{3B}\frac{\partial B}{\partial \varphi} \right\} \mathrm{d}t' \tag{1}$$

where r and φ are plane polar coordinates, and where

$$\Omega = \frac{3\mu}{er^2}. \tag{2}$$

(Ω is the angular velocity of azimuthal drift in the magnetic dipole field.) The fields entering into the integrand are to be evaluated at the instantaneous position of the particle. If the fields were known in sufficient detail, it would in principle be a trivial matter to calculate $r(t)$. However, the problem to be considered here is that of the average long-term behaviour of particles moving in fields of whose time variations only some general statistical properties are known. This amounts to a problem of diffusion by continuous movements, and the aim is to express the diffusion coefficient in terms of the statistical properties of the fields. The Fokker-Planck-type equations that have generally been used for describing the transport and calculating the particle distributions, involve two coefficients. However, it has been pointed out by DUNGEY (1965) that for the kind of transport process considered here, a single diffusion coefficient is enough for characterizing the transport, which means that one of the two coefficients is derivable from the other. The transport equation for equatorial particles may then be written in the form

$$\frac{\partial n^*}{\partial t} = \frac{\partial}{\partial r}\left(\frac{D}{r^2}\frac{\partial}{\partial r}(r^2 n^*)\right) + \text{collision terms} \tag{3}$$

(cf. FÄLTHAMMAR, 1966), where D is the diffusion coefficient and $n^* \mathrm{d}r\,\mathrm{d}\mu\,\mathrm{d}J$ is the number of particles between r and $r+\mathrm{d}r$ having magnetic moments between μ and $\mu+\mathrm{d}\mu$ and longitudinal invariants between J and $J+\mathrm{d}J$. (J is small because we consider particles that stay near the equatorial plane.)

In the derivation of the diffusion coefficient D we represent the azimuthal electric field by a Fourier sum of the form

$$E_\varphi(r,t) = \sum_{n=0}^{n_0} E_n(r,t)\cos(n\varphi + \gamma_n(r,t) \tag{4}$$

and the equatorial magnetic field by the expression

$$B(r,t) = B_0(r) + S(t) + A(t)\, r \cos\varphi \tag{5}$$

Here $B_0(r)$ is the magnetic dipole field, and S and A represent symmetrical and non-symmetrical parts of the disturbance in the magnetosphere. From theoretical models of the magnetosphere, S and A can be related to the instantaneous position, $r_m(t)$, of the magnetopause. For example, MEAD's (1964) model of the magnetosphere gives (in MKSA units)

$$S = 0.816(\mu_0 M/4\pi)/r_m^3(t)\,; \qquad A = 0.637(\mu_0 M/4\pi)/r_m^4(t) \tag{6}$$

Like in the previously mentioned studies of special cases we have to assume the disturbances to be small in a certain sense so as to allow expansions in powers of the disturbance strength. Under assumptions discussed in detail elsewhere (FÄLTHAMMAR, 1965, 1966) the following results are obtained;

1. *Electric potential fields.* ($\partial \mathbf{B}/\partial t \equiv 0$.)

$$D_E = \frac{1}{8B_0^2} \sum_{n=1}^{n_0} P_n(r, n/T_a) \tag{7}$$

where $P_n(r, \nu)$ is the spectral power density of the quantity $E_n(r, t)$ at the frequency ν. T_a is the azimuthal drift period, which depends on location and particle energy.

Notice that in (7) the spectral density of the nth Fourier component is to be evaluated at the nth harmonic of the azimuthal drift frequency. The drift frequency and its harmonics play the role of resonance frequencies analogous to the cyclotron frequency in a cyclotron. The particles respond most strongly to the frequencies near the resonance frequencies. The effects of other frequencies tend to average out in the long run, and the longer time scales we consider, the narrower the relevant frequency intervals are, until, in the limit, we have the result (7). (There is actually a close analogy between particle motion in a so-called "stochastic synchrocyclotron" (see e.g. LIVINGOOD, 1961, pp. 319–373) and the kind of magnetospheric particle motion considered here.)

2. *Magnetic disturbances and their associated electric induction fields.* ($\partial \mathbf{B}/\partial t \not\equiv 0$)

$$D_B = \frac{\pi^2}{2}\left(\frac{5}{7}\right)^2 \frac{r^4 P_A(1/T_a)}{B_0^2 T_a^2} \tag{8}$$

where $P_A(\nu)$ is the spectral power density, at the frequency ν, of the quantity $A(t)$ defined in (5).

The formulas (7) and (8) reduce the problem of computing diffusion coefficients to the mathematically simple problem of a spectral analysis. The relevant frequency for a given location and particle energy is $1/T_a$ (and its harmonics). A survey diagram of T_a is given in Figure 1.

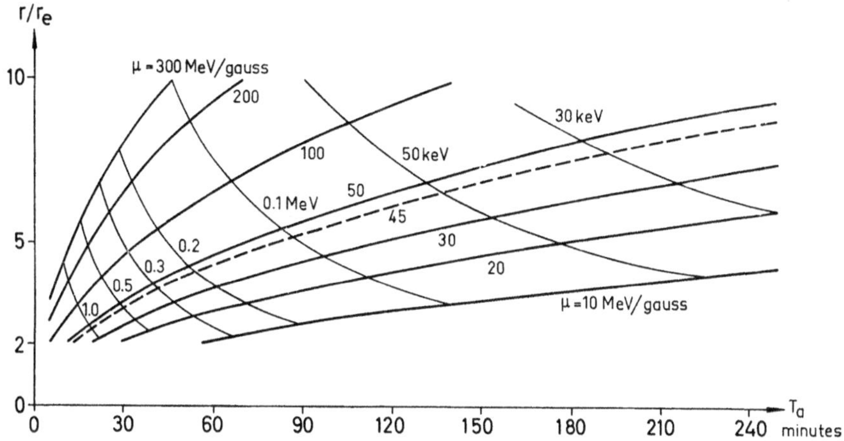

Fig. 1. Values of the azimuthal drift period T_a in the equatorial plane of the Earth's magnetic dipole field. The dashed line represents a characteristic μ-value for the exponential energy spectrum of the soft outer-belt protons.

As a simple illustration we may apply the formula (8) to magnetic impulses of amplitude ΔA occurring at random with a probability N per second ($N \ll 1/T_a$). If the rise time is very short and the decay time very long (on the time scale of T_a), the power spectrum is simply $P_A = 2N(\Delta A)^2/(2\pi\nu)^2$ (cf. e.g. RICE, 1964, p. 137). Using this expression and the relations, (6) characterizing MEAD's (1964) magnetosphere model, we immediately obtain

$$D_B = \frac{N}{2}\frac{25}{98}(0.637)^2 r^{10}\left(\frac{1}{r_{m,\min}^4} - \frac{1}{r_{mo}^4}\right)^2 \tag{9}$$

in agreement with NAKADA and MEAD (1965). However, by means of (8) it is a straight-forward matter to compute D_B for any distribution of impulse shapes and amplitudes. By considering impulses of finite length one can show that the tenth-power radial dependence characterizing the expression (9) and other diffusion coefficients calculated before, is a consequence of the assumption of a very long decay time. Impulses of short duration (on the time scale of T_a) give a contribution to the diffusion coefficient that varies as the sixth power of r (FÄLTHAMMAR, 1966).

What really matters, according to (8), is the *total* magnetic disturbance spectrum. This is due not only to impulses and initial phases but also to various other phenomena, among them the kind of world-wide events studied by NISHIDA and JACOBS (1962) and NISHIDA and CAHILL (1964), and the long-period regular magnetic variations discovered in the Explorer XIV data (PATEL, 1966). What shape the total magnetic disturbance spectrum has in the magnetosphere remains to be determined by direct observations in the magnetosphere. Here we may notice, by mere inspection of (8), that if the spectrum is of the form ν^{-n}, the diffusion coefficient varies with radius as r^{6+2n}.

Electric potential fields in the magnetosphere are now generally realized to be important for particles of auroral energies (as originally suggested in Alfvén's theory of auroras and magnetic storms). Time variations of these fields can also cause a diffusion of trapped particles, the diffusion coefficient being given by (7). Although the lack of observational data makes numerical estimates uncertain, there are indications that this contribution to the diffusion may be very important (DUNGEY, 1965; FÄLTHAMMAR, 1966).

References

ALFVÉN, H. and FÄLTHAMMAR, C.-G.: 1963, *Cosmical Electrodynamics, Fundamental Principles,* Clarendon Press.

BOSTRÖM, R.: 1964, *J. Geophys. Res.* **69**, 4983–4999.

DAVIS, L. Jr., and CHANG, D. B.: 1962, *J. Geophys. Res.* **67**, 2169–2179.

DAVIS, L. R., HOFFMAN, R. A., and WILLIAMSON, J. M.: 1964, *Trans. Am. Geophys. Union* **45**, 84.

DAVIS, L. R. and WILLIAMSON, J. M.: 1963, *Space Res.* **3**, 365–375.

DUNGEY, J. W.: 1965, *Space Sci. Rev.* **4**, 199–222.

DUNGEY, J. W.: 1966, see this volume p. 389.

DUNGEY, J. W., HESS, W. N., and NAKADA, M. P.: 1965, *Space Res.* **5**, 399–403.

FÄLTHAMMAR, C.-G.: 1965, *J. Geophys. Res.* **70**, 2503–2516. Cf. also *Space Res.* **6** (to be published).

FÄLTHAMMAR, C.-G.: 1966, On the Transport of Trapped Particles in the Outer Magnetosphere, *J. Geophys. Res.* (in press).

HESS, W. N.: 1966, this volume.

KELLOGG, P. J.: 1959, *Nature* **183**, 1295–1297.

LIVINGOOD, J. J.: 1961, *Principles of Cyclic Particle Accelerators,* Van Nostrand.

MEAD, G. D.: 1964, *J. Geophys. Res.* **69**, 1181–1195.

NAKADA, M. P. and MEAD, G. D.: 1965, *J. Geophys. Res.* **70**, 4777–4791.

NAKADA, M. P. and MEAD, G. D.: 1966, this volume p. 469.

NISHIDA, A. and CAHILL, L. J. Jr.: 1964, *J. Geophys. Res.* **69**, 2243–2255.

NISHIDA, A. and JACOBS, J. A.: 1962, *J. Geophys. Res.* **67**, 525–540.

PARKER, E. N.: 1960, *J. Geophys. Res.* **65**, 3117–3130.

PATEL, V. L.: 1966, *Space Res.* **6** (to be published).

RICE, S. O.: 1954, in *Selected Papers on Noise and Stochastic Processes,* pp 133–294, Dover.

TVERSKOY, B. A.: 1964, *Geomagnetism and Aeronomy,* **3**, 351–366.

ELECTRON LOSS FROM THE VAN ALLEN ZONES DUE TO PITCH ANGLE SCATTERING BY ELECTROMAGNETIC DISTURBANCES

CHARLES S. ROBERTS

Bell Telephone Laboratories, Inc., Murray Hill, New Jersey

Abstract. Experimental observations indicate that relativistic electrons in the Van Allen zones are perturbed by physical processes that violate one or more adiabatic invariants of the electron's motion. This paper discusses one such possible process, that of pitch angle or mirror point diffusion. A qualitative explanation of pitch angle diffusion and how this process can lead to loss of trapped particles is presented, and then a Fokker-Planck type diffusion equation is introduced to describe the process mathematically. This diffusion equation is used to treat the loss of electrons from trapped orbits due to pitch angle diffusion that results from cyclotron resonant scattering of electrons by whistlers. Predictions are compared with satellite data on the loss from the Van Allen zones of electrons artificially injected by a nuclear explosion. It is found that the data indicate that pitch angle diffusion does indeed play an important role in the loss process, but the dominant mechanism producing the pitch angle scattering must be something other than cyclotron resonant scattering by whistlers since this mechanism leads to several predictions that are in disagreement with the data. These include prediction of the electron pitch angle distribution, the variation of the pitch angle distribution with L-shell and energy, the variation of the loss rate with energy, and the magnitude of the loss rate. Further investigation reveals that a process that, unlike whistlers, scatters electrons more or less uniformly along their spiral path would produce a pitch angle diffusion mechanism in much better accord with observation. Interaction with some type of wide band electromagnetic noise is suggested as such a process which might be of importance.

1. Introduction

The simplest picture of charged particles trapped in the earth's magnetic field is provided by the adiabatic theory of charged particle motion (NORTHROP and TELLER, 1960; NORTHROP, 1963). In this picture a particle once trapped in the geomagnetic field would remain so forever, since its motion is such as to conserve the three adiabatic invariants. Since the time of the initial discovery of the trapped radiation zones (VAN ALLEN *et al.*, 1958) a wealth of experimental evidence has been accumulated which indicates that such a picture is inadequate to describe the behavior of the electrons trapped in the Van Allen zones (BROWN *et al.*, 1963; BROWN, 1965; FRANK *et al.*, 1964; FRANK, 1965; MCILWAIN, 1963; O'BRIEN, 1962, 1964; VAN ALLEN *et al.*, 1959; VAN ALLEN, 1964; WILLIAMS and SMITH, 1965; also many of the papers in this *Proceedings* volume). These measurements all indicate that instead of being conserved one or more of a trapped electron's adiabatic invariants changes, even during periods of low magnetic activity, by a significant amount in times of the order of a few hours to a few years, depending upon the energy of the electron and its location in the trapping region. One result of this is that electrons do not remain trapped in the earth's field forever, but instead are eventually lost due to these nonadiabatic processes. This paper is concerned with such loss of electrons from the Van Allen zones as a result of one particular type of non-adiabatic process, that of pitch angle diffusion.

WALT and MAC DONALD (1964) have shown that for L-shells (MCILWAIN, 1961) below 1.25 small angle coulomb scattering of electrons by the atmospheric constituents is the dominant loss mechanism. Above $L=1.25$ the atmospheric scattering mechanism becomes too weak to be responsible for the observed rates of electron loss. We are thus led to the hypothesis that above $L=1.25$ the loss rate of electrons must be governed by electromagnetic disturbances in the temperate, tenuous plasma which permeates this region. This hypothesis must still be regarded as just an educated guess, however, since experiments have yet to support its validity or to shed much light upon the characteristics of electromagnetic disturbances which may actually exist. DUNGEY (1964) has reviewed some possible effects of such disturbances upon trapped particles and has concluded that cross L-shell diffusion, energy diffusion, and pitch angle diffusion can all result. Thus a model of electron loss based upon pitch angle diffusion alone cannot be expected to describe the actual state of affairs in the Van Allen zones unless it is known beforehand that cross L-shell diffusion and energy diffusion are negligible.

There is experimental evidence that in the region below about $L=2.5$ cross L-shell diffusion is negligible compared with other processes, at least for electrons above about 0.5 MeV in energy. In this region the observed rate of cross L-shell diffusion (FRANK, 1965) is too low to be of much importance in times as short as the typical lifetimes for these electrons. Verification of this fact was provided by electrons artificially injected into narrow L-ranges of the trapping zone by nuclear detonations. These narrow shells of electrons created by the Argus tests (VAN ALLEN *et al.*, 1959) and the Russian test of November 1, 1962 (BROWN, 1965) were all lost with only a very minor amount of cross L-shell movement or spreading. Thus if interest is confined to relativistic electrons trapped below about $L=2.5$, the neglect of cross L-shell diffusion processes seems justified.

The case for the neglect of processes which change electron energy is not so clear cut however. Fluctuating electric fields produce changes in both energy and pitch angle. Pure pitch angle diffusion with no energy diffusion can be produced by perturbing electron orbits with magnetic fields, since a magnetic field alone can never change the energy of a charged particle. But since Maxwell's equations require that any disturbance containing a fluctuating magnetic field must also contain a fluctuating electric field, it is clear that this electric field will always cause some energy diffusion to accompany pitch angle diffusion. The relative importance of the energy diffusion will depend upon E/B, the ratio of the magnitude of the fluctuating electric field in the disturbance to that of the fluctuating magnetic field, and also upon v/c, the ratio of the velocity of the particles being diffused to that of light. The case of relativistic particles being perturbed by a disturbance with E/B small is the one in which energy diffusion would be expected to be negligible compared to pitch angle diffusion. This is the case for the scattering of greater than 0.5 MeV electrons by disturbances propagating in the natural whistler mode, where $E/B \approx 0.1$, a case which we shall want to consider in detail later in this paper. The relative importance of energy diffusion for the processes which actually affect electrons in the Van Allen zones is, unfortunately, still unknown.

2. Qualitative Description of Pitch Angle Diffusion

A charged particle trapped in the earth's magnetic field moves in a spiral path around a line of force between two conjugate mirror points. The location of a mirror point on a line of force may be specified by any of several coordinates such as B, the local magnetic field strength, or λ, the magnetic latitude. However, for the considerations of this paper it is simpler to use the x coordinate described by ROBERTS (1965), which equals zero at the minimum B point or magnetic equator of the line of force and increases monotonically towards unity as one moves along the line towards the earth. As an adiabatically trapped particle travels between mirror points the value of its local pitch angle α, which is the angle between the local magnetic field vector and the particle velocity vector, oscillates between a minimum value α_0 and a maximum value $\pi - \alpha_0$, the extremum values α_0 and $\pi - \alpha_0$ being obtained as the particle crosses the magnetic equator and the value $\pi/2$ being obtained at the mirror points. The value of a particle's pitch angle at some point along its orbit is directly related to the position of its mirror points by the so called mirror equation

$$\mu^2(x) = (x_m^2 - x^2)/(1 - x^2) \tag{1}$$

where $\mu = \cos\alpha$, x is the instantaneous position of the particle, and x_m is the position of the mirror point. Thus specifying a pitch angle for a particle at a given point is equivalent to specifying its mirror position and *vice versa*. In particular, the particle's pitch angle as it crosses the magnetic equator is related to its mirror position by the simple relationship $x_m = \mu_0$, where $\mu_0 = \cos\alpha_0$. Thus descriptions of a particle's orbit in terms of the mirror point position or the equatorial pitch angle are equivalent.

Suppose now that instead of travelling adiabatically between mirror points, the trapped particle interacts with some electromagnetic disturbance along its orbit, which changes the direction of its velocity vector by a small amount. Another way of describing this is to say that the pitch angle of the particle has undergone a scattering by the disturbance. Since such a change in local pitch angle must according to (1) result in a corresponding change in mirror point position, we see that a pitch angle scattering is equivalent to a mirror point scattering, and *vice versa*. Similarly, since $x_m = \mu_0$ a scattering of the mirror point must be related to a corresponding scattering of the particle's equatorial pitch angle. Relating all scatterings occurring anywhere along a particle's orbit to the corresponding change in the equatorial pitch angle is a very convenient way to keep track of the effects of many small scatterings and is the technique which will be used in this paper.

If a population of many trapped particles is subjected to a series of small pitch angle scatterings that are random in both direction and size, the individual particles of the population will undergo a sort of random walk in their equatorial pitch angle, and a diffusion process in equatorial pitch angle space results. Such a diffusion process will result in loss of trapped particles because of the exponentially increasing atmospheric density near the earth and the fact that diffusion of pitch angles also means diffusion of mirror positions. The presence of the atmosphere creates a "loss cone" for particle pitch angles since particles having pitch angles smaller than some critical value will have mirror positions close enough to the earth so that it becomes overwhelmingly probable that loss by interaction with the atmospheric constituents will occur. Thus particles whose pitch angles diffuse into the loss cone will be removed from the trapped particle population. It is interesting that even for pitch angle scattering, a so called nonatmospheric loss mechanism, it is the atmosphere which is responsible for the actual loss of the particles.

3. Fokker-Planck Equation for Pitch Angle Diffusion

MAC DONALD and WALT (1961) used a Fokker-Planck equation to treat the simultaneous pitch angle and energy diffusion of a population of trapped particles caused by interactions with the atmospheric constituents. The Fokker-Planck equation for pure pitch angle diffusion can be obtained from their equation by setting the energy diffusion coefficients equal to zero. Doing this, we obtain

$$\frac{\partial f(\mu_0, t)}{\partial t} = -\frac{\partial}{\partial \mu_0} f \langle (\Delta\mu_0) \rangle + \frac{1}{2} \frac{\partial^2}{\partial \mu_0^2} f \langle (\Delta\mu_0)^2 \rangle + Q(\mu_0) \tag{2}$$

Here t is the time, $\mu_0 = \cos \alpha_0$ is the cosine of the particle's equatorial pitch angle, and $f(\mu_0, t)$ is the MAC DONALD and WALT (1961) distribution function for the particles under consideration, which will be of a specified kinetic energy and type. If, for example, we are considering electrons with energies lying in a small energy interval centered around 0.5 MeV, $f(\mu_0,t)\,\mathrm{d}\mu_0\,\mathrm{d}A_0$ would represent the number of these particles which would cross the equator with pitch angle cosine between μ_0 and $\mu_0+\mathrm{d}\mu_0$ trapped at time t in a tube of force of equatorial cross sectional area $\mathrm{d}A_0$. The distribution function f is related to the equatorial unidirectional flux by the expression

$$j(\mu_0,t) = f(\mu_0,t)/4\pi\mu_0 t_b \tag{3}$$

where t_b is the time for the trapped particle to travel from one mirror point to the other. In (2) the term $Q(\mu_0)$ is a source term, and $Q(\mu_0)\,\mathrm{d}\mu_0\,\mathrm{d}A_0$ equals the number of particles per unit time injected into a tube of force of equatorial cross sectional area $\mathrm{d}A_0$ with equatorial pitch angle cosines between μ_0 and $\mu_0+\mathrm{d}\mu_0$. The diffusion coefficients $\langle(\Delta\mu_0)\rangle$ and $\langle(\Delta\mu_0)^2\rangle$ in (2) depend on the nature of the process producing the pitch angle diffusion and represent the average, over all scatterings, of $(\Delta\mu_0)$ and $(\Delta\mu_0)^2$ multiplied by the scattering frequency. The quantity $(\Delta\mu_0)$ represents the change in one scattering of μ_0. It should be noted that for scatterings which occur off the magnetic equator the change $(\Delta\mu_0)$ in equatorial pitch angle cosine must be computed from the change in local pitch angle cosine with the formula

$$(\Delta\mu_0) = (\Delta\mu)\left(\frac{\partial\mu_0}{\partial\mu}\right) + \tfrac{1}{2}(\Delta\mu)^2\left(\frac{\partial^2\mu_0}{\partial\mu^2}\right) \tag{4}$$

The second term on the right hand side of (4) is necessary since the Fokker-Planck formalism is one in which all second order terms are retained. The derivatives in (4) are computed by differentiating the mirror equation,

$$\mu_0^2 = \mu^2(1-x^2) + x^2 \tag{5}$$

holding the value of x constant. It is apparent from (4) that even if the average of $(\Delta\mu)$ taken over all scatterings is zero, the average of $(\Delta\mu_0)$ can be nonzero.

Before working with (2) it is convenient to make a few mathematical substitutions. We let

$$D(\mu_0) = \tfrac{1}{2}\langle(\Delta\mu_0)^2\rangle, \qquad D'(\mu_0) = \tfrac{1}{2}\langle(\Delta\mu_0)^2\rangle - \mu_0\langle(\Delta\mu_0)\rangle, \qquad r(\mu_0) = D'/D \tag{6}$$

$$w(\mu_0,t) = D(\mu_0)f(\mu_0,t)/8\pi\mu_0 L r_e \tag{7}$$

where L is McIlwain's (1961) parameter for the shell of interest and r_e is the average radius of the earth $\approx$6371 km. Using (6) and (7), Equation (2) now becomes

$$\frac{1}{D(\mu_0)}\frac{\partial w}{\partial t}=\frac{1}{\mu_0}\frac{\partial}{\partial\mu_0}\left[r(\mu_0)w+\mu_0\frac{\partial w}{\partial\mu_0}\right]+s(\mu_0)\bar{q}(\mu_0) \tag{8}$$

where $\bar{q}(\mu_0)$ is the average over the particle's spiral path of the source strength in particles/cm^3 sec ster, and $2Lr_e s(\mu_0)$ is the distance along the spiral path between mirror points. Evaluation of integrals over a particle's spiral path, such as $\bar{q}(\mu_0)$ and $s(\mu_0)$, have been previously discussed by Lenchek *et al.* (1961) and Lenchek and Singer (1963) as well as by several other authors. For a dipole magnetic field Lenchek *et al.* (1961) give the approximation for $s(\mu_0)$

$$s(\mu_0)=1.38-0.32\left[(1-\mu_0^2)^{1/2}+(1-\mu_0^2)^{1/4}\right] \tag{9}$$

In terms of $w(\mu_0, t)$, the equatorial unidirectional flux is

$$j(\mu_0,t)=vw(\mu_0,t)/D(\mu_0)s(\mu_0) \tag{10}$$

4. Normal Mode Solution of the Fokker-Planck Equation

Equation (8) is a partial differential equation in two variables, the solutions of which give the behavior of a population of particles undergoing pitch angle diffusion. As discussed in Section 2, the presence of the earth's atmosphere produces a loss cone for particle pitch angles, and therefore to account for this (8) must be solved subject to the boundary condition $w(\mu_{c0}, t)=0$, where μ_{c0} is the cosine of the equatorial loss cone angle. Equation (8) is a separable partial differential equation, and its general solution may be written

$$w(\mu_0,t)=w_e(\mu_0)+\sum_{l=1}^{\infty}w_l(\mu_0)\exp(-k_l t) \tag{11}$$

where $w_e(\mu_0)$ satisfies the equation

$$\frac{1}{\mu_0}\frac{d}{d\mu_0}\left[rw_e+\mu_0\frac{dw_e}{d\mu_0}\right]=-s\bar{q} \tag{12}$$

The $w_l(\mu_0)$ are the eigenfunctions and k_l the eigenvalues of the equation

$$\frac{1}{\mu_0}\frac{d}{d\mu_0}\left[rw_l+\mu_0\frac{dw_l}{d\mu_0}\right]=-\frac{k_l}{D}w_l \tag{13}$$

subject to the boundary condition $w_l(\mu_{c0})=0$. Equation (11) is called the normal mode solution and the $w_l(\mu_0)$ the normal modes of the diffusion equation in analogy to the similar type of general solution that occurs in the theory of vibrating strings.

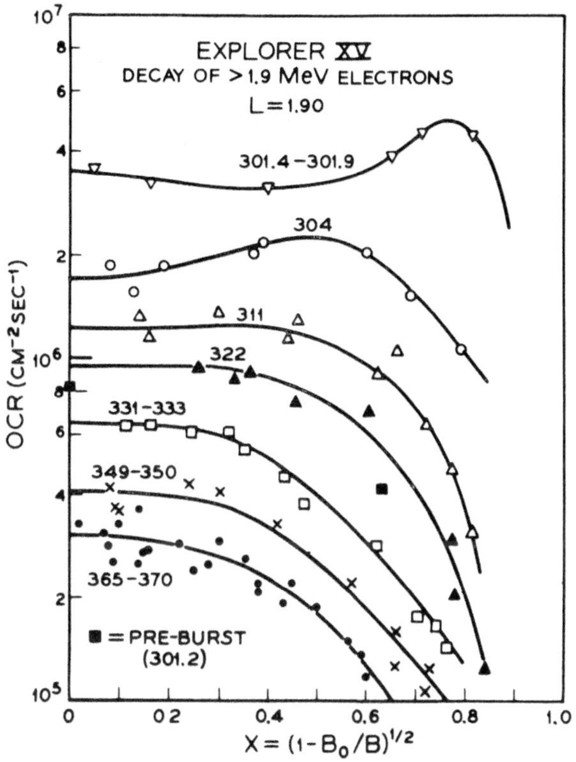

Fig. 1. Counting rate of the Bell Telephone Laboratories >1.9 MeV electron detector aboard Explorer XV after the U.S.S.R. nuclear detonation of October 28 (day 301), 1962. The data points and associated curves show the omnidirectional counting rate on $L=1.9$ plotted as a function of position on the L-shell measured with the x-coordinate. The various curves and data point symbols are for the different time periods indicated (in days of 1962). The sequence of curves shows how a severely perturbed particle distribution decays back to the "normal shape", as represented by the data after about day 330.

Several interesting properties of a population of particles undergoing pitch angle diffusion can be understood from the normal mode solution. First, there will be a smallest eigenvalue, k_1, for Equation (13), which in practice usually turns out to be about a factor of 3–6 smaller than k_2. This means that after a sufficiently long time the $l \geqslant 2$ terms in the summation of (11) will become negligible compared to the $l=1$ term, and the time dependence will thereafter be a simple exponential decay. Thus the decay of an artificially injected population of particles via pitch angle diffusion can be divided into an initial stage, where the time dependence of the flux is not exponential and the pitch angle distribution changes with time, and a later stage, where $(w-w_e) \approx w_1(\mu_0) \exp(-k_1 t)$. Second, Equations (12) and (13) differ only in the term on the right hand side of the equal sign, and it turns out that the solutions for $w_e(\mu_0)$ and $w_1(\mu_0)$ are rather insensitive to the μ_0 dependence of this term. This results in the following interesting features. The μ_0 dependence of $w_e(\mu_0)$ is insensitive to the μ_0 dependence of $\bar{q}(\mu_0)$, and thus one can tell almost nothing about the pitch angle dependence of the source by looking at the equilibrium pitch angle distribution. Also, both $w_e(\mu_0)$ and $w_1(\mu_0)$ will have very nearly the same μ_0 dependence so that in the later stage of the decay, where $l \geqslant 2$ terms in (11) can be neglected, $w(\mu_0, t) \approx w_e(\mu_0)\,[1 + C \exp(-k_1 t)]$, where C is some constant, and the pitch angle dependence remains approximately constant with time. Thus if one plots flux of particles on a logarithmic scale versus μ_0 at various times during this later stage of decay, the shape of the curves will remain approximately constant.

Data showing the decay of electrons artificially injected into the trapping region by the U.S. high altitude nuclear test, Starfish, of July 9, 1962 and the U.S.S.R. tests of October 22, October 28, and November 1, 1962 seem to be in accord with the idea that pitch angle diffusion plays an important part in the loss mechanism for these electrons. An illustration of this is shown in Figure 1, which shows the decay of >1.9 MeV electrons following the second U.S.S.R. test. The data points were obtained by the Bell Telephone Laboratories' experiment aboard the Explorer XV satellite, and the omnidirectional counting rate of the electron detector is plotted versus the x coordinate at various times after the nuclear burst. The omnidirectional counting rate of the detector is proportional to J, the omnidirectional flux of electrons >1.9 MeV, and the function $J(x,t)$ is in turn related to $j(\mu_0, t)$ and $w(\mu_0, t)$ by an integral transformation (Roberts, 1965). The curves show that up until around day 320–330, the dependence of the counting rate upon x changes with time, while after that the main change is a downward displacement of the same shape curve. The rate of this downward displacement can be seen in Figure 2 where the equatorial counting rate of the Explorer XV electron detectors is plotted versus time. The points for the >1.9 MeV electrons show nearly exponential decay between about day 320 and day 360. It thus appears that the period before day 320–330 can be identified with the initial stage of the decay, discussed in connection with the normal mode solution (11), while the period from day

330–360 can be identified as the later stage of the decay where only the $l=1$ term in (11) need be considered. The deviations of the data from this simple picture may be attributable to several things. First, the solution (11) gives the flux of electrons at one particular kinetic energy, while the electron detector counts electrons greater than 1.9 MeV. Since k_1 may be a strong function of

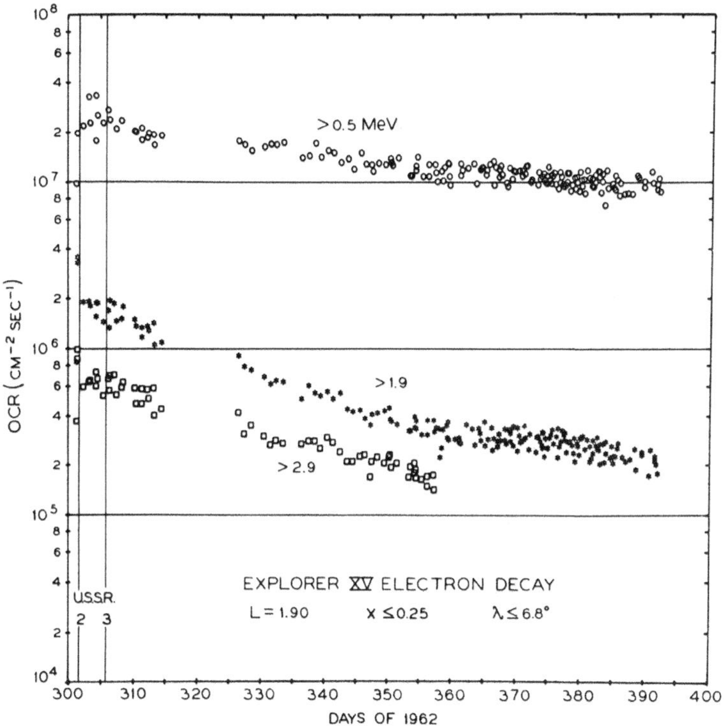

Fig. 2. Counting rate of the Bell Telephone Laboratories' electron detectors aboard Explorer XV for the near equatorial region, $x \leqslant 0.25$, on $L=1.9$. The omnidirectional counting rate is plotted versus time, and the times of the second and third U.S.S.R. nuclear tests on October 28 (day 301) and November 1 (day 305) are marked with vertical lines.

electron energy one would really like to compare with the counting rate of a detector measuring electrons in a narrow energy range. Secondly, energy diffusion may be of some importance, and therefore a treatment based upon pure pitch angle diffusion would be incomplete. Finally, the diffusion coefficients $D(\mu_0)$ and $D'(\mu_0)$, which depend upon the number of electromagnetic disturbances present, are probably not really independent of time. Instead their value is probably correlated with magnetic activity (McIlwain, 1963). Despite these complications, the overall behavior of the data obtained after artificial injection of electrons is in general accord with the idea that pitch angle diffusion plays at least an important role in the subsequent loss of these electrons. The paper by West in this *Proceedings* volume contains further experimental verification of this.

5. Loss of Electrons Due to Interaction with Whistlers

Dungey (1963a) and Cornwall (1964) have suggested that pitch angle scattering due to cyclotron resonant interactions with whistlers might be responsible for the loss of electrons from the Van Allen zones above $L \approx 1.7$. Equation (8) will now be used to compute the properties of such a process, and these will then be compared with existing experimental knowledge concerning electrons in this region.

In order to apply (8) to the scattering of electrons by whistlers the diffusion coefficients $D(\mu_0)$ and $D'(\mu_0)$, defined in (6), must be evaluated on the basis of the dynamics of the electron-whistler interaction. Although specific formulas and details will be left for another paper (Roberts, 1966), a sketch of the important ideas involved will be given here. This should suffice to enable the reader to understand physically the qualitative results which ensue.

Computation of the diffusion coefficients involves determining the points at which cyclotron resonance occurs between a whistler and an electron. A whistler is a type of disturbance that has its origin in the electromagnetic energy released in a lightning bolt discharge below the ionosphere (Helliwell, 1965), and for our purposes here it may be thought of as an electromagnetic wave whose frequency at a given point in space decreases with time. For simplicity, we take the direction of the whistler propagation vector **k** to be parallel to the earth's magnetic field. The condition for cyclotron resonance is

$$(\omega + kv_{\parallel} - \Omega) = 0 \tag{14}$$

where $\omega/2\pi$ is the frequency of the whistler, $v_{\parallel}$ the electron's velocity component parallel to the earth's field, $2\pi/k$ the wavelength of the whistler, and $\Omega/2\pi$ the electron cyclotron frequency. Equation (14) simply states that resonance occurs when the instantaneous frequency of the whistler doppler shifted by the electron's motion is equal to the local electron cyclotron frequency. Since the quantities in (14) all change with time as the electron travels along its orbit, it is clear that cyclotron resonance can only be achieved instantaneously, unless very fortuitous conditions are obtained. Using the fact that the cyclotron frequency of a relativistic electron varies inversely as the total energy of the electron, (14) can be rewritten

$$\omega = \Omega_r \frac{T_r}{(T_r + T)(1 + n\mu\beta)} \tag{15}$$

where $\Omega_r/2\pi$ is the cyclotron frequency for a nonrelativistic electron, T_r is the rest energy of an electron $=0.51$ MeV, T is the kinetic energy of the electron, $n = ck/\omega$ is the index of refraction for whistler mode propagation at frequency ω, $\mu = \cos\alpha = v_{\parallel}/v$, and $\beta = v/c$. Equation (15) shows how the whistler frequency necessary to resonate with an electron varies with local magnetic field intensity,

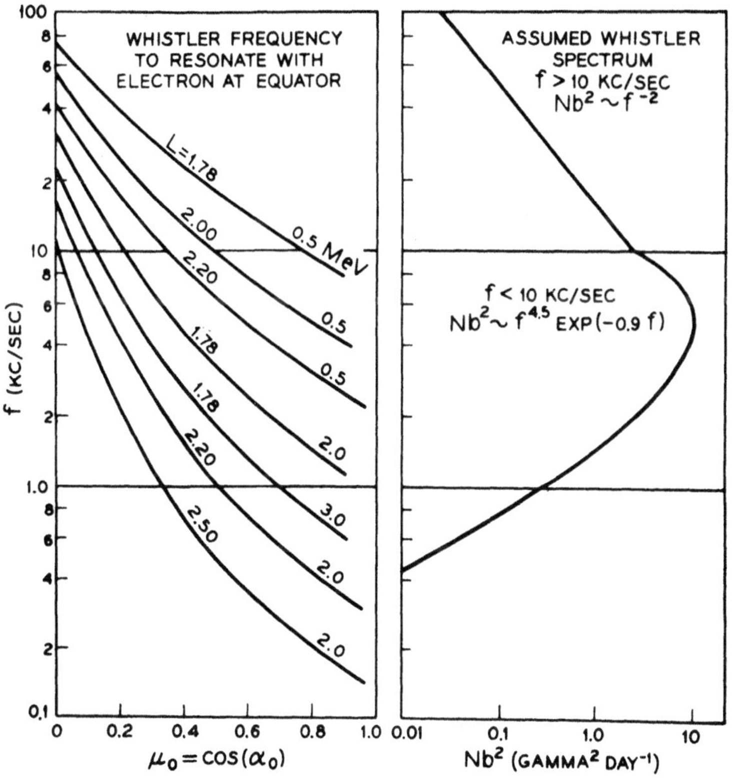

Fig. 3. Whistler frequency needed to produce cyclotron resonance with an electron crossing the magnetic equator plotted versus μ_0 for various L-shells and electron energies. The right graph shows the whistler power spectrum that was assumed in computing the diffusion coefficients for pitch angle scattering by whistlers. Comparison of the left and right graphs gives an estimate of the relative amount of pitch angle scattering that will occur at various energies, L-shells, and equatorial pitch angles.

electron energy, and electron pitch angle, and this is displayed graphically on the left side of Figure 3, which shows the whistler frequency necessary to resonate with an electron passing through the magnetic equator plotted as a function of the cosine of its equatorial pitch angle. Curves are drawn for several L values and electron energies so that one may also see the dependence upon these factors. Resonance, is of course, possible off the magnetic equator, and similar curves could be drawn for those cases. Such curves would necessarily lie above the corresponding equatorial curves since as (15) will verify an electron resonates with the lowest possible frequency when it crosses the equator.

For each resonant whistler-electron interaction it is necessary to compute the perturbation produced in the electron's local pitch angle by the fields of the whistler wave. The motion of an electron in resonance with a circularly polarized wave has been solved exactly (Roberts and Buchsbaum, 1964), but the perturbation technique of integrating over the zero order helix of the electron (Dungey, 1963b) is sufficient to yield $(\Delta\mu)$, the change in local pitch angle cosine as a result of the interaction. The value of $(\Delta\mu)$ is of course dependent upon the strength of the whistler at the frequency necessary to produce cyclotron resonance, and thus the diffusion coefficients $D(\mu_0)$ and $D'(\mu_0)$ turn out to be dependent upon the whistler frequency spectrum. It is most convenient to express this spectrum in terms of $Nb^2(\omega)$, which is the average sum at the frequency ω over all whistlers per day of b^2, the square of the whistler magnetic field intensity.

Unfortunately, few if any experimental measurements are yet available on the spectrum of whistlers above the ionosphere above about 10 kc/sec. However measurements have been made of the spectrum of lightning discharges which produce whistlers (Helliwell, 1965; Horner, 1961). These indicate that b^2 peaks around a frequency of 5 kc/sec and drops about a factor of 4 at 2 kc/sec and 10 kc/sec, and that above 10 kc/sec, b^2 decreases approximately as the inverse square of the frequency. Since the spectrum of whistlers should resemble the spectrum of the lightning which produces them, a hypothetical whistler spectrum based upon the preceding facts was constructed and used to compute the diffusion coefficients. This spectrum is shown in the right hand graph of Figure 3. The value at the peak of 10 gamma^2/day seems to be a reasonable guess based upon satellite measurements below 10 kc/sec (Cain *et al.*, 1961; Gurnett and O'Brien, 1964).

Simultaneous examination of the two graphs of Figure 3, shows that the rate of pitch angle diffusion should be highly dependent upon L, energy, and pitch angle. Scattering will be greatest when the resonance can occur with a whistler frequency of about 5 kc/sec, the frequency at which the whistler spectrum peaks. Scattering of electrons which mirror near the equator, $\mu_0 \approx 0$, will be small since whistlers have only a small amount of energy at the high frequencies necessary to rescnant with these electrons. Typical results of these effects can be seen in Figure 4 which shows the diffusion coefficient $D(\mu_0)$ as computed for diffusion by whistlers on $L=1.9$. The curves are for electron

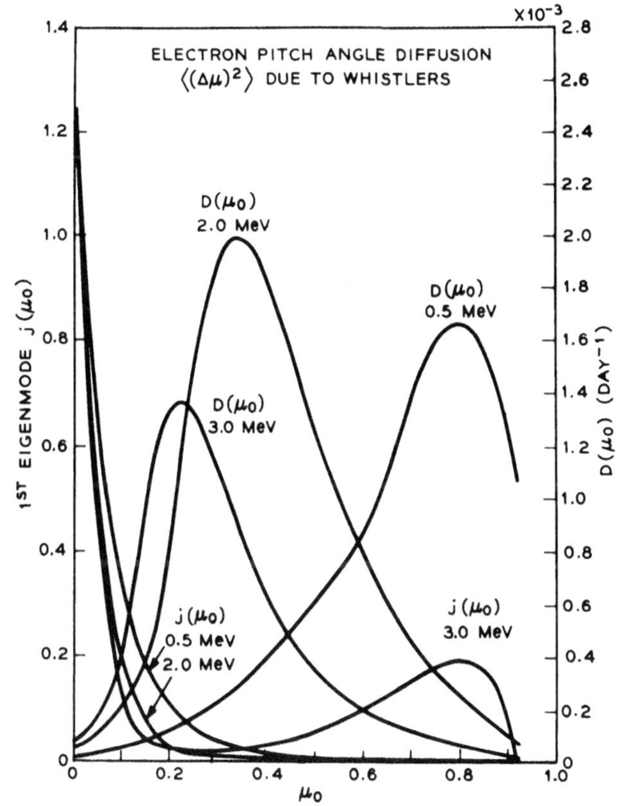

Fig. 4. Some results of the calculation of electron pitch angle diffusion on $L=1.9$ produced by cyclotron resonant scattering of the electrons by whistlers. The diffusion coefficient $D(\mu_0)$, defined by Equation (6), and the equatorial unidirectional flux given by the $l=1$ solution of Equation (13) are plotted as a function of $\mu_0 = \cos \alpha_0$ for 0.5, 2.0, and 3.0 MeV electrons.

energies 0.5, 2.0, and 3.0 MeV, and the $l=1$ eigenmode unidirectional flux $j(\mu_0)$ as calculated with (13) and (10) is also shown. The large peaks in $D(\mu_0)$are typical of the whistler mechanism, and result from the peak in the whistler spectrum.

When the results of this pitch angle diffusion calculation are compared with experiment, severe discrepancies are found to exist. The $l=1$ eigenmode solution of (13) and equilibrium solution of (12) with any reasonable source $\bar{q}(\mu_0)$ both have about the same μ_0 dependence, as previously pointed out in

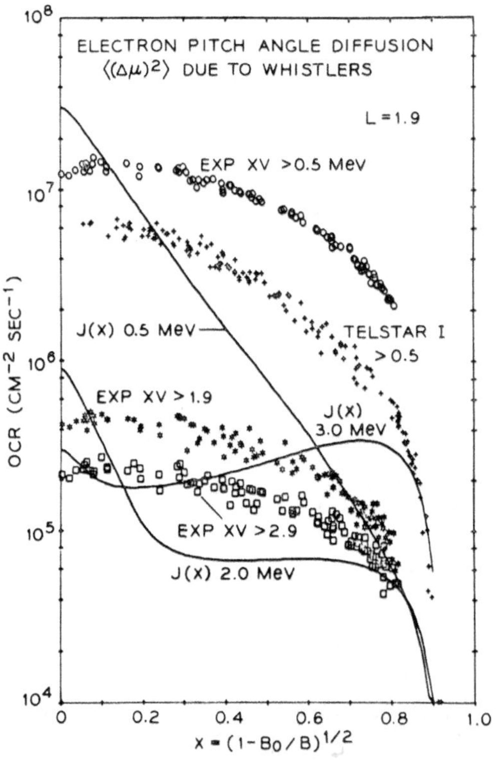

Fig. 5. Omnidirectional flux J of electrons on $L=1.9$, as predicted by the whistler pitch angle diffusion mechanism, plotted versus the x-coordinate. The theoretical $J(x)$ curves for the 0.5, 2.0 and 3.0 MeV electrons were computed from the corresponding equatorial unidirectional fluxes $j(\mu_0)$ shown in Figure 4, and all the $J(x)$ curves may be displaced vertically on the log scale by an arbitrary amount. The shapes of the theoretical curves are in severe disagreement with the shapes of the sets of data shown, thus indicating that something other than whistlers is responsible for the pitch angle diffusion of these electrons.

Section 4, and they both predict a unidirectional flux $j(\mu_0)$ that disagrees violently with that observed experimentally. An example of this is given in Figure 5, which shows the dependence of the omnidirectional counting rate upon the x coordinate for >0.5, >1.9, and >2.9 MeV electrons on $L=1.9$.

The data were obtained by the Bell Telephone Laboratories' experiments aboard the Explorer XV and Telstar I satellites, as marked, and the x dependence of the data points shown is typical for periods when the x dependence does not vary with time. The solid curves in Figure 5 show the omnidirectional flux $J(x)$ as computed from the theoretical $l=1$ eigenmode $j(\mu_0)$ shown in Figure 4. These $J(x)$ curves may be displaced vertically by any desired amount, since an eigenmode may always be multiplied by an arbitrary constant. The point however is that the x dependence of the computed curves is in violent disagreement with the x dependence of the data, i.e. the shapes are different. This failure of the whistler mechanism to predict correct x or μ_0 dependences applies also to other L values and energies. The computed pitch angle distributions typically have a large peak at $\mu_0=0$, due to the relatively small amount of whistler energy present at the high frequencies necessary to resonate with electrons mirroring near the equator, and also display bumps and dips whose position varies with energy and L. None of these features are in agreement with observation since time independent pitch angle distributions for electrons of any energy above a few hundred keV are generally qualitatively similar to those represented by the data points in Figure 5.

The failure of the whistler mechanism to predict correct pitch angle distributions does not depend critically upon the details of the whistler spectrum that was assumed. Rather it depends only on the fact that the spectrum falls rapidly above 10 kc/sec and that a maximum exists at some frequency below 10 kc/sec. On the basis of the information already presented about the spectrum of lightning that causes whistlers, it is difficult to see how the actual whistler spectrum could fail to possess these features. The true whistler spectrum should actually fall even more rapidly above 10 kc/sec than the lightning spectrum in Figure 3, since higher frequencies are attenuated more in passing through the ionosphere.

Another prediction of the whistler mechanism that disagrees with observation is the dependence of the loss rate on electron energy. For instance on $L=1.9$, the decay constant k_1 associated with the $l=1$ eigenmodes shown in Figure 4 are 0.65×10^{-3}, 2.6×10^{-3}, and 3.0×10^{-3} day^{-1} respectively for 0.5, 2.0, and 3.0 MeV electrons. Thus the whistler mechanism predicts that 3 MeV electrons should decay more rapidly than 2 MeV electrons. Experimentally, the Bell Telephone Laboratories' experiment aboard Explorer XV measured decay rates for 1.9 MeV electrons which were higher than those for 2.9 MeV electrons, as is evident in Figure 2, in disagreement with the whistler predictions. Similar disagreements exist at other energies and L values, and are particularly serious if one considers electrons of about 0.25 MeV. Since the frequencies necessary to resonate with these electrons on $L=1.9$ would lie in the very weak portion of the whistler spectrum, their loss rate would be expected to be very low.

Instead, Telstar I observed loss rates for such electrons comparable to those for 0.5 MeV electrons. Finally, one should note that the absolute value of the decay rates as predicted by whistlers are generally too low. For instance on $L=1.9$, the decay constants given above for 0.5, 2.0, and 3.0 MeV electrons

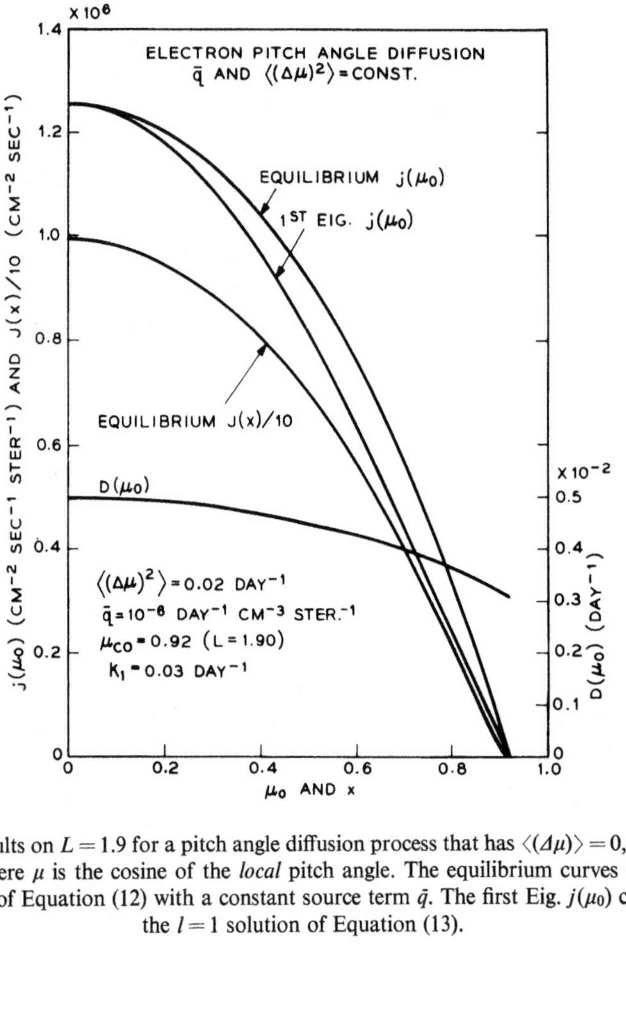

Fig. 6. Results on $L=1.9$ for a pitch angle diffusion process that has $\langle(\Delta\mu)\rangle=0$, $\langle(\Delta\mu)^2\rangle=$ constant, where μ is the cosine of the *local* pitch angle. The equilibrium curves come from the solution of Equation (12) with a constant source term $\bar{q}$. The first Eig. $j(\mu_0)$ comes from the $l=1$ solution of Equation (13).

are a factor of 30, 10, and 5 lower respectively than the decay constants measured by Explorer XV. This disagreement in absolute value is, however, directly dependent upon the value of 10 gamma^2/day taken as the peak of whistler frequency spectrum in Figure 3. Since this number is apt to be in error, either high or low, one should attach minimum significance to absolute values of predicted decay rates. However, on the basis of the qualitative disagreements one must conclude that resonant pitch angle scattering of electrons by whistlers is not the dominant mechanism responsible for loss of electrons in the Van Allen zones.

6. Pitch Angle Diffusion with Constant $\langle(\Delta\mu)^2\rangle$

In order to get a feel for what type of a pitch angle scattering process would predict unidirectional flux functions $j(\mu_0)$ that agree with experiment, a scattering process for which $\langle(\Delta\mu)\rangle=0$ and $\langle(\Delta\mu)^2\rangle=$ constant was investigated. The quantity $\langle(\Delta\mu)\rangle$ is the average of the change in *local* pitch angle cosine,

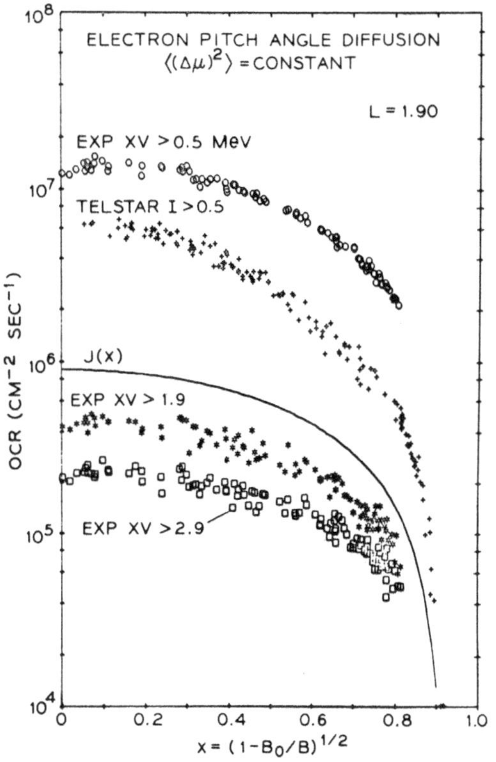

Fig. 7. The omnidirectional flux function $J(x)$ corresponding to the first Eig. $j(\mu_0)$ shown in Figure 6. The curve may be displaced vertically on the log scale by an arbitrary amount. The data shown is the same data shown in Figure 5 with the $J(x)$ results based on the whistler mechanism. The shape of the $J(x)$ curve in Figure 7 clearly resembles the shapes of the sets of data better than do the $J(x)$ curves in Figure 5, indicating that the pitch angle diffusion mechanism represented by Figures 6 and 7 must be closer to the truth than the whistler mechanism.

and $\langle(\Delta\mu)^2\rangle$ is the average of the square of this quantity. Physically, this would correspond to a scattering process which produces perturbations in local pitch angle cosine that are random in direction and independent of where along the particle's spiral path the scattering occurs. Figures 6 and 7 show the results of the diffusion calculation for such a process operating on $L=1.9$. In Figure 7 the predicted omnidirectional flux for the $l=1$ eigenmode is plotted along with the data previously shown in Figure 5. The curve in Figure 7 can be displaced vertically by any desired amount, corresponding again to the multiplication of an eigenmode by an arbitrary constant. However, unlike the whistler case shown in Figure 5, there is general agreement between the shape of the predicted curve in Figure 7 and the shape of the data. Thus pitch angle scattering by a process which operates more or less uniformly along the electron's spiral path yields a physical model which is in much better agreement with observation than pitch angle scattering by whistlers. Scattering by whistlers does not of course operate uniformly along the spiral path since the whistler frequency necessary to produce resonance changes along the spiral path, and the spectrum of whistlers varies with frequency. Instead, one would guess that pitch angle scattering produced by interaction with some sort of wide band noise would yield a mechanism that operates more or less uniformly along the electron's spiral path. The wide band noise need not be whistler mode noise; it could be noise propagating in any of the other possible plasma modes including the longitudinal modes. It appears that electromagnetic noise sources and their possible effects on trapped electrons is a subject worthy of future investigation, both experimental and theoretical.

7. Conclusion

It seems likely that pitch angle diffusion is at least an important part of the loss mechanism for relativistic electrons in the Van Allen zones in the region $L \approx 1.7$ to 2.5 and possibly also elsewhere. On the basis of pitch angle diffusion one can explain several of the features of the loss of electrons artificially injected into the trapping region by nuclear explosion. Among these are the fact that the electrons move only a very small amount in L during their lifetime, that the rate of loss is faster when the shape of the electron pitch angle distribution is perturbed severely from its normal or equilibrium shape, that a "shape independent" stage exists in the decay when the shape of the pitch angle distribution on a log scale no longer changes much with time and the flux at all pitch angles decreases with time approximately as $\sim \exp(-kt)$, and that the pitch angle distribution seems finally to settle down after a severe perturbation to a shape which is very similar to that during the shape independent stage of the decay. What remains unanswered, however, is the key question: what physical process is responsible for producing the electron pitch angle diffusion? Pitch angle diffusion produced by cyclotron resonant interactions between electrons and whistlers leads to several predictions that disagree with observations, and so this mechanism is ruled out as the dominant one. Some other physical process that scatters the local pitch angle more or less uniformly all along the electron's spiral path would lead to better agreement with the data. It is not yet clear, however, just what the nature of such a process might be. Interaction between electrons and some kind of wide band electromagnetic noise seems at present to be a likely possibility.

References

Brown, W. L., Hess, W. N., and Van Allen, J. A.: 1963a *J. Geophys. Res.* **68**, 605.
Brown, W. L., Gabbe, J. D., and Rosenzweig, W.: 1963b, *Bell System Tech. J.* **42**, 1505.
Brown, W. L.: 1965, *Proceedings of the Plasma Space Science Symposium,* D. Reidel Publishing Co., p. 189.
Cain, J. C., Shapiro, I. R., Stolarik, J. D., and Heppner, J. P.: 1961, *J. Geophys. Res.* **66**, 2677.
Cornwall, J. M.: 1964, *J. Geophys. Res.* **69**, 1251.
Dungey, J. W.: 1963a, *Planet. Space Sci.* **11**, 591.
Dungey, J. W.: 1963b, *J. Fluid Mech.* **15**, 74.
Dungey, J. W.: 1964, *Space Sci. Rev.* **4**, 199.
Frank, L. A., Van Allen, J. A., and Hills, H. K.: 1964, *J. Geophys. Res.* **69**, 2171.
Frank, L. A.: 1965, *J. Geophys. Res.* **70**, 3533.
Gurnett, D. A. and O'Brien, B. J.: 1964, *J. Geophys. Res.* **69**, 65.
Helliwell, R. A.: 1965, *Whistlers and Related Ionospheric Phenomena,* Stanford University Press, p. 121.
Horner, F.: 1961, *J. Atmos. Terr. Phys.* **21**, 13.
Lenchek, A. M., Singer, S. F., and Wentworth, R. C.: 1961, *J. Geophys. Res.* **66**, 4027.
Lenchek, A. M. and Singer, S. F.: 1963, *Planet. Space Sci.* **11**, 1151.
MacDonald, W. M. and Walt, M.: 1961, *Ann. Phys. (N.Y.)* **15**, 44.
McIlwain, C. E.: 1961, *J. Geophys. Res.* **66**, 3681.
McIlwain, C. E.: 1963, *Science* **142**, 355.
Northrop, T. G. and Teller, E.: 1960, *Phys. Rev.* **117**, 215.
Northrop, T. G.: 1963, *The Adiabatic Motion of Charged Particles,* Interscience Publishers.
O'Brien, B. J.: 1962a, *J. Geophys. Res.* **67**, 1227.
O'Brien, B. J.: 1962b, *J. Geophys. Res.* **67**, 3687.
O'Brien, B. J.: 1964, *J. Geophys. Res.* **69**, 13.
Roberts, C. S. and Buchsbaum, S. J.: 1964, *Phys. Rev.* **135**, A381.
Roberts, C. S.: 1965, *J. Geophys. Res.* **70**, 2517.
Roberts, C. S.: 1966, *J. Geophys. Res.*, to be published.
Van Allen, J. A., Ludwig, G. H., Ray, E. C., and McIlwain, C. E.: 1958, *Jet Propulsion* **28**, 588.
Van Allen, J. A., McIlwain, C. E., and Ludwig, G. H.: 1959, *J. Geophys. Res.* **64**, 877.
Van Allen, J. A.: 1964, *Nature* **203**, 1006.
Walt, M., and MacDonald, W. M.: 1964, *Rev. Geophys.* **2**, 543.
Williams, D. J., and Smith, A. M.: 1965, *J. Geophys. Res.* **70**, 541.

OBSERVATIONS OF MAGNETOSPHERIC BOUNDARY PHENOMENA

L. A. FRANK

Department of Physics and Astronomy, University of Iowa, Iowa City, Iowa

1. Introduction

From early observations of auroras (STÖRMER, 1955) and of cometary tails (BIERMANN, 1951, 1957) the earth was thought to be immersed in a plasma streaming radially outward from the sun. On the basis of observations of cometary tails and of the solar corona a hydrodynamic model of the solar corona was developed (PARKER, 1958) which later proved to be in agreement with direct measurements (SNYDER, NEUGEBAUER, and RAO, 1963) of the solar corpuscular radiation, or solar wind. The advent of the artificial earth satellite and the interplanetary probe provided *in situ* measurements of the environment of the earth in the interplanetary medium: a continual solar plasma directed radially outward from the sun with a bulk velocity of 300–900 km $(\text{sec})^{-1}$ (corresponding proton kinetic energy of ~ 1 keV), a density of ~ 1 to 20 protons $(\text{cm})^{-3}$, and a characteristic temperature of $\sim 10^5$–10^6 °K (~ 100 eV) (SNYDER *et al.*, 1963; BRIDGE *et al.*, 1965), threaded by a weak interplanetary field of $\sim 5\gamma (1\gamma = 10^{-5}$ gauss) forming a general Archimedean spiral structure stretching outward from the sun (NESS and WILCOX, 1965). The motion of the solar wind past the earth is impeded by the presence of the earth's magnetic field. This interaction of the solar plasma with the geomagnetic field is presumed to be directly related, for example, with the auroras, the geomagnetic storm, and the energetic charged particles of the earth's radiation zones. The solar wind confines the earth's magnetic field to a cavity, or magnetosphere, which is terminated in the solar direction at ~ 10 R_E (R_E = earth radius) geocentric radial distance (CAHILL and AMAZEEN, 1963; NESS *et al.*, 1964) and is drawn into a magnetic tail extending beyond 30 R_E in the antisolar direction (NESS, 1965) as suggested qualitatively by earlier theoretical models (cf. PIDDINGTON, 1959, 1960; JOHNSON, 1960; AXFORD and HINES, 1961). Since the propagation of disturbances (Alfvén waves) in the interplanetary medium can proceed at only a fraction ($\sim 1/5$) of the bulk velocity of the solar plasma, a collisionless magnetohydrodynamic shock should develop on the sunward side of the magnetosphere (cf. AXFORD, 1962; KELLOGG, 1962) and has been directly observed (NESS *et al.*, 1964; FREEMAN, 1964; BRIDGE *et al.*, 1965). Measurements of charged particles near and in the outer magnetosphere have revealed reservoirs of charged particles energized by 'local' acceleration mechanisms driven by the solar wind, for example in the transition region between the shock wave and the sunward magnetospheric boundary and also in the tail of the magnetosphere, which are indicative of charged particle sources for auroral phenomena. The energy flux supplied by the solar wind to the area of the magnetosphere projected toward the sun and available to drive magnetospheric and upper atmospheric phenomena is $\sim 10^{20}$ ergs $(\text{sec})^{-1}$, and is sufficient for supplying the average corpuscular energy flux $\sim 10^{18}$ ergs $(\text{sec})^{-1}$ precipitated into the earth's upper atmosphere at auroral latitudes (cf. O'BRIEN, 1964; FRANK and VAN ALLEN, 1964a). In the following presentation the major results of an already substantial and interesting collection of experimental data concerning phenomena in the earth's outer magnetosphere (a complex and variable structure with a volume of at least $10^5 R_E^3$) and its immediate vicinity will be discussed.

2. Coordinate Systems

A coordinate system for the description of the geomagnetic field direction and magnitude and the spatial and angular distributions of charged particles within the magnetosphere is necessary for the understanding and concise summary of experimental data. Examples of extremely useful coordinates for the summary and analysis of charged particle distributions in the magnetosphere are the familiar B and L coordinates (MCILWAIN, 1961) based upon the adiabatic invariants of charged particle motion in a magnetic field and measurements of the geomagnetic field at the surface of the earth. This coordinate system provides a two-dimensional map of a three-dimensional spatial distribution of charged particles, and temporal changes and longitude nondegeneracies, for example, in the B-L map can be interpreted in terms of the violation of the adiabatic invariants and of sources and sinks of charged particles. Since the flow of solar plasma perturbs the distant geomagnetic field so that an extrapolation of the surface measurements of the geomagnetic field (assuming $\nabla \times \mathbf{B} = 0$ everywhere external to the earth's surface) becomes inaccurate it is to be expected that this surface-derived B and L coordinate system becomes invalid for the organization of particle intensities in the distant magnetosphere. This situation also becomes more acute as predominant magnetic control of the motion of the charged particle is subdued in the distant magnetosphere by electric drifts, by rapid radial diffusion, and by loss and source mechanisms effective in periods less than a longitudinal drift period. Realistic models of the distorted geomagnetic field have been introduced (cf. HONES, 1963; MEAD, 1964) and the motions of charged particles in these models (MALVILLE, 1960; FAIRFIELD, 1964) are studied by utilizing the adiabatic invariants in order to separate, for example, the effects of gradient drifts from electric drifts (HONES, 1963; TAYLOR and HONES, 1965;

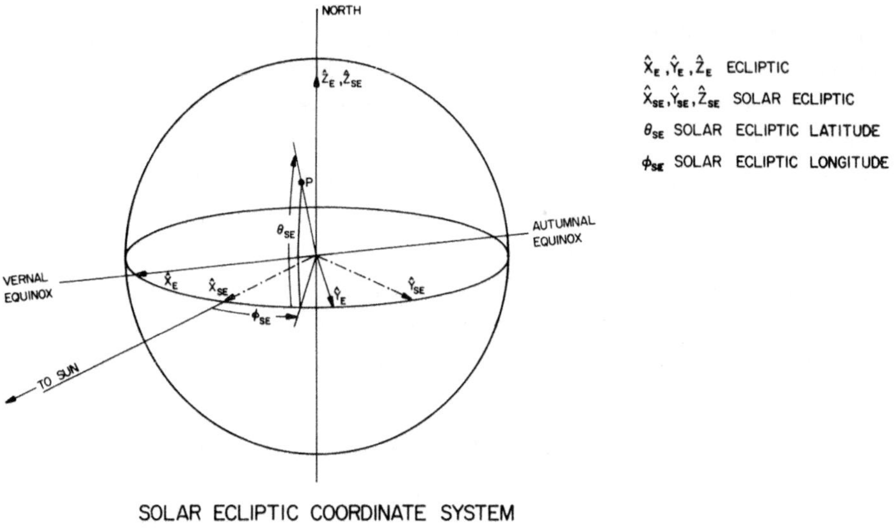

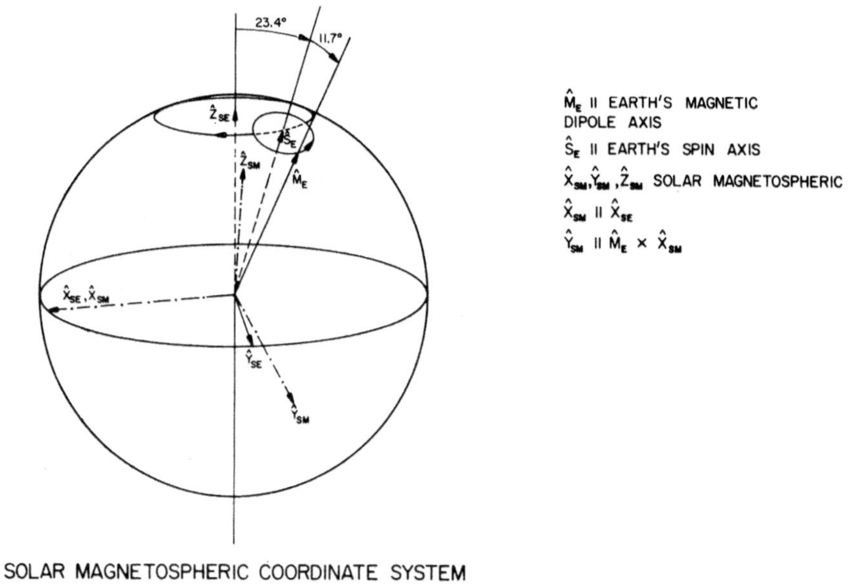

Fig. 1. Illustrations of the solar ecliptic coordinate system (upper diagram) and of the solar magnetospheric coordinate system (lower diagram).

TAYLOR, 1965). Significant perturbation of the observed geomagnetic field from the field as derived from a geomagnetic potential based upon surface measurements begins at 6 to 7 R_E in the magnetic equatorial plane (CAHILL and AMAZEEN, 1963; NESS *et al.*, 1965). A significant perturbation of the geomagnetic field by the solar wind beyond $L \simeq 6$ has also been reflected in the measurements of energetic electrons $E \sim 1.5$ MeV by showing that (a) a nondegeneracy of intensities at a given L as a function of local time (FRANK, 1965b) and (b) an asymmetry in observed omnidirectional intensities about the magnetic equator (FRANK, 1965a) exist when the surface-derived geomagnetic field is used for the determination of B and L of the positions of the observations. Since the solar wind is responsible for the formation of the magnetosphere (terrestrial ring currents, closed or open, and current sheets will also perturb the distant geomagnetic field) a coordinate system based upon the solar direction immediately appears appropriate. Such coordinates were first used in the evaluation of Explorer 10 observations of plasma (BONETTI *et al.*, 1963) and magnetic fields (HEPPNER *et al.*, 1963) and were designated as *solar ecliptic coordinates*. In the top diagram of Figure 1 is an illustration of this right-handed Cartesian coordinate system (X_{SE}, Y_{SE}, Z_{SE}): $\mathbf{X}_{SE}$ is parallel to the earth-to-sun direction and $\mathbf{Z}_{SE}$ is directed toward the North ecliptic pole, θ_{SE} and ϕ_{SE} are the solar ecliptic latitude and longitude, respectively, of point P. For example, in an earth-centered solar ecliptic coordinate system the solar wind vector is directed toward $\theta_{SE} = 0°$, $\phi_{SE} = 180°$ neglecting the aberration effect of $\sim 5°$ due to the earth's revolution about the sun. A more recent coordinate system based upon IMP 1 magnetic measurements in the tail of the magnetosphere has been developed (NESS, 1965) which recognizes the role of the direction of the earth's magnetic moment as well as the solar direction. These coordinates are referred to as solar magnetospheric coordinates (X_{SM}, Y_{SM}, Z_{SM}) (refer to bottom, Figure 1) and are defined with $\mathbf{X}_{SM}$ parallel to $\mathbf{X}_{SE}$ and $\mathbf{Y}_{SM}$ parallel to $\mathbf{M}_E \times \mathbf{X}_{SM}$ where $\mathbf{M}_E$ is parallel to the earth's magnetic dipole axis. In an earth-centered solar magnetospheric coordinate system, for example, a point fixed in celestial coordinates will move in an annual cyclic motion due to the earth's revolution about the sun upon which is superimposed a diurnal motion due to the rotation of the earth. Extensive surveys of magnetic fields and particle distributions in the tail of the magnetosphere are currently the indicators as to whether magnetic, solar ecliptic, or solar magnetospheric coordinates provide the highest degree of coherency in the organization of the data. A summary of the major experimental observations directed toward this topic is as follows:

(1) The neutral sheet in the earth's magnetic tail is more precisely defined by the solar magnetospheric equatorial plane (NESS, 1965).

(2) Vela satellite measurements of electrons $E > 50$ keV at ~ 17 R_E show that the occurrence of measurable intensities is preferentially near the magnetic

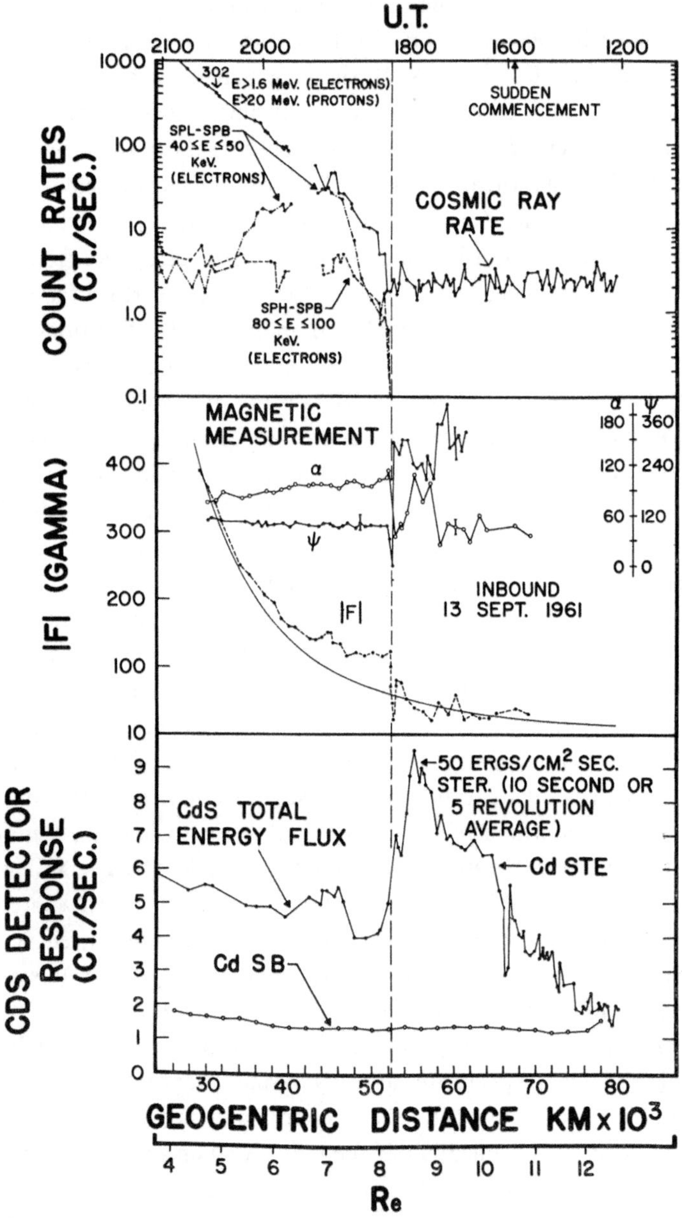
U.T.
2100
2000
1800
1600
1200
1000
100
10
1.0
0.1
COUNT RATES (CT./SEC.)
302
E>1.6 MeV. (ELECTRONS)
E>20 MeV. (PROTONS)
SUDDEN COMMENCEMENT
SPL-SPB 40 ≤ E ≤ 50 KeV. (ELECTRONS)
COSMIC RAY RATE
SPH-SPB 80 ≤ E ≤ 100 KeV. (ELECTRONS)
MAGNETIC MEASUREMENT
|F| (GAMMA)
400
300
200
100
10
α
ψ
|F|
180
360
120
240
60
120
0
0
INBOUND
13 SEPT. 1961
CDS DETECTOR RESPONSE (CT./SEC.)
CdS TOTAL ENERGY FLUX
50 ERGS/CM.² SEC. STER. (10 SECOND OR 5 REVOLUTION AVERAGE)
Cd STE
Cd SB
GEOCENTRIC DISTANCE KM x 10³
30
40
50
60
70
80
Re
4
5
6
7
8
9
10
11
12

Fig. 2

equator rather than near the ecliptic plane or the solar magnetospheric plane (ASBRIDGE *et al.*, 1965; MONTGOMERY *et al.*, 1965).

(3) Explorer XIV measurements of the omnidirectional intensities of electrons $E > 40$ keV near the local midnight meridian indicated a tail of electron intensities lying more nearly in the ecliptic plane than near the geomagnetic equator beyond 10 R_E. In fact measurements during a relatively quiescent period in magnetic activity and in intensity temporal variations at similar geomagnetic latitudes but at different ecliptic latitudes repeatedly indicated that geomagnetic latitude was inadequate for organizing the observed intensities (FRANK, 1965a).

(4) The further sophistication of the solar magnetospheric coordinates has not yet proved to be necessary for the organization of shock wave and magnetopause positional data. Solar ecliptic coordinates continue in practice to be adequate for the summary of these data (cf., NESS *et al.*, 1964; BRIDGE *et al.*, 1965).

It must be remembered in consideration of the above results that the observation of electron intensities $E > 50$ keV in the tail region (for example, (2) above) is not necessarily indicative of the coincident presence of the neutral sheet (1). Further surveys of the electrons and protons in this region of the magnetosphere are necessary in order to pursue the interrelationships of the above results.

3. Measurements of the Distant Geomagnetic Field

It is easily recognized that the interpretation of charged particle observations in the earth's magnetosphere and its vicinity depends heavily upon the character of the associated magnetic fields. Major experimental evidences concerning the topology of these magnetic fields are summarized here.

MAGNETOPAUSE-TRANSITION-SHOCK REGIONS

One of the early direct observations of the interaction of the solar wind plasma with the geomagnetic field is shown in Figure 2 as Explorer XII passed through the magnetospheric boundary, or magnetopause, near the earth-to-sun direction (FREEMAN, VAN ALLEN, and CAHILL, 1963). The scalar value |F| of the observed

---

Fig. 2. Particle and magnetic field measurements with Explorer XII for the inbound pass on 13 September 1961. The CdSB (magnet in aperture) detector count-rate has been normalized to the energy scale of the CdSTE (no magnet in aperture) detector. The counting rates of both CdS detectors are nearly linear with energy flux. Both spectrometer channels (SpL and SpH) have been corrected for background counts by the subtraction of the counting rate of the background detector SpB. The CdS optical monitor (not shown) indicated that during this pass the CdS detectors did not have any bright objects within their field of view. F denotes the scalar magnetic field strength; α the angle between the F-vector and the spin-axis of the satellite; and ψ the dihedral angle between the plane containing the F-vector and the spin axis and the plane containing the spin axis and satellite-sun line (after FREEMAN *et al.* (1964)).

magnetic field (center, Figure 2) monotonically increases over the FINCH and LEATON (1955) extrapolated field with increasing geocentric radial distance while remaining relatively constant in direction. At 52000 km (8.2 R_E) with $|\mathbf{F}|$ approximately a factor of 2 larger than the theoretical field a sharp decrease in the magnitude and change of character of the magnetic field occurred and highly

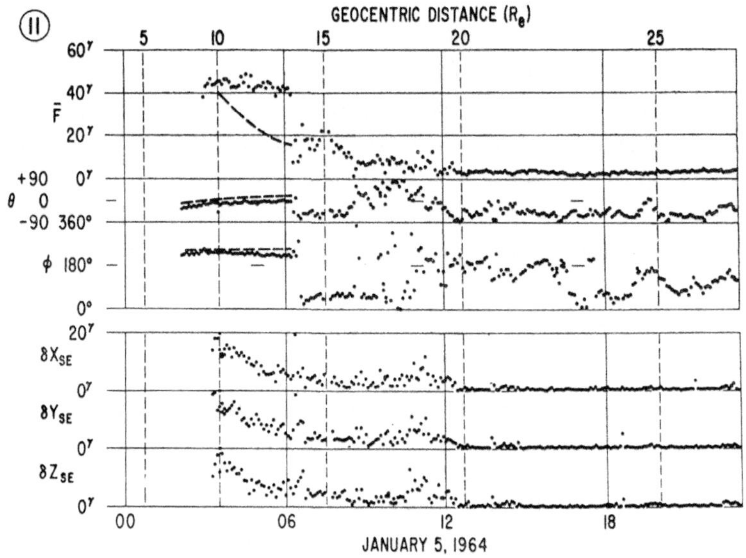

Fig. 3. Magnetic field data (IMP 1) from orbit No. 11 January 5, 1964, illustrating outbound traversal of magnetopause boundary at 13.6 R_E and shock wave boundary at 19.7 R_E (after NESS *et al.* (1964)).

disordered magnetic fields with magnitudes of tens of gammas were observed at larger radial distances. Simultaneous observations of charged particles displayed (1) a severe decrease of trapped electron 40 keV $<E<$ 50 keV fluxes (SPL-SPB, top, Figure 2) and (2) a large increase of low-energy electrons $\sim$ a few keV of energy with increasing radial distance concurrent with the above change in the character of the magnetic field. These measurements were interpreted as (1) the termination of durable geomagnetic trapping of charged particles at 8.2 R_E (2) concurrent with the termination of a perturbed but smooth geomagnetic field at the magnetopause and (3) the peak of quasi-thermalized plasma beyond the magnetopause supplying the plasma pressure necessary to balance the magnetic pressure due to the compressed geomagnetic field just inside the magnetopause. These early observations, although not the most detailed available at present, provide a valuable guide to the gross characteristics of the sunward magnetopause region. Although the radial thickness of the persistent quasithermalized plasma observed with Explorer XII was typically 2 to 3 R_E in depth (FREEMAN, 1964) in agreement with theoretical predictions of the stand-off distance of the bow shock (KELLOGG, 1962; AXFORD, 1962), no observations with the Explorer XII magnetometer of a change in the character of the magnetic field designating the position of the shock wave were reported. Direct observations of a collisionless magnetohydrodynamic shock wave with a magnetometer were first reported by NESS *et al.* (1964) with the IMP 1 satellite. An example of these observations is given in Figure 3 which displays the magnetopause at 13.6 R_E and the disordered magnetic field at greater radial distances in the region between the magnetopause and the shock wave. The coordinates θ, ϕ, X_{SE}, Y_{SE}, and Z_{SE} are satellite-centered solar ecliptic coordinates. At a radial distance of 19.7 R_E and beyond the variances in the magnetic field components of discrete sets of measurements (δX_{SE}, δY_{SE}, δZ_{SE}) become small relative to those within the transition region (or 'magnetosheath') at smaller radial distances and a similar effect is noticeable in the averaged scalar field F. The change in the character of the magnetic field at 19.7 R_E which is a persistent feature of the IMP 1 measurements beyond the sunward side of the magnetosphere, although variable in position, is interpreted by NESS *et al.* (1964) as delineating the shock wave boundary. NESS *et al.* (1964) have summarized their crossings of the magnetopause and shock wave in the ecliptic plane by rotating the observations in a meridian plane containing the Z_{SE} axis and the position of the observation; this summary is reproduced in Figure 4. These data have also been normalized, or rectified, to a geomagnetic latitude of the subsolar point $\chi_{SS}=0°$. In the ecliptic plane at the subsolar point the average positions of the magnetopause and the shock wave are at ~ 10 R_E and 13.5 R_E and flare out to ~ 13 R_E and 22 R_E, respectively, at $\phi_{SE}=270°$. The solid lines indicate the result of a theoretical model calculation of the shock position with the magnetospheric boundary used as the blunt body (SPREITER and JONES, 1963). Calculations of the shape of the sunward magnetopause (cf. BEARD, 1960; MIDGELY and DAVIS, 1962; SPREITER and BRIGGS, 1962a, b; MEAD and BEARD, 1964) are in substantial agreement with these observations. Explorer X observations of the magnetopause toward the tail of the magnetosphere as determined from simultaneous plasma (BONETTI *et al.*, 1963) and magnetic field (HEPPNER *et al.*, 1963) measurements have been rotated into the ecliptic plane for comparison and to strengthen the experimental evidence that no closure of the geomagnetic tail is discernible contrary to closures predicted due to transverse plasma pressures (cf. JOHNSON, 1960; BEARD, 1960).

Motions of the magnetospheric boundary and of the shock wave are implicit in the variations of the measurements summarized in Figure 4 (NESS *et al.*, 1964) and are manifestations of temporal variations in the solar wind densities, velocities, and associated interplanetary magnetic fields. Explorer X observations in the evening flank of the magnetosphere over the radial distance

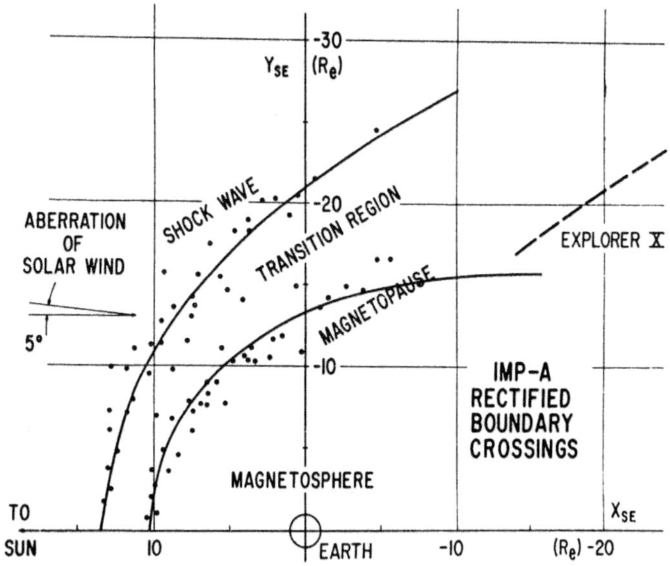

Fig. 4. Summary of location of rectified shock wave and magnetopause boundary crossings and comparison with theoretical predictions using analogy with supersonic fluid flow (after NESS *et al.* (1964)).

range of ~22 R_E to 42 R_E of alternate periods of no plasma and magnetic field directed radially from the sun and of plasma moving generally from the direction of the sun and disordered magnetic fields were interpreted as the motion of the magnetopause across the satellite's position. FREEMAN (1964) has shown that the correlation of D_{ST} values with the position of the sunward magnetopause as determined by Explorer XII observations indicates that the magnetosphere is compressed during the initial phase of a magnetic storm presumably by enhanced solar plasma flux and is characterized by a general expansion during the recovery phase. A single Explorer XII observation of a transient presence of trapped protons near the magnetopause has given an impulsive, outward radial

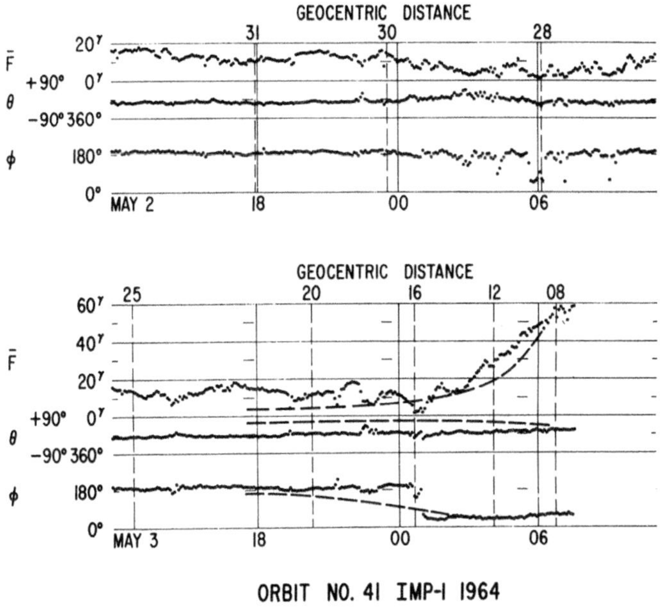

Fig. 5. Magnetic field data from inbound orbit No. 41, May 2 through May 4, 1964. Throughout most of this time interval the magnetic field is pointed antisolar although at a geocentric distance of 16 R_E the magnetic field abruptly reverses direction at the same time that it becomes very weak or zero. This spatially limited region is identified as a neutral sheet in the earth's magnetic tail (after NESS (1965)).

velocity of the boundary position exceeding 150 km $(sec)^{-1}$ (KONRADI and KAUFMANN, 1965).

GEOMAGNETIC TAIL REGION

The form of the distant geomagnetic field in the anti-solar direction has been of theoretical interest before definitive measurements were available (cf. PIDDINGTON, 1959; BEARD, 1960; AXFORD and HINES, 1961); active investigations concerning the linear dimensions of the geomagnetic tail in the anti-solar direction are presently continuing (DESSLER, 1964; DUNGEY, 1965; MICHEL, 1965; PIDDINGTON, 1965). Explorer X (HEPPNER *et al.*, 1963) provided the first direct experimental evidence indicating the nature of the magnetic field topology in the tail of the magnetosphere. Beyond approximately 8 R_E to 22 R_E a magnetic field directed radially outward (the probe was at southerly latitudes during the measurements) from the sun and considerably larger in magnitude over the theoretical values was detected. Beyond 22 R_E the repeated transversal of the magnetopause across the satellite's position occurred. Measurements again at

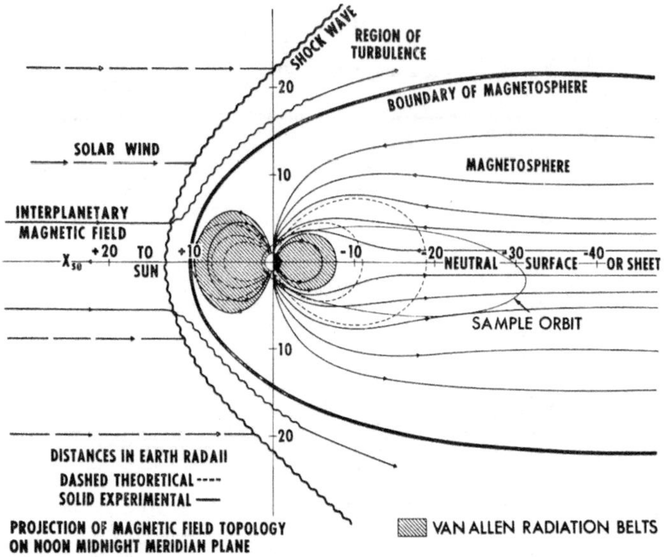

Fig. 6. Illustration of the interpretation of the IMP 1 magnetic field data field topology within the noon-midnight meridian plane. The relative position of the neutral surface or sheet in the earth's magnetic tail and the corotating magnetic field lines supporting trapped particle motion are indicated. These include the classical Van Allen radiation belts. The collisionless shock and magnetosphere boundaries are extrapolated to the polar regions indicating a depression but the relative position of a polar neutral point and the size of the boundary are not experimentally determined. Cylindrical symmetry about the earth-sun line has been assumed for the boundary of the earth's magnetic tail in this presentation (after NESS (1965)).

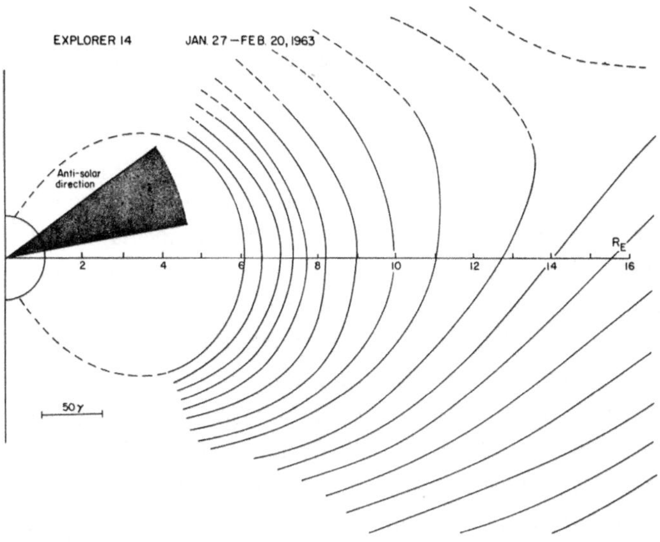

Fig. 7. A schematic representation of the magnetic lines of force projected upon a magnetic meridional plane as observed by Explorer XIV near local midnight (after CAHILL (1965)).

southerly ecliptic latitudes and to a radial distance of ~16 R_E with Explorer XIV (CAHILL, 1964) provided confirmation of the above Explorer X results and indicated a depression of the magnetic field below theoretical values at 7 to 8 R_E. Similar depressions of the geomagnetic field were observed during the period near solar maximum with Explorer VI (August 1959) at altitudes as low as 5 R_E and beyond (SONETT *et al.*, 1960 a, b; see also SMITH, (1963) for a review of Pioneers I and V, Lunik I and II, and Explorers VI and X measurements of the distant magnetic field and the possibility of an associated terrestrial ring current). The most comprehensive measurements of the magnetic fields in the magnetospheric tail region reported thus far have been obtained by NESS (1965) with IMP 1 (apogee 31.4 R_E). A representative inbound pass (NESS, 1965) through the tail region is reproduced in Figure 5; the magnitude $\bar{F}$ and the direction in solar ecliptic coordinates of the measured field are compared with the usual theoretical field (dashed lines). Ness concludes that (1) the large magnetic field of 10 to 30 gammas and characterized by regularity in direction observed to apogee distances (~half the distance to the lunar orbit) indicate

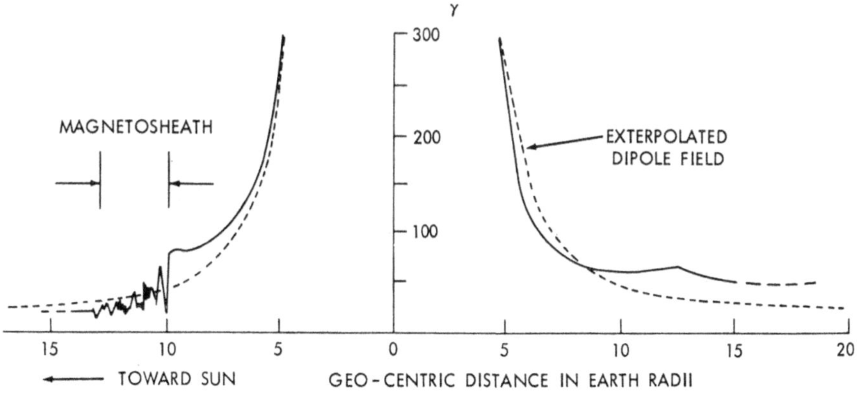

Fig. 8. Schematic representation of the outer geomagnetic field in the equatorial plane along the sun-earth line (after OBAYASHI (1965)).

that the earth's magnetic tail extends to at least the lunar orbit and (2) the simultaneous observations of a practically null field and a reversal from anti-solar direction to solar direction of the field at ~16 R_E is indicative (supported by similar observations over 18 consecutive orbits) of a neutral sheet in the tail region with a thickness of a fraction of an earth radius. Hence the earth's polar cap lines of force are pulled to the night side of the earth in the magnetic tail with lines of force from the Northern and Southern Hemispheres directed toward the solar and anti-solar directions respectively, and with these two flux tubes separated by a neutral sheet ($B \simeq 0$) containing sufficient plasma to provide a pressure to balance the magnetic pressure of the adjacent tubes of force (see AXFORD *et al.*, (1964) for a theoretical discussion). A summary due to NESS (1965) of the magnetic field topology in the noon-midnight meridional plane is shown in Figure 6. CAHILL (1965) has recently reported similar results near the midnight meridian which are summarized in Figure 7. The shape of the magnetic field has been projected into a magnetic meridian plane and significant stretching of the field lines in the antisolar direction appears beyond $\sim 10\ R_E$; observation of a sharp field reversal indicative of the neutral sheet found by Ness was also reported. A coarse comparison (OBAYASHI, 1965) of the measured magnetic field magnitudes and the extrapolated field along the earth-to-sun line (X_{SE} axis) is given in Figure 8 and illustrates several of the experimental measurements discussed above. The magnetic field topology at high latitudes and large radial distance yet awaits experimental survey.

4. Observations of Charged Particles in the Distant Geomagnetic Field

The spatial distributions of charged particles surrounding and within the distant geomagnetic field are highly dependent upon the energy and species of charged particles under investigation. An attempt to summarize the major reported evidences provides insight into the complexity of the magnetospheric structure. Other summaries pertaining to the distributions of charged particles within the earth's magnetosphere have been given, for example, by VAN ALLEN (1961), FARLEY (1963), O'BRIEN (1963), FRANK and VAN ALLEN (1964a), and OBAYASHI (1965).

LOW-ENERGY CHARGED PARTICLES

An example of observations of a quasi-thermalized plasma just beyond the magnetopause and compressing the sunward geomagnetic field has been given previously in Figure 2 (FREEMAN *et al.*, 1963). The plasma energy flux was of the order of tens of ergs $(cm^2\ sec\ sr)^{-1}$ and was inferred from rudimentary dynamical considerations to be electrons of energy $\sim$ few kilovolts and intensities $J_0 \simeq 10^9 - 10^{10} (cm^2\ sec)^{-1}$. Measurements of the proton and electron intensities in this portion of the magnetosphere were continued with a plasma cup on IMP 1 by BRIDGE and his associates (1964, 1965) and a typical pass is reproduced here in Figure 9. Shown are the average proton energy V, $\log_{10}$ of the total positive ion flux and of the electron flux 65-210 eV looking toward and away from the sun and a measure, *T*, of the width of the proton distribution. Beyond

15 R_E the satellite is in the interplanetary medium and a anisotropic flux of protons $J_0 \simeq 10^8$ (cm^2 sec)$^{-1}$ is directed away from the sun; the electron intensity 65 eV to 210 eV was reported to be at approximately the noise level of the instrument 6×10^6 (cm^2 sec)$^{-1}$. In the transition region between ~ 10 R_E and 15 R_E approximately isotropic but variable intensities of protons *and*

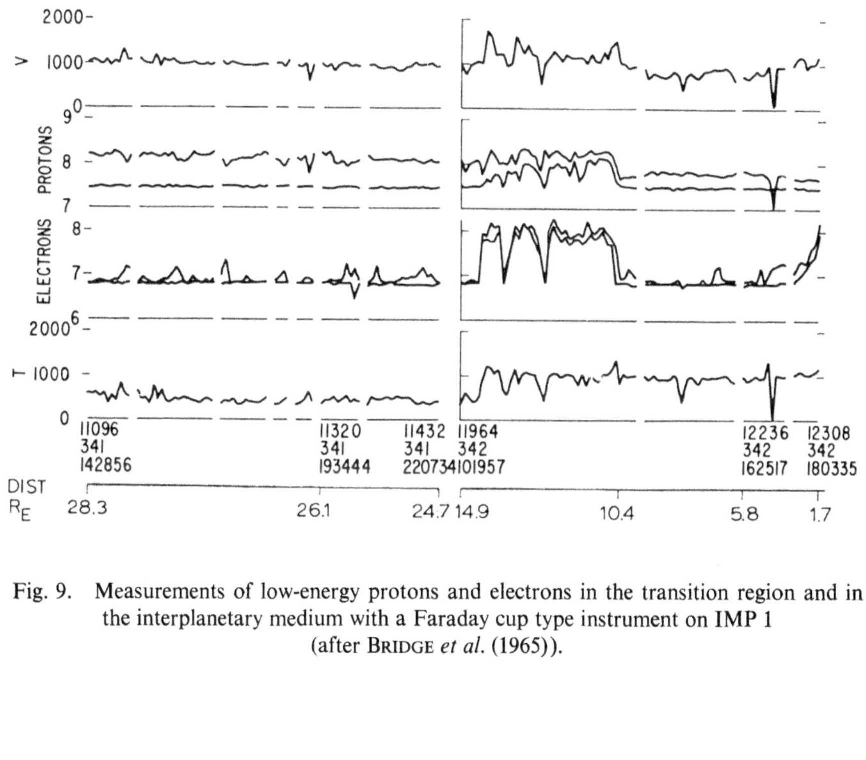

Fig. 9. Measurements of low-energy protons and electrons in the transition region and in the interplanetary medium with a Faraday cup type instrument on IMP 1 (after BRIDGE *et al.* (1965)).

electrons $J_0 \simeq 10^9$ (cm^2 sec)$^{-1}$, a marked increase in ion temperature (measurable intensities in all channels 45 eV through 5400 eV), and a sharp decrease in the intensities of protons and electrons at the magnetopause near 10.5 R_E were observed. These measurements indicate that the proton density in the transition region exceeds the solar wind proton density by a factor of 8 or 10. A summary of the portions of the IMP 1 trajectory along which a hot, approximately isotropic proton flux was observed is shown in Figure 10 (BRIDGE *et al.*, 1964, 1965); the transition regions as determined by these plasma observations and by simultaneous magnetic field measurements are coincident (cf. Figure 4 (NESS *et al.*, 1964) of this paper). WOLFE and SILVA (1964) generally confirm the above results using an electrostatic analyzer also on IMP 1 and find that the proton angular distributions in the transition region indicate a highly turbulent flow but statistically directed away from the sun. Measurements of low-energy electron intensities utilizing a retarding potential analyzer (also IMP 1, one outbound pass) have been obtained by SERBU (1965) and are shown here in

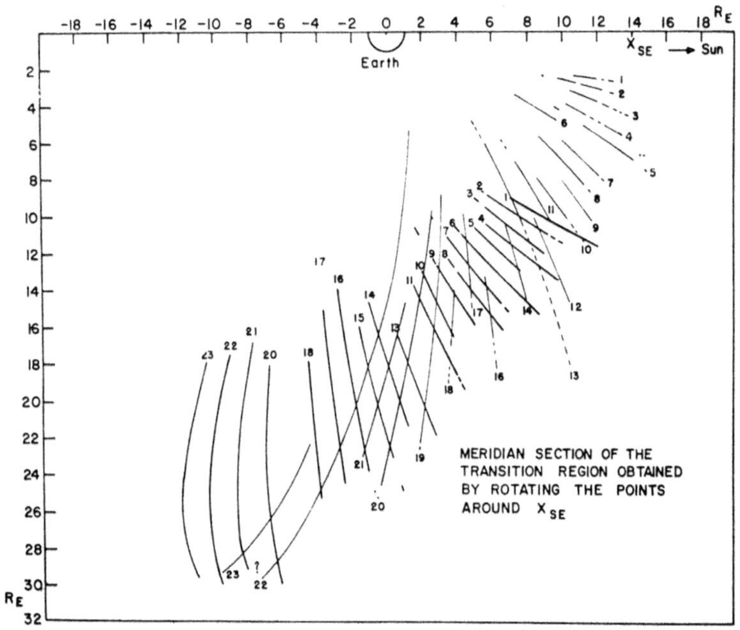

Fig. 10. Summary of portions of IMP 1 trajectory during which a hot, nearly isotropic plasma flux was observed (after BRIDGE *et al.* (1965)).

Figure 11. The omnidirectional intensity of electrons $E > 100$ eV is $\sim 2 \times 10^8$ $(\text{cm}^{-2}\,\text{sec})^{-1}$ from 8 R_E to the position of the shock front at 16 R_E as determined by the magnetometer measurements. Intensities of these electrons decrease sharply at the shock boundary but no change in the observed intensities, increase or decrease, is seen at the magnetopause at ~ 11 R_E. The radial profiles of the omnidirectional intensities of electrons $0 < E < 5$ eV and $5 < E < 10$ eV appear to enjoy a similar independence of the magnetospheric boundary.

A summary of the observations of the low-energy charged particle distributions near the magnetopause as discussed above and in the tail of the magnetosphere is given in Figure 12. The ecliptic plane has been selected here for mapping and comparing the results. Explorer X plasma observations (BONETTI *et al.*, 1964) indicate that the plasma resumes supersonic flow along the flanks of the magnetosphere. The results of GRINGAUZ and his collaborators

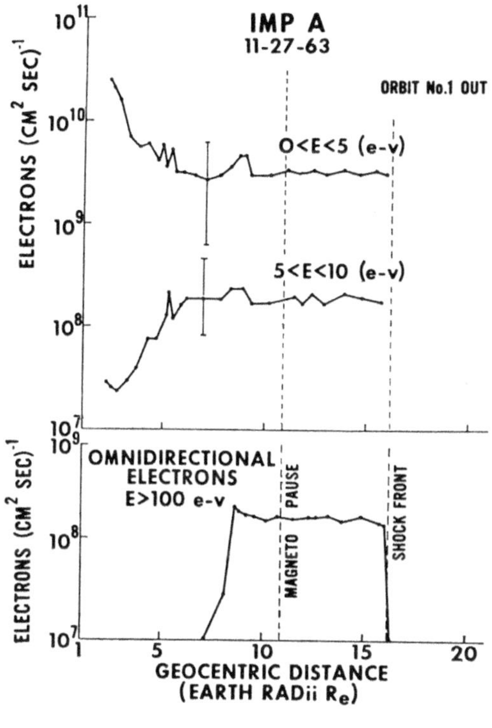

Fig. 11. Plot of low-energy electron fluxes as a function of geocentric radial distance (after SERBU (1965)).

(1964) at high latitudes ($\lambda_m \sim 50°$) (for coordinates, see VAKULOV *et al.* (1964)) have been roughly translated into the ecliptic plane by direct L-shell mapping. Observations of low-energy electron intensities $E \gtrsim 100$ eV by Lunik II (GRINGAUZ *et al.*, 1961), Explorer XII (FREEMAN, 1963), and Mars I (GRINGAUZ *et al.*, 1964) beyond $\sim 8\ R_E$ are in coarse agreement although there is an obvious need for more definitive measurements. The position of these large electron intensities coincides with the onset of significant distortion of the tail field in the anti-solar direction measured with IMP 1 and Explorer XIV as reviewed above. The observed electron energy flux of several tens of ergs (cm^2 sec)$^{-1}$ (energy density $\sim 10^{-8}$ ergs $(\mathrm{cm})^{-3}$ if $E \sim$ a few kilovolts) appears sufficient to support the neutral sheet discovered by NESS (1965) if $B \sim 30$ gammas for the adjacent fields. Large intensities of electrons of similar energies at low altitudes and high latitudes $L \gtrsim 8$ have been observed during local night (SHARP *et al.*, 1965; FRITZ and GURNETT, 1965) and further suggest the tail region as at least a partial reser-

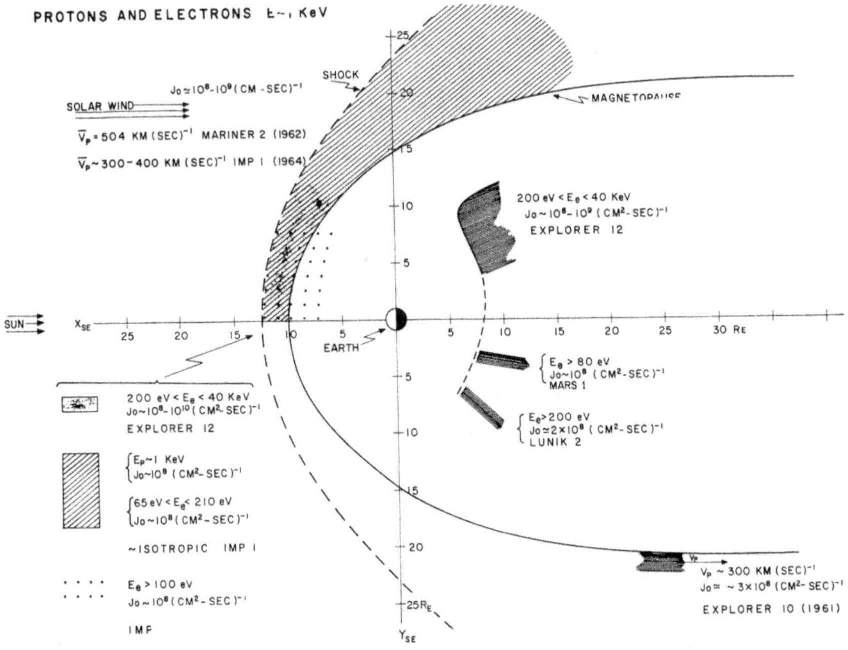

Fig. 12. Coarse summary of low-energy charged particle observations. Average positions of the magnetopause and the shock as given by NESS *et al.* (1964) have been included in this diagram.

voir and/or accelerating region for auroral particles. The spatial connection of the two regions characterized by large intensities of electrons $E > 100$ eV near the sunward magnetopause and in the geomagnetic tail has not been determined (cf. GRINGAUZ, 1964; VAN ALLEN, 1964) and further surveys in the dawn, evening, and tail portions of the magnetosphere are necessary to establish the relationship.

MEDIUM-ENERGY CHARGED PARTICLES

Simultaneous measurements of magnetic fields and electron $E \sim 50$ keV intensities with Explorer XII revealed that a large decrease of electron intensities occurred concurrently with the disappearance at the sunward magnetopause of the uniform, although perturbed, geomagnetic field necessary for the durable trapping of charged particles (FREEMAN *et al.*, 1963) (refer to present Figure 2). Another example of the typical behavior of the radial profiles of electron $E > 40$ keV, $E > 230$ keV, and $E > 1.6$ MeV intensities near the sunward magneto-

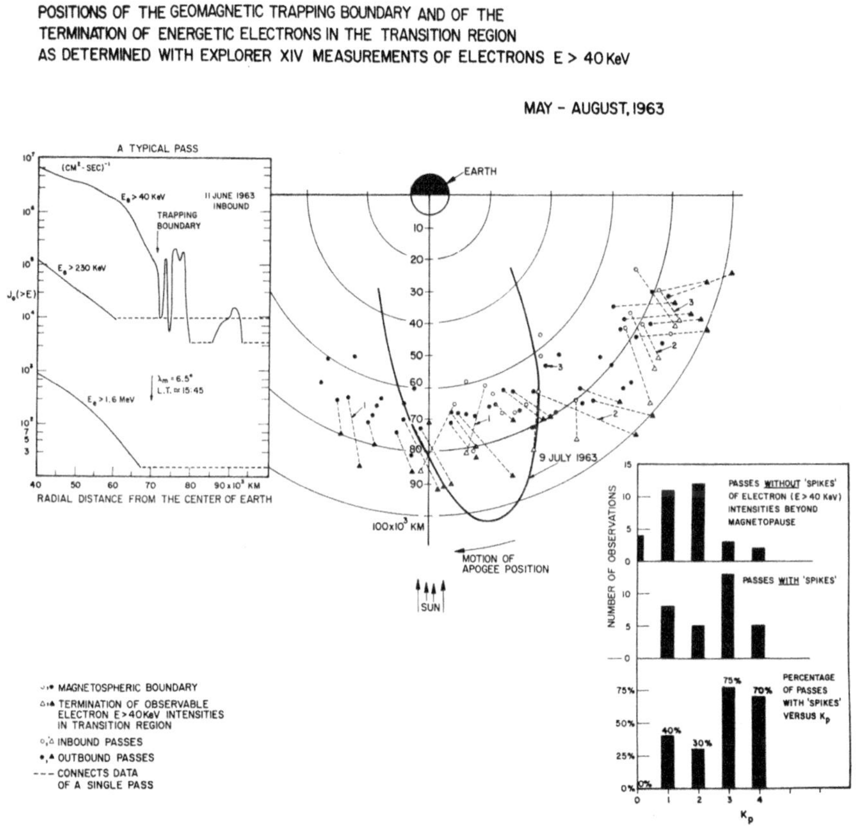

Fig. 13. Summary of Explorer XIV observations of the sharp termination of electron $E > 40$ keV intensities at the sunward magnetopause and of the intensity 'spikes' in and near the transition region (after FRANK and VAN ALLEN (1964b)).

spheric boundary with Explorer XIV (FRANK and VAN ALLEN, 1964b) is given in the left-hand inset of Figure 13. The "spikes" of electron $E > 40$ keV intensities beyond the position of the trapping boundary at ~70000 km and extending in this example to ~93000 km in the general vicinity of the transition region have been observed by ANDERSON *et al.* (1965), FRANK and VAN ALLEN (1964b), and FAN *et al.* (1964) and are suggestive of a high energy tail of the electron spectrums in the transition region $(J_0(E>40\text{ keV})/J_0(E\gtrsim 1\text{ keV}) \sim 10^{-5})$ (FRANK and VAN ALLEN, 1964b). Theoretical hypotheses for the acceleration of these energetic electrons have been given by SCARF *et al.* (1965) and JOKIPII and DAVIS (1964). A coarse positive correlation of the occurrence of these intensity 'spikes' with the planetary magnetic index K_p (right-hand inset Figure 13) was found (FRANK and VAN ALLEN, 1964b) with average K_p values $\simeq 2.5$ and $\simeq 1.5$ for passes with and without intensity spikes beyond the magnetopause, respectively. The percentage of passes with observations of spikes while the satellite was in the transition region and K_p was 3 or 4 (~ 20 passes) was $\sim 75\%$. Figure 13 also displays the 'flaring' of the magnetospheric boundary as deduced from the characteristic sharp termination intensities in general agreement with IMP 1 magnetic field measurements (NESS *et al.*, 1964) and the increasing thickness of the regions with extra-magnetospheric spikes with angular displacement from the subsolar point (FRANK *et al.*, 1963; FAN *et al.*, 1964; FRANK and VAN ALLEN, 1964b; ANDERSON *et al.*, 1965). Direct correlation of the position of the shock as determined by simultaneous magnetic field measurements with the positions of these intensity spikes shows that the intensity spikes are not coincident with, but are also sometimes positioned upstream from, the position of the shock wave (ANDERSON *et al.*, 1965; JOKIPII and DAVIS, 1964). The interpretation due to ANDERSON *et al.* (1965) of small intensity spikes ($\sim 10^3$ to 10^4 (cm^2 sec)$^{-1}$ observed at large distances in front of the shock as being of terrestrial origin may be complicated by the recent discovery (VAN ALLEN and KRIMIGIS, 1965) of solar electron $E \sim 40$ keV events with Mariner IV data.

A summary of several gross features of the spatial distributions of electrons $E \sim 40$ keV is given in Figure 14. A rough upper limit of the average electron $E > 40$ keV intensity in the interplanetary medium is ~ 10 (cm^2 sec)$^{-1}$ as determined with Mariner IV (cf. VAN ALLEN, 1965). Surrounding the earth and extending to the magnetopause in the solar direction ($\sim 10\ R_E$) and to $\sim 8\ R_E$ in the antisolar direction is a 'hard core' of electron $E > 40$ keV intensities $\sim 10^7$–10^8 (cm^2 sec)$^{-1}$ characterized by relatively weak temporal and spatial variations when compared to outlying regions of the magnetosphere (FRANK, 1965a). Within $\sim 45°$ of longitude of the subsolar point at low latitudes these intensities are terminated coincident with the magnetopause. In the dawn and evening 'skirts' of the spatial distribution (i.e., at larger angular displacements from the subsolar point) tortuous temporal and spatial variations are observed (FRANK *et al.*, 1963; ANDERSON *et al.*, 1965; FRANK, 1965a) and their association with the magnetopause in terms of spatially correlated decreases or increases in intensity weakens and disappears toward the flanks of the magnetosphere

(ANDERSON *et al.*, 1965). MONTGOMERY *et al.* (1965) have noted a large asymmetry in the dawn and evening skirts with the dawn skirt having generally more intense and frequent electron $E > 50$ keV intensity spikes when compared with

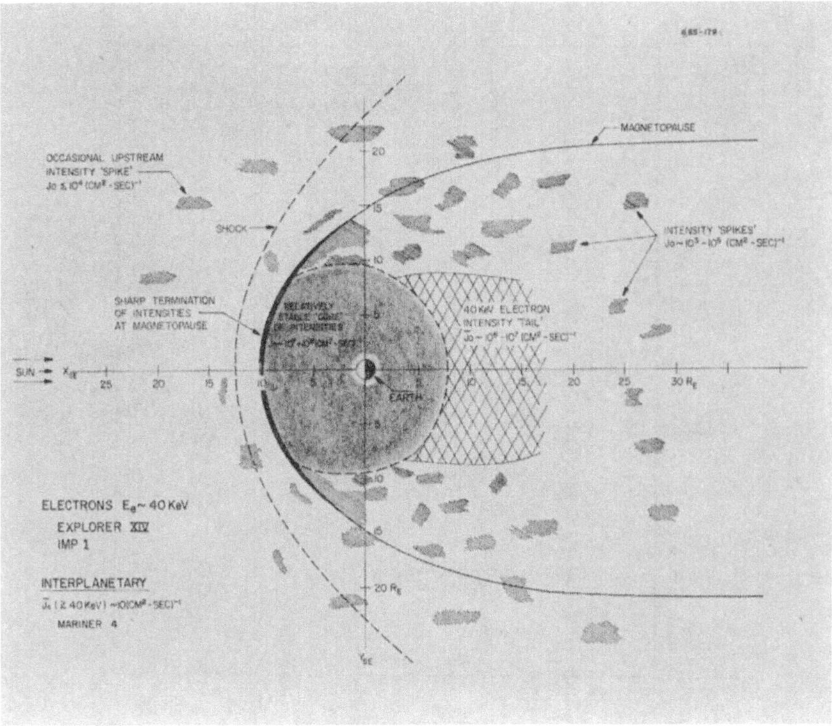

Fig. 14. Coarse summary of the observations of electrons $E > 40$ keV near the ecliptic plane.

observations in the local evening skirt. Observations with Explorer XIV displayed no significant dawn-evening asymmetry during relatively quiescent magnetic conditions ($\sum K_p \leqslant 25$) (FRANK, 1965a) but the nearly circular Vela orbits (MONTGOMERY *et al.*, 1965) are more favorable for observing such an asymmetry than the highly eccentric Explorer XIV orbit. In the anti-solar direction a 'tail' of electron $E > 40$ keV intensities has been found (FRANK, 1965a; ANDERSON *et al.*, 1965) characterized by much weaker temporal and spatial dependences when compared to those of the skirt regions and a shallow latitude dependence compared to that of the 'core' of intensities. The position of this 'tail' of electron intensities presumably is coincident with the neutral sheet. ANDERSON (1965) has found 'spikes' of intensities with fast risetimes (~a few minutes) with approximately exponential decays (~a few minutes to hours) in

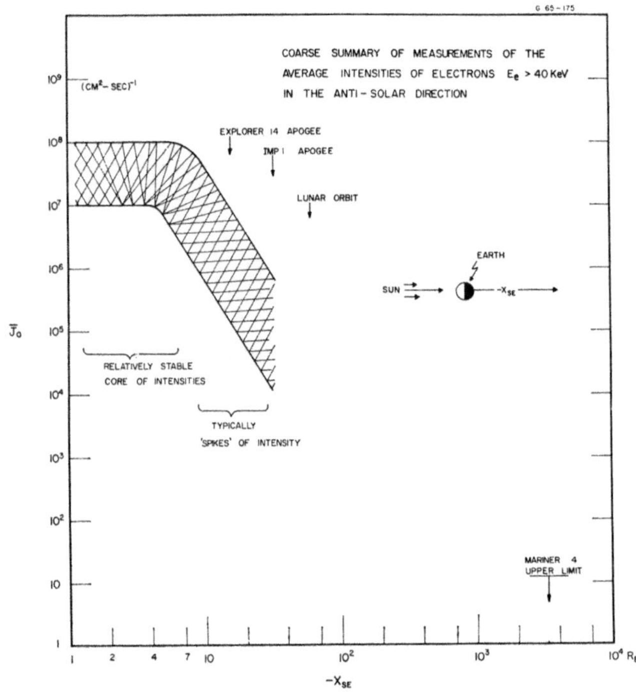

Fig. 15. Summary of the observations of electron $E > 40$ keV intensities as a function of radial distance in the anti-solar direction.

the anti-solar direction to at least IMP 1 apogee position of $\sim 32\ R_E$ with a strong decrease in frequency of occurrence with increasing radial distance and infers that these 'islands' (typically several R_E in linear dimensions) are impulsively injected into the tail region. ASBRIDGE *et al.* (1965) have observed electrons of similar energies streaming in the anti-solar direction in the geomagnetic tail at $\sim 17\ R_E$. Recently intensity spikes of low-energy protons $E > 125$ keV, $j \simeq 10^3$ to 10^4 (cm² sec sr)$^{-1}$ in these regions have been reported (KONRADI, 1965). Observations of rapid increases of intensities during the onsets of several intensity 'spikes' in the evening skirt of intensities with OGO 1 allowed FRANK *et al.* (1965) to set a lower limit on the bulk velocity of the irregularities of 50 km $(\text{sec})^{-1}$. A coarse summary of the average omnidirectional intensities of electrons $E > 40$ keV as a function of geocentric radial distance in the anti-solar direction is shown in Figure 15 which includes the Explorer XIV and IMP 1 data discussed above and a single determination at 3300 R_E with Mariner IV (VAN ALLEN, 1965); average intensities at the lunar orbit in the anti-solar direction might reasonably be expected to be $\sim 10^3$ $(\text{cm}^2 \text{ sec})^{-1}$.

Periodic variations of the intensities of electrons $E > 40$ keV with periods ~ 6 minutes and persisting for several hours in the skirt and tail regions of the magnetosphere have been recently reported (ANDERSON, 1965). These variations are presumably related to periodic variations in the magnetic field (cyclic acceleration and deceleration by hydromagnetic waves) such as the long-period hydromagnetic waves observed by PATEL (1965) (for theoretical discussions of the generation and propagation of magnetospheric hydromagnetic waves, refer to DESSLER (1958, 1961), DESSLER and PARKER (1959), DESSLER and WALTERS (1964), and MACDONALD (1961)).

ENERGETIC ELECTRONS

The spatial distributions of electron $E \gtrsim 1.5$ MeV intensities in the outer magnetosphere and beyond are summarized in Figure 16. The isopleths of constant intensities $J_0 = 10^2, 10^3, 10^4$ $(\text{cm}^2 \text{ sec})^{-1}$ as measured with Explorer XIV near the magnetic equatorial plane exhibit a large local time dependence in radial position (for example, the isopleth $J_0 = 10^2$ $(\text{cm}^2 \text{ sec})^{-1}$ is positioned at 10 R_E in the solar direction and at 8 R_E in the anti-solar direction) and the spatial distribution of these energetic electrons lies within the relatively stable 'core' of 40 keV electron intensities (see Figure 14) (FRANK, 1965b); this region as defined by the outermost contour is probably indicative of the region in which complete longitudinal drift paths around the earth are possible. Several intensity spikes of electrons $E > 1.5$ MeV ($J_0 \sim 10$ $(\text{cm}^2 \text{ sec})^{-1}$) were observed in the dawn skirt and transition region and none in the corresponding evening portions of the magnetosphere but the statistics are poor with only several events of observable intensity discovered in the data. Arrows on the isopleths indicate excursions (temporal variations) in the position of the contour $J_0 = 10^2$ $(\text{cm}^2 \text{ sec})^{-1}$ observed with Explorer XIV. Upper limits on the average interplanetary omnidirectional intensities are $\bar{J}_0(> 3 \text{ MeV}) \lesssim 0.1$ $(\text{cm}^2 \text{ sec})^{-1}$ (CLINE *et al.*, 1964) and $\bar{J}_0$ $(> 1.5 \text{ MeV}) \lesssim 1$ $(\text{cm}^2 \text{ sec})^{-1}$ (FRANK, private communication) with IMP 1 and Explorer XIV data, respectively.

The diurnal variation of electron $E > 280$ keV intensities observed at low altitudes $L \gtrsim 6$ (WILLIAMS and PALMER, 1965) is in qualitative agreement with the observations of electrons $E > 1.5$ MeV in the magnetic equatorial plane (FRANK,

1965b) and is consistent with the motion of charged particles in models of the perturbed geomagnetic field (cf. HONES, 1963).

Determinations of the isopleths of electron $E > 1.5$ MeV intensities (FRANK, 1965b) in the outer magnetosphere indicate that the axis of symmetry of the

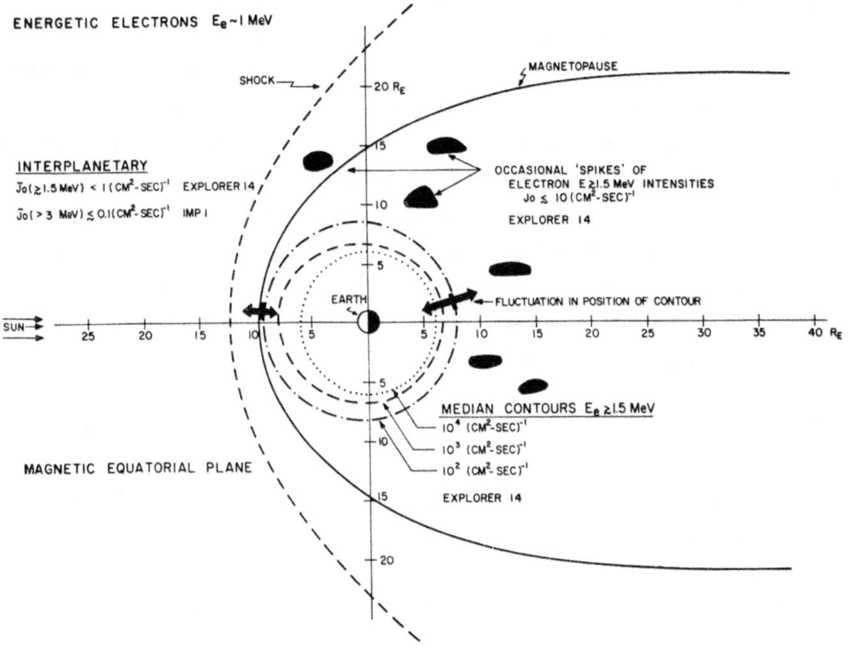

Fig. 16. Coarse summary of the observations of electrons $E > 1.6$ MeV in the magnetic equatorial plane.

local time distribution of charged particles may not be the earth-to-sun direction but may lie closer to the 11.00–23.00 (local time) meridional plane. Although this single result is inconclusive, a similar effect has been observed in the diurnal variations of electron $E > 40$ keV intensities at low altitudes with Injun 3 data (FRANK *et al.*, 1964).

Acknowledgements

The research was supported in part by the National Aeronautics and Space Administration under Grant NsG-233-62 and by the Office of Naval Research under Contract Nonr 1509(06).

References

ANDERSON, K. A.: 1965, *J. Geophys. Res.* **70** (to be published).

ANDERSON, K. A., HARRIS, H. K., and PAOLI, R. J.: 1965, *J. Geophys. Res.* **70,** 1039.

ASBRIDGE, J. R., BAME, S. J., FELTHAUSER, H. E., OLSON, R. H., and STRONG, I. B.: 1965, *Trans. Am. Geophys. Union* **46,** (1) 142.

AXFORD, W. I.: 1962, *J. Geophys. Res.* **67,** 3791.

AXFORD, W. I., and HINES, C. O.: 1961, *Canadian J. Phys.* **39,** 1322.

AXFORD, W. I., PETSCHEK, H. E., and SISCOE, G. L.: 1965, *J. Geophys. Res.* **70,** 1231–1236.

BEARD, D. B.: 1960, *J. Geophys. Res.* **65,** 3559.

BIERMANN, L.: 1951, *Z. Astrophys.* **29,** 274.

BIERMANN, L.: 1957, *Observatory* **77,** (898), 109.

BONETTI, A., BRIDGE, H. S., LAZARUS, A. J., ROSSI, B., and SCHERB, F.: *J. Geophys. Res.* **68,** 4017.

BRIDGE, H., EGIDI, A., LAZARUS, A., LYON, E., and JACOBSON, L.: *Space Res.* **5,** 969.

BRIDGE, H. S., EGIDI, A., LAZARUS, A., LYON, E. F., JACOBSON, L., and BAKER, R. H.: 1965, *Trans. Am. Geophys. Union,* **45** (3), 515.

CAHILL, L. J.: 1964, *Trans. Am. Geophys. Union* **45,** (1), 231.

CAHILL, L. J.: 1965, *Univ. of New Hampshire Res. Rep.* 65–4.

CAHILL, L. J. and AMAZEEN, P. G.: 1963, *J. Geophys. Res.* **68,** 1835.

CLINE, T. L., LUDWIG, G. H., and MCDONALD, F. B.: 1964, *Phys. Rev. Letters* **13** (26), 786.

DESSLER, A. J.: 1958, *J. Geophys. Res.* **63,** 507.

DESSLER, A. J.: 1961, *J. Geophys. Res.* **66,** 3587.

DESSLER, A. J.: 1964, *J. Geophys. Res.* **69,** 3913.

DESSLER, A. J. and PARKER, E. N.: 1959, *J. Geophys. Res.* **64,** 2239.

DESSLER, A. J. and WALTERS, G. K.: 1964, *Planet. Space Sci.* **12,** 227.

DUNGEY, J. W.: 1965, *J. Geophys. Res.* **70,** 1753.

FAN, C. Y., GLOECKLER, G., and SIMPSON, J. A.: 1964, *Phys. Rev. Letters* **13,** 149.

FARLEY, T. A.: 1963, *Rev. Geophys.* **1,** 3.

FAIRFIELD, D. H.: 1964, *J. Geoph. Res.* **69,** 3919.

FRANK, L. A.: 1965a, *J. Geophys. Res.* **70,** 1593.

FRANK, L. A.: 1965b, *J. Geophys. Res.* **70** (to be published).

FRANK, L. A. and VAN ALLEN, J. A.: 1964a, *Res. Geophys.* **1** *(Sun, Upper Atmosphere and Space,* ed. by H. ODISHAW), 161.

FRANK, L. A. and VAN ALLEN, J. A.: 1964, *J. Geophys. Res.* **69,** 4923.

FRANK, L. A., VAN ALLEN, J. A., and CRAVEN, J. D.: 1964, *J. Geophys. Res.* **69,** 3155.

FRANK, L. A., VAN ALLEN, J. A., and MACAGNO, E.: 1963, *J. Geophys. Res.* **68,** 3543.

FRANK, L. A., VAN ALLEN, J. A., HILLS, H. K., and FILLIUS, R. W.: 1965, *Trans. Am. Geophys. Union,* **46** (1), 124.

FREEMAN, John W., Jr.: 1964, *J. Geophys. Res.* **69,** 1691.

FREEMAN, J. W., VAN ALLEN, J. A., and CAHILL, Jr., L. J.: 1963, *J. Geophys. Res.* **68,** 2121.

FRITZ, T. A. and GURNETT, D. A.: 1965, *J. Geophys. Res.* **70,** 2485.

GRINGAUZ, K. I.: 1964, *J. Geophys. Res.* **69,** 1007.

GRINGAUZ, K. I., BEZRUKIKH, V. V., MUSATOV, L. S., RYBCHINSKY, R. E., and SHERONOVA, S. M.: 1964, *Space Res.* **4,** 621.

GRINGAUZ, K. I., KURT, V. G., MOROZ, V. I., and SHKLOVSKII, I. S.: 1960, *Astron. Ž.* **37,** 716 (translation in: *Soviet Astronomy – A.J.* **4,** (1961) 680).

HEPPNER, J. P., NESS. N. F., SCEARCE, C. S., and SKILLMAN, T. L.: 1963, *J. Geophys. Res.* **68,** 1.

HONES, E. W.: 1963, *J. Geophys. Res.* **68,** 1209.

JOKIPII, J. R. and DAVIS, Jr., L.: 1964, *Phys. Rev. Letters* **13,** 739.

JOHNSON, F. S.: 1960, *J. Geophys. Res.* **65,** 3049.

KELLOGG, P. J.: 1962, *J. Geophys. Res.* **67,** 3805.

KONRADI, A.: 1965, *Trans. Am. Geophys. Union* **46,** (1), 124.

KONRADI, A. and KAUFMANN, R. L.: 1965, *J. Geophys. Res.* **70**, 1627.
MACDONALD, G. J. F.: 1961, *J. Geophys. Res.* **66**, 3639.
MCILWAIN, C. E.: 1961, *J. Geophys. Res.* **66**, 3681.
MALVILLE, J. M.: 1960, *J. Geophys. Res.* **65**, 3008.
MEAD, G. D.: *J. Geophys. Res.* **69**, 1181.
MEAD, G. D. and BEARD, D. B.: *J. Geophys. Res.* **69**, 1169.
MICHEL, F. C.: 1965, *J. Geophys. Res.* **70**, 1.
MIDGLEY, J. E. and DAVIS, L.: 1962, *J. Geophys. Res.* **67**, 499.
MONTGOMERY, M. D., SINGER, S., CONNER, J. P., and STOGSDILL, E. E.: 1965, *Phys. Rev. Letters* **14**, 209.
NESS, N. F.: 1965, *J. Geophys. Res.* **70**, 2989.
NESS, N. F., SCEARCE, C. S., and SEEK, J. B.: 1964, *J. Geophys. Res.* **69**, 3531.
NESS, N. F. and WILCOX, J. M.: 1964, *Phys. Rev. Letters* **13**, 461.
OBAYASHI, T.: 1965, *NASA Tech. Note* NASA TN D-2789.
O'BRIEN, B. J.: 1962–1963, *Space Sci. Rev.* **1**, 415.
O'BRIEN, B. J.: 1964, *J. Geophys. Res.* **69**, 13.
PATEL, V. L.: 1965, *Univ. of New Hampshire Res. Rep.* 65-3.
PARKER, E. N.: 1958, *Astrophys. J.* **128**, 664.
PIDDINGTON, J. H.: *Geophys. J.*, **2**, 173., 1959.
PIDDINGTON, J. H.: 1960, *J. Geophys. Res.* **65**, 93.
PIDDINGTON, J. H.: 1965, *Planet. Space Sci.* **13**, 281.
SCARF, F. L., BERNSTEIN, W., and FREDRICKS, R. W.: 1965, *J. Geophys. Res.* **70**, 9.
SERBU, G. P.: 1965, *Space Res.* **5**, 564.
SHARP, R. D., EVANS, J. E., IMHOF, W. L., JOHNSON, R. G., REAGAN, J. B., and SMITH, R. V.: *J. Geophys. Res.* **69**, 2721.
SMITH, E. J.: 1963, *Space Science* (ed. by D. P. LEGALLEY). J. Wiley and Sons, New York, p. 316.
SNYDER, C. W., NEUGEBAUER, M., and RAO, U. R.: *J. Geophys. Res.* **68**, 6361.
SONETT, C. P., SMITH, E. J., JUDGE, D. L., and COLEMAN, P. J.: *Phys. Rev. Letters* **4**, 160.
SONETT, C. P., SMITH, E. J., and SIMS, A. R.: 1960, *Space Res.* **1**, (ed. by H. K. KALLMAN-BIJL), 982.
SPREITER, J. R. and BRIGGS, B. R.: 1962a, *J. Geophys. Res.* **67**, 2983.
SPREITER, J. R. and BRIGGS, B. R.: 1962b, *J. Geophys. Res.* **67**, 2985.
SPREITER, J. R. and JONES, W. P.: 1963, *J. Geophys. Res.* **68**, 3555.
STÖRMER, C.: 1955, *The Polar Aurora,* The Clarendon Press, London.
TAYLOR, H. E.: 1965, Fifth Western Am. Geophys. Union Meeting (abstract), Dallas, Texas, 1965.
TAYLOR, H. E. and HONES, E. W.: *Univ. of Iowa Research Report* 65–10, 1965.
VAKULOV, P. V., VERNOV, S. N., GORCHAKOV, E. B., LOGACHEV, Yu. I., CHARAKHCHYAN, A. N., CHARAKHCHYAN, T. N., and CHUDAKOV, A. E.: *Investigation of cosmic rays, Space Res.* **4**, 26.
VAN ALLEN, J. A.: 1962, *Trans. Intern. Astron. Union,* **11B**, 99.
VAN ALLEN, J. A.: 1964, *J. Geophys. Res.* **69**, 1011.
VAN ALLEN, J. A.: 1965, *J. Geophys. Res.* **70**, 4731.
VAN ALLEN, J. A. and KRIMIGIS, S. M.: 1965, *Univ. of Iowa Research Report* 65–28; *J. Geophys. Res.* (submitted for publication, August, 1965).
WILLIAMS, D. J. and PALMER, W. F.: 1965, *J. Geophys. Res.* **70**, 557.
WOLFE, J. H. and SILVA, R. W.: *Trans. Am. Geophys. Union* **45**, (3), 512.

MAGNETOSPHERIC MODELS

A. J. DESSLER and F. C. MICHEL
Department of Space Science, Rice University, Houston, Texas

Abstract. Several magnetospheric models are compared and checked against available experimental data. No evidence is found to support the hypothesis of fast magnetic merging (or annihilation of magnetic energy) and neutral line formation. It is concluded that the concept of fast magnetic merging plays no significant role either in determining the configuration of the magnetosphere or in the acceleration of charged particles within the magnetosphere.

Introduction

It is now well established that the magnetosphere is deformed asymmetrically by the solar wind (e.g., CAHILL, 1964; NESS, 1965; BONETTI *et al.*, 1963). The earth's magnetic field is deformed into a comet-like shape with the tail pointing generally away from the sun. (Interestingly, this turns out to be quite an old idea. The earliest reference that we have been able to find is to the work of FITZGERALD (1892, 1900), of the Lorentz-FitzGerald contraction fame, and LODGE (1900), who suggested that magnetic storms were due to "a torrent or flying cloud of charged atoms or ions"; that auroras were caused by "the cathode ray constituents... as they graze past the polar regions"; that comet tails could not be accounted for by solar electromagnetic radiation pressure but could be accounted for by particle radiation emanating from sunspots "like a comet's tail" and "projected from the sun" with a velocity of about 500 km/sec; and, finally, that "there seems to be some evidence from auroras and magnetic storms that the earth has a minute tail like that of a comet directed away from the sun".) Although most of the present theories agree as to the gross configuration, a significant disagreement exists in an important aspect of the theoretical formulations. This disagreement centers on the rate at which magnetic field lines from the interplanetary medium can merge with the geomagnetic field. The differing views may be summarized as follows:

DUNGEY (1961, 1962, 1963) favors the model shown in Figure 1a where the outer portion of the geomagnetic field is merged with a southward (i.e., nearly perpendicular to the plane of the ecliptic) directed interplanetary field. The field lines in the polar cap all connect directly to the interplanetary magnetic field within $\sim$0.04 AU of the earth (DUNGEY, 1965). This model has been elaborated and made more quantitative by LEVY *et al.* (1964) and AXFORD *et al.* (1965) by using the neutral line magnetic merging mechanism outlined by DUNGEY (1958)

McCormac (ed.), Radiation Trapped.

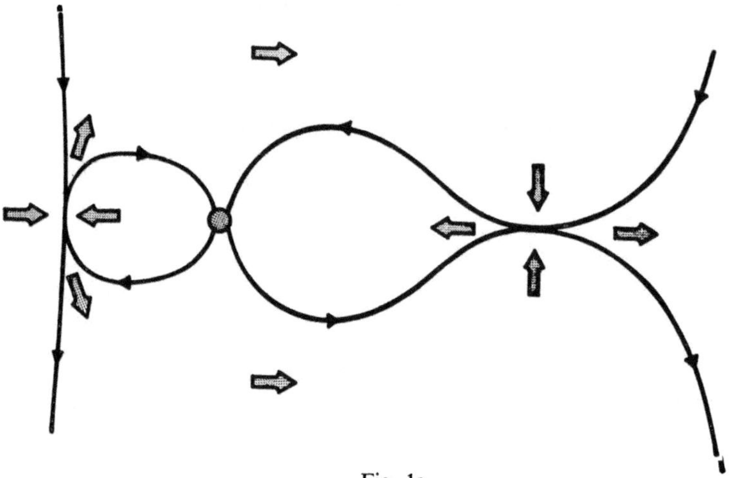

Fig. 1a

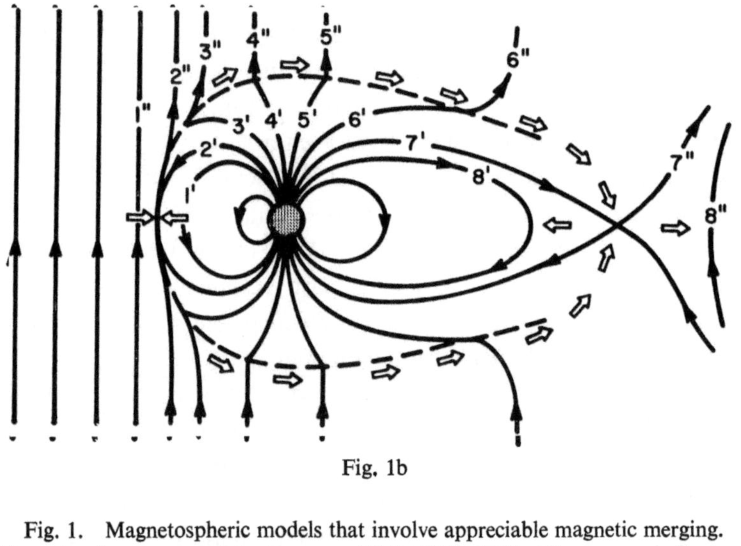

Fig. 1b

Fig. 1. Magnetospheric models that involve appreciable magnetic merging.
(a) Magnetosphere proposed by DUNGEY (1961, 1962, 1963). The arrows indicate plasma flow. Although not shown explicitly, the magnetic field lines from the polar caps connect directly to the interplanetary magnetic field.
(b) A more detailed drawing of Dungey's magnetosphere by LEVY *et al.* (1964) showing the direct connection of field lines coming out of the polar caps. Note that beyond the neutral line at the back of the magnetosphere, the field tends to return to interplanetary conditions as indicated by field line 8″.
(c) A magnetospheric model due to AXFORD *et al.* (1965). Although not shown explicitly, the very existence of a neutral line in the antisolar direction implicitly requires that the polar magnetic flux be connected directly to the interplanetary magnetic field as shown in Figure 1b, if the merging mechanism that created the neutral line also works at the magnetopause. The "neutral sheet" in this figure (as in Figures 1a and 1b) is not really neutral but has a weak perpendicular field passing through it and is actually a volume since its thickness is much greater than the local ion cyclotron radii.

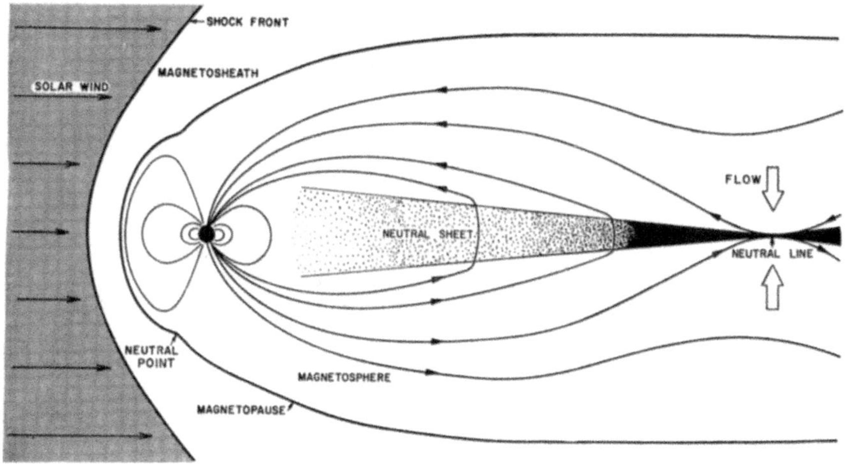

Fig. 1c

and derived formally by PETSCHEK (1964). The fields configuration proposed by LEVY *et al.* (1964) and AXFORD *et al.* (1965) are shown in Figure 1b and 1c, respectively. Some details of such magnetospheric configurations have also been discussed by SONNERUP (1965).

An alternate model in which magnetic merging is negligible has been proposed by DESSLER (1964). The magnetospheric tail in this model extends 20 to 50 AU to a postulated charge-exchange boundary shell (AXFORD *et al.*, 1963; PATTERSON *et al.*, 1963). The near-earth portion of this magnetospheric field configuration is shown in Figure 2. The principal differences between this model and the one of DUNGEY (1961, 1962, 1963), LEVY *et al.* (1964), and AXFORD *et al.* (1965) are that in Figure 2 there is a neutral sheet rather than a neutral line (note that the "neutral sheet" in Figure 1c is not really neutral), and that there is no quantitatively significant direct connection of geomagnetic field lines with interplanetary magnetic field lines, although such connection cannot be entirely absent.

Fast Magnetic Merging

The concept of merging oppositely directed lines of force that are frozen into a highly conducting plasma is of significant geophysical interest. This concept has thus far found its greatest application to the problem of solar flares where, it is generally thought, the energy of a solar flare is stored as magnetic energy and somehow suddenly released by some process such as fast merging. For further details on this subject, the reader is referred to a collection of papers on

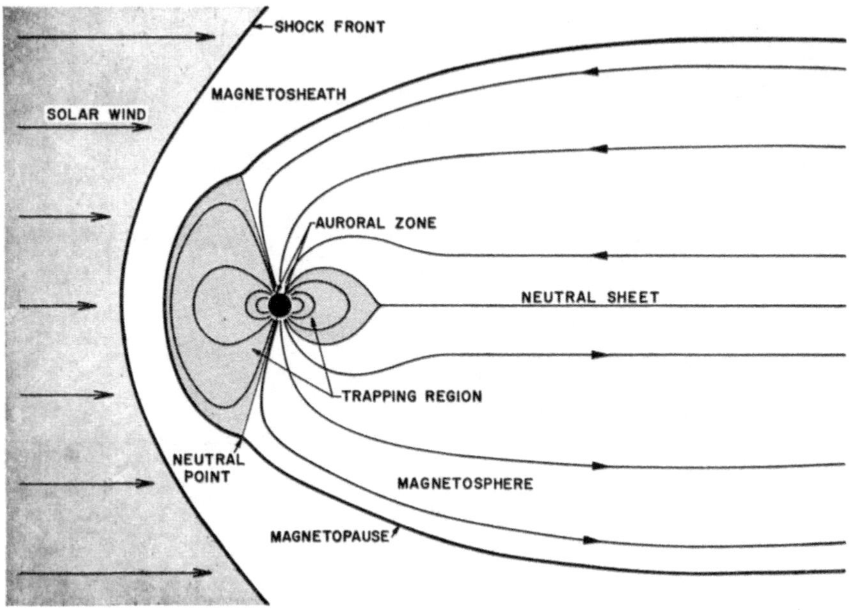

Fig. 2. Magnetospheric model in which magnetic merging is negligible (after DESSLER (1964); and DESSLER and O'BRIEN (1965)). In this model there is a true neutral sheet. Arguments are presented against any significant merging so that neutral lines do not form; hence there is negligible direct connection between the geomagnetic and interplanetary magnetic fields or across the neutral plane.

the physics of solar flares (HESS, 1964). The fastest of the various merging mechanisms is that due to PETSCHEK (1964). Petschek concludes that the merging is quite insensitive to the value of the conductivity since it appears only as the argument of a logarithm. This insensitivity to conductivity allows this particular fast-merging scheme to be applied to the earth's magnetic tail to yield the configuration shown in Figure 1c, while the general concept of fast merging leads to the similar configurations shown in Figure 1a and 1b.

In this paper, the concept of fast merging, as it applies to magnetospheric theories (and to the state of the interplanetary medium), will be checked against available experimental data. No detailed theoretical discussion of magnetic merging will be given. However, we would point out that Petschek, in the derivation of his fast-merging mechanism, assumed (in his equation (4)) that the conductivity was a scalar (rather than a tensor with large imaginary components), and, in addition, near the neutral line he allowed diffusion to proceed faster than wave motion. These points do not mean that the analysis is necessarily invalid, but certainly provide grounds for further examination.

Fast Merging and the Magnetosphere

There are two regions where the process of fast merging involving the magnetosphere might occur: in the neutral sheet between oppositely directed field lines in the tail (as shown in Figure 1c), and at the magnetopause between the interplanetary and the geomagnetic fields (as shown in Figures 1a and 1b).

We will first discuss the experimental evidence against fast merging in the neutral sheet.

The neutral sheet has been observed repeatedly with the magnetometers on board the IMP 1 satellite (NESS, 1965). Ness states that the neutral sheet has a thickness of approximately 600 km – approximately the cyclotron radius of a 1 keV proton in the tail field. He further states that such a thin sheet is representative of several traversals of the neutral sheet. Thus, it appears that a thick neutral volume as shown in Figure 1c is not a normal feature of the magnetospheric tail.

An argument against merging in the neutral plane is quoted here from an earlier paper by DESSLER (1964).

"It might be thought that the oppositely directed field lines coming from the two polar caps might be able to merge in a relatively short distance from the earth to form a closed tail... However, an examination of the merging process and the conductivity of a plasma leads us to the conclusion that the fields cannot be merged in a time shorter than several times the ion-ion or ion-atom collision period. One way of seeing this is to note that in the complete expression for conductivity in a plasma (cf. HANSON, 1961) the collision frequency ν appears in quadrature with the driving frequency ω (i.e., $\nu - i\omega$). Thus, when $\omega > \nu$, which would correspond to an event taking place in a time short compared with the collision period, the conductivity will be principally imaginary. There can be no energy dissipation with an imaginary conductivity.* Since merging of magnetic fields involves the dissipation of magnetic energy, only the real component of the conductivity can be effective. Thus, the merging time must be of the order of $1/\nu$ or greater. For conditions that prevail at say, 50 R_E, the proton-hydrogen charge-exchange collision is probably the fastest process. If there is 1 neutral hydrogen atom/cm^3, the charge-exchange collision period is about 4 months. The corresponding merging time would be several years. (Thus) merging or diffusion of magnetic field lines in the tail can little affect the length of the tail."

We will now discuss experimental evidence against fast merging at the magnetopause.

* Diagonal conductivity

$$\boldsymbol{\sigma} = \begin{Bmatrix} \sigma_1 & \sigma_2 & 0 \\ \sigma_2^* & \sigma_1 & 0 \\ 0 & 0 & \sigma_0 \end{Bmatrix};$$

σ_2 is real but non-diagonal — if $\boldsymbol{\sigma}$ is *diagonalized*, then all terms are imaginary for $\nu = 0$.

Polar cap absorption (PCA) events are caused by the bombardment of the atmosphere over the polar caps by solar flare cosmic rays. The morphology of PCA events is controlled in large part by the configuration of the magnetospheric tail. A magnetosphere whose polar cap field lines are directly connected to the interplanetary field, as in Figures 1a and 1b, should result in rather homogeneous PCA events with irregularities and inhomogeneities lasting only a few minutes. For a magnetospheric model in which there is a magnetospheric tail longer than about 1 AU and little merging between geomagnetic and interplanetary magnetic field lines, the low-energy ($\sim$5 MeV) PCA events should take several hours to become homogeneous over the polar caps.

If the geomagnetic field lines coming out of the polar caps merge with the interplanetary magnetic field within 0.04 AU, as calculated by DUNGEY (1965), then low-energy solar flare particles have access to the polar caps directly along magnetic lines of force. A 5 MeV proton can traverse 0.04 AU in about 3 minutes. Thus, if rapid magnetic merging at the magnetopause were to cause field configuration as shown in Figure 1, with a magnetospheric tail as short as 0.04 AU, one would expect on the basis of Liouville's theorem that the polar caps would be uniformly illuminated by soft cosmic rays within a few minutes following the onset of a PCA. Furthermore, a fluctuation in particle intensity at one point on the polar caps would be duplicated at other points within this same time – a few minutes. We are here assuming that there is no strong gradient in solar cosmic rays across the tail; such a gradient is unlikely since the tail diameter is only 3×10^{-3} AU (i.e., $\sim 70\ R_E$) (DESSLER, 1964). It would be difficult to sustain a difference in particle flux over such small scales – for example, the cyclotron diameter of a 5 MeV proton in a 5γ interplanetary field is approximately 20 R_E or about 0.3 tail diameters. Bathing the tail uniformly in cosmic ray flux should therefore lead to a uniform cosmic ray flux into the polar caps. Thus, it is our opinion that such a short magnetospheric tail with field lines that connect to the interplanetary magnetic field should result in relatively uniform PCA behavior over the polar caps with few spatial inhomogeneities (i.e., differences in particle flux from one part of the polar cap to another) that last longer than a few minutes.

In contrast to the above expectation, the long (>1 AU) magnetospheric tail with negligible magnetic merging (Figure 2) should lead, as will be discussed, to spatial inhomogeneities over the polar caps over time periods about two orders of magnitude larger than expected for the short (0.04 AU) magnetospheric tail. The manner of entry of low-energy cosmic rays into this magnetospheric tail and thence into the polar cap is by diffusion via scattering with magnetic field irregularities. MICHEL (1965) has argued that for diffusion into the tail to be effective, the tail must be longer than about 1 AU. The direct travel time for a 5 MeV particle over a distance of 1 AU is approximately 1 hour. The diffusion time for this distance is ~ 3 hours (MICHEL and DESSLER, 1965). As long as entry into the tail is within 1 AU of the earth, the total time for 5 MeV proton PCA flux to cover the polar caps uniformly will be controlled by the time for diffusion onto the polar cap field lines, hence about 3 hours. Including the transit time for particles to reach to 1 AU behind the earth then gives a time constant of about 4 hours.

With the long tail model shown in Figure 2, the development of a PCA event should, on the average, show characteristic patterns of development. Since the particles enter the magnetosphere along the magnetospheric boundary (the magnetopause) and then diffuse in toward the center of the tube of flux from each polar cap, the PCA events for low-energy particles (~ 5 MeV protons) should start first in the auroral zones and progress smoothly toward the magnetic poles. We assume that beyond a relatively short distance, $\sim 10^2$ R_E, the polar tubes of flux form the figure 8 or double cylinder pattern proposed by DESSLER and JUDAY (1965), and therefore the entire surface of the tube is exposed to the solar protons. Of course, the field configuration is probably more irregular and complex than suggested by this simple model. Any irregularities due to splitting, distortions, or fluting within the tail will result in deviations from the ideal behavior described above.

Experimental evidence bearing on the time behavior and spatial characteristics of low-energy (~ 5 MeV) solar flare cosmic rays is sparse. However, data that are available (OBAYASHI and HAKURA, 1960; HAKURA, 1964; BAILEY, 1964) seem to indicate that PCA events are not simultaneous and homogeneous across the polar cap; significant spatial inhomogeneities across the polar cap are usually evident in the early part of PCA events. The most detailed study known to us is due to HAKURA (1964), who demonstrates that there are irregularities and inhomogeneities in the early stages of a PCA that have time constants of 3 to 6 hours. These times are quite consistent with the estimate of 4 hours derived by MICHEL and DESSLER (1965) on the basis of a long tail model with insignificant magnetic merging. We are not aware of any experimental study supporting the 3-minute time constant, the latter being derived from the short magnetospheric tail with direct connection between the geomagnetic and interplanetary magnetic fields. Axford has suggested (private communication) that anisotropy in the solar flare cosmic rays may account for the observed inhomogeneities even if the field configuration is as shown in Figures 1a and 1b.

As a final point we will examine merging in the interplanetary medium.

Regions of oppositely directed field lines are created in the interplanetary medium by the solar wind as it pulls out magnetic field lines from adjacent photospheric regions of opposite magnetic polarity. These oppositely directly magnetic field regions move from the sun to the earth in about 3 days. If fast merging were universally applicable, large neutral volumes should develop (like the "neutral sheet" shown in Figure 1c).

Experimental evidence in the interplanetary medium (NESS and WILCOX, 1965) shows that the interplanetary field is sharply divided into regions of opposite magnetic polarity. There is, once again, no evidence of magnetic merging.

Conclusions

The phenomenon of rapid magnetic merging and connection between geomagnetic and interplanetary magnetic fields or in the neutral sheet of the earth's magnetic tail is not supported by space probe or cosmic ray data. There are no direct satellite or space probe magnetometer measurements that can differentiate between the different magnetopause configurations of Figures 1a and 1b and Figure 2. However, polar cap absorption (PCA) data, obtainable at ground level, can answer the above questions regarding fast merging and the configuration of the geomagnetic field lines. We conclude that available PCA data indicate a magnetospheric tail longer than about 1 AU with negligible connection between the geomagnetic and interplanetary magnetic fields.

Similarly, space probe data obtained in the interplanetary medium and in the earth's magnetic tail show large longlived regions of oppositely directed magnetic field with no evidence for neutral regions to appear.

Thus, we conclude that fast magnetic merging is not important in the interplanetary medium or in the earth's magnetic tail, and that the magnetospheric model shown in Figure 2 is most consistent with theory and experiment.

As a corollary to this conclusion we note that if merging proceeds slowly, it is difficult to see how it can, by itself, be important in acceleration of particles or in other effects of possible geophysical interest.

Acknowledgement

This work was supported in part by the U.S. Air Force Cambridge Research Laboratories under Contract AF 19(628)-3858.

References

AXFORD, W. I., DESSLER, A. J., and GOTTLIEB, B.: 1963, *Astrophys. J.* **137**, 1268.
AXFORD, W. I., PETSCHEK, H. E., and SISCOE, G. L.: 1965, *J. Geophys. Res.* **70**, 1231.
BAILEY, D. K.: 1964, *Planet. Space Sci.* **12**, 495.
BONETTI, A., BRIDGE, H. S., LAZARUS, A. J., ROSSI, B., and SCHERB, F.: 1963, *J. Geophys. Res.* **68**, 4017.
CAHILL, L. J., Jr.: 1964, *Trans. Am. Geophys. Union* **45**, 231.
DESSLER, A. J.: 1964, *J. Geophys. Res.* **69**, 3913.
DESSLER, A. J. and JUDAY, R. D.: 1965, *Planet. Space Sci.* **13**, 63.
DESSLER, A. J. and O'BRIEN, B. J.: 1965, in *Satellite Environment Handbook* 2nd ed., (ed. by F. S. JOHNSON), Stanford University Press, Stanford, California, p. 64.

DUNGEY, J. W.: 1958, *Cosmic Electrodynamics*, Cambridge University Press, London, p. 98.
DUNGEY, J. W.: 1961, *Phys. Rev. Letters* **6**, 47.
DUNGEY, J. W.: 1962, *J. Phys. Soc., Japan* **17**, Suppl. A–II, 15.
DUNGEY, J. W.: 1963, *Planet. Space Sci.* **10**, 233.
DUNGEY, J. W.: 1965, *J. Geophys. Res.* **70**, 1753.
FITZGERALD, G. F.: 1892, *The Electrician* **30**, 48.
FITZGERALD, G. F.: 1900, *The Electrician*, **46**, 287.
HAKURA, Y.: 1964, *J. Radio Res. Lab.* **11**, 273.
HANSON, W. B.: 1961, *Satellite Environment Handbook* (ed. by F. S. JOHNSON), Stanford University Press, Stanford, California, Ch. 2.
HESS, W. N., Ed.: 1964, *AAAS-NASA Symposium on the Physics of Solar Flares*, NASA SP-50, Washington, D.C.
LEVY, R. H., PETSCHEK, H. E., and SISCOE, G. L.: 1964, *AIAA Journal* **2**, 2065.
LODGE, Oliver: 1900, *The Electrician* **46**, 249.
MICHEL, F. C. and DESSLER, A. J.: 1964, *J. Geophys. Res.* **70**, 4305.
MICHEL, F. C.: 1965, *Planet. Space Sci.* **13**, 753.
NESS, N. F.: 1965, *J. Geophys. Res.* **70**, 2989.
NESS, N. F. and WILCOX, J. M.: 1965, *Science* **148**, 1592.
OBAYASHI, T. and HAKURA, Y.: 1960, *J. Geophys. Res.*, **65**, 3131.
PATTERSON, T. N. L., JOHNSON, F. S., and HANSON, W. B.: 1963, *Planet. Space Sci.* **11**, 767.
PETSCHEK, H. E.: 1964, *AAAS-NASA Symposium on the Physics of Solar Flares* (ed. by W. N. HESS) NASA SP-50, Washington, D. C.
SONNERUP, B. U. Ö.: 1965, *J. Geophys. Res.* **70**, 1051–1060.

Discussion

After Dessler presented his paper there was much interest in trying to understand the difference between DESSLER's model and DUNGEY's model and how experimental data may support the various models.

This distinction between models is important in looking at polar cap events. On the point of the development pattern for polar-cap-absorption events, VAN ALLEN said that his satellite data showed essentially uniform precipitation of solar cosmic rays over the polar cap – distinctly contrary to the prediction made by Michel and Dessler, and in agreement with what one would expect from Dungey's magnetospheric model. Van Allen invited a comment from STOLOV, who stated that most of the PCA events he had studied started at the pole and swept toward the auroral zone. Dessler pointed out that this particular pattern of development gave comfort to neither of the magnetospheric models, although the several-hour time delay of PCA occurrence from one part of the polar cap to another was in agreement with a diffusion process. Dessler also pointed out that Stolov's statement was in disagreement with Van Allen's experimental findings of essentially instantaneous coverage of the polar cap by solar cosmic rays.

WILLIAMS reported preliminary data that showed solar cosmic rays arriving outside the magnetosphere several hours before they were detected by a satellite well within the polar cap. Unfortunately, the threshold sensitivity of the detector outside the magnetosphere was about 15 MeV, while the one over the polar cap was about 5 MeV. Van Allen argued that it is characteristic of solar-flare cosmic-ray propagation that low energy particles should arrive at the earth after the high-energy component. Thus, this particular time delay might well have nothing to do with diffusion into the magnetospheric tail.

Van Allen further argued that Mariner IV, which was within 4 degrees of the earth-sun line at 3300 earth radii for about one week, did not detect any electrons that are normally associated with the magnetosphere. Dessler pointed out that since magnetospheric electrons could be expected to drift and diffuse out of the tail, the absence of electrons that could be detected with a 40 keV threshold geiger counter was not surprising. This loss of energetic electrons within the tail is supported by Anderson's IMP I findings that 45 keV electrons are almost absent at even 30 earth radii. Dessler stated that he expected the magnetospheric tail to be filled with solar-wind plasma at such large distances and thus be undetectable with a geiger counter. Van Allen concluded that, "If you can't see it, it's not there."

NATURE OF THE OUTER MAGNETOSPHERE AND THE MAGNETOPAUSE THEORY

DAVID BEARD
University of Kansas, Lawrence, Kansas

In an attempt to explain the correlation between solar flares and magnetic storms observed on the surface of the earth, CHAPMAN and FERRARO (1931, 1932, 1933, 1940; FERRARO, 1952) proposed that streams of plasma emitted from the sun were compressing the geomagnetic field. They recognized that

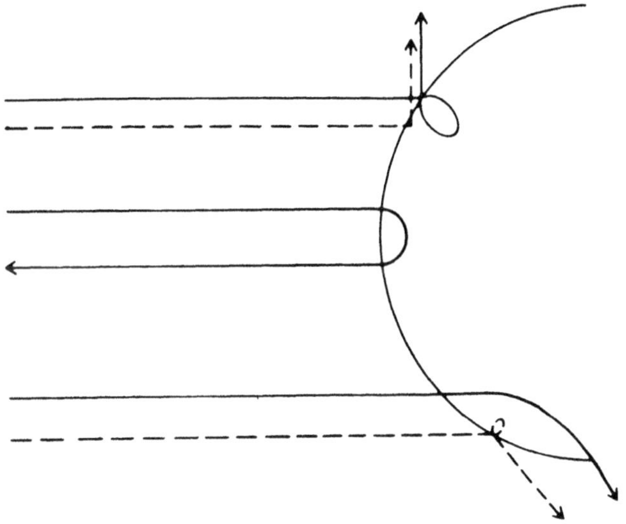

Fig. 1. Naïve charged particle trajectories (solid lines positive ions, dashed lines electrons) incident perpendicular to a magnetic dipole projected on the equatorial plane.

the diamagnetic behavior of the plasma would shut out the geomagnetic field, and there would be a sharp boundary between the plasma and the magnetic field. Figure 1 illustrates a naive picture of the particle trajectories at the boundary. Since the boundary is at most a proton gyroradius thick (≈ 50 km) and the radius of curvature of the surface is about ten earth radii (≈ 60000 km), the fields on the exit path of the particles duplicate the fields on the entrance path. Hence the exit path is the mirror image of the entrance path; the reflection is specular and electrons and protons are reflected from the surface in the same way.

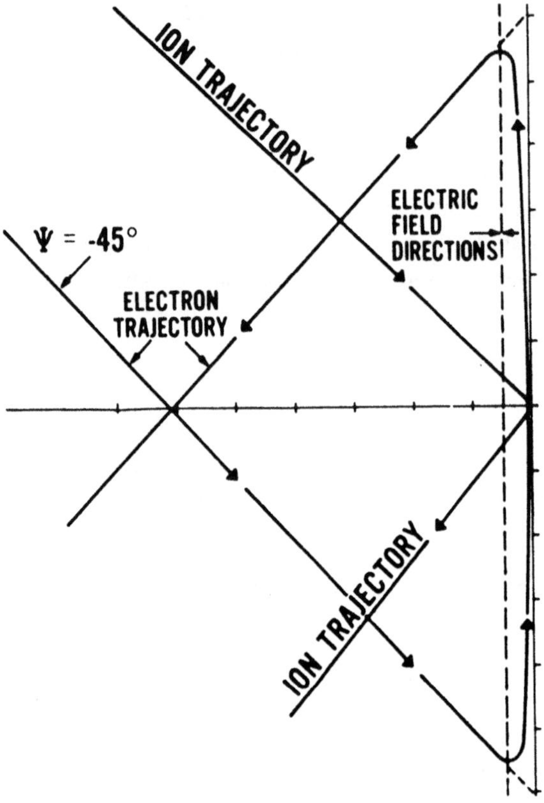

Fig. 2. Charged particle trajectories in a boundary layer between a plasma and a magnetic field as computed by MacMahon. The magnetic field is at the right and perpendicular to the plane of the figure.

The actual situation is quite different from that illustrated in the figure. Since the protons penetrate more deeply than the electrons, an electric field due to a charge separation layer will be created. The situation is further complicated by the presence of an interplanetary magnetic field which we have not yet included. DUNGEY (1958) and ROSENBLUTH (1957) have analyzed the charge layer sheath and shown that the magnetic field decreases with depth, X, into the plasma as $B \approx B_0 e^{-X/X_0}$ where $X_0 = 3.8/\sqrt{N_0}$ km and N_0 is the number of electrons per cubic centimeter in the plasma (See for example BEARD (1960,

1964a)). Mac Mahon has computed the trajectories of the charges which are illustrated in Figure 2. Note that the electrons which are accelerated by the electric field but do not penetrate as deeply into the magnetic field, travel a long distance along the boundary and carry the electrical current in the boundary which terminates the geomagnetic field. The protons penetrate more deeply into the magnetic field and have a shorter effective radius of curvature.

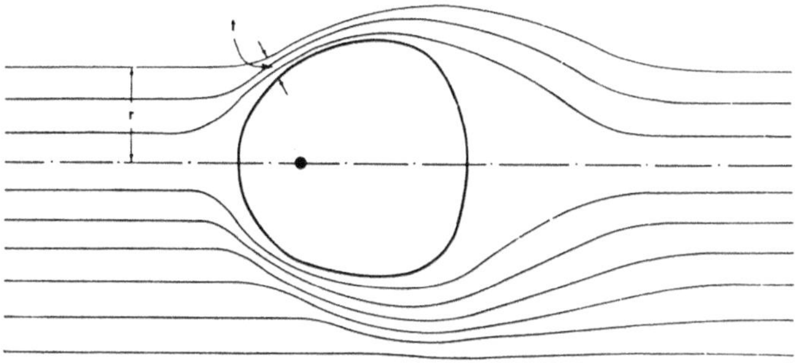

Fig. 3. Illustration of deflection of interplanetary field lines by the magnetosphere when the lines are parallel to the solar wind velocity.

The presence of an interplanetary magnetic field, however weak, ties the plasma together so that specular reflection does not take place. If the magnetic field were aligned along the stream velocity vector the magnetic field would diverge around the earth's magnetic field – the magnetosphere – as illustrated in Figure 3 (Beard, 1964a, b). A more relevant and interesting case, however, is that for which the magnetic field has a significant component perpendicular to the stream velocity vector.

Consider, for a moment, a gas made up of positrons and electrons containing a weak magnetic field traveling with velocity, v_0, perpendicular to a sharp boundary containing a stationary magnetic field. The magnetic pressure in the stream is much less than the forward stream pressure and the stream magnetic field is carried along with the stream (Figure 4). The field is perpendicular to the figure. On entering the stationary field the particle is turned around. The particle density behind the boundary is twice the free stream particle density since the pressures are equal and only half the particles are moving towards the boundary in the stationary field (Beard, 1965a). Since in a collisionless plasma the particle density is proportional to the magnetic field intensity the field intensity is twice the free stream field intensity also. Therefore if particle conservation is invoked, we find the velocity of the boundary to the left is simply $v_0/2$. That is, the flux of particles entering the boundary from the left is n_0v_0 and the number of particles piling up behind the boundary per second as it moves to the left must be $(2n_0)(v_0/2)$. The important point is that by the time the particle is turned around, $\pi\omega_c$, where ω_c is the cyclotron frequency, $e2H_0/mc$, the boundary has moved to the left a distance $(v_0/2)(\pi/\omega_c)$ which is $\pi/2$ gyroradii. That is, the boundary moves to the left faster than the particle is turned around and the particle is trapped into a stationary circular orbit.

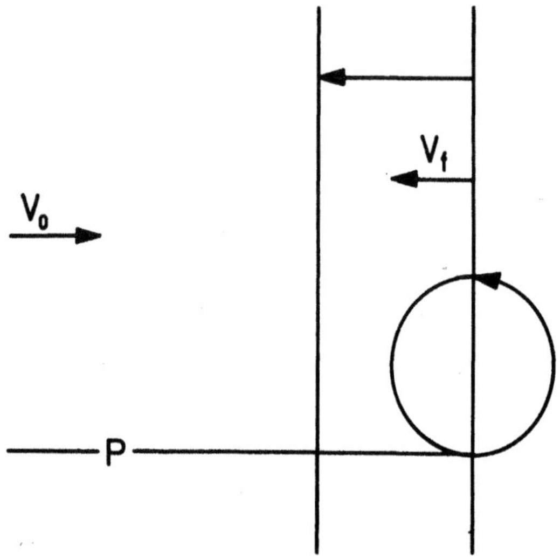

Fig. 4. Illustration of particle orbit when a weak magnetic field is imbedded in the plasma incident on a strong stationary magnetic field. The incident plasma and weak trapped field are compressed and stopped on the surface and the surface advances to the left with velocity V_f.

The magnetosphere, however, is a blunt curved surface and not an infinite plane. The plasma is not held stationary but is pushed around any finite obstacle as is illustrated in Figure 5 where again the magnetic field trapped in the plasma is perpendicular to the figure. R is the radius of the stationary cylindrical permanent magnetic field chosen to represent the magnetosphere.

Where the change of momentum of a specularly reflected particle was $2mv_0 \cos\psi$, ψ being the angle between the surface normal and the stream velocity vector, the change of momentum of a particle entering the surface is $mv_0 \cos\psi$ for the particle change in momentum is still perpendicular to the boundary between the free stream and the transition sheath (BEARD,

1965a). As a result the particle pressure on the outer boundary is $(n_0 v_0 \cos\psi)$ $(mv_0 \cos\psi) = n_0 m v_0^2 \cos^2\psi$, not $2n_0 m v_0 \cos^2\psi$ as in specular reflection. This is the external particle pressure confining the particle and enhanced magnetic field pressure in the sheath and it acts perpendicular to the sheath. There is no change in particle center of motion momentum component parallel to the boundary which would result in a momentum change factor not linearly dependent on $\cos\psi$.

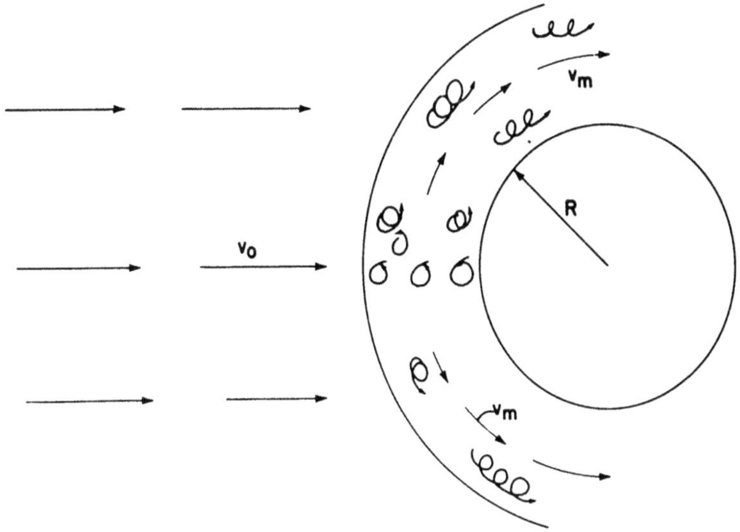

Fig. 5. Illustration of particle orbits when the surface is a cylinder instead of the plane discussed in Figure 4. V_m represents the velocity of magnetic field lines after they are slowed from their initial velocity v_0.

The magnetic field lines are shown in another projection in Figures 6 and 7. In Figure 7 the plane of projection is perpendicular to the earth-sun line and the field lines are shown slipping around the magnetosphere.

Since the particle pressure is still proportional to $\cos^2\psi$ as in the case of specular reflection, we are justified, therefore, in making the same calculation as is described in fuller detail in CHAPMAN and FERRARO (1931, 1932, 1933, 1940); FERRARO (1952); DUNGEY (1958) or BEARD (1960, 1964a), for example. The pressure communicated to the front surface of the magnetosphere is

$$n_0 m v_0^2 \cos^2 \Psi \equiv p_0 \cos^2 \Psi = p_0 (\hat{\mathbf{n}}_s \cdot \hat{\mathbf{v}})^2 \tag{1}$$

where $\hat{\mathbf{n}}_s$ is the unit normal to the surface and $\hat{\mathbf{v}}_0$ is a unit vector in the direction of the stream velocity. We have neglected the magnetic pressure of the interplanetary magnetic field since it is only one percent of the stream pressure at the stagnation point, but around well to the back of the magnetosphere it should be included in a complete calculation since the particle pressure is very small there.

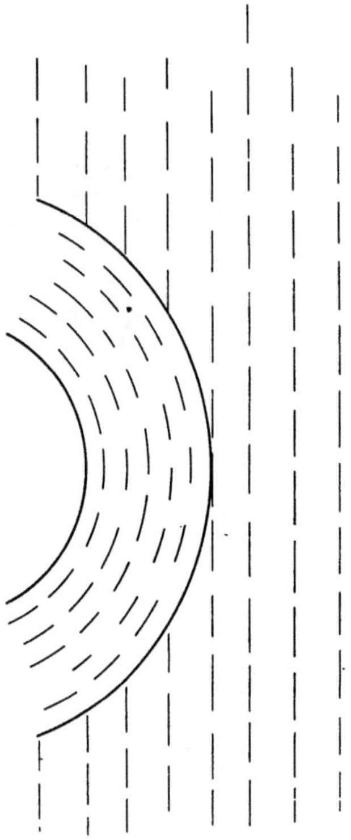

Fig. 6. Magnetic field lines in free interplanetary space as they are being compressed in the magnetosheath.

A conducting fluid moving in a magnetic field will experience a body force expressed by the familiar magnetohydrodynamic equation

$$\operatorname{grad} p = \mathbf{j} \times \mathbf{B}/c \tag{2}$$

where $\mathbf{j}$ is the electrical current density in the fluid and p is the particle pressure. Maxwell's equation

$$\operatorname{curl} \mathbf{B} = 4\pi \mathbf{j} \tag{3}$$

can be substituted into (2) to obtain

$$\operatorname{grad} p = (1/4\pi)(\mathbf{B}\cdot\operatorname{grad})\mathbf{B} - (1/8\pi)\operatorname{grad} B^2 \tag{4}$$

The change in the magnetic field in the direction of the magnetic field is negligible compared to the change in the field perpendicular to the boundary

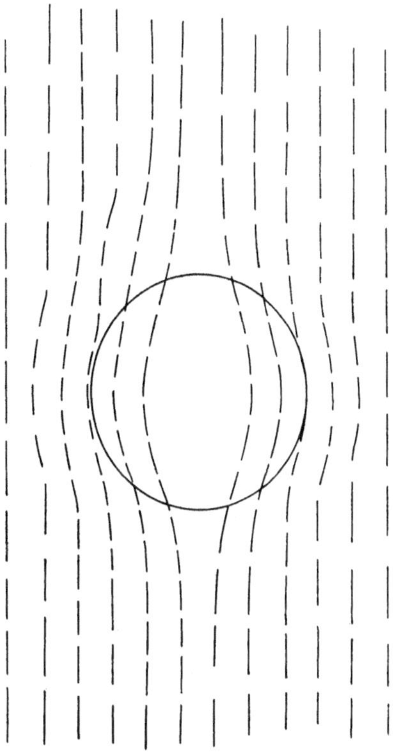

Fig. 7. Magnetic field lines projected on a plane perpendicular to the earth sun line showing how they slip around the magnetosphere.

in the direction of **p** except very near the two null points on the magnetosphere where the field reverses direction. Therefore the first term on the right hand side can be neglected and the equation integrated to yield the well-known diamagnetic approximation in magnetohydrodynamics

$$p + B^2/8\pi = \text{constant} \tag{5}$$

Letting the constant be the free space value of the stream pressure, we obtain the pressure equation relating the particle pressure outside the boundary to the magnetic pressure inside the boundary. This equation determines the surface

$$p_0(\hat{\mathbf{n}}_s \cdot \hat{\mathbf{v}})^2 = B^2/8\pi$$

$$B = \pm (8\pi p_0)^{\frac{1}{2}} \hat{\mathbf{n}}_s \cdot \hat{\mathbf{v}}$$

Since the normal component of **B** is continuous across the surface in this description and must be zero outside the surface, the magnetic field inside and close to the surface is everywhere tangential to the surface. Therefore we can write instead

$$|\hat{\mathbf{n}}_s \times \mathbf{B}| = -(8\pi p_0)^{\frac{1}{2}} \hat{\mathbf{n}}_s \cdot \hat{\mathbf{v}} \tag{6}$$

Since $\hat{\mathbf{n}}_s \cdot \hat{\mathbf{v}}$ must be negative everywhere on the surface for the outside of the surface to be exposed to the solar stream, the minus sign is used in (6).

The surface is defined by a function $F(r, \theta, \varphi) = \text{constant}$. Let the constant be zero and the function be

$$F(r, \theta, \varphi) = r - R(\theta, \varphi) = 0 \tag{7}$$

Then the normal to the surface is written

$$\hat{\mathbf{n}}_s = \operatorname{grad} F / |\operatorname{grad} F| = a\left(\hat{\mathbf{r}} - \frac{1}{r}\frac{\partial R}{\partial \theta}\hat{\boldsymbol{\theta}} - \frac{1}{r \sin\theta}\frac{\partial R}{\partial \varphi}\hat{\boldsymbol{\varphi}}\right) \tag{8}$$

where a is a normalizing constant. In the polar coordinate system illustrated in Figure 8 the unit stream velocity vector is

$$\begin{aligned}\hat{\mathbf{v}} = &-(\cos\theta \sin\lambda + \cos\varphi \sin\theta \cos\lambda)\hat{\mathbf{r}} \\ &-(\cos\varphi \cos\theta \cos\lambda - \sin\theta \sin\lambda)\hat{\boldsymbol{\theta}} \\ &+ \sin\varphi \cos\lambda\, \hat{\boldsymbol{\varphi}}\end{aligned} \tag{9}$$

Equations 8 and 9 are then substituted into the surface equation, Equation 6 which must be solved in order to obtain the shape of the magnetopause, the thin outer boundary of the magnetosphere.

Approximate Methods for Calculating the Surface

THE SELF-CONSISTENT FIELD

In order to calculate the surface, that is, in order to calculate $\hat{\mathbf{n}}_s$ everywhere from Equation 6 we have to know the magnetic field, **B**. However **B** which includes the effect of the surface currents cannot be evaluated until the surface is known. The self-consistent approximation consists of (i) assuming a reasonable value for B (ii), calculating a surface, and then after (iii) computing a more correct value for **B** based on this surface, (iv) substituting the new value of **B** in Equation 6, and (v) recomputing the surface. In all of the several different plasma-magnetic-field-boundaries that have been computed all of the reiterated surfaces after the second are indistinguishable from one another.

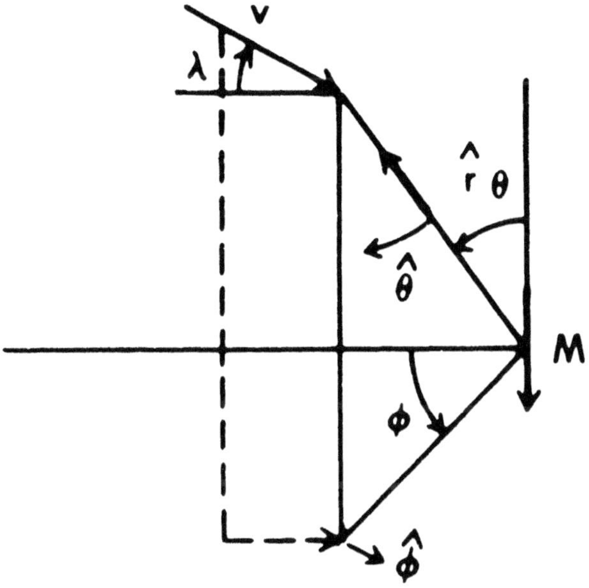

Fig. 8. Coordinate system used in the calculations. **M** is the geomagnetic dipole, **V** is the solar wind velocity and λ is the geomagnetic latitude position of the sun.

A reasonable approximation to the magnetic field is suggested by the large radius of curvature of the surface. If the surface were a plane the field due to the surface currents would be equal and opposite on both sides of the surface. Since the field outside is zero and the geomagnetic dipole field is the same on both sides of the surface, the field due to the surface currents must be equal to the tangential component of the geomagnetic field. Therefore in first approximation we need only double the tangential component of the geomagnetic field, $\hat{\mathbf{n}}_s \times \mathbf{B}$, where

$$\mathbf{B}_g = -(M/r^3)(2\cos\theta\,\hat{\mathbf{r}} + \sin\theta\,\hat{\boldsymbol{\theta}}) \tag{10}$$

M being the dipole moment of the earth. When this approximate field is substituted into Equation 6 a first approximation surface can be computed. A solution may not exist, however, for all angles θ and φ if the correct value of $\mathbf{B}$ is not used. *Indeed, the solution for all values of θ and φ on the antisolar side does not exist in first approximation*, at least for an inappropriate choice of coordinate system. In this case the corrected magnetic field used in higher approximation may be computed by completing the surface with a reasonable extrapolation of the part of the approximate surface which can be computed.

In the meridian plane $\mathrm{d}R/\mathrm{d}\varphi=0$, and if we let $R_0=(M^2/2\pi p_0)^{1/6}$ and let r be measured in units of R_0 then Equation 6 takes a particularly elementary form in the meridian plane.

$$\frac{\mathrm{d}R}{\mathrm{d}\theta} = r\tan\theta\,(r^3-1)/(r^3+2) \tag{11}$$

whose solution is

$$r = 1 \tag{12}$$

or

$$\frac{r_0^3-1}{r_0^2}\,\frac{r^2}{r^3-1} = \cos\theta \tag{13}$$

where the constant r_0 occurring in the integration constant is the value of r at $\theta=0$. Equation 13 yields a surface for which r is either zero or infinite at the equator ($\theta=90°$). Only equation 12 for a circle yields an acceptable surface.

Since $\mathbf{B}$ changes sign at the poles and the absolute value of $\hat{\mathbf{n}}_s\times\mathbf{B}$ is required, the equation for the surface in the meridian plane on the "dark" side of the pole becomes

$$(\mathrm{d}R/\mathrm{d}\theta) = r\tan\theta\,(r^3+1)/(r^3-2) \tag{14}$$

whose solution is

$$kr^3/(1+r^3) = r\cos\theta \tag{15}$$

The first approximation surface calculated for the entire solar side in 1960 has since been very well confirmed by satellite observations.

Unfortunately all the boundary conditions are not satisfied in the first approximation. Nowhere is this more apparent than near the poles where the surface current drops to zero and the magnetic field due to the surface curvature cannot be neglected as this approximation assumes. As a result matching the first approximation solutions (12) and (15) presents minor difficulties which have been arbitrarely treated in various ways as illustrated in Figure 9. The middle curve is one chosen by SPREITER and BRIGGS (1962) who obtained a meridian plane solution for non-perpendicular incidence of the stream on the solar side surface

$$r^3\cos(\theta+\lambda) - \cos\theta = Kr^2 \tag{16}$$

and on the anti-solar side surface

$$r^3 \cos(\theta + \lambda) + \cos\theta = Kr^2 \tag{17}$$

These solutions are illustrated in Figure 10.

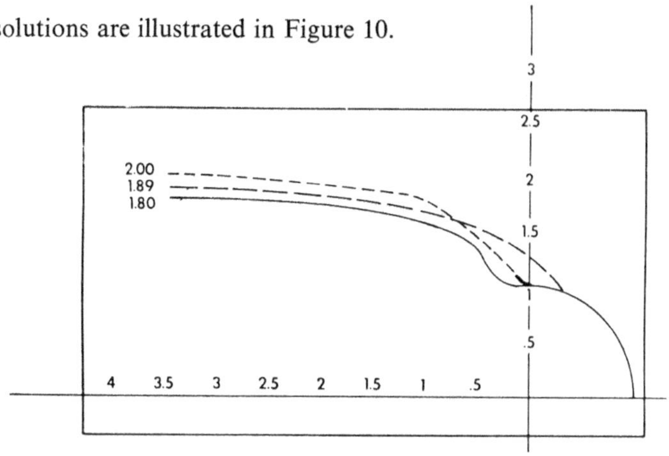

Fig. 9. First approximation surfaces in the meridian plane with various arbitrary choices of day-time night-time solution matching at the poles.

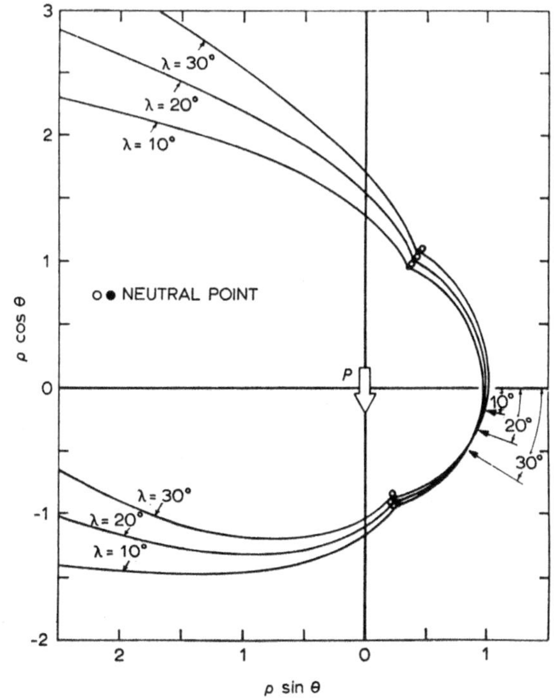

Fig. 10. Meridian plane first approximation solutions for various angles of inclination of the solar wind vector to the magnetic equatorial plane.

Having obtained a first-order approximation surface, we are now able to integrate over the surface and obtain the correction to the magnetic field due to the curvature of the surface, which was assumed to be a plane in first order. The current density on the surface is given by

$$\mathbf{j}/c = \mathbf{B} \times \operatorname{grad} p/B^2 \tag{18}$$

where grad p is proportional to $\hat{\mathbf{n}}_s \cdot \hat{\mathbf{v}}$ and is in the direction of $\hat{\mathbf{n}}_s$. The approximate vector **B** used in the computation of the surface is used in Equation 18.

The magnetic field due to the surface currents is computed from Laplace's equation

$$\mathbf{B}_{sc} = \int \frac{\mathbf{J} \times \hat{\mathbf{r}}}{cr^2} \mathrm{d}S \tag{19}$$

where **r** is the vector from the surface element dS to the point where **B** is evaluated. Of course the integrand has a singularity at $\mathbf{r}=0$ and depends on whether it is evaluated on one side of the infinitesimally thin surface or at the center. The center of the current sheath has been chosen in order to compute only the correction due to the curvature of the surface. The integral has been evaluated numerically with a five degree mesh. When the correction field Equation 19 is substituted into Equation 6 the altitude of the subsolar point is found to be raised

$$r(90°) = (1 + B_{sc})^{-1/3}$$

since B_{sc} (90°) adds to the geomagnetic field at the subsolar point.

The first approximation surface does not exist everywhere because there can be no solution in some regions if the first approximation magnetic field is used in the calculation. This is especially true around the meridian plane on the antisolar side. An arbitrary surface must be constructed in order to carry out higher order approximations which give a sufficiently accurate magnetic field for a surface to be computable everywhere.

The fourth and higher reiterated surfaces are indistinguishable from one another (Mead and Beard 1964). The final surface is illustrated in Figures 11–14. The proof of the pudding is that the magnetic field outside the surface

---

Fig. 11. Noon meridian and equatorial intersections of the first and fourth self-consistent approximations to the magnetosphere boundary. Midgley and Davis' solution is shown for comparison. Their solution is interior to the other solutions at the subsolar point.

Fig. 12. Intersections of the fourth order self-consistent approximation to the magnetospheric boundary with planes differing by 15° in rotation about the earth-sun line.

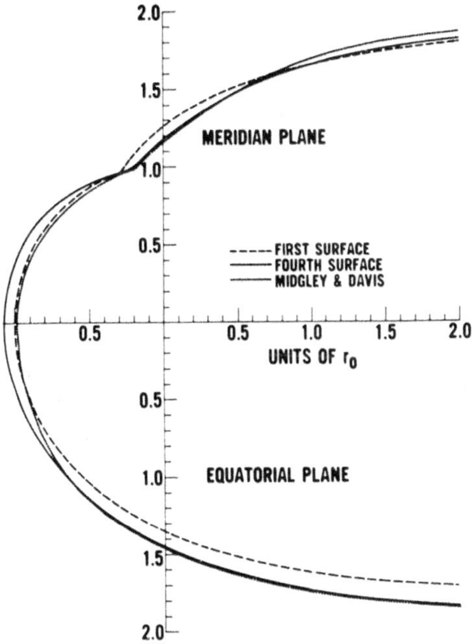

MERIDIAN PLANE
FIRST SURFACE
FOURTH SURFACE
MIDGLEY & DAVIS
EQUATORIAL PLANE
UNITS OF r_0
2.0
1.5
1.0
0.5

Fig. 11

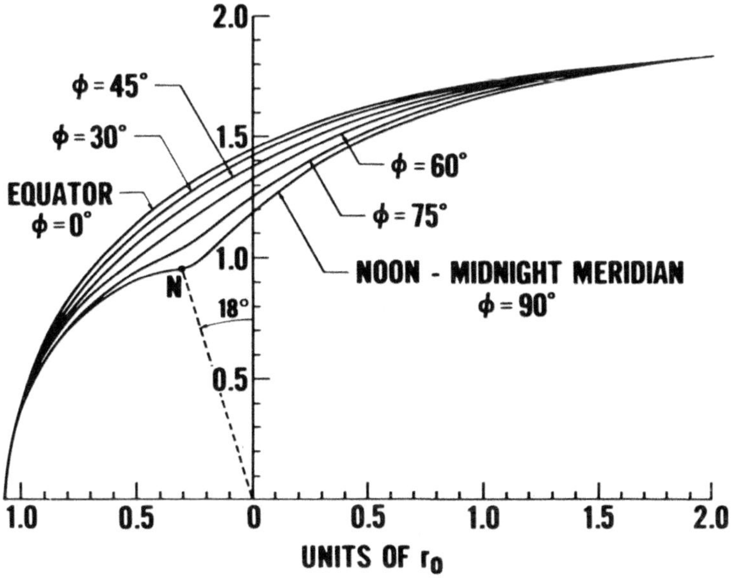

$\phi = 45°$
$\phi = 30°$
EQUATOR
$\phi = 0°$
$\phi = 60°$
$\phi = 75°$
NOON - MIDNIGHT MERIDIAN
$\phi = 90°$
N
18°
UNITS OF r_0

Fig. 12

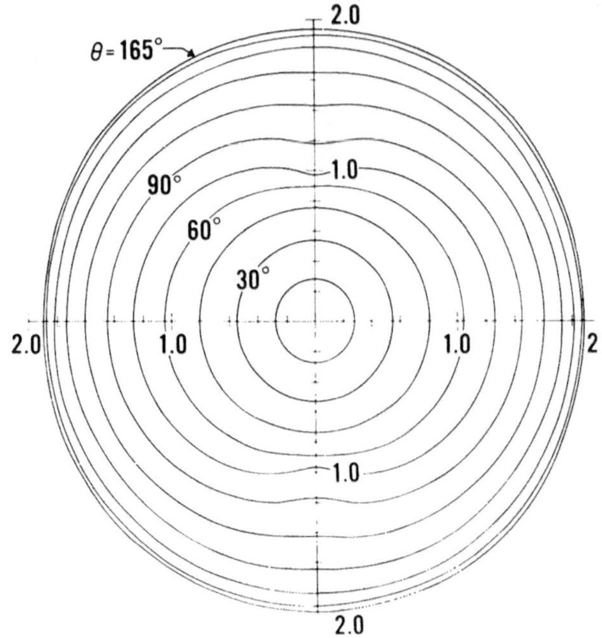

Fig. 13

Fig. 14

cannot be zero. The ratio of the total magnetic field to the undisturbed dipole field is less than one per cent everywhere exterior to the surface. This is sufficiently close to the theoretical value of zero that a more precise numerical computation of the surface correction field is not necessary. The altitude of the final surface at the subsolar point is seven per cent higher than for the first approximation surface, R_0. The neutral point, where the surface is tangential to the stream velocity vector, is at $r = 1.00\ R_0$ and 18° from the dipole axis. At the antisolar point the surface is nearly circular in cross-section, with a radius of $1.90\ R_0$ at the equator and $1.94\ R_0$ at the meridian. The neutral sheet confirmed by Ness will expand the tail to a larger vertical dimension and lengthen the tail considerably.

The method has been used for several other simpler problems for which solutions exist and converges to the correct answer by the second reiteration (Baker *et al.*, 1964).

Midgley and Davis (1962, 1963) have worked with the boundary condition that the magnetic field in the plasma region must be identically zero. Before the approximations were worked out on a computer, Prof. Davis was disturbed by the failure of the first approximation to satisfy Maxwell's equations ($\nabla \cdot \mathbf{B} = 0$) requiring the field in the plasma to be zero. In treating the case of a uniform plasma pressure surrounding a magnetic dipole, they represented the surface by a polynomial expansion in even powers of latitude angle α

$$\varrho = C\left[1 - \sum_{s=1}^{N} C_s \alpha^{2s}\right] \tag{20}$$

The cocfficients were then determined by requiring the multipole moments of the magnetic field resulting from such a surface to vanish in all lower orders higher than the dipole moment. The vector potential at point r_2 outside the surface is given by an integral over the surface current

$$A(r_2) = \int \frac{J(r)\,\mathrm{d}s}{|r_2 - r|} \tag{21}$$

A may be expanded in Legendre polynomials which after integration become

$$A(r_2) = \hat{\varphi} \sum_{n=1}^{\infty} (2\pi p)^{\frac{1}{2}} \frac{R_0'^{n+2} P_n'(\sin\alpha_2)}{n(n+1)\, r_2^{n+1}} I_n \tag{22}$$

Fig. 13. Contours of the fourth order self-consistent approximation to the magnetospheric boundary projected on a plane perpendicular to the earth sun line.

Fig. 14. Pictorial view of the magnetospheric boundary in fourth order self-consistent approximation.

where

$$(R_0')^3 = M(8\pi p_0)^{-\frac{1}{2}}$$

and

$$I_n = \int_{-\pi/2}^{\pi/2} \varrho^{n+1}\left[\varrho^2 + \left(\frac{\mathrm{d}\varrho}{\mathrm{d}\alpha}\right)^2\right]^{\frac{1}{2}} P_n'(\sin\alpha)\cos\alpha\,\mathrm{d}\alpha \tag{23}$$

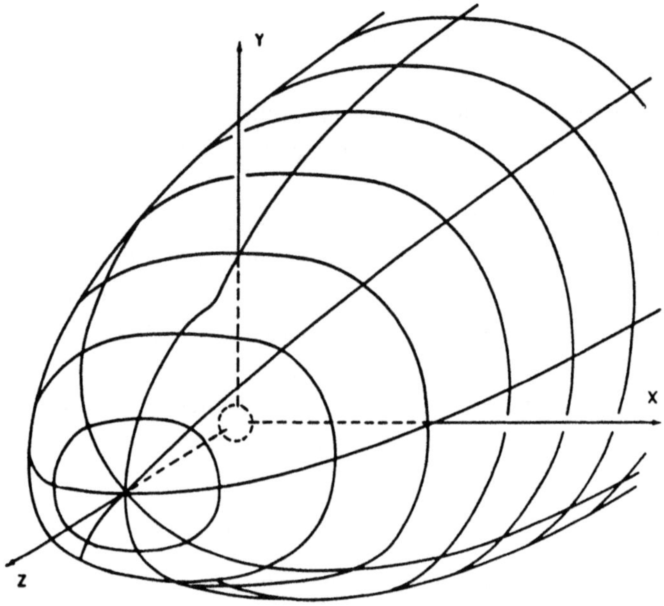

Fig. 15. Pictorial view of the magnetospheric boundary as computed by Midgley and Davis using a moment technique.

Only odd numbered n are permitted because of symmetry considerations. The I_n are numerically computed, and the c_s are adjusted so that all I_n up to $n=15$ are zero except I_1, the dipole term. The vector potential of a dipole is $M(R_0')^{-2}P_1(\sin\alpha_2)$, and therefore in order for the surface current field to cancel the geomagnetic dipole at all $r_2>\rho$, $I_1=4$, and thus C is determined.

Midgley and Davis thus obtained an exceptionally good approximate solution to the particular problem of a dipole immersed in a plasma exerting uniform pressure.

The extension of Midgley and Davis' multipole expansion method to the solar wind problem is very much more complicated, not only because the vector potential must be expanded in associated spherical harmonics rather than Legendre polynomials, but because the surface current is not constant in this problem. By using a generalized Newton's method, they have obtained a surface for which the lower undesired moments were made zero. Their results (MIDGLEY and DAVIS, 1963) are illustrated in Figures 11 and 15. The magnetic ratios outside the surface are very much inferior to (that is, larger than) the simpler problem they solved, being typically 10–100 percent in the meridian plane.

FREE SURFACE SOLUTION OF SLUTZ

SLUTZ (1962) attacked the uniform plasma pressure problem at the same time as Midgley and Davis by a different technique. He used a method developed for hydrodynamic problems by TREFFTZ (1916) in finding a surface free to move in seeking equilibrium between pressures on its two sides. He described the magnetic field interior to the surface by a scalar potential

$$B = -\operatorname{grad} \Psi \tag{24}$$

If (24) is integrated along a meridian line and the magnetic pressure interior to the surface is equated to the particle pressure outside, the result is

$$\Psi = \int (8\pi p_0)^{\frac{1}{2}} \mathrm{d}\zeta = (8\pi p_0)^{\frac{1}{2}} \zeta \tag{25}$$

since B is constant and tangential to the surface. Using Green's theorem for the potential at any interior point

$$\Psi = \frac{M \cos\theta}{r^2} + \frac{1}{4\pi} \int_s \left[\frac{1}{r} \frac{\partial \Psi}{\partial n} \Big|_s - \Psi \Big|_s \frac{\partial}{\partial n} \left(\frac{1}{r} \right) \right] \mathrm{d}s \tag{26}$$

where the first term is the dipole source term inside the surface, r is the distance from t to the surface point s, and n is the outward normal direction to the surface. Since B is tangential to the surface, the first term in the integrand is zero, and by definition of the solid angle

$$(\partial/\partial n)(1/r)\,\mathrm{d}s = -\,\mathrm{d}\Omega$$

A cusp appearing over the magnetic pole creates a difficulty in integrating (26) as t is caused to approach the surface which is surmounted in the following way. t is surrounded with a sphere of radius ε (which approaches zero in the limit), and the integral is separated into that part of the surface outside the sphere S' and that part of the surface S'' inside the sphere. As t approaches infinitesimally close to the surface, and $\varepsilon \to 0$, ψ can be taken outside the S'' integral, since it is constant in this infinitesimal region. Therefore

$$\Psi = \frac{M \cos\theta}{r^2} + \frac{1}{4\pi} \int\limits_{S'} \Psi \mathrm{d}\Omega + \frac{1}{4\pi} \Psi \int\limits_{S''} \mathrm{d}\Omega$$

$$= \frac{M \cos\theta}{r^2} + \frac{1}{4\pi} \int\limits_{S'} \Psi \mathrm{d}\Omega + \frac{1}{4\pi} \Psi \left[4\pi - \int\limits_{S'} \mathrm{d}\Omega \right]$$

We finally obtain

$$\frac{M \cos\theta}{r^2} + \frac{1}{4\pi} \int\limits_{S'} \Psi \mathrm{d}\Omega - \frac{1}{4\pi} \Psi \int\limits_{S'} \mathrm{d}\Omega = 0 \tag{27}$$

If all distances are put in units of $R'_0 = (M/\sqrt{8\pi p_0})^{1/3}$, Equation 24 and 27 become

$$\frac{\cos\theta}{r^2} + \frac{1}{4\pi} \int\limits_{S'} \Psi \mathrm{d}\Omega - \frac{1}{4\pi} \Psi \int\limits_{S'} \mathrm{d}\Omega = 0 \tag{28}$$

$$\Psi = \zeta \tag{29}$$

The surface was divided into areas subtending equal arcs at the origin, and (28) and (29) were solved by averaging the potential over these areas. If an arbitrary surface shape is assumed, (29) will yield the potential at points along the meridian, which can then be substituted into (28) as a test of how good a surface shape was assumed. The process was made to converge to a satisfactory surface by a steepest descent method in that the changes in (28) were studied as the assumed surface points were varied. A surface was obtained which differed from Midgley and Davis' solution by as little as 1 per cent at the pole and only 3 per cent at the equator. Slutz has also applied his method to the solar wind problem, but it was necessary for him to limit the surface on the antisolar side by adding a small uniform plasma pressure. His present result is illustrated in Figure 16.

The calculations discussed above are only of the magnetosphere boundary, that is, of the boundary of the geomagnetic field with its trapped particles and attendant phenomena. The calculations neglect the side pressures arising from the compressed interplanetary magnetic field and the thermal motion of the incident solar wind plasma. Except for the tail, however, the calculations as given are correct, since, as discussed at the beginning, the pressure with or without an interplanetary magnetic field is the same function on the front surface and only differs around in the back where the stream pressure is only one or two percent of its stagnation point value.

I would like to return to my earlier discussion in the introduction of this paper. The important difference which the interplanetary magnetic field makes, is to tie the plasma together in a manner similar to but by no means identical to the manner in which collisions couple the particles together in ordinary hydrodynamics. AXFORD (1962), KELLOGG (1962), SPREITER and JONES (1963), and LEES (1964) have, with increasing sophistication, applied conventional

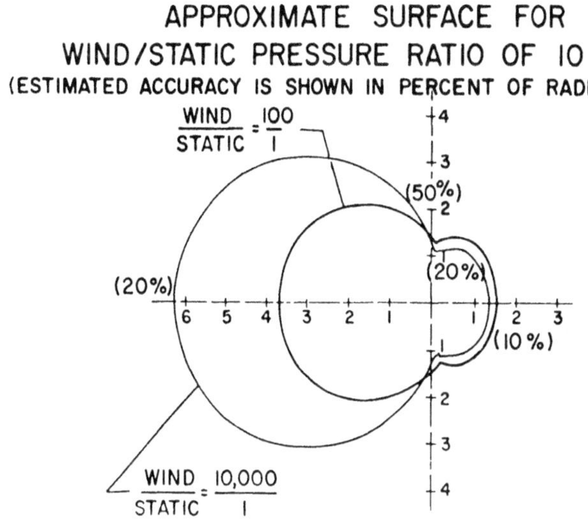

Fig. 16. Cross-sections in the noon meridian of the magnetosphere as computed by Slutz free surface method. A small constant static pressure has been added to the solar wind pressure.

shock theory to this problem and have very successfully predicted the shape of the magnetosheath, the transition layer of compressed magnetic field and plasma on the surface of the magnetosphere which is generally referred to as a shock layer or shock structure. Figure 17 shows a frequently quoted result of SPREITER and JONES (1963) for the magnetosheath; this is a refinement of an earlier calculation by KELLOGG (1962). Kellogg used a hemisphere and cylinder for the magnetosphere instead of the calculated shape and treated the solar wind as an incompressible fluid. Spreiter and Jones obtained a somewhat better result by improving on both of these assumptions.

Many very interesting suggestions have been made about the magnetosheath by DESSLER (1964), Petschek and his collaborators at Avco (LEVY *et al.*, 1964), and by DUNGEY (1963) to mention only a few; some of which you have heard here yesterday. Instead of pursuing this subject further, I would like to conclude by limiting myself to one additional very important process which I believe, in addition to possible turbulence, wave motion, and other effects, must occur in the outer shock boundary (BEARD, 1965a). This concerns the creation of energetic electrons within the magnetosheath. FERRARO (1952), and especially DUNGEY (1958) and ROSENBLUTH (1957), followed by BEARD

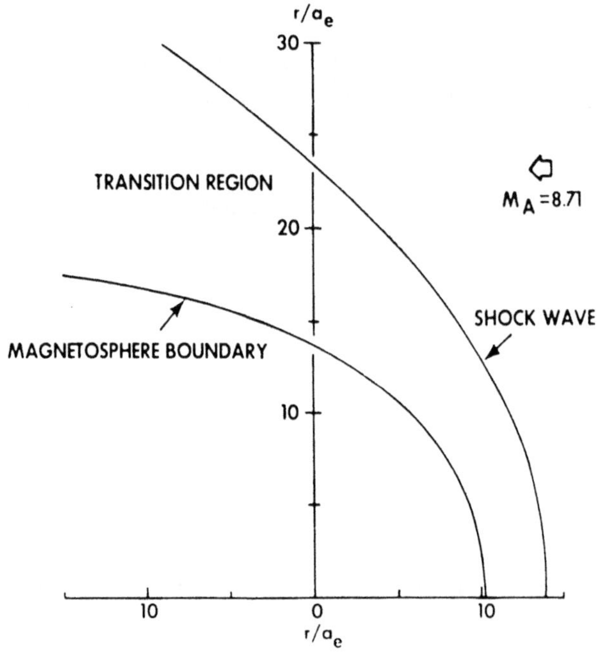

Fig. 17. Traces of the rotated first approximation equatorial magnetospheric boundary and the gas hydrodynamic shock wave on the surface (after SPREITER and JONES, 1963).

(1960), GRAD (1961), SESTERO (1964), and others, have all considered the electric field due to charge separation whenever a plasma is excluded by a magnetic field in motion relative to the plasma velocity frame. In the case of the magnetosheath the plasma moves into the compressed field region whose velocity frame differs from the plasma stream. We could then anticipate that the plasma would move through an electric field due to charge separation which would accelerate the electrons and decelerate the protons.

Figure 18 illustrates the coordinate system I wish to use in discussing a simplified problem of an entrapped field in a moving plasma in which the entrapped field is perpendicular to the stream motion and the normal to the boundary.

The way the field trapped in the plasma is maintained is, of course, by the small gyromotion of the plasma particles in the velocity frame of the stream. In a stationary frame the particles have a wavelike trajectory of small amplitude.

The equations of motion of the particles crossing the boundary are:

$$\ddot{x} = \frac{q}{m} E(x) + \frac{q}{mc} H\dot{y} \tag{30}$$

$$\ddot{y} = -\frac{q}{mc} H(\dot{x} - v_m) \tag{31}$$

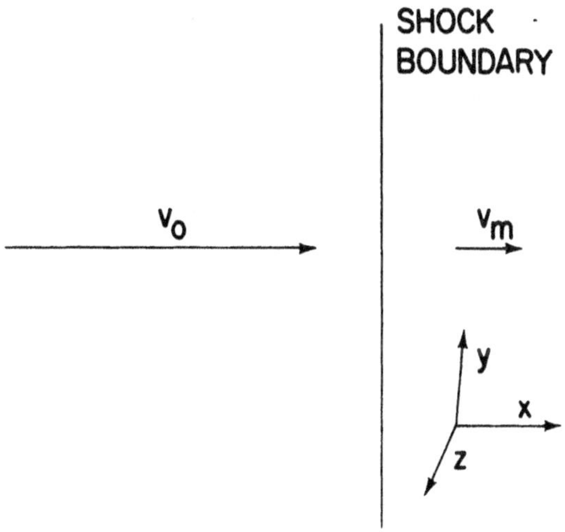

Fig. 18. The coordinate system used in analyzing particle motion in a magnetic field parallel to the Z axis whose field lines travel with a velocity v_0 to the left of the shock boundary and with velocity v_m to the right of the shock boundary.

q and m are the charge and mass of the particle, E is the electric field perpendicular to the boundary, H is the magnetic field, and v_m is the velocity of the guiding centers (the magnetic lines of force). We easily integrate Equation 31 to obtain

$$\dot{y} = \frac{-q}{mc} \int_{-\infty}^{t} H(\dot{x} - v_m)\, dt' \tag{32}$$

Hence the y component of momentum is the same for both electrons and protons. *That is, the component of the particles' momentum perpendicular to the stream velocity is always equal.*

The electric potential drop across the boundary if deceleration were neglected would be much larger than the kinetic energy of the proton stream. The problem may be solved by expanding in powers of m_e/m_i, the ratio of

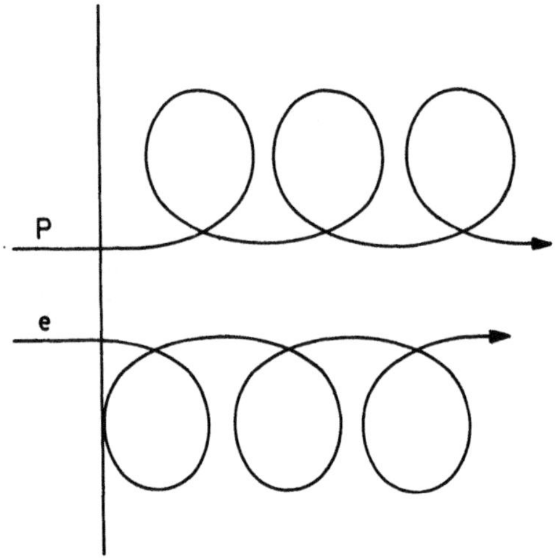

Fig. 19. Electron and proton trajectories initially travelling with a moving magnetic field which is slowed down and compressed downstream from the shock boundary. The magnetic field is perpendicular to the plane of the figure.

electron mass to proton mass, which means that the electric field is a higher order effect. In zero order the acceleration in the x-direction of the electrons and protons due to the combined electric and magnetic fields is zero so that in first order

$$E(x) = \frac{e}{m_e c^2} H \int_{-\infty}^{t} H(\dot{x} - v_m)\,\mathrm{d}t' \tag{33}$$

and

$$\dot{x}_i^2 = v_0^2 - \frac{2e}{m_i} \int_{-\infty}^{x} E(x')\,\mathrm{d}x' \tag{34}$$

Differentiating this last expression with respect to x, we obtain

$$eE(x) = \frac{\mathrm{d}}{\mathrm{d}x}\left(\tfrac{1}{2} m_i \dot{x}_i^2\right) \tag{35}$$

A change in stream kinetic energy is always accompanied by an electric field due to charge separation such that the stream kinetic energy plus the electron kinetic energy perpendicular to the field lines is a constant. If X_m is the position at which the incident proton first reaches the final guiding center velocity, V_m,

$$\tfrac{1}{2}m_i(v_0^2 - V_m^2) = e \int_{-\infty}^{x_m} E(x')\,\mathrm{d}x' \tag{36}$$

A free hand drawing of the particle motion across the boundary is illustrated in Figure 19.

Note that in Figure 5 the plasma in the magnetosheath is shown being accelerated away from the stagnation point. As a result, there is an electric field throughout the magnetosheath which decelerates the electrons. The potential drop along a path within the magnetosheath is proportional to the square of the guiding center velocity, $v_0^2 \sin^2\psi$. Since the field on the outer boundary, like the pressure, is proportional to $\cos^2\psi$, the line integral of the electric field around any closed path is zero as it must be. Near the stagnation point there should be a large flux ($\sim 10^{10}/\text{cm}^2$ sec) of near kilovolt electrons. The same phenomena should occur in the heads of large comets, and it appears very likely that ionization by electron impact is the dominant process occurring in the creation of type 1 comet tails (Beard, 1964c, 1965b).

Bibliography

Axford, W. I.: 1962, *J. Geophys. Res.* **67**, 3791.

Baker, J. C., Beard, D. B., and Young, J. C.: 1964, *Phys. Fluids* **7**, 504.

Beard, D. B.: 1960, *J. Geophys. Res.* **65**, 3559.

Beard, D. B.: 1964a, *Rev. Geophys.* **2**, 335.

Beard, D. B.: 1964b, *J. Geophys. Res.* **69**, 1159.

Beard, D. B.: 1964c, Conference on the Solar Wind, April 1–4 Pasadena California, sponsored by the Calif. Inst. of Tech. and the Jet Propulsion Lab, proceedings to be published.

Beard, D. B.: 1965a, *J. Geophys. Res.* **70**, 4181.

Beard, D. B.: 1965b, to appear in *Planet. Space Sci.*

Chapman, S. and Ferraro, V. C. A.: 1931, 1932, 1933, 1940, *Terrest. Magnetism Atmos. Elec.* **36**, 77, 171; **37**, 147, 421; **38**, 79; and **45**, 245.

Dessler, A. V.: 1964, *J. Geophys. Res.* **69**, 3913.

Dungey, J. W.: 1958, *Cosmic Electrodynamics,* Cambr. Univ. Press, London.

Dungey, J. W.: 1963, *Planet. Space Sci.* **10**, 233.

Ferraro, V. C. A.: 1952, *J. Geophys. Res.* **57**, 15.

Grad, H.: 1961, *Phys. Fluids* **4**, 1366.

Kellogg, P. J.: 1962, *J. Geophys. Res.* **67**, 3805.

Lees, L.: 1964, Proc. paper presented at AIAA New York meeting, Jan.

Levy, R. H., Petschek, H. E., and Siscoe, G. L.: 1964, *AIAA Journal* **2**, 2065.

Mead, G. D. and Beard, D. B.: 1964, *J. Geophys. Res.* **69**, 7.

Midgley, J. and Davis, L. Jr.: 1962, *J. Geophys. Res.* **67**, 499.

Midgley, J. and Davis, L. Jr.: 1963, *J. Geophys. Res.* **68**, 5111.

ROSENBLUTH, M. N.: 1957, *Magnetohydrodynamics* (ed. by R. LANDSHOFF), Stanford Univ. Press, Palo Alto, California.
SESTERO, A.: 1964, *Phys. Fluids* **7**, 44.
SLUTZ, R. J.: 1962, *J. Geophys. Res.* **67**, 505.
SPREITER, J. R. and BRIGGS, B. R.: 1962, *J. Geophys. Res.* **67**, 37.
SPREITER, J. R. and JONES, W. P.: 1963, *J. Geophys. Res.* **68**, 3555.
TREFFT, E.: 1916, *Z. Math. Phys.* **64**, 34.

Discussion

DESSLER and others discussed experimental measurements that should be made, e.g., he desires that the temperature of the ions and electrons be measured just behind the shock on the sub-solar side. In the Dessler-Michel model there is thermodynamic equilibrium giving the protons and electrons equal energy. BEARD stated that the electrons would have a higher temperature than the protons. COLGATE stated that the ratio of electron to proton energy is sensitive to the Mach number, that there would not be thermodynamic equilibrium, and emphasized that space is much better than the laboratory for making measurements on collisionless shocks. Therefore, it is necessary to measure the ratio of energies at many locations.

DUNGEY pointed out that spatial and temporal variations can probably not be separated without 2 or more satellites separated by reasonable distances, certainly less than 1000 km.

A "fluid" picture of the magnetosheath treated in terms of standard hydrodynamic shock theory was presented by J. SPREITER in contrast to the particle model which was given by Beard. Spreiter gave velocity profiles and particle temperatures throughout the magnetosheath on the solar side. The particle thermalization which warrants the use of standard aerodynamics is assumed to arise from unspecified randomizing processes such as various kinds of plasma instabilities. VAN ALLEN asked if the fluid model provided for particle injections on the day side, and in fact no model seems to provide for particle injection. DUNGEY felt that if you have a magnetic field model you can compute backwards and see if the particle comes from the boundary.

G. MEAD presented a very good summary of the relation between the stable trapping boundary, closed field lines, and aurora phenomena, which is reproduced below.

THE MOTION OF TRAPPED PARTICLES IN A DISTORTED FIELD

GILBERT D. MEAD
Goddard Space Flight Center, Greenbelt, Maryland

Abstract. The motion of charged particles in a geomagnetic field distorted by the solar wind is examined, using adiabatic theory. It is shown that the drift motion of particles trapped on high L-shells depends strongly on their equatorial pitch angle. Those mirroring near the equator move in on the night side, while those mirroring at low altitudes move out, thus producing a significant splitting of the L-shell with pitch angle. One consequence is that the very high latitude lines on the day side may contain trapped particles near the equator, but none near the ends of the lines. The opening up of field lines in the geomagnetic tail may be the controlling factor governing the high-latitude limit of trapping on both the day side and the night side.

1. Introduction

The theory of adiabatic invariants enables one to determine the motion of trapped particles, so long as a complete description of the magnetic field is available. For the inner radiation belt, i.e., for $L \lesssim 2$, the internal earth's field dominates, and very accurate ordering of the data has been possible by using a spherical harmonic expansion of the earth's field with internal source terms only. In fact, the accuracy in this region seems to be limited only by the uncertainty in the values of the coefficients in the expansion. As more accurate magnetic field data are obtained, the resulting improvement in the field model is expected to increase even further our ability to order the trapped particle data in the inner belt.

However, at higher L values, the effect of external current sources becomes more and more important. Experimenters have found that beyond $L \approx 5$, a two-dimensional coordinate system based on a field model with internal field sources only is no longer able to order the data. With polar-orbiting satellites, particle fluxes are strongly dependent upon local time, in addition to the usual B and L coordinates. Satellites in highly eccentric orbits find large time variations in particle fluxes which depend upon magnetic activity, in addition to variations with local time which occur even during magnetically quiet times.

In this paper we shall examine the consequences of adiabatic theory in regions of the magnetosphere where external current sources produce significant distortions of the geomagnetic field. The variation of particle fluxes with local time will be investigated. We shall also demonstrate the substantial splitting of L shells as a function of equatorial pitch angle. That is, it will be shown that particles at large L which mirror near the equator (i.e., with large equatorial pitch angles) follow substantially different drift paths from those particles which mirror near the earth (i.e., with small equatorial pitch angles). One consequence of this is that field lines at large L on the day side of the earth may be only partially loaded with trapped particles. The absence of trapped particles at the ends of these lines, near the earth, is related to the inability of these particles to drift completely around the earth without entering the geomagnetic tail, where they no longer remain trapped.

2. Model of the Distorted Field

The distortions of the geomagnetic field at large altitudes are primarily due to two sources: currents at the magnetosphere boundary and currents in the tail

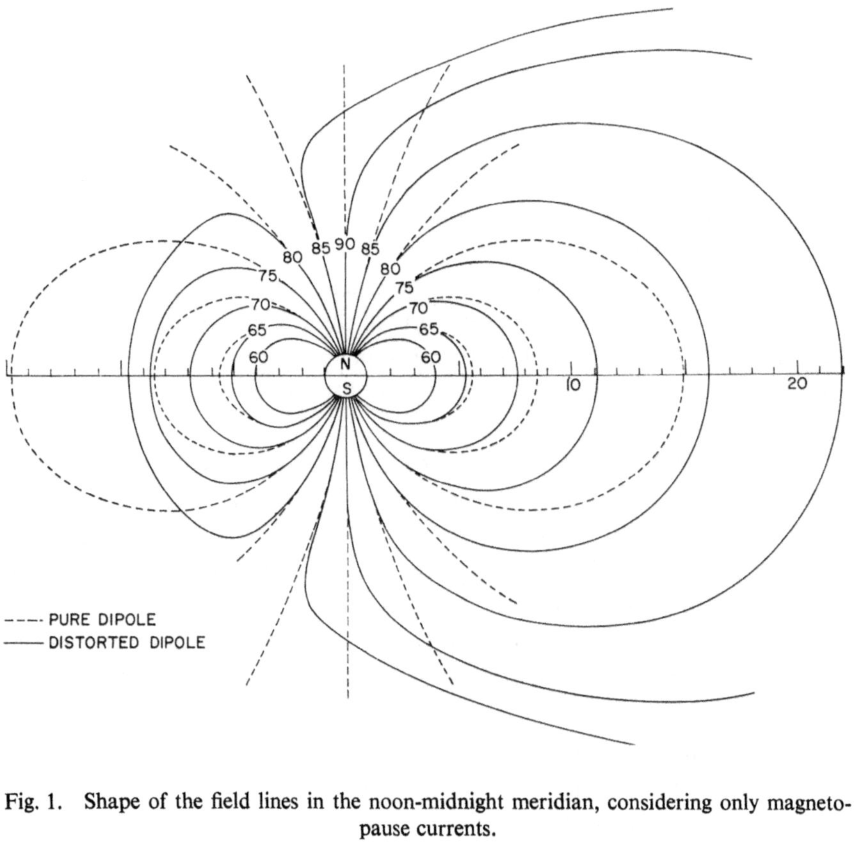

Fig. 1. Shape of the field lines in the noon-midnight meridian, considering only magnetopause currents.

of the magnetosphere. We shall ignore here currents due to the drift motions of the trapped particles themselves ("ring currents"), which are generally small during magnetically quiet times, and ionosphere currents, which in this region of space represent internal sources whose field falls off rapidly along with the earth's main field.

The effect of magnetopause currents has been calculated in some detail (MEAD, 1964). Basically, the coefficients of a spherical harmonic expansion with external source terms have been determined, using the solution to the Chapman-Ferraro problem as given by MEAD and BEARD (1964). This expansion is then used to calculate field topologies, field line distortions, and the effect of changes in the boundary position. The topology of the distorted field in the noon-midnight meridian plane is shown in Figure 1, and Figure 2 shows a comparison of this distorted field with the experimental measurements made by Cahill on a sample pass of Explorer XII. It is seen that the predicted distortions are in reasonable agreement with the measurements.

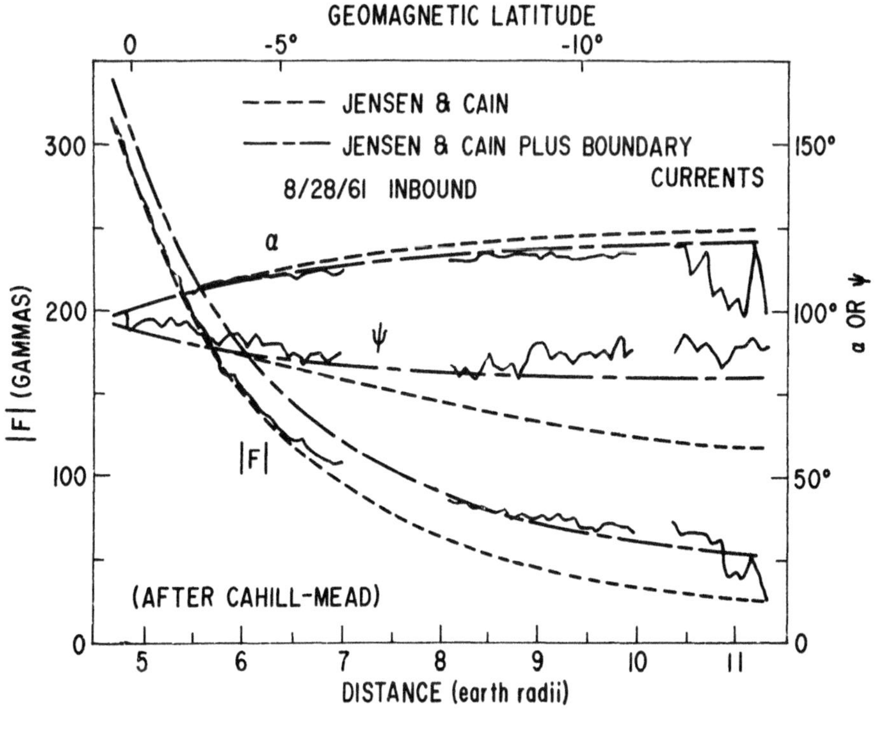

Fig. 2. Comparison of model of distorted field with experimental data from Explorer XII. $|F|$ is the field magnitude, and α and ψ determine its direction in spacecraft coordinates (CAHILL and AMAZEEN, 1963). This pass is near the noon meridian.

The expression for the distorted field in the equatorial plane is particularly simple if we restrict ourselves to the first two terms in the external source expansion, which turn out to be the most important ones:

$$B_{\mathrm{eq}} = \frac{M}{r^3} + \frac{a_1}{r_{\mathrm{b}}^3} + \frac{a_2}{r_{\mathrm{b}}^4} r \cos \varphi \tag{1}$$

Here $M=0.31$ gauss (the equatorial value of the dipole field at the earth's surface), r_{b} is the distance to the magnetopause in the solar direction, φ is the azimuthal angle as measured from the noon meridian, and a_1 and a_2 are the first two non-vanishing external coefficients (note that φ and a_2 are defined somewhat differently in MEAD (1964)). Similar but more complicated expressions give the three components of the field off the equator. The values of the coefficients are shown in Table I for Mead's model and for an image dipole model (PARKER, 1960).

TABLE I

Values of the coefficients a_1 and a_2 in the external source expansion of the distorted field

	a_1	a_2
Image Dipole (PARKER, 1960)	$\frac{M}{8}=0.04$ gauss	$\frac{3M}{16}=0.06$ gauss
MEAD (1964)	$0.816\ M=0.25$ gauss	$0.673\ M=0.21$ gauss

The expansion of the field near the earth is the same in both models, except for the values of the coefficients. In the image dipole model, r_{b} is the distance to the infinite plane surface. Table I indicates that for a given value of r_{b}, the image dipole model greatly underestimates the magnitude of the distortion near the earth.

From Equation (1) the ratio of external field to internal field is given by

$$\frac{\langle B_{\mathrm{ext}} \rangle}{B_{\mathrm{int}}} = \frac{a_1}{M} \frac{r^3}{r_{\mathrm{b}}^3} \approx 0.8 \frac{r^3}{r_{\mathrm{b}}^3} \tag{2}$$

where $\langle B_{\mathrm{ext}} \rangle$ is the external field averaged over local time. This ratio is about

0.1% at the earth's surface, 2.5% at $3R_E$, 10% at $5R_E$, and 40% at $8R_E$, assuming a boundary at $10R_E$. Thus the external terms rapidly become important as the outer regions of the magnetosphere are approached.

Recent experimental measurements (NESS, 1965) have shown that this model is incapable of accurately describing the topology of the field in the tail region.

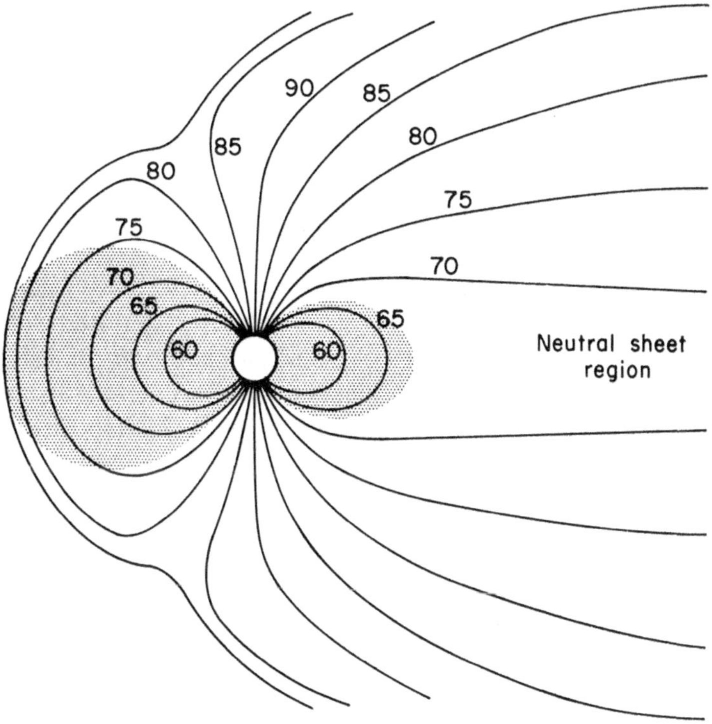

Fig. 3. Model of distorted field including a current sheet in the tail. The shaded region represents the region of stable trapping.

In the region beyond about $10R_E$ near the midnight meridian, a "neutral sheet" has been found which separates antisolar directed fields in the southern hemisphere from solar directed fields in the northern hemisphere. The presence of this neutral sheet implies the existence of strong plasma currents inside the magnetosphere. Since the Chapman-Ferraro problem assumes that no plasma exists inside the boundary, the original model must be modified to include effects in the magnetosphere tail. This can be done in a somewhat ad hoc manner by calculating the field due to a current sheet in the tail and vectorially adding this to the field obtained from the Chapman-Ferraro problem, as was done by WILLIAMS and MEAD (1965). The general effect is to stretch out the lines of force on the night side. Lines of force at high latitudes are no longer closed, but stretch out into the tail. The configuration in the noon-midnight meridian is shown in Figure 3. With the strength of the tail field used in this calculation, all lines above approximately 67° at the earth's surface in the midnight meridian are open, and thus cannot hold stably trapped particles. The critical latitude separating open field lines from closed field lines will decrease if the tail field increases. Such increases have been observed by Ness during magnetically disturbed periods. Thus one would expect the upper trapping boundary on the night side to move to lower latitudes during magnetic disturbances.

3. Motion of Trapped Particles in a Distorted Field

In order to study the motion of particles trapped in a distorted field, we shall assume that adiabatic theory is valid and that the energy of the particles remains constant. This is equivalent to assuming a static magnetic field and the absence of electric fields. This situation is expected to be valid for relatively energetic particles during magnetically quiet times. Any electric fields present will affect primarily the lower-energy particles, i.e., those with energies up to 50 or 100 keV. Above this energy, electric fields do not seem to be an important factor.

We shall consider two kinds of particles: those whose mirror points are near the magnetic equator and those which mirror near the ends of the field lines at low altitudes. The drift motion of these two classes of particles will be substantially different.

A consequence of the conservation of the first invariant, μ, is that a particle will always mirror at the same value of magnetic field strength, B_m. For particles that mirror near the equator, therefore, the second invariant $J \approx 0$, and the drift paths will follow contours of constant field strength along the equator. Since the field is compressed by the solar wind on the day side, at a given radial distance the field is stronger here than on the night side. Thus in order to remain on contours of constant field strength, particles must drift closer to the earth on the night side. An estimate of the magnitude of this effect is obtained by equating the field magnitude (in gammas) given by Equation (1) in the subsolar and antisolar direction:

$$\frac{31\,000}{r_d^3} + 2.1\, r_d = \frac{31\,000}{r_n^3} - 2.1\, r_n \tag{3}$$

where r_d and r_n are the dayside and nightside distances, r_b is assumed to be $10R_E$, and Mead's value for a_2 is used. Some typical values of r_n versus r_d are given in Table II, showing that the effect is strongly dependent upon radial distance.

TABLE II

Regions of constant magnetic field intensity in the solar direction (r_d) and antisolar direction (r_n)

r_d	r_n	Δr
10	7.7	2.3
8	7.0	1.0
6	5.7	0.3
4	3.96	0.04

Particles trapped near the magnetopause on the solar side will move several earth radii closer to the earth as they drift around to the night side. FRANK *et al.* (1963) have found that contours of constant omnidirectional fluxes of electrons >40 keV are displaced towards the sun by distances corresponding roughly to those given in Table II. That is, fluxes fall off more rapidly on the night side than on the day side, in accordance with what one might expect from considerations of particle drifts.

Particles on high L-shells with small equatorial pitch angle, i.e., those that mirror near the earth, will follow substantially different drift paths. They will mirror at approximately constant altitude as they drift around the earth, since the external source terms produce only negligible change in the field magnitude at low altitudes. Assuming constant energy, the second invariant requires that

$$I = \int_{M}^{M^*} \left(1 - \frac{B}{B_m}\right)^{\frac{1}{2}} ds \tag{4}$$

remain constant. B_m is the field magnitude at the mirror point, and the line integral is taken along the field line between mirror points. For high-latitude lines, the value of the integrand is essentially unity over almost the entire length of the line (for particles mirroring at 1000 km, $(1-B/B_m)^{\frac{1}{2}}=0.90$ at $2R_E$ and

0.99 at $3.5 R_E$). Therefore, at high latitude I is essentially a measure of the length of the field line between mirror points. Thus we reach the conclusion that *particles on high L-shells that mirror near the earth drift to field lines with constant length between mirror points.*

A necessary consequence of this is that the equatorial crossing point of these particles moves *out* on the night side, rather than in. This can be seen by examining the field lines in Figure 4. One is a pure dipole line at high latitude. The

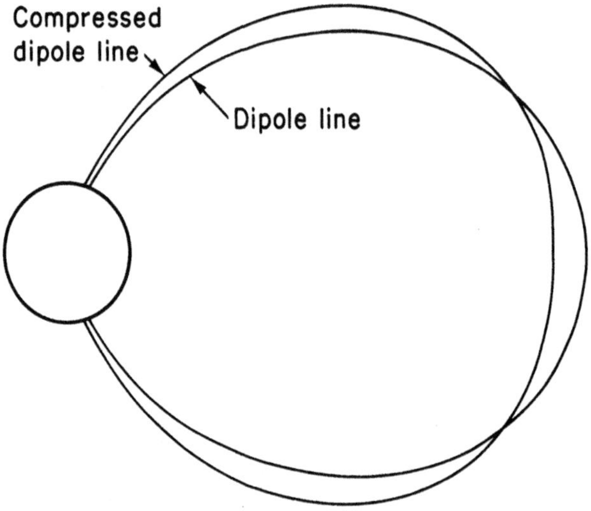

Fig. 4. Shapes of pure dipole and compressed dipole lines having approximately the same length between low-altitude mirror points and same value of I. The compressed line crosses the equator closer to the earth, but enters the earth at a higher latitude.

other is compressed in a manner similar to that expected on the day side. Both lines have approximately the same length and the same value of I for particles mirroring near the earth. The shapes of the two lines, however, are different. A line which is compressed enters the earth at a higher latitude, but crosses the equator at a lower altitude, than a pure dipole line of the same length. The opposite holds for lines which may be extended. The change in latitude at the earth's surface with local time is quite comparable to that actually observed with polar-orbiting satellites (WILLIAMS and MEAD, 1965).

Thus the drift paths of particles trapped on high-latitude field lines depend strongly on their equatorial pitch angle. Those with large pitch angle, i.e., those that mirror near the equator, drift to lower altitudes on the night side, the difference in altitude amounting to as much as several earth radii. Those with small equatorial pitch angles mirroring near the earth move out on the night side. This is equivalent to a splitting of L shells with pitch angle. The splitting appears to be much more than the maximum of about 1% that McIlwain (1961) found using internal source terms only for the earth's field.

We turn now to another question: what controls the high-latitude limit of trapping on the day side? For the most part, trapped particles are not observed with polar-orbiting satellites beyond an invariant latitude of about 70–75° on the day side, the exact upper limit being somewhat variable and dependent upon the threshold energy of the detector. It has often been assumed that this latitude corresponds to that separating the field lines crossing the equator on the day side from those which pass over the pole and are drawn back into the tail. However, Table III shows that this cannot be the case.

TABLE III

Latitude where field line enters earth vs. equatorial crossing distance on the noon meridian for a pure dipole and a distorted dipole with a magnetosphere boundary at 10 R_E

Equatorial Crossing Distance	*Latitude at Earth's Surface*	
	Pure Dipole	*Distorted Dipole*
4	60.0	60.4
5	63.4	64.4
6	65.9	67.6
7	67.8	70.3
8	69.3	73.0
9	70.5	76.5
10	71.6	82.0

An undistorted dipole field line crossing the equator at $10R_E$ would enter the earth at a latitude of 71.5°. On the day side the lines are strongly compressed, and the line crossing at $10R_E$ actually enters the earth at a much higher latitude, most likely above 80° (Mead, 1964). Thus there exists a bundle of field lines between approximately 70° and 80° on the day side which contain trapped particles in the equatorial regions, but which seem to be devoid of particles near the earth.

The most likely explanation of this is found in the phenomenon of L-shell splitting. Those particles trapped on this bundle of field lines whose mirror points are near the equator will drift closer to the earth on the night side and stay in the region of the magnetosphere where the field lines remain closed. Those which mirror near the earth, however, move out on the dark side and find themselves in the region of the tail, where the field lines are no longer closed. Since complete drift paths around the earth are not available to these particles, they become lost and this region is relatively free of trapped radiation.

It thus appears that the geomagnetic tail may be the controlling factor governing the high-latitude limit of trapping. On the night side particles are trapped on all field lines up to about 65–70°. Beyond this latitude field lines are drawn back into the tail. The upper limit is several degrees higher on the day side, the difference corresponding to the shift in latitude as particles drift around the earth. The trapping region in the noon-midnight meridian according to this picture is shown in Figure 3. During periods of magnetic activity, the tail field increases, more field lines open up on the night side, and the upper trapping limit moves to lower latitudes on both the day side and the night side (NESS and WILLIAMS, 1965).

Finally, we note that the auroral zone is located at or near the upper limit of trapping, both on the day side and the night side. On the night side, this upper limit is in just that region where field lines are beginning to open up into the tail. However, on the day side, auroras and electron precipitation are often observed in latitudes of 65–70°. The field lines here extend well inside the trapping region, crossing the equator on the day side between 5 and 7 earth radii (see Table III). They are definitely not drawn back over the poles into the geomagnetic tail. Any theory of auroras or electron precipitation on the day side must take this into account.

References

CAHILL, L. J. and AMAZEEN, P. G.: 1963, *J. Geophys. Res.* **68**, 1835–1843.

FRANK, L. A., VAN ALLEN, J. A., and MACAGNO, E.: 1963, *J. Geophys. Res.* **68**, 3543–3554.

MCILWAIN, C. E.: 1961, *J. Geophys. Res.* **66**, 3681–3691.

MEAD, G. D.: 1964, *J. Geophys. Res.* **69**, 1181–1195.

MEAD, G. D. and BEARD, D. B.: 1964, *J. Geophys. Res.* **69**, 1169–1179.

NESS, N. F.: 1965, *J. Geophys. Res.* **70**, 2989–3005.

NESS, N. F. and WILLIAMS, D. J.: 1966 *J. Geophys. Res.* **71**, 322–325.

PARKER, E. N.: 1960, *J. Geophys. Res.*, **65**, 3117–3130.

WILLIAMS, D. J. and MEAD, G. D.: 1965, *J. Geophys. Res.* **70**, 3017–3029.

SOME PLASMA INSTABILITIES OF THE MAGNETOSPHERE

DAVID B. CHANG

General Atomic Division of General Dynamics Corporation, John Jay Hopkins Laboratory for Pure and Applied Science, San Diego, California

Abstract. Four plasma instabilities which may operate in planetary magnetospheres are described: (1) the low frequency Rayleigh-Taylor-type flute instability as affected by a resistive ionosphere, a dipole field geometry and by resonant particle-wave interactions, (2) a whistler instability, with the concomitant particle diffusion, which occurs in a plasma containing a small component of flat-helix energetic electrons, (3) a nonconvective high frequency two energy component electrostatic loss-cone instability, and (4) the resistive instabilities occurring in sheared magnetic fields, as affected by the Chapman and Cowling collisional and magnetic viscosities and Hall currents.

1. Introduction

A frightening number of plasma instabilities has been uncovered in the last few years in connection with the work on controlled fusion, and it seems likely that some of these could operate in planetary magnetospheres. I'd like to describe four candidates for magnetospheric instability that M. N. Rosenbluth, L. D. Pearlstein and I have been studying.

(1) The first is the low frequency Rayleigh-Taylor-type flute instability which may occur when a plasma is supported by a magnetic field against some destabilizing force. At first glance it would seem that a planetary radiation belt should fall victim to such an instability since a particle confined in a dipole-like field experiences an outwardly directed inertial force associated with its motion in the curved field. On the other hand, the presence of a conducting ionosphere tends to short out the electric fields which drive the instability. In addition, the geometry of the field itself can play a stabilizing role. We were interested in studying the competition between these various effects to see if a flute instability could ever occur fast enough to cause diffusion of the energetic belt particles.

(2) The second instability concerns disturbances of a much higher frequency, in fact frequencies high enough to resonantly violate the magnetic moments of the energetic trapped particles. Study of this instability was prompted by observations of the sporadic circularly polarized decameter radiation which comes from the planet Jupiter. The sporadic nature of the radiation suggests that it might be associated with some instability, and the circular polarization suggests that it might be an instability of electromagnetic waves propagating along a magnetic field. Other radio observations of Jupiter in the shorter

decimeter wavelengths indicate that the plasma in Jupiter's belts has an energetic electron component comprised mostly of particles with very flat helices. What we have done, then, is to see if a plasma, the bulk of which is thermalized and of low temperature but which contains a small number of quite flat helix energetic particles, can exhibit maser action. It's interesting also to look at what effect this energy exchange between the waves and particles has on the particle distribution, and a quasilinear diffusion equation has been constructed to look at this.

(3) Recently, ROSENBLUTH and POST (1965) have pointed out that there are some high frequency electrostatic plasma instabilities which are peculiar to "loss-cone" particle distributions such as exist in mirror confinement geometries. They treated the case in which there is only one energy component of each species present, and found that in a uniform plasma the instability was convected ($k_{\|} \neq 0$) along the magnetic field. Rosenbluth and Post have pointed out that gradients of the density could lead to nonconvective ($k_{\|} = 0$) instabilities. With the Van Allen belt in mind, we have looked at a loss-cone instability in a plasma containing both high and low energy components, and it turns out that the presence of two components also leads to nonconvective instabilities of rather large growth rates.

(4) The fourth instability studied is a low frequency resistive type which occurs in sheared magnetic fields. This type may have some chance of existing in long twisted magnetospheric tails, and if it does it could play some role in the annihilation and recombination of field lines. The instability owes its existence to the local relaxation of the constraint that the plasma must remain attached to a magnetic field. In the instability, finite conductivity permits the system to go to a lower potential energy by allowing lines of force which are initially distinct to link up. FURTH *et al.* (1963), and JOHNSON *et al.* (1963) have studied these instabilities with hydromagnetic equations which contain a scalar resistivity. For these resistive instabilities, the hydromagnetic equations are only directly applicable when the ion collision frequency is larger than the ion gyro-frequency (when the plasma and magnetic pressures are comparable). Since the mean free paths in magnetospheres are quite long and the plasma and magnetic pressures in the tails may be comparable, such a description would not be directly applicable. We have tried to analyze the instabilities with the Chapman and Cowling plasma equations to see how they behave under a broader range of conditions.

2. Low Frequency Flute

GOLD (1959) and SONNERUP and LAIRD (1963) have looked at the energetics of interchanges in a dipole field, and treating the plasma as a fluid have defined circumstances under which the energy change would permit an interchange instability to occur. By looking at the dynamics of the instability, a dispersion relation can be constructed to see if an interchange can occur fast enough to affect the energetic trapped particles. It appears that the ionosphere could play an important role in this instability, as it would strongly affect any driving electric fields. The ionosphere would behave as a resistive medium for rapidly occurring instability and as a reactive medium for interchanges occurring slowly enough for the neutral ionospheric particles to take part in the motion. (Here we do not consider those situations in which the change in ionospheric gravitational energy is important (See SONNERUP and LAIRD, 1963). The character of the instability would be different in the two cases: Thus, hydromagnetically a reactive ionosphere might be expected to give the usual Rayleigh-Taylor-type dispersion relation in which growth rates are proportional to the square-root of the product of the destabilizing force and the density gradient of the energetic particles which drive the instability. The growth rate in this case would be smaller; the smaller the ratio of the number of energetic driving particles to the number of driven ionospheric particles, and stability of the energetic particles might be expected if this ratio were small enough. A resistive ionosphere, on the other hand, would give a dispersion relation with hydromagnetic growth rates proportional to the effective collision frequency of the resistivity and to the density gradient of the driving energetic particles. Again, stability of the energetic particles might be expected if the ratio of the number of energetic particles to the number of ionospheric particles were small enough.

We have studied the resistive situation in the electrostatic approximation to see if diffusion of the energetic particles could ever result (CHANG *et al.*, 1965). The driving currents of the energetic particles have been obtained from the description of NORTHRUP and TELLER (1960) of particle motion in a mirror system when the magnetic moment and mirror actions are invariant. The polarization and resistive currents have been described by a dielectric tensor of a zero temperature three fluid (ions, electrons, neutrals) plasma model. In this situation, currents due to higher order drifts are not important: even though the frequencies of interest are of the order of the drift frequencies of the energetic particles – so that their finite Larmor radius and polarization drifts are of the same order, their polarization currents are already negligible compared with that of the low energy belt component, and especially compared with the ionospheric currents.

The resulting dispersion relation shows geometry itself to play a stabilizing role, associated with the fact that the flux tubes in a dipole field increase in volume with increasing radial distance. The lowest order solution to the dispersion relation (i.e., the solution obtained by treating the growth rate as being much larger than the azimuthal drift frequencies – the latter being obtained from the time required for a particle to drift through a wavelength of the disturbance) yields the familiar interchange energy $\delta^2 W > 0$ sufficiency – condition for stability (TAYLOR, 1964).* If this condition is violated, or if one is interested in instabilities in which the growth rates are not treated as being much larger than the azimuthal drift frequencies – e.g., those in which resonant interactions between the disturbances and the azimuthal drift velocities play a role, then a Nyquist analysis of the dispersion relation gives another sufficiency condition for stability of the energetic particles which takes into account the effect of a resistive ionosphere. This more general condition is essentially the requirement that the growth rates be less than the azimuthal drift frequencies of the energetic particles.

3. Whistler Maser

To date, three different kinds of radio emission have been detected from Jupiter (ROBERTS, 1963). Below wavelengths of 3 cm, the radiation is thermal, following closely the λ^{-2} dependence of the Rayleigh-Jeans law. In the decimeter region beyond 3 cm, the flux behavior differs considerably from a λ^{-2} law, remaining quite constant between 3 cm and 70 cm. Observations of the intensity, polarization and spatial extent of the decimeter radiation strongly suggest that the decimeter radiation might be synchrotron radiation from energetic electrons in Jovian Van Allen belts. To explain the high degree of linear polarization that is observed, it's necessary to postulate electrons with very flat helices. The third kind of radiation occurs in the decameter band, and is quite different from the other two in that it is quite sporadic. It is observed in the 8 to 80 meter range, occurs in bursts, has features which are strongly correlated with longitude on Jupiter, and 90% of the time is very strongly RH circularly polarized. A variety of mechanisms has been suggested over the years for the origin of this sporadic radiation. The storms have been attributed to thunderstorm activity, atmospheric chemical explosions, plasma oscillations, spin-flip, the focusing by Jupiter's magnetosphere of radio waves from distant sources, magnetic bremsstrahlung from electrons dumped into the Jovian magnetic polar region, and coherent emission by relativistic electrons in Jupiter's Van Allen belts. An instability of whistlers may have some relevance to the last two explanations (CHANG, 1963).

The whistler instability calculation consists first of evaluating an absorption coefficient from a Vlasov equation, and then obtaining the corresponding emissivity by Einstein coefficient arguments. For a plasma the bulk of which is thermalized but which also contains a small number of energetic electrons with very flat helices, it is found that efficient amplification is possible under

* Dungey, Hess and Nakada (see discussion following Van Allen's Aug. 23 talk) have pointed out that this condition seems to be satisfied over a large portion of the earth's magnetosphere. G. Davidson and M. Watt have reached similar conclusions (private communication).

two quite different conditions, when the flat-helix relativistic electrons have a net drift along the field. Amplification occurs for waves travelling in the same direction as the drift. The two conditions are: (1) $\omega_p^2 \gg \omega_c \omega \gg \omega^2$, and (2) $\omega \approx \omega_c$, $\omega_p^2 \gg \omega_c \omega - \omega^2 > 0$. Here, ω, ω_c and ω_p are the whistler frequency, (non-relativistic) electron gyrofrequency, and plasma frequency, respectively. It appears that case (1) may exist in Jupiter's ionosphere. ZHELEZNIAKOV (1958) has estimated from photoionization considerations that the electron density there might be 10^7 cm^{-3}, and it appears that higher densities may be obtained on taking into account the ionizing particle flux from the sun and particles dumped from the belts. For some numbers, suppose the field in the ionosphere is 16 gauss, the whistler frequency desired is 10 Mc/s and the electron density is 10^8 cm^{-3}, say: then $\omega_p^2/\omega\omega_c \approx 13$, and case (1) is satisfied. Beams of particles travelling down the field might also be important in this case. Case (2), on the other hand, might easily exist in the belts themselves. The decameter radiation has been observed between frequencies of 5 and 39 Mc/s, and these correspond to gyrofrequencies for fields between 1.8 and 14 gauss. Fields of this order have already been required to provide the decimeter radiation.

If amplified whistlers are the source of the decameter radiation, then it appears reasonable to associate the sporadic nature of the radiation with the only occasional occurrence of net drift motions along the magnetic field of the relativistic electrons (and possibly in case (1) with the only occasional occurrence of the required plasma densities). A net drift motion might exist, for instance, when increased solar wind activity causes dumping of radiation belt electrons, as might be suggested by the correlation between the Jovian decameter activity and solar events. The predominant presence of right-handed polarization may be attributed to a combination of the preference in drift velocity directions – for instance, dumping would correspond to drifts from the belts toward the planet – and to geometry. The geometrical aspects of the problem have been discussed in detail by WARWICK (1963), FIELD (1963), and CARR (1963). Insertion of reasonable Jovian parameters in the expressions give acceptable emissivities and growth rates, so it seems possible that some such process might supply the decameter radiation.

It is interesting to ask for the effect of these circularly polarized waves on the particles with which they resonantly interact. A Fokker-Planck equation may be constructed to describe the behavior of a group of particles in the presence of damped or growing waves (CHANG, 1964). In this way it can be seen that the behavior of particles which resonate with a wave differs considerably from that of particles which are far from resonance. The nonresonant particles have wave-induced motions which increase or decrease as the amplitude of a wave increases or decreases: they are described by a "diffusion coefficient" which can be either positive or negative. The resonant particles, on the other hand, are exposed to an almost constant electric field which carries them rapidly away from resonance: a real (positive) diffusion coefficient results.

Alternatively, the diffusive behavior of the particles may be studied directly from the Vlasov equation by constructing a quasilinear diffusion equation following the prescription of DRUMMOND and PINES (1962), directing attention to resonant interactions in which some doppler-shifted wave frequency is equal to a particle's gyrofrequency (CHANG and PEARLSTEIN, 1965).

If we assume that the dispersion relation of the waves has the form $\omega \sim k^b$ – e.g., for whistlers $b=2$ and for Alfvén waves, $b=1$, and if we assume that the energy spectrum of the waves goes like $k^{-\alpha}$ – for whistlers resulting from lightning discharges, Dungey has suggested an α close to unity, whereas Alvfén waves might have this form if they result from some turbulence spectrum, and if we make the approximations that the particles can be treated non-relativistically, that the phase velocities are much less than the particle velocities, and that only particles in flat helices are to be considered; then the diffusion equation takes a very simple form in which energy changes may be ignored compared to changes in the helix angles. If particles are inserted initially at one helix angle, then this equation shows the peak of the distribution to move to flatter helices while at the same time the particles leak off to steeper helices where they disappear into a loss cone. Eventually the simple equation doesn't apply since the derivatives with respect to energy cannot be ignored compared to the derivatives with respect to helix angle when the latter become small. The steady state diffusion equation keeping both the energy and angle derivatives has as possible solutions some very flat helix distributions. These may have interesting properties, e.g., it's possible to have distributions in which energy and helix angle are strongly correlated.

4. Two-Component Electrostatic Loss-Cone

The dispersion relation of the high frequency flute-like two energy component loss-cone instability in a uniform magnetic field is shown below in dimensionless variables (PEARLSTEIN *et al.*, 1965)

$$-\left(\frac{K}{1-\Delta}+\frac{1}{1-y^2}\right)\frac{\tan \pi y}{y}=\frac{\Delta}{1-\Delta}\frac{1}{|kR|^3} \tag{1}$$

For a situation in which there are n_H+n_L low energy electrons, n_L low energy ions and n_H energetic ions in a loss cone distribution,

$$y=\frac{\omega}{\Omega_i} \qquad K=\frac{m_e}{m_i}+\frac{\Omega_i^2}{\omega_{pi}^2} \tag{2}$$

$$R=\frac{\bar{V}_i}{\Omega_i} \qquad \Delta=\frac{n_H}{n_H+n_L}$$

Here, ω is the frequency, k is the wave number, Ω_i is the ion cyclotron frequency, ω_{pi} is the ion plasma frequency, m_e and m_i are the electron and ion masses, and $\bar{V}_i^{-3}$ is obtained from the perpendicular velocity $V_\perp$ as the average value of $V_\perp^{-3}$ for the energetic ions. For a situation in which there are n_L low energy electrons and n_H high energy electrons in a loss cone distribution with the ions treated as unperturbed, the quantities are

$$\begin{aligned} y &= \frac{\omega}{\Omega_e} \qquad K = \frac{\Omega_e^2}{\omega_{pe}^2} \\ R &= \frac{\bar{V}_e}{\Omega_e} \qquad \Delta = \frac{n_H}{n_H + n_L} \end{aligned} \tag{3}$$

Here the subscript e denotes electron quantities.

To determine stability properties, we examine the dispersion relation for real ω and look for values of the parameters at which real solutions disappear. Except for a special resonant case, the condition for instability found in this way is

$$\frac{2}{\pi}\left(\frac{K}{1-\Delta}\right)^2 \approx \frac{2}{\pi[1-(N+1/2)^2]^2} \lesssim \left(\frac{\Delta}{1-\Delta}\right)\frac{1}{(kR)^3} \lesssim \left(\frac{\Delta}{1-\Delta}\right)\frac{1}{N^3} \lesssim \frac{\pi}{2} \tag{4}$$

where $N=2, 3, \ldots$. For the $\Delta \ll 1$ case, we have as the instability condition

$$\Delta \gtrsim \frac{2}{\pi}\sqrt{K} \tag{5}$$

and as a solution

$$y \approx \pm\frac{1}{\sqrt{K}} - i\frac{\Delta}{2K^2(kR)^3} \tag{6}$$

For electrons, Equations (5) and (6) become

$$(n_H/n_L) \gtrsim (2/\pi)(\Omega_e/\omega_{pe}) \tag{7}$$

$$\omega \approx \omega_{pe} - i\Omega_e \tag{8}$$

whereas for ions, they become

$$(n_H/n_L) \gtrsim (2/\pi)[(m_e/m_i) + (\Omega_i^2/\omega_{pi}^2)]^{1/2} \tag{9}$$

$$\omega \approx \Omega_i[(m_e/m_i) + (\Omega_i^2/\omega_{pi}^2)]^{-1/2} - i\Omega_i \tag{10}$$

For a magnetic field of 1/24 gauss and $n = 10^4$ cm^{-3}, say, $\Omega_e/\omega_{pe} \approx 1$; so that the electron case wouldn't be satisfied except at rather large L in the earth's field. In the ion case, the requirement for these numbers is that $n_H/n_L \gtrsim 1/65$. If the instabilities do occur, the growth rates are quite large (on the order of the cyclotron frequencies): on the other hand, the wavelengths are quite small so that turbulence, enhanced resistivity and diffusion rather than mass motion would be expected.

Finally, a special resonance case occurs when

$$\frac{1-\Delta}{K} = N^2 - 1\,, \quad \Delta < \frac{2}{\pi}\sqrt{K} \tag{11}$$

In that case, we have for the electrons,

$$(\omega_{p_e}/\Omega_e)^2 = N^2 - 1 \tag{12}$$

$$\omega \approx N\Omega_e - i\Omega_e\left[(n_H/n_L)(N/2\pi)\right]^{1/2} \tag{13}$$

and for the ions

$$\left[(m_e/m_i) + (\Omega_i/\omega_{p_i})^2\right]^{-1} = N^2 - 1 \tag{14}$$

$$\omega \approx N\Omega_i - i\Omega_i\left[(n_H/n_L)(N/2\pi)\right]^{1/2} \tag{15}$$

We have not yet looked into the effect of nonlinearities on these instabilities, but these might be expected to be important in quenching the instabilities by putting enough structure on the distribution function to reduce the contribution of the energetic portion.

5. Resistive Instabilities

Finally, we come to the resistive instabilities which, it has been suggested (NESS, 1965), might occur in magnetospheric tails. One of the interesting features of these instabilities is that the purely resistive part of the instabilities is very strongly localized in the high conductivity limit. The finite conductivity actually only allows the plasma to decouple from the magnetic field in a very small region, about where the perturbation practically lies along a field line. The longer the mean free path, the smaller is the size of this decoupled region. The very large gradients associated with this narrow decoupled-flow region then give rise to the possibility that viscosity and finite Larmor radius effects might be important. With the long mean free paths in magnetospheric tails, this is likely to be the case there. To see how these effects modify the finite resistivity hydromagnetic results, we have been looking at the instabilities with the Chapman and Cowling equations (CHANG and ROSENBLUTH, 1964). As the ratio β of plasma pressure to magnetic pressure might not be too small in the tails, we have treated β to be of order unity.

When temperature gradients may be ignored, the Chapman and Cowling equations differ from the hydromagnetic equations principally at two points. In the momentum equation, the hydromagnetic approximation has a scalar pressure. In the Chapman and Cowling momentum equation the scalar pressure is accompanied by a pressure tensor which accounts for the viscosity which arises from ion-ion collisions and for the current which arises when the ions because of their larger gyroradii sample a different environment than the electrons. The second place where the equations differ is in the Ohm's law. In the hydromagnetic approximation, the Ohm's law relates the electric field in the fluid frame of reference to the current through a scalar resistivity. The Chapman and Cowling Ohm's law states that the electron-ion collisions should relate the current to the electric field in the frame of reference drifting with the electrons. In addition, the scalar resistivity of the hydromagnetic approximation is replaced by an anisotropic resistivity which results from the presence of a magnetic field.

At low temperatures, when the ion gyrofrequency is less than the ion collision frequency these equations give the same results as the finite resistivity hydromagnetic equations: there are again tearing and gravitational modes. As the temperature is increased, and the ion gyrofrequency becomes larger than the collision frequency, the first correction to the growth rates comes from the ion-ion collision contribution to the pressure tensor. In the curl of the momentum equation, the ratio of the viscosity term to the inertial term is of the order of β, the ratio of the plasma and magnetic pressures, times the square root of the ratio of the ion to electron mass, times the quantity $1/p\varepsilon^2$. "p" is the growth rate expressed in terms of the resistive diffusion time, and ε is the ratio of the width of the decoupled-flow region to the width of the original current layer. For the tearing mode, $p\varepsilon^2$ is of order ε. At any rate, with the square root of the mass ratio occurring here too, unless β is very small, the viscosity term will dominate the inertial term. This then gives an ε that is inversely proportional to the cube root of the magnetic Reynold's number. But otherwise, the results are qualitatively the same as the hydromagnetic results: the only difference is that viscosity replaces inertia here, confirming the conclusions of FURTH *et al.* (1963).

As the temperature is increased further, the phases and magnitudes of the quantities are such that the next correction to the growth rates arises from the Hall current term in Ohm's law. The Hall current corrections do not become appreciable, though, until $r\varepsilon^\delta = 0(1)$, where r is the ratio of the electron gyrofrequency to the electron collision frequency; and δ is equal to 2 for the gravitational mode and for the tearing mode which occurs where the pressure is maximum. With ε determined by viscosity, the quantity $r\varepsilon^2$ is just equal to three times the ratio of the ion Larmor radius to the size of the decoupled flow region. This ratio has been assumed to be small in the Chapman and Cowling expansion, so these Hall current corrections to the growth rate do not become important until the validity of the Chapman and Cowling plasma equations itself becomes somewhat questionable. The finite Larmor radius terms in the momentum equation's pressure tensor do not contribute sizeable corrections to the growth rate until even higher temperatures, when $r\varepsilon^{\delta}$ is of the order of the square root of the ratio of the ion mass to the electron mass. This is quite outside the range where the Chapman and Cowling equations are known to be valid.

As some representative numbers, if we take a typical length of the shear field to be 10^9 cm, look at perturbation wavelengths of the same order, assume a density of 10 cm^{-3}, set $\beta=1/10$ and assume a temperature of 1 ev (so that the magnetic field is on the order of 10 gamma), the Chapman and Cowling equations show that viscosity dominates the behavior. In this case, ε is on the order of 10^{-3}, and for the tearing mode the growth time is on the order of 10^{-3} times the resistive diffusion time, i.e., on the order of a year. Smaller lengths would give faster growth times, but then these equations would become suspect.

Appendix

SUMMARY OF INSTABILITY EQUATIONS

(1) Two stability sufficiency criteria are given for the electrostatic low frequency flute, the first (1a) applying only to gross hydromagnetic interchanges, and the second (1b) including those in which resonant interactions between waves and particles are important.

(1a) Specifically, the sufficiency condition for stability against gross hydromagnetic interchanges is:

$$\int\int \mathrm{d}\mu\,\mathrm{d}J\,\frac{\partial F_0}{\partial \alpha}\frac{\partial K_0}{\partial \alpha} < 0 \tag{A1}$$

where $F_0(\mu, J, \alpha)\mathrm{d}\mu\mathrm{d}J\mathrm{d}\alpha$ is the number of particles with magnetic moment and bounce actions in the intervals $(\mu, \mu+\mathrm{d}\mu)$ and $(J, J+\mathrm{d}J)$, located in the magnetic flux interval $(\alpha, \alpha+\mathrm{d}\alpha)$.*

$K_0(\mu, J, \alpha)$ is the relativistic energy of a particle.

(1b) The second sufficiency condition takes a simple approximate form for disturbances with large azimuthal (longitudinal) variation when the ionosphere behaves resistively:

$$m^2\oint \frac{4\pi\sigma_1}{R^2B}\mathrm{d}s > \omega\,\mathscr{I}m(-\mathscr{M}) \tag{A2}$$

where

$$\mathscr{M} = 8\pi\sum\int\int \mathrm{d}\mu\,\mathrm{d}J\,\frac{ecm\dfrac{\partial F_0}{\partial \alpha}}{\omega+\dfrac{cm}{e}\dfrac{\partial K_0}{\partial \alpha}} \tag{A3}$$

* For a dipole field of moment M, a line of force with an equatorial distance of r_{E} has $\alpha=-M/r_{\mathrm{E}}$.

Here ω and m are the frequency and azimuthal wave number (disturbances of the form $e^{i(\omega t+m\theta)}$ are assumed, where θ is the longitude), δ_1 is the Pederson conductivity, B is the magnetic field intensity, R is the distance from the dipole axis, the integral $\oint ds$ is over the line of force whose stability is in question, c and e are the velocity of light and electron charge, and the sum is over the different particle species.

(2) The whistler instability condition may be expressed in terms of the sign of an integral. Specifically, in a plasma which is comprised mostly of low energy, thermal particles but which also contains a small number of relativistic electrons, the condition for the instability of a whistler of frequency ω propagating along the magnetic field is:

$$I(p_1, p_2) > 0 \tag{A4}$$

where, approximately,

$$I(p_1, p_2) = -2\int_{p_1}^{p_2} \mathrm{d}p\, pF_0(p, \theta)\left[1 - \left(1 - \frac{\xi}{(1+p^2)^{1/2}}\right)\left(\frac{\xi - 1}{\xi + \chi^2 - 1}\right)\right]$$
$$\pm \frac{\chi^2}{[(\xi + \chi^2 - 1)(\xi - 1)]^{1/2}}\int_{p_1}^{p_2} \mathrm{d}p \frac{p^2 \sin\theta}{(1+p^2)^{1/2}}\frac{\partial F_0}{\partial\theta} \tag{A5}$$

(and where here the mass of the electron and the velocity of light have been set equal to unity). The plus sign is used when $\xi > 1$ and the minus sign is used when $\xi < 1$. Here, $\chi = \omega_p/\omega$ and $\xi = \omega_c/\omega$, with ω_p and ω_c the plasma frequency and (nonrelativistic) cyclotron frequency. $F_0(p, \theta)\mathrm{d}p\,\mathrm{d}\theta$ is the number of electrons with relativistic momentum p in the interval $(p, p+\mathrm{d}p)$ and helix angle in the interval $(\theta, \theta+\mathrm{d}\theta)$. In the integrands, θ is determined from

$$\cos\theta = -\frac{\omega}{kp}\left[(1+p^2)^{1/2} - \frac{\omega_c}{\omega}\right] \tag{A6}$$

$$\left(\frac{k}{\omega}\right)^2 = 1 - \frac{\chi^2}{1-\xi} \tag{A7}$$

and p_1 and p_2 are the smallest and largest momenta for which $|\cos\theta| = 1$. When the relativistic electrons are in flat helices ($\theta \approx \pi/2$), the condition $I(p_1, p_2) > 0$ can be easily satisfied for slight drift motions of the relativistic electrons along the field under two quite different conditions of low phase velocity:

(1) $$\omega_p^2 \gg \omega_c\omega \gg \omega^2 \tag{A8}$$

(2) $$\omega \approx \omega_c \quad \text{and} \quad \omega_p^2 \gg \omega_c\omega - \omega^2 > 0\,. \tag{A9}$$

(3) The two component electrostatic loss-cone instability condition is given in the general nonresonant case for electron and ion frequencies by Equations (7) and (9) of the text, respectively. For the resonant case, the instability condition is given by Equations (12) and (14) for electron and ion frequencies, respectively.

(4) For the resistive instabilities occurring for comparable plasma and magnetic pressures and governed by viscosity in the absence of temperature gradients, the Chapman and Cowling equations apply when all of the following quantities are less than unity:

$$r_I/\varepsilon a \approx 10^7 [\beta^{1/6}\alpha^{-2/3}a^{-2/3}(\log \Lambda)^{-1/3}] T^{2/3}n^{-1/2} \tag{A10}$$

$$k_{\|}v_\theta/v_+ \approx 6 \times 10^{13} [\beta^{1/3}\alpha^{2/3}a^{-4/3}(\log \Lambda)^{-2/3}] T^{4/3}n^{-1} \tag{A11}$$

$$\omega/v_+ \approx 5 \times 10^{13} [\beta^{-2/3}\alpha^{2/3}a^{-4/3}(\log \Lambda)^{-2/3}] p\varepsilon^2 T^{4/3}n^{-1} \tag{A12}$$

$$\omega/\Omega_+ \approx 10^7 [\beta^{-1/6}\alpha^{2/3}a^{-4/3}(\log \Lambda)^{1/3}] p\varepsilon^2 T^{-2/3}n^{-1/2} \tag{A13}$$

$$v_+/\Omega_+ \approx 2 \times 10^{-6} [\beta^{1/2} \log \Lambda] T^{-2}n^{1/2} \tag{A14}$$

$$\omega\tau_T = 0[p\varepsilon^2\beta^{-1}(m_-/m_+)^{1/2}] \tag{A15}$$

$$\varepsilon \approx \beta^{1/3}(a\alpha)^{-1/3}(\log \Lambda)^{1/3}T^{-2/3} \tag{A16}$$

$$S^{-1} \approx (1/6)[\beta^{1/2}a^{-1} \log \Lambda] T^{-2} \tag{A17}$$

Here β is the ratio of the plasma to the magnetic pressure, a is the width of the equilibrium current layer expressed in cm, the dimensionless quantity α is the product of a and the perturbation wave number k, T is the temperature in eV, n is the density in cm^{-3}, r_I is the ion gyroradius, ε is the ratio of the width of the decoupled flow region to a, $k_{\|}$ is a measure of the projection of k onto the magnetic field in the decoupled flow region, v_θ is the ion thermal velocity, ω is the growth rate, v_+ is the ion-ion collision frequency, Ω_+ is the ion gyrofrequency, τ_T is the temperature relaxation time, S is the magnetic Reynold's number, p is the product of the growth rate and the resistive diffusion time, m_- and m_+ are the electron and ion masses, and Λ is the ratio of the maximum to minimum impact parameter for coulomb collisions.

When these quantities are less than unity, then the growth rate may be obtained from

$$p\varepsilon = 0(1) \tag{A18}$$

for the tearing mode (the mode which is driven by magnetic pressure) and from

$$p\varepsilon^2 = 0(1) \tag{A19}$$

for the gravitational mode (the mode which is driven by a destabilizing force, such as the inertial force of acceleration).

References

CARR, T. D.: 1963, Proceedings of the NASA Conference on the Planet Jupiter.
CHANG, D. B.: 1963, *Astrophys. J.* **138,** 1231.
CHANG, D. B.: 1964, *Phys. Fluids* **7,** 1980.
CHANG, D. B. and PEARLSTEIN, L. D.: 1965, *J. Geophys. Res.* **70,** 3075.
CHANG, D. B. and ROSENBLUTH, M. N.: 1964, *Bull. Phys. Soc.* **9**. 308 (1964).
CHANG, D. B., PEARLSTEIN, L. D., and ROSENBLUTH, M. N.: 1965, *J. Geophys. Res.* **70,** 3085; CHANG, D. B., PEARLSTEIN, L. D., and ROSENBLUTH, M. N.: to be submitted to *J. Geophys. Res.* as Correction and Addition to preceding paper.
DRUMMOND, W. E. and PINES, D.: 1962, *Nucl. Fusion, Suppl.* **2,** 1049.
FIELD, G. B.: 1963, Proceedings of the NASA Conference on the Planet Jupiter.
FURTH, H. P., KILLEEN, J., and ROSENBLUTH, M. N.: 1963, *Phys. Fluids* **6,** 459.
GOLD, T.: 1959, *J. Geophys. Res.* **64,** 1219.
JOHNSON, J. L., GREENE, J. M., and COPPI, B.: 1963, *Phys. Fluids* **6,** 1169.
NESS, N. F.: 1965, *J. Geophys. Res.* **70,** 2989.
NORTHRUP, T. G. and TELLER, E.: 1960, *Phys. Rev.* **117,** 215.
PEARLSTEIN, L. D., ROSENBLUTH, M. N., and CHANG, D. B.: 1965, to be submitted to *Phys. Fluids.*
ROBERTS, J. A.: 1963, *Planet. and Space Sci.* **11,** 221.
ROSENBLUTH, M. N. and POST, R. F.: 1965, *Phys. Fluids* **8,** 547.
SONNERUP, B. U. O. and LAIRD, M. J.: 1963, *J. Geophys. Res.* **68,** 131.
TAYLOR, J. B.: 1964, *Phys. Fluids* **7,** 767.
WARWICK, J. W.: 1963, *Astrophys. J.* **137,** 41.
ZHELEZNIAKOV, V. V.: 1958, *Soviet Astronomy-A.J.* **2,** 206.

AURORAL PHENOMENA ASSOCIATED WITH PARTICLE PRECIPITATIONS

BERNT MÆHLUM
Norwegian Defence Research Establishment, Kjeller, Norway

1. Introduction

About 65 years ago it was shown by BIRKELAND (1901) that the aurora could be produced through impact of energetic, charged particles from space, and for many years the different phenomena in the auroral ionosphere were the main indicators of the particle fluxes in space. Research workers have compiled numerous observations on auroral phenomena, and some information on the primary particles was gained from these observations. The main theoretical problems in auroral physics in recent years have been:

(1) What can the observations of the various auroral phenomena tell us about the upper atmosphere?

(2) How can the observations yield information on the energy, directional, spatial and charge distributions of the primary particles?

(3) Where do the primary particles come from, and how do they gain energy?

(1) Detailed analysis of the auroral spectrum has shown (CHAMBERLAIN, 1961) that all spectral lines can be referred to upper atmosphere constituencies, except the Balmer lines. Temperatures at altitudes above 100 km may also be derived from spectral observations and by various radio wave propagation techniques. With present days optical techniques an auroral spectrum may be obtained within a minute or less (OMHOLT *et al.*, 1962).

(2) Information on the ionizing particle fluxes may be inferred from ground based observations in various ways:

(a) From the vertical- and horizontal spectral line profiles of the Balmer lines the energy and angular distribution of the fast protons may be derived directly, although the interpretation of the observational results has proven difficult (CHAMBERLAIN, 1961).

(b) The altitude distributions of (i) density of free electrons, or (ii) luminosity within an auroral display, are directly related to the distributions of ionizing particles. If the energy dissipation processes of low energy electrons (and protons) were accurately known, it should be possible to derive the energy- and angular distributions of the primary particles from the height profiles, although

McCormac (ed.), Radiation Trapped.

not unambiguously. Rigid theories are, however, not available, particularly not in the low energy regions for electrons. Recent works by SPENCER (1959), WALT *et al.* (1962, 1964) and MAEDA (1965) may bring the problem nearer to a solution.

(3) Theories on the origin and dynamics of the auroral particles are manifold. The three "classical" theories by STÖRMER (e.g. 1955), CHAPMAN and FERRARO (1931–1933) and ALFVÉN (1939, 1940) have recently found competition by theories which take into account satellite observations of charged particles in and outside the magnetosphere. The most fully developed auroral particle theory from recent years is probably the one proposed by AXFORD and HINES (1961), which in some respects is a modification and development of Chapman and Ferraro's model.

Theories on the dynamics of the auroral particles have now become a biproduct of the more general problem of the dynamics in the magnetosphere during periods when the geomagnetic field is distorted by the pressure of solar plasma streams.

Some of the problems associated with auroral phenomena have reached a temporary solution through observations from satellites. There are, however, still a large number of problems to be investigated, some of which can hardly be solved by the present day techniques used in space vehicle borne experiments. These methods have at present two serious limitations, which are important when a transient aurora is to be observed: (i) the satellite passes an active auroral curtain within a fraction of a second, and a very high time resolution in the counters and satellite telemetry is required for detailed studies of this phenomenon, (ii) the satellite borne counters can obviously only take a sample at one point at a time. In order to obtain a full picture of the dynamics of the auroral display one needs synchronous observations at closely spaced points.

The importance of the present day ground based activity in auroral physics for space problems can be summarized in the following points:

(1) Synoptic studies of the structure and motion of auroral displays provide indirect information on the particle dynamics outside the earth's atmosphere.

(2) Studies of special features of the auroral phenomena (low intensity aurorae, sharply localized phenomena, proton aurorae and others) provide information on the charged particles which has not or cannot be obtained from satellite borne counters.

(3) By comparing quantitatively observations of particles by satellite borne counters with ground based observations of atmospheric ionization and excitation one may possibly arrive at a complete solution to the problem of particle/atmospheric interaction. If this were possible, the aurora might be used with confidence as a "charged particle spectrometer" in the future.

This paper is mainly concerned with the problems reviewed in the present paragraph. Special emphasis is placed on the problems in which a coordinated study by satellite borne and ground based experiments may be most fruitful. It is not the intention to give full credit to all works done on auroral phenomena. For more detailed discussions the reader is referred to standard text-books (MITRA, 1948; HARANG, 1951; STÖRMER, 1955; RATCLIFFE, 1960; CHAMBERLAIN, 1961; CHAPMAN and AKASOFU, 1964 and HULTQVIST, 1964).

General problems related to visual aurorae and ionization phenomena are discussed in Sections 2 and 3, respectively, whereas Section 4 is devoted to indirect particle studies from ground based observations. Sections 5 and 6 give results from auroral structure/drift observations and observations at geomagnetically conjugate points.

2. Visual Aurora, General Description

In the present and the following section some general properties of the visual aurora and the auroral ionization are reviewed. More detailed discussions on some special problems related to these phenomena are given in Sections 5 and 6.

A. AURORAL TYPES

The visual aurora shows a large variety of forms and structures, which have been the object of much classification work. According to the IQSY agreement the aurora is arranged in a complex system based on six qualifying parameters (JACKA and PATON, 1963). For the present paper a simpler system seems adequate (ELVEY, 1964):

(1) *Ordinary displays*, which are by far the most commonly observed types at auroral latitudes, with a lower limit between 95 and 115 km.

(2) *Intermediate red arc*, which is a rare phenomenon observed near the auroral zone at about 195 km altitude.

(3) *Red M-arc*, which is mainly observed on the equatorial side of the auroral zone at altitudes above 400 km (MOORE and ODENKRANTZ, 1961; ROACH and MAROWICH, 1959).

(4) *Polar glow*, which is observed over the polar caps in conjunction with influx of cosmic protons in the energy range 10–100 MeV (SANDFORD, 1961, 1963).

(5) *Discrete polar cap aurora*, which is observed over the polar caps, mainly during quiet geomagnetic conditions (DAVIS, 1963).

To this list should be added a type which by some workers are referred to as:

(6) *Proton glow*. This is observed mainly near the auroral zone. Due to the sensitivity of the eye at these wavelengths this glow can hardly be seen, and no estimates of the height of the layer have been made (OMHOLT, 1963).

Most visual auroral observations made refer to type (1), and the "statistical auroral zone" is defined by the occurrence of this type.

It is beyond the scope of this paper to go into a detailed discussion of the various types. Only a few comments on the vertical and horizontal extent of the ordinary displays seem appropriate.

The auroral arcs are generally very limited in the north-south direction, whereas the east-west extent may be very large. Arc thicknesses less than 1 km

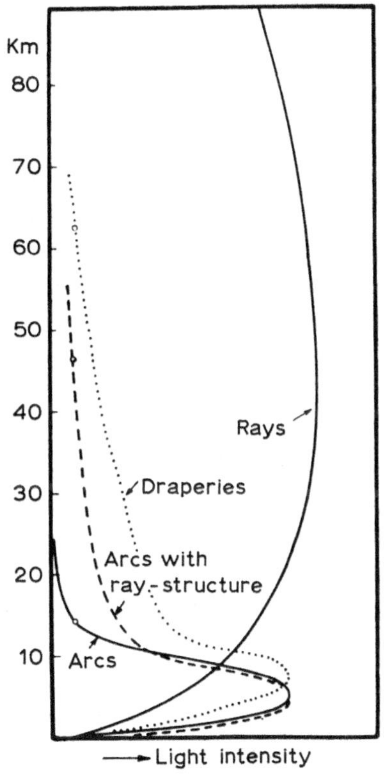

Fig. 1. Luminosity height profiles of various auroral forms.

have been reported (ELVEY, 1957 and AKASOFU, 1961), and synoptic studies indicate that an arc may occur simultaneously over an east-west range of 4000 km or more (AKASOFU, 1963b).

The height distributions of the curtainlike aurorae and auroral rays were originally deduced by VEGARD and KROGNESS (1920), and these two types of visual aurorae show a distinct difference in the luminosity profile (Figure 1).

This may imply that the energy distributions of the primary particles responsible for the two types differ significantly, or that the rays are generated by a different source than the arcs (WEBSTER, 1957).

B. GEOGRAPHICAL DISTRIBUTION OF AURORAL INCIDENCE

The visual aurorae (ordinary display type) occur most frequently in two zones, centered approximately around the northern and southern geomagnetic poles, respectively. These two "statistical auroral zones" are deduced from more or less systematical, visual observations of all aurorae above the threshold of visibility.

Until recently very little information has been presented in publications on the latitudinal distribution of the intensity of visual aurorae. Some observations indicate that the more intense auroral phenomena occur outside of the "statistical auroral zone". Hence, a comparison between the latitudinal variation of the occurrence of visual aurorae and the average flux of precipitated electrons of energy ≈ 10 keV should not necessarily give a maximum at the same latitude.

The occurrence of visual aurorae in the polar regions is the matter of some disagreement. According to ALFVÉN (1955) one should expect to find an "inner auroral zone" about 5–10° from the "center of symmetry". Some workers report the existence of a second zone (LASSEN, 1959), whereas others oppose this finding (HATHERTON, 1960; DENHOLM, 1961). It seems clear that if this zone is present, it is located at a larger polar distance than is predicted by Alfvén (DAVIS, 1962).

The "statistical auroral zone" is, by definition, fixed in latitude. For detailed, quantitative studies of the correspondence between particles and excitation this zone is of minor importance. AKASOFU (1964) has introduced the term "auroral belt" for the instantaneous position of auroral phenomena. This concept is adopted in the following discussion in order to avoid confusion.

Long term variations in the latitude of maximum occurrence of visual aurorae have been observed. The maximum is situated 1–2° nearer to the poles during sunspot minimum than during maximum solar activity (DAVIES, 1947; GARTLEIN, 1959).

Both the ordinary aurorae and the "proton aurorae" are situated at lower latitudes near local midnight than near sunset (e.g. AXFORD and HINES, 1961; REID and REES, 1959).

The occurrences of different types of auroral displays show different latitudinal distributions. The auroral rays are generally observed 2° nearer the equator than the auroral arcs (GARTLEIN, 1959). Furthermore, the position of the "proton auroral belt" differs from the "active auroral belt" (STOFFREGEN and DERBLOM, 1962). Before midnight the hydrogen emission is regularly observed at lower latitudes than the active aurora, whereas the positions are reversed later in the night. None of these observations have been verified by direct particle observations.

The mysterious M-arcs observed at lower latitudes are restricted to a very narrow latitude range. Over North America 50% of all events takes place between 53.2° and 53.5° geom. lat. (MAROWICH and ROACH, 1963). The east-west extent of these arcs may, however, be many hundred kilometers (MOORE and ODENKRANTZ, 1961).

C. SOURCES OF EXCITATION

It is generally agreed that low energy electrons ($E_e < 20$ keV) are the main contributors to auroral excitation, as these electrons are absorbed in the appropriate altitude range. This has also been confirmed by rocket observations (MCILWAIN, 1960). Higher energy electrons ($E_e = 100$ keV) may, however, occasionally take part in the excitation near the lower border of auroral arcs (MALVILLE, 1959).

The auroral spectrum is dominated by nitrogen and oxygen bands, due to the abundance of these elements in the upper atmosphere (CHAMBERLAIN, 1961). In addition, H_α and H_β lines are occasionally observed in the auroral zone (VEGARD, 1939; MEINEL, 1950a, b) as well as inside the zone (DENHOLM, 1961). The H_α line is regularly broadened and the zenith profile is Doppler shifted towards the violet (VEGARD, 1948), indicating a rapidly moving source. Occasionally the H_α-line is very narrow (MALVILLE, 1960), but the significance of this observation is uncertain.

The importance of energetic protons ($E_p > 30$ keV) as contributors to auroral excitation has not been appreciated until recently (OMHOLT, 1963), and only a few reliable rocket observations on primary protons have as yet been made (MEREDITH *et al.*, 1958).

The intensity distribution of the common yellow-green auroral line 5577 Å has been studied in College, Alaska (ROACH and REES, 1960). 80 per cent of the observations lie between 0.75 and 11.5 kR, corresponding to an International Brightness Coefficient of I to II. Only during 0.4 per cent of the time does the intensity exceed 100 kR. The energy input per cm^2 required to maintain an aurora of intensity 100 kR is about 300 ergs/sec (OMHOLT, 1963). Such high fluxes have never been observed above the auroral zone.

The excitation efficiency of the primary particles has been debated. For regular auroral lines the efficiency ranges from 0.2% (MCILWAIN, 1960) to 5% (CHAMBERLAIN, 1961). Recent satellite measurements (O'BRIEN and TAYLOR, 1964) indicate a value "more nearly 1% than 0.1%", which is in fair agreement with values given by KRASSOVSKY (1965). The efficiency is a function of the altitude (KRASSOVSKY, 1965), and this may possibly explain the apparent discrepancies between various observations.

One type of auroral excitation which is not fully explained is the one giving rise to the red oxygen doublet 6300/6364 Å at medium latitudes (FAN and SCHULTE, 1954; BARBIER, 1958). The excitation levels of these lines are about 2 eV, and the excitation level of the 5577 Å line in the oxygen atom is about 4 eV. As the 5577 Å line is not present in the M-arcs, one can hardly imagine that the energy of the exciting particles exceeds 4 eV (ROACH and ROACH, 1963). Hence, it is not likely that the fast magnetospheric electrons take part in the excitation. The present day belief is that the excitation is caused by heated, atmospheric electrons or that some photochemical reaction occurs (e.g. ROACH and ROACH, 1963; MEGILL and CARLETON, 1964).

3. Auroral Ionization, General Description

The auroral particles will regularly cause enhanced ionization in the lower part of the auroral ionosphere. These irregular layers of ionization, known as the auroral sporadic E(Es)- and D-layers, have been studied by various ground based radio propagation techniques and in situ measurements from rockets.

In the present section the two layers are discussed separately, as the main problems associated with the layers and the methods used for observing the two layers differ.

A. SPORADIC E IONIZATION

The sporadic E ionization in the auroral zone occurs exclusively during the nighttime (SMITH, 1957). This ionization is mainly investigated by the ionosonde technique, and it may be separated into two types: (i) a layer which causes group retardation of HF radio signals near the critical frequency and (ii) a layer which causes no group retardation near the critical frequency. The former type (Es type *r*) consists of a fairly thick, opaque layer, whereas the latter type (Es types *f*, *l*, *h*, *n*, *a*) is generally semitransparent for radio frequencies just below the critical frequency, and consists probably either of one or several thin sheets of ionization, or the reflection is simply caused by steep gradients in the "background" ionization.

The "statistical sporadic E auroral zone" (MATSUSHITA, 1959) is presented in Figure 2. The Es layer occurs most frequently at about 69° geom. lat. over North America. During disturbed conditions, however, the maximum Es ionization may be shifted to much lower latitudes, as will be shown in Section 5.

The electron density in the E region varies proportionally with the square root of the ionization rate, and the relaxation time of the electron density is some minutes or more. One can therefore hardly imagine a rate of change in the electron density comparable with the rate of change in the visual aurora. It is, however, known that rapid changes in the electron density at these altitudes sometimes occur (JACKSON and SEDDON, 1959). Typical time variations in the maximum frequencies of reflection from the sporadic E layer (*f*Es) as observed during medium disturbances by three spaced ionosondes in Northern Norway are shown in Figure 3 (MÆHLUM, 1962). The horizontal distances between the stations were about 100 km, and the time variations of the retardation type

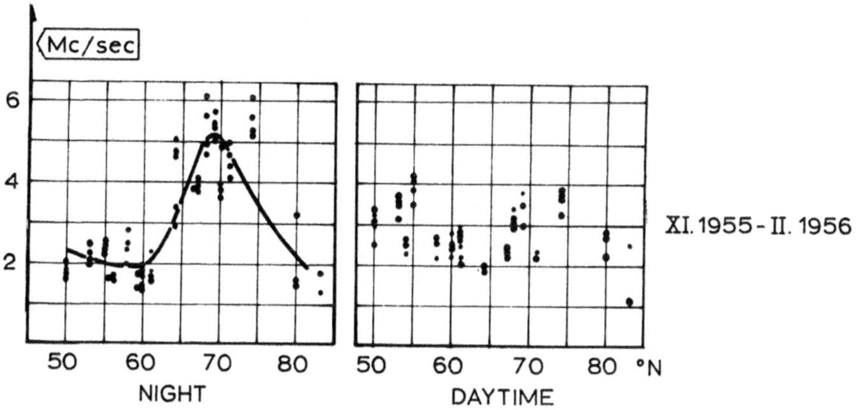

Fig. 2. Average *f*Es-values observed over North America (MATSUSHITA, 1959).

(*r*) and non-retardation type (*n*) of Es reflections are shown separately for the three stations. The *f*Es type *r* varies smoothly during the night, and the correlation between the three stations is nearly unity. The ionization in the Es layer type *n* varies apparently much more randomly, and the correlation between observations at different points is negligible. It has been suggested that the two Es types *r* and *n* are produced by fast protons and fast electrons, respectively, (SHAW, 1959; OGUTI, 1962) but a detailed investigation of the problem remains to be performed.

A fair correlation is found between the maximum frequency of high frequency radio wave reflection from Es ionization and the luminosity of visual aurora (OMHOLT, 1955; KNECHT, 1956; ULWICK *et al.*, 1964). Particularly high values of *f*Es are observed when an auroral band is present in zenith (HEPPNER *et al.*, 1952; HUNSUCKER and OWREN, 1962).

The critical frequencies of the Es type *r* observed in Northern Norway by an ionosonde indicate that the peak electron density in the layer seldom exceeds 5×10^5 cm^{-3}. Electron densities up to 8×10^5 cm^{-3} have, however, occasionally been observed by rockets (JESPERSEN *et al.*, 1965).

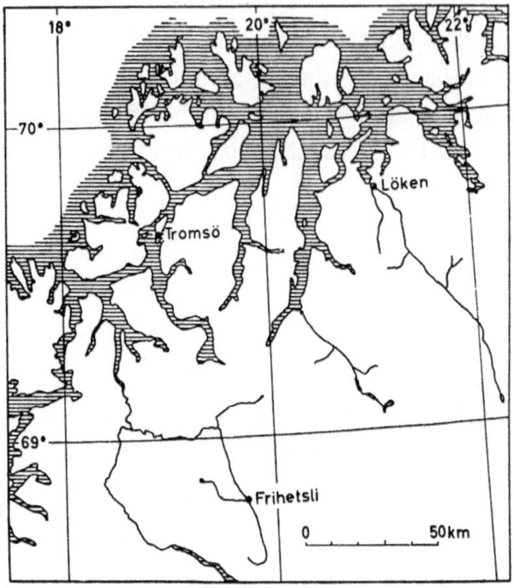

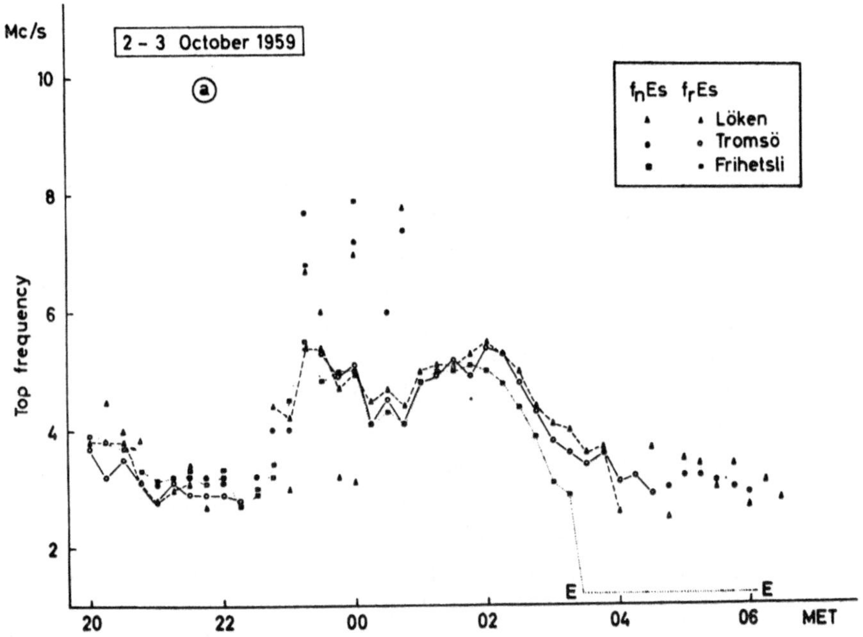

Fig. 3. Typical time variations of *f*Es-type *r* and *f*Es-type *n* during medium disturbances at high latitudes as observed by three spaced ionosondes in Northern Norway.

B. AURORAL D IONIZATION

The enhanced ionization below the E-region associated with disturbed conditions in the auroral zone was first observed by APPLETON *et al.* (1933), and this ionization has been the subject of intense investigation during recent years (e.g. LITTLE, 1954; LITTLE and LEINBACH, 1958; REID and COLLINS, 1959; HOLT *et al.*, 1962).

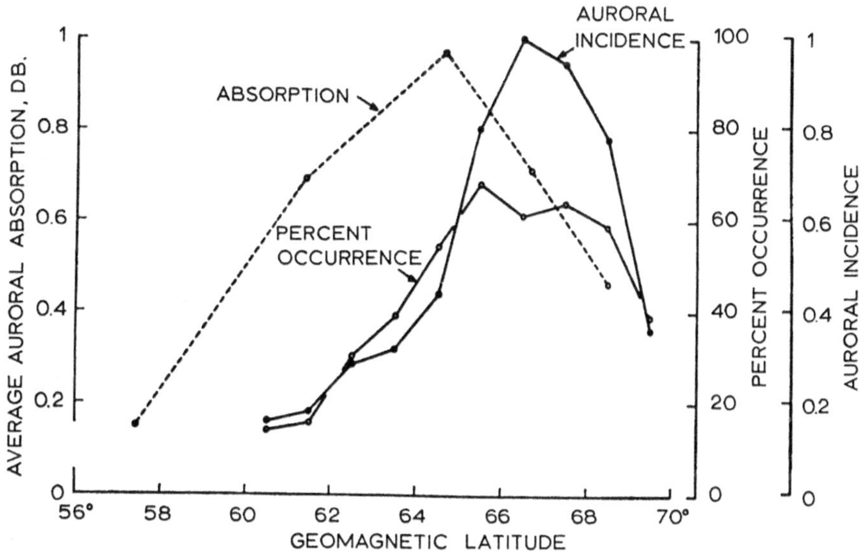

Fig. 4. Comparison of the auroral absorption zone with the visual auroral zone over North America (BASLER, 1963).

Observations indicate that the main contributors to the formation of this ionization are energetic electrons (McDIARMID *et al.*, 1961), and the secondary X-rays may play a certain role in the very lower part of the layer (AIKIN and MAIER, 1964).

The location of the "statistical auroral D-layer zone" over North America is shown in Figure 4 (After BASLER, 1963). It is found that the average auroral absorption peaks 2° south of the visual auroral zone.

The actual height distribution of free electrons during different degrees of auroral activity has been the matter of discussion during recent years. (For a review see BROWN, 1964 and HULTQVIST, 1964a, b). Most of the ground based observational techniques used for investigating the ionospheric D-region are not well suited for an exact determination of the height where the high frequency radio wave absorption really takes place.

Recent rocket measurements of the electron density in the lower ionosphere during small and medium disturbances have given a fairly consistent picture of typical day- and night profiles (JESPERSEN *et al.*, 1965) (Figure 5a). If these profiles are normalized to simplify the comparison (Figure 5b) one finds that the shape of the profiles is similar in the upper portion. Below 80 km there is, however, a clear difference between the day and the night profiles. This difference has been ascribed to the day/night difference in the density of negative ions at these altitudes.

From Figure 5 one might expect a clear diurnal variation in the auroral absorption. Such variations have been observed (CHAPMAN and LITTLE, 1957), but the actual value of the day/night ratio in absorption is debated (BROWN, 1964; HULTQVIST, 1964a, b). The solution of this problem is obscured by the fact that the day/night ratio in the auroral absorption may partly be ascribed to diurnal variations in the source of ionization (BASLER, 1963).

The horizontal extent of the "auroral absorption D-layer" is on an average much larger than 100 km (HOLT *et al.*, 1962). It is, however, clear that the layer is not homogeneous (LEONARD, 1962), and on some occasions the ionization is very localized (PARTHASARTHY and BERKEY, 1965).

As the energetic electrons responsible for the auroral D-layer ionization differ in energy from the ones responsible for the auroral excitation, one should not a priori expect a one-to-one correspondance between the two phenomena. Observations indicate a general correlation between auroral luminosity and auroral absorption (HOLT and OMHOLT, 1962; JOHANSEN, 1964), particularly if only periods with pulsating surfaces are selected (MURCRAY, 1959). There are, however, clear cases when there is a lack of correlation between the two phenomena (JOHANSEN, 1964; ANSARI, 1962), which indicates interesting variations in the primary electron spectrum.

4. Indirect Particle Studies from Observations of Ionization and Light Phenomena in the Auroral Ionosphere

In spite of the large number of high latitude satellites launched during recent years very little detailed information on the time and spatial variations in the flux of the "auroral particles" is yet available. It is therefore believed that ground-based studies of the phenomena in the upper auroral atmosphere may still yield information on the particle precipitation which, if properly interpreted, may fill gaps in our present knowledge.

In the present section some of the results obtained from such studies are briefly reviewed. Most of these results have not yet been verified by particle observations by satellites.

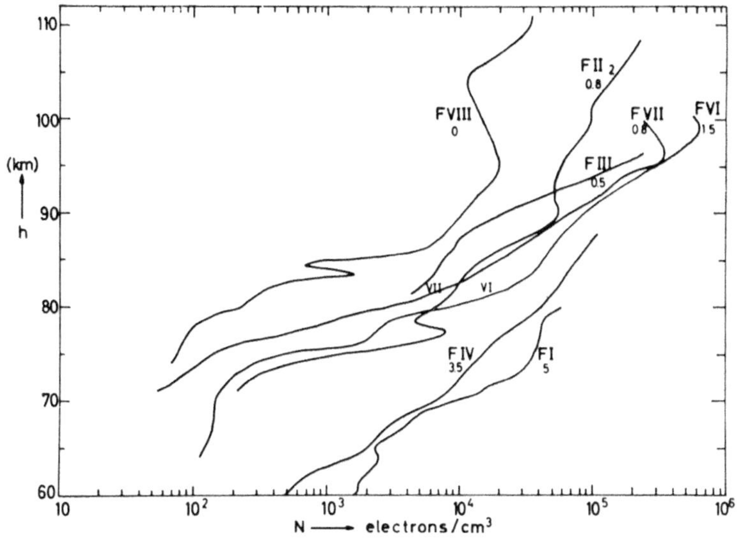

Fig. a

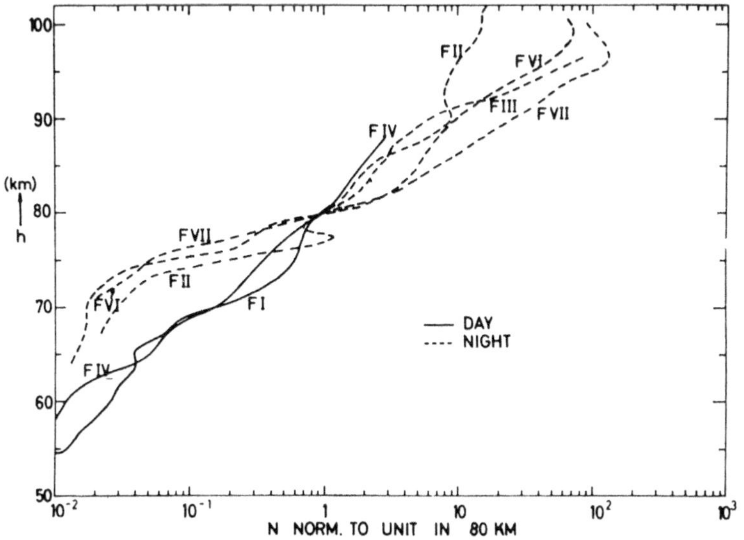

Fig. b

Fig. 5. (a) Unnormalized and (b) normalized electron density profiles observed by sounding rockets over Northern Norway during various degree of geomagnetic activity.

A. INDIRECT OBSERVATIONS OF PRECIPITATED ELECTRONS

The spectrum of precipitated electrons has been studied semiquantitatively from ground based observations in three different ways: (i) by analysing the height profile of luminosity of visual aurora, (ii) by analysing the height profile of ionization and (iii) by comparing visual phenomena and radio wave absorption phenomena. These three techniques give information on the electron spectrum in different energy ranges, and they are therefore not redundant.

Studies of the electron density profile in the lower ionosphere indicate that the auroral electron energy spectrum in the 30–300 keV range may be approximated by an exponential law $i \sim \exp(-E/E_a)$, where $E_a < 10$ keV (BROWN, 1964) *or* by a power law spectrum $i \sim E^{-\gamma}$, where $\gamma \geqq 4$ (MÆHLUM, unpublished results). From comparative studies of electron density profiles and luminosity profiles one may extrapolate the spectrum down to 5–10 keV. If one single model spectrum were to cover the entire energy range 5–300 keV the exponential spectrum seems to be in best accordance with upper atmosphere observations (REES, 1963; BROWN, 1964).

It has been found by some workers that the primary electron energy spectrum is softer within an auroral display than the "normal" spectrum (MCILWAIN, 1960), whereas other studies indicate that there is no correlation between the intensity of particle precipitation and the spectral shape (MCDIARMID *et al.*, 1964). A contribution to this discussion has recently been presented by LERFALD *et al.* (1964), based on studies of multifrequency riometer observations (Figure 6). From the frequency dependence of the VHF cosmic radio wave absorption these authors derive the electron density profiles, characterized by a parameter *n*. For high *n*-values the electron density is low in the lower D-region. This implies that high *n*-values correspond to steep primary electron energy spectra. Figure 6b clearly demonstrates that the spectrum of primary electrons becomes steeper as the flux of precipitated electrons increases. When the cosmic radio noise absorption on a frequency of 10 Mc/s exceeds 16 db, all but one values of *n* exceed 1.8.

The correlation between the auroral luminosity and auroral absorption varies with time. Late in the night after break up, the two phenomena are particularly highly correlated (BROWN and BARCUS, 1964). Correlation studies of bremsstrahlung X-rays in the 20–260 keV energy range and the luminosity of overhead, visual aurorae (BHAVSAR, 1961) support the conclusion that the electron fluxes between 40–200 keV and between 1–20 keV vary grossly in phase during the early morning hours. Before break up the correlation is less pronounced.

Detailed indirect studies of the time variations of the electron energy spectra have revealed important information, which should be investigated by satellite

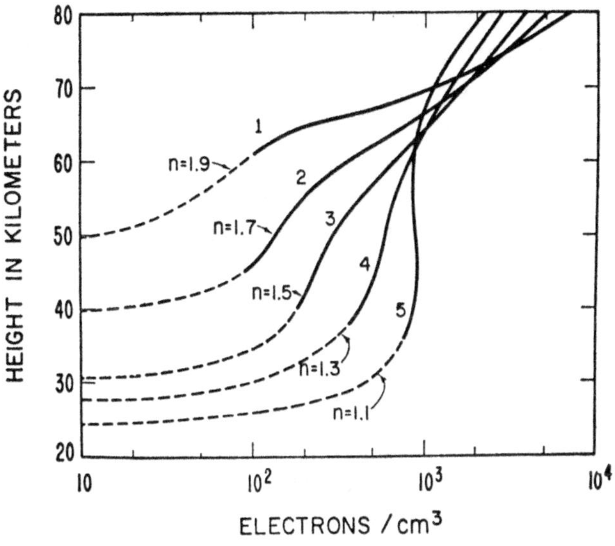

Fig. a

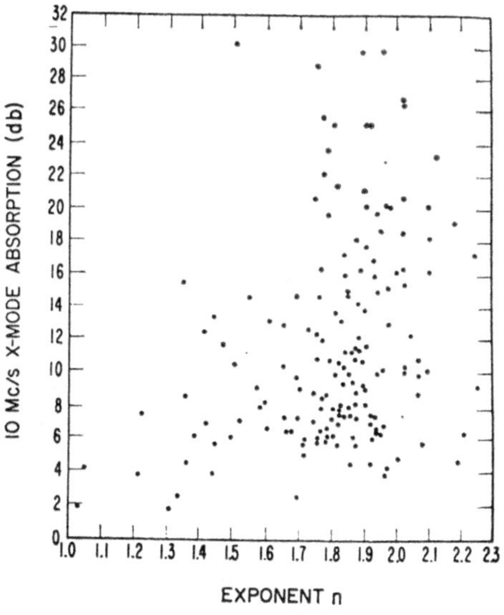

Fig. b

Fig. 6. (a) Electron density profiles defining a parameter n and (b) observed distribution of n as function of riometer absorption.

measurements. According to JOHANSEN (1964) the primary spectrum of electrons during periods when homogeneous auroral forms cover most of the sky remains constant for hours, interrupted by rapid changes. Statistically, the energy spectrum is much harder after midnight than earlier in the night. The shift to a harder spectrum occurs between 2000 and 0100 L.T. (JOHANSEN, 1964). According to ANSARI (1962) the shift in spectrum occurs near break-up.

Another important observation by JOHANSEN (1964) is the large variations in the auroral luminosity which occasionally take place during periods of low auroral absorption. This indicates that the precipitation of low energy electrons ($E_e < 10$ keV) and electrons of higher energies ($50 < E_e < 500$ keV) at times are uncorrelated.

Johansen's observations refer to visual aurorae which more or less cover the entire sky. PARTHASARATHY and BERKY (1965) have investigated particular absorption events which are fairly limited in horizontal extent. The primary electron energy spectra obtained during such events are essentially harder than "typical" auroral events. It is, however, not known whether there is any correlation between the electron spectrum and the horizontal structure of particle precipitation.

The "fine structure" of the particle precipitation, associated with the formation of auroral rays has not yet been paid much attention. According to some authors (GARTLEIN, 1959; AKASOFU, 1963), the rays are generated by wave formations in the auroral arcs, due to space charges in the impinging particle flux. In the rays there is a net excess of positive particles, which causes the arc to "curl up" anticlockwise (WEBSTER, 1957). Two questions related to this model have, however, to be answered: (i) How is it possible to accumulate the positive space charge required to account for the effect? (ii) How do the large differences in the height luminosity profiles of the rays and the arcs (Figure 1) compare with this model? It has been suggested that the auroral rays are produced by electrons which have a wide energy distribution, quite different from the spectrum of the electrons responsible for the auroral arcs (CHAMBERLAIN *et al.*, 1960). This may account for the difference in vertical extent of the two phenomena, but the dynamics of the ray formation (AKASOFU, 1963) has to be explained by some additional effect.

Occasionally "waves of light" which move rapidly upward along the geomagnetical field lines are observed within auroral displays. Apparent speeds up to 1000 km/sec are reported by Carleton (see OMHOLT, 1963). The following model has been forwarded to explain the observation: If the particles are accelerated at a certain (localized) area in space, the more energetic particles will reach the atmosphere rapidly, and excite the air atoms in the lower atmosphere, whereas the less energetic particles responsible for excitation in the higher atmosphere will reach the atmosphere after a certain time (Δt). From the height dependence of Δt Carleton estimates the distance to the acceleration region. Some of the observations can be explained by assuming that the excitation is produced by electron bunches accelerated in a region ≈ 2500 km from the visual aurora.

B. INDIRECT OBSERVATIONS OF PRECIPITATED PROTONS

Two specific phenomena in the auroral zone have been ascribed to protons: (a) the visual "proton aurora" (OMHOLT, 1963) and (b) the sporadic E ionization type "*r*" (SHAW, 1959; OGUTI, 1962). The precipitation of fast protons into the high latitude atmosphere has not been thoroughly studied from space vehicles, and we feel that indirect studies of the proton fluxes from ground-based observations may therefore be particularly fruitful.

Only a few observations of the detailed structure and time variations in the sporadic E layer, type *r*, in the auroral zone are presented in the literature. Observations in Northern Norway show that this layer is very uniform in a horizontal direction and the ionization is normally restricted to a certain latitude interval.

The peak electron density in the Es region seldom exceeds 5×10^5 electrons/cm^3 (MÆHLUM, 1962). From this value we infer that the fluxes of protons of energies about 100–200 keV infrequently exceed $(0.5\text{–}1) \times 10^7$ cm^{-2} sec^{-1}. Direct proton observations above the auroral zone are sparse, but some investigations by rocket borne counters indicate that the proton flux generally is less than some 10^6 cm^{-2} sec^{-1} in the 100 keV energy range, in agreement with our indirect findings (L. R. DAVIS, Private communication).

No studies have yet been done on possible relationships between Es type *r* and the "proton aurora". The "proton aurora", which is observed during periods when the Balmer lines are present in the auroral spectrum, is barely visible to the eye, but it may be observed by high sensitivity spectrometers (OMHOLT *et al.*, 1962). The close relationship between the occurrence of this glow and enhanced H_α-H_β-emission strongly suggests a proton source of excitation.

Correlation studies of electron produced (ordinary) aurorae and the "proton aurorae" indicate that these two phenomena "live their own life" irrespective of each other (OMHOLT, private communication). Statistically, the Balmer lines are most frequently observed during periods when quiet visual auroral forms (arcs and glow) are present (ROMIC and ELVEY, 1958). Most authors agree that the H_α and H_β emission disappears abruptly near break-up (FAN, 1958; GALPERIN, 1959; DAHLSTROM and HUNTEN, 1951; ROMIC and ELVEY, 1958).

The variable degree of correlation between proton aurorae and ordinary (active) aurorae has been ascribed to differences in the dynamics of the two phenomena. It seems well established that the zone of proton precipitation is situated at lower latitudes near midnight than during the day (REID and REES, 1959; AKASOFU, 1960), and detailed studies of the "electron auroral belt" and the "proton auroral belt" indicate that the proton precipitation takes place south of the electron precipitation before midnight in the northern hemisphere (OMHOLT, 1963). Later in the night the two zones of precipitation merge together, and some research workers claim (STOFFREGEN and DERBLOM, 1962) that their observations indicate a reversal in the positions of the two zones in the morning hours. These results, if verified by more direct measurements, are of extreme importance for a full understanding of the particle dynamics.

Studies of the energy- and angular distributions of the primary protons have been performed from analysis of the zenithal and horizontal profiles of the H_α-line (CHAMBERLAIN, 1961). It has, however, proved difficult to obtain exact values of the proton distribution by this method. The observations indicate typical differential energy spectra in the energy range 1–300 keV which may be approximated by a power law (OMHOLT, 1963):

$$i(\mathrm{E_p}, \theta)\,\mathrm{d}E_\mathrm{p} \sim E_\mathrm{p}^{-1.8} \cos^6 \theta \,\mathrm{d}E_\mathrm{p},$$

(where θ refers to the zenith angle of arrival). During selected events the relative abundance of low energy protons is much higher than indicated by this equation (MALVILLE, 1960). Furthermore, the angular distribution varies with time.

The total influx of protons of energy $E_\mathrm{p} \approx 100$ keV during a strong event of "proton aurora" has been estimated to 10^7 cm^{-2} sec^{-1} (OMHOLT *et al.*, 1962) or 10^7–10^8 cm^{-2} sec^{-1} (MONTALBETTI, 1959), which is in fair agreement with the values obtained from Es studies (MÆHLUM, 1962).

5. Morphology and Motion of Auroral Ionization

The complex over-all picture of auroral morphology and motion prohibits any detailed discussion of the various phenomena in this brief presentation. Furthermore, the "average" behaviour of auroral phenomena as described by various authors differs significantly. This makes the picture even more confusing.

Ordinary displays of visual aurorae occur in three main forms (CHAPMAN and AKASOFU, 1964): (i) quiet arcs, which have a latitudinal extent of a few kilometers, (ii) rayed arcs, whose latitudinal extent is as small as 1 km or less, and (iii) homogeneous surfaces or patches. Most studies of auroral morphology refer to the two first groups of auroral displays, as the exact shape and motions of the latter can hardly be investigated.

The discussion of auroral motions is generally separated into two groups: (i) North-South displacements of the auroral pattern as a whole, and (ii) drifts (mainly East-West) of individual forms within the pattern. These two modes of motions, although not unrelated, are most conveniently discussed separately.

A. AURORAL MORPHOLOGY

Studies of the auroral morphology in recent years have centered on two problems: (i) the time variations in the average orientation of auroral arcs during the night and during a polar substorm, and (ii) "typical" development of the auroral pattern during a polar substorm in general.

The average orientation of an auroral arc is in a direction about normal to the horizontal component of the earth's magnetic field (VEGARD and KROGNESS, 1920), nearly parallel to the geomagnetic latitude circles (EVANS and THOMAS, 1959). The instantaneous orientation of the auroral arcs differs, however, frequently from this average direction, and the time variations in the orientation of auroral arcs may elucidate some of the problems associated with particle acceleration and distortions of the magnetosphere.

The main contributions to the investigation of the auroral morphology during recent years have been presented by two research groups in College, Alaska; supervised by T. N. Davis and S.-I. Akasofu, respectively. These two groups have based their studies on different approaches, and reached conclusions which differ significantly.

According to Akasofu and colaborators (AKASOFU, 1963, 1964; AKASOFU and KIMBALL, 1964; AKASOFU *et al.*, 1965) the aurora may consist of faint arcs extending along the entire auroral zone that is in darkness. When a polar magnetic sub-storm occurs, sudden changes are observed at all longitudes where aurora is present (see Figure 7). The type of activity observed depends on the local hour at the observer and the intensity of the substorm. The largest disruption in the arc generally takes place on the morning side of the earth or near the midnight meridian. West of this "break up" region waves and folds are generated, which move rapidly toward the West. Within and East of this region similar waves move toward the East. During especially strong events the "break up" region may extend over a large part of the night side. Hence, according to Akasofu's model, which is based on synoptical studies of individual auroral storms, the auroral activity peaks in the break-up region, from which different types of waves are transmitted along the auroral arc. Sometimes the break up region may, according to this model, expand into the polar cap.

Davis and colaborators (DAVIS, 1960, 1962a, b; DAVIS and KIMBALL, 1962) deduce the average direction of auroral forms as a function of the local hour. The orientation of the auroral forms inside the auroral zone is generally in a direction parallel to the sun-earth line. Hence, the auroral forms over the poles rotate 360° relative to the earth during a day. Furthermore, many morphological and drift features in the visual aurora are known to be local time dependent.

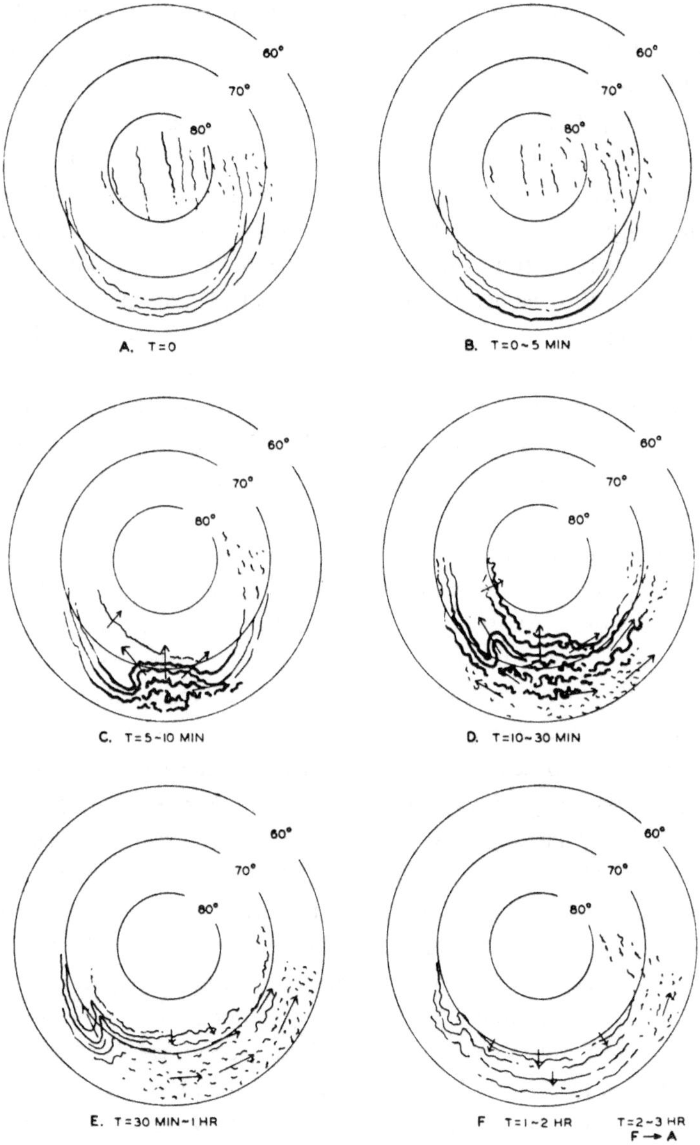

Fig. 7. Typical development of auroral forms during geomagnetic storms (AKASOFU, 1964).

This has led Davis to propose a model of the auroral morphology in which the auroral pattern remains fixed in orientation relative to the sun-earth line, and the average orientation of forms is almost parallel to the idealized S_D-current system (CHAPMAN and BARTHELS, 1940). The model auroral pattern proposed by Davis is shown in Figure 8. As the earth rotates under this pattern different

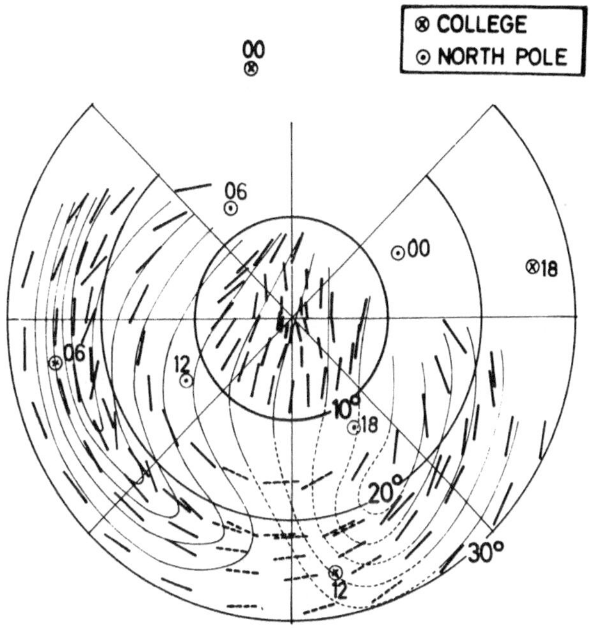

Fig. 8. Model of average auroral pattern (DAVIS, 1962b).

observing sites on the earth move along circles centered approximately on the geomagnetic pole. For illustration the apparent motions of College, Alaska and the geographical north pole are shown. Thule, Greenland is situated close to the center of symmetry. The main conclusions drawn from DAVIS' model (1962b) are:

(1) The auroral forms are generally aligned along the lines shown in Figure 8. An arc is never extending indefinitely from the evening side to the morning side of the earth, but is bent toward the north just West of the break-up region (indicated by the discontinuous lines).

(2) The motion of auroral forms is clockwise in the evening loop array, i.e. toward the West in the southernmost part of the pattern and toward the East further North. East of the break-up region the general drift is toward the East.

One may infer from this pattern that the reversal of drift velocity takes place later in the night at the southermost stations than further North, which is in agreement with observations (STOFFREGEN, 1961).

At present the all-sky camera coverage in the auroral zone is inadequate for a full investigation of the two models. The most significant difference between the models is found in the region west of the break-up, where Davis' model predicts closed "stream lines" between the auroral zone and the polar cap aurora, whereas Akasofu claims that the arcs are more or less continuous throughout the night. It is well known that in the region West of the break-up the auroral arc is heavily perturbed, and Akasofu claims that the "loop array" suggested by Davis in this region is a statistical result only, related to the westward traveling folds. Maybe future satellite borne cameras can distinguish between the two models.

In order to track the dynamics of the auroral particles one should try to project the auroral forms along the geomagnetic field lines into the equatorial plane. This has not yet been done with any degree of confidence, as the geomagnetic field configuration in space is not fully known.

B. NORTH-SOUTH DISPLACEMENTS OF THE AURORAL BELT

The problem of latitudinal shifts of the auroral belt during different geophysical conditions is an important piece in the auroral puzzle, and some observational results will be discussed in this section.

The geomagnetical storm dependence of the latitude of the auroral belt was pointed out by STÖRMER (1910), who explained the latitudinal shift of the belt during geomagnetic storms by assuming (a) time variations in the energy of the primary corpuscular beam, or (b) time variations in the earth's magnetic field (Dst-component). It is still not clear whether the auroral belt generally moves toward lower latitudes during disturbances, or whether the occurrence of auroral phenomena at low latitudes during disturbed conditions is due to an expansion of the auroral belt in the north- and southward direction.

AKASOFU and CHAPMAN (1963) have reported that the latitude of the southern limit of quiet auroral arcs is correlated with the Dst component of the geomagnetic field (see Figure 9). Recent detailed investigations of the auroral belt (STRINGER *et al.*, 1965) indicate that the "average auroral display" shifts from $\approx 78°$ (dipole latitude) during periods of complete geomagnetical calmness to $\approx 67°$ when the local K-index is 1. During heavily disturbed conditions (Feb. 11, 1958) the auroral belt may be shifted as far south as $L = 3.5$, which corresponds to an invariant latitude 57°–58° (AKASOFU and CHAPMAN, 1960).

Similar time variations have been observed in the latitude of the maximum sporadic E ionization and maximum high frequency radio wave absorption zone (BELLCHAMBERS and PIGGOT, 1960; HOLT *et al.*, 1962; NAGATA, 1950; MÆHLUM, 1962; BASLER, 1963).

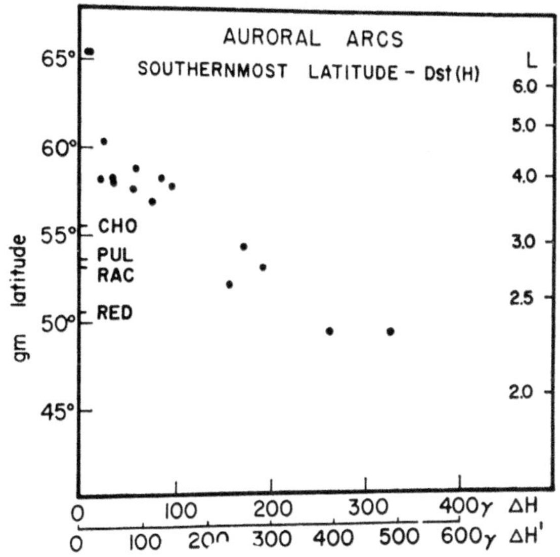

Fig. a

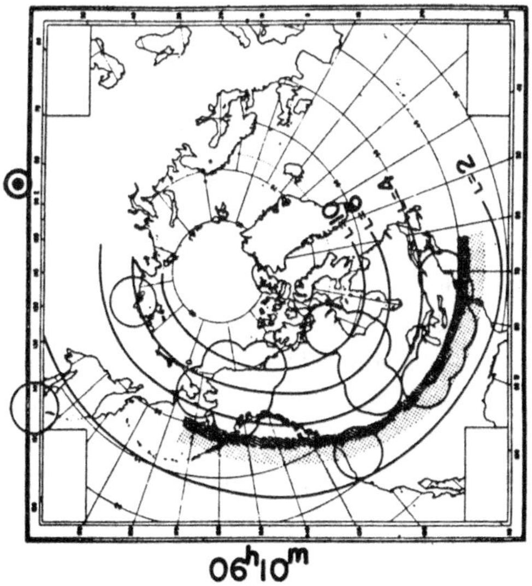

Fig. b

Fig. 9. (a) Relationship between southern limit of auroral arcs and the Dst (H) component (Akasofu and Chapman, 1963). (b) Position of the auroral arc during the February 1, 1958 storm (Akasofu and Chapman, 1960).

It is assumed that the auroral phenomena are in some way related to the outer boundary of the magnetosphere, which is known to be moving to and from the earth during different degrees of solar activity (FRANK and VAN ALLEN, 1964). Stormtime variations in the boundary of trapped electrons have also been observed at low altitudes in the auroral zone (MÆHLUM and O'BRIEN, 1962), but the physical relationship between the different phenomena is still unknown.

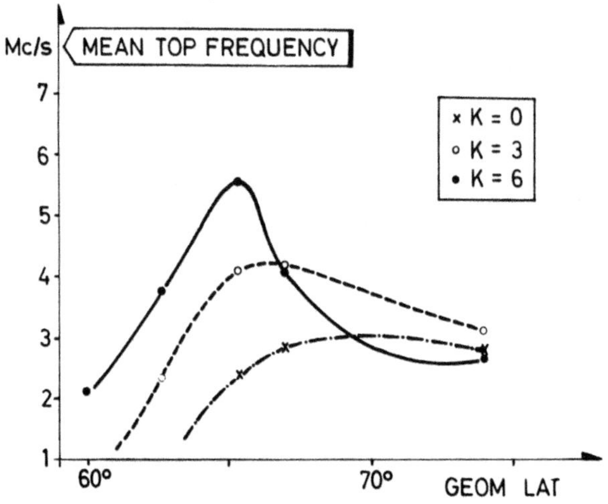

Fig. 10. Average latitudinal variation of fEs during different storm condition. (MÆLUM, 1962a.)

The average ionization in the auroral E-region during different degrees of disturbances is indicated in Figure 10. It has been found (BELLCHAMBERS and PIGGOT, 1960) that the sporadic E layer of type *r* disappears at high latitudes during disturbed periods. From this observation one might infer that the proton precipitation region is shifted toward lower latitudes during geomagnetic disturbances. Such variations have not been reported from satellite observations.

The diurnal variations of the different upper atmosphere phenomena in the auroral zone have been considered by a large number of workers. (For a review see OGUTI, 1962, 1963). The visual aurora is generally observed $\approx 5°$ nearer the equator around midnight than during the morning and evening hours. From a more detailed study of the local time dependence of the phenomena the picture becomes more complex.

It was initially shown by NIKOL'SKII (1947) that the maximum occurrences of polar blackouts on high frequency radio waves took place much earlier at low latitudes than at higher latitudes, and the hours of maximum frequency of occurrence were located on a spiral (see Figure 11). Corresponding studies of

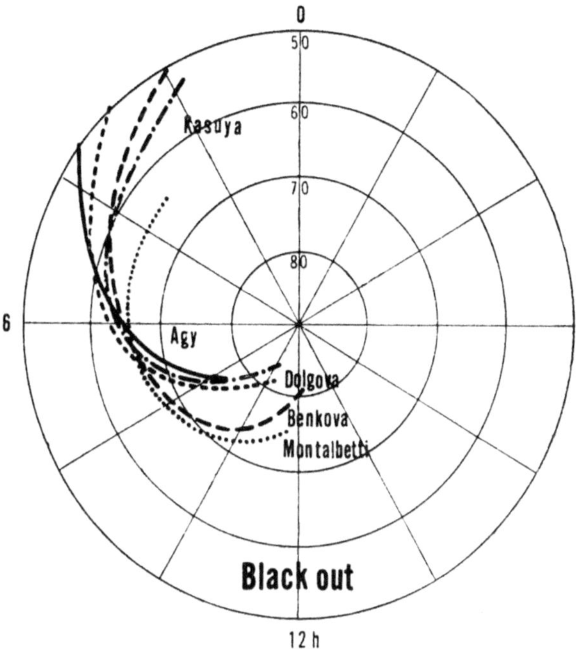

Fig. 11. Hour of maximum auroral absorption on HF radio waves at various latitudes (OGUTI, 1962).

other auroral zone phenomena have unveiled a series of spirals, some of which are curved to the right, others to the left, and it has been suggested that the difference in curvature implies different polarity of the ionizing particles (OGUTI, 1962).

The local hours of maximum occurrence of sporadic E and auroral absorption are indicated in Figure 12 (compiled by AXFORD and HINES, 1961). From this illustration one can hardly deduce whether the sporadic E maxima and the auroral absorption maxima form two separate spirals with opposite curvature, or whether the two curves together form a complete ellipse centered at about 75° latitude on the night side of the earth. We realize that there are no large basic differences between the two phenomena, as only a minor change in

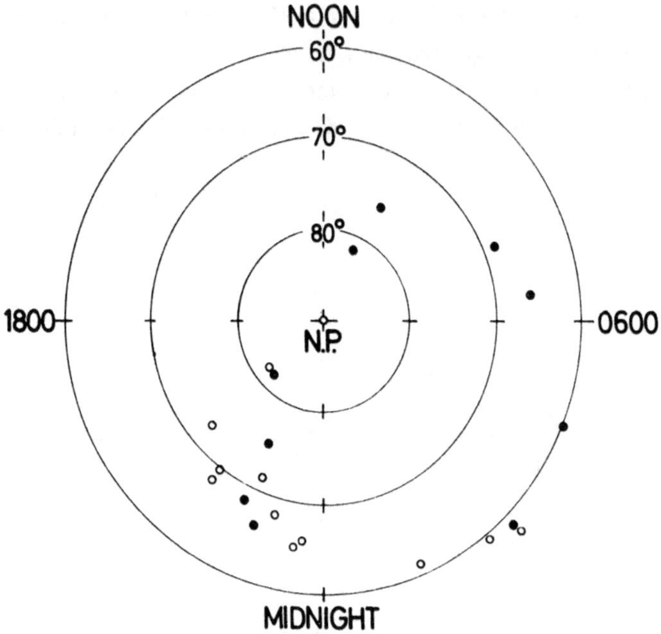

Fig. 12. Hour of maximum Es ionization (open circles) and auroral absorption (filled circles) at selected latitudes (AXFORD and HINES, 1961).

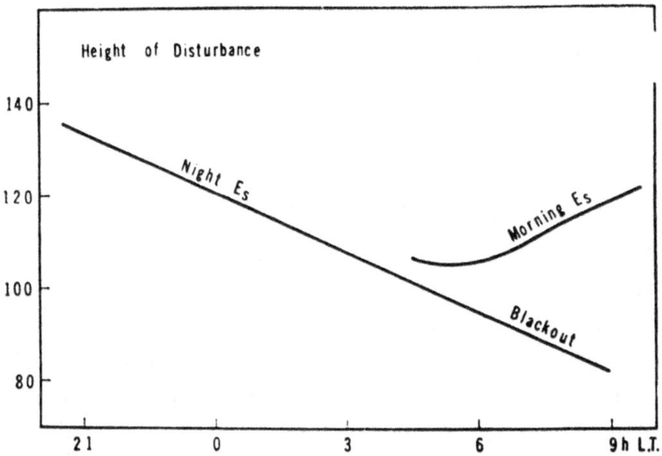

Fig. 13. Time variations in the height of sporadic ionization in the lower ionosphere (model) (OGUTI, 1963).

the ionizing electron spectrum is required for a transition from one phenomenon to the other, the harder spectrum being responsible for an auroral absorption. As earlier shown, the visual auroral observations indicate that the electron energy spectrum becomes harder in the middle of the night, and the nighttime Es layer may gradually develop into a D-layer (Figure 13, after OGUTI, 1963). Hence, the two spirals shown in Figure 12 may possibly reflect the general southward shift in the electron precipitation zone during the night (O'BRIEN, 1963), combined with a spectral hardening late in the night.

C. EAST-WEST DRIFTS OF AURORAL PHENOMENA

As the auroral zone is oriented exactly in an East-West direction at a few longitudes only, there are no physical reasons for separating the motion of auroral phenomena into pure meridional and zonal components. In fact, most observations indicate clearly that the main directions of drifts are along the auroral arcs. However, the observational results on auroral drift are regularly given in terms of North-South and East-West components instead of drift speeds along and normal to the auroral forms.

Basically, there are three different observational techniques for studying irregularities in the auroral ionosphere:

1. *Optical techniques*

(a) *Short time interval all-sky camera snapshots.* By comparing subsequent observations the angular velocity of visual auroral forms are easily deduced. Motion of auroral forms as well as inhomogeneities in the luminosity distribution (rays, patches) may be studied by this technique.

(b) Different *photometric techniques* give information on the speed of inhomogeneities in the luminosity (OMHOLT, 1962).

2. *HF radio wave techniques*

(a) "*Cambridge method*". From statistical studies of the fading characteristics of the high frequency radio wave signals reflected from the ionosphere deduction may be drawn on the scattering ionospheric irregularities.

(b) By comparing *time variations in the height and maximum ionization* in the ionospheric layers above spaced observatories information is obtained on the large scale drifts.

3. *VHF radio wave techniques*

(a) Studies of *scintillation of radio stars* on very high radio frequencies give information on small scale irregularities in the upper atmospheric ionization.

(b) The *Doppler shift* of an ionospheric reflected very high frequency radio signal is directly proportional to the line-of-sight velocity of the irregularity.

There are wide divergencies in the results obtained by the various techniques. The average speeds of the irregularities measured differ from one method to the other. Observations of visual aurorae show that the auroral forms never overtake one another, and extended forms (rays) may move as a whole without changing form (COLE, 1963), whereas the irregularities in the ionization generally show large irregular drifts superimposed on the average motion (MÆHLUM, 1962b).

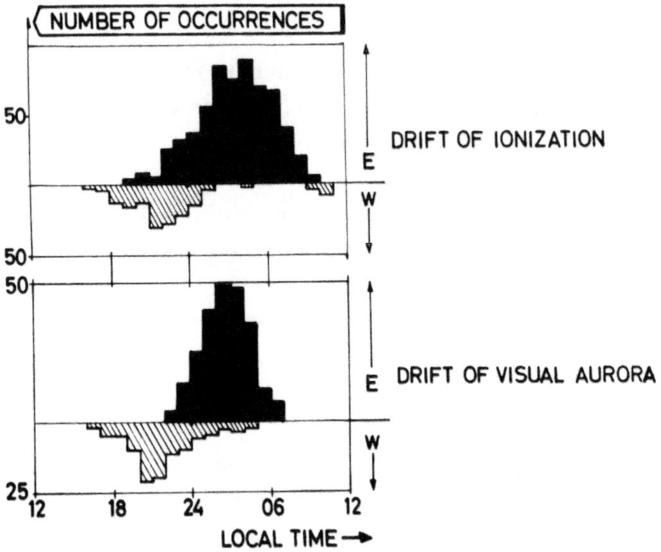

Fig. 14. Number of easterly and westerly drift observations of (a) auroral ionization (UNWIN, 1959) and (b) visual aurora (STOFFREGEN, 1961).

Most observations of auroral motions show that the irregularities tend to drift toward the west in and south of the auroral zone during the evening hours, shifting to easterly drift later in the night (BULLOUGH and KAISER, 1955; MEINEL and SCHULTE, 1953; KAISER, 1955; BLESS *et al.*, 1955; HARANG and TRØIM, 1961; STOFFREGEN, 1961; UNWIN, 1959). The close agreement between drift directions of visual aurorae and irregularities in the ionization is shown in Figure 14. The drift directions at higher latitudes are shifted 180° relative to the drift further south (STOFFREGEN, 1961; DAVIS, 1962). The exact hour of drift reversal has been debated, but it occurs near midnight. The reversal is masked by the presence of irregular East-West drifts all through the night (KIM and CURRIE, 1958; BLESS *et al.*, 1955).

The drift speeds of visual aurorae range from 200–500 m/sec for loop types to 4–6 km/sec (KIM and CURRIE, 1958) for inhomogeneities in the luminosity. Very active rays may move with velocities up to 30 km/sec (OMHOLT, 1962). Highest velocity ever observed in aurorae is 125 km/sec (DAVIS and HICKS, 1963).

The drift speeds of irregularities in the ionization are generally much less than the speeds of visual aurorae. VHF radio wave observations indicate horizontal (East-West) speeds in the 500–4000 m/sec range (BULLOUGH and KAISER, 1954, 1955; KAISER, 1955; UNWIN, 1959), whereas ionospheric irregularities which give rise to HF radio wave fading in the sporadic E region have

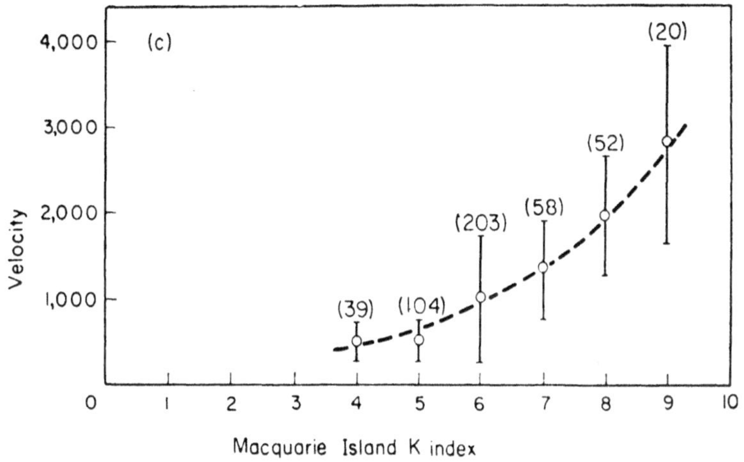

Fig. 15. Relationship between drift speed of auroral ionization and geomagnetic activity (UNWIN, 1959).

East-West drifts between 150 and 500 m/sec (MÆHLUM, 1962). Drift speeds up to 1000 m/sec have been reported for the Es clouds (HAGG and HANSON, 1954).

The direction and magnitude of ionospheric drifts in the auroral zone depend on the geomagnetic activity (NICHOLS, 1959). During periods when the horizontal component of the geomagnetic field is higher than normal the drift of ionization in the auroral zone is preferably toward the East (HEPPNER, 1954), whereas negative geomagnetic storms are accompanied by drifts toward the West (MÆHLUM, 1962). The variation of the drift speed with magnetic activity is indicated in Figure 15. The drift speed increases by a factor of 5 or 6 as the local K index increases from 4 to 9. Similar (qualitative) results are obtained from radio star scintillation studies (LITTLE and MAXWELL, 1952).

As there apparently is a whole spectrum of drift speeds one would expect to find a series of drift sources. The lower speeds measured are in the same range as the wind speeds of the neutral atmosphere (ELFORD, 1959) or the velocity of sound at these altitudes. Hence, one may postulate that the irregularities are drifting with the neutral air, or that the irregularities are produced by some pressure wave phenomena (HINES, 1964). External electric fields or moving sources of ionization cannot, however, be excluded as an explanation of low speed drifts. Observations of bremsstrahlung X-rays indicate that the primary electron beam ($E_e > 50$ keV) occasionally moves horizontally in an East-West direction with a speed of 200–300 m/sec (BROWN, 1962), and AKASOFU (1960) has shown that an electrical field required to account for the observed drift in the ionization is ≈ 20 mV/m, which is in fair agreement with observations of electric fields within aurorae (KAVADAS and JOHNSSON, 1965).

The higher drift speeds (i.e. $v > 1$ km/sec) cannot be ascribed to neutral atmosphere phenomena. Most probably these drifts reflect presence of electric fields or horizontal motion of the source of ionization (HEPPNER, 1954).

For a detailed investigation of the drift phenomena one needs information on (i) electric fields in and above the visual auroral forms, (ii) longitudinal motions of the electron precipitation zones, and (iii) time variations in neutral atmospheric winds and the coupling between the neutral and ionized atmosphere.

6. Auroral Phenomena at Geomagnetically Conjugate Points

For indirect studies of particle dynamics and acceleration in the magnetosphere, recent investigations of auroral phenomena in geomagnetically conjugate areas may prove useful. The results obtained from such studies are at present very qualitative, and no firm conclusions may yet be drawn.

A general agreement exists between the auroral phenomena in the northern and southern hemisphere (LEINBACH and BASLER, 1963; DE WITT, 1962; HOOK, 1962). In fact, the correlation is higher between conjugate aurorae than between auroral phenomena observed in the same hemisphere at two stations separated by 2° latitude interval (LEINBACH and BASLER, 1963). Hence, reliable quantitative studies of this relationship can only be performed from observations at exactly conjugate points.

As the conjugacy breaks down during periods when the earth's magnetic field is distorted (LITTLE *et al.*, 1965), one needs a series of observatories in order to keep track on the motion of the conjugate point for a given station. Another problem in the interpretation of observations is the difference in external condition for the two conjugate points due to (i) an asymmetric geomagnetic field, (ii) difference in local time, (iii) difference in altitude of mirror points of particles above the stations.

Up to now only a general agreement between auroral phenomena, such as break-up and drifts, can be derived. Future research will, it is hoped, throw light on the significance of certain discrepancies in the behaviour of the auroral phenomena in the northern and southern auroral zones (DAVIS and DE WITT, 1963).

7. Concluding Remarks

Ground based observations of upper atmosphere phenomena at high latitudes associated with particle precipitation have been reviewed in this paper. The main effects of these particles in the earth's atmosphere which may be observed from the ground are: (i) ionization, (ii) excitation and (iii) heating of the air atoms/molecules. In the present paper the two former of these effects have been covered, mainly. From the height profile of the free electron density and auroral luminosity semiquantitative results on the energy distribution of the primary particles may be drawn. The horizontal distribution of auroral phenomena, on the other hand, provides information on the spatial distribution of the particles.

Upper atmospheric heating due to bombardment of fast particles has not yet been thoroughly studied. In a bright aurora each air molecule gains a thermal energy of some 20 μeV/sec (CHAMBERLAIN, 1961) which is sufficient for a significant enhancement of the temperature within a few minutes. Sporadic temperature enhancements of many hundred degrees have been observed in the upper atmosphere at high latitudes (BLAMONT and CHANIN-LORY, 1965; KRASSOVSKY, 1965). It is still an open question whether these increases in the air temperature may be ascribed to direct particle bombardment, or whether Joule heating due to auroral associated current systems is the main source (KATO, 1962; COLE, 1962). It is, however, clear that temperature measurements are very important for investigations of the auroral phenomena and for the particle physics generally.

In order to obtain a more complete understanding of the various phenomena which occur as a consequence of the particle/atmospheric interaction it is imperative to take all the available observations into consideration. Surprisingly, only a few co-ordinated ground-based and satellite borne experiments have been performed up to now. In the future such joint studies may prove very useful.

Acknowledgement

I wish to express appreciation to Professors L. Harang and A. Omholt of the University of Oslo and Dr. O. Holt of the NDRE for informative discussions. Professor A. Omholt also kindly furnished me with an unpublished manuscript of lectures on aurora presented at the Alpbach Summer School 1963, which was of great help in the preparation of the present paper.

References

AIKIN, A. C. and MAIER, E. J.: 1964, in *Electron Density Distribution in Ionosphere and Exosphere,* North-Holland Publ. Co., Amsterdam.
AKASOFU, S.-I.: 1960, *J. Atm. Terr. Phys.* **19,** 19.
AKASOFU, S.-I. and CHAPMAN, S.: 1960, *J. Atm. Terr. Phys.* **24,** 740.
AKASOFU, S.-I.: 1961, *J. Atm. Terr. Phys.* **21,** 287.
AKASOFU, S.-I. and CHAPMAN, S.: 1963, *J. Atm. Terr. Phys.* **25,** 9.
AKASOFU, S.-I.: 1963a, *J. Atm. Terr. Phys.* **25,** 163.
AKASOFU, S.-I.: 1963b, *J. Geophys. Res.* **68,** 1667.
AKASOFU, S.-I.: 1964a, *J. Atm. Terr. Phys.* **26,** 1167.
AKASOFU, S.-I.: 1964b, *Plan. Space Sci.* **12,** 273.
AKASOFU, S.-I. and KIMBALL, D. S.: 1964, *J. Atm. Terr. Phys.* **26,** 205.
AKASOFU, S.-I., KIMBALL, D. S., and MENG. C.-I.: 1965, *J. Atm. Terr. Phys.* **27,** 173.
ALFVÉN, H.: 1939–1940, *Kgl. Svenska Veterskapsakad. Handl.* **18,** (3, 9).
ALFVÉN, H.: 1955, *Tellus* **7,** 50.
ANSARI, Z. A.: 1962, *The spatial and temporal variations in high latitude cosmic noise absorption and their relation to luminous aurora,* PhD thesis, Univ. Alaska.
ANSARI, Z. A.: 1964, *J. Geophys. Res.* **69,** 4493.
APPLETON, E. V., NAISMITH, R., and BUILDER, G.: 1933, *Nature* **132,** 340.
AXFORD, W. I. and HINES, C. O.: 1961, *Canadian J. Phys.* **39,** 1433.
BARBIER, D.: 1958, *Ann. Géophys.* **14,** 334.
BASLER, R. P.: 1963, *J. Geophys. Res.* **68,** 4665.
BELLCHAMBERS, W. H. and PIGGOTT, W. R.: 1960, *Proc. Roy. Soc.* **A256,** 200.
BHAVSAR, P. D.: 1961, *J. Geophys. Res.* **66,** 679.
BIRKELAND, K.: 1901, *Vid. Selsk. Skr. Math. Nat. Klasse,* No 1.
BLAMONT, J. E. and CHANIN-LORY, M. L.: 1965, *Space Res.* **5,** 313.
BLESS, R. C., GARTLEIN, C. W., and KIMBALL, D. S.: 1955, *Astrophys. J.* **122,** 205.
BROWN, R. R.: 1962, *J. Geophys. Res.* **67,** 31.
BROWN, R. R.: 1964, *Arkiv Geofysik* **4,** 405.
BROWN, R. R. and BARCUS, J. R.: 1964, *Arkiv Geofysik* **4,** 395.
BULLOUGH, K. and KAISER, T. R.: 1954, *J. Atm. Terr. Phys.* **5,** 189.
BULLOUGH, K. and KAISER, T. R.: 1955, *J. Atm. Terr. Phys.* **6,** 198.
CHAMBERLAIN, J. W., KERN, J., and VESTINE, E. H.: 1960, *J. Geophys. Res.* **65,** 2535.
CHAMBERLAIN, J. W.: 1961, *Physics of the Aurora and Airglow, Academic Press.*
CHAPMAN, S. and AKASOFU, S.-I.: In Research in Geophysics, (ed. H. ODISHAW), The M.I.T. Press, 1964.
CHAPMAN, S. and FERRARO, V. C. A.: 1931–1933, *Terr. Mag.* **36,** 77 and 171; **37,** 147 and 421; **38,** 79.
CHAPMAN, S. and BARTELS, J.: 1940, *Geomagnetism,* Clarendon Press, Oxford.
CHAPMAN, S. and LITTLE, C. G.: 1957, *J. Atm. Terr. Phys.* **10,** 20.
COLE, K. D.: 1962, *Australian J. Phys.* **15,** 223.
COLE, K. D.: 1963, *Plan. Space Sci.* **10,** 129.
DAHLSTROM, C. E. and HUNTEN, D. M.: 1951, *Phys. Rev.* **84,** 378.
DAVIES, F. T.: 1948, *ATME Bull.* **13.**
DAVIS, T. N.: 1960, *J. Geophys. Res.* **65,** 3497.
DAVIS, T. N.: 1962a, *J. Geophys. Res.* **67,** 59.
DAVIS, T. N.: 1962b, *J. Geophys. Res.* **67,** 75.
DAVIS, T. N. and KIMBALL, D. S.: 1962, *Univ. Alaska, Science Report,* No. 2.
DAVIS, T. N.: 1963, *J. Geophys. Res.* **68,** 4447.
DAVIS, T. N. and DEWITT, R. N.: 1963, *J. Geophys. Res.* **68,** 6237.
DAVIS, T. N. and HICKS, G. T.: 1963, Paper presented at AGU Meeting, Boulder.
DENHOLM, J. V.: 1961, *J. Geophys. Res.* **66,** 2105.
DEWITT, R. N.: 1962, *J. Geophys. Res.* **67,** 1347.

ELFORD, W. G.: 1959, *Proc. Antarctic Symp.* Buenos Aires.
ELVEY, C. T.: 1957, *Proc. Nat. Acad. Sci. U.S.* **43,** 63.
ELVEY, C. T.: 1964, *Auroral Morphology, Lockheed Symposium on Aurorae.*
EVANS, S. and THOMAS, G. M.: 1959, *J. Geophys. Res.* **64,** 1381.
FAN, C. Y. and SCHULTE, D. H.: 1954, *Astrophys. J.*, **120,** 563.
FAN, C. Y.: 1958, *Astrophys. J.*, **128,** 420.
FRANK, L. A., VAN ALLEN, J. A., and CRAVEN, J. D.: 1964, *J. Geophys. Res.* **69,** 3155.
GALPERIN, G. I.: 1959, *Plan. Space Sci.* **1,** 57.
GARTLEIN, C. W.: 1959, *Ann. Géophys.* **15,** 31.
HAGG, E. L. and HANSON, G. H.: 1954, *Canadian J. Phys.* **32,** 790.
HARANG, L.: 1951, *The Aurorae,* John Wiley and Sons.
HARANG, L. and TRØIM, J.: 1961, *Plan. Space Sci.* **5,** 33.
HATHERTON, T.: 1960, *Nature* **186,** 288.
HEPPNER, J. P., BYRNE, E. C., and BELON, A. E.: 1952, *J. Geophys. Res.* **57,** 121.
HEPPNER, J. P.: 1954, *J. Geophys. Res.* **59,** 329.
HINES, C. O.: 1964, *Res. Geophysics* **1** (ed. by H. ODISHAW).
HOLT, O., LANDMARK, B., and LIED, F.: 1962, In: *Radio Wave Absorption in the Ionosphere* (ed. by N. C. GERSON), Pergamon Press.
HOLT, O. and OMHOLT, A.: 1962, *J. Atm. Terr. Phys.* **24,** 467.
HOOK, J. H.: 1962, *J. Geophys. Res.* **67,** 115.
HULTQVIST, B.: 1964, *Plan. Space Sci.* **12,** 579 and 1035.
HULTQVIST, B.: 1964, *Aurora,* Goddard Energetic Particle Preprint Series, X-611-64-97.
HUNSUCKER, R. D. and OWREN, L.: 1962, *J. Res. Natl. Bur. Std.* **D66,** 581.
JACKA, F. and PATON, J.: 1963, *IQSY Instruction Manual,* No. **3,** CIG-IQSY Committee, London, 1963.
JACKSON, J. E. and SEDDON, J. C.: 1959, *IGY Bulletin, No.* **28.**
JESPERSEN, M., HAUG, A., and LANDMARK, B.: 1965, Paper presented at the NATO advanced study institute, Finse, Norway.
JOHANSEN, O. E.: 1965, *Plan. Space Sci.* **13,** 225.
KAISER, T. R.: 1955, *Proc. of Belfast Symp. on the Aurora and Airglow,* Pergamon Press, London.
KATO, Susumu: 1962, *Plan. Space Sci.* **9,** 939.
KAVADAS, A. and JOHNSON, D. W.: 1965, *Space Res* **5,** 312.
KIM, J. S. and CURRIE, B. W.: 1958, *Canadian J. Phys.* **36,** 160.
KNECHT, R. W.: 1956, *J. Geophys. Res.* **61,** 59.
KRASSOVSKY, V. I.: 1965, *Space Res.* **5,** 71.
LASSEN, K.: 1959, *Nature,* **184,** 1375.
LEINBACH, H. and BASLER, R. P.: 1963, *J. Geophys. Res.* **68,** 3375.
LEONARD, R. S.: 1962, *J. Geophys. Res.* **67,** 939.
LERFALD, G. M., LITTLE, C. G., and PATHASARATHY, R.: 1964, *J. Geophys. Res.* **69,** 2857.
LITTLE, C. G. and MAXWELL, A.: 1952, *J. Atm. Terr. Phys.* **2,** 356.
LITTLE, C. G.: 1954, *Proc. IRE* **42,** 1700.
LITTLE, C. G. and LEINBACH, H.: 1958, *Proc. IRE* **46,** 334.
LITTLE, C. G., SCHIFFMACHER, E. R., CHIVERS, H. J. A., and SULLIVAN, K. W.: 1965, *J. Geophys. Res.* **70,** 639.
MAEDA, Kaichi: 1965, *NASA Technical Note,* NASA TN-D-2612.
MÆHLUM, B.: 1962a, *Geophys. Publ.* **23,** No. 1.
MÆHLUM, B.: 1962b, *Geophys. Publ.* **23,** No. 2.
MÆHLUM, B. and O'BRIEN, B. J.: 1963, *J. Geophys. Res.* **68,** 997.
MALVILLE, J. M.: 1959, *J. Atm. Terr. Phys.* **16,** 59.
MALVILLE, J. M.: 1960, *Plan. Space Sci.* **2,** 130.
MALVILLE J. M.: 1964, *J. Geophys. Res.* **69,** 1285.
MAROVICH, E. and ROACH, F. E.: 1963, *J. Geophys. Res.* **68,** 1885.
MATSUSHITA, S.: 1959, *J. Atm. Terr. Phys.* **15,** 68.
MCDIARMID, I. B., ROSE, D. C., and BUDZINSKI, E.: 1961, *Canadian J. Phys.* **39,** 1888.

McDiarmid, I. B. and Burrows, J. R.: 1964, *Canadian J. Phys.* **42,** 616.
McIlwain, C. E.: 1960, *J. Geophys. Res.* **65,** 2727.
Megill, L. R. and Carleton, N. P.: 1964, *J. Geophys. Res.* **69,** 101.
Meinel, A. B.: 1950a, *Science,* **112,** 590.
Meinel, A. B.: 1950b, *Phys. Rev.* **80,** 1096.
Meinel, A. B. and Schulte, D. H.: 1953, *Astrophys. J.* **117,** 454.
Meredith, J. H., Davis, L. R., Heppner, J. P., and Berg, O. E.: 1958, *IGY Rocket Report Series,* No. 1.
Mitra, S. K.: 1952, *The Upper Atmosphere, The Asiatic Society,* Calcutta.
Montalbetti, R.: 1959, *J. Atm. Terr. Phys.* **14,** 200.
Moore, J. G. and Odencrantz, F. K.: 1961, *J. Geophys. Res.* **66,** 2101.
Murcray, W. B.: 1959, *J. Geophys. Res.* **64,** 955.
Nagata, T.: 1950, *J. Geophys. Res.* **55,** 127.
Nichols, B.: 1959, *Proc. IRE,* **47,** 245.
Nikolskii, A. P.: 1957, *Dokl. Akad. Nauk SSSR* **115,** 84.
O'Brien, B. J.: 1963, *J. Geophys. Res.* **68,** 989.
O'Brien, B. J. and Taylor, H.: 1964, *J. Geophys. Res.* **69,** 45.
Oguti, T.: 1962, *Rep. on Ionosph. and Space Res. in Japan* **16,** 363.
Oguti, T.: 1963a, *Rep. on Ionosph. and Space Res. in Japan* **17,** 291.
Oguti, T.: 1963b, Polar Section Nat. Sci. Museum, *Sci. Rep. Ser. A,* No. 1.
Omholt, A.: 1955, *J. Atm. Terr. Phys.* **7,** 73.
Omholt, A.: 1962, *Plan. Space Sci.* **9,** 285.
Omholt, A., Stoffregen, W., and Derblom, H.: 1962, *J. Atm. Terr. Phys.* **24,** 203.
Omholt, A.: 1963, *Plan. Space Sci.* **10,** 247.
Parthasarathy, R. and Berkey, F. T.: 1965, *J. Geophys. Res.* **70,** 89.
Ratcliffe, J. A.: 1960, *Physics of the Upper Atmosphere,* Academic Press.
Rees, M. H.: 1963, *Plan. Space Sci.* **11,** 1209.
Reid, G. C. and Collins, C.: 1959, *J. Atm. Terr. Phys.* **14,** 63.
Reid, G. C. and Rees, M. H.: 1959, *Nature* **184,** 539.
Roach, F. E. and Marovich, E.: 1959, *J. Res. Natl. Bur. Std.* **D63,** 297.
Roach, F. E. and Rees, M. H.: 1960, *J. Geophys. Res.* **65,** 1489.
Roach, F. E. and Roach, J. R.: 1963, *Plan. Space Sci.* **11,** 523.
Romic, G. J. and Elvey, C. T.: 1958, *J. Atm. Terr. Phys.* **12,** 283.
Sandford, B. P.: 1961, *Nature* **190,** 245.
Sandford, B. P.: 1963, *Plan. Space Sci.* **10,** 195.
Shaw, J. E.: 1959, *Plan. Space Sci.* **2,** 49.
Smith, E. K. jr.: 1957, *Natl. Bur. Std. Circ.* **582.**
Spencer, L. V.: 1959, *Nat. Bur. Std. Monograph* **1.**
Stoffregen, W.: 1961, *J. Atm. Terr. Phys.* **21,** 257.
Stoffregen, W. and Derblom, H.: 1962, *Plan. Space Sci.* **9,** 711.
Stringer, W. J., Belon, A. E., and Akasofu, S.-I.: 'The Latitude of Auroral Activity During Periods of Zero and Very Weak Magnetic Disturbance'. Preprint, 1965.
Störmer, C.: 1910, *Acad. Sci.* **151,** 736.
Störmer, C.: 1955, *The Polar Aurora,* Clarendon Press, 1955.
Ulwick, J. C., Pfister, W., Haycock, O. C., and Balen, K. D.: 1964, *Environmental Research Paper,* No. **56,** (AFCRL–1964).
Unwin, R. S.: 1959, *Nature* **183,** 1044.
Vegard, L. and Krogness, O.: 1920, *Geophys. Publ.* **1,** No. 1.
Vegard, L.: 1939, *Nature* **144,** 1089.
Vegard, L.: 1948, URSI, Mixed Commission on the Ionosphere, Brussels.
Walt, M. and Mac Donald, W. M.: 1962, *J. Geophys. Res.* **67,** 5013.
Walt, M., Crane, G. E., and MacDonald, W. M.: 1964, *Space Res.* **4,** (ed. by P. Muller), 631.
Webster, H. F.: 1957, *J. Appl. Phys.* **28,** 1388.

Discussion

H. STOLOV presented some preliminary attempts at correlating f_{min} measurements in the auroral region with satellite particle measurements. These are described in the following brief.

IONOSPHERIC RESPONSE TO PARTICLE PRECIPITATION BY f_{min}-BLACKOUT TECHNIQUE

HAROLD L. STOLOV*

Goddard Institute for Space Studies, National Aeronautics and Space Administration, New York, N. Y. 10027

Abstract, Using the f_{min}-blackout method of studying auroral absorption, data from over 50 ionosonde stations in the northern hemisphere are analyzed during the geomagnetic storm with SC 0810 UT on 28 October 1961. Injun I energetic particle measurements are available during the main phase. It is demonstrated that auroral absorption is present on both sides of the high-latitude boundary of the trapping region for electrons greater than 40 keV. The increased auroral absorption observed during the morning hours is apparently associated with the increased precipitation of energetic electrons during the morning hours reported by several investigators. Large fluxes of electrons with energies between 10 keV and 40 keV, reported north of the 40 keV boundary, and the associated auroral absorption observed north of the boundary suggest that the high-latitude trapping boundary for 10 keV electrons might be at a greater latitude than the high-latitude trapping boundary for 40 keV electrons. Ionospheric responses to isolated high-latitude spikes of 40 keV electrons have not yet been detected.

Introduction

Satellite measurements of the dumping or precipitation of energetic electrons into the atmosphere have been extensive (O'BRIEN, 1962, 1964; FRANK *et al.*, 1964; MCDIARMID *et al.*, 1963). Detailed studies of the absorption of HF and VHF radio waves by ionosonde and riometer techniques have investigated the ionization of the lower auroral ionosphere (OBAYASHI, 1959; OBAYASHI and HAKURA, 1960; BASLER, 1963; ANSARI, 1964). To the best of our knowledge, no attempts have been made to correlate the particle precipitation and auroral absorption measurements on a global scale. However, two limited studies have been conducted. MÆHLUM and O'BRIEN (1963) calculate the percentage of ionograms that show no ionospheric echo (blackout) for each 6-hour interval, using the quarter-hourly observations of 5 ionosonde stations in North America. From Injun I data for the same period, the northern boundary (L_n) of trapped electrons is determined. Their conclusion is that the northern boundary of the auroral absorption is the northern boundary of energetic electrons ($E > 40$ keV). JELLY *et al.* (1964) affirm a positive statistical correlation between auroral absorption and precipitated electrons of energies greater than 40 keV using individu-

* On leave from Department of Physics of the City College of the City University of New York on a National Academy of Sciences – National Research Council senior postdoctoral research associateship.

al satellite passes of Alouette I selected within 5° longitude of Canadian riometer stations. However, the correlation is shown to be quite weak on occasion.

The general purpose of this preliminary study is to assist in understanding the acceleration and precipitation processes within the magnetosphere by correlating the ionospheric and particle measurements on a global scale. The following questions will be investigated:

1. Is the auroral absorption inside the high-latitude boundary of the trapping region, as suggested by MÆHLUM and O'BRIEN (1963), or might it be outside or straddling the boundary?

2. Is the morning-afternoon asymmetry in the distribution of precipitated electrons (FRANK *et al.*, 1964; McDIARMID and BURROWS, 1964) visible in the measurements of auroral absorption?

3. Are the high-latitude boundaries of the trapping region for 40 keV electrons and for 10 keV electrons the same?

Method of Analysis

Due to the extreme variability of auroral zone phenomena, detailed studies of the morphology of individual events must be undertaken to examine the conclusions suggested by the statistical investigations. In this paper, ionosphere data from over 50 stations in the northern hemisphere and Injun I measurements of trapped particles are examined for the geomagnetically disturbed period following the sudden commencement at 0810 UT on 28 October 1961. The ionosonde probes the ionosphere at vertical incidence with a pulsed radio transmitter, sweeping from 1–20 Mc/sec with data taken at 15 minute intervals. The minimum frequency returned to ground is f_{min}. Since absorption is inversely proportional to frequency squared, the lower frequencies are absorbed and f_{min} is increased when there is unusual ionization below 100 km. The minimum frequency depends on local time (maximum at noon), season, and solar cycle. Therefore, in order to study anomalous ionization produced by particle precipitation, the Δf_{min} value is used. This is the departure of f_{min} from the median value for that hour for that month. OBAYASHI and HAKURA (1960) have pioneered the use of the f_{min} technique for the study of ionosphere storms both of the polar cap and auroral zone variety. It is important that the limitations of the f_{min} method be recognized as well as the advantages.

Limitations: (a) Since critical frequencies and equivalent heights are of major interest in ionospheric studies, the sensitivity of the ionosonde is usually adjusted to obtain the best recording of these parameters. Changes of transmitted power, radio interference, and low critical frequency of the reflecting region, all affect f_{min}.

(b) The sensitivity of the ionosonde is very high, and for strong absorption events total blackout is recorded at an early stage.

As a result of the limitations mentioned above, detailed quantitative studies are seldom attempted by the f_{min} method.

Advantages: (a) The second limitation is an advantage as well, since the f_{min} technique will detect weak absorption events that are missed by riometers. In fact, many weak PCA (polar cap absorption) events were detected in this way (JELLY and COLLINS, 1962).

(b) There is available a complete geographical coverage with ionosonde stations operating on a routine basis at all times. The distribution of riometers is rather sparse.

Since we do not even understand qualitatively the factors involved in acceleration and precipitation, the difficulties with the quantitative aspects of the f_{min} technique are far from pressing at this time.

Ionospheric Observations

The Δf_{min} values for over 50 ionosonde stations are plotted on a northern hemisphere map in geomagnetic coordinates. Contours are constructed for Δf_{min} values of 1 Mc/sec, 2 Mc/sec, and 3 Mc/sec. Values of Δf_{min} greater than 3 Mc/sec are rare, and are contained within the hatched blackout region. The position of the sun is indicated on the map margin.

Satellite Observations

The State University of Iowa satellite Injun I was aloft during the period in an orbit with initial apogee 1010 km, perigee 890 km, period 104 minutes and inclination 67°. Since the satellite did not orient with respect to the local field vector as expected, the 213 G-M directional detector for electrons greater than 40 keV made varying angles with the magnetic field vector, measuring trapped particles at times and dumped particles at other times. During the geomagnetic storm being investigated, the satellite made only four passes across the northern hemisphere, and the 213 counter was measuring trapped electrons in each case. No solar protons were observed at high latitudes during this period that could contaminate the measurements. Figures 1 to 4 show the four passes plotted on the corresponding ionospheric map. The abscissa gives the time and location of the satellite. The ordinate gives the 213 G-M counting rate in counts/sec on a log scale.

Analysis

Figure 1 shows the auroral absorption south of the high-latitude boundary of the trapping region exactly as indicated by MÆHLUM and O'BRIEN (1963). However, Figure 2 indicates, just as clearly, the auroral absorption north of the high-

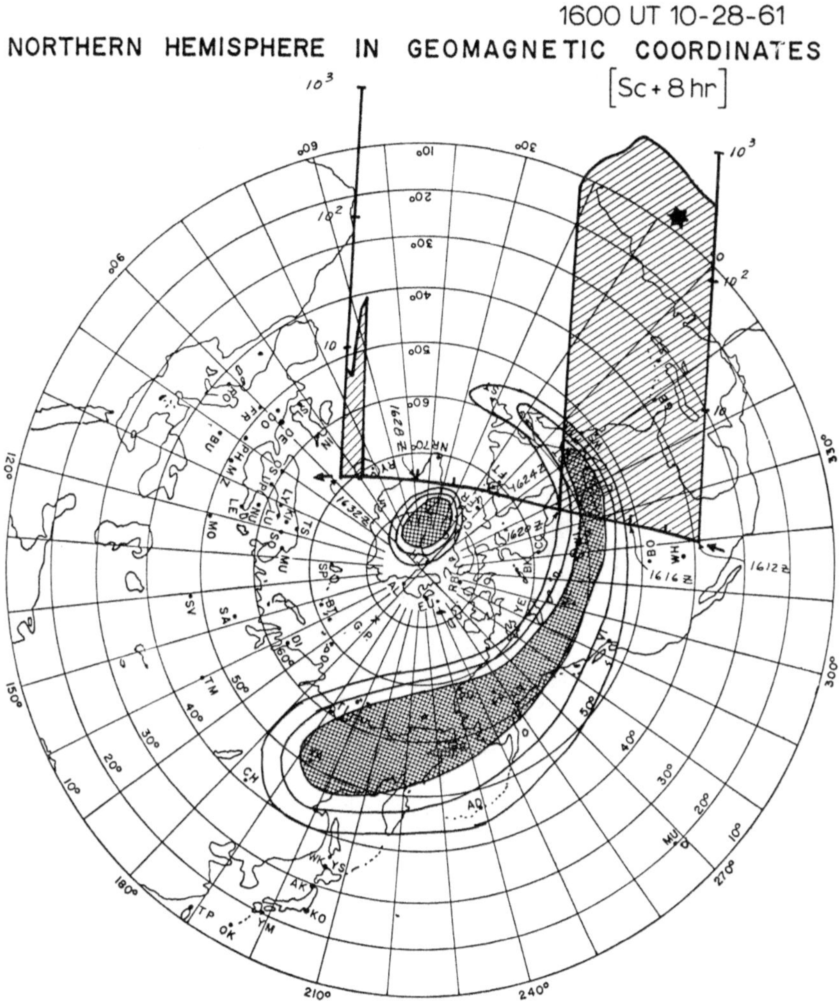

Fig. 1. Global pattern of $\Delta f_{\min}$ at 1600 UT on 28 October 1961 and Injun I pass no. 1685 showing 213 G-M counting rate in counts/sec on a log scale.

latitude boundary. The synchronization of satellite and ionospheric measurements is almost perfect in Figure 2 with the boundary of trapping measured at approximately 1803 UT and ionospheric measurements taken at 1800 UT. It is highly unlikely that a sudden impulsive shift of the boundary could be invoked to explain Figure 2, since the k_p values were consistently high during the period. Figure 3 would have presented another opportunity to observe the relationship between the boundary and the auroral absorption, but unfortunately data were

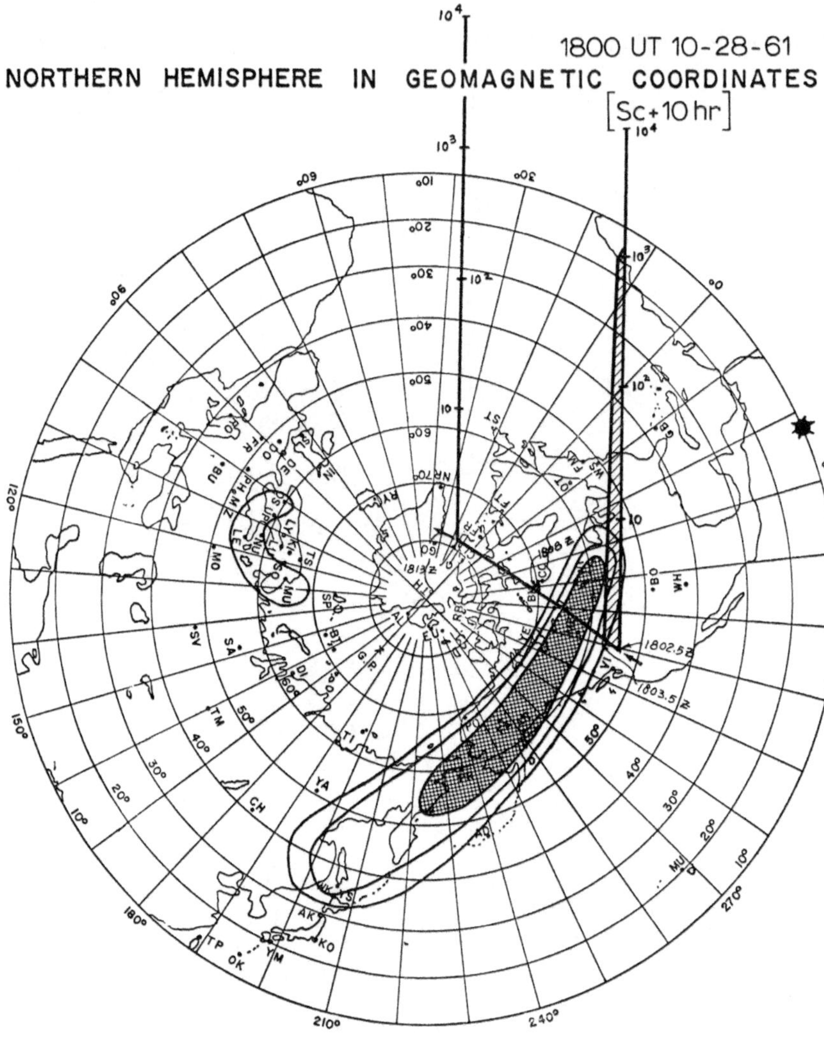

Fig. 2. Global pattern of $\Delta f_{\min}$ at 1800 UT on 28 October 1961 and Injun I pass no. 1686 showing 213 G-M counting rate in counts/sec on a log scale.

not available from the 213 counter for the period from 1951.5 UT to 1954 UT, as Injun I crossed the blackout region. Although a single case study of this nature can hardly be regarded as conclusive, it indicates that auroral absorption is present on both sides of the high-latitude boundary of the trapping region for 40 keV electrons and implies the presence of fluxes of precipitating electrons of energies less than 40 keV north of the high-latitude trapping boundary for

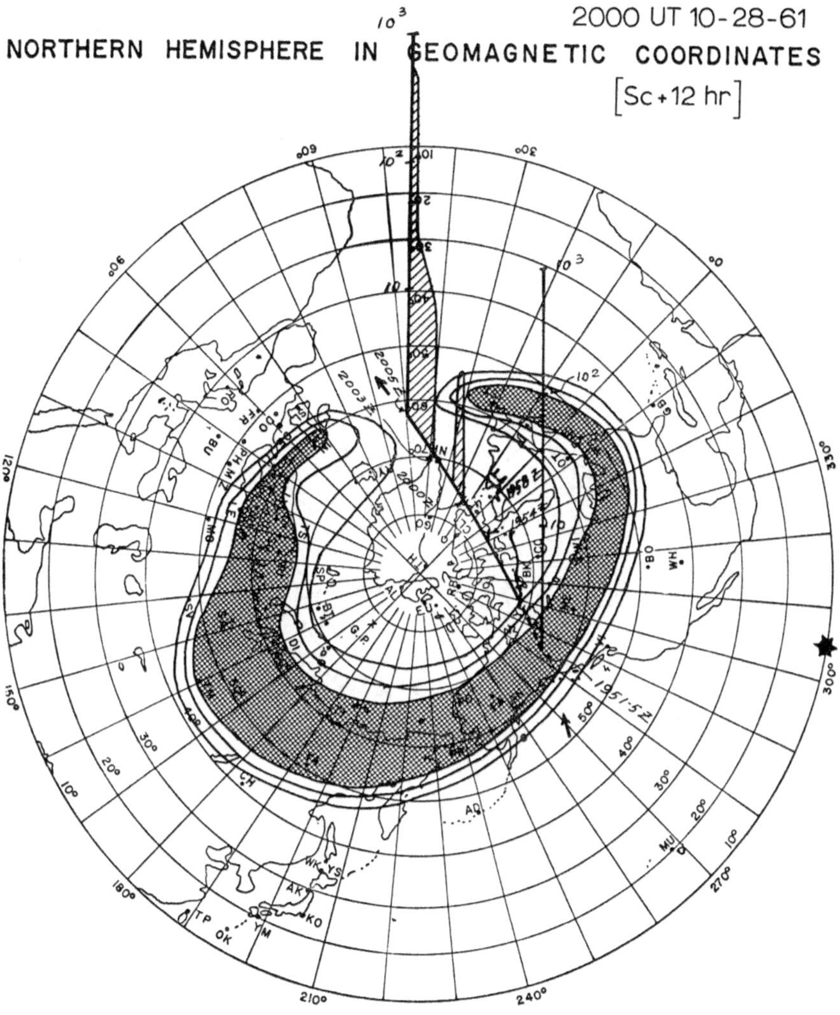

Fig. 3. Global pattern of Δf_{min} at 2000 UT on 28 October 1961 and Injun I pass no. 1687 showing 213 G-M counting rate in counts/sec on a log scale.

40 keV electrons. Support for these conclusions may be found in other investigations.

Fritz and Gurnett (1965) report a flat energy spectrum between 10 keV and 40 keV south of the high-latitude trapping boundary for 40 keV electrons. However, north of the boundary the spectrum becomes extremely steep, and a soft electron spectrum with large fluxes of electrons between 10 keV and 40 keV,

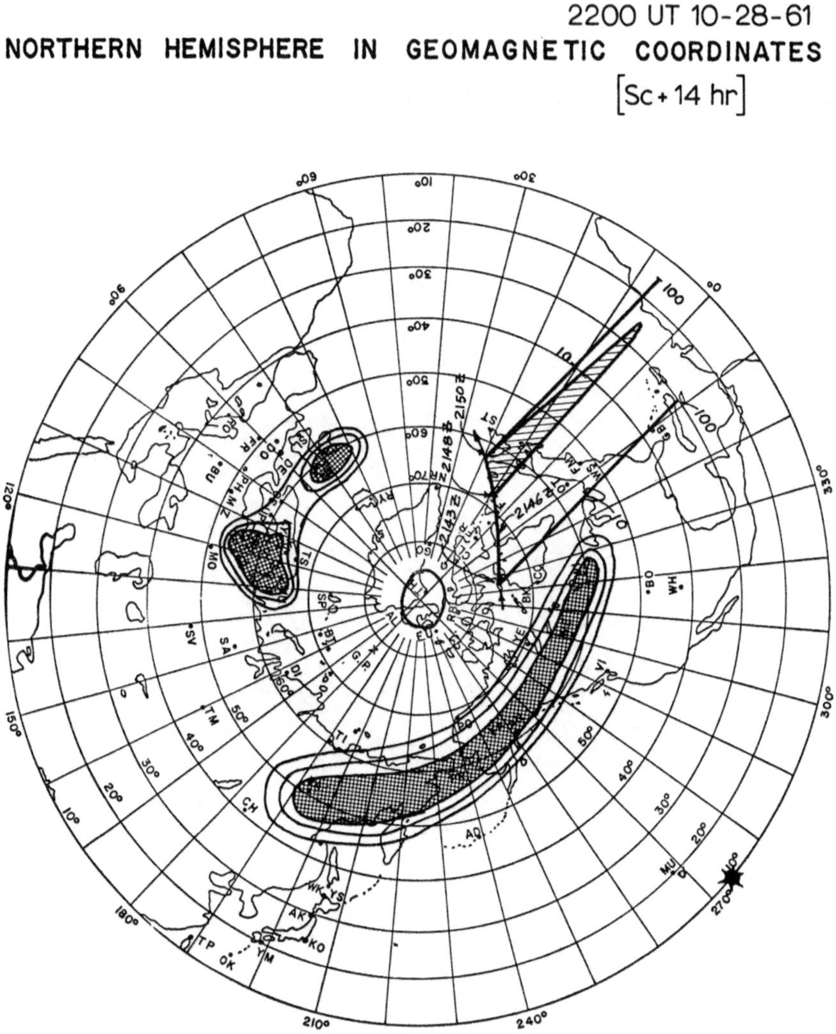

Fig. 4. Global pattern of $\Delta f_{\min}$ at 2200 UT on 28 October 1961 and Injun I pass no. 1688 showing 213 G-M counting rate in counts/sec on a log scale.

is observed during local night. These intense fluxes of low energy electrons occur predominantly during periods of high geomagnetic activity.

SHARP *et al.* (1964, 1965), measuring the total energy of precipitating electrons of energy greater than 80 eV in the auroral regions, find values higher by a factor of 5 on the night side than on the day side. Spectra are harder on the day side than on the night side.

O'BRIEN and TAYLOR (1964) indicate that weak auroral activity has been found to extend to higher latitudes than the precipitation of electrons with energies greater than 40 keV, indicating that electrons of $E<40$ keV are being precipitated at these higher latitudes.

It is of some interest to see if the morning-afternoon asymmetry in the precipitation of electrons greater than 40 keV reported by FRANK *et al.* (1964) and MCDIARMID and BURROWS (1964) can be found in the synoptic ionospheric analysis. The ionospheric maps verify to some extent the reported increased precipitation during the morning hours. However, Figure 3, for the most disturbed period shows the precipitation to be essentially world-wide. Maxima of auroral absorption during the morning hours have also been demonstrated in the riometer statistical studies of HOLT *et al.* (1962) and HARTZ *et al.* (1963).

Another item of interest, worthy of mention, is the isolated high-latitude spike of intensity of electrons greater than 40 keV in the Injun 1 measurements seen in Figure 3. High-latitude spikes of this variety, well removed from the 40 keV trapping boundary, have been studied in detail by MCDIARMID and BURROWS (1965) with Alouette I particle detectors. In all probability, the spikes are related to the "islands" of energetic electron fluxes observed by ANDERSON *et al.* (1965) and ANDERSON (1965) using detectors on Imp I. The relationships of spikes and islands to neutral sheet processes (NESS, 1965; AXFORD *et al.*, 1965) in the magnetospheric tail are clearly items for further investigation. The limited spatial extent of the high-latitude spike in Figure 3 and the absence of a nearby ionosonde station made it impossible to study the radio wave absorption associated with the event. Future studies of ionospheric response with riometers should contribute to our understanding of the morphology of these high-latitude phenomena.

Summary

The correlation of ionospheric and particle measurements on a global scale leads to the following conclusions:

1. Auroral absorption is present on both sides of the high-latitude boundary of the trapping region for electrons greater than 40 keV.

2. The increased auroral absorption observed during the morning hours is apparently associated with the increased precipitation of energetic electrons during the morning hours.

3. There are indications that the high-latitude boundary for 10 keV electrons might be at a greater latitude than the high-latitude trapping boundary for 40 keV electrons. This is implied by the large fluxes of electrons with energies between 10 keV and 40 keV observed north of the 40 keV boundary and by the presence of auroral absorption north of the 40 keV boundary. Further study will have to be undertaken with Injun 3 data, to decide conclusively the question

of the high-latitude trapping boundary locations for 10 keV and 40 keV electrons.

Further details

A complete synoptic study of the auroral absorption associated with the geomagnetic storm with SC at 0810 UT on 28 October 1961, containing hourly ionospheric maps during the height of the storm, is submitted to the *Journal of Atmospheric and Terrestrial Physics* (1966).

Acknowledgements

It is a pleasure to thank Drs. J. A. Van Allen and L. A. Frank for their assistance in supplying the Injun I data and for helpful suggestions on the manuscript. I especially wish to thank Dr. Robert Jastrow for his hospitality while I held a National Academy of Sciences – National Research Council senior postdoctoral research associateship at the Institute for Space Studies.

References

Anderson, K. A.: 1965, *J. Geophys. Res.* **70,** 4741.
Anderson, K. A., Harris, H. K., and Paoli, R. J.: 1965, *J. Geophys. Res.* **70,** 1039.
Ansari, Z. A.: 1964, *J. Geophys. Res.* **69,** 4493.
Axford, W. I., Petschek, H. E., and Siscoe, G. L.: 1965, *J. Geophys. Res.*, **70,** 1231.
Basler, R. P.: 1963, *J. Geophys. Res.* **68,** 4665.
Frank, L. A., Van Allen, J. A., and Craven, J. D.: 1964, *J. Geophys. Res.* **69,** 3155.
Fritz, T. A. and Gurnett, D. A.: 1965, *J. Geophys. Res.* **70,** 2485.
Hartz, R. R., Montbriand, L. E., and Vogan, E. L.: 1963, *Can. J. Phys.* **41,** 581.
Holt, C., Landmark, B., and Lied, F.: 1962, *J. Atmos Terr. Phys.* **23,** 229.
Jelly, D. H. and Collins, C.: 1962, *Can. J. Phys.* **40,** 706.
Jelly, D. H., McDiarmid, I. B., and Burrows, J. R.: 1964, *Can. J. Phys.* **42,** 2411.
McDiarmid, I. B., Burrows, J. R., Budzinski, E. E., and Wilson, M. D.: 1963 *Can. J. Phys.* **41,** 2064.
McDiarmid, I. B. and Burrows, J. R.: 1964, *Can. J. Phys.* **42,** 1135.
McDiarmid, I. B. and Burrows, J. R.: 1965, *J. Geophys. Res.* **70,** 3031.
Mæhlum, B., and O'Brien, B. J.: 1963, *J. Geophys. Res.* **68,** 997.
Ness, N. F.: 1965, *J. Geophys. Res.* **70,** 2989.
Obayashi, T.: 1959, *J, Radio Res. Lab.* **6,** 375.
Obayashi, T. and Hakura, Y.: 1960, *J. Atmos. Terr. Phys.* **18,** 101.
O'Brien, B. J.: 1962, *J. Geophys. Res.* **67,** 3687.
O'Brien, B. J.: 1964, *J. Geophys. Res.* **69,** 13.
O'Brien, B. J. and Taylor, H.: 1964, *J. Geophys. Res.* **69,** 45.
Sharp, R. D., Evans, J. E., Imhof, W. L., Johnson, R. G., Reagan, J. B., and Smith, R. V.: 1964, *J. Geophys. Res.* **69,** 2721.
Sharp, R. D., Evans, J. E., Johnson, R. G., and Reagan, J. B.: 1965, *Space Res.* **5,** 282.

THE TAIL OF THE MAGNETOSPHERE AND PARTICLE ACCELERATION

F. C. HOH
Boeing Scientific Research Laboratories, Seattle, Wash.

This paper is a shorter version of an earlier paper (HOH, 1965) but emphasizes some of the questions which arose during the Institute.

The IMP satellite has measured a magnetic field configuration on the night side of the earth as shown in Figure 1, where the directions are given (NESS, 1965). The field is directed along the z axis and has no steady variations in the y-z plane over large distances. It changes sign at $x=0$ and has a value of about 15γ far away from the $x=0$ plane. The thickness of the neutral sheet 2λ is ~ 600 km, the width L in the y direction is $\sim 10^5$ km, and the length l in the z direction is greater than 10^5 km.

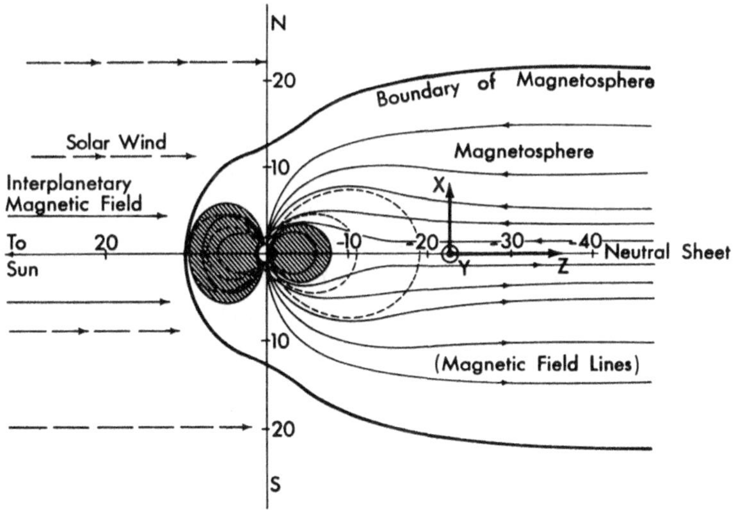

Fig. 1. Magnetic field configuration measured by the IMP satellite.

1. Sheet Pinch Model

When we look for a model for the neutral sheet, the question of the existence of a time independent electric field in the y direction E_y was often raised. As we can see in Figure 1, an E_y field will produce an $\mathbf{E}_y \times \mathbf{B}_z$ drift in the x direction that collects (or deletes) plasma to the neutral surface from both its sides. In steady state, the plasma so piled up on the surface must flow away along the y or z axis. SPEISER (1965) thinks that the plasma, after acceleration along the y axis by E_y, is in some way deflected and made to flow along the field lines (z axis) towards the earth and cause aurora. However, his theory is not derived self-consistently. In particular, such a flow of plasma would build up gradients of plasma flux nV_z along the z axis. If the flow velocity V_z is somewhat constant, the density n must increase monotonically towards the earth. If the particle temperature perpendicular to the magnetic field $T_\perp$ is constant along z or does not decrease towards the earth, the particle pressure $nkT_\perp$ also must increase towards the earth. This pressure is balanced by the magnetic pressure $B_{max}^2/2\mu_0$. Since observations do not see a systematic increase of B_{max} towards the earth, we conclude that this plasma flow model is quite unlikely. This criticism also applies to the suggestion that plasma flows along the y axis instead of the z axis.

A more natural model for the tail is the sheet pinch model, which is the planar version of the BENNETT (1934) cylindrical pinch. It describes the equilibrium of a plasma sheet which carries a current flowing along the y axis. The current produces a magnetic field which in turn confines the plasma itself.

The plasma is collisionless and has no variation along the y and the z axis (i.e. $E_y = -\partial\phi/\partial y = 0$). The sheet pinch model is constructed on the assumption that the distribution function f is of the Maxwell-Boltzmann type:

$$f \sim \exp - \{W + UP_y\}, \tag{1}$$

$$W = \frac{mv^2}{2} + q\phi(x), \qquad P_y = mv_y + qA_y(x) \tag{2}$$

where W is the total energy and P_y the canonical momentum of the particle, m the particle mass, v the velocity, q the charge, ϕ the electro-static potential, A_y the vector potential in the y direction, and U a constant. Inserting this distribution into the Maxwell equations (see HOH, 1965), one obtains

$$B_z = -[2\mu_0 n_0 K(T_e + T_i)] \tanh(x/\lambda), \tag{3}$$

$$n(x) = n_0 \cosh^{-2}(x/\lambda), \tag{4}$$

where λ is related to the electron and ion temperatures T_e and T_i, to the density n_0 at $x=0$, and to U_i. The spatial distributions of (3) and (4) fit the measurements rather well. We assume $T_e+T_i \approx 600$ eV (3) gives $n_0 \approx 1/\text{cm}^3$ for a maximum field of 15 gamma.

It is shown that the above results are also valid for relativistic particle energies. Further, they do not change significantly if we replace v^2 in (2) by some function which makes f in (1) non-Maxwellian so as to contain more high energy particles.

2. Particle Acceleration

The first acceleration mechanism utilizes particle orbits shown in Figure 2. In the field of (3), there are two types of particle motions. One of them does not cross the neutral surface but gyrates and drifts in the inhomogeneous magnetic field. The other type crosses the neutral surface and performs a sinusoidal type of orbit as is shown in Figure 2. If we follow the motion of all the particles which cross the $x=0$ surface at a certain time, about 130/180 of them will subsequently move along the positive y axis. These particles are most easily amenable for acceleration.

Although our model does not allow a steady E_y, one can show that a small

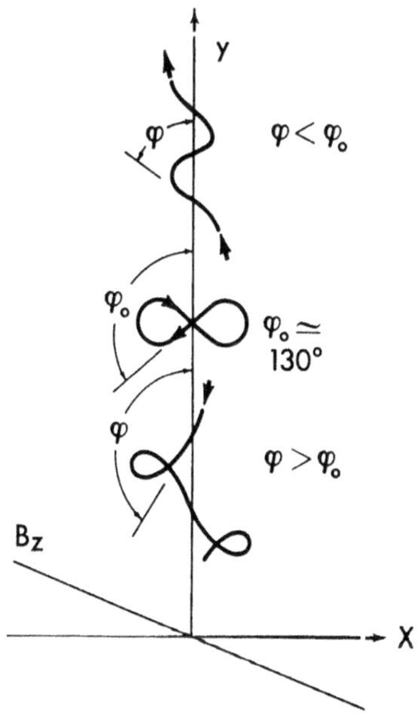

Fig. 2. Type of particle motion, according to Equation 3. It crosses the neutral surface and performs a sinusoidal type of orbit.

and slowly oscillating $E_y = E_0 \sin\omega_a t$ can be allowed. In such a field, the plasma sheet will be thinner and thicker with frequency ω_a. If we prescribe that the fluctuation amplitude is a small quantity of order $\varepsilon < 0.2$, one gets a criterion that $E_0/\omega_a = \varepsilon B_{\max}\lambda/2 < 4.5 \times 10^4$ (mks). Assuming $\omega_a \approx 0.1/s$ (which is much less than the cyclotron frequency), then $E_0 = 4.5 \times 10^{-5}\ v/m$, and the total voltage drop across the sheet in the y direction $E_0 L$ is about 4.5 keV.

The average crossing electrons, having an energy of about 300 eV, would traverse the sheet in a time of about 20 sec. Since this time is less than $2\pi/\omega_a$, the electrons will see a constant $E_y(\approx E_0)$ field during their traversal and pick a maximum energy of a few keV. Thus, the neutral sheet with its finite size, can act as an *acceleration source for the electrons*. The average ion traversal time is, however, much longer than $2\pi/\omega_a$ so the mechanism is not effective for ions.

For aurora purposes, however, these few keV electrons have large energies perpendicular to the field lines and have great difficulty in reaching the atmosphere because of the huge mirror ratio present.

The second acceleration mechanism is based upon instability of the sheet. The sheet pinch is unstable microscopically against a perturbation of the form $g(x)\sin kz$. One can conclude that a coupled Alfvén and electro-static wave will propagate along the z axis. If the unstable amplitude is again of the order of $\varepsilon \approx 0.2$, the electro-static potential trough can be of 5 keV high. Ions and electrons trapped by this trough will be accelerated to that energy. This may have some relevance to the measurements by K. A. Andersson (1965).

References

Anderson, K. A.: 1965, *Phys. Rev. Letters* **14**, 888.

Hoh, F. C.: 1965, Boeing Scientific Research Laboratories Report No. D1-82-0457; submitted to *J. Geophys. Res.*

Ness, N. F.: 1965, to be published in *J. Geophys. Res.*

Speiser, T. W.: 1965, *Goddard Space Flight Center Report* X-640-65-165, April (1965).

A FLUID MODEL FOR THE INTERACTION OF THE SOLAR WIND AND THE GEOMAGNETIC FIELD

JOHN R. SPREITER, AUDREY L. SUMMERS, and ALBERTA Y. ALKSNE
Ames Research Center, NASA, Moffett Field, California

Data obtained in space, particularly that from IMP-1, have established that many of the gross features of the interaction of the solar wind and the geomagnetic field can be described by the continuum theory of fluid flow. The purpose of this paper is twofold: (a) to indicate briefly how one can arrive at results for the magnetosphere boundary, the bow wave, and the associated flow field from a purely fluid point of view without resort to a mixture of particle and fluid models as usually employed and (b) to present numerical results for the flow parameters in the shock layer between the bow wave and the magnetosphere boundary. The objective is not to get a complicated theory that encompasses all phenomena of importance, but rather the simplest body of analysis that appears able to describe the average bulk properties of the solar wind as it flows steadily through the bow wave and around the forward portion of the magnetosphere. A more complete account of this study is given by SPREITER *et al.* (1966).

The fundamental assumption is that the flow can be described adequately by the standard magnetohydrodynamic equations for steady flow of a dissipationless perfect gas. No attempt will be made here to justify this assumption in detail, except to remark that the presence of a weak and irregular magnetic field in the incident solar wind plasma seems sufficient to couple the motions of the particles even in the absence of collisions. The differential equations are thus as follows (see, e.g., LANDAU and LIFSHITZ, 1960).

$$\left.\begin{aligned} &\nabla\cdot\rho\mathbf{v}=0\\ &\rho(\mathbf{v}\cdot\nabla)\mathbf{v}+\nabla p=-\frac{1}{4\pi}\mathbf{H}\times\mathbf{curl}\,\mathbf{H}=-\frac{1}{8\pi}\nabla H^2+\frac{1}{4\pi}(\mathbf{H}\cdot\nabla)\mathbf{H}\\ &\mathbf{curl}(\mathbf{H}\times\mathbf{v})=0,\qquad \operatorname{div}\mathbf{H}=0\\ &(\mathbf{v}\cdot\nabla)S=0,\qquad p=e^{S/c_v}\rho^{\gamma}\end{aligned}\right\}\qquad(1)$$

where ρ, p, S, and v refer to the density, pressure, entropy, and velocity of the gas, $\mathbf{H}$ refers to the magnetic field, $\gamma=c_p/c_v$, and c_p and c_v are constants representing the specific heats at constant pressure and constant volume. Important auxiliary relations for γ, temperature T, speed of sound a, internal energy e, and enthalpy h are

$$\left.\begin{aligned} &\gamma = (N+2)/N\,, \qquad p = \rho RT/\mu = nkT \\ &a = (\partial p/\partial \rho)^{1/2} = (\gamma p/\rho)^{1/2} = (\gamma RT/\mu)^{1/2} \\ &e = c_v T\,, \qquad h = c_p T = e + p/\rho \end{aligned}\right\} \tag{2}$$

where N represents the number of degrees of freedom, $R = (c_p - c_v)\mu = 8.314 \times 10^7$ erg/°K, μ = mean molecular weight $= \frac{1}{2}$ for fully ionized hydrogen plasma, n = number of particles/cm^3 $= 2n_p$ where n_p is the number of protons/cm^3, and k = Boltzmann's constant $= 1.38 \times 10^{-16}$ erg/°K.

Although only first derivatives appear in this system of equations, the neglected dissipative terms are described by second derivatives. Their neglect requires that the gradients be small. In magnetohydrodynamics, as in gasdynamics, however, compressions tend to coalesce and steepen into shock waves of such small thickness that they can be considered as virtual discontinuities. In addition, attraction between like currents tends to cause current distributions to collapse into thin sheaths, across which the magnetic field can be considered in the same sense to be nearly discontinuous. Mathematically, continuous solutions of the dissipationless differential equations cease to exist, and the flow is no longer governed solely by the equations given above. Mass, momentum, magnetic flux, and energy must still be conserved, however, and the following relations must hold between quantities on the two sides of any such discontinuity:

$$\left.\begin{aligned} [\rho v_n] &= 0 \\ \left[\rho v_n \mathbf{v} + \left(p + \frac{H^2}{8\pi}\right)\hat{n} - \frac{1}{4\pi} H_n \mathbf{H}\right] &= 0 \\ [H_n \mathbf{v}_t - \mathbf{H}_t v_n] &= 0\,, \qquad [H_n] = 0 \\ \left[\rho v_n \left(\tfrac{1}{2} v_n^2 + h\right) + v_n \frac{H^2}{4\pi} - \frac{1}{4\pi} H_n \mathbf{v}\cdot\mathbf{H}\right] &= 0 \end{aligned}\right\} \tag{3}$$

The subscripts n and t refer to components normal and tangential to the discontinuity surface and $[Q] = Q_1 - Q_0$ where subscipts 0 and 1 refer to conditions on the upstream and downstream sides of the discontinuity.

Five classes of discontinuities are described by Equation (3). Following Landau and Lifshitz (1960), discontinuities which lie along streamlines ($v_n = 0$) are called tangential discontinuities or contact discontinuities according to whether or not the normal component of the magnetic field H_n vanishes. Discontinuities across which there is flow ($v_n \neq 0$) are divided into three categories called rotational discontinuities, and fast and slow shock waves. Some proper-

ties that distinguish the various discontinuities are summarized below:
Tangential discontinuities

$$v_n = H_n = 0; \quad [\mathbf{v}_t] \neq 0, \quad [\mathbf{H}_t] \neq 0, \quad [\rho] \neq 0, \quad [p + H^2/8\pi] = 0 \tag{4}$$

Contact discontinuities

$$v_n = 0, \quad H_n \neq 0; \quad [\mathbf{v}] = [\mathbf{H}] = [p] = 0, \quad [\rho] \neq 0 \tag{5}$$

Rotational discontinuities

$$\left. \begin{array}{l} v_n = \pm H_n/(4\pi\rho)^{1/2}, \quad [\mathbf{v}_t] = [\mathbf{H}_t]/(4\pi\rho)^{1/2}, \\ [\rho] = [p] = [v_n] = [H_n] = [H^2] = 0 \end{array} \right\} \tag{6}$$

Shock waves, fast and slow

$$\left. \begin{array}{l} v_n \neq 0, \quad [\rho] > 0, \quad [H_n] = 0, \\ (\rho v_n)_{\text{fast}} \geqslant (\rho v_n)_{\text{rot}} = H_n(\rho/4\pi)^{1/2} \geqslant (\rho v_n)_{\text{slow}} \\ \mathbf{H}_t \text{ and } H^2 \begin{pmatrix} \text{increase} \\ \text{decrease} \end{pmatrix} \text{ through } \begin{pmatrix} \text{fast} \\ \text{slow} \end{pmatrix} \text{ shock waves} \end{array} \right\} \tag{7}$$

Of these, only the tangential discontinuity has properties compatible with those required to describe the boundary of the geomagnetic field. As in the classical Chapman-Ferraro theory based on particle concepts, the condition $H_n = 0$ holds and requires that there is no connectivity between the geomagnetic and interplanetary fields. Slight differences stem from the additional assumptions in the Chapman-Ferraro theory that the incident plasma is free of magnetic field and the outer magnetosphere is free of plasma so that $p_0 = H_1^2/8\pi$. Although neither of these statements is strictly true, estimates of the magnitudes of the gas pressure p and the magnetic pressure $H^2/8\pi$ lead to the conclusion that $p \ll H^2/8\pi$ in the outer magnetosphere and that $p \gg H^2/8\pi$ in the shock layer. So far as the flow outside the magnetosphere is concerned, the discontinuities at the magnetosphere boundary may thus be approximated satisfactorily by those of the limiting case of a tangential discontinuity in which there is a vacuum on one side, and negligible magnetic pressure on the other side. It is demonstrated further by Spreiter *et al.* (1966) that the gas pressure on the magnetosphere boundary is adequately approximated by the simple Newtonian formula $p = K\rho_\infty v_\infty^2 \cos^2\psi$ where the subscript ∞ refers to the values in the undisturbed incident stream, ψ is the angle between the normal to the boundary and the velocity vector of the undisturbed stream, and K is a constant usually taken to be unity but which yields better accuracy if equated to 0.84 for $\gamma = 2$

and 0.88 for $\gamma = 5/3$. In this way, the calculation of the shape of the magnetosphere boundary is decoupled from the detailed analysis of the surrounding flow and reduced to the identical mathematical problem described by the classical Chapman-Ferraro theory.

Examination of conditions typical of the solar wind shows that the mass flux ρv_n through the nose of the bow wave is characteristically about an order of magnitude greater than that through a rotational discontinuity. The bow wave must therefore be a fast shock wave.

Although it has been necessary to invoke the presence of a magnetic field to provide a mechanism for providing interactions between the particles, and to consider the complications of hydromagnetic theory to provide an adequate description of the magnetosphere boundary as a tangential discontinuity and the bow wave as a fast hydromagnetic shock wave, the magnetic field strengths typically encountered in the solar wind are sufficiently small as otherwise to produce only secondary effects on the flow. On this basis, one can justify the neglect of all the terms containing the magnetic field H in Equations (1) and (3) when calculating the flow. The fluid motion then calculated is just exactly that indicated by ordinary gasdynamic theory. Deformations of the magnetic field are determined subsequently using the equations

$$\operatorname{curl}(\mathbf{H} \times \mathbf{v}) = 0, \quad \operatorname{div} \mathbf{H} = 0, \quad [H_n] = 0 \tag{8}$$

which are frequently interpreted as indicating the magnetic field lines move with the fluid.

In this way, the exterior magnetosphere flow problem is reduced to a purely gasdynamic problem of supersonic flow past a given body, the shape of which is determined by solving the standard Chapman-Ferraro problem. Difficulties associated with the nonlinear and mixed elliptic-hyperbolic character of the governing partial differential equations are such, however, that it is necessary at the present time to approximate the magnetosphere boundary with an axisymmetric shape. The precise shape selected is that obtained by rotating the equatorial trace of the boundary given by the approximate solution of SPREITER and BRIGGS (1962) about an axis through the center of the earth that extends parallel to the velocity vector of the undisturbed incident stream.

The principal gasdynamic result presented by SPREITER and JONES (1963) was a plot of the calculated position of the bow shock wave associated with the simplified axisymmetric magnetosphere described in the preceding paragraph. The calculations were performed for a ratio of specific heats γ of 2, and a free-stream Mach number of 8.71. The latter was identified with the free-stream Alfvén Mach number associated with a representative choice of values for the density, velocity, and magnetic field of the incident solar wind. It is implicit in the analysis presented above, however, that the gasdynamic Mach number should more properly be identified with the free-stream Mach number $M_\infty = v_\infty / a_\infty$ than the Alfvén Mach number $M_{A_\infty} = v_\infty (4\pi\rho / H^2)^{\frac{1}{2}}$ if M_{A_∞} is much greater than unity, as is indeed generally the case in the present applications. Since a value of 8.71 is also a reasonable choice for M_∞, we turn now to a presentation of further details of the flow field for the same set of conditions as employed by SPREITER and JONES (1963).

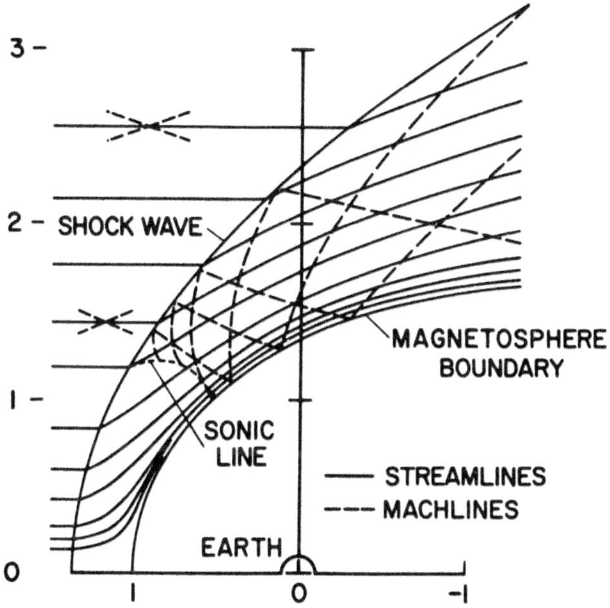

Fig. 1. Streamlines and wave patterns for supersonic flow past the magnetosphere; $M_\infty = 8.71$, $\gamma = 2$.

Figure 1 shows a plot of the magnetosphere boundary and shock wave position for $\gamma = 2$ and $M_\infty = 8.71$ in terms of a length scale in which the distance D from the center of the earth to the magnetosphere nose is unity. D is generally of the order of 10 earth radii, and fluctuates in response to variations in the incident stream in accordance with the expression

$$D = a_e H_{po}^{1/3} / (2\pi K \rho_\infty v_\infty^2)^{1/6}$$

where $a_e = 6.37 \times 10^8$ cm is the radius of the earth and $H_{po} = 0.312$ gauss is the intensity of the earth's magnetic field at the equator. Also included on this

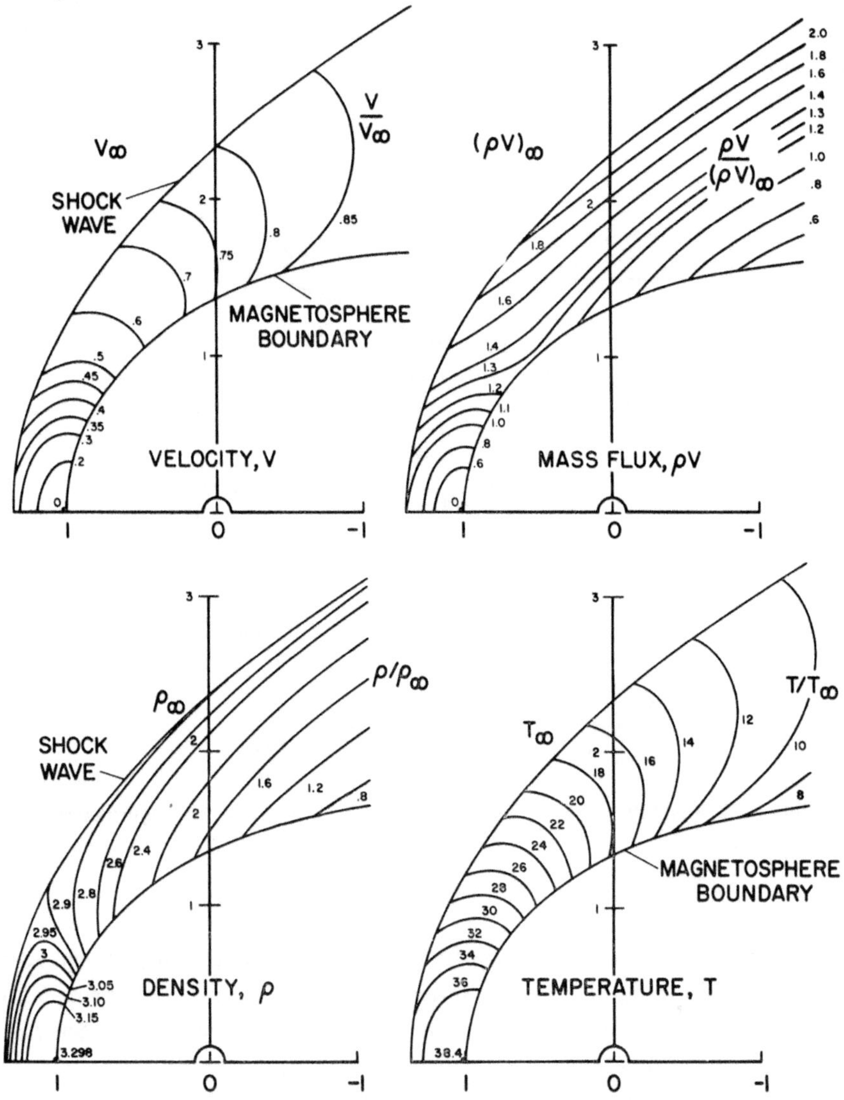

Fig. 2. Contour maps of constant velocity, mass flux, density, and temperature for supersonic flow past the magnetosphere; $M_\infty = 8.71$, $\gamma = 2$.

figure are several solid lines representing streamlines, and broken lines representing characteristic or Mach lines of the flow. The latter correspond to standing compression or expansion waves of infinitesimal amplitude. They cross the streamlines at such angles that the local velocity component normal to the wave is exactly equal to the local sound speed. Mach lines thus exist only where the flow is supersonic; they are absent from the vicinity of the magnetosphere nose because the flow there is subsonic.

Contour maps showing lines of constant density, velocity, temperature, and mass flux, each normalized by dividing by the corresponding quantity in the incident stream are presented in Figure 2. The results show that the density ratio ρ/ρ_∞ remains close to the maximum value $(\gamma+1)/(\gamma-1)=3$ for a strong shock wave in a gas with $\gamma=2$ along nearly the entire length of the portion of the bow wave shown. The gas undergoes a small additional compression as it approaches the stagnation point at the magnetosphere nose and then expands to less than free-stream density as it flows around the magnetosphere. The velocity remains less than in the free stream, however, throughout the same region. The temperature T/T_∞ is closely related to the velocity ratio through the expression

$$\frac{T}{T_\infty} = 1 + \frac{(\gamma - 1)M_\infty^2}{2}\left(1 - \frac{v^2}{v_\infty^2}\right) \tag{9}$$

derived from Equations (1) and (2) with $\mathbf{H}=0$. Of particular interest is the tremendous increase in temperature of the solar wind as it passes through the bow shock wave. If, for example, the temperature of the incident solar wind is 50000 °K, the temperature at the magnetosphere nose is indicated to be nearly 2000000 °K. This value is consistent with the temperature possessed by the gas in the solar corona before it is accelerated to the high velocities characteristic of the solar wind in the vicinity of the earth's orbit.

As noted above, calculation of the deformation of the magnetic field in the flow around the magnetosphere can be accomplished directly, once the flow field is determined, either by direct integration of Equation (8) or by considering the field lines to move with the fluid. Although the field lines are, in general, spatial curves, simplicity may be achieved at the expense of completeness by confining attention to the plane containing the velocity and magnetic field vectors in the incident stream. Since the magnetosphere has been approximated by an axisymmetric shape, it follows that the resulting field lines for this case are also confined entirely to the same plane. Results of two such calculations are shown in Figure 3. The magnetic field in the incident stream is inclined at 90° to the direction of the velocity vector in the left portion of Figure 3, and 45° in the right portion. The corresponding results for 0° inclination are not presented, but can be visualized easily because the field lines for that case are aligned everywhere with the streamlines, and the field strength ratio H/H_∞ is proportional to $\rho v/\rho_\infty v_\infty$.

These results clearly show how the magnetic field lines bend discontinuously as they pass through the bow wave at any angle except a right angle, and then

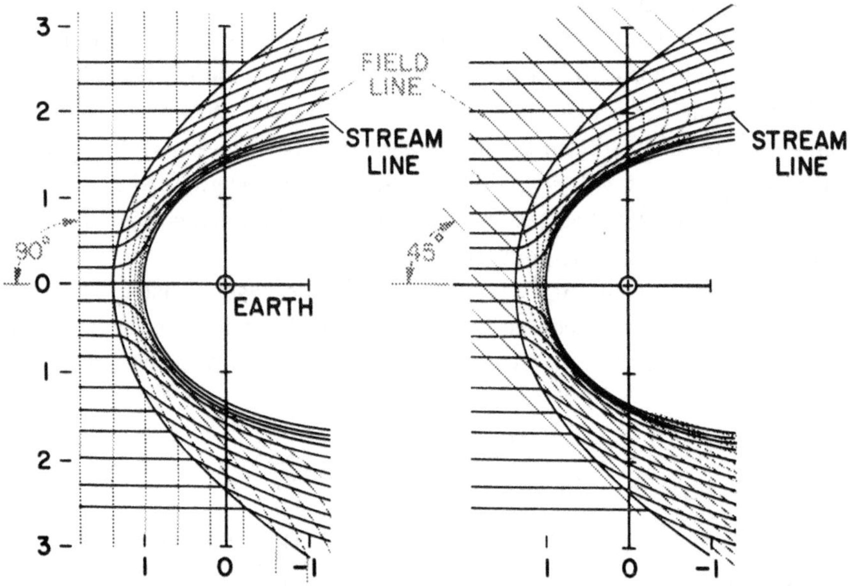

Fig. 3. Magnetic field lines in plane of free-stream velocity and magnetic field vectors for supersonic flow past the magnetosphere; $M_\infty = 8.71$, $\gamma = 2$.

curve in a continuous manner throughout the entire region between the bow wave and the magnetosphere. The discontinuous bend at the shock wave is, moreover, always in the direction that preserves the sign of the tangential component of the field, as is required for all physically relevant hydromagnetic shock waves. It may be seen that the field lines illustrated in Figure 3 are all draped around the nose of the magnetosphere. Outside of this plane, however, the field lines drift past the nose with the flow and deform into three-dimensional curves. The strong constraint imposed on the magnetic field by the stagnation point is thus greatly reduced, and the field lines may be anticipated to remain much straighter than illustrated for the plane of symmetry.

The choice of the value 2 for the ratio of specific heats employed in calculating Figures 2 and 3 is usually justified by reference to the presumed two-degree-of-freedom nature of the interactions of charged particles in a magnetic

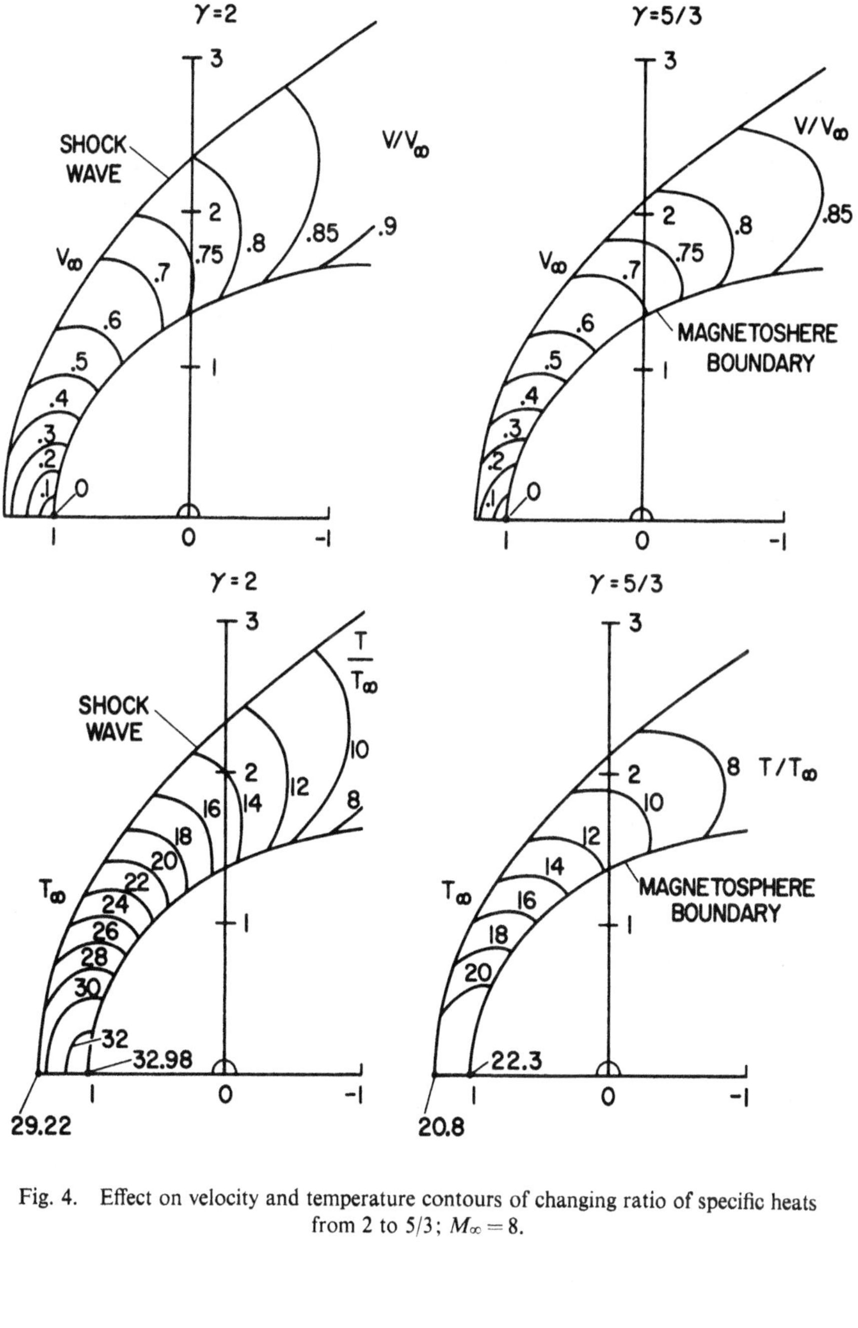

Fig. 4. Effect on velocity and temperature contours of changing ratio of specific heats from 2 to 5/3; $M_\infty = 8$.

field. This argument weakens, however, when consideration is given to the irregular character of the magnetic fields observed in space, particularly downstream of the bow shock wave. In fact, the whole concept of applying hydromagnetic theory to the flow of solar plasma around the magnetosphere involving as it does the assumption of an isotropic pressure appears more consistent internally if the particles are considered to behave as if they have three rather than two degrees of freedom and γ is equated to 5/3. The effects on the velocity and temperature contours of changing γ from 2 to 5/3 are shown in Figure 4.

Although the values of 8 and 8.71 employed in Figures 2, 3, and 4 for the free-stream Mach number are well centered in the range of values to be expected in the incident solar wind, considerable variation is known to occur. In order to illustrate the magnitude of the effects to be anticipated, results are presented in Figure 5 for Mach numbers of 5 and 12 for a gas with $\gamma = 5/3$. They show that the bow wave recedes from the magnetosphere as the Mach number diminishes and as γ increases. The change is small in the ranges covered, however, as is the change in the velocity and density contours. Temperatures in the flow field depend strongly on both Mach number and γ, however, with higher values associated with higher Mach number and larger γ.

A useful quantity for characterizing the location of the bow shock wave is the standoff distance Δ at the nose of the magnetosphere. This distance has been shown in an aerodynamic context by Seiff (1962) and Inouye (1965) to be very nearly proportional to the density ratio across the nose of the bow wave for a wide range of values for γ and M_∞. The degree to which this simple empirical result is able to represent the standoff distance is illustrated in Figure 6.

In conclusion, we have presented a sample of the type of results that can be obtained using the continuum equations of magnetohydrodynamics and gasdynamics. They augment the results for the location of the bow wave given by Spreiter and Jones (1963) which has been widely used in the interpretation of data from IMP-1 satellite and Mariner II and IV space probes. It has been reported in this symposium by Coon that data from the Vela satellites are in at least qualitative agreement with these results. With respect to such comparisons, it should be observed that our calculations only yield information regarding the average bulk properties of the flow, whereas the usual plasma probes and magnetometers observe a much more microscopic and instantaneous view of the plasma. With respect to the latter, it should also be recalled that substantial quantitative discrepancies still remain between the values for bulk parameters deduced from readings of plasma probes of different design even when flown on the same spacecraft. These differences are certain to be resolved as better understanding develops, and we present these results at this time in order that they be available for comparison with observational data as the latter becomes increasingly available.

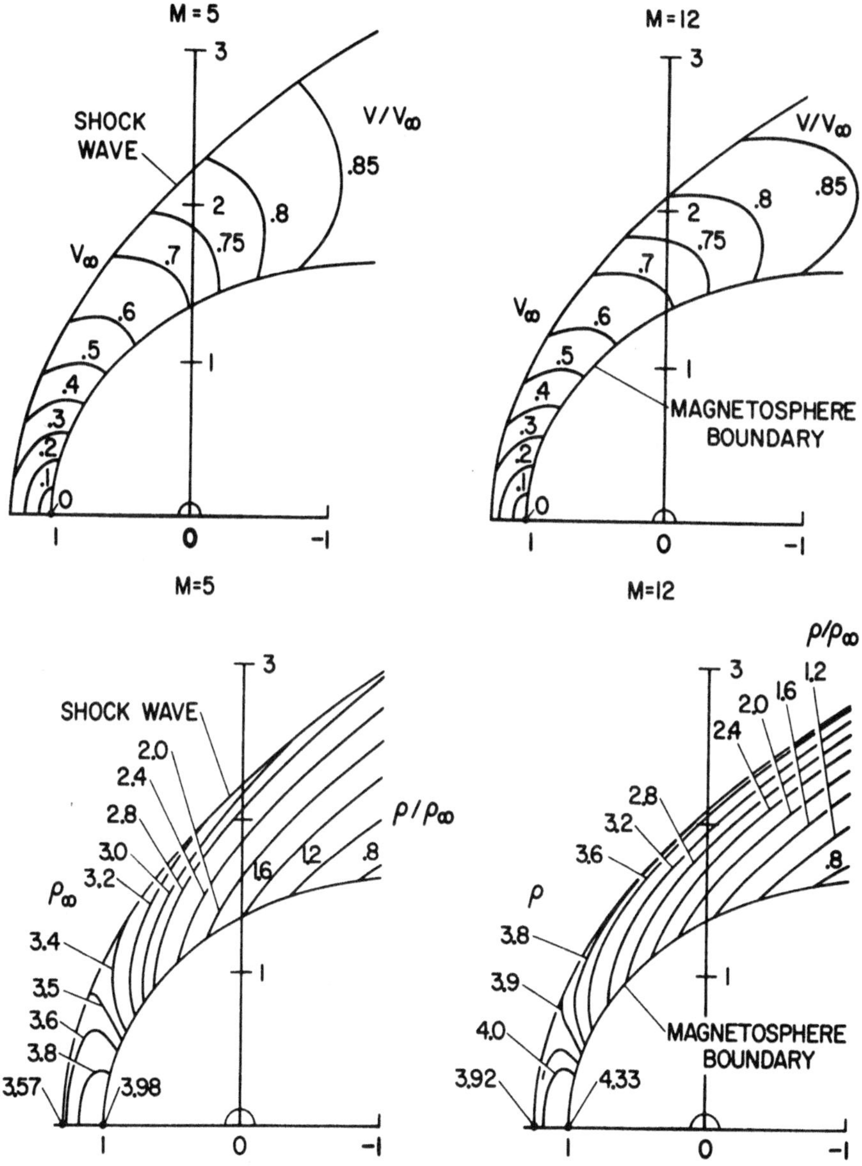

Fig. 5. Effect on velocity and density contours of changing Mach number to 5 and 12; $\gamma = 5/3$.

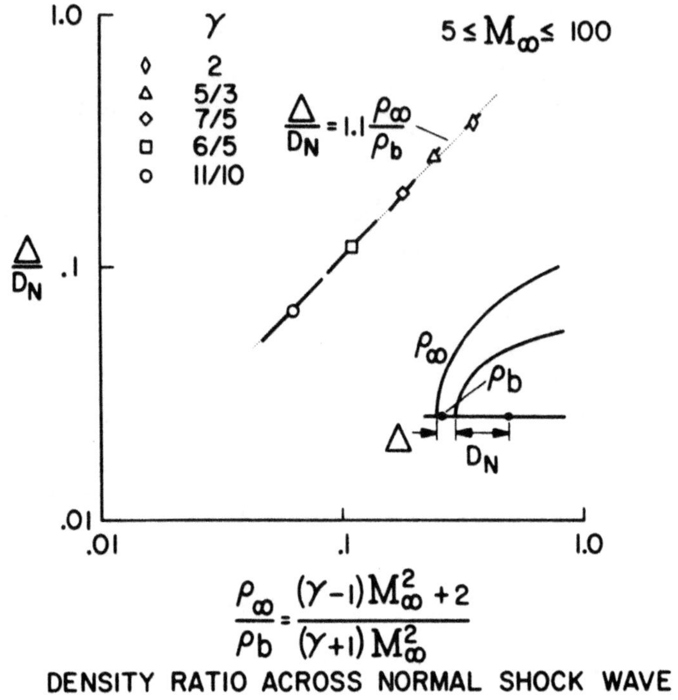

Fig. 6. Variation of standoff distance with density ratio across the nose of the bow shock wave.

References

INOUYE, M.: 1965, *AIAA J.* **3**, 172.

LANDAU, L. D. and LIFSHITZ, E. M.: 1960, *Electrodynamics of Continuous Media,* Pergamon Press, Oxford-New York.

SEIFF, A.: 1962, *Gasdynamics in Space Exploration,* NASA SP-24.

SPREITER, J. R. and BRIGGS, B. R.: 1962, *J. Geophys. Res.* **67**, 37.

SPREITER, J. R. and JONES, W. P.: 1963, *J. Geophys. Res.* **68**, 3555.

SPREITER, J. R., SUMMERS, A. L., and ALKSNE, A. Y.: 1966, *Plan. Space Sci.* (in press).

SESSION VI

ARTIFICIAL INJECTED RADIATION

Chairman

N. C. CHRISTOFILOS

Lawrence Radiation Laboratory

SOURCES OF ARTIFICIAL RADIATION BELTS *

N. C. CHRISTOFILOS

Lawrence Radiation Laboratory, University of California, Livermore, California

1. Introduction

With the advent of the satellite era the thoughts of the scientists have extended to outer space, since means have become available for measurements and experimentation in the region of the upper atmosphere. Two classes of experiments are possible with satellite-carried instruments: (1) To measure natural phenomena; (2) To create artificial effects under controlled conditions.

This class of experiments is most interesting as the initial conditions are known. Hence physical quantities can be measured under completely different conditions than those which the natural phenomena allow.

Just after the first Sputnik was launched I was attracted by outer space problems. The second class of experiments appeared very intriguing to me and I started thinking about possible experiments in outer space. The first experiment which occurred to me stemmed out of my main field of endeavor in the last five years, namely, the Astron thermonuclear device. In this device a cylindrical layer of relativistic electrons is established between two mirror coils. This electron layer then will establish in turn a magnetic bottle for plasma confinement when the trapped electron current reaches a predetermined value. Thereafter the relativistic electrons will eventually subsequently heat this plasma by Coulomb collisions up to fusion temperature. In October 1957 it occurred to me to extrapolate this idea of the electron layer to global dimensions, inasmuch as the earth's magnetic field provides a natural mirror effect. The trapping phenomenon was known theoretically for many years from the work of Störmer, Alfvén, and others, namely, that charged particles released at one point will spiral about a magnetic line being reflected at two points along this line. These points came to be known as mirror points. In addition to this motion the electrons would drift eastward due to the fact that the earth's field is decreasing as one moves away from the earth. Consequently, it was obvious that if sufficient electrons would be released at one point, these electrons would spread all around the earth creating a layer of rather small thickness but of global dimensions.

* Work performed under the auspices of the U.S. Atomic Energy Commission.

The question then arose whether or not these electrons will survive long enough to complete several circuits around the earth. The time for one circuit around the earth is approximately inversely proportional to the electron energy. The lifetime has been calculated. The conclusions from these calculations was that most of the known losses are due to Coulomb scattering with air atoms. Since the density of the atmosphere decreases exponentially with altitude it turned out that most of the losses occur near the mirror points. The result of the scattering is that the mirror points move down toward denser atmosphere. The electrons are practically lost after the mirror points have moved downward by about one scale height. The lifetime is proportional to the air density at the mirror points and approximately to the square of the electron energy, and is proportional to the length of the magnetic line. The conclusion was that electrons of one to two MeV energy injected at an altitude of a few hundred miles will survive several hours or more, depending on the air density at the mirror points (CHRISTOFILOS, 1959).

The energy loss ($\dot{E}$) of the electrons by Coulomb scattering is equal to the energy loss ($\dot{E}_0$) corresponding to the air density (N_0) at the mirror points divided by the length (ψ) of the magnetic line from the mirror point to the equatorial plane, measured in units of earth radii, times the square root of the ratio of the earth's radius (r_0) to the scale height (h). Namely

$$\dot{E} = \dot{E}_0/(\psi\sqrt{r_0/h}) \tag{I-1}$$

Electrons in the few MeV range are lost by scattering without appreciable energy loss. In this case the lifetime is approximately

$$T = 1.7 \cdot 10^6 \frac{\psi\gamma^2}{N_0}\sqrt{\frac{h}{r_0}}\ \text{days} \tag{I-2}$$

where γ is the electron energy in rest mass units.

For the following parameters

$$(r_0/h)^{\frac{1}{2}} = 8 \qquad N_0 = 10^4$$
$$\psi = 1.5\ (\text{latitude } 45°) \qquad \gamma = 3$$
$$T = 2.86\ \text{days}$$

These results pointed toward a convenient source of a large quantity of electrons, namely, to an A-bomb. This source is so plentiful that even one megaton of fission would create an electron layer so dense as to constitute a radiation hazard in outer space.

Fortunately a much smaller yield, namely in the kiloton range, is sufficient to yield detectable quantities without creating any radiation hazard at all.

It is obvious that the scientific aspect of such experiments is extremely important. Consequently, as soon as I suggested this experiment about the end of October, 1957, the matter was discussed with interest within the Radiation Laboratory for several weeks. In January, 1958, the matter was further discussed in a meeting with scientists from other laboratories. Finally, a group of scientists was appointed to evaluate the proposal, and it was recommended to proceed with a test employing a small yield A-bomb in the kiloton range. Further an altitude was selected such so that the effect would not last more than a few days.

In March 1958 the natural radiation belt was discovered by Dr. James Van Allen. This discovery was partial proof that the trapping effect actually exists as postulated, so that about the end of April, 1958, it was decided to proceed with the Argus test.

Argus was the code name given to the experiment. The most pertinent question then was the selection of the location of the tests. Since the magnetic axis of the earth is tilted and the center of the dipole is displaced toward the Pacific Ocean, it is obvious that the trace of the mirror points will have their closest distance from the surface of the earth over the South Atlantic Ocean. Hence a location in the South Atlantic was finally selected for the test.

2. The Argus Experiment

The over-all responsibility for the preparation and conduct of the experiment was vested in the Advanced Research Projects Agency (ARPA), then just established. It required only four months from the time it was decided to proceed with the tests until the first bomb was exploded. This is a remarkable achievement in organization, and if we were to single out the one man as most responsible for this accomplishment it would be Dr. Herbert York, who was at that time Chief Scientist of the ARPA. During this short time a task force including the rocket launching ship, Norton Sound, was organized, and special instrumentation on satellites was prepared by Dr. James Van Allen. Many other activities were coordinated with the help of dozens of scientists.

In addition to the trapping effect the creation of an artificial aurora was predicted. The nontrapped electrons would be guided along the magnetic lines of force (which pass through the burst point) toward the upper atmosphere, producing strong ionization along these magnetic lines. Hence in both points where these lines re-enter the atmosphere auroral luminescence was expected.

The persons responsible for organizing the observations of auroral phenomena and magnetic disturbances as well as some of the minitrack stations were Dr. Philip Newman (AFCRC) and Dr. Allen Peterson (SRI). Besides the satellite observations, some rocket launchings were scheduled for observations at the very beginning of the formation of the Argus shell. This responsibility was assumed by a group of scientists in the AFSWC.

The experiment was successful in all respects except that Explorer V was not placed in orbit. However, the one satellite available, Explorer IV, yielded most of the important information although valuable information resulted from other observations.

The bursts occurrred on the 27th and the 30th of August in the early morning hours and on the 6th of September shortly before midnight, Greenwich time. The locations were, respectively, 38 °S, 12 °W; 50 °S, 8 °W; and 50 °S, 10 °W, approximately.

A very interesting sequence of observations was obtained. The initial flash of the burst was followed by an auroral luminescence extending upward and downward along the magnetic line where the burst occurred. Simultaneously at the point where the same magnetic line returns to the earth's atmosphere (the so-called conjugate point) in the North Atlantic another auroral luminescence appeared in the night sky near the Azores Islands.

The instruments of Explorer IV recorded and reported to ground stations, around the earth, the electron density in the shell, as well as the position and its thickness. The satellite continued to penetrate the shell several times a day at various altitudes, so that the density as a function of altitude and time was measured. The apogee of the electrons was about an earth's radius out in space or 4000 miles approximately. Thus, for the first time the earth's magnetic field was being plotted experimentally. The conjugate points were a few hundred miles farther away than the theoretical calculations indicated. However, this is rather a small error considering that all the measurements of the earth's field were thus far taken at the earth's surface.

When the location in the South Atlantic was selected, the data on the natural radiation belt was rather meager. However, it turned out that the Argus shell was located at a region where the intensity of the natural radiation passes through a minimum. If the location were further south or north of the one selected, the Argus shell effects would have disappeared in the background much earlier. For the first time an experiment was conducted in outer space on a global scale where all the measured quantities were related to a known cause, namely, to the trapping in the earth's field of a known number of electrons of a known energy, injected at a known location at a known time.

An important observation was that (within the limit of observational error) neither did the electron shell move at all across the magnetic lines nor did the electrons diffuse across these lines.

In connection with the position stability of the shell a conclusion can be drawn concerning hydromagnetic instabilities. According to hydromagnetic theory, an ionized plasma either permanently existing in the area where the Argus shell was located or temporarily trapped as a result of the sun's activity could create motions characteristic of unstable magnetic line interchange. In such a case the Argus shell would have followed the motion of the magnetic lines and be deformed and change location. Since no such motion has been observed this is a good indication that such instabilities are not occurring in the earth's field at least at the latitudes of the Argus shell or lower.

3. Other Sources of Artificial Radiation Belts

The first Argus experiment provided us with very important information. A variety of experiments is possible as well as a variety of methods of injecting electrons into the geomagnetic field. We can classify the means of injection into three categories: (a) Injection through atomic explosions as in the Argus experiment; (b) Injection from a satellite carrying a payload of radioactive β-decay material or alpha emitters; (c) Injection from a satellite-borne electron accelerator.

Since nuclear explosions in outer space have been banned, we shall discuss the two other possible sources of artificial radiation belts. The absolute value of the particle density in the belt is not of importance as long as the particles can be detected. Therefore, by forming a radiation belt of particles not occurring in the natural belts it is possible to detect the particles at densities much lower than we can detect electrons injected as a result of nuclear explosions. In this respect, if a positron (HONES, JR., 1964) or alpha particle belt is formed, the required density may be one thousand or more times lower than the density of an artificial electron belt. A kiloton of fission yields approximately $5 \cdot 10^{23}$ electrons. A megacurie of a positron emitter with half-life of several hours will yield 10^{21} positrons. Comparing an artificial belt formed with 10^{21} positrons with another of the same volume formed with the electrons of a kiloton nuclear explosion we observe that the positron belt will be much easier to detect. A nuclear explosion of higher yield than one kiloton might distort the magnetic field and complicate the measurements of its properties especially in the turbulent regions. Consequently, it appears, however surprising it may be, that a positron belt constitutes for its own merit a much more controlled experiment than a belt produced by nuclear explosions.

Similar reasoning applies for an alpha particle belt which will simulate the proton belts. The energy of the alpha particles of most of the α-emitters is approximately 5 MeV; thus the momentum is approximately 100 MeV/C. Therefore the Larmor radius of alpha particles injected at $L=5$ to 7 will exceed 1000 kilometers allowing the study of non-adiabatic effects. Solid state detectors can discriminate protons from alpha particles of the same energy; hence the required flux is not very high. Possible sources of alpha particles and the configuration of the source are discussed in Section 5.

4. Sources of Positrons

There are two alternative sources of positrons. Electron accelerators of a few MeV energy converting the electrons into X-rays and then to electron pairs, or radioactive isotopes. The only economical method to produce isotopes is irradiation in fission reactors. Therefore, we shall confine ourselves to isotopes produced by n, γ reactions. In addition, the isotope must have a half-life of several hours allowing for transportation from the reactor to outer space where the positrons will be injected to create the radiation belt. A search in the table of isotopes has shown that four isotopes, positron emitters, can be produced by n, γ reactions. The four isotopes are listed in Table I. It is also shown the natural abundance of the irradiated isotope (column 2) and its thermal neutron cross-section.

TABLE I

The Four Isotopes Producible by n, γ reactions

Irradiated isotope	*% Natural abundance*	$\sigma_{n,\gamma}$ *(barns)*	*Radioactive isotope*	*Half-life*	E_{max} (MeV)
Cu^{63}	69.09	4.5	Cu^{64} *	12.8 h	0.65
Zn^{64}	48.87	0.47	Zn^{65}	247 d	0.32
Kr^{78}	0.354	2.0	Kr^{79}	34.5 h	0.595
Cd^{106}	1.22	1.0	Cd^{107}	6.9 h	0.32

* The branching ratio of Cu^{64} is: β^{+} 19%, β^{-} 38% and K 43%.

The lifetime of Zn^{65} is rather long thus committing at a given orbit in space a radiation source for a time which may be much longer than desired. The isotope Cd^{107} is difficult to produce because of the presence of Cd^{113} which is a strong neutron absorber with a cross-section of 20000 barns. Thus, its use involves an expensive isotope separation to remove practically every trace of Cd^{113}. The krypton isotope appears attractive because it is in gaseous form. It involves an isotope separation to avoid absorption of neutrons by Kr^{83} (250 barns) but not as meticulous as is the case of Cd. The half-life of the isotope Kr^{85} is 9.4 years however, thus, introducing a contamination of a beta-emitter with a very long lifetime.

Therefore, it appears that Cu^{64} is the most attractive of the four isotopes. The half-life of the isotope produced by the neutron irradiation of Cu^{65}

is only 5.2 minutes; thus no isotope separation is required. Therefore, we may conclude that Cu^{64} is the most convenient isotope to employ to create an artificial positron radiation belt. Four megacuries of Cu^{64} will yield $2 \cdot 10^{21}$ positrons.

The problems associated with the source are its geometric configuration to avoid self-absorption of the positrons, cooling and shielding before and during the first phase of launching.

The copper to be irradiated can be coated in aluminum tubes approximately 1 cm diameter, 100 cm long and a fraction of a millimeter thick. The layer of copper is 25 microns thick so that the energy loss of the positrons is not excessive. One-half of the positrons are emitted outward, the other half are absorbed in the aluminum tube.

A total of 3000 tubes are required, containing 20 000 grams of Cu to provide a positron source of one megacurie six hours after they have been removed from the reactor. The rods are placed in a container which is filled with water.

The heat released by the radioactivity of the source evaporates approximately 10 kg of water per hour. The total weight of the rods, cooling water and container is approximately 300 kg, excluding shielding.

The container is placed in a shielding approximately 15000 pounds for personnel protection before the launching of the rocket and during the boost phase.* Thereafter, the shielding can be ejected; thus the payload which should be placed in orbit is reduced to a few hundred pounds. After the payload is placed in the desired orbit the rods are ejected from the container. At this phase radiation cooling is adequate. The ejection velocity must be small enough that in the next 20–30 hours the spread of the source is smaller than the cyclotron radius of the positrons in the earth's field. In this way the source can be treated as "point source". If the positrons are injected in the equatorial plane the thickness of the belt depends on the eccentricity of the orbit and the earth's magnetic dipole. At an orbital distance of ten earth's radii and a belt thickness of 1000 km at the equatorial plane the expected density is approximately 10^{-7} positrons/cm^3 or a flux of 1000 positrons/cm^2, sec which should be detectable in the presence of the cosmic ray or other radiation background. An accurate evaluation of the background radiation is required however, as well as of possible detector designs before a conclusion on the lowest detectable positron density.

5. Sources of Alpha Particles

There is a large number of alpha emitters from Polonium to Transuranic elements. The required half-life is from several hours to several days. Most of

* Considering the possibility of accident which may require abortion of the launching.

the alpha emitters, however, are the products of nuclear reactions with high energy deuterons or alpha particles. The quantity of the isotopes produced in particle accelerators is inherently limited. Therefore for large quantities we should find an isotope produced in fission reactors by n, γ reactions. The isotope Po^{210} can be produced by the following reaction

$$Bi^{209}(n,\gamma)\,Bi^{210}$$
$$Bi^{210},\, e^- \rightarrow Po^{210}$$

The half-life of Bi^{210} is 5 days. The half-life of Po^{210} is 138.3 days which is somewhat longer than required for artificial belt experiments. The energy of the alpha particles is 5.3 MeV. A megacurie of Po^{210} yields $6\cdot10^{23}$ alpha particles. The required weight of the isotope is 210 grams. Its cost is negligible in comparison to the cost of the launching rocket.

Approximately 0.1% of the isotope yield γ's of 0.77 MeV. Thus the γ activity of a megacurie of Po^{210} is 1000 curies, while a positron emitter of one megacurie is associated with four megacuries of beta and γ emitters. The low level of γ activity of Po^{210} simplifies the shielding problems in that, under certain precautions, a shielding may not be required at all during the launching phase.

Since all the alpha particles are of the same energy, it will be very useful to inject the isotope in orbit in a gaseous compound to avoid any energy loss by ionization. A spherical container 50 cm. diameter can contain the gas at high temperature but at a reasonable pressure. The external surface of the container is 0.8 m^2. The energy release of one megacurie of Po^{210} is 32000 watts. The temperature of the container required to radiate this energy is approximately 700 °C. There are many alloys which can safely contain the radioactive gas at this temperature. When the container is placed at the desired orbit the gas can be released through a tube of small diameter and adequate length to allow for cooling of the gas thus minimizing the ejection velocity. The total payload should not exceed 200–300 lbs.; thus an Atlas booster can place it on a 24 hour orbit.

As an illustrative example we shall calculate the alpha particle density in an artificial radiation belt created at $L=6.6$ (24 hour orbit). The thickness of the belt depends on the Larmor radius of the particles and the orbit perturbation due to the thermal velocity of the gas. If the gas is ejected at room temperature, its thermal velocity is approximately 100 meters/sec. The corresponding orbit perturbation is ±3000 km. The Larmor radius of $L=6.6$ is 3200 km or one half earth radius. Thus the thickness of the artificial belt at the equator will extend from $L=5.1$ to $L=8.1$. The mean time of circling the earth (at the equator) is approximately 200 seconds.

The alpha particle flux at the equator is approximately:

$$F = 500(1 - e^{-t/T})/\text{cm}^2, \text{sec} \tag{II-1}$$

where

$$T = 195 \text{ days}.$$

We observe that even 20 days after injection the flux will be at least 50 counts/cm^2, sec., which is much above the background count of solid state alpha counters. Since the life time at this large L value may be much shorter than 6 months, the flux may reach a peak much earlier. In this way the life time of the trapped particles can be measured.

During magnetic storms three effects can be studied:

(a) Abnormal losses which will be observed as an anomalous reduction in the flux.

(b) Acceleration or deceleration of the alpha particles which can be observed by providing the detectors with a pulse height analyser thus measuring the change of energy of the alpha particles. This observation, obviously, is not possible with a positron belt.

(c) Part of the belt will disappear altogether because of the inward motion of the magnetosphere.

These observations will shed a lot of light in the phenomena associated with the natural radiation belts. Thus both the origin and losses of natural radiation belts will be elucidated.

The technical developments associated with the practical configuration of the source should not present any major difficulties. Therefore in my opinion an artificial alpha particle radiation belt can be created before the next maximum of the sunspot cycle.

6. Discussion

Of the three possible sources which can be employed to create artificial radiation belts, it appears that positrons and alpha particles are much more useful for they do not occur in the natural belts in quantities comparable to the trapped electron density. Furthermore positrons or alpha particles injected from a radioactive source or a particle accelerator would not create any distortion of the magnetic field. In this respect a large nuclear explosion is undesirable in spite of the fact that it can provide copious quantities of electrons.

An electron accelerator in orbit may be feasible but in comparison to a positron source of one megacurie it appears impractical, because of the large weight of the required power supply. A much smaller accelerator is required, however, if the injection is limited within a narrow range of magnetic moments. Such an application will be useful to study diffusion of the pitch angle of the trapped particles.

A positron belt will be more useful to simulate the trapped electrons of the natural belts. An alpha particle belt can simulate the trapped protons. A unique property of the alpha particles, however, is that they are monoenergetic. Therefore they offer the possibility of studying the energy change of the trapped particles during magnetic storms.

References

Christofilos, N. C.: 1959, *Proc. Nat. Acad. of Sci. U.S.* **45**, 000.
Hones, Jr., Edward W.: 1964, *J. Geophys. Res.* **69**, 182.

Discussion

Roberts pointed out that whistler interactions with positrons behave differently than with electrons. Therefore, if whistlers are an important diffusion mechanism in the outer belt, positrons will not provide the correct data.

An objection was made by Hynds and others to the injection of artificial alpha particles until the natural alpha particle energy and flux as a function of time are measured. This may require a number of years.

There was no objection to the use of artificial electrons as these have already been injected from a number of high altitude nuclear tests. In fact, several people pointed out that an international experiment using small nuclear bursts at different locations in *B-L* space would be very useful. There was no reaction from the French attendees to the specific suggestion that France could aid many scientific investigations by conducting such high altitude nuclear tests.

SPATIAL DISTRIBUTION AND TIME DECAY OF THE INTENSITIES OF GEOMAGNETICALLY TRAPPED ELECTRONS FROM THE HIGH ALTITUDE NUCLEAR BURST OF JULY 1962

J. A. VAN ALLEN

Department of Physics and Astronomy, University of Iowa, Iowa City, Iowa

Abstract, A one-year observational study of the artificial radiation belt produced by the nuclear burst Starfish on 9 July 1962 is reported. It is estimated that 1.3×10^{25} electrons from radioactive fission products, or some 2.6% of the total yield, were present in geomagnetically trapped orbits at 10 hours after the burst. These electrons disappeared in the manner expected from the atmospheric loss theory of Walt for values of the magnetic shell parameter $L < 1.25$ earth radii. At increasing values of L the rate of disappearance was progressively more rapid than expected by this theory. The maximum observed value of the apparent mean lifetime of ~ 2 MeV electrons in the time range $4000 < \Delta t < 10\,000$ hours was 2 years at $L = 1.5$. About 15% of the initially injected electrons (or 0.4 of 1% of the total) survived the first 5½ months, about 10% the first year.

Detailed spatial distribution and temporal data are given in a series of plots.

Introduction

Störmer (1907) discussed the dynamical motion of an isolated, electrically-charged particle in the field of a point magnetic dipole and showed that, under specified conditions, it is possible for a particle to be trapped in such a field with its spiraling, drifting trajectory forever confined within a toroidal region which is a figure of revolution about the axis of the dipole.

In October 1957 Christofilos (1959) suggested the possibility of artificially populating a Störmer region around the earth with beta-rays from the radioactive fission products of a nuclear bomb, exploded at high altitude. Such a populated region is now known as a "radiation belt". Following the discovery in early 1958 (Van Allen, 1958) that the region around the earth already contained large intensities of charged particles from natural sources, it was decided to proceed with a series of experiments of the type envisioned by Christofilos. These experiments were conducted in August and September 1958 and resulted in the successful production of artificial radiation belts. Extended publication of the results has been made ('Scientific Effects of Artificially Introduced Radiations at High Altitudes', *J. Geophys. Res.* **64** (1959) 865–938).

In the period 1958–1962, there was created a total of nine artificial radiation belts which remained detectable for periods greater than one day in the presence of the natural radiation belts of the earth. All of these were produced by high-altitude nuclear bombs, six by the United States and three by Russia, and all have been observed by satellite equipment prepared in the author's laboratory as well as by an increasing variety of satellite and other equipment prepared in other laboratories in the United States, Canada, Russia, England, and elsewhere. Table I gives an abridged summary.

The purpose of the present paper is to summarize the observations by satellites Injun I, Injun III, and Explorer XIV of the Starfish radiation belt which was produced on 9 July 1962. Comprehensive review of the literature is not attempted in this brief paper, though references to other work are given wherever directly pertinent.

Data on the Starfish Nuclear Burst

Publicly available data on the Starfish nuclear burst are:

(a) That it occurred at 09:00:05 UT on 9 July 1962 at an altitude of about 400 kilometers;

(b) That the geographic position was near Johnston Island (geographic coordinates 16.7° N, 190.5° E);

(c) That, from the above data, the position in trapped particle coordinates was $B = 0.29$ gauss, $L = 1.12$ earth radii;

(d) That the explosive yield was 1.4 megatons TNT equivalent; and

(e) That the nominal total yield of fission-decay electrons was 5×10^{26}.

Observations with Injun I, Injun III, and Explorer XIV

Three previous papers have been published (O'BRIEN, *et al.*, 1962; VAN ALLEN, *et al.*, 1963; VAN ALLEN, 1964) and several further University of Iowa research reports have been circulated privately.

The University of Iowa-Office of Naval Research satellite Injun I was launched on 29 June 1961 into an orbit with perigee altitude 890 km, apogee altitude 1010 km, inclination 67°, and period 104 minutes. Useful data were received from launch until latter August 1962. Thus a special virtue of the Injun I observations is that extensive pre-Starfish measurements were available from the same detectors in the same satellite as were post-Starfish measurements for over a month after the burst. Hence, it is possible to study the incremental effect of the burst with a confident knowledge of the pre-burst "background" due to natural radiation.

The series of observations was resumed with the launching of the University of Iowa-ONR satellite Injun III on 13 December 1962 into an orbit having the following initial parameters: perigee altitude 237 km, apogee altitude 2785 km, inclination 70.4°, and period 116.2 minutes. This satellite carried a variety of

TABLE I

Artificial Radiation Belts 1958–1962

Designation	*Date of burst*	*Altitude* (km)	*Nominal Explosive yield TNT equivalent*	*Maximum omnidirectional intensity at $t=0$,* $(cm^2 sec)^{-1}$	*L value of burst*	*Apparent mean lifetime (approx.)*
Teak	1 August 1958	$\sim$ 75	10 megatons	10^3	1.1	Few days
Orange	12 August 1958	$\sim$ 45	10 megatons	10^3	1.1	Few days
Argus I	27 August 1958	$\sim$200	1.4 kilotons	10^5	1.7	3 weeks
Argus II	30 August 1958	$\sim$250	1.4 kilotons	10^5	2.1	3 weeks
Argus III	6 September 1958	$\gtrsim$480	1.4 kilotons	10^6	2.0	1 month
Starfish	9 July 1962	$\sim$400	1.4 megatons	10^9	1.12	1.5 years
U.S.S.R. I	22 October 1962	Unknown	"Submegaton"	10^7	1.9	1 month
U.S.S.R. II	28 October 1962	Unknown	"Submegaton"	10^7	2.0	1 month
U.S.S.R. III	1 November 1962	Unknown	"Megaton"	10^7	1.8	1 month

particle detectors including one nominally identical to one in Injun I. Observations with Injun III continued to early October 1963.

Additional data of value were obtained with University of Iowa equipment on the NASA satellite Explorer XIV, launched on 2 October 1962 into a very eccentric orbit having the following initial parameters: perigee altitude 282 km, apogee altitude 98,533 km, inclination 33°, and period 36.4 hours. Useful data were obtained over a ten-month period.

A further continuation of the long term study of the decay of the Starfish radiation belt was begun on 21 November 1964 with Injun IV and is continuing over a year later. Results from Injun IV will be reported at a later date.

Detectors

Detectors designated SpB on Injun I and Injun III were small Geiger tubes having an effective cross-section for penetrating particles (galactic cosmic rays) of $\sim 0.12\ cm^2$, averaged over direction. The shielding of the counting volume was complex in detail but over some 3π steradians is estimated to be effectively as follows: 0.3 g cm^{-2} of magnesium, silicon and glass, 3.5 g cm^{-2} of lead, and 0.4 g cm^{-2} of steel in the order given starting from the outside. The remaining solid angle was shielded by a miscellany of equipment totalling some tens of g cm^{-2}.

The SpB detector counts fission-product decay-electrons via the bremsstrahlung which is produced when they are stopped in the shield (PETSCHEK, 1963; VAN ALLEN *et al.*, 1963). The ratio ε of the counting rate of the tube R in counts sec^{-1} (corrected for dead time when necessary) to the omnidirectional intensity J_0 in $cm^{-2}\ sec^{-1}$ has been calculated (FRANK, private communication) for isotropic exposure to monoenergetic electrons and is plotted as a function of electron energy E in Figure 1. Also shown in Figure 1 is the differential energy spectrum of electrons from the thermal neutron fission of U^{235}, measured in approximate secular equilibrium by CARTER *et al.* (1959), designated dN/dE; and the product $\varepsilon\ dN/dE$. Thus, it is clear that the sensitivity of SpB to a fission electron-spectrum is concentrated in the approximate energy range 1 to 5 MeV.

The calculated mean value of ε for the fission spectrum (from the data of Figure 1) is 6×10^{-6}. Three different experimental determinations give somewhat higher values. The direct inflight determination and a Sr^{90} calibration are described by VAN ALLEN *et al.* (1963). A subsequent laboratory calibration using machine-generated beams of monoenergetic electrons has been made over the energy range 0.4 to 2.4 MeV by B. J. Farmer of the Ling-Temco-Vought Company. On the basis of these various determinations I have adopted the relation:

$$\varepsilon = \frac{R}{J_0} = 1.0 \times 10^{-5}$$

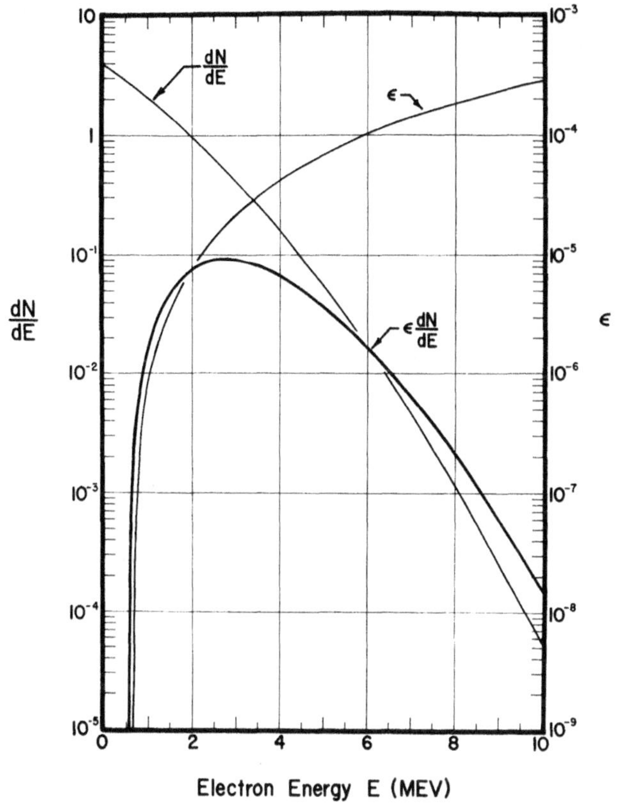

Fig. 1.

for the fission spectrum. The estimated uncertainty in ε is some 50%. The form of the dependence of ε on E as shown in Figure 1 has been used for estimates on spectra which evolve as atmospheric losses modify the initial spectrum.

Most of the curves and data of the present paper are in terms of the true counting rate R of the SpB.

During the flight period of Injun III a larger, lightly shielded Geiger tube, designated 302, became useful. Its detection efficiency curve is approximated by zero efficiency below 1.8 MeV and unit efficiency with an effective area of 0.5 cm^2 above 1.8 MeV for the fission spectrum (direct penetrations).

Observations

Observed data were segmented into time eras of length varying from a day early in the history of the artificial belt to three months later on. Counting rates

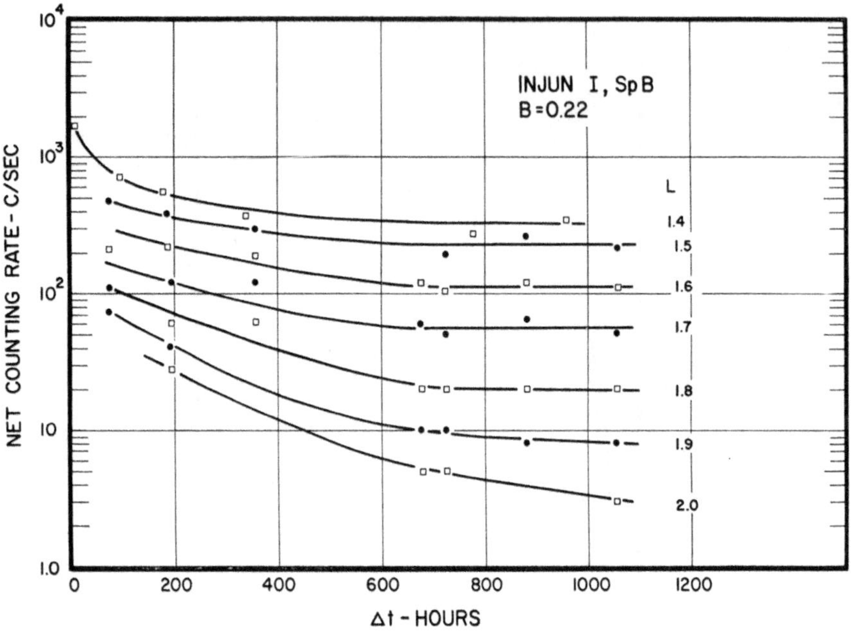
INJUN I, SpB
B=0.22
L
1.4
1.5
1.6
1.7
1.8
1.9
2.0
NET COUNTING RATE - C/SEC
Δt - HOURS

Fig. 2.

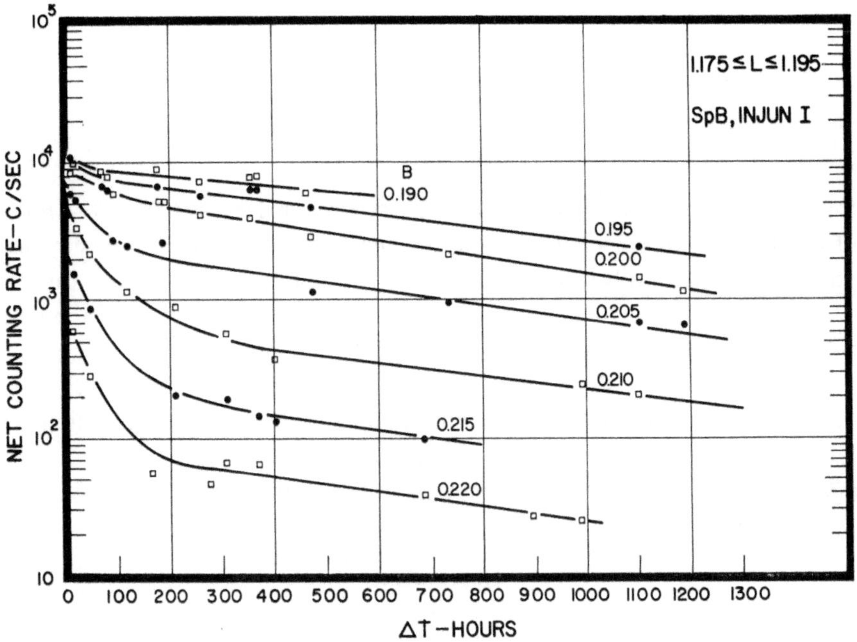
1.175 ≤ L ≤ 1.195
SpB, INJUN I
B
0.190
0.195
0.200
0.205
0.210
0.215
0.220
NET COUNTING RATE - C/SEC
ΔT - HOURS

Fig. 3.

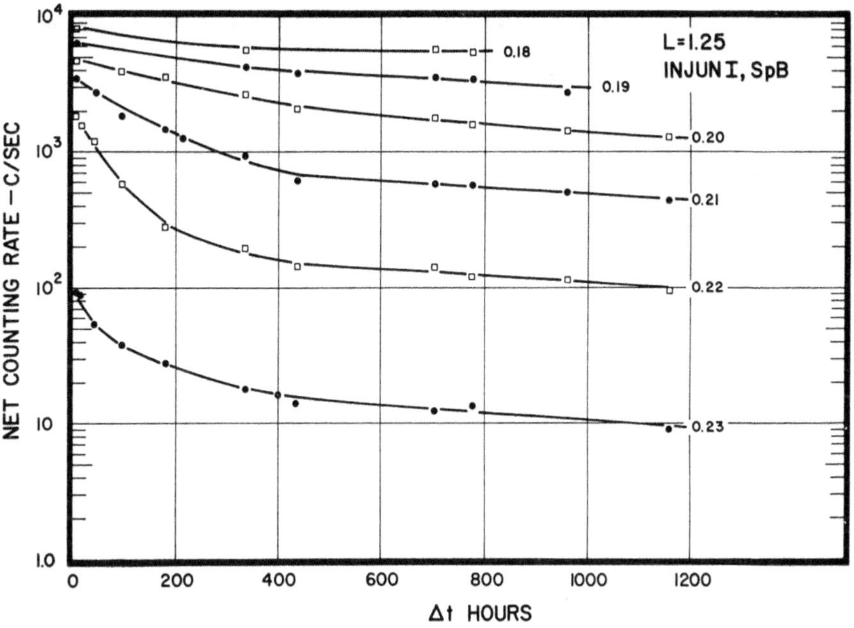
L=1.25
INJUN I, SpB
0.18
0.19
0.20
0.21
0.22
0.23
NET COUNTING RATE – C/SEC
Δt HOURS

Fig. 4.

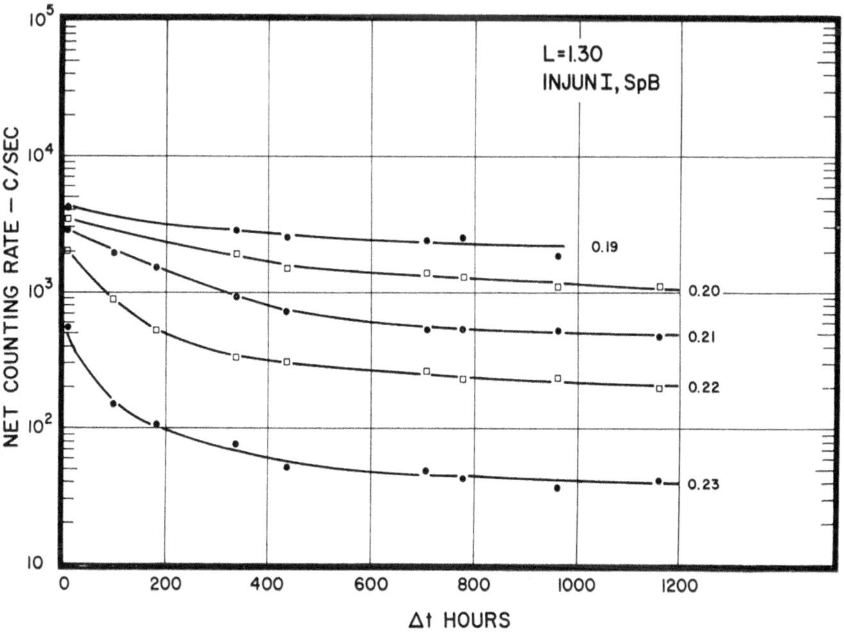
L=1.30
INJUN I, SpB
0.19
0.20
0.21
0.22
0.23
NET COUNTING RATE – C/SEC
Δt HOURS

Fig. 5.

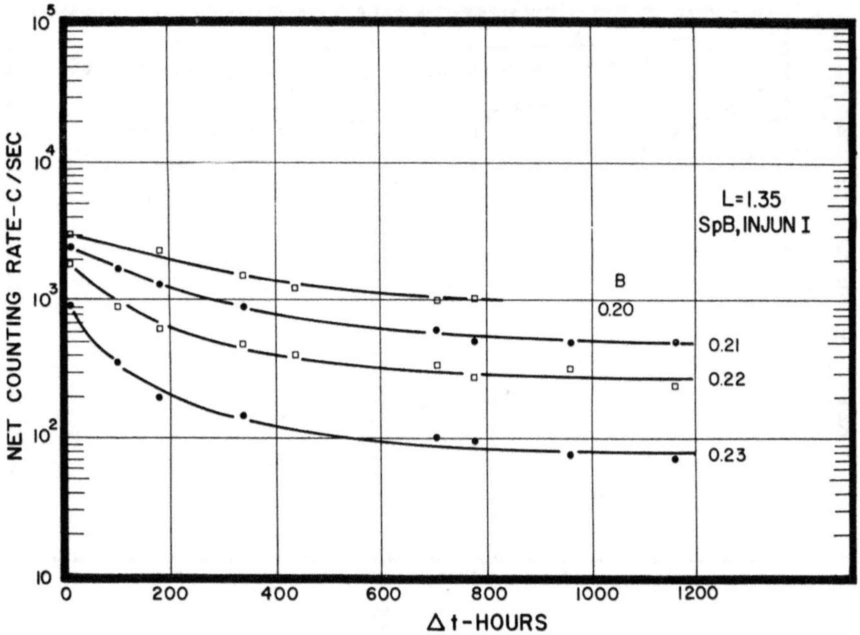
NET COUNTING RATE-C/SEC
Δt-HOURS
L=1.35
SpB, INJUN I
B
0.20
0.21
0.22
0.23

Fig. 6.

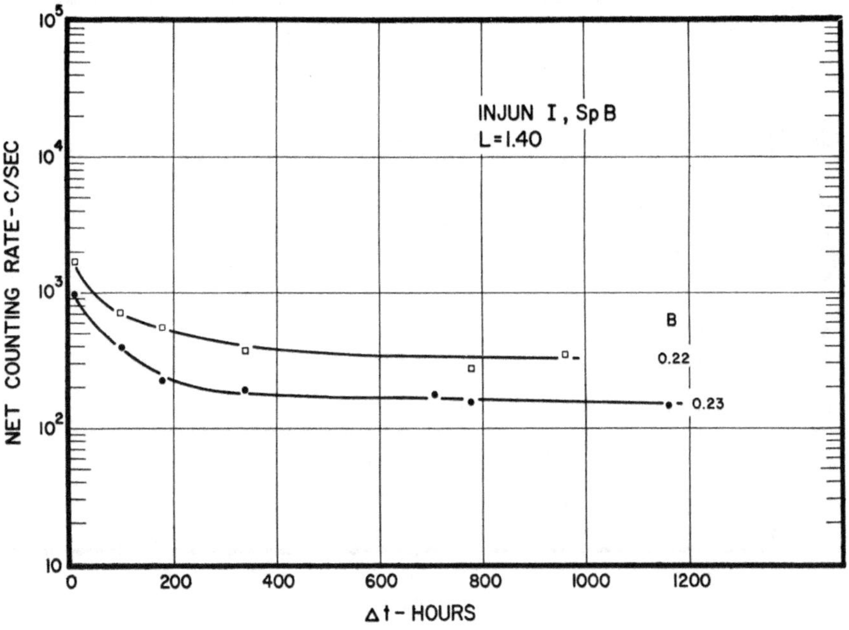
NET COUNTING RATE-C/SEC
Δt-HOURS
INJUN I, Sp B
L=1.40
B
0.22
0.23

Fig. 7.

for each era were entered on a large sheet of rectangular graph paper at points having the calculated values of magnetic shell coordinates B and L at the positions in space at which they were observed. Smooth contours of constant counting rate were drawn by hand through the array of data and were taken to represent the situation for the era.

From these basic graphs a number of other presentations of the data were derived:

(a) Counting rates versus B at selected values of L for each era.

(b) Counting rates versus L at selected values of B for each era.

(c) Counting rates versus time at selected values of B and L.

(d) Contours of constant counting rate in R, λ space.

In the figures which follow Δt is the elapsed time measured from 09:00 UT on 9 July 1962.

Figures 2, 3, 4, 5, 6, and 7 give time decay data from Injun I. The pre-burst background has been subtracted. Open squares and solid circles are used alternately to distinguish adjacent curves. According to FILZ and HOLEMAN (1965) there was a significant impulsive increase in the intensity of 55 MeV trapped protons at low altitudes over the South American magnetic anomaly ($L \sim 1.4$) at the time of the Starfish burst. A correction for this effect (not shown here) will result in a steeper rate of decay of the curves of Figures 2–7, the correction being significant for the highest B-value member of each of the families of curves in Figures 3, 4, and 5 and probably negligible elsewhere.

Since the spectrum of trapped electrons from an impulsive source varies with time, the characteristic of the SpB detector (Figure 1) should be borne in mind in the interpretation of the time decay curves.

The time history of the artificial radiation belt is continued in Figures 8, 9, 10, 11, 12, and 13 in which data from Injun I and Injun III have been plotted on the same graphs assuming that the absolute sensitivities of the respective SpB detectors are identical. The effective omnidirectional factor of the detector on Injun III (for counting penetrating particles) was found to be 1.5 to 1.9 times as great as that of the one on Injun I by comparing galactic cosmic ray counting rates at high latitudes. Thus it is probable that normalization of the two sets of data would result in somewhat steeper decays during the period $1200 < \Delta t < 4000$ hours than those shown. I have not made this normalization on the grounds (a) of its uncertain applicability to the counting of non-penetrating electrons via their bremsstrahlung and (b) of other undetermined uncertainties of the order of unity due to slightly different instrumental environments of the respective detectors on the two different satellites.

In Figures 8, 9, 10, 11, 12, and 13 total counting rates are plotted and the pre-Starfish Injun I background rates are shown by horizontal bars at the right hand side of the graphs.

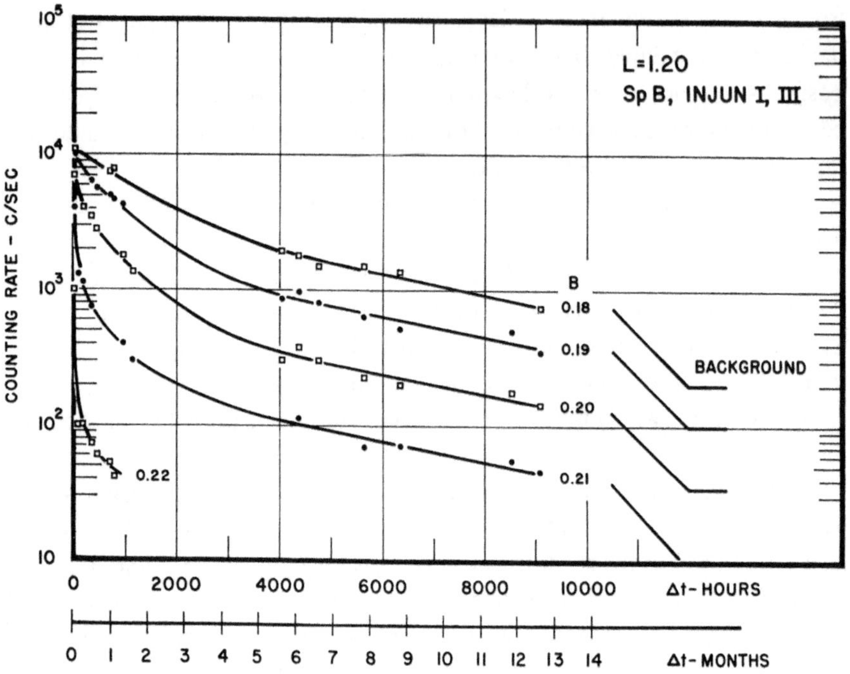
L=1.20
Sp B, INJUN I, III
B
0.18
0.19
0.20
0.21
0.22
BACKGROUND
COUNTING RATE - C/SEC
10
10²
10³
10⁴
10⁵
0
2000
4000
6000
8000
10000
Δt-HOURS
0 1 2 3 4 5 6 7 8 9 10 11 12 13 14
Δt-MONTHS

Fig. 8.

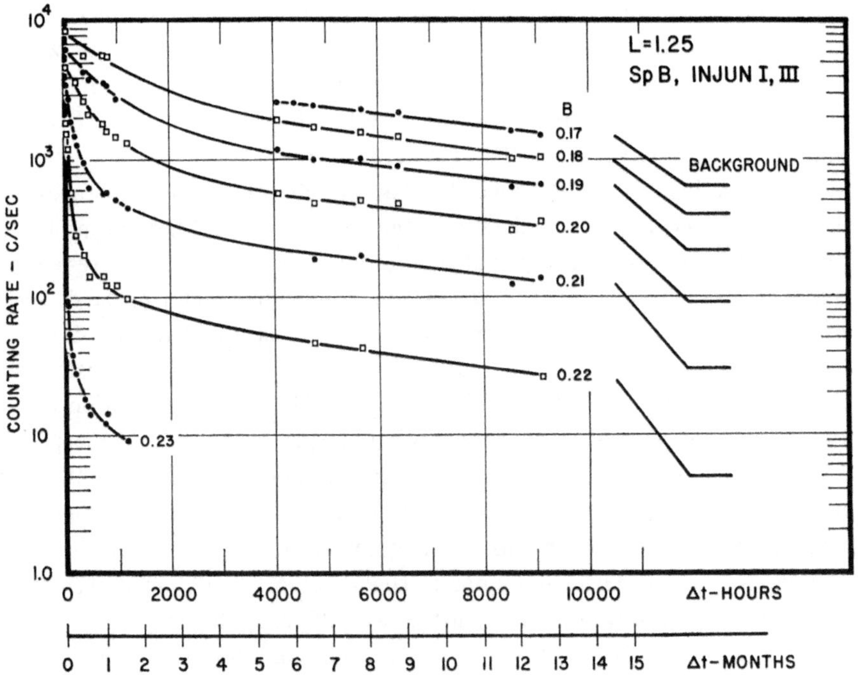
L=1.25
Sp B, INJUN I, III
B
0.17
0.18
0.19
0.20
0.21
0.22
0.23
BACKGROUND
COUNTING RATE - C/SEC
1.0
10
10²
10³
10⁴
0
2000
4000
6000
8000
10000
Δt-HOURS
0 1 2 3 4 5 6 7 8 9 10 11 12 13 14 15
Δt-MONTHS

Fig. 9.

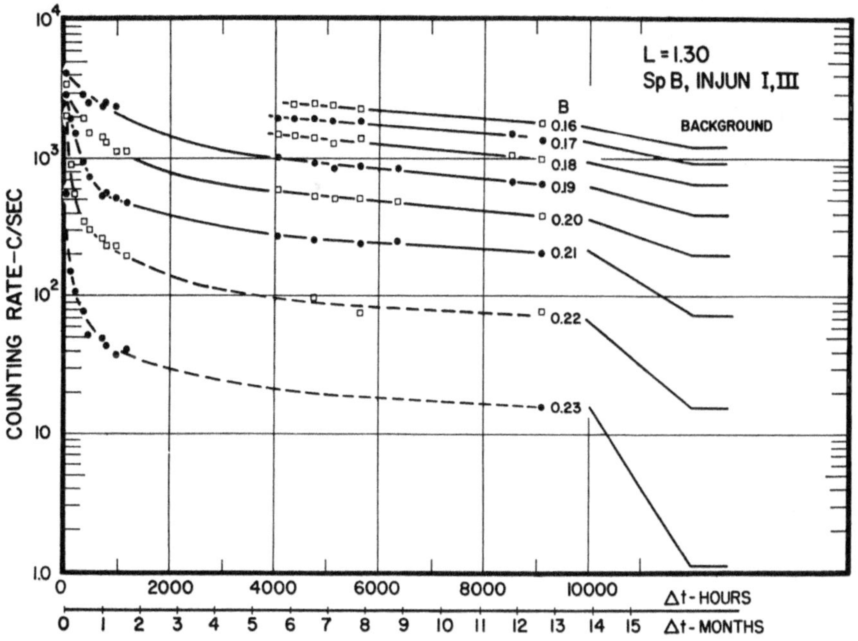
L = 1.30
Sp B, INJUN I, III
B
BACKGROUND
0.16
0.17
0.18
0.19
0.20
0.21
0.22
0.23
COUNTING RATE – C/SEC
Δt - HOURS
Δt - MONTHS

Fig. 10.

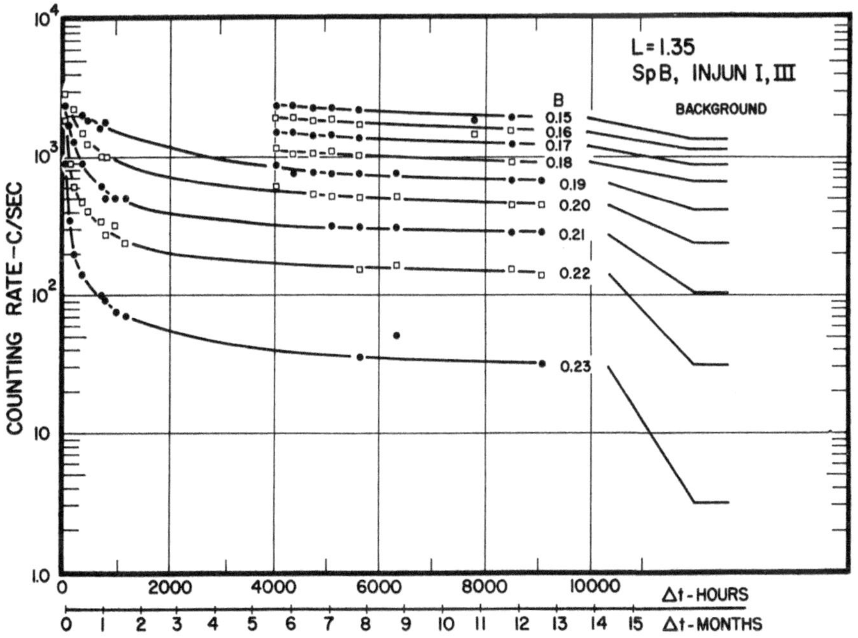
L = 1.35
Sp B, INJUN I, III
B
BACKGROUND
0.15
0.16
0.17
0.18
0.19
0.20
0.21
0.22
0.23
COUNTING RATE – C/SEC
Δt - HOURS
Δt - MONTHS

Fig. 11.

In Figure 14 are shown iso-counting-rate contours on a *B-L* diagram from Injun I for the year preceding the Starfish burst (plus an extension of the 800 count/sec contour with Pioneer IV, March 1959 data) and from Injun III at $\Delta t \sim 4050$ hours. Two other sets of contours from Injun III data are shown in Figures 15 and 16.

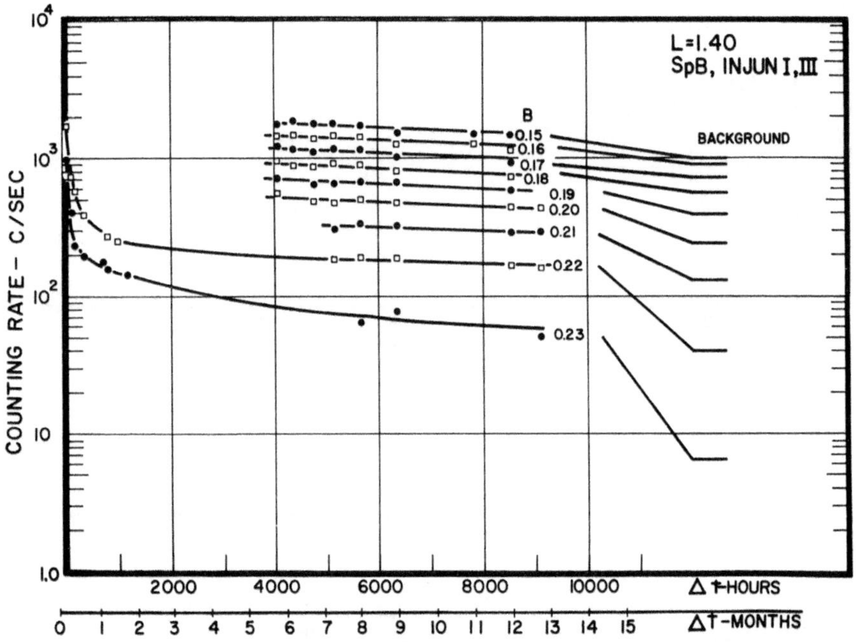

Fig. 12.

Summary

(a) In Figure 17 are shown meridian cross-sections of the Starfish radiation belt for two different eras, in R, λ coordinates. The diagrams are obtained by coordinate transformation from the basic *B-L* plots to equivalent polar co-ordinate R, λ space. The solid portions of the contours are those observed directly; the dashed portions of the contours in the left hand diagram are obtained by extrapolation to the equator in *B-L* space. As viewed in Figure 17, the extrapolation process appears to permit considerable uncertainty in the outer portion of the belt. However, in consideration of the fact that electrons from radioactive fission products must be injected isotropically at the point of decay, there is a calculable dependence of J_0 on B for L=constant corresponding to any assumed distribution of decays (VAN ALLEN, 1962). On these considerations it is found to be effectively impossible to obtain a substantially different outer boundary for the $\Delta t = 10$ hour distribution by extrapolation of Injun I

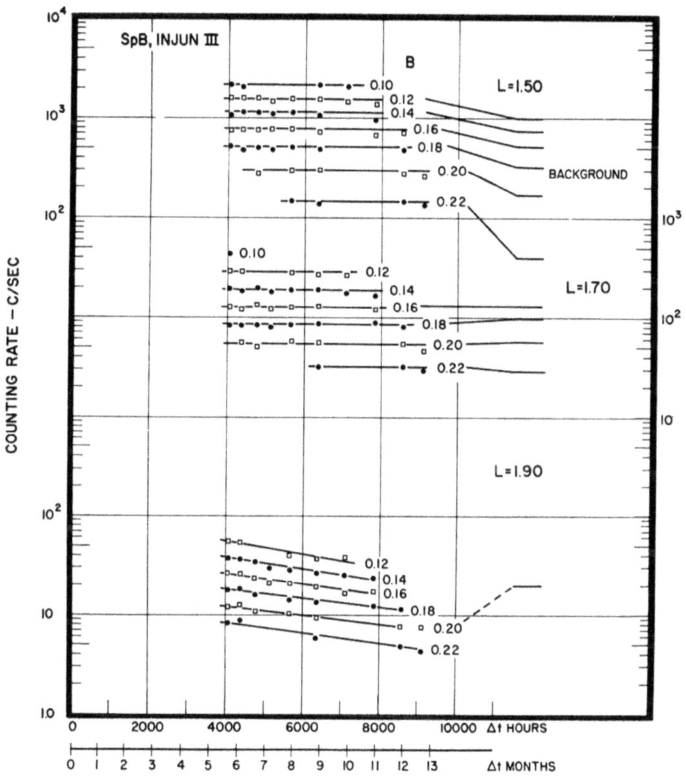

Fig. 13.

contours in B-L space. Hence our results are thoroughly irreconcilable with the widely published diagram of HESS and NAKADA (1962) (cf. VAN ALLEN *et al.*, 1963), insofar as the latter was represented as showing the distribution of fission-decay electrons. BROWN and GABBE (1964) have, however, shown persuasive evidence for an outer fringe (higher L values) of the basic belt of "hard" electrons (Figure 17) caused by the burst in some secondary manner and comprised of electrons having a much softer energy spectrum than that of fission decay electrons.

(b) The volume integration of the left hand diagram of Figure 17 gives a total inventory of fission-decay electrons of

$$1.3 \times 10^{25}$$

in trapped orbits at $\Delta t = 10$ hours. This volume integration corresponds to an "injection efficiency" of 2.6%. The overall uncertainties in this estimate are

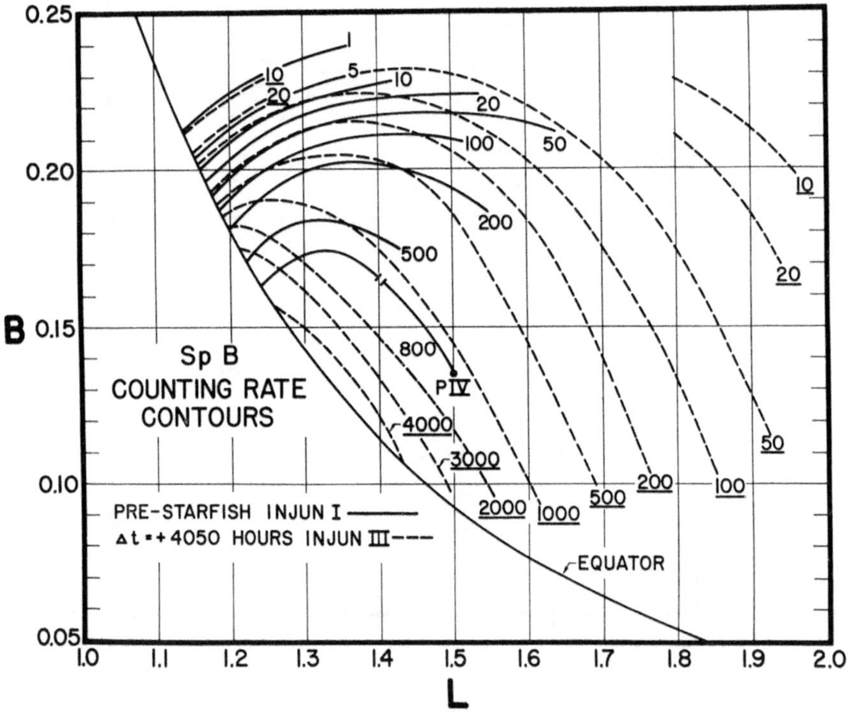

Fig. 14.

such as to make it quite unlikely that the true value lies outside of the range 2 to 5%.

(c) The "apparent mean lifetime" τ of ~2 MeV trapped electrons has been taken from the linear portions of decay curves similar to Figures 8, 9, 10, 11, and 12 for $1.15 < L < 1.4$ for the time period $4000 < \Delta t < 10000$ hours. Additional values at $L = 1.4$ and $L = 1.5$ have been derived from the 302 data of Injun III. Note that these values of τ are independent of the uncertainty in relative sensitivities of the SpB detectors in Injun I and Injun III. In the range $2.1 < L < 3.6$ a series of comparable values of "apparent mean lifetime" are available from observations with Explorer IV of the artificial radiation belt of Argus III

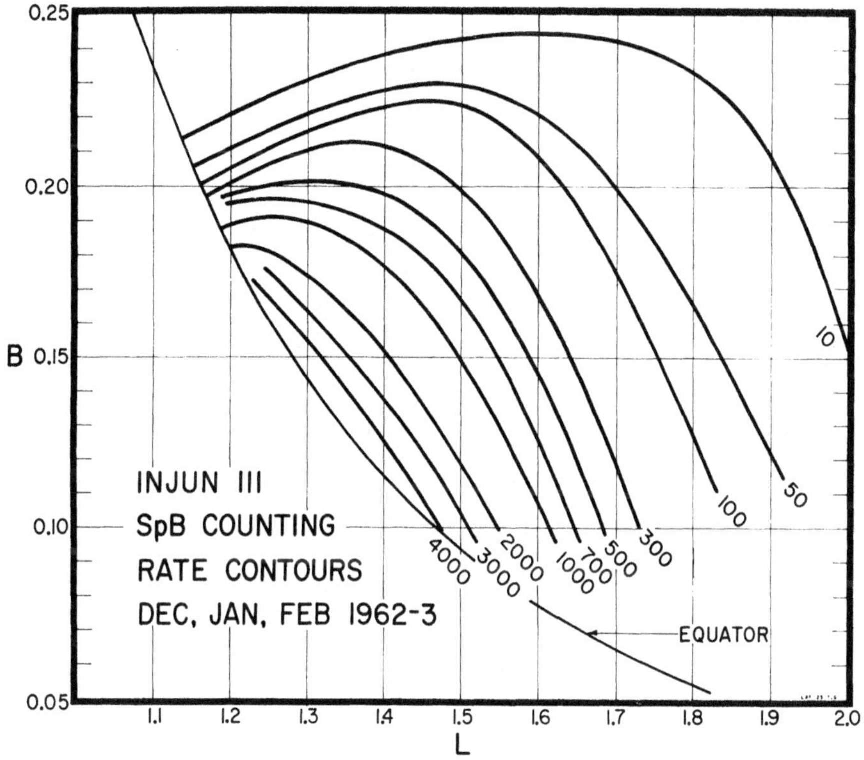

Fig. 15.

(VAN ALLEN *et al.*, 1959) and from observations with Explorer XIV of the three artificial radiation belts produced by Russian high altitude nuclear bursts in October-November 1962 (FRANK *et al.*, 1964). A summary plot of all of these values of τ is given in Figure 18 (VAN ALLEN, 1964). A maximum value of about 2 years occurs at $L \sim 1.5$.

(d) WALT (1964) has achieved an essentially complete understanding of the time decay data shown herein for $L < 1.25$ on the basis of elementary processes of scattering and energy loss of electrons in the quiescent atmosphere.

(e) For L increasing above 1.25 the observed values of τ fall below the expected theoretical curve of Walt in a progressively more dramatic way (Figure 18), a result well known for many years from observation of the time variations of the natural outer radiation belt of the earth.

(f) The maximum value of the observed lifetime is about fifty times less than that predicted by Lovell, Hess, and others in the late summer of 1962.

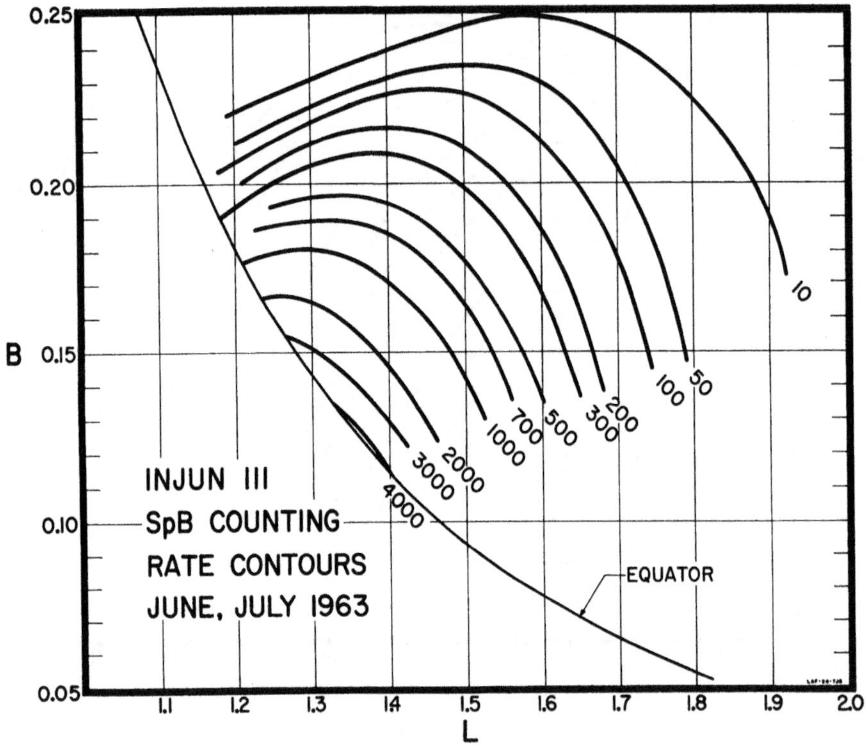

Fig. 16.

(g) Of the some 2.6% (2 to 5%) of the total number of fission decay electrons from the Starfish burst which were trapped at $\Delta t \sim 10$ hours, about 15% (or 0.4 of 1% of the total) survived the first $5\frac{1}{2}$ months and about 10% the first year.

Acknowledgement

This work was supported primarily by the U. S. Office of Naval Research under contracts N9onr 93803 and Nonr 1509(06).

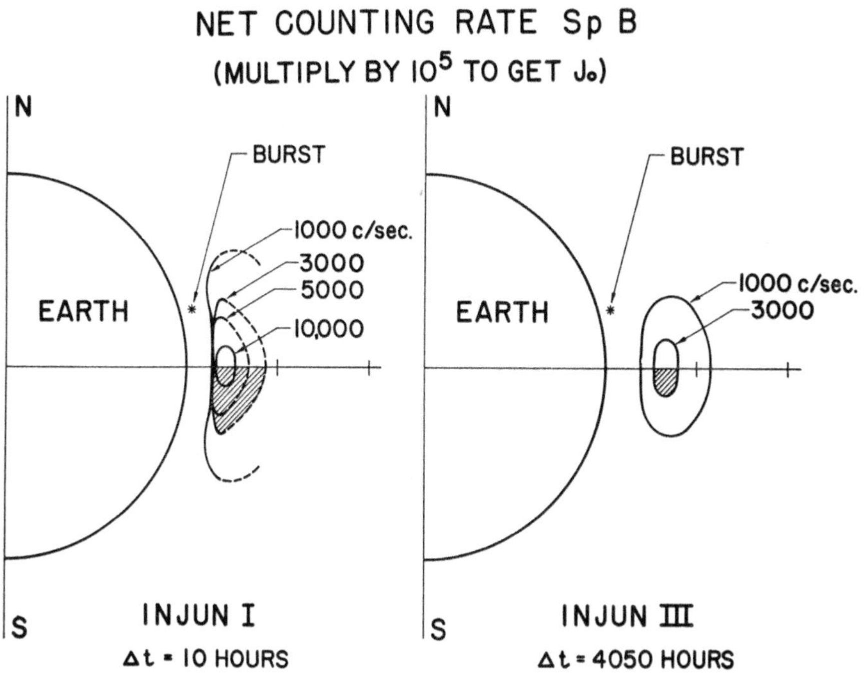
NET COUNTING RATE Sp B
(MULTIPLY BY 10^5 TO GET J₀)
N
BURST
1000 c/sec.
3000
5000
10,000
EARTH
S
INJUN I
Δt = 10 HOURS
N
BURST
1000 c/sec.
3000
EARTH
S
INJUN III
Δt = 4050 HOURS

Fig. 17.

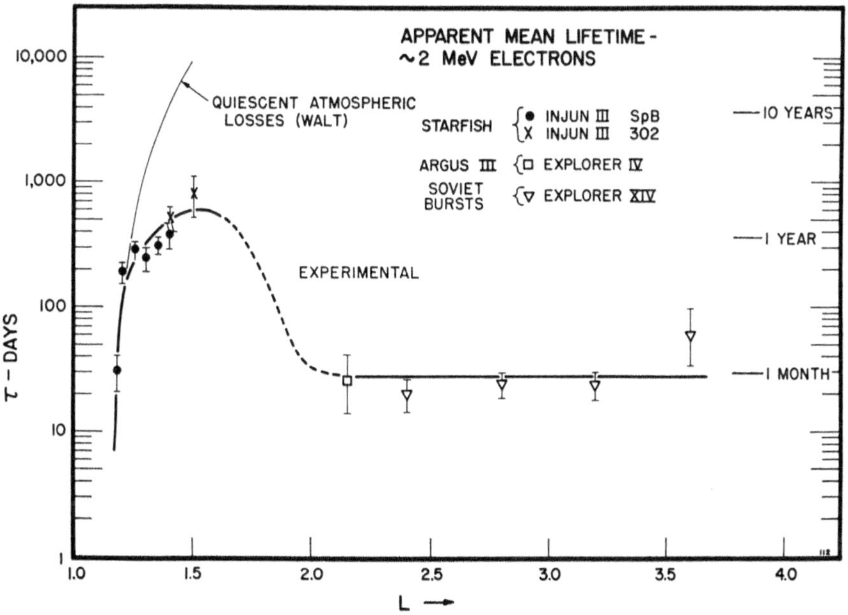
APPARENT MEAN LIFETIME -
~2 MeV ELECTRONS
QUIESCENT ATMOSPHERIC
LOSSES (WALT)
STARFISH
● INJUN III SpB
X INJUN III 302
ARGUS III
□ EXPLORER IV
SOVIET BURSTS
▽ EXPLORER XIV
10 YEARS
1 YEAR
1 MONTH
EXPERIMENTAL
10,000
1,000
100
10
1
τ - DAYS
1.0
1.5
2.0
2.5
3.0
3.5
4.0
L →

Fig. 18.

References

CARTER, R. E., REINES, F., WAGNER, J. J., and WYMAN, M. E.: 1959, *Phys. Rev.* **113,** 280–286.
CHRISTOFILOS, N. C.: 1959, *J. Geophys. Res.* **64,** 869–875.
FILZ, R. C. and HOLEMAN, E.: 1965, *J. Geophys. Res.* **70,** 5807–5822.
HESS, W. N. and NAKADA, PAUL: 1962, *Science* **138,** 53–54 and illustration on the cover.
O'BRIEN, B. J., LAUGHLIN, C. D., and VAN ALLEN, J. A.: 1962, *Nature* **195,** 939–943.
PETSCHEK, A. G.: 1963, *J. Geophys. Res.* **68,** 663–665.
STÖRMER, C.: 1907, *Arch. Sci. phys. et naturelles* **24,** 317–364.
VAN ALLEN, J. A.: 1958, Special Joint Meeting of National Academy of Sciences and American Physical Society, Washington, D. C., May 1, 1958.
VAN ALLEN, J. A.: 1961, Verbatim transcript of lecture published in IGY Satellite Report No. 13 January 1961, IGY World Data Center A, Rockets and Satellites, National Academy of Sciences-National Research Council, Washington 25, D. C.
VAN ALLEN, J. A.: 1962, *Trans. IAU* **X1B,** 99–136.
VAN ALLEN, J. A.: 1964, *Nature* **203,** 1006–1007.
VAN ALLEN, J. A., FRANK, L. A., and O'BRIEN, B. J.: 1963, *J. Geophys. Res.* **68,** 619–627.
VAN ALLEN, J. A., MCILWAIN, C. E., and LUDWIG, G. H.: *J. Geophys. Res.* **64,** 877–891.
WALT, MARTIN: 1964, *J. Geophys. Res.* **69,** 3947–3958.

MEASUREMENTS OF TRAPPED ELECTRON INTENSITIES MADE BY THE EXPLORER XV SATELLITE

C. E. McILWAIN

University of California, San Diego, La Jolla, California

Abstract. The spatial distributions of three categories of trapped particles were observed for three months by two detectors on the Explorer XV satellite. The decay time constants for electrons with energies greater than 0.5 MeV were generally greater than 6 months in the region $L < 1.7$ earth radii. The corresponding time constants for electrons with energies greater than 5 MeV were unobservably long (greater than one year). On November 5, 1962, over one percent of the electrons available from the July 9, 1962 event were still present.

Preface

The following paper was written in the fall of 1963. Since a number of improvements in the methods of analyzing the data were in progress, it was decided to withhold publication until the improved results became available. The principal result of the new methods of analysis has been to show that the earlier methods were in fact quite accurate. The "final" analysis should be completed in the near future, but there appears to be no reason to withhold publication of the present paper.

1. Introduction

This paper is principally based upon the observations of two charged particle detectors on the Explorer XV satellite. The orbit for this satellite was chosen such that the region with 4 earth radii and ± 30 degrees magnetic latitude was regularly traversed in a way that permits a complete determination of the particle fluxes within this region with as little as a one week segment of data.

Some of the results obtained by these detectors have been published elsewhere (McILWAIN, 1963) and therefore will not be discussed at length.

The determination of the detailed behavior of the electrons injected by the detonations ascribed to the Soviet Union which occurred shortly before and after the launch of Explorer XV requires the development of analytic techniques which have not yet been completed. The explicit use of this data will therefore be deferred until a later paper.

2. Coordinates

The magnetic coordinates B and L used in this paper have been discussed elsewhere (McIlwain, 1961; Stone, 1963). The relationships between B and L and the other subsidiary coordinates R, λ, and B/B_0 are given by the equations:

$$\begin{aligned} B &= (0.311653/R^3)(4 - 3R/L)^{\frac{1}{2}} \\ R &= L\cos^2\lambda \\ B/B_0 &= (L/R)^3(4 - 3R/L)^{\frac{1}{2}} \end{aligned}$$

The Jensen-Cain (1962) coefficients for the epoch 1960 were used to compute B and L.

3. Detectors

The University of California directional electron detector on Explorer XV is shown in Figure 1. The 0.25 cm diameter plastic scintillator is partially surrounded by platinum which serves three purposes: (1) it provides a final area defining aperture 0.13 cm in diameter; (2) it provides shielding against penetrating particles; (3) it increases the detection efficiency by scattering some of the electrons emerging from the scintillator back into the active volume. The

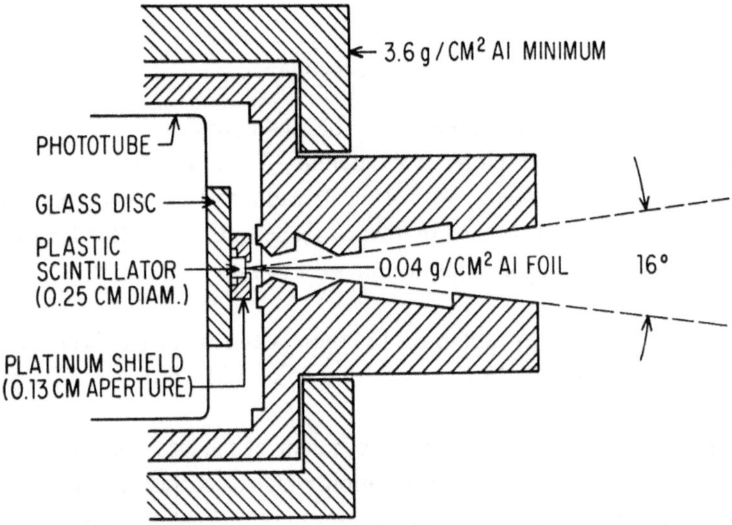

Fig. 1. Cross section of the directional detector used to detect electrons with energies greater than 0.5 MeV.

aluminum solid angle defining aperture is designed such that the surface available for scattering occupies less than 10 percent of the solid angle (the ± 8 degree cone) available without scattering. The rates of pulses corresponding to electron energy losses of 280 keV and 660 keV are counted. The 49 milligram/cm^2 absorber raises the proton thresholds to 5.0 and 5.2 MeV respectively. Since most electrons entering the scintillator with more than 660 keV energy leave the active volume before losing 660 keV, the efficiency for counting electrons in the second channel is small. By direct measurement the first channel is about 5 times as efficient as the second channel for energetic electrons. The proton contribution to the counting rate of the first channel can therefore be removed by subtracting 1.1 times the second channel without appreciably changing the character of the electron contribution. The factor of 1.1 corresponds to a typical ratio of proton counting rates in the two channels. The minimum of 3.6 g/cm^2 of shielding is sufficient to reduce the counting rate due to penetrating particles to less than 10 percent of the total rate. The electron efficiency versus energy was measured with the Lockheed beta ray spectrometer and the Bell Telephone Laboratory electron Van de Graff over the range of 200 to 1800 keV. The predicted response at higher energies was checked with the fission electron source at Los Alamos. The electron efficiency rises from near zero at 400 keV to about 0.5 at 600 keV and about 0.7 at 1000 keV (after subtraction of the second channel).

For convenience, the efficiency versus energy is approximated by a step function at 500 keV. This approximation introduces an error of less than ± 20 percent over a wide range of electron energy spectra. The uncertainty in the absolute geometrical factor and efficiency adds an additional possible error in the derived electron fluxes of about -30 to $+10$ percent.

The directional detector is directed perpendicular to the satellite spin axis. The 9.3 second accumulations telemetered correspond to averages over about 7.6 revolutions. The scatter introduced by differences in the initial azimuthal angles is typically less than 5 percent. The method used for converting the spin averages into omnidirectional intensities is illustrated in Figure 2. The spin averages (data) corrected for the system dead time of 3.05 ± 0.2 microseconds are shown at the top of the figure as a function of the scalar magnetic field along the line of force $L = 1.5$ earth radii. These spin averages are approximately converted into omnidirectional intensities by multiplying by the function $(0.0308 + Q)/(0.0524 + 0.632\,Q)$ where $Q = (\theta/90)^4$ and θ is the angle between the magnetic field vector and the satellite spin vector. An initial least squares fit is made to the resulting data points as illustrated by the curve labelled J_0' in Figure 2. This curve is then used to obtain an initial function representing the unidirectional intensity $(j_\perp')$ perpendicular to magnetic field vector by the method given by Ray (1960) and by Farley and Sanders (1962). This function can then be used to obtain the angular distribution at any point along the field line by way of the relation $B_1 \sin^2 \alpha_2 = B_2 \sin^2 \alpha_1$. A good approximation to the true ratio of the omnidirectional to spin average intensity can therefore be

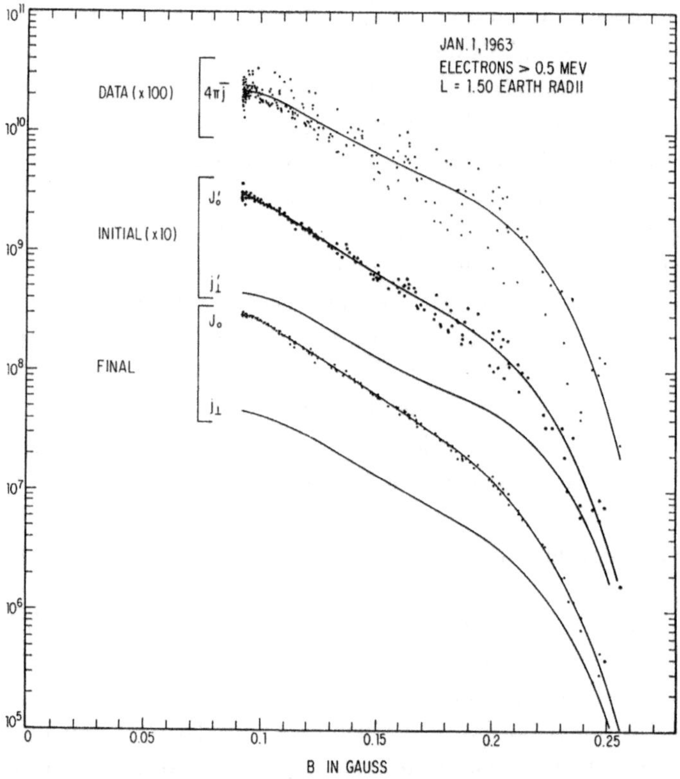

Fig. 2. Electron intensities along the line of force $L = 1.5$ earth radii illustrating the method of data analysis. The top curve (data) corresponds to the spin average counting rates, the next curve (J_0') corresponds to the omnidirectional intensity obtained from the spin averages using a typical function of the angle between the spin and magnetic field vectors. The third curve ($j_\perp'$) is the unidirectional intensity derived from J_0'. The points around the curve labelled J_0 were obtained from the original data using the curve $j_\perp'$ to determine the ratio between the omnidirectional intensity and the spin average intensity for the different satellite orientations with respect to the magnetic field.

obtained for each data point by appropriate integrations over the initial angular distribution given by the function $j_\perp'(B)$. The integration for the spin average also includes the instantaneous effect of the system dead time so that initial approximate correction for dead time is effectively replaced by the true correction factor. In the present set of data, the dead time correction is never more than 30 percent so that systematic errors introduced by the uncertainty in the dead time should be less than 2 percent. The data multiplied by the calculated ratios of omnidirectional to spin average intensities can be seen in Figure 2 to scatter by an average of less than 3 percent from the final fit the omnidirectional intensity (J_0). The final unidirectional intensity ($j_\perp$) is obtained from the omnidirectional intensity as before. It should be noted that $j_\perp$ depends to first order

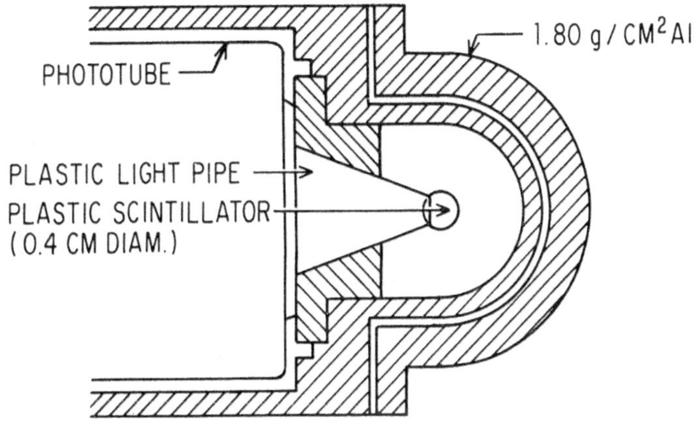

Fig. 3. Cross section of the omnidirectional detector used to detect electrons with energies greater than 5 MeV and protons with energies between 40 and 110 MeV.

upon the derivative of J_0 with respect to B so that the slope of $j_\perp$ depends strongly upon the curvature of J_0. A second iteration with this procedure is not performed because the error in the desired ratio obtained with the initial angular distribution is less than the scatter expected from other causes. The average scatter of less than 3 percent for example corresponds to an average error in L of less than 0.3 percent. The known discrepancies between the Jensen-Cain magnetic field representation and the true earth's magnetic field predict an average error in L of about this magnitude. It might be noted that this is a case in which the use of $L_0 \equiv (M/B_{\text{min}})^{1/3}$ instead of L would introduce a significant amount of additional scatter in the data.

The University of California omnidirectional detector on Explorer XV is shown in Figure 3. The 0.4 cm diameter sphere of plastic scintillator is shielded by a uniform 1.80 g/cm^2 of aluminum over the front 2π steradians of the solid angle. In the other direction, the minimum absorber is about 2.6 g/cm^2 and is greater than 3.5 g/cm^2 over at least 80 percent of this back 2π steradian solid angle. The rates of pulses corresponding to electron energy losses greater than 0.8 and 3.0 MeV are counted. Since the average electron energy loss in the scintillator is less than 0.8 MeV, the first channel has a low efficiency for counting electrons and the second channel has a near zero efficiency for electrons. Pileup of electron produced pulses can produce pulses large enough to trigger the second channel but delay line clipping of the pulses down to a 0.25 microsecond length prevents this from happening at electron counting rates in the first channel of less than 10^6/sec. In flight, the highest rates obtained by the first channel were less than 10^5/sec. The efficiency of the second channel versus energy for protons has been calculated taking into account the distribution of path lengths in the scintillator; the scintillator saturation effects at high $\mathrm{d}E/\mathrm{d}X$ values, and the distribution of absorber over the full 4π steradian solid angle. The relative efficiency for protons rises abruptly from 0.0 to 1.0 at 40 MeV. Using the University of Minnesota linear proton accelerator with a similar detector which had a 1.3 g/cm^2 absorber (instead of the present 1.8 g/cm^2) the rise from zero to near full efficiency occurred over less than a 2 MeV range at 34 MeV. The calculated relative efficiency of the present detector remains between 0.75 and 1.35 from about 40 to 95 MeV and falls to 0.1 and 0.01 at about 130 and 160 MeV respectively. By integrating the calculated efficiency versus energy over a wide range proton energy spectra it was found that considering the efficiency to be constant over the range 40 to 110 MeV and zero at other energies introduces an error typically less than 15 percent. The calculated value of the efficiency times the geometrical factor ($=1/18$ cm^2) is probably in error by less than 15 percent.

The discrimination level of 0.8 MeV for the first channel is sufficiently high that the efficiency for γ rays and therefore for electron bremsstrahlung is very low. The calculated relative electron efficiency for the first channel rises from less than 10^{-2} at 4.3 MeV to about 0.2 at 5 MeV and up to about 0.8 at 8 MeV. The electron spectra encountered were apparently decreasing rapidly above 5 MeV, therefore over 70 percent of the counts registered in the first channel were probably due to electrons in the energy range between 4.8 and 6.5 MeV. The efficiency above 5 MeV is therefore taken to be the caclulated efficiency at 5.6 MeV and the efficiency for energies below 5 MeV to be zero. The calculated value was experimentally confirmed to within 30 percent with the fission electron source at Los Alamos. The expected low efficiency for 1 to 2 MeV electrons was confirmed by noting that a 15 millicurie Sr^{90} source placed directly on the detector produced less than 0.3 counts/sec. The counts due to protons in the first channel are effectively eliminated by subtracting 1.4 times the counts in the second channel.

4. Spatial Dependence of the Observed Fluxes

Plots of the contours of constant omnidirectional intensity in the R, λ coordinates on November 10, 1962 for the 40 to 110 MeV protons, the electrons greater than 0.5 MeV and the electrons greater than 5 MeV are shown in Figures 4, 5, and 6 respectively. The corresponding plots for the latter two in B, L

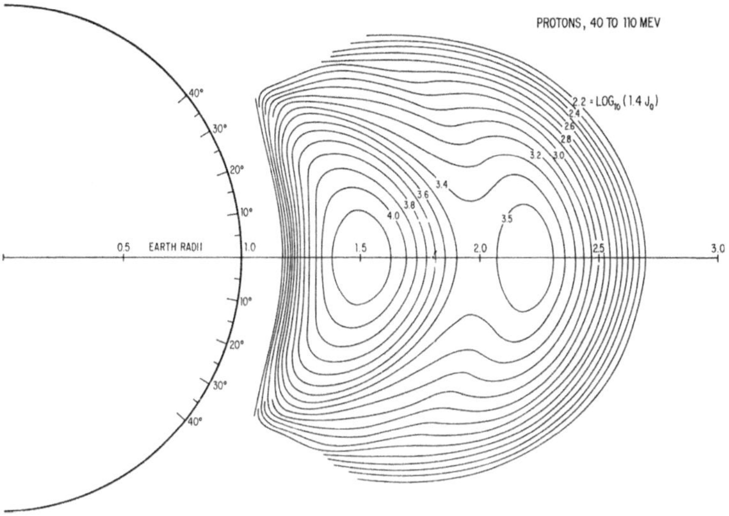

Fig. 4. Contours of constant intensity of 40 to 110 MeV protons on November 10, 1962, plotted in R, λ space. The horizontal line represents the magnetic equator.

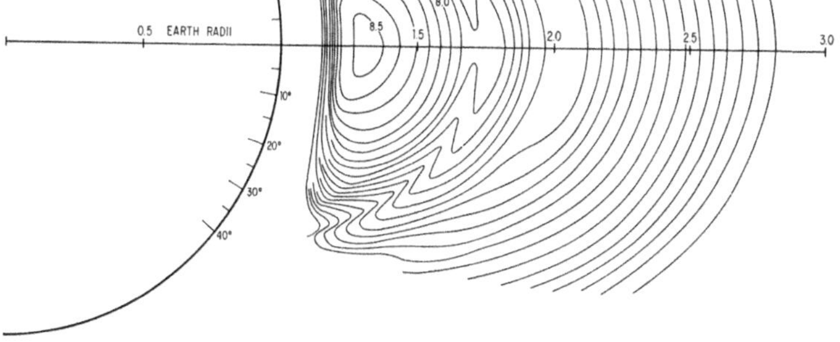

Fig. 5. Same as Figure 4 except for electrons with energies greater than 0.5 MeV.

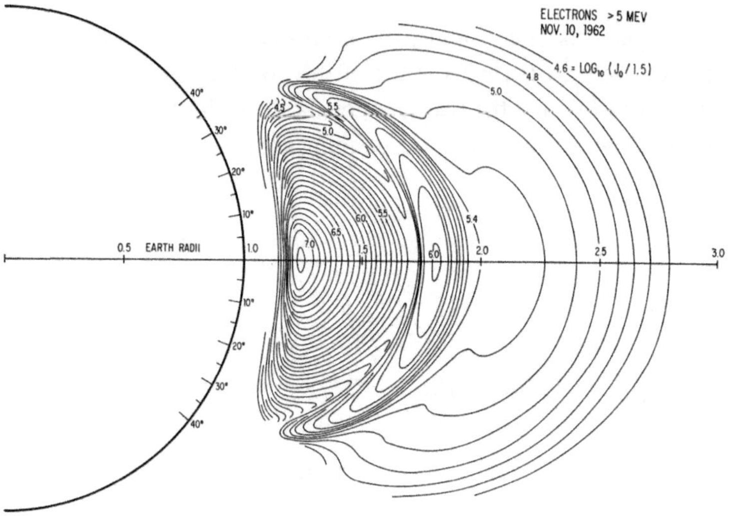

Fig. 6. Same as Figure 4 except for electrons with energies greater than 5 MeV.

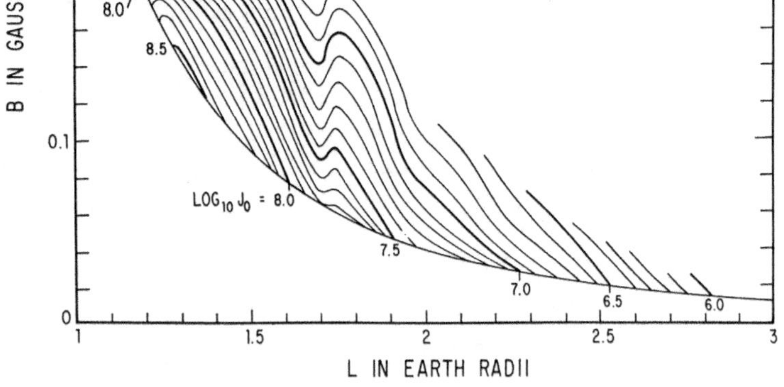

Fig. 7. The contours of Figure 5 mapped into B, L space.

coordinates are shown in Figures 7 and 8. These contours were obtained by first interpolating the data to a set of 55 different lines of force. The fit to the intensities versus B for each line of force was then searched to find the B values which correspond to each contour. These particular plots (Figures 4–8) were produced with an early set of fits to the data and contain some systematic errors which are typically less than 20 percent but go up to almost a factor of two at the extreme low altitude edge.

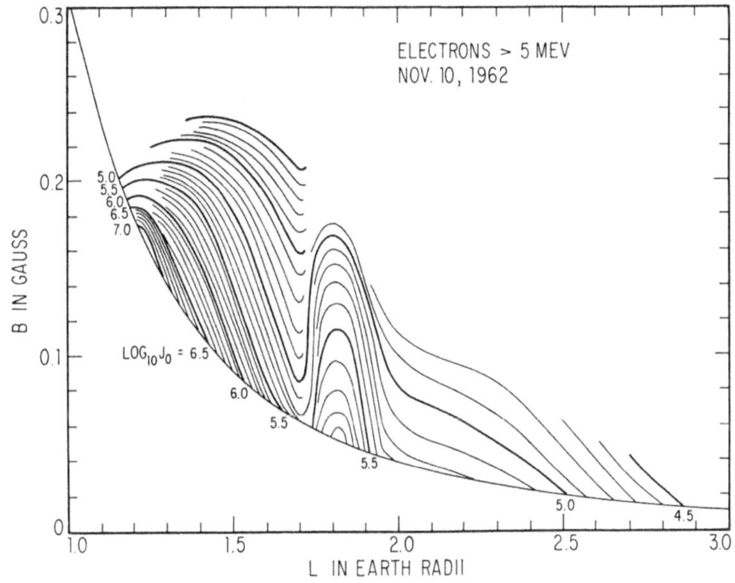

Fig. 8. The contours of Figure 6 mapped into B, L space.

The electrons in the region around 1.4 earth radii were presumably injected by the July 9, 1962 event (Starfish). The extra electrons in the $L = 1.75$ to 2.0 region were presumably injected by Soviet events in October and early November 1962.

The approximate distributions on January 1, 1963 can be seen in Figure 9. The extra low electrons around $L = 1.9$ at this time are almost completely masked by the Starfish electrons and possibly electrons of natural origin.

The most recent fits to the omnidirectional intensities and the derived unidirectional intensities as functions of B along the different lines of force are shown in Figures 10–14. The unidirectional intensities of electrons with energies greater

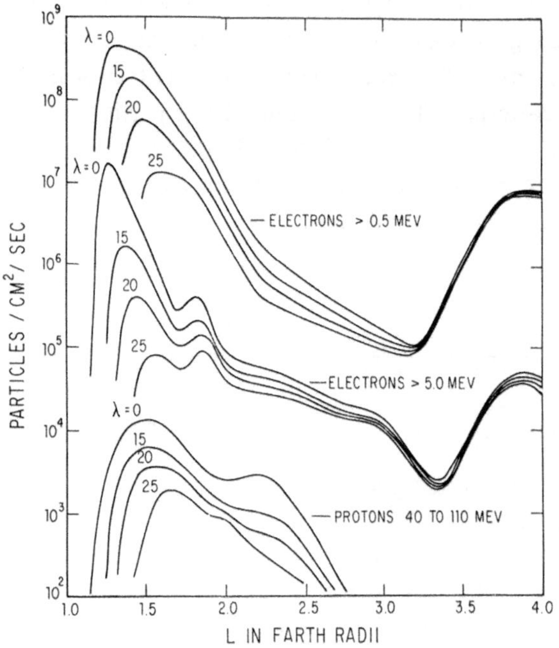

Fig. 9. The radial dependence of the omnidirectional intensities at four different magnetic latitudes on January 1, 1963.

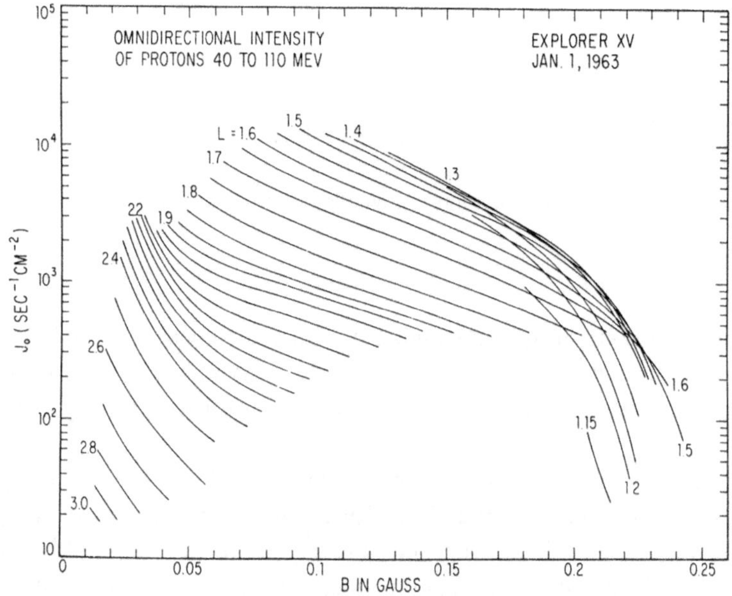

Fig. 10. Omnidirectional intensities of the 40 to 110 MeV protons on January 1, 1963 versus B along different lines of force.

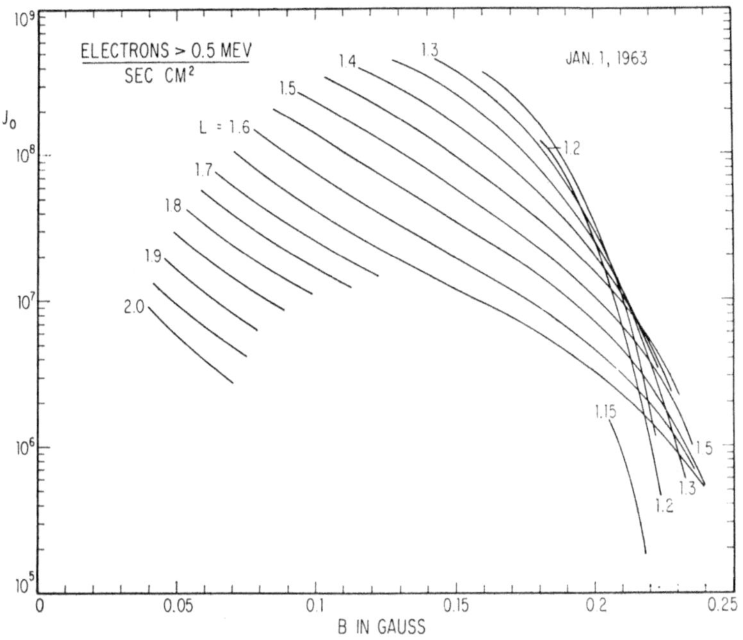

Fig. 11. Omnidirectional intensities of electrons with energies greater than 0.5 MeV on January 1, 1963 versus B along lines of force.

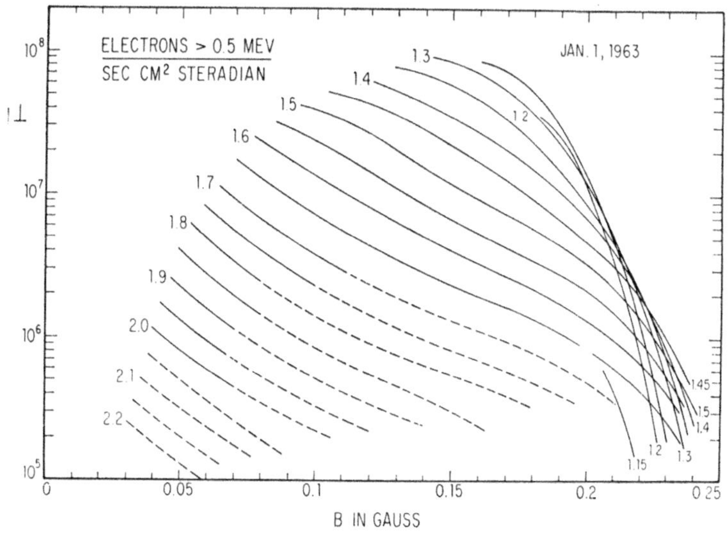

Fig. 12. Unidirectional intensities derived from the curves shown in Figure 11.

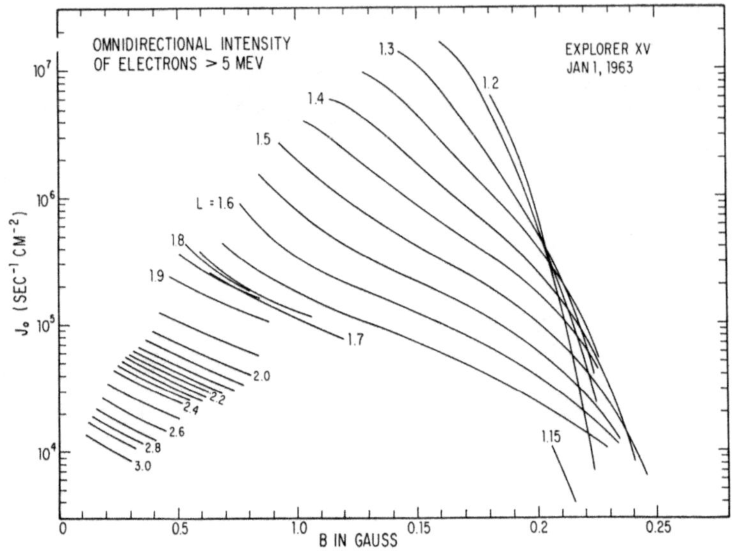

Fig. 13. Omnidirectional intensities of electrons with energies greater than 5 MeV on January 1, 1963 versus B along lines of force.

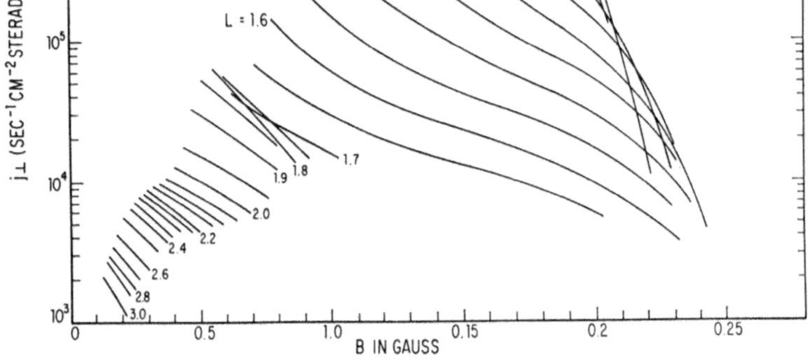

Fig. 14. Unidirectional intensities derived from the curves shown in Figure 13.

than 0.5 MeV have been supplemented with the data obtained by a similar detector on the Relay I satellite.

5. Electron Inventory

On November 5, 1962 the total number of electrons with energies greater than 0.5 and 5 MeV trapped on all magnetic shells labelled by $L \leqslant 1.72$ earth radii were 8.8×10^{24} and 9.8×10^{22} respectively. The probable error introduced by the data fitting and volume integration procedures is less than 10 percent. If the total number of electrons available as a result of the Starfish event was 10^{27} electrons (VAN ALLEN *et al.*, 1963) and if these electrons had an energy spectrum similar to that of a U^{235} fission source in time equilibrium then the present measurements indicate that 1.3 and 1.2 percent of the available electrons with greater than 0.5 and 5 MeV energy respectively were still trapped 120 days after the event. Comparisons with pre-event measurements indicate that less than 10 percent of the electrons trapped in the region $L \leqslant 1.7$ on November 5, 1962 were of natural origin.

6. Electron Energy Spectrum

As can be seen in the various figures, the spatial distributions of the two categories of electrons are not the same. This is shown explicitly in Figure 15 where the ratios of omnidirectional intensities, $J_0(>0.5)/J_0(>5)$ on January 1, 1963 are plotted versus L for $B/B_0 = 1.0 (\lambda = 0)$ and $B/B_0 = 1.6 (\lambda = 19°)$. The ratio for a fission spectrum would be about 80. In the region labelled "B" the spectrum is "harder" than a fission spectrum as might be predicted by the more rapid loss rate of low energy electrons. In the lowest region labelled "A" the spectrum is

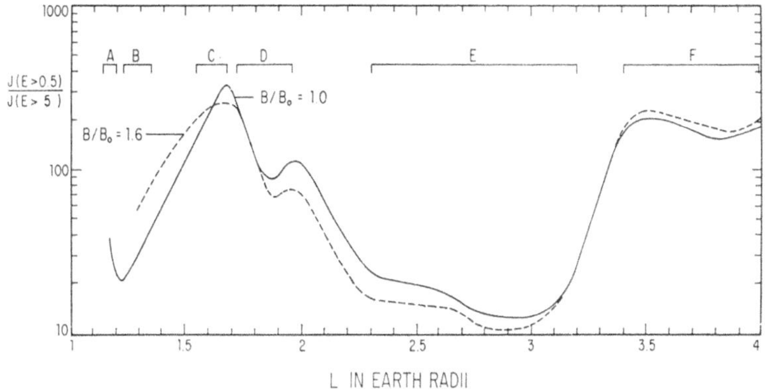

Fig. 15. Ratios of the omnidirectional intensities of electrons with energies greater than 0.5 MeV and 5 MeV.

starting to become "softer" possibly due to the increasing proportion of naturally injected electrons as the artificially injected electrons are rapidly removed by the dense atmosphere. In the region labelled "C", there is a relative absence of high energy electrons and may reflect a difference in the injected distributions of high and low energy electrons. The region labelled "D" corresponds to the regions in which the Soviet events injected electrons. At the high altitudes of the present measurements, these electrons appeared to have a fission like energy spectrum except at L values greater than 2 earth radii where the second Soviet event appeared to preferentially inject low energy electrons. In the region labelled "E" the electron spectrum had become "hard" by January 1, 1963 due to the more rapid decrease in the low energy fluxes with time. In region labelled "F", the intensity of high energy electrons had decreased drastically by December 18, 1962 but had increased by January 1, 1963 to produce the ratio indicated.

7. Time Variations

The time variations observed by the present detectors at high L values has been previously published (McIlwain, 1963). The general results are similar to those of many other experimenters: for $L>3$ the time variations are large and are correlated with magnetic disturbances; for $2>L>3$ the time variations are not as large (except after an impulsive injection) but still appear to be correlated with magnetic disturbances.

The time constants for decreasing a factor of $e=2.72$ for the electrons with energies greater than 0.5 MeV are shown in Figure 16. The statistical probable

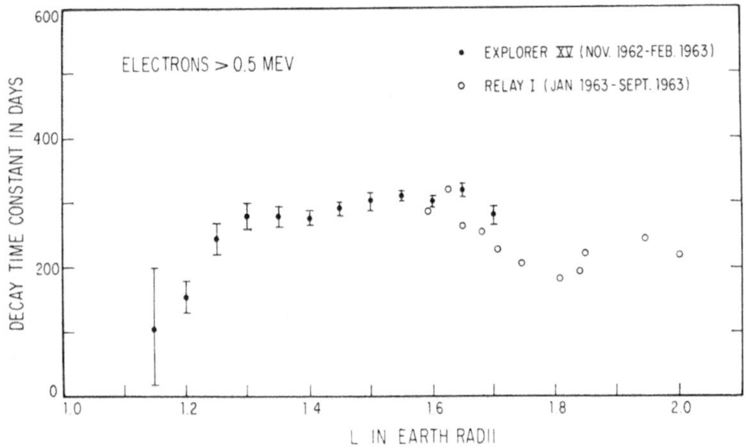

Fig. 16. Average decay time constants of electrons with energies greater than 0.5 MeV.

errors are shown for the Explorer XV data. The open circles were obtained by comparison of the January 1963 Explorer XV data with the September 1963 Relay I data. It should be noted that the rapid decreases in intensity which were observed in the region $L=1.75$ to 2.3 after impulsive injections (BROWN *et al.*,

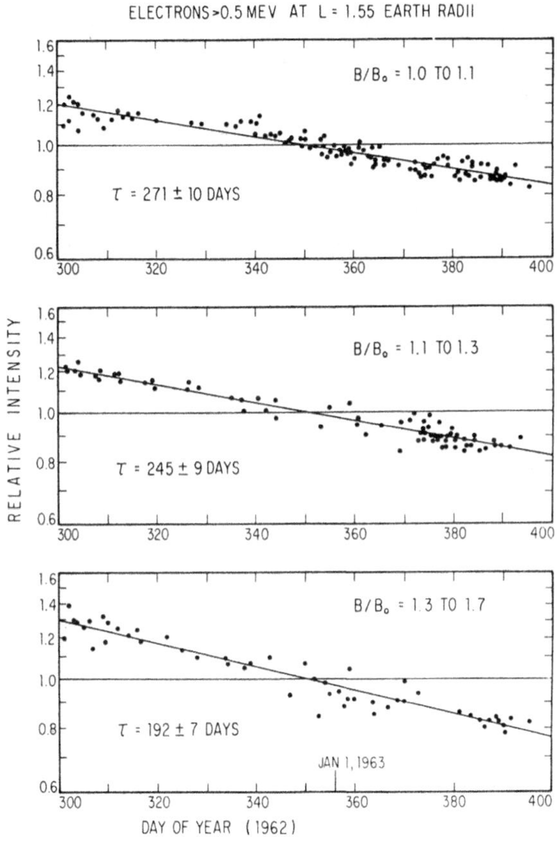

Fig. 17. Relative time dependences of the intensities of electrons with energies greater than 0.5 MeV at three locations along the line of force $L=1.55$.

1963; MCILWAIN, 1963) did not continue into 1963. This may be due to two effects: (1) the increase in decay time constant as an equilibrium distribution is approached (WELCH *et al.*, 1963); (2) the approach to the intensity of the naturally present electrons.

The time constants for the electrons with energies greater than 5 MeV typically correspond to a slowly increasing intensity in the region $1.30 < L < 1.7$. This may be an instrumental effect since the counting rate of this channel is somewhat sensitive to gain changes.

Conclusion

The distributions in space and time of three categories of trapped particles were measured from October 27, 1962 to January 30, 1963. Analytic functions have been obtained which represent the omnidirectional and unidirectional intensities along different lines of force.

Within $L = 1.7$ earth radii, it was found that over one percent of the electrons assumed to be available from the July 9 event were still trapped in the earth's magnetic field 120 days after the event. Over most of this region, the electron fluxes changed by less than 40 percent during the period 110 to 206 days after the event.

Within $L < 1.7$ the electron energy spectrum was found to depend strongly on L and to some extent upon B. At all higher L values, the spectrum depended strongly on both L and time.

The slow rate of change in some of the artificially injected electron fluxes indicates that it will be possible to follow their behavior for a number of years before they become hidden in the naturally occurring electron fluxes.

Acknowledgements

I thank J. Valerio of the University of California, San Diego, J. Townsend, Jr. of the Goddard Space Flight Center, Greenbelt, Maryland, and G. Frohwein and D. Enemark of the State University of Iowa for their work associated with the Explorer XV satellite. The particle sources used in detector calibration were provided through the cooperation of R. Smith of the Lockheed Missiles and Space Division, W. Brown of the Bell Telephone Laboratory, R. Featherstone of the University of Minnesota, and H. Motz of Los Alamos.

This research was supported in part by NASA contracts NAS5-3063 and NASr-116.

References

Brown, W. L., Gabbe, J. D., and Rosenzweig, W.: 1963, *Bell System Tech. J.* **42,** 1505.
Farley, J. A. and Sanders, N. L.: 1962, *J. Geophys. Res.* **67,** 2159.
McIlwain, C. E.: 1961, *J. Geophys. Res.* **66,** 3681.
McIlwain, C. E.: 1963, *Science* **142,** 355.

Ray, E. C.: 1960, *J. Geophys. Res.* **65,** 1125.
Stone, E. C.: 1963, *J. Geophys. Res.* **68,** 4157.
Van Allen, J. A., Frank, L. A., and O'Brien, B. J.: 1963, *J. Geophys. Res.* **68,** 619.
Welch, Jr., J. A., Kaufmann, R. L., and Hess, W. N.: *J. Geophys. Res.* **68,** 685.
Jensen, D. C. and Cain, J. C.: 1962, *J. Geophys. Res.* **67,** 3568 (a).

OBSERVATIONS OF THE TRANSIENT BEHAVIOR OF ELECTRONS IN THE ARTIFICIAL RADIATION BELTS

W. L. BROWN
Bell Telephone Laboratories, Inc.,
Murray Hill, New Jersey

1. Introduction

The sudden artificial alteration of the electron distribution in space by high altitude nuclear tests has provided a unique opportunity to study the relaxation processes or loss mechanisms that are effective in controlling the natural particle distribution. Observations of the artificial belts of energetic electrons and their transient decay following the U.S. and U.S.S.R. nuclear tests of 1962 have been made by instruments on the Telstar I and II, Relay I and II, and Explorer XV and XXVI satellites. The most dramatic results from these experiments have been obtained on Telstar I and Explorer XV which were advantageously in orbit at times of major activity during the testing period. Descriptions of their observations are contained in Sections 2 and 3 and comprise the primary part of this paper. Some observations made by Relay II long after the creation of the artificial radiation belts will be discussed in Section 4.

All the measurements described in this paper were made with the same type of particle detector, a silicon diffused p-n junction device (BUCK *et al.*, 1964) with a sensitive volume about 2 millimeters in diameter and 0.4 millimeters in thickness. The geometrical arrangement of the detector, its entrance aperture and absorber, varied among the different experiments as did also the electronics that processed the pulses produced in a proportional way by charged particles in the sensitive volume. Because of the relatively small size of this sensitive volume the measurement of high energy electrons required entrance absorbers to remove the lower energy portion of the electron flux distribution. In the Telstar I configuration with an open entrance aperture the detector had a principle sensitivity to electrons of less than 1 MeV and primarily measured electrons above about 0.5 MeV. In the Explorer XV experiments a series of detectors were used with different entrance absorbers. Results from three of these devices measuring electrons above 0.5 MeV, above 1.9 MeV, and above 2.9 MeV are described in Section 3.

2. Telstar I, Starfish and the U.S.S.R. Tests

2.1 *Electron distributions following the Starfish test*

Telstar I was launched on June 10, 1962 (day 191), one day after the U.S. high altitude nuclear test Starfish. The order of these events has clouded the evaluation of the effect of the nuclear explosion because of the meager information concerning the natural electron distribution present before the test in

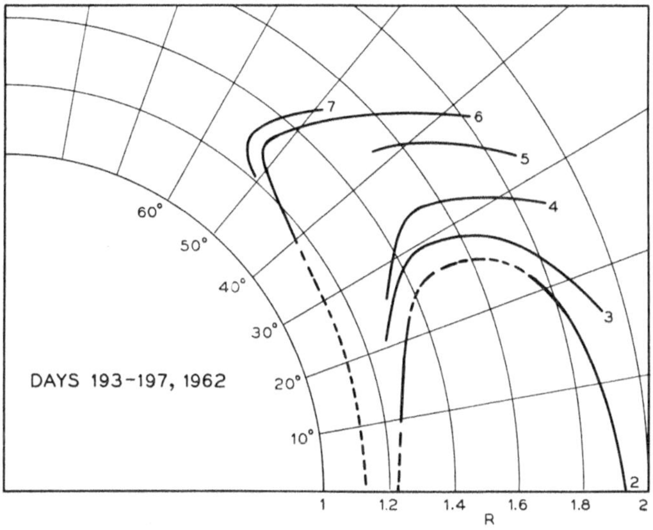

Fig. 1. Contours of equal omnidirectional counting rate for electrons from Telstar I in the time period from Day 193–197, 1962. The contours are two to a decade as follows: $2 - 6.6 \times 10^7$/ cm^2 sec; $3 - 2.1 \times 10^7$; $4 - 6.6 \times 10^6$, etc. The omnidirectional counting rate is related to the omnidirectional flux of particles by the efficiency factor of the detector.

most of the region that Telstar I actively explored. The changes in the distribution observed by Telstar subsequent to July 10 lead to the conclusion that Starfish introduced a high flux of electrons over a very extensive region of space. These electrons may have originated from fission beta decay (CARTER *et al.*, 1959) or alternatively they may have been thermal electrons heated by a shock wave created by the explosion as discussed by COLGATE (1966).

The earliest electron distribution that could be well determined from Telstar I data (BROWN *et al.*, 1963a; BROWN and GABBE, 1963) is shown in Figure 1. The plot is in R-λ space where the coordinates are derived from the McIlwain parameter L (MCILWAIN, 1961) and from the magnetic field strength B as given by the 48 term

Jensen and Cain expansion for the earth's field (McIlwain, 1962). The figure spans a five-day time period starting on July 12 (day 193) three days after Starfish. A period of five days was required to gather enough data to construct a satisfactory contour plot. The distribution is changing radically in time as will be seen in subsequent figures, and thus the contours of Figure 1 are averages over the indicated time period. The contours are of equal omnidirectional counting rate and were hand drawn through sets of measured points corresponding to a given

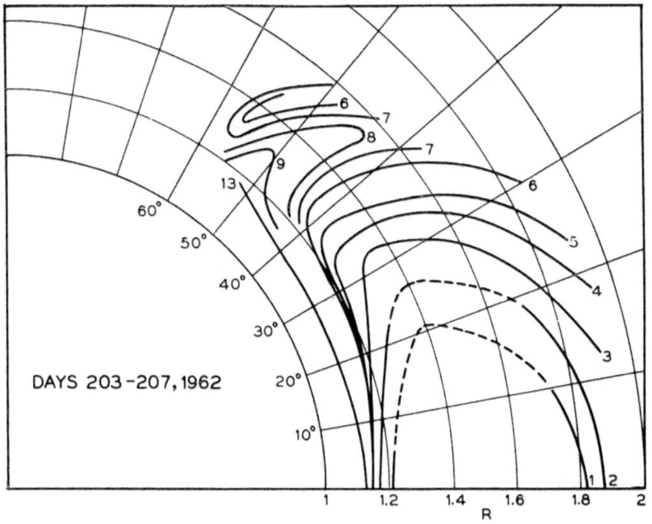

Fig. 2. Contours of equal omnidirectional counting rates for Days 203–207, 1962. The contour labelling is the same as in Figure 7 except that Contour 1 with an omnidirectional couning rate of 1.5×10^8 is now in evidence. This contour is slightly out of the normal two-to-a-decade order.

range of values of the omnidirectional counting rate (Brown *et al.*, 1963a). The contours are two to a decade and are dashed where data is missing. Contour 1, for the highest electron flux density measured, does not show in the distribution of Figure 1 because of the satellite orbit. The contour number has been saved for use on later figures. Contour 2, the highest in Figure 1, corresponds to an omnidirectional counting rate of about $7 \times 10^7/\text{cm}^2$ sec. Notice that the contours of Figure 1 show only a single maximum in electron flux even up to latitudes above 50 degrees where the outer electron belt would be expected to be seen.

Figure 2 illustrates the electron distribution about two weeks after Starfish and 10 days later than Figure 1. Contour 1 is now observed at low latitudes, near the equator at both high and low altitudes. The satellite orbit has not yet precessed sufficiently to complete the contour as indicated by the dashed portion of the curve. Contour 1 has an omnidirectional counting rate of about 1.5×10^8 counts/cm^2 sec. Because of the detector sensitivity as a function of electron energy, the overall efficiency of the detector in the presence of an electron spectrum produced by fission fragment beta decay is approximately 0.2 (Brown *et al.*, 1963a). The highest flux contour thus represents a total electron flux for

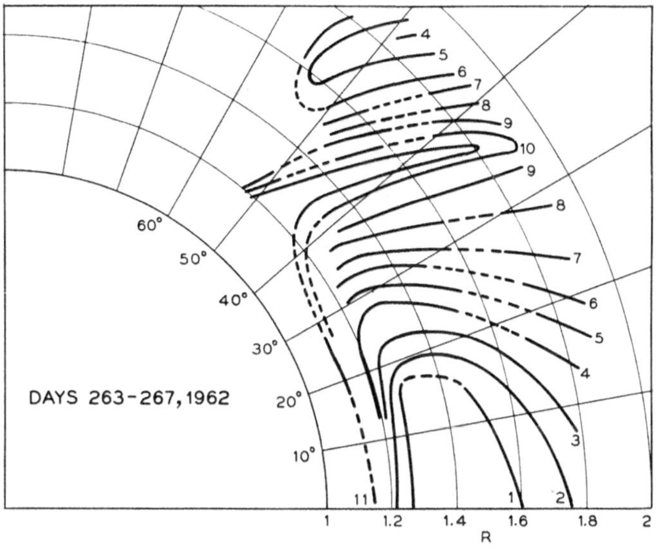

Fig. 3. Contours of equal omnidirectional counting rates for Days 263–267, 1962. The contour designations are the same as in Figures 1 and 2.

that assumed spectrum of approximately 8×10^8/cm^2 sec. If the spectrum is substantially softer than a fission beta decay spectrum, as it probably is in much of the figure, the total electron flux corresponding to Contour 1 might decrease by as much as a factor of 1.5. The contours in Figure 2 have changed very substantially in comparison with those in Figure 1. A second maximum in the distribution now appears as a result of a decrease in the electron flux that is particularly pronounced in the region between 35 and 50 degrees latitude. This second maximum is the tip of the outer belt which was indistinguishable from the inner belt because of the anomalous population of electrons that had earlier filled in "the slot" between the two. Notice also that where Contour 2 had crossed the $R=1.8$ earth radii line at about $\lambda=18$ degrees in Figure 1, Contour 3 occupies a similar position in Figure 2. There has been a general decay in the electron population.

Decreases in the electron flux distribution over the whole region of Telstar I's orbit continued throughout the summer. The situation in mid-September is shown in Figure 3. Contour 5 now crosses $R=1.8$ at 18 degrees, a decrease by a factor of 30 compared with Figure 1. The slot between the inner and outer belt has become very deep. The omnidirectional counting rate at $R=1.6$ and $\lambda=40$ degrees is given by Contour 11, $2\times10^3/\text{cm}^2$ sec., a thousand times less than the intensity at the same position in Figure 1.

2.2 *Decay times for the Starfish electrons*

Figure 4 shows the omnidirectional counting rate on several L lines as a function of time between the launch of Telstar I and the day before the first of the U.S.S.R. high altitude nuclear tests on October 22 (day 295). The figure shows in a more continuous way the electron flux decay discussed in connection with

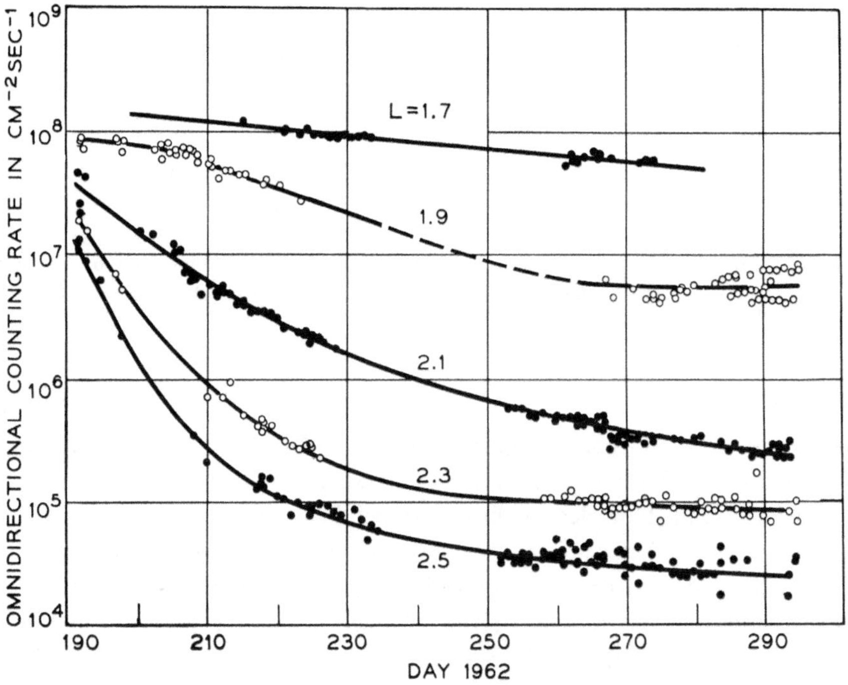

Fig. 4. The time dependence of the counting rate in narrow lamda ranges on several L lines.

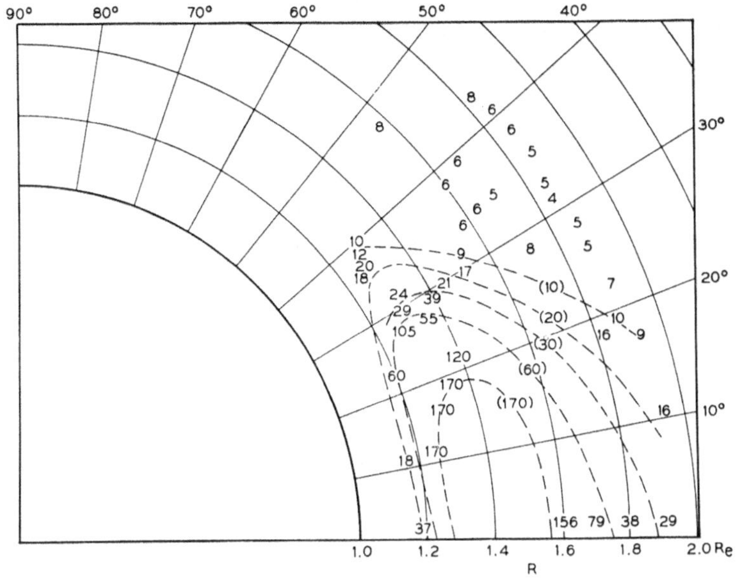

Fig. 5. The initial time constants from data such as in Figure 4 displayed in R, λ space. The decay times are in days. The dashed lines connect regions of equal decay times as indicated by the numbers in parentheses.

Figures 1 through 3. The points are taken in each case in a narrow range of λ on an individual L line. The λ intervals are taken near the equator for $L=1.7$ and 1.9 and at increasingly large λ values for larger values of L. The data points occur in clusters because of the precession of the satellite orbit. On $L=2.5$ the decay is observed to be fast and, overall, very large. This region is in the heart of the slot where the contour plots show the decay of a factor of about 1000. On $L=1.7$ the decay is slow and only amounts to about a factor of two over the three month period. There are indications that in the last month of the time interval the electron flux on the higher L lines tends toward a steady-state. This may quite possibly be at values determined by natural source and loss processes.

The decay curves of Figure 4 are not exponential. If they were, the curves would be straight lines on the semilogarithmic plot and would approach a steady-state value with a rather sharp corner. However, assuming the initial decay is approximately exponential, a time constant has been associated with the initial slope of these curves and others like them throughout the space in which Telstar I collected data. The time constants are themselves plotted in R-λ space in Figure 5. At the lowest altitudes the decay times are short as noted by Van Allen (1964). At intermediate altitudes near the equator the decay time is relatively long and at the highest altitudes and in the slot the time is once again short. The lightly dashed lines in the figure attempt to connect regions of approximately equal decay time. The shape of these lines is very roughly equivalent to the shape of the inner belt itself as seen in Figure 3. Clearly the rates of decay of the particles are closely related to the number of particles to be found in different regions. This is to be expected in a steady-state case and in a case of a still decaying transient following a broad injection of new particles. The decay times over the slot region are the order of a week and vary very little with L. Decay times of this same order are observed in the outer belt region after natural increases in the normally unstable outer belt flux (BOSTROM and WILLIAMS, 1965).

The question of the loss mechanisms responsible for the results of Figure 5 is of evident importance. At low altitudes (on L lines below about 1.3 earth radii) there seems to be no question but that loss is produced by interaction of the electrons with the earth's atmosphere. This correspondence is very beautifully displayed in the results of VAN ALLEN (1964) from the Injun satellite and the calculations that WALT (1964) has made of atmospheric scattering processes. The Telstar results are entirely consistent with this work but are much less extensive because of differences in the orbital properties of the Telstar and Injun satellites. The lifetime for atmospheric scattering increases very rapidly with increasing altitude and is incapable of explaining lifetimes the order of a year between $L=1.3$ and 1.6, much less lifetimes the order of a week on still higher L lines that fall in slot. It has been proposed by DUNGEY (1963) and by CORNWALL (1964) that resonant interaction of circularly polarized electromagnetic whistlers with the high energy trapped electrons might provide the new loss mechanism needed. Figure 6 shows lifetimes from Figure 5 replotted against L and compared with Dungey's initial calculations. Decay times measured by Explorer XV for electrons with energy >1.9 MeV have been added to the >0.5 MeV results of Telstar I in the region between $L=2.0$ and 2.6. The higher energy particles also have decay times in this region which are of the order of one to two weeks. Figure 6 shows an extremely appealing correspondence with the observed L dependence of τ on the high altitude side of the inner belt. However, ROBERTS (1966) has shown in a more complete calculation of the whistler model that there are very substantial failures of this model when an attempt is made to explain the observations of the L, λ, and energy dependence of the electron decay times.

2.3 *New particles from the U.S.S.R. tests*

A series of three high altitude nuclear tests on October 22 (day 295), October 28 and November 1 again grossly disturbed the electron population in the region of the slot and the high altitude side of the still slowly decaying Starfish belt.

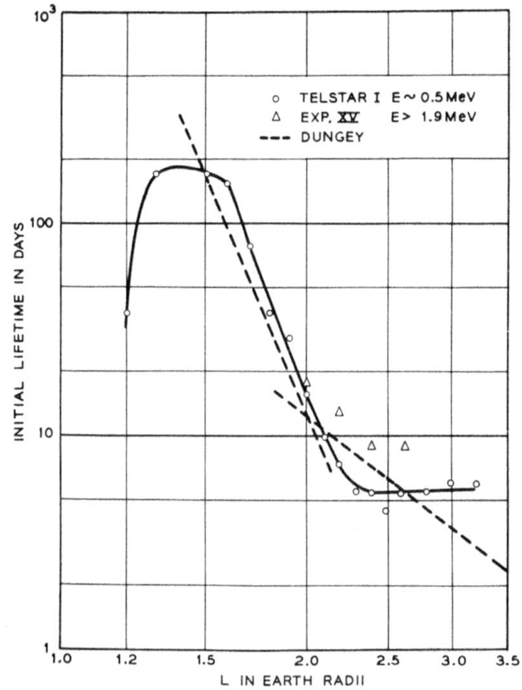

Fig. 6. The L dependence of the initial decay time for electrons compared with Whistler mode electron loss according to DUNGEY (1963).

No new electrons were detected on L lines below 1.7 in any of these three tests. Because of the character of the new particle distributions and the more extensive satellite instrumentation which was in space to observe them (McILWAIN, 1963, 1966; KATZ, 1966; IMHOF and SMITH, 1965 and WEST, 1966) these events added significantly in support and extension of information gained from the Starfish test. Telstar observations are shown in Figure 7 for three L lines in the slot region. The figure shows the long period of decay following Starfish and the sudden increase in flux on day 295, almost a factor of 100 on $L=2.2$. A rapid decay (decay time of the order of three days on $L=2.2$) is observed immediately afterward. It is followed by another upsurge in electron flux on day 301. The decay following the second U.S.S.R. test is not quite so rapid as that following the first, but still the characteristic time constants are the order of a week or less. A third U.S.S.R. test on November 1 added no electrons to any of the L lines shown in Figure 7. As will be seen in the results of Explorer XV, the third test created a very narrow band of electrons on an L shell of approximately 1.76.

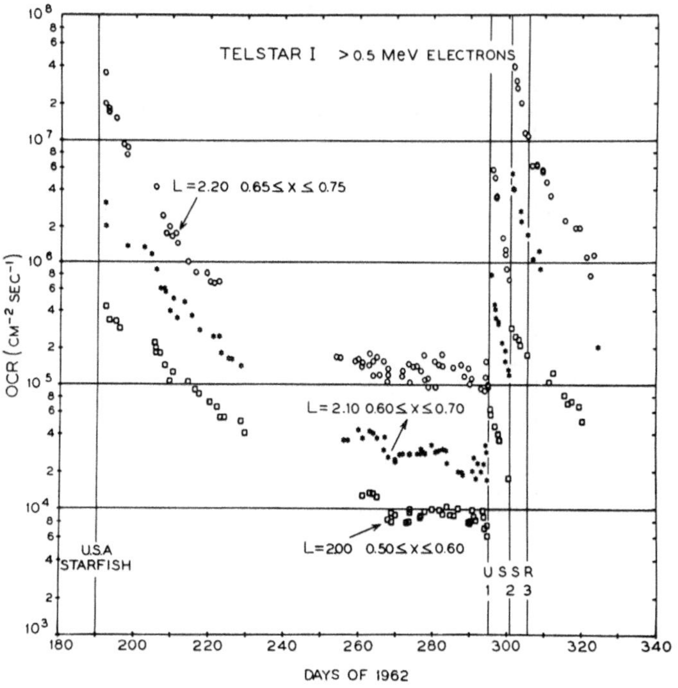

Fig. 7. Telstar observations of the U.S.S.R. tests on October 22, October 28, and November 1 following long term decay of electrons from Starfish. In the figure $x = \sqrt{1 - B_0/B}$.

3. Explorer XV, the U.S.S.R. Tests

3.1 *New particles from the second and third U.S.S.R. tests*

Explorer XV was launched into a low inclination orbit about five hours before the second U.S.S.R. test on October 28. Fortuitously it was thus able to obtain data on one complete orbit before new electrons were added to the distribution. The data for the early orbits on October 28 are shown in Figure 8 for an electron detector measuring electrons >0.5 MeV. Orbit 1′ is the nearly equatorial returning half of the first orbit and shows the inner edge of the outer belt, the slot and the rise toward the inner belt maximum. The slot is not as empty of electrons as it was a week earlier because of particles added in the first U.S.S.R. test (see Figure 7). At 0407 hours on October 28 the satellite crossed $L = 2.0$. When it crossed outward bound about an hour later (0518) new particles had been added above $L = 1.8$. The initial transient of particles in their longitudinal distribution around the earth is not yet complete at this time as evidenced by

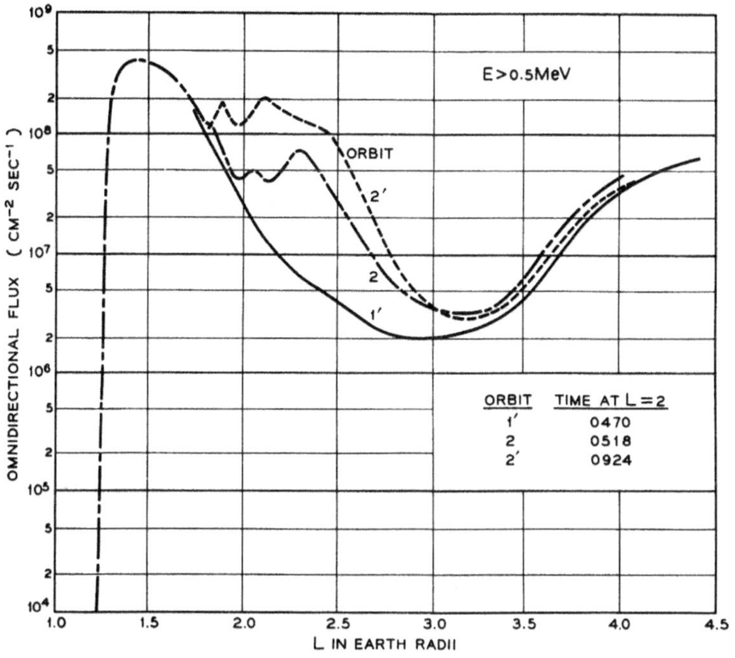

Fig. 8. The omnidirectional flux of electrons of > 0.5 MeV for Explorer XV for the early passes on October 28, 1962. The satellite is nearly equatorial and the figure thus gives a nearly equatorial trace through the particle distribution.

the further increases in flux seen on the returning half of the second orbit some four hours later, curve 2′. By this time electrons with energies >0.5 MeV will have drifted at least five times around the earth and the distribution in various L shells should be longitudinally reasonably uniform. The observation of the transient in longitude permits a rough determination of the initial time of injection of new particles at 0440 ± 10 minutes (Brown, 1965).

Figure 9 shows the early results of a second detector on Explorer XV which measures electrons above 1.9 MeV. The two small spikes on orbit 1′ at $L = 1.85$ and 2.0 are probably remnants from the first U.S.S.R. test as is most of the flux in the slot region. In this higher energy detector a higher flux of electrons is observed in orbit 2 at $L \approx 1.8$ then in orbit 2′. This might indicate electron decay, but it is more likely to be due to a clump of newly injected high energy electrons not yet dispersed in longitude on orbit 2. By orbit 2′ the clump will have disappeared. The structure of the added electrons is quite complex as observed in both the low and high energy electron detectors. It is related in a complicated

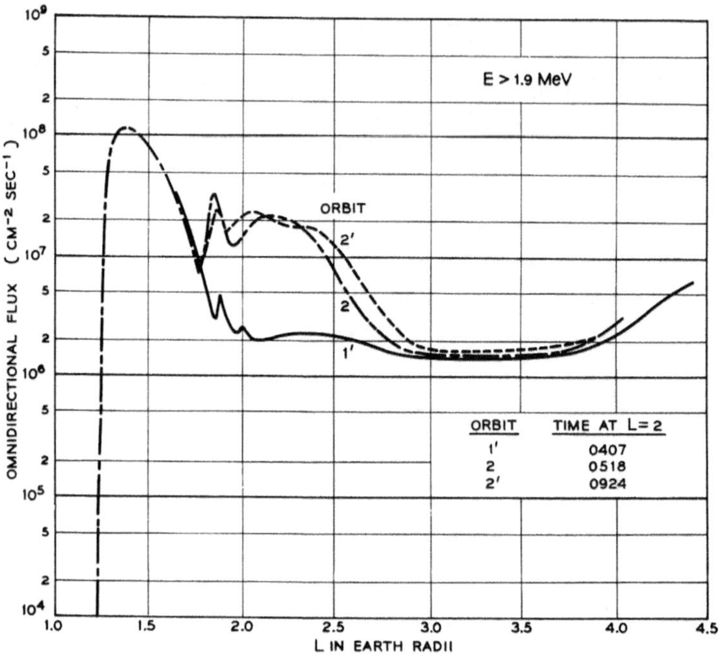

Fig. 9. The omnidirectional flux of electrons of > 1.9 MeV from Explorer XV for the same period as that in Figure 8.

way to the motion of the radioactive fragments carried in the expanding plasma of the nuclear explosion and to the geometry of shock waves created by the explosion.

A qualitative picture of the spectral distribution of the electrons can be obtained by combining the results from the 0.5 MeV and 1.9 MeV detectors with results from a third detector measuring electrons with energies greater than 2.9 MeV. Figure 10 shows these three curves for orbit 1′, just before the second U.S.S.R. test. The vertical spacings between these curves on the semilogarithmic plot give the relative spectral hardness of the electron distribution. If the spectrum is that of equilibrium fission beta decay the ratios between the curves should be 1:0.3:0.09. Such ratios are very nearly observed in the peak of the inner belt (not shown in the figure) at an L value of about 1.4. It is clear from Figure 10 that in the center of the slot the spectrum is much harder than a fission beta spectrum; it is much softer in the outer belt, and it is substantially softer on the high-L side of the inner belt. There is no particular reason why the spectrum of particles should be appropriate to fission beta decay in any of these

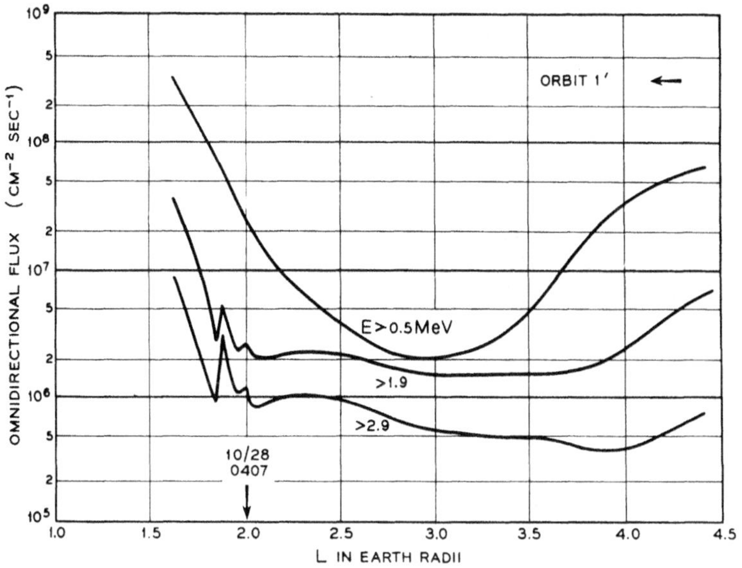

Fig. 10. The omnidirectional flux from three electron detectors on Explorer XV for the first returning pass from apogee on October 28. The variation in the vertical spacing of the curves in the semilogarithmic plot indicates the change in the spectral distribution of the electrons.

regions. The outer belt is composed primarily of natural electrons. The slot contains a residue of electrons from the first U.S.S.R. explosion degraded by loss mechanisms over a period of six days. The high-L side of the inner belt contains a substantial residue of electrons from the Starfish test as well as electrons from the first U.S.S.R. test. The curve spacings of Figure 10 may now be compared with those of Figure 11 which show the results for the same three detectors for orbit 2′, after the added electrons have had time to disperse longitudinally. The spectrum in the slot is substantially softer than it was before at $L=3.0$. A greater number of 0.5 MeV electrons have been added to the population in comparison to the higher energy electrons than would be the case if the added electrons had a fission spectrum. A similar thing is true to varying degrees throughout the region affected by the second U.S.S.R. test. These spectral characteristics are not understood. They either represent degradation of an initial fission beta spectrum or they represent particles that are associated with the explosion but do not originate from the fission beta process at all. Spectral differences of this kind and the difference in the detector response characteristics of the Telstar I and Injun I detectors are doubtless responsible for the apparent discrepancy between the Telstar I and Injun I pictures of the Starfish electron distribution in space (Brown *et al.*, 1963b).

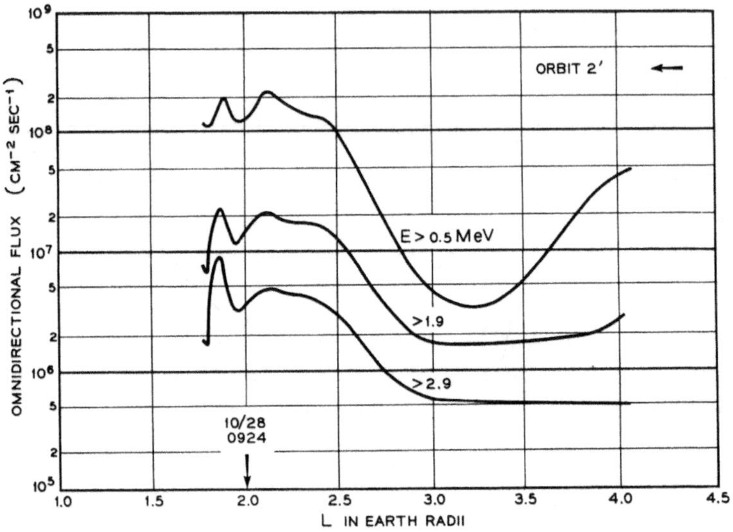

Fig. 11. The omnidirectional flux from the electron detectors of Figure 10 for the second returning pass of Explorer XV. The newly injected electrons have now had time to disperse uniformly in longitude.

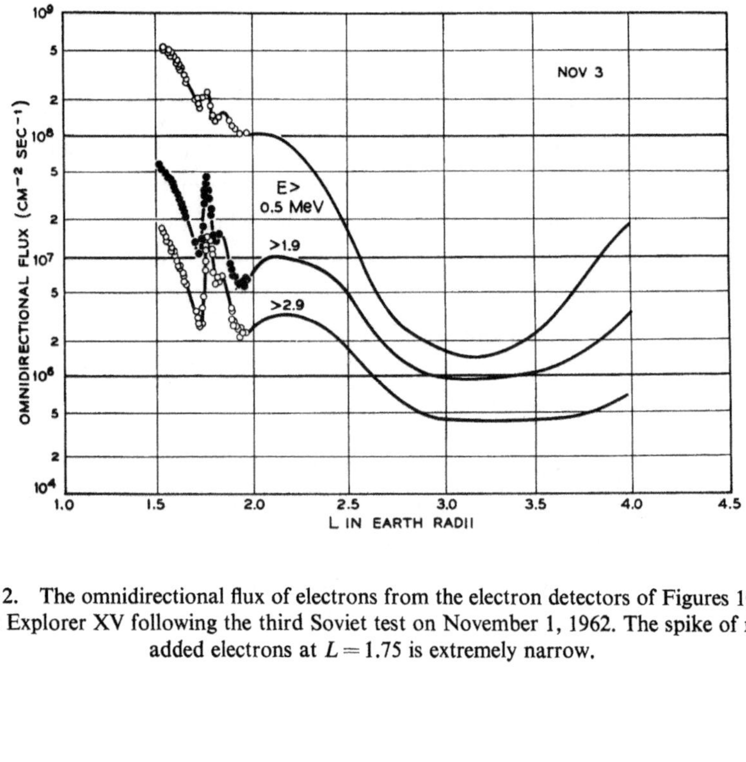

Fig. 12. The omnidirectional flux of electrons from the electron detectors of Figures 10 and 11 on Explorer XV following the third Soviet test on November 1, 1962. The spike of newly added electrons at $L = 1.75$ is extremely narrow.

Figure 12 shows the influence of the third U.S.S.R. test of November 1 as observed on November 3. The only electrons added by this test are contained in a narrow spike at an L value of about 1.76. The second peak at about 1.83 is one of the peaks created by the second U.S.S.R. test. The actual data points are shown on the curves in the low L region to indicate the resolution of the measurements which is important to determination of possible cross-L diffusion of electrons from the new spike. (See Section 3.3.) The spectrum of added particles in this new spike is very nearly that appropriate to fission beta decay.

3.2 *The transient flux distribution along field lines*

The changes in the flux distribution along particular field lines during the general decay of the particle flux following a nuclear test has not been considered explicitly in earlier sections of this paper. Three cases of this kind have been selected from the >1.9 MeV data of Explorer XV, and are shown in Figures 13, 14, and 15. Figure 13 is for $L = 1.75$. Each plotted value is obtained by interpolation between actual data points measured along an individual satellite pass.

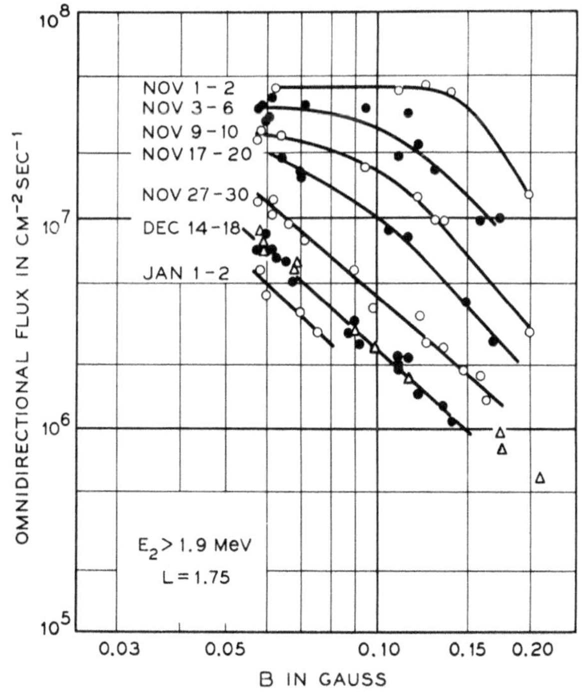

Fig. 13. The B-variation of the electron distribution along the $L = 1.75$ line as it changes with time. The triangles are points obtained before the third U.S.S.R. test on November 1.

The triangles in the figure are points obtained prior to the third U.S.S.R. test. There were no particles added at an L value as low as $L=1.75$ by the first or second tests. The triangles define a reasonably straight line in this log flux-log B plot. That is, the distribution is essentially an inverse power law in B, in this

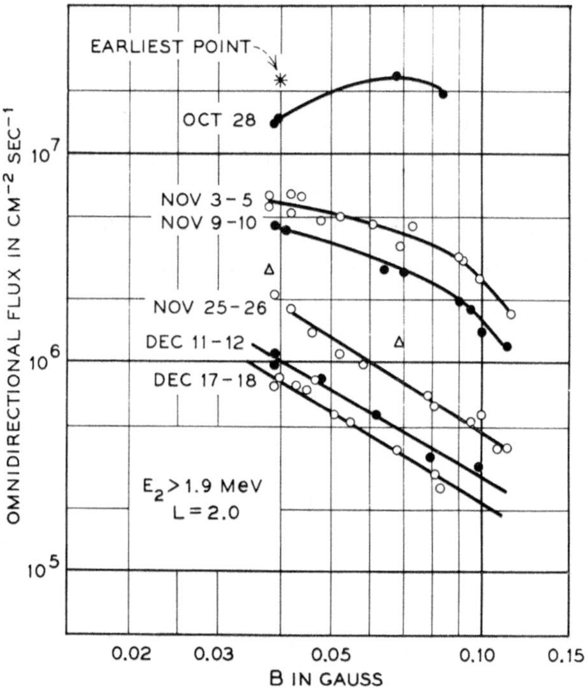

Fig. 14. The B-variation of the electron distribution along the $L=2.0$ as it changes with time. The triangles are the two points obtained before the second U.S.S.R. test obtained on October 28. The star is the earliest point following the shot as in Figure 9.

case with a power of approximately two. The lowest values of B in the figure correspond to the equator on $L=1.75$, and the region of data coverage extends out to approximately $B/B_0 \approx 3$, or $\lambda \approx 30$ degrees.

On November 1 the flux jumps up suddenly by a factor of about five at the equator and a factor of about 20 at $B \approx 0.2$ gauss. The distribution is altered drastically and is no longer nearly a simple power law in B. Over the next few weeks the anomalous new shape of the distribution gradually disappears. By the end of November the distribution is again nearly an inverse power law in

B and at later times the shape is maintained while the distribution as a whole decreases in intensity. More rapid disappearance of particles at large values of B in an abnormal distribution along the field line, and the decay of the distribution as a whole at long times are to be expected in processes controlled by

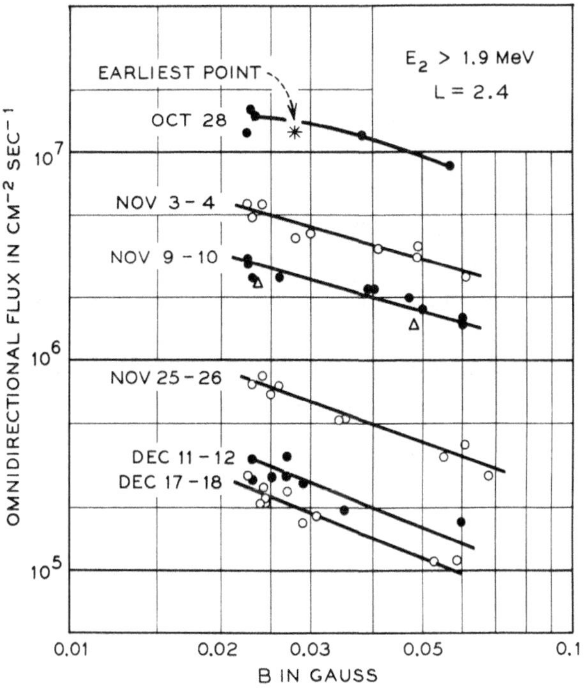

Fig. 15. The B-variation of the electron distribution along the $L = 2.4$ line as it changes with time. The triangles and the star are the same as in Figure 14.

diffusion of particle mirror points along field lines with ultimate particle loss in the atmosphere. This is a situation found by Welch *et al.* (1963) and by Walt (1964) for lower L values where the atmosphere controls the whole process. It is also the situation discussed by Roberts (1966) for a general case of pitch angle diffusion. The fact that the flux in January in Figure 13 has dropped below the data of the triangles for late October is presumably due to a continuing high energy decay of electrons from the Starfish test.

Figure 14 is anologous to Figure 13 but for $L = 2.0$. There are now only two data points (triangles) before the second U.S.S.R. test on October 28. One of these corresponds to the outgoing, the other to the incoming pass of the first orbit of Explorer XV. The equatorial point is that for orbit 1′ of Figure 9. The star marked "earliest point" in Figure 14 is for orbit 2, at which time the high energy electrons had not yet dispersed in longitude. At $L=2.0$ the flux-B variation shows a maximum off the equator, an even more anomalous distribution than that seen on $L=1.75$. Such a new distribution results from injection of new particles far off the equator (WEST, 1966). This grossly distorted shape rapidly disappears and the decay process carries past the triangles of October 28 because of continuing decay of electrons from the first U.S.S.R. test. The slope of the power law distribution which characterizes the steady-state shape during the later stages of decay has a value on $L=2.0$ of only approximately 1.35.

Results for a third field line $L=2.4$, are shown in Figure 15. On this field line the shape of the electron distribution following injection is much less anomalous and a steady-state power law dependence is rapidly reassumed. The slope of this line is approximately 1.05.

The form of the decay illustrated in Figures 13, 14, and 15 is characteristic of all the field lines examined. The decay is in essence by normal modes. Higher modes, corresponding to more extreme deviations from the steady-state shape, decay rapidly. However, there are systematic L dependent trends in the parameters of the system. First, the decay is more rapid going to higher L values from

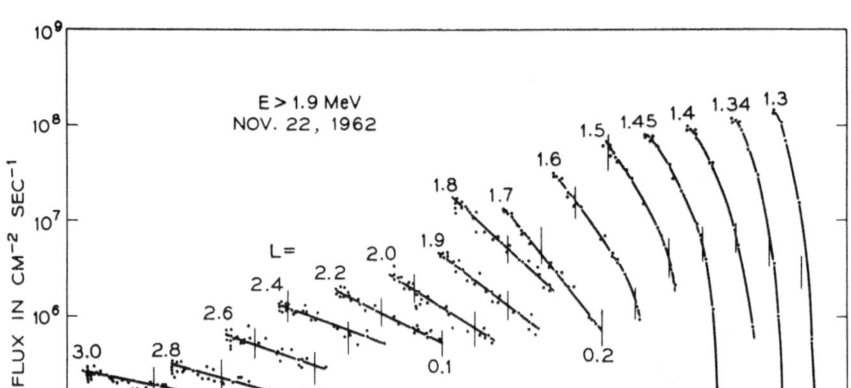

Fig. 16. The B-variation of the electron flux along a number of different L lines as observed by Explorer XV on November 22. Note the decrease in the slope of the log flux vs log B lines as L increases.

$L=1.7$. This is in reasonable quantitative agreement with the lower energy electron results of Telstar I following the Starfish test as shown in Figure 6. Second, the steady-state shape of the log flux-log B distribution is progressively more shallow on higher L shells. This feature is illustrated more completely in Figure 16 which shows the distributions on November 22 over a wide range of L values. The curves in the figure have been shifted horizontally with respect to one another to avoid overlapping. Since the observed particle loss rate increases in going to L values greater than 1.7, and since the gradient of the distribution at the same time decreases, then a pitch angle diffusion mechanism which may be stirring up the particles on a given field line and bringing them at random into the atmospheric loss cone, must operate with increasing effectiveness on higher L lines. That is, the pitch angle diffusion coefficient must increase with L.

3.3 *The width of the electron distribution from the third U.S.S.R. test – Cross-L diffusion*

As was observed in Figure 12, the U.S.S.R. test on November 1 introduced a very narrow band of high energy electrons at $L\approx 1.76$, below the electron distributions added by the two earlier tests. The width of this band as a function of time provides a sensitive measure of cross-L diffusion in this region of space. Figure 17 shows snapshots of the >1.9 MeV electrons in this band observed in four passes of Explorer XV during the three weeks following the creation of this new perturbation in the trapped particle distribution. A total of 18 such passes have been examined in comparable detail. Each peak in Figure 17 is marked with the L value of the center of the peak and with it the corresponding time and value of B. In each case examined the center falls at $L=1.765\pm 0.010$. This variability is due to limitations in the calculated value of the L parameter arising from uncertainties in the magnetic field. The flux observed in the peak is a function of both B and time as noted in Figure 13. During November the distribution is decaying toward the steady-state shape that characterizes its lowest normal mode. In each part of Figure 17 the data shown include a portion of the high-L side of the inner belt at L values below the new peak. The >1.9 MeV electrons in this region are those left from the Starfish test of July. They are continuing to decay slowly. The passes also show fragments of the data at L values above the new peak. These electrons were those created in the first and second U.S.S.R. tests. They are decaying more rapidly than the Starfish electrons at $L=1.7$ and below but not much more rapidly than the new peak itself. There are breaks in the data which occur because of the time sharing mode in which this part of the experiment in the satellite is being operated.

The full width at half maximum of each peak in Figure 17 is indicated. It has been deduced assuming a Gaussian shape and is obtained by subtraction of an extrapolated background of the Starfish electrons. The fit has made use

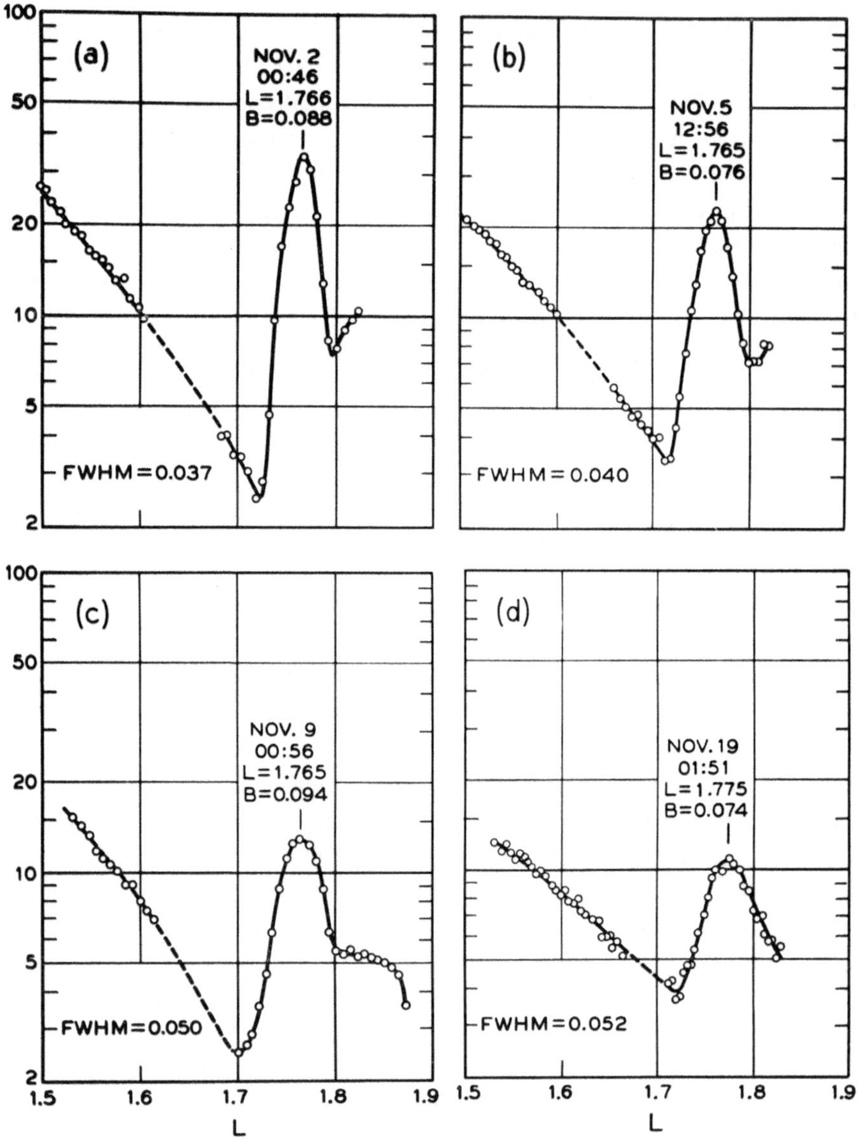

Fig. 17. Details of the narrow peak produced by the third U.S.S.R. test as observed on four passes in November. The date, time and *B* value of each peak center are noted as is also the full width at half maximum of a Gaussian fitted to the peak.

of the high-L side of the peak only for determination of the peak's center location. The width has been determined from the low-L side where the background subtraction is more easily carried out. The widths observed are all quite narrow, varying from 0.037 earth radii for part (a) to 0.052 for part (d) of the figure. The electrons are thus located in a shell about 250 kilometers thick, a dimension still large compared with the approximately one kilometer radius of the electron spiral in the magnetic field.

Figure 18 is a compilation of the line widths measured on 18 different satellite passes. The square of the full width at half maximum is plotted against time.

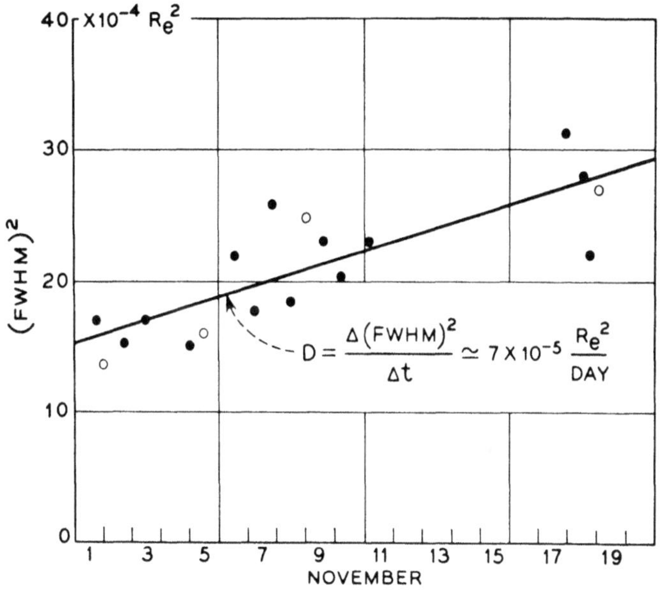

Fig. 18. Composite results of the peak width from 18 different passes of the satellite. The four passes included in Figure 17 are shown as open circles.

The scatter of the determinations is rather large, due primarily to the difficulties of peak fitting. The peak width nevertheless clearly increases with time. The square of the width is satisfactorily represented as a linear function of time as would be expected if the initially created width were broadening by diffusion. The diffusion coefficient corresponding to the line drawn in the figure is $7 \times 10^{-5} R_e^2$/day, a very small diffusion coefficient. There does not seem to be any systematic shift in the peak position over the time of measurement. Certainly any such linear diffusion (FRANK, 1966) is less than $2 \times 10^{-4} R_e$/day.

The broadening of the peak contributes very little to the decrease in the electron flux trapped in it. In the B and time range of the measurements in Figure 17 the peak flux decreased by a factor of about four, much more than the 30 percent decrease that would be expected if the same electrons were simply confined in a broader region of trapping.

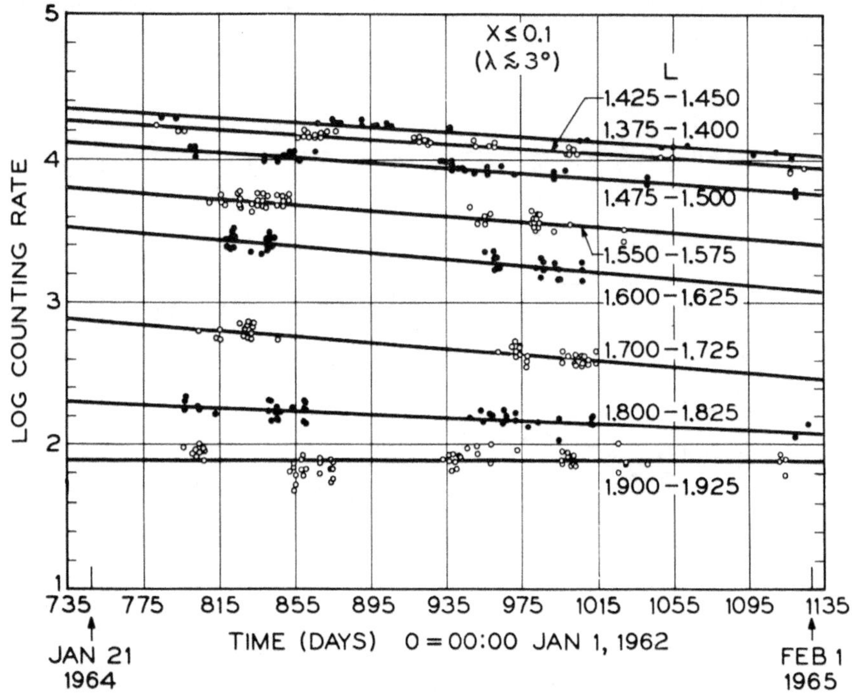

Fig. 19. The decay of electrons >0.75 MeV as observed by Relay II during 1964. The lines are exponentials, least square fit to the points.

4. Relay II, the Artificial Radiation Belt at Long Times

Relay II has provided a measure of the continuing decay of the electron flux in the higher altitude portions of the artificial belt during 1964 and 1965. The detector in this case is counting electrons >0.75 MeV in the directional flux perpendicular to the magnetic field. The results for near the magnetic equator ($x=\sqrt{1-B_0/B}<0.1$ or $\lambda<3$ degrees) for a number of narrow L slices are shown in Figure 19. The figure starts about a year and a half after the Starfish test. The measured points are in clusters because of the precession of the satellite orbit. The lines are least square fits of exponentials to the data. They show gradual particle decays over the year of measurement.

The values of the decay constants determined from fits of the kind shown in Figure 19 have been plotted against L in Figure 20, not only for the near equatorial measurements, but also for three other regions in x further along the field lines. The near equatorial values have vertical bars which indicate the confidence limits of the fits. There seems to be a tendancy for decay constants at higher values of x to be greater than the near equatorial values, but the tendency is not far outside the accuracy of the determinations. The results also suggest decay constants slightly higher between $L=1.55$ and 1.7 than between $L=1.35$ and 1.55, but this L variation of K is quite small. The decay constant definitely decreases as one goes above $L=1.7$ and actually passes through zero at an L of about 1.9. For still higher L values the negative values of K imply small net increases in the electron flux over the one year period.

The results of Figure 20 are quite different from those determined from measurements soon after the nuclear tests, for example as shown in Figure 6. In Figure 6 the decay time is a maximum of about half a year between $L=1.3$ and 1.6 and decreases on higher L lines into the slot. However, in Figure 20 the decay constants, which are reciprocals of decay times, imply minimum decay times of the order of a year between $L=1.35$ and 1.7 and longer decay times at higher values of L. The two sets of results do not, however, seem to be in essential conflict. It is only in those regions where the number of residual excess electrons is large compared with the steady-state distribution of natural particles that the fitted decay constants will be a simple measure of the decay properties of the excess particles. In the region of $L=1.7$ or 1.8, if the measured initial decay times, the order of 40 to 60 days, had persisted, a flux decrease of between 10^4 and 10^6 would have been expected over the intervening two years. Such a decrease would almost certainly have brought the >0.75 MeV electrons to their steady-state distribution and a measurement of the decay constant would yield a value of zero. In the region where substantial numbers of excess particles still exist between $L=1.35$ and 1.7, decay processes with time constants the order of one or two years are operating. For decay times this long, particle loss might be proceeding by cross-L diffusion to regions where loss processes are normally more active.

Such diffusion could also account for the non-zero values of decay constant observed on L-lines above $L=1.7$, since the central maximum in the particle distribution could be feeding electrons onto these higher lines. However, if the cross-L diffusion coefficient discussed in Section 3.3 for $L=1.76$ is applicable over the $L=1.35$ to $L=1.7$ region and at the different electron energy of the two sets of measurements, it is too small to account for the Relay II observations. This whole question deserves more detailed consideration.

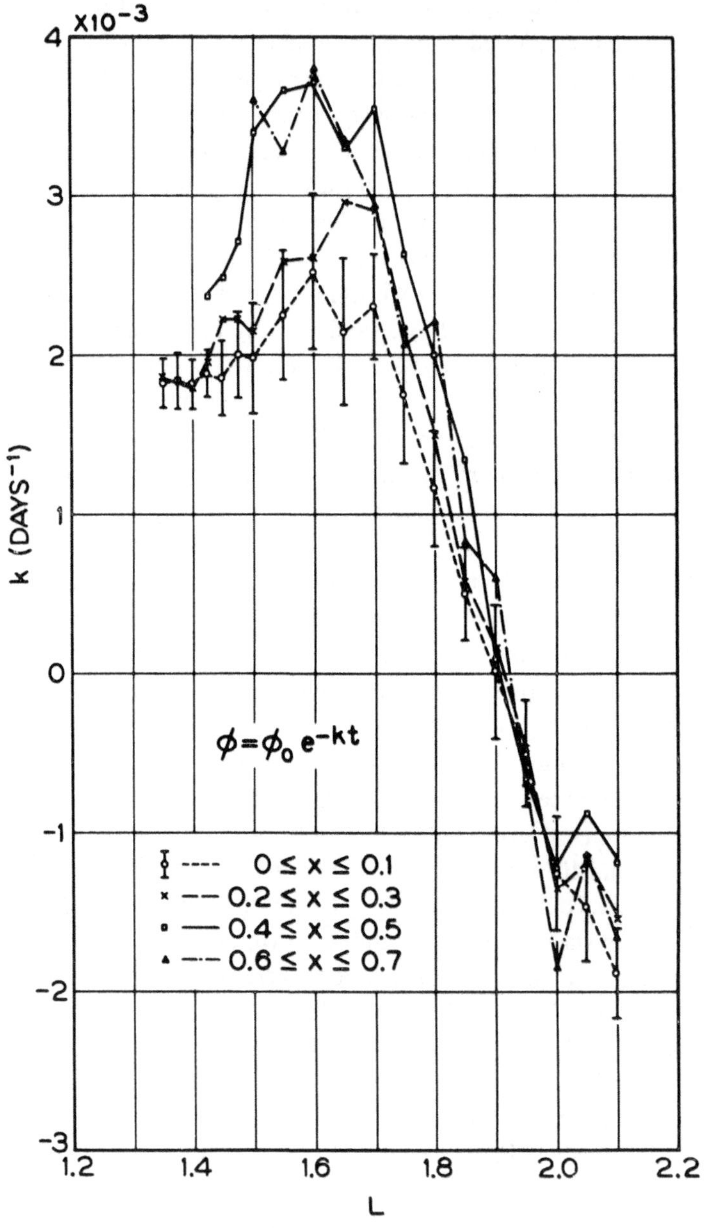

Fig. 20. The decay constant K corresponding to the exponential fits of electron decay curves such as those of Figure 19.

5. Acknowledgements

The results described in this paper have been obtained with the indispensable collaboration of a number of colleagues: J. D. Gabbe who carried out the analysis of the Telstar I data, Mrs. W. L. Mammel who managed the data reduction on Explorer XV, Miss L. W. Davidson who had a similar role for the Relay results, and C. S. Roberts who has given detailed consideration to the transient decay of electrons by pitch angle scattering. Their contributions to this work are gratefully acknowledged.

References

Bostrom, C. O. and Williams, D. J.: 1965, *J. Geophys. Res.* **70**, 240.
Brown, W. L.: 1965, *Proc. Plasma Space Sci. Symp.*, D. Reidel Publ. Comp., Dordrecht, p. 189.
Brown, W. L. and Gabbe, J. D.: 1963, *J. Geophys. Res.* **68**, 607.
Brown, W. L., Gabbe, J. D., and Rosenzweig, W.: 1963a, *Bell System Tech. J.* **62**, 1505.
Brown, W. L., Ness, W. N., and Van Allen, J. A.: 1963b, *J. Geophys. Res.* **68**, 605.
Buck, T. M., Wheatley, G. H., and Rogers, J. W.: 1964, *IEEE Trans. Nucl. Sci.* **11**, 294.
Carter, R. E., Reines, F., Wagner, J. J., and Wyman, M. E.: 1959, *Phys. Rev.* **113**, 280.
Colgate, S. A.: 1966, this volume, pp. 693.
Cornwall, J. M.: 1964, *J. Geophys. Res.* **62**, 1251.
Dungey, J. W.: 1963, *Planet. Space Sci.* **11**, 591.
Frank, L. A.: 1966, this volume, p. 422.
Imhof, W. L. and Smith, R. V.: 1965, *J. Geophys. Res.* **70**, 569.
Katz, L.: 1966, this volume, p. 129.
McIlwain, C. E.: 1961, *J. Geophys. Res.* **68**, 607.
McIlwain, C. E.: 1962, private communication.
McIlwain, C. E.: 1963, *Science* **142**, 355.
McIlwain, C. E.: 1966, this volume.
Roberts, C. S.: 1966, this volume, p. 403.
Van Allen, J. A.: 1964, *Nature* **203**, 1006.
Walt, M.: 1964, *J. Geophys. Res.* **69**, 3947.
Welch, Jr., J. A., Kaufman, R. L., and Hess, W. N.: *J. Geophys. Res.* **68**, 685.
West, H. I.: 1966, this volume.

SOME OBSERVATIONS OF THE TRAPPED ELECTRONS PRODUCED BY THE RUSSIAN HIGH-ALTITUDE NUCLEAR DETONATION OF OCTOBER 28, 1962*

HARRY I. WEST, Jr.
Lawrence Radiation Laboratory, University of California, Livermore, California

Abstract. The trapped electrons produced by the Russian high altitude nuclear detonation of October 28, 1962, were observed by a five-channel magnetic spectrometer covering the range of 325 to 3250 keV. The results are presented in terms of energy spectra and pitch angle distributions. Mirror-point density distributions are derived which give an insight into the details of the injection of the trapped electrons into the earth's geomagnetosphere.

Introduction

The Air Force satellite** 1962 βK (STARAD) was launched October 26, 1962, for the specific purpose of measuring the trapped electrons produced by the United States high altitude nuclear detonation Starfish. An analysis of these results has already been published (West *et al.*, 1965). On October 28 at 04:41:18 ± 5 G.M.T. (Selzer, 1964, as quoted by Burrows and McDiarmid, 1964) the U.S.S.R. detonated a nuclear device above the atmosphere. The satellite measurements place the geographic latitude at about 48° North. If we are to keep the detonation over Russia, we find the longitude to be in the range of 54° to 80° East. The yield has been reported in the Nuclear Effects Handbook as an intermediate yield that is in the range of 20 to 999 kilotons. The following is a study of the trapped electrons produced from the detonation. There were other high-altitude nuclear detonations in this general period of time by both the United States and the U.S.S.R. but their results were not nearly as spectacular as the detonation of October 28, and they will be discussed only in connection with that detonation. The data were obtained with a five-channel magnetic spectrometer covering the range of 325 to 3250 keV. The results are presented in terms of energy spectra and pitch angle distributions. Part of the aim of the present paper is to attempt to understand just how and where the electrons from the detonation were trapped in the earth's geo-

* This work was performed under the auspices of the U. S. Atomic Energy Commission.
** Efforts were coordinated by Mr. Ludwig Katz and Mr. Donald Smart of the Air Force Cambridge Research Laboratories.

magnetosphere. Unfortunately we do not know the details of the Russian detonation and hence are somewhat hampered in the final analysis of the data.

Technical Details

The pertinent parameters of the satellite at early times were apogee 5580 km, perigee 198 km, orbital inclination 71°, orbital period 148 min, and apsidal rate 0.8° per day. The pitch period was 127 seconds and the roll period 53 seconds.

The spectrometer aperture looked out the aft end of the satellite inclined at 30° with the major axis of the satellite. The data acquisition rate was about once per second for each data channel. Aspect information was obtained from a three-axis magnetometer. Conditions were fairly good for obtaining pitch angle distributions.

The electron spectrometer had five differential energy channels. The primary energy selection was by means of a uniform magnetic field. Secondary energy selection was by means of solid state detectors. Each channel was read out via four logarithmic rate meters of overlapping range and graduated response time covering the range of ≈2 to 150000 counts per second. Two background channels were provided and were correlated with the minima in roll modulation of count rate to determine the background. The properties of the spectrometer are shown in Table I.

TABLE I

Properties of the spectrometer

Energy (keV)	*Effective Channel Width* (keV)	*Discriminator Bias*	*Detector Type*	*Depletion Depth* (cm)	*Geometrical Factor* (cm^2-keV-sr)
325	137	213	PN	0.2	0.147
955	162	680	LiD	1.2	0.301
1630	184	1120	LiD	2.5	0.276
2400	214	850	LiD	2.5	0.360
3250	224	850	LiD	2.5	0.377

The spectrometer was calibrated by the use of extended sources of beta emitters. A recalibration was prompted by a comparison of our Starfish data with that of McIlwain (1963). It was found that our primary calibration standard $Sr^{90}-Y^{90}$ was 50% more intense than the value we had used earlier. The geometrical factor in Table I reflects this change which brings our results into better agreement with those of McIlwain (1963).

Many of the results in this paper came from pitch angle distributions. Only selected portions of an orbit were suitable for reliable pitch-angle distributions. This occurred when the view angle of the spectrometer made a smooth rapid

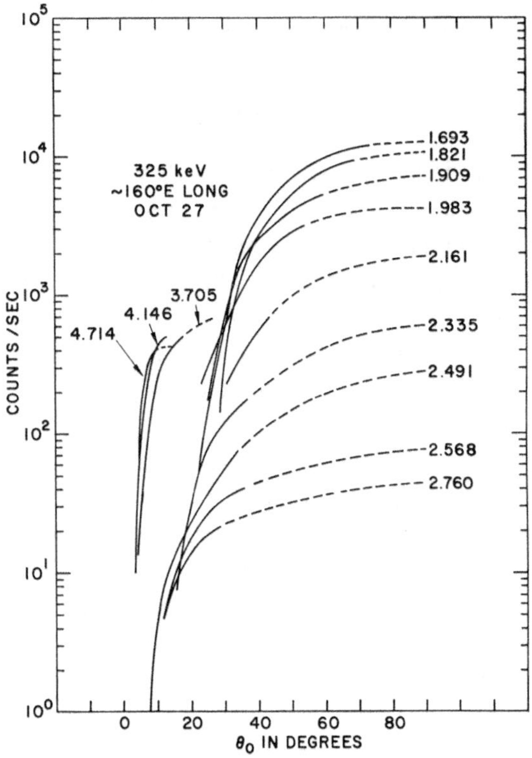

Fig. 1. The pitch angle distributions of 325 keV electrons on October 27 referred to the equatorial pitch angle.

sweep from near zero degrees to near 180 degrees. The magnetic field of the spectrometer produced first order focusing of the electrons. In the direction of focus the acceptance angle was about 17 degrees. In the direction perpendicular to focus the acceptance angle was narrow, and it was this narrow acceptance angle which came into play when the aperture of the spectrometer made the rapid sweep described above. The acceptance angle in the narrow direction varied from about 8 degrees for the lowest energy channel to about 5 degrees for the highest energy channel. The angular response was almost square.

Observations

To put one in the right frame of reference, we start with what must be called one of the final results of the analysis, that is, plots of mirror-point density. From satellite passes at 7.3 and 41 hours after the detonation especially good

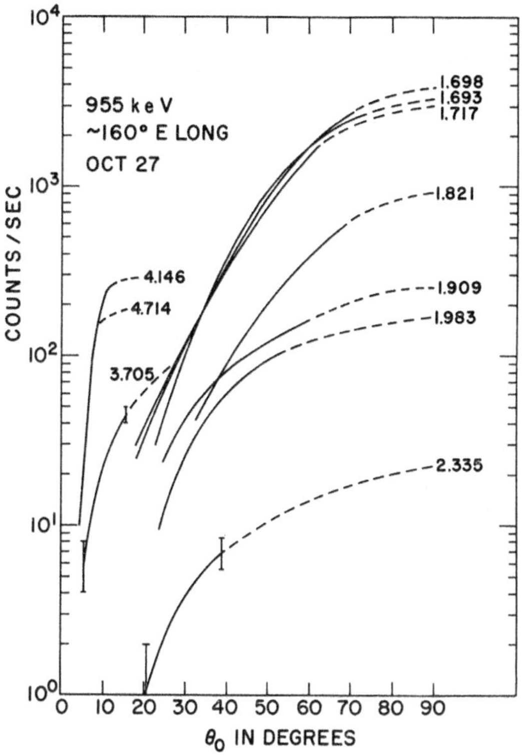

Fig. 2. The pitch angle distributions of 955-keV electrons on October 27 referred to the equatorial pitch angle.

data were obtained. Because of the pitch angle distributions we were able in a single pass to obtain many of the details of the trapped radiation at that time. Decay rates at early times were so very high that it was important to obtain a view of the trapped radiation from data covering a short time span, in these cases about 30 minutes.

The data from the 955-keV channel of the spectrometer gave especially good pitch angle distributions in that the statistical fluctuations due to counting rate were low and the natural background (predetonation electron fluxes as well as bremsstrahlung and proton detection) was low. The background of "natural radiation" was evaluated from a pass prior to the detonation. The pitch angle distributions of the predetonation backgrounds are shown in Figures 1 and 2. An attempt was made to remove the effect of the finite angular resolution of the spectrometer by computer. The correction was most important for the data at low pitch angles where the counting rate varied rapidly with pitch angle. The correction usually amounted to a shift of about 3° towards larger angles at counting rates of about 2% of the peak counting rates.

The corrected pitch angle distributions have been organized into a plot of mirror-point density. Mirror-point density has the advantage of being rather similar in distribution to that of injection density. The relation between the two has been given by (WELCH and WHITTAKER, 1959) in the case of isotropic injection as

$$\omega(r) = \frac{1}{2B(r)} \frac{dB(r)}{dl} \int \frac{n(r')\,dl'}{\sqrt{1 - B'/B}}$$

in which $n(r')$ is the injection density. Here we restrict ourselves to a given guiding line of center as characterized by constant L. Here l is the arc length along the guiding line of center, the primed values refer to points of injection, and the unprimed to the point of interest. The integral is to be evaluated from mirror point to mirror point for each value of $\omega(r)$. Determining $n(r)$ from this integral is quite tedious and has not been done. Indeed all that we can arrive at is the modified injection density as it is rather difficult to assess the interactions which have occurred between injection time and the time of first good observation. It was felt to be sufficient to present mirror-point density which to first approximation is similar in shape to that of injection density.

The equations for mirror point density have been given by FARLEY and SANDERS (1959). Briefly we ask where do the particles with a given pitch angle mirror on a given guiding line of center. Choosing the equator for reference we have the number of particles in the solid angle $d\Omega_0$ which mirror in the volume element dV to be

$$T(\theta_0) \frac{dj(\theta_0)}{dE} 2\pi \sin\theta_0 \, d\theta_0 \cos\theta_0 \, dA_0 = \omega \, dA \, dl \, .$$

This equation defines the mirror-point density ω. In this equation $T(\theta_0)$ is the total mirror period, $dj(\theta_0)/dE$ is the unidirectional flux as a function of energy (electrons/cm^2-sec-keV-sr), θ_0 is the equatorial pitch angle, A_0 is the cross section of a flux tube at the equator, A the cross section of a flux tube at the mirror-point, and l the arc length. This equation through proper manipulations becomes

$$\omega(\lambda) = \frac{3\pi}{r_0} T(\lambda) \frac{\sin\lambda(8-5\cos^2\lambda)}{(4-3\cos^2\lambda)^{3/2}\cos^2\lambda} \frac{dj(\lambda)_\perp}{dE} \text{ particles/unit volume,}$$

where r_0 = radius of the earth, λ = invariant geomagnetic latitude, $dj(\lambda)_\perp/dE = n(\lambda)/g$, in which $n(\lambda)$ follows from $n(\theta)$, the counting rate, by determining the mirroring latitude for electrons observed at an angle (θ), and g is the geometrical factor. The above equation is just one half of that given by Farley and Sanders. Through a conceptual error they count the particles twice in their derivation. For practical purposes (exactly true only for a dipole field) we can write

$$T(\lambda) = 8.5 \times 10^{-2} L \frac{(1+\varepsilon)}{\sqrt{\varepsilon(\varepsilon+2)}} t(\lambda) \text{ seconds}$$

in which $t(\lambda) = 1.380 - 0.640 \cos^{2.7} \lambda$. This is a re-expression of a result given by HAMLIN *et al.* (1961). Here ε is the electron kinetic energy in rest mass units and $t(\lambda)$ follows empirically from an analytic solution. For plotting purposes we have used the latitude-dependent terms $n(\lambda) f(\lambda) t(\lambda)$ in which

$$f(\lambda) \equiv [\sin \lambda (8\text{–}5 \cos^2 \lambda)]/(4\text{–}3 \cos^2 \lambda)^{3/2} \cos^2 \lambda .$$

One should note that $f(0°) = 0$ and $f(90°) = \infty$.

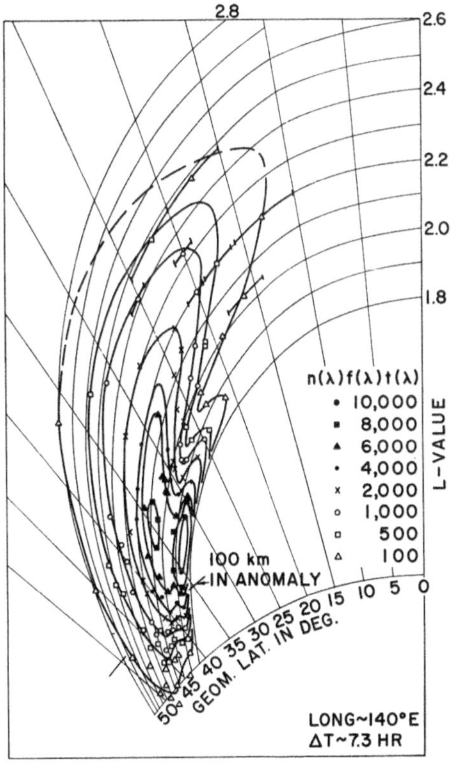

Fig. 3. A contour map of relative mirror-point density for 955-keV electron ≈ 7.3 hours after the detonation. The terms $n(\lambda) f(\lambda) t(\lambda)$ represent λ dependent terms and can be transformed to electrons/cm³ by equations in the text.

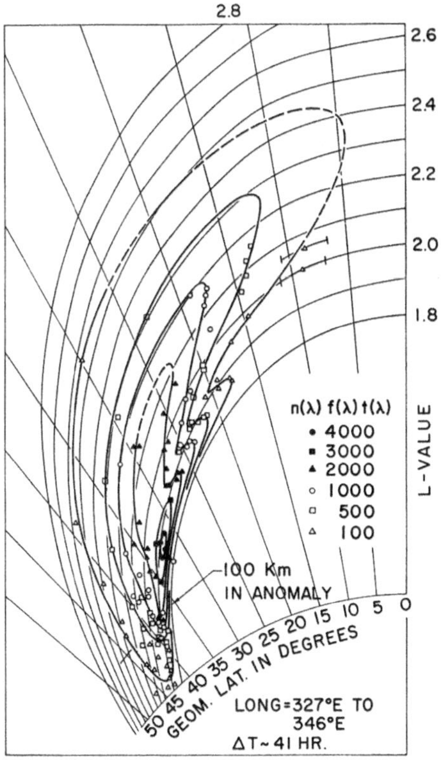

Fig. 4. Contour map of relative mirror-point density for 955-keV electrons ≈ 41 hours after the detonation.

Figures 3 and 4 are contour maps plotted in terms of $n(\lambda)f(\lambda)t(\lambda)$. Here we note the "injection spike" in the region of $L=1.8$ to 1.9 and then the "blowout region" from $L=1.9$ to ≈ 2.9. The explanation of these distributions will be left until later. The purpose at present is to provide a frame of reference for presenting the experimental data in an understandable manner. The path of the satellite may be noted by the uniform set of data points in Figure 3 starting at 37°, 2.8 (λ L) moving across to 19°, 2.0; and 46°, 2.8 moving across to 25°, 1.85; and in Figure 4, 30°, 2.56 to 8°, 2.0 and 39°, 2.57 to 17°, 1.87. These points representing the ascending and descending portion of the orbit come from the counting rate at 90° pitch angles. These are the most precise data. The data at smaller λ are estimates from the trend of the data and the dotted curves represent an educated guess. The error bars are believed to be liberal. The data at λ greater than the points of observation come from the lower angle data of the pitch angle distributions. The determination of the important region at high latitude in Figure 3 is believed to be good to within $\pm$ 2 degrees in that it is based upon data taken at a nearby latitude. These then represent the mirror point density of $\approx$955-keV electrons at $\approx$7.3 and 41 hours after the detonation. The other energy groups only approximately follow these profiles as will be seen later. We digress for a moment, still using the above injection profile as a frame of reference.

Early Observations

The first detection of the detonation occurred at $\approx$52 minutes after the detonation at $B=0.2801$, $L=1.796$, and 60.8° East longitude as a sharp rise in the background. A few moments later the data shown in Figure 5 was obtained. All data channels except the 1630-keV channels were counting at near the predetonation level. This observation is a consequence of the energy dependence of the eastward drift rate of the electrons. The eastward drift period for a dipole is given in minutes as

$$t = 85.7 \frac{1+\varepsilon}{\varepsilon(2+\varepsilon)} \frac{1}{L} \frac{1}{F(\lambda)},$$

in which $F(\lambda) \approx 3.00 - \cos^2(\lambda + 0.001\lambda^2)$ with λ in degrees (Hamlin *et al.*, 1961). Analysis shows that we are observing $\approx$1630-keV electrons on approximately their second eastward drift around the earth. Walt (1965) has calculated the correction to the dipole drift rate per drift around the earth at $L=1.9$ and B mirror$=0.26$ gauss and finds it to be $+5\%$. The drift rate at $L=1.87$, $B=0.26$, and $E=1630$ keV is then 27.2 min/2π radians. Correction for B from 0.279 to 0.26 is probably not justified since 100 km in the anomaly comes at $\approx$0.274. Probably a major component of the counting rate is due to scattering down from higher mirror points (justified by later observations). It appears that we are counting electrons at a predominant energy of 1630±100 keV; this transforms to 1.915±0.159 revolutions or an injection point of 92±57 degrees East longitude. The next observation was at $L=2.131$, $B=0.267$, and $\Delta T=53.1$

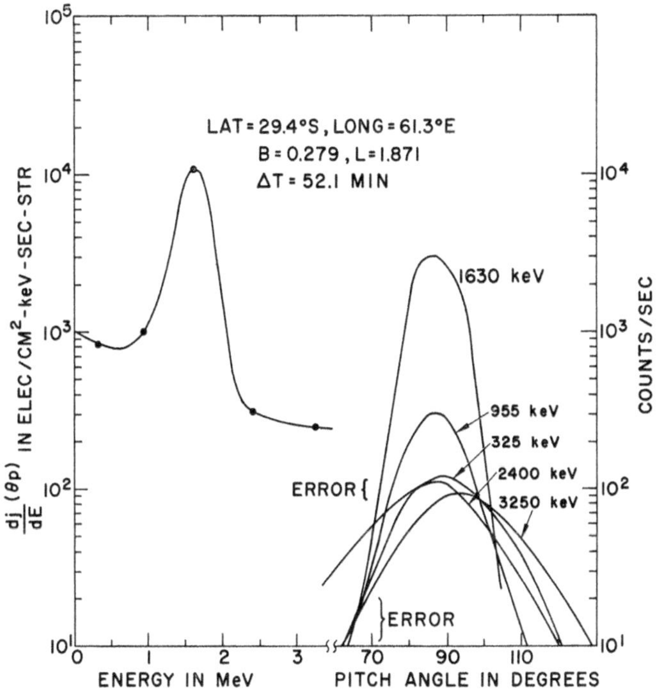

Fig. 5. Early time-of-flight data. The spectrum is given in terms of the peak unidirectional intensity. The peaking of the spectrum at ≈1600 keV is consistent with time-of-flight considerations.

min (ΔT refers to the time after the detonation) as shown in Figure 6. As before, the spectrum is plotted in terms of peak unidirectional flux. An analysis of the data shows that we are observing electrons rather close in energy to the 1630-, 2400-, and 3250-keV channels on their 2, 3, and 4 drift periods around the earth. The drift period for 325- and 955-keV electrons is 94 and 38 min respectively (reduced by 5% over the dipole calculations). The 325-keV channel is counting at essentially preshot fluxes. The 955-keV electrons are being observed on what would be ≈1.4 drift periods. The energy for one drift period is 650 keV which would be completely impossible to detect in this channel. The observation strongly suggests delayed injection from decaying fission fragments.

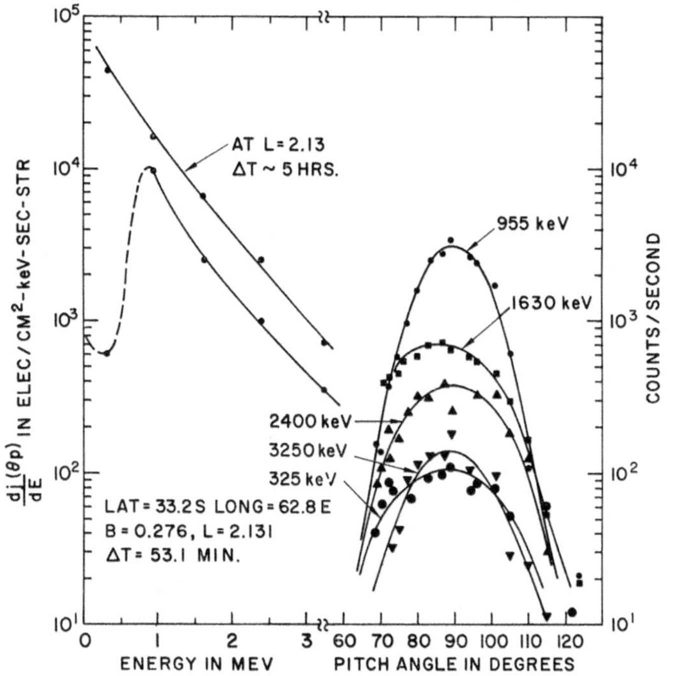

Fig. 6. Early time-of-flight data. The top three energy channels detected electrons on their 2nd, 3rd, and 4th pass around the earth. The 955-keV channel is detecting electrons on their 1.4th drift (nominal) around the earth. This is suggestive of delayed injection.

Pitch Angle Distributions

Later we will find that the pitch angle distributions are quite energy dependent. The question could then be asked: was there an energy dependence in the injection profile? Early pitch angle distributions can help resolve this problem. Unfortunately much of the data obtained at about 2 hours was garbled through poor telemetry. A reasonable distribution from this pass is shown in Figure 7. For the 3250-keV data the signal-to-background ratio was $\sim 1/1$. Since the measured falloff of counting rate at low angle is background dependent there may be a significant error in the peak position of this energy group. The lower energy groups, however, should be correct as far as relative position is concerned.

Comparison of these data with data obtained on an excellent pass at $\approx$7.3 hours as shown in Figures 8–11 shows that indeed the various energy groups could have been injected uniformly. Figures 12 and 13 show pitch angle distributions obtained at $\approx$41 hours after the detonation. In all pitch angle distributions the predetonation level has been subtracted using data of the type

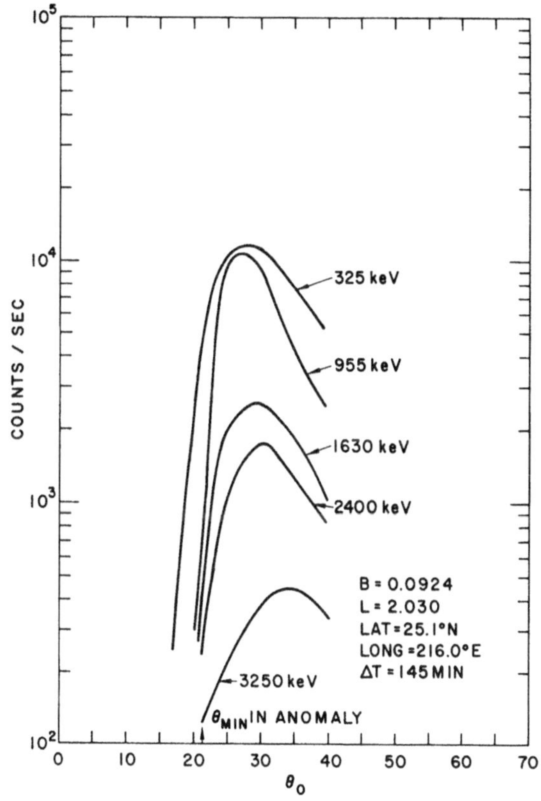

Fig. 7. A pitch angle distribution at early time. The pitch angle is referred to the equator.

shown in Figures 1 and 2. The data at 41 hours show even further energy dependence in the pitch angle distributions and also show an increased flux of electrons at larger pitch angles. These data were taken near the equator and hence the data at low pitch angles are not of the highest accuracy. The large apparent flux in the loss cone in this region of the anomaly is believed due to experimental error. The data taken at $\sim$7.3 hours, however, were taken at low latitude and the fluxes in the loss cone are believed to be of reasonable accuracy.

We were fortunate in obtaining some low altitude data east of the anomaly at $\Delta T \approx 6$ hours. These data showed an almost square pitch angle distribution. The data for 955-keV electrons as a function of θ_0 are contrasted with data at $\Delta T \approx 7.3$ hours taken at $\approx 145°$ East longitude in Figure 14. The relative absence of electrons in the loss cone for the earlier data is noted. The data at ≈ 7.5 hours

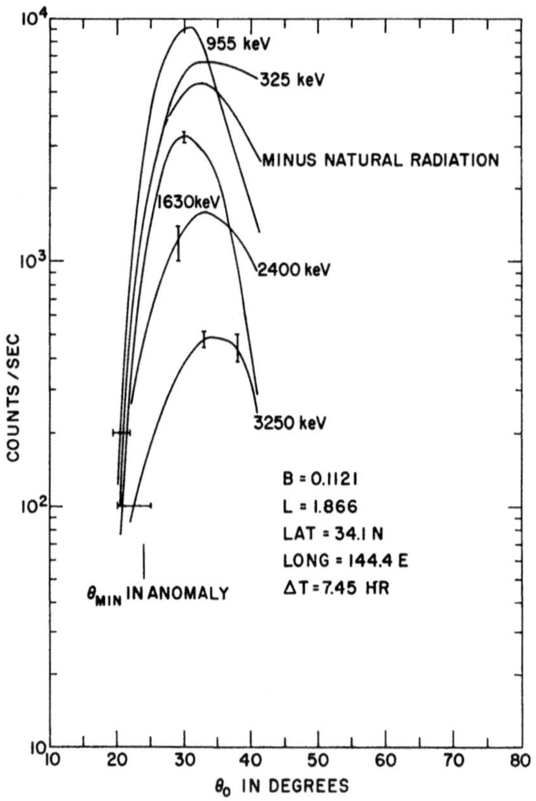

Fig. 8. A pitch angle distribution at 7.45 hours after the detonation.

as shown in Figures 8–10 has been checked for loss rate and it is found that the amount in the loss cone is somewhat less than might be expected from the longitude and the peak unidirectional decay rates (discussed later). Loss of counting rate in the peak unidirectional flux does not correspond exactly to loss from the trapping region since there is pitch angle diffusion both up and down the field lines. In general it is believed that at least for the 955-keV electrons, for which the most accurate measurements exist, the general picture of scattering towards both higher and lower pitch angles at a roughly uniform rate as a function of longitude is correct.

Some interesting pitch angle distributions were obtained on the later passes in the region of $L \approx 1.9$ to ≈ 2. What appears to be a secondary maximum in the pitch angle distribution showed up when the natural background was sub-

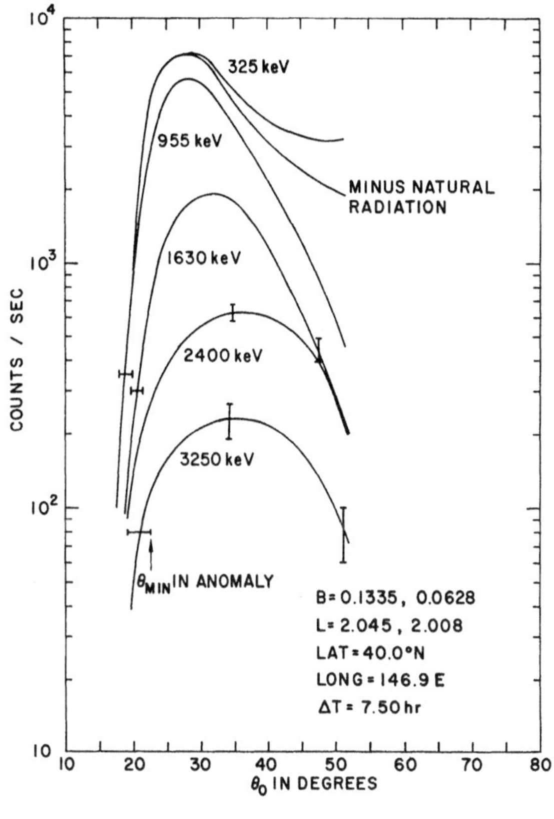

Fig. 9. A pitch angle distribution at 7.5 hours after the detonation. This distribution is a composite using off-equator data to give the results at the low pitch angles.

tracted from the 325 keV distributions. The 955-keV distributions show a less pronounced effect. An example obtained for the 325 keV distribution at ≈ 41 hours is shown in Figure 15. This figure shows the most pronounced effect observed. In these observations there was no effect at $L = 1.89$, there was the maximum depression in the region 1.92 to 1.93, and a filling up of the depression in the transition to 2.0. One should note that the errors are large, especially in the region of the depression, but it is believed that the general aspects are true. It is not known if the effect was present at early times since these passes did not traverse the right region of space to check this. The effect was present, however, in a pass at $\Delta T \approx 13$ hours.

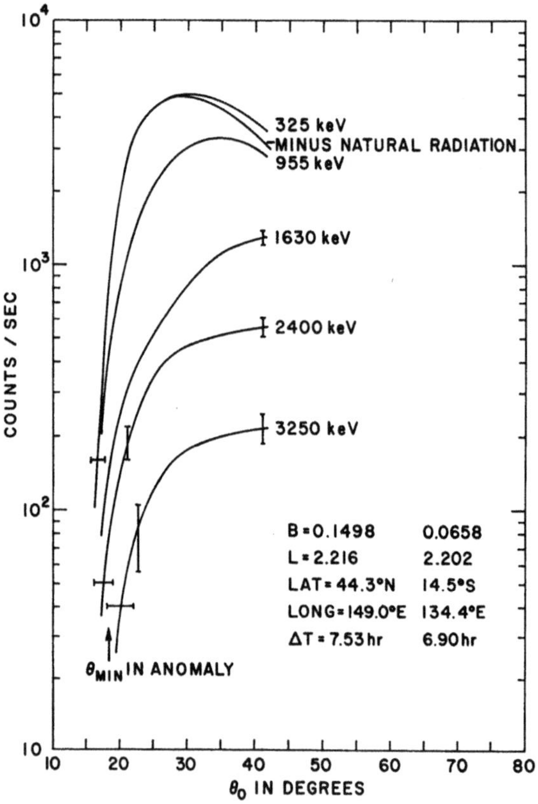

Fig. 10. A pitch angle distribution at ≈ 7.3 hours after the detonation. The results are a composite of two distributions taken at widely different latitudes.

Decay Rates

The ideal presentation of decay rates would come from equatorial pitch angle distributions as a function of time and energy. The satellite orbit made this impossible. The parameter that we could obtain with the greatest degree of accuracy was the change in the peak unidirectional decay rate. This takes advantage of the fact that unidirectional flux on a given L-shell transforms as $\mathrm{d}j(\theta_1)/\mathrm{d}E = \mathrm{d}j(\theta_2)/\mathrm{d}E$, in which $\sin^2\theta_1/B_1 = \sin^2\theta_2/B_2$, hence $\mathrm{d}j(\theta_P)/\mathrm{d}E =$ constant on a given L-shell as long as $\theta_P < 90$ degrees. A plot of the peak unidirectional counting rates including the values when θ_P was at 90° is shown in Figure 16 for the ascending and descending portion of the orbit at $\Delta T \approx 7.5$

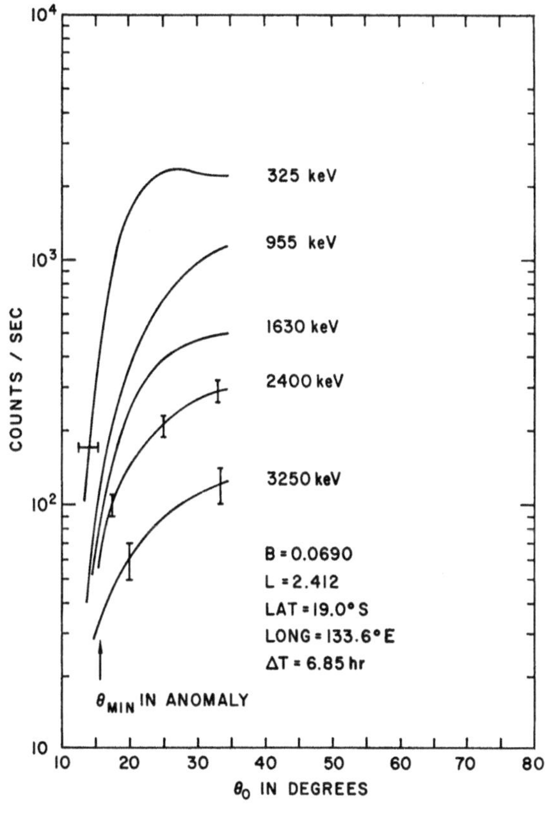

Fig. 11. A pitch angle distribution at 6.85 hours after the detonation.

hours. In all cases the peak counting rate was at <90 degrees out to $L \approx 2.1$ and for the upper curve to $L \approx 2.3$. One notes that for the 325 keV data there is no separation in the peak counting rates out to $L = 2.5$. This may be indicative of the injection profile or difference in decay rates.

The decay of the various energy groups obtained from plots of peak counting rate versus L is shown for $L = 1.885$ in Figure 17. The data points at about 10^4 seconds show the effect of the time dependence in the eastward drift.

One will note that the peak decay rate at these early times quite accurately follows a power law. It should be noted, however, that due to pitch angle diffusion up the field lines the loss rate of electrons from the L-shell is somewhat less than this. Such a power law decay was also observed by BURROWS and MCDIARMID (1964) on the Alouette satellite. These measurements with an

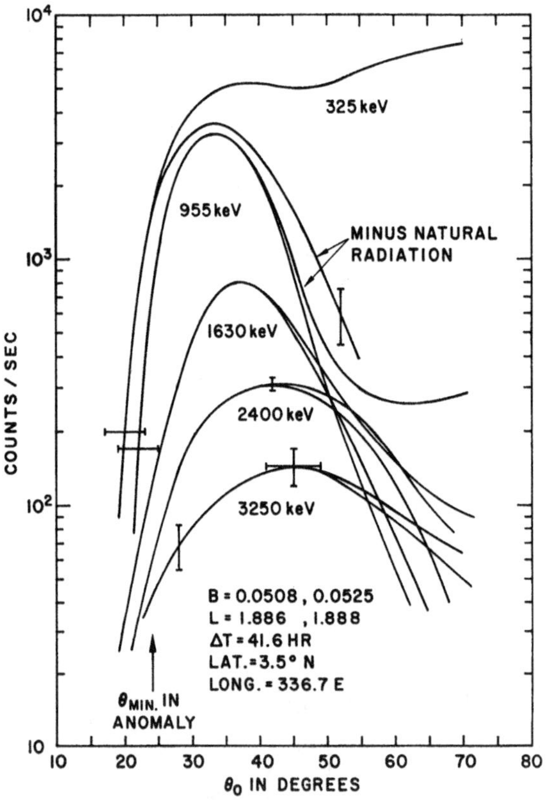

Fig. 12. A pitch angle distribution at 41.6 hours after the detonation. Two distributions were averaged for these results.

omnidirectional detector with a threshold at 3.9 MeV showed a power law decay of exponent $(n) \approx 1.2$ in which it is assumed that the intensity varies as $I = I_0 t^{-n}$. The exponents we obtained are shown in Table II. Since the Alouette orbit was low (≈ 1000 km circular) Burrows and McDiarmid were not measuring the peak omnidirectional flux but rather what might be referred to as resulting from the "tails" of the mirror-point density distribution, hence differences in decay rate should be expected.

TABLE II

Power law decay exponents

Energy in keV	*Values of exponent n for L =*		
	1.885	2.1	2.3
325	0.25 ± 0.03	0.25 ± 0.04	0.25 ± 0.05
955	0.52 ± 0.03	0.52 ± 0.04	0.52 ± 0.05
1630	0.84 ± 0.05	0.83 ± 0.06	0.53 ± 0.06
2400	0.93 ± 0.15	0.80 ± 0.15	0.28 ± 0.07
3250	0.69 ± 0.15	0.63 ± 0.15	0.29 ± 0.08

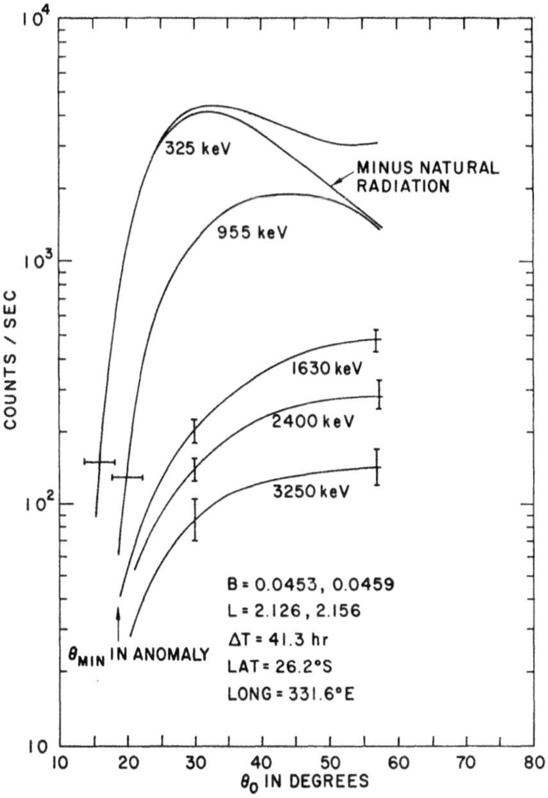

Fig. 13. A pitch angle distribution at 41.3 hours after the detonation. This represents the average of two distributions.

WALT (private communication) has used our pitch angle distributions as the input to his atmospheric model of electron decay. The results conclusively show that the decay as observed here is not dominated by the atmosphere.

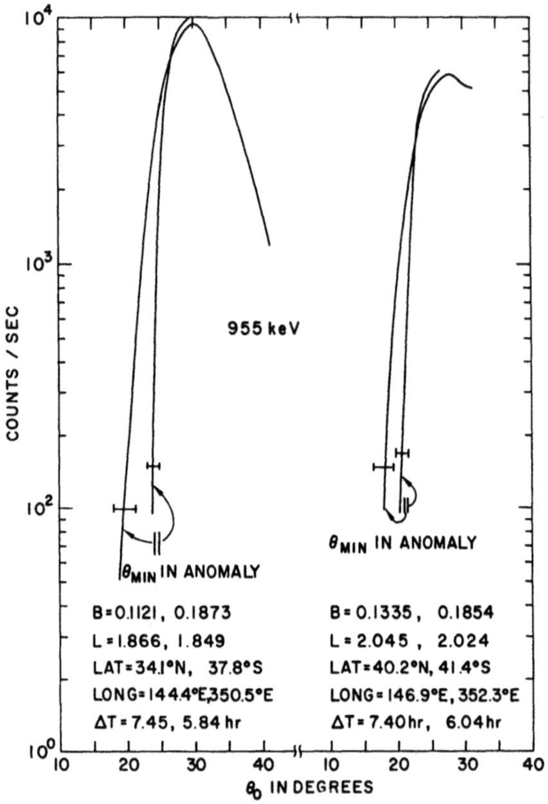

Fig. 14. Pitch angle distributions for electrons measured just east of the anomaly compared with distributions far east of the anomaly.

Energy Spectra

Omnidirectional energy spectra have been obtained at the point of observation from the usual relation

$$\frac{\mathrm{d}J}{\mathrm{d}E} = 2\int_0^{\pi/2} \frac{\mathrm{d}j(\theta)}{\mathrm{d}E} 2\pi \sin\theta \,\mathrm{d}\theta \,.$$

The predetonation spectra have been subtracted in all but the indicated cases. Representative spectra are shown in Figures 18 and 19 for ≈7.5 hours and in Figures 20 and 21 for ≈41 hours after the detonation. One of the most prolific producers of electrons in nuclear detonations results from the decay of fission fragments which beta decay as they spiral along field lines. Hence we must

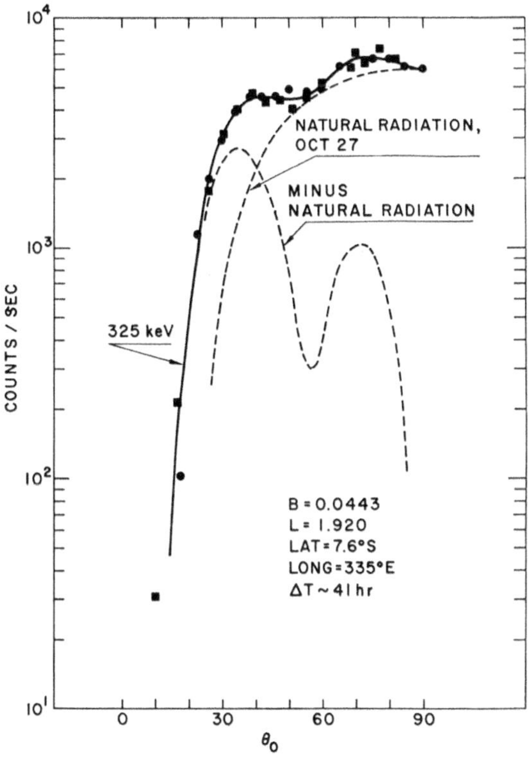

Fig. 15. Low-energy pitch angle distribution taken near the equator at about 41 hours after the detonation. In this approximate region of space the 325 keV pitch angle distributions showed a weak secondary peak when predetonation fluxes were substracted. An effect showed up in the 955-keV distributions as a marked widening of the distribution towards 90° but with no evidence of a secondary peak.

compare the measured spectra to the spectrum produced by fission. CARTER *et al.* (1959) have given the beta spectrum produced by fission fragments decaying in equilibrium with their formation, which is just the spectrum produced by the total decay of a group of fission fragments. This spectrum corrected for the velocity dependence in mirror period is shown in Figures 18, 20, and 21. Making a comparison with this spectrum is extremely subjective as will be discussed later. Thermonuclear reactions can also produce energetic electrons up to a few MeV but at a much lower yield per kiloton than fission and this source cannot be completely eliminated. Speculation in this area is extremely

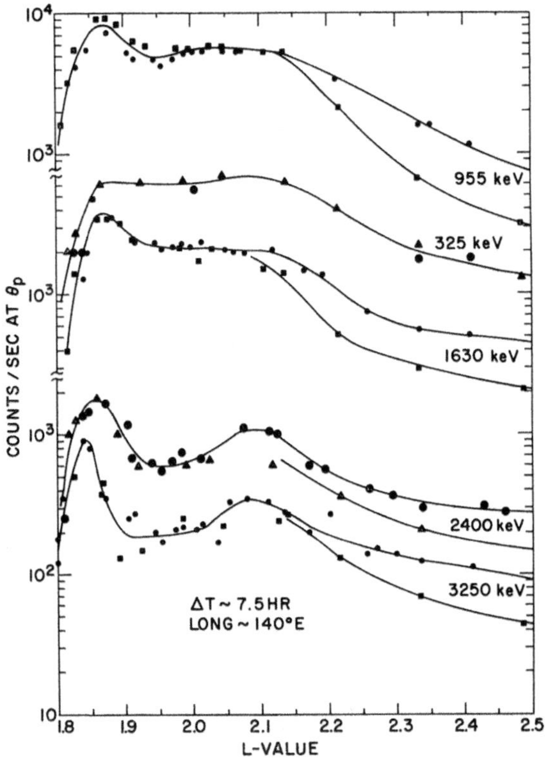

Fig. 16. The counting rates at the peak of the pitch angle distributions plotted vs *L*-value. The data were taken from the ascending and descending portions of the orbit about 7.5 hours after the detonation.

uncertain and the approach has been to assume that the majority of the trapped electrons are of fission origin. Figures 18 and 19 show omnidirectional spectra obtained at about 7.5 hours after the detonation. For spectra numbers 1, 2, 3, and 4 in Figure 18 the point of observation was between the equator and the point of maximum mirror-point density, hence the observational point did not adversely affect the observation. The point to be made is that the pitch angle distributions are energy dependent and a simple observation of spectra can be very misleading, the most misleading result coming from unidirectional spectra since the decay rates are energy dependent.

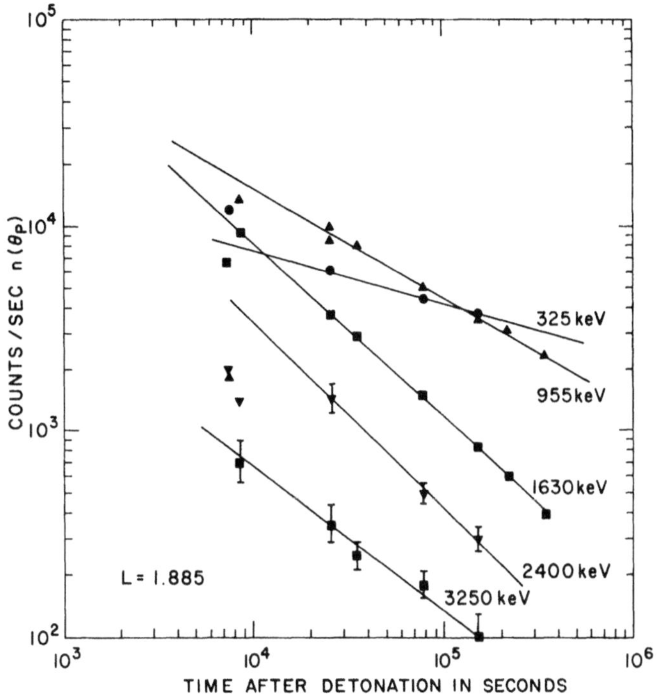

Fig. 17. Time decay of the peak unidirectional counting at $L = 1.885$. The data at $\approx 10^4$ seconds show the effects of the eastward drift in that the given L-shell was not uniformly filled by all energy groups at this time of observation.

Spectra 5, 6, 7, 8, and 9 in Figure 18 show omnidirectional spectra obtained well off of the equator and we know from Figure 3 that there was considerable injection above the point of observation. One notes in these spectra a development of an upturn in the low energy spectra with increasing L-value. In all cases the high energy portion of the spectrum is similar in shape to that of a fission spectrum with a tendency to be somewhat softer. Figure 19 contrasts data taken on similar L-shells but at widely different latitudes. Since the major contribution to the omnidirectional flux comes from nearby mirror-points (below the point of observation) is it obvious that the mirror-point density profile of the two lowest energy groups is considerably different from that of the higher groups at this time of observation.

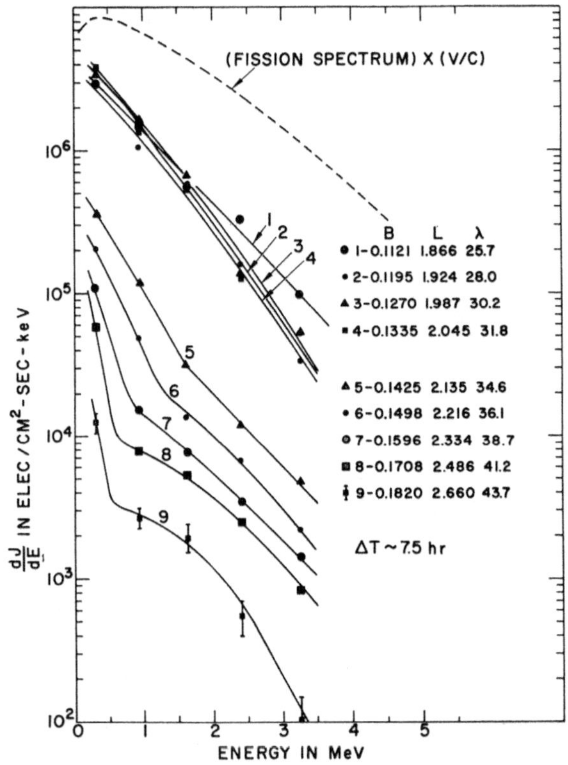

Fig. 18. Electron spectra at ≈ 7.5 hours after the detonation. The results are presented as omnidirectional fluxes.

Figures 20 and 21 show spectra obtained at about 41 hours after the detonation. Again as we go further off of the equator and to higher *L*-values, we note the upturn of the spectrum at lower energies. The high energy portion of the spectrum is similar to that expected from fission. A more complete understanding of the spectra must come out of a study of the decay rates of the various energy groups and the effects of delayed injection and decay of fission fragments.

Electron Inventory

We can do an inventory of the trapped electrons from two equivalent approaches. One is to use the original pitch angle distributions and the second is to work from plots of mirror-point density. In the first case we have for N the total

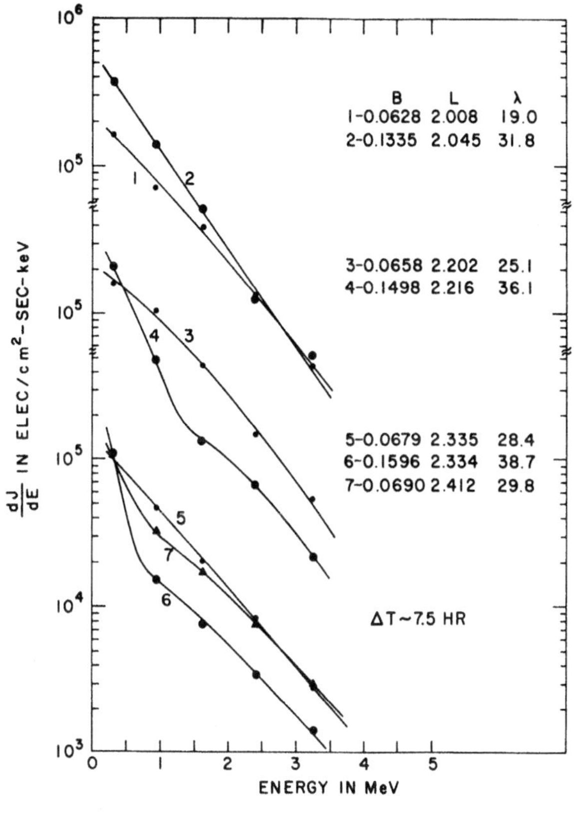

Fig. 19. Electron spectra at ≈ 7.5 hours after the detonation. The presentation is intended to show the λ dependence in the spectra.

number of electrons

$$N = \int T_m(\theta_0, E, L) \frac{dj}{dE}(\theta_0, E, L) \cos\theta_0 2\pi \sin\theta_0 \, d\theta_0 2\pi r(L) \, dr(L) \, dE .$$

$T_m(\theta_0, E, L)$ is the total bounce period as a function of the equatorial pitch angle, energy, and L, and r is the average equatorial radius of the respective

L-shell. The limits of integration are $\theta_0=0$ to $\pi/2$, $r_1(L)$ to $r_2(L)$ determined from the range of L-values for which trapping occurred, and E_1 to E_2 determined from the range of energies examined.

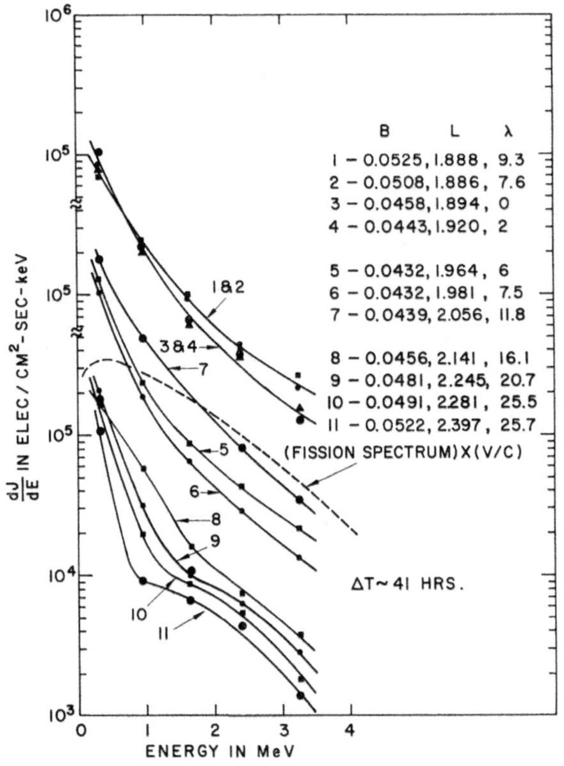

Fig. 20. Electron spectra at ≈ 41 hours after the detonation.

For the mirror-point density plot we have

$$N = \int N(E)\,\mathrm{d}E$$

in which

$$N(E) = \int \omega(\lambda, L, E)\,\mathrm{d}V.$$

Now $\mathrm{d}V = \mathrm{d}l\,\mathrm{d}A$, in which using cylindrical coordinates we find

$$\mathrm{d}l = r_0 \cos^7 \lambda \left(\frac{B}{B_0}\right) \mathrm{d}\lambda,$$

and from the flux invariant $dA = B_0/B\, dA_0$. Here $r_0 = 6371$ km, the mean radius of the earth. We also originally expressed $\omega(\lambda, L, E)$ as

$$\omega(\lambda, L, E) = \frac{3\pi}{r_0} T(\lambda) f(\lambda) \frac{n(\lambda, L, E)}{g}.$$

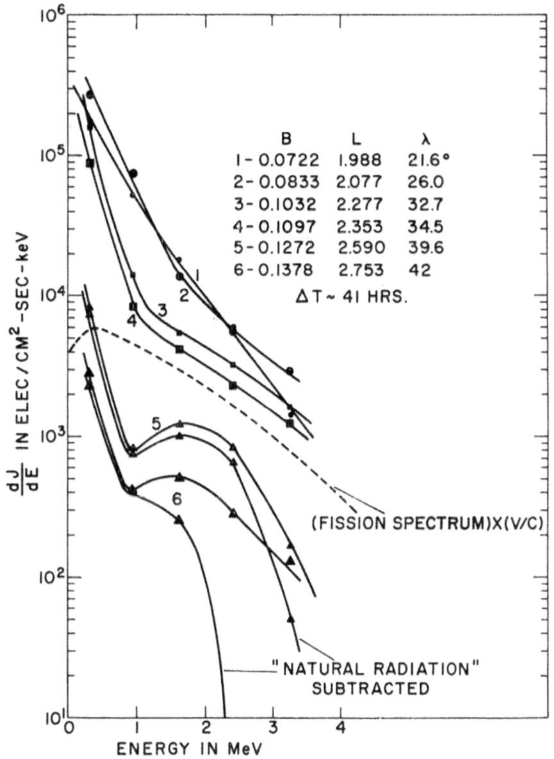

Fig. 21. Electron spectra at ≈41 hours after the detonation. Spectra 5 and 6 show the effect of subtracting the preshot fluxes. In *all* other spectra the preshot fluxes have already been subtracted.

Using the originally stated expression for $T(\lambda)$ and letting $dA_0 = 2\pi r_0^2 L\, dL$, the cross-sectional area of an equatorial L-shell we obtain

$$N(E) = 6\pi^2 r_0^2 \times 8.5 \times 10^{-2} \frac{(1+\varepsilon)}{\sqrt{\varepsilon(2+\varepsilon)}} \frac{1}{g} \int_0^{\lambda max} \int_{L_1}^{L_2} n(\lambda, L, E) t(\lambda) f(\lambda) \cos^7 \lambda L^2 \, d\lambda\, dL,$$

from which we define

$$\frac{N(L,E)}{L^2} \equiv 6\pi^2 r_0^2 \times 8.5 \times 10^{-2} \frac{1}{g} \frac{(1+\varepsilon)}{\sqrt{\varepsilon(2+\varepsilon)}} \frac{1}{g} \int_0^{\lambda \max} n(\lambda, L, E)\, t(\lambda) f(\lambda) \cos^7 \lambda \, d\lambda .$$

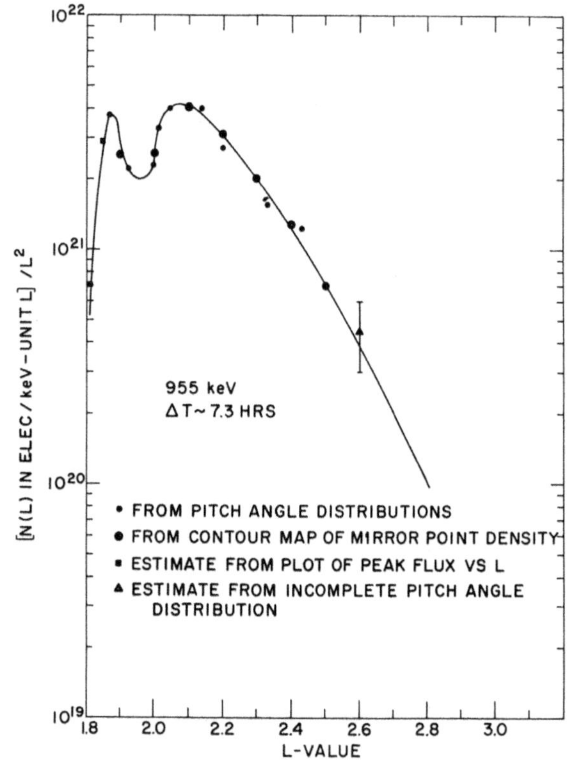

Fig. 22. The electron population profile for 955-keV electrons as a function of L at ≈7.3 hours after the detonation.

The function $N(L, 955)/L^2$ was determined for the data from the 955-keV channel and is plotted in Figure 22. Using these procedures we can arrive at an approximate electron inventory of $(2.0 \pm 1) \times 10^{25}$ for $E > 200$ keV, $(1.2 \pm 0.5) \times 10^{25}$ for $E > 500$ keV, and $(5.9 \pm 2) \times 10^{24}$ for $E > 1$ MeV at 7.3 hours after the detonation; that is, at 12.0 hours G.M.T., October 28, 1962. The last figure compares most favorably with the inventory of Imhof *et al.* (1964) at the same time for which they obtained 5×10^{24} electrons in the region $L = 1.9$ to 2.9 for $E > 1.2$ MeV.

Discussion

Plots of relative mirror point density for the 955-keV energy group are shown in Figures 3 and 4 for the times 7.3 and 41 hours respectively after the detonation. We may assume that the data in Figure 3 is roughly indicative of the injection profile although it is obvious from the decay rates that a considerable decay has occured prior to obtaining this first plot. It should be emphasized also that it takes several hours for electrons of about 1 MeV to reach transient equilibrium in their eastward drift around the earth. Obtaining the injection profile from nonequilibrium data is a difficult task even if good data exists, as it does not in this case because of telemetry malfunctions.

From examination of the pitch angle distributions and energy spectra at $\Delta T \approx 7.3$ hours we deduce that the profile for the 325 keV group should be weighted over the distribution for the 955-keV group in the direction of an increase in density at higher latitudes and conversely for the 1630-, 2400-, and 3250-keV groups. It is believed that some of the difference is due to the energy dependence in the decay rates since those few pitch angle distributions at about 2 hours after the detonation, that we have been able to examine, indicate a markedly reduced energy dependence in the pitch angle distribution.

In interpreting the data in Figure 3 we assume first that the detonation occurred well above the atmosphere. It would appear that the detonation occurred on an L-shell at about $L = 1.85$, possibly as low as $L = 1.8$. The expanding diamagnetic bubble of plasma which had been formed excluded the earth's magnetic field. The field was strongest near the earth, hence the resultant forces were upwards and along the constraining field lines. The bubble expanded, elongated, and rose up to $\lambda \approx 32$ degrees (probably at a somewhat smaller angle since the injection density profile must be weighted towards smaller angles relative to the mirror-point density profile) at which point instabilities set in on the outward side radially away from the earth. The jetting plasma was then in a region of much weaker magnetic field. The final phase probably occurred from the breakup of increasingly smaller plasma bubbles. It does not appear that the injected fission fragments retained enough kinetic energy at the completion of plasma breakup to pass over the geopotential barrier. In the region of $L = 1.8$ to 1.9 we can be quite sure of this. In the "blowout" region this is not as certain. The ionized fission fragments were semitrapped; that is, the earth's field controlled the motion. The fragments spiraled slowly down the constraining field lines on a time scale from tens of minutes to days depending on their initial velocity and injection angle.

We must consider how the beta spectrum* from decaying fission fragments

* The author's attention was refocused on this effect by the presentation of a paper by COLGATE (1945).

changes with time as a step in understanding the electron spectra. At delayed times after fission the spectrum softens as measured by WEST (1960) and calculated by a number of people. The author has experimentally checked some of the calculations of NELMS and COOPER (1959) (from 31.2 min to later times) and finds the spectral shape in agreement with experiment, hence we use these calculations. For the beta decay rate we use the tabulation of KNABE and PUTNAM (1958). We must assume that there has been a time delay (t) before a fission fragment decays, and in addition we must assume that there is no contribution to the spectrum after a given time. This later time has been arbitrarily set at 10^5 seconds or about 1 day. We then obtain the spectrum for time t to 10^5 seconds, weighting the instantaneous spectrum for each interval of time. An exponential beta spectrum was assumed in the range of interest of this experiment of 1 to 3.5 MeV. This is characteristic of only a limited range of energy. The results are shown in Table III. Accuracy is $\approx \pm 10\%$.

TABLE III

Characteristics of the beta decay of fission fragments as a function of time after fission.

t(sec)	*Spectral Characteristics for* 1 *to* 3.5 MeV *Instantaneous e-fold rate*	 $t \to 10^5$ sec *e-fold rate*	$t \to \infty$ *fraction of decay for all energies*
0		≈1200 keV	1.00
1	2000 keV	1000	0.95
2	1790	960	0.92
5	1460	920	0.86
10	1250	890	0.81
50	940	820	0.69
100	920	790	0.64
500	870	730	0.54
1000	840	700	0.50
5000	740	630	0.41
10 000	670	600	0.38

From the data in Table III we see how the spectrum softens with time after fission. One is tempted to obtain an approximate injection time for the fission fragments by a comparison of the measured spectra with the data in Table III. Indeed we note that a fairly hard spectrum, rather similar in shape to an equilibrium spectrum, can be obtained by restricting the decay to the 10-to 50-second interval. Also we can obtain a spectrum similar to number 5 in Figure 18 by adding up the decay from fission fragments initially injected into the region of observation plus the decay from fragments that were injected above the point of observation and had subsequently slowly spiraled down the constraining field lines. The contribution from fission fragments that were injected well above the point of their decay may be responsible for much of the apparent nonuniformity in the injection of the various energy groups.

We can estimate the effect of the energy-dependent decay in the spectra by utilizing our decay data. It is to be noted that the decay data is for the decrease in the peak unidirectional intensity, whereas the spectra are of omnidirectional intensity. Interpreting the spectra using the decay data is extremely subjective and we can only establish the trend.

Initial injection of the fission fragments probably occurred on a time scale of tens of seconds to a few hundred seconds, and the intensity of the trapped electrons built up from that time to a peak in the range of a few hundred to a few thousand seconds. A peak must occur due to the competing action of injection of electrons from decaying fission fragments and their natural decay. We might expect that the decay rate of electrons injected at t to $t+\Delta t$ could be determined from an extrapolation of the experimental decay rates back to that period of time. We could then weight the decay during the various time intervals of fission fragment decay and obtain the spectra as determined at $\Delta T=7.3$ hours. Such a procedure is not completely satisfactory. It does, however, generate roughly some of the shapes shown in spectra 5–9 of Figure 18. Unfortunately we do not have an adequate knowledge of the decay rates. The rates given are for only a particular region of space; also, there has been pitch angle diffusion as exemplified by the pitch angle distributions and mirror-point density maps. The diffusion alters the decay rates so as to make them λ dependent as well as L dependent. The only place we can hope for agreement is in the region $L\approx 1.8$ to 1.9. Here agreement can be obtained if we reduce the power law exponents for the 1630 and 2400 keV groups by about 20%. The general conclusions, however, are that the spectra could be explained through a detailed knowledge of the electron decay and the details of the fission fragment injection. This is not necessarily the whole picture as we have not considered other effects such as shock acceleration, thermonuclear sources of electrons, and the loss of energy of the electrons in the "plasma bubble". There is also the question as to what happened to the electrons from fission decay occurring while the "plasma bubble" was rising. It is believed that most of these electrons leaked out of the "bubble" at an altitude that would allow them to be quickly lost in the anomaly. There should be a strong energy dependence in the containment of energetic electrons by the "bubble" and its encompassing magnetic field. Evidence that such a fractionation occurred is shown in Figure 16 showing the change in peak unidirectional flux as a function of L. The effect is most pronounced for the two extreme energy groups at $L\approx 1.8$ to 1.9.

It is evident that we have been dealing with an extraordinarily complicated situation. Unfortunately we have been unduely handicapped by a lack of knowledge of the details of the nuclear detonation and have had to speculate at times.

Nevertheless we have been able to put together a reasonable picture of the situation and have gained insight into many aspects of trapped radiation.

References

Burrows, J. R., McDiarmid, I. B.: 1964, *Can. J. Phys.* **42**, 1529.
Carter, R. E., Reines, F., Wagner, J. J., and Wyman, M. E.: 1959, *Phys. Rev.* **113**, 280.
Colgate, S. A.: 1965, this volume, p. 693.
Hamlin, D. A., Karplus, R., Vik, R. C., and Watson, K. M.: 1961, *J. Geophys. Res.* **66**, 1.
Imhof, W. L., Gaines, E. E., King, D. R., Nakano, G. H., Smith, R. V., Vaughn, F. J.: 1964, *DASA* 1540.
Knabe, W. E., Putnam, G. E.: 1958, *General Electric Report* APEX-448.
McIlwain, C. E.: 1963, *Science* **142**, 355, and private communication.
Nelms, A. T. and Cooper, J. W.: 1959, *Health Physics* **1**, 427.
Walt, Martin, private communication.
Welch, Jasper A. Jr. and Whittaker, William A.: 1959, *J. Geophys. Res.* **64**, 909.
West, H. I., Jr.: 1960, *Trans. Am. Nucl. Soc.* **3**, No. 2, 537, and unpublished data.
West, H. I., Jr., Mann, L. G., and Bloom, S. D.: 1965, *Space Research* **5**, 423.

THE TRAPPED ELECTRON SPECTRA FROM THE STARFISH DETONATION AND FROM THE OUTER BELT IN THE FALL OF 1962*

HARRY I. WEST, Jr.
Lawrence Radiation Laboratory, University of California
Livermore, California

Abstract. A brief summary of some of the electron spectra measured in the fall of 1962 is presented. Measurements were made with a five-channel magnetic electron spectrometer covering the range of 325 to 3250 keV.

The following is a brief summary of some of the electron spectra of the radiation belts in the fall of 1962. A complete discussion of the results has been published elsewhere (WEST *et al.*, 1965) and the interested reader is referred to that paper. The instrument was a five-channel magnetic electron spectrometer with differential energy channels at 325, 955, 1630, 2400, and 3250 keV. The omnidirectional spectra presented in Figures 1 and 2 came from integrating the pitch angle distributions obtained. Normally such a pitch angle distribution consisted of about 30 data points obtained at an acquisition rate of one/second. In many cases where the data were not statistically different, more than one spectrum was averaged, and this is indicated by the ranges in the coordinates. The relative accuracy of the data points unless otherwise indicated is within $\pm$ 20%. A calibration error (same for all energies) was discovered recently and the spectra presented here have been properly corrected.

Spectra 1–7 of Figure 1 show the remnants of the trapped electrons resulting from the Starfish detonation. The upturn of the spectra at low energies is believed to be of natural origin. Spectra 1 and 2 show the effects of preferential loss of low energy electrons in agreement with the electron decay calculations of WALT (1964) and ANDERSON *et al.* (1964) utilizing the atmospheric model. Most of the energetic electrons are believed to have resulted from the decay of fission products. Such a spectrum characterizing the decay from time zero to infinity and corrected for the velocity dependence in observing the spectra is shown in Figure 1. One notes that the high energy tails for spectra obtained near the equator and for $L < 1.3$ indeed strongly resemble the fission spectrum. As we go to higher L-shells the spectrum softens as one would expect from a

* This work was performed under the auspices of the U. S. Atomic Energy Commission.

fission spectrum measured at delayed time after fission. The inventory of electrons >500 keV for $L<1.7$ in early December of 1962 was $(6.6\pm 2)\times 10^{24}$.

Part of the high energy tails of spectra 8, 9, and 10 of Figure 1 and of spectra 1 and 2 of Figure 2 are believed to have resulted from the Russian

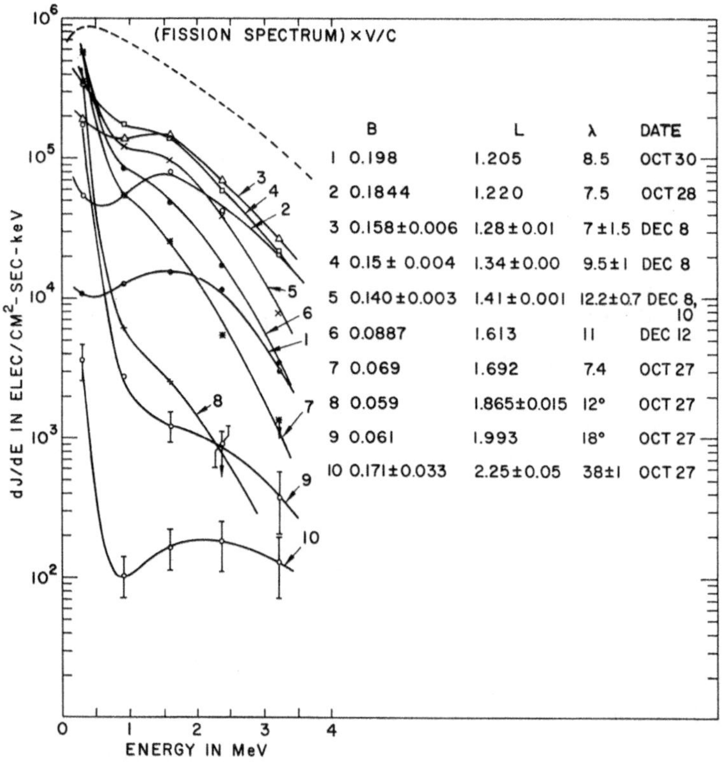

Fig. 1. Electron spectra in the range $L\approx 1.2$ to ≈ 2. The range of values on the coordinates indicates that two or more spectra have been averaged and also indicates the approximate range of validity.

detonation of October 22, 1962. Spectra 3, 4, and 5 of Figure 2 are typical of the heart of the outer belt. Generally as one goes to higher L-values the spectrum softens and decreases in intensity. Electrons of 325 keV energy were usually observed out to $L\approx 7$.

The B-dependence of the spectra can be inferred from the pitch angle distributions. Generally one finds changes in the spectra with B on a given

L-shell only when near the limits of the trapping regions. An exception occurred in the region $L \approx 1.7$ to 1.9 in which the 325-keV distributions were on the order of 30 to 40% wider than the higher energy groups.

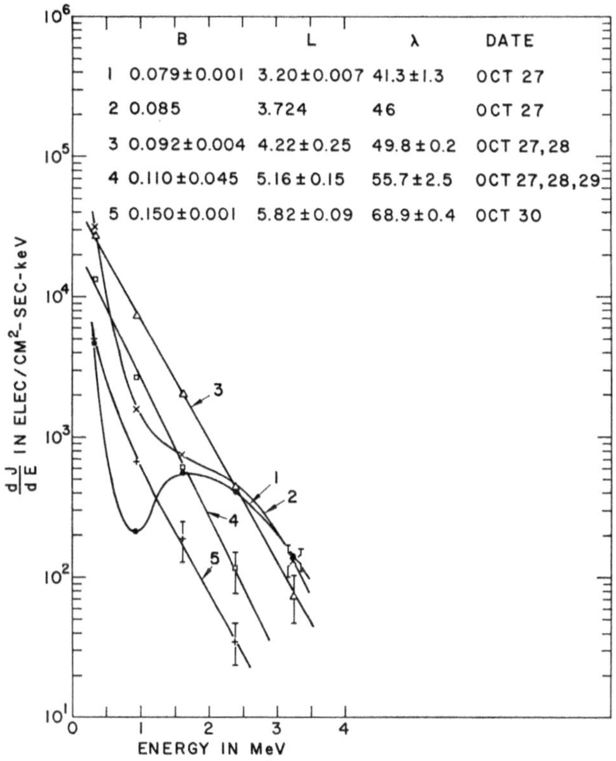

Fig. 2. Electron spectra in the outer belt region. Curve 4 is the average of 4 sets of data covering a range of about 2/1 in intensity.

References

ANDERSON, A. D., CRANE, G. E., FRANCIS, W. E., NEWKIRK, L. L., and WALT, M.: 1964, *Lockheed Missiles and Space Company, LMSL*-895355.
WALT, M.: 1964, *J. Geophys. Res.* **69**, 3947.
WEST, H. I., Jr., MANN, L. G., and BLOOM, S. D.: 1965, *Space Research* **5**, 423.

OVERALL ANALYSIS OF EXPERIMENTS ON ARTIFICIAL RADIATION BELTS

MARTIN WALT

Lockheed Missiles and Space Company, Research Laboratories, Palo Alto, California

1. Introduction

The analysis of experiments on the artificial radiation belts has been a difficult task. Invariably enough data are available to stimulate one's interest and to suggest mechanisms and processes of geophysical importance, but the data are seldom extensive or exact enough to provide adequate tests for the theories they suggest. However, in spite of these drawbacks, experiments with artificial radiation belts have led to a number of important discoveries concerning trapped radiation, and additional study of the available experimental information should further improve our understanding of magnetospheric physics.

One should bear in mind that the conclusions applicable to the artificial belts may not be valid for the natural electron belts, much less the proton belt. In the first place, the energies of the bulk of the electrons in the naturally occurring belts are lower than the energy of fission fragment beta decay particles. Furthermore, the energy densities and the spatial distributions of artificial belts often differ markedly from the natural ones, and these differences may lead to collective effects which are not normally present. Nevertheless, a study of magnetospheric phenomena induced by unusual conditions may still be valuable, since these unusual conditions may increase the detectability of processes which might be unnoticed in the unperturbed environment.

As the preceding speakers have indicated, there are many uncertainties in our knowledge of the actual phenomena following nuclear detonations and even greater uncertainties in the theoretical interpretation. It is not possible to settle these questions now. The purpose of the present paper is to summarize the data obtained from the detonations of 1958 and 1962 and to bring out what has been learned about trapped radiation from these tests.

2. Argus Explosions (1958)

The first tests which furnished definitive information on trapping were the low-yield, Argus explosions in 1958 (Christofilos, 1959; Van Allen *et al.*, 1959; Cladis and Walt, 1962). These detonations occurred at L values of about 1.7, 2.1, and 2.2. In all three cases, trapped electrons were observed following the detonations, although the spatial coverage of the observations was limited and the energy resolution of the measurements was not as high as desired. Insofar as the physics of trapping is concerned, the chief results of these tests were: (1) The electron shells remained in position throughout the detectable life of the belts, indicating that large latitude drifts did not occur. (2) The decay rate of the trapped electron flux was too rapid for the loss to be attributed to collisions with the atmosphere. (3) The energy spectrum was that expected from fission product beta decay, although the precision of this measurement was not high. (4) The position of the shells at various longitudes was useful in testing coordinate systems for trapped radiation. Some unexplained features of these tests were the "wings" of trapped electrons which certainly occur on the high latitude side of the central peak and probably occur on the low latitude side. Whether these are produced by the migration of fission fragments before their decay or by the transport or diffusion of electrons across shells after beta decay is not known. Any theory of their occurrence must account for the fact that the spectra in the wings and peak are the same within the limits of measurement, the wings and central peak decay at the same rate, and the wings appeared within 30 minutes of the detonation.

In view of the recent interest in the diffusion of trapped particles across L shells, the Argus tests of 1958 take on renewed importance. In these experiments trapped electrons were observed to leak out the ends of the magnetic trap while the position of the shells remained unchanged. Theories of electron transport from one magnetic shell to another must take into account the fact that in the Argus test, leakage by pitch angle change occurred at a far greater rate than transport across L shells.

2. Starfish (1962)

The belt produced in 1962 by the Starfish detonation (see collected papers of Hess, 1963) is the most extensive and longest lived of the artificial radiation belts. Consequently, a considerable amount of data is available concerning its characteristics. However, even in this case there are serious deficiencies in our knowledge of the actual facts. The data obtained in the first few hours are limited in energy discrimination and spatial extent, and the data taken at later times are difficult to interpret due to uncertainties in the background radiation. These later data, taken some months after the explosion, do not, of course, include much of the early and interesting phenomena.

From the available data, it seems clear that fission electrons were injected to at least $L \approx 2.0$ and radiation effects were noted considerably beyond this value. The broad spread in latitude of the injected electrons has made it impossible to deduce the stability of L shells as was done for the Argus explosions. However, the broad coverage has enabled definitive values of electron flux decay rates to be obtained over an extended portion of the inner belt region. Analysis of this data has led to a knowledge of the region in which the atmosphere causes the loss of electrons, and has shown that above $L > 1.25$ some other loss mechanism must be involved (WALT, 1964). These observed loss rates will

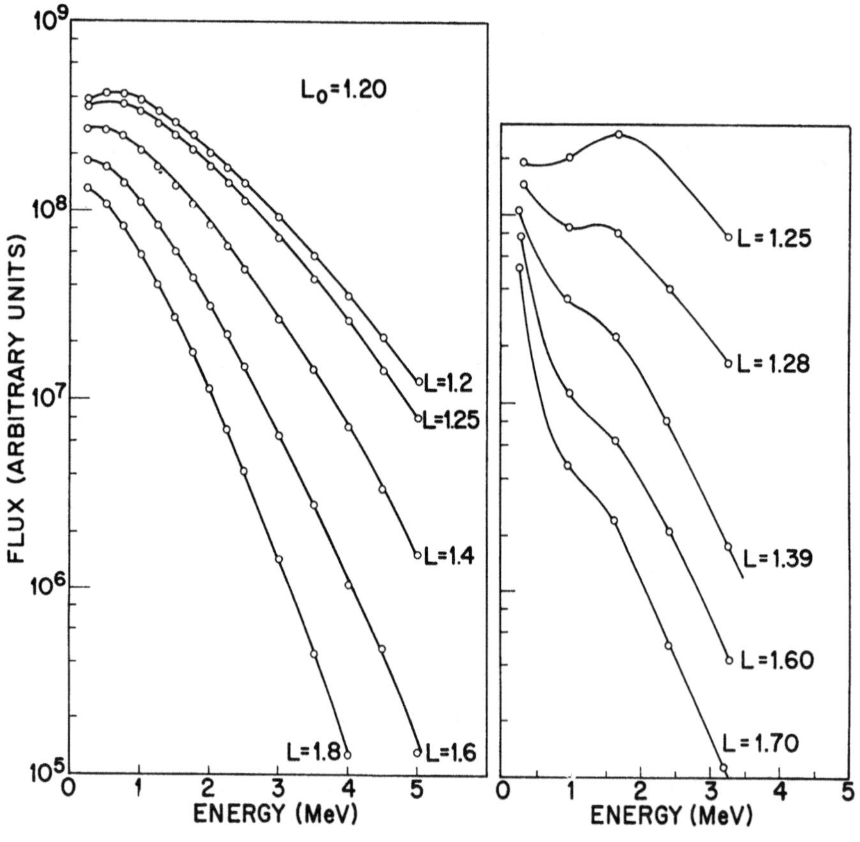

Fig. 1. Experimental and theoretical spectra of radiation belt electrons following Starfish. The left hand figure gives theoretical spectra obtained by assuming that the electrons moved across field lines conserving the magnetic moment and the integral invariant. Right hand figure shows experimental values of WEST *et al.* (1964).

be valuable in testing theories of this loss mechanism. Another important feature learned from Starfish was the degree of validity of the *B-L* coordinate system, which received extensive tests in the organization of electron measurements. A recently learned result from Starfish was reported by IMHOF and

SMITH (1965) who found that, even at L values as low as 1.15, electrons are affected by fluctuations in the geomagnetic field.

Perhaps the most serious controversy concerning Starfish is the discrepancy between the interpretation of Telstar and Injun 1 data. These two results can be brought into agreement only with difficulty by postulating an extensive region of bomb-induced electrons with energies lower than those of fission electrons. However, to date the means of producing this energy distribution are in question. Possible mechanisms include the reduction in energy of fission electrons due to multiple collisions from the sides of the expanding bubble, formation of an energetic band of shock heated electrons, or the diffusion of electrons across magnetic fields by collective effects. If fission electrons are moved to higher L values by a process which conserves the magnetic moment and integral invariant, then the energy will be reduced, leading to spectral changes as shown in Figure 1. Also shown in Figure 1 are some spectra measured by West *et al.*, which indicate a softening at high L values in fair agreement with theory. However, the experimental data are undoubtedly contaminated with naturally occurring electrons, the contamination increasing with increasing L. Hence, quantitative comparisons will require a more elaborate effort to evaluate the background, and in view of the observed time variation in the background, may not be possible.

Other features of Starfish which merit further study are the geophysical effects such as the high latitude riometer absorption and the enhanced ELF and VLF emissions which persisted many days after the actual detonation.

3. Russian Detonations (1962)

The series of Russian tests in October and November occurred at a time when a number of payloads designed to measure the residual radiation from Starfish were in operation. Thus, a considerable amount of information was obtained on these intense but short lived belts, much of the data being obtained with better energy and pitch angle resolution than was available at the time of the Starfish detonation. The decay rates of these shells have provided definitive information on the loss rates of electrons at L shells between 2.5 and 3.5, and the observed diffusion in pitch angle offers an opportunity to test various loss mechanisms postulated for this region. The Russian tests also gave indications of a non-fission energy distribution of electrons on the outer fringes of the artificial belts, although closer examination of the experimental data and a careful study of the natural background will be needed to clarify this finding.

4. Areas of Future Study

As mentioned above, additional information can undoubtedly be gleaned from study of existing data on artificial radiation belts and from further measurements on the long-lived portion of Starfish. The variation of the Starfish belt decay rate with solar activity will be important additional information and will complement the study of the variation of proton belt intensities with the solar cycle.

There are a number of experimental "facts" concerning artificial radiation belts for which adequate theoretical explanations are not available at present. These features are summarized below.

(1) Observed decay rate of electron fluxes at $L > 1.25$. A possible mechanism for these losses is the interaction of electrons with external electric or magnetic fields such as whistlers, geomagnetic micropulsations, or magnetic fluctuation induced by solar wind variations. On the other hand, the observed losses may result from self-induced plasma instabilities. In this regard the experimental data are quite extensive and the contraints on the possible theoretical explanations will be serve. Not only should the overall loss rates be obtained, but the diffusion in pitch angles observed by West *et al.* (1964) must be reproduced by theory.

(2) Transfer of electrons across L shells. The observations by Imhof and Smith (1965) strongly suggest the transfer of electrons across L shells by magnetic variations. Sufficient experimental data are available on the electron flux and on magnetic field variations that analysis of this low latitude feature of Starfish should be fruitful.

(3) The generation of an outer fringe of non-fission spectrum electrons by nuclear explosions. A number of mechanisms for this feature have been proposed, but up to date a completely convincing explanation is not available.

(4) Perturbation of existing belt populations by nuclear explosions. Some (Filz and Holeman, 1965) of the experimental results obtained following Starfish suggest that the natural radiation belts were perturbed by the detonation. While confirmation of these results is highly desirable, the possibility that the natural belts were altered substantially by nuclear explosions is an important area for further study.

References

Christofilos, N. C.: 1959, *J. Geophys. Res.*, **64**, 869.
Cladis, J. B. and Walt, M.: 1962, *J. Geophys. Res.* **67**, 5035.
Filz, R. C. and Holeman, Ernes : 1965, *J. Geophys. Res.* **70**, 5807.
Hess, W. N.: 1963, *J. Geophys. Res.* **68**, 605.
Imhof, W. L. and Smith, R. V.: 1965, *Phys. Rev. Letters* **14**, 885.
Van Allen, J. A., McIlwain, C. E., Ludwig, G. H.: 1959, *J. Geophys. Res.* **64**, 877.
Walt, M.: 1964, *J. Geophys. Res.* **69**, 3947.
West, H. I., Mann, L. G., and Bloom, S. D.: 1964, *University of California report* UCRL-7659, TID-4500 (29th Ed.).

THE MOTION OF BOMB DEBRIS FOLLOWING THE STARFISH TEST

JOHN ZINN, HERMAN HOERLIN, and ALBERT G. PETSCHEK

Los Alamos Scientific Laboratory, Los Alamos, New Mexico

Abstract. Theoretical computations are presented which identify some of the more important processes affecting the debris expansion following the Starfish explosion. A MHD shock wave is formed, through which the moving debris and air plasma is coupled to stationary plasma. The major component of the stationary plasma is produced by the explosion through ionization of neutral air by X rays and by collisions. The detailed behavior of a model MHD shock system with collisional effects is studied by means of a Lagrangian computer program, and the computed results are compared with photographic data. The data and computations are in good qualitative agreement as to the behavior of the downward and east-west expansion. The upward expansion is slowed partly by the processes mentioned, but it is ultimately stopped and reversed through the action of conduction currents flowing in the F1 layer in a region connected to the moving plasma by way of the field lines.

The observed motion of the plasma along the field lines is not correctly predicted by the machine code in its present form. The major source of electron injection at high altitudes was through a jet of plasma directed initially along the field lines toward the south. This jet was concentrated to an extent sufficient to permit it to move in a straight line across the bending *B* field without deflection, though leaving a trail of ionized material which was guided by the field to the southern conjugate and other points further south. The trail of the jet is quite clearly seen in the photographs. Its observed motion upward and to the west is interpreted in accordance with well known principles governing the motion of a dilute plasma in a nonuniform field.

The Starfish explosion was fired on July 9, 1962 at an altitude of 400 km above Johnston Island. The nominal yield was 1.5 Mt. Visual effects of the explosion were observable at extremely large distances, extending to points in New Zealand 4600 miles to the south. A very appreciable number of high energy electrons were trapped in the magnetosphere at altitudes far above the burst. The present paper describes LASL and EGG photographic observations of the event, along with a theoretical study of the bomb plasma expansion.

The expanding ionized bomb material from an explosion at 400 km altitude is coupled to the ionized component of the atmosphere through the magnetic field, and a magnetohydrodynamic (MHD) shock wave is formed. The shock is decelerated through the act of sweeping up stationary air ions. Out of the aggregate of stationary ions to be swept up, a part is present before the explosion. Another part is created as the result of charge exchange processes occuring between the moving ions in the shock and stationary neutral air molecules. Another arises out of ionizing collisions with neutral air. Still another part arises as the result of absorption of X rays emitted by the bomb case.

At a relatively late stage the expanding shock becomes coupled to neutral air at low altitudes through the flow of conduction currents along the field lines. This gives rise to a strong decelerative force which ultimately stops the expansion. Each of these processes will be considered in detail later.

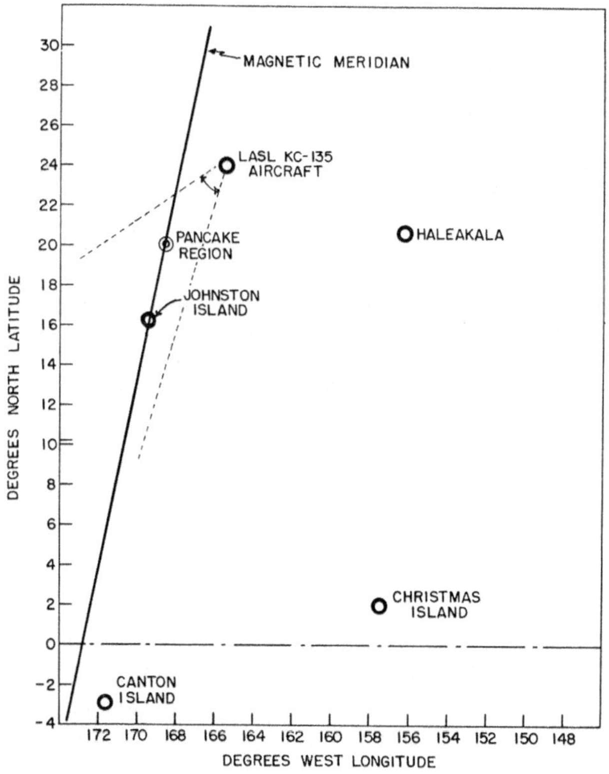

Fig. 1. Map of station locations.

Figure 1 is a map showing some of the observing stations and the locations of certain points related to the phenomenology.

Figures 2a, b, c and d are from a sequence of motion picture photographs taken from Maui. They show an early-to-intermediate stage in the expansion of the bomb debris. Figure 3 is a computed projection of field lines in the burst magnetic meridian plane as seen by the camera. The photographs show the downward and horizontal progress of the debris-MHD shock expansion and

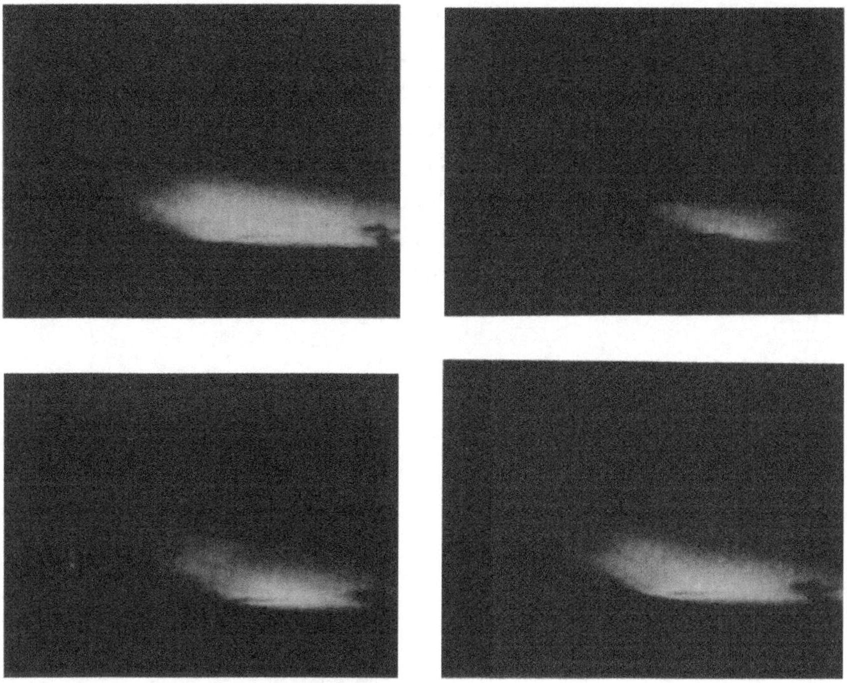

Fig. 2. Early Bomb Debris Expansion From Maui; (a) 0.06 seconds, (b) 0.09 seconds (c) 0.12 seconds, (d) 0.15 seconds.

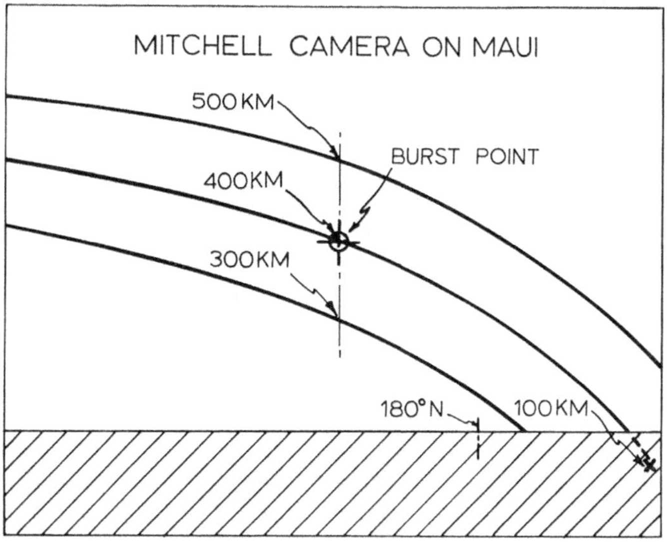

Fig. 3. Maui Mitchell B field projection.

a limited portion of the beta-ray-excited region near the northern magnetic conjugate. (Beta excitation occurred mainly at altitudes which were below the horizon for observers on Maui.) In the first photograph (Figure 2a) the expansion still retains a degree of symmetry about the vertical axis. In the later photographs this symmetry has disappeared, and the motion is largely collimated along the direction of the *B* field. The region below 220 km altitude above Johnston Island is obscured by the cloud bank. Horizontal dimensions can be identified out to a radius of about 200 km, and beyond this distance it is difficult to distinguish horizontal motion from motion along the *B* field.

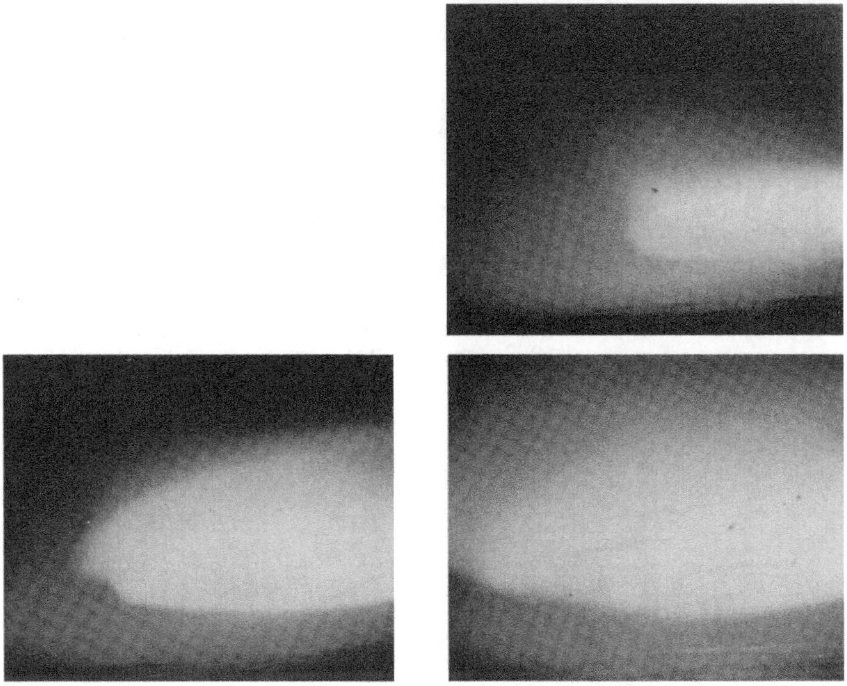

Fig. 4. View of Northern Conjugate Region from Aircraft; (a) 0.1 seconds, (b) 0.15 seconds, (c) 0.2 seconds.

Figures 4a, 4b, and 4c are early photographs taken from the LASL aircraft north of the conjugate region. Figure 5 is the corresponding *B* field projection. The expansion of the bomb debris itself is largely masked by the more intense excited air at the conjugate, which is in the foreground. The excitation at the conjugate was due initially to beta electrons which spiralled along the field lines at approximately the speed of light. The shape of the beta-excited region is defined by the projection of the radiating debris cloud onto the magnetic field, and thus provides a means of determining the shape of the debris cloud. The

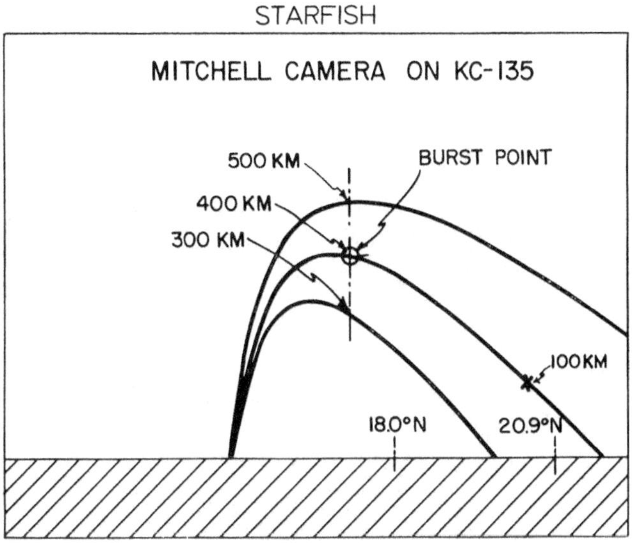

Fig. 5. Aircraft Mitchell B field projection.

altitude of maximum beta excitation was approximately 88 km – somewhat higher than would be predicted on the basis of single-particle interactions. Dimensions inferred from the size of the beta patch are in good agreement with those obtained from the Maui photographs. The maximum diameter attained by the beta patch was somewhat less than 1000 km. At approximately 0.2 sec after burst a very marked increase in brightness occurred, corresponding to the arrival of bomb debris in the conjugate region. A corresponding brightening occurred in the southern conjugate region at about 2.0 sec, coincident with the arrival of radioactive debris (D'Arcy and Colgate, 1965).

The data thus lead to a picture of the debris expansion which sets limiting radii of somewhat less than 500 km for the horizontal east-west directions and slightly more than 180 km for the downward direction. The expansion north and south directly along the B field was not inhibited at all.

The very tight collimation along field lines is dramatically illustrated in photographs taken from Canton Island looking north. Figure 6 contains two photographs taken from Canton Island at approximately 30 sec and 60 sec after burst, and kindly furnished to us by EG+G, Boston. Because of the proximity of Canton Island to the burst magnetic meridian, the pertinent field lines pass almost directly overhead. The burst point was 0.6° above the geometric horizon, but obscured by clouds. The width of the luminous region was approximately 30 km – close to the dimensions of an ion cyclotron orbit; moreover this

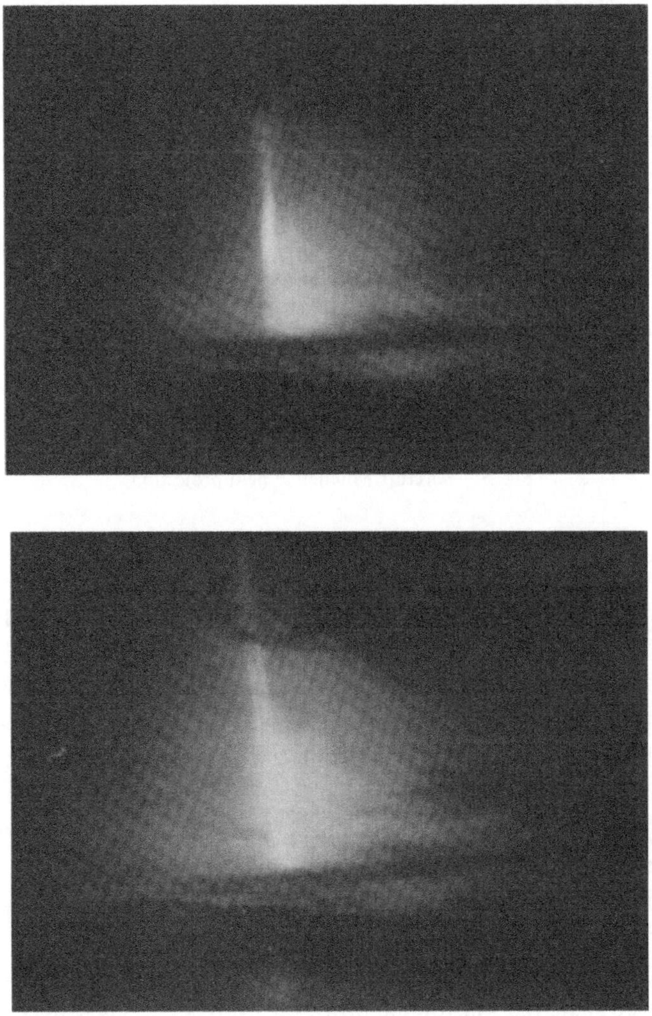

Fig. 6. Canton Island; (a) 30 seconds, (b) 60 seconds.

Fig. 7. Christmas Island at 10 seconds.

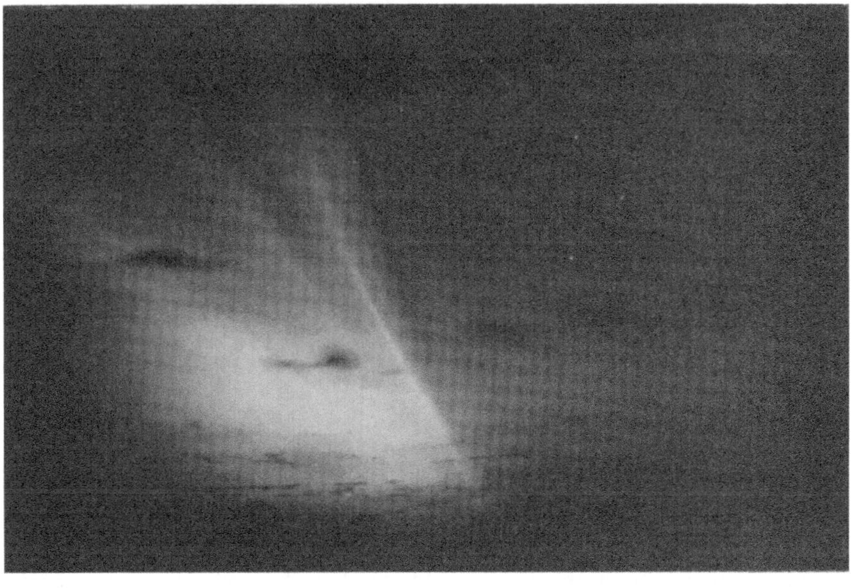

Fig. 8. Christmas Island at 60 seconds.

width was essentially constant down to the visible horizon – i.e. to within 300 km of the burst point. The luminous band was observed to drift toward the west with a velocity of about 0.5 km per second.

Figures 7, 8, and 9 are photographs of about the same region in space, but taken from Christmas Island. The times are approximately 10, 60 and 180 sec

Fig. 9. Christmas Island at 180 seconds.

after burst. Figure 10 is the magnetic field projection. The burst point was 1.3° above the geometric horizon, but again obscured by clouds. We shall need to elaborate on these photographs later, but for the moment note only that the luminous region lies entirely below and to the left of a well defined straight line which is tangent to the magnetic field direction at a point 200 km above the burst.

The formation and westward motion of the narrow streamers was also observable directly overhead from Johnston Island. Spectra showed that the yellow luminescence was due to excited air. The westward motion still continued one minute after detonation, giving proof that ionization persisted for that long.

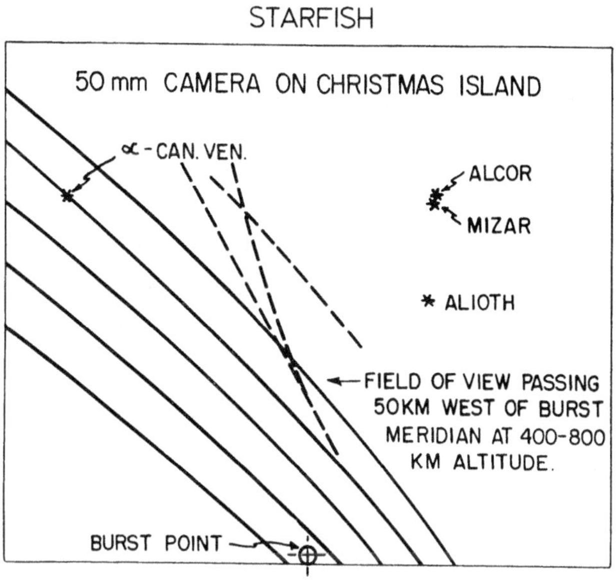

Fig. 10. Christmas Island B field projection.

Let us return to the mechanics of the debris expansion. The calculations will be based on a fictitious bomb, which will be assumed to be spherically symmetric, with a mass of 10^6 g, and made entirely of iron. The debris expansion will be taken to be shell-like, with an initial velocity of 1500 km per sec, giving a kinetic energy of 1.1×10^{22} ergs.

The air nighttime density at 400 km altitude is about 4.2×10^{-15} g/cc with a scale height of 61 km. The magnetic field strength is 0.3 G and varies inversely as R^3, with R being the distance from the center of the earth. The ambient nighttime ion density at 400 km altitude is about 2×10^5 per cm^3, corresponding to a mass density of ions of 3×10^{-18} g/cc. The ion density is maximum at 400 km altitude.

An iron ion ($A = 26$, $M = 56$) moving at 1500 km per sec has a kinetic energy of 650 keV, and in a B field of 0.3 G it has a cyclotron radius of $29/Z$ km if Z is the charge. An electron accompanying the ion has a drift kinetic energy of

6.4 eV and a cyclotron radius of 29 cm. The stopping range for a 650 keV iron atom in neutral air at 4.2×10^{-15} g/cc comes out to be 8×10^5 km. Therefore collisions *alone* would be ineffective in stopping the debris (except in the direction parallel to B). Collisions are nevertheless important because they produce slow-moving ions which are coupled to the MHD shock by electromagnetic forces, causing it to slow down. Cross sections for charge exchange collisions are typically in the range 5×10^{-16} to 2×10^{-15} cm^2, so that the corresponding mean free paths at 400 km altitude are 35 to 150 km. Some details on the assumed cross sections are given in Appendix I.

In the early stages when the expansion velocity is larger than the Alfvén speed (about 380 km/sec for an ion density of 2×10^5 per cc), a good approximation to the motion of the MHD shock is obtained with a so-called "snow-plow" model, which assumes that all of the ions in the shock move with the same velocity , v. If M is the mass per unit solid angle moving with the MHD shock in some given direction, and if $-\dot{P}$ represents the rate of loss of ion momentum due to neutralization of fast ions in charge-exchange collisions, then we can write

$$\frac{\mathrm{d}(Mv)}{\mathrm{d}t} = M\frac{\mathrm{d}v}{\mathrm{d}t} + v\frac{\mathrm{d}M}{\mathrm{d}t} = -\dot{P}$$

The mass M changes as the result of sweeping up (1) ions present in the stationary air, and (2) ions produced in ionizing collisions, and as the result of exchanging heavy ions for lighter ones in charge exchange collisions. Let σ_x be the effective charge exchange cross section and σ_i be the collisional ionization cross section. Let M_n be the average molecular weight of neutrals ahead of the shock and m_i the molecular weight of ions in the shock. n_n and n_i are the number densities of neutrals and of ions in the shock, and ρ_p is the mass density of ions ahead of the shock. Then the shock motion can be described approx imately by the coupled equations:

$$M\frac{\mathrm{d}v}{\mathrm{d}t} = -v\frac{\mathrm{d}M}{\mathrm{d}t} - \sigma_x n_n v M v$$

$$\frac{\mathrm{d}M}{\mathrm{d}t} = R^2 v\rho_p + (\sigma_i n_n v) M\frac{m_n}{m_i} + (\sigma_x n_n v) M\frac{(m_n - m_i)}{m_i}$$

$$\frac{\mathrm{d}N}{\mathrm{d}t} = (\sigma_i n_n v) N + 6 \times 10^{23} R^2 v\rho_p/m_n$$

$$R = R_0 + \int_0^t v\,\mathrm{d}t \qquad m_i = 6 \times 10^{23} M/N$$

with the initial conditions $M = M_0$ (the mass of the bomb), and $N = 6 \times 10^{23}$ $M/56$ the number of atoms in the bomb, and $v = v_0$ the given initial velocity.

The air ion density ahead of the shock is greater than the ambient plasma density as a result of the absorption of x rays produced in the explosion. It will be shown that this is unimportant at altitudes above 300 km, but very important at low altitudes. The mass absorption coefficient of air above the K edge ($\sim$400 eV) is approximately $4.3 \times 10^3/(hv)^3$ cm^2/g if hv is the quantum energy in keV. Then if it is assumed that the thermal radiation from the bomb case can be represented by a Planck function characterized by a constant temperature T, it can be shown that the energy deposited in the air per unit mass is

$$E \cong 53 \cdot \frac{W}{R^2 (kT)^3} \left[\frac{0.4}{kT} - \log_e (e^{0.4/kT} - 1) \right] \text{ergs/g}$$

where W is the total energy radiated, R is the distance in centimeters, and kt is the radiating temperature in keV. (This equation assumes that $kT > 400$ eV and R is less than a typical X-ray mean free path.) If the radiating temperature exceeds 1 keV, no significant attenuation occurs above 200 km altitude; the energy deposited per megaton of energy radiated is everywhere less than 2.5×10^{24} ρ_N/R^2 ergs per cc, and the X-ray produced ion density is less than 5×10^{34} ρ_N/R^2 ions (or electrons) per cc. Above 300 km altitude the mean free path for production of secondary electrons by the primary X-ray photoelectrons is very long, and the primaries simply escape along the field lines without increasing the local ion density. (The first primary photoelectrons from Starfish were observed to arrive in the northern conjugate region at $H + 60$ msec.) Therefore the X-ray-produced enhancement of ion density above 300 km altitude is negligible.

Below 250 km altitude the X-ray-produced ionization is definititely not negligible, and is an important factor in stopping the downward moving MHD shock. The ion density just prior to the arrival of the shock is determined not only by the density of energy deposition by the X-rays, but also by the amount of electron-ion recombination which will have occurred. If the electron density produced by the X rays at $t = 0$ is n_0 per cc, then after time t it is $n = n_0/(1 + n_0 \alpha t)$, where α is the recombination coefficient, equal to about 5×10^{-7} cm^3/sec. At 200 km altitude, n_0 should be of the order of 5×10^7 ions per cc, and by 0.2 sec it should have decreased to about 1×10^7 – still about 5000 times the ambient value. At 250 km the picture is rather different. The X-ray produced electrons have mean free paths of 130 km and longer, and the number of ions produced locally per X-ray photon absorbed is rather small. We shall assume $n \sim 1 \times 10^6$ ions per cc.

The moving debris and air plasma necessarily develops an electric field $\mathbf{E} = -\mathbf{v} \times \mathbf{B}$ (in electromagnetic units). Since $\mathbf{v}$ is radially outward, $\mathbf{E}$ is in the azimuthal direction. Depending on the degree of azimuthal symmetry of the expansion this E field may be associated with local concentrations of polarization charges, or else simply with the changing magnetic field via the Maxwell equation

$$\nabla \times \mathbf{E} = -\frac{\partial B}{\partial t}.$$

In the frame of the moving plasma the force due to the E field exactly cancels the Lorentz $\mathbf{v} \times \mathbf{B}$ force. This situation is disturbed if there are collisions with stationary neutral molecules. Upon being stopped by a collision an electron or ion feels only the E field, whereupon it falls down the E field until it reacquires the drift energy. The particles falling along the E field constitute a conduction current. Without collisions there was no net force on the particles, and, therefore, no conduction current in the usual sense. The interaction of the conduction currents j_c with the B field gives rise to a $j \times B$ force which retards the moving plasma. If one uses the familiar Pedersen expression for the conductivity, viz.

$$\sigma = ne^2 \left[\frac{1}{m_+ \nu_+} \left(\frac{\nu_+^2}{\nu_+^2 + \omega_+^2} \right) + \frac{1}{m_e \nu_e} \left(\frac{\nu_e^2}{\nu_e^2 + \omega_e^2} \right) \right]$$

(where ν_+ and ν_e are the frequencies of collisions of ions and electrons respectively with neutral molecules and ω_+ and ω_e are the cyclotron frequencies) then it can be shown that when $\omega_+ \gg \nu_+$ and $\omega_e \gg \nu_e$ the $j_c \times B$ retarding force is identical to the momentum loss by the plasma per cubic centimeter per second due to the collisions.

Since the air density decreases with increasing altitude, the upward expansion naturally is decelerated less rapidly than the downward expansion. By 0.25 sec the computed radius of upward-moving debris is 250 km, and the velocity is 800 km per sec. The downward debris at the same time has progressed to 170 km from the burst point and the velocity has dropped to about 60 km per sec. Because of the greater velocity in the upward direction, the debris and shock system must then tend to become polarized, with positive charges concentrating on the west side and electrons on the east. This polarization is necessary for the existence of the E field $\mathbf{E} = -\mathbf{v} \times \mathbf{B}$, but it tends to be shorted out by polarization and conduction currents flowing in the lower hemisphere. The currents produce a large viscous-like drag on the upward motion. Through the same process the downward motion in the lower hemisphere tends to be reversed, and the center of gravity of the entire system dragged upward. The details of these processes seem unfortunately to be too complicated to calculate. Many parallel current paths are involved, and the currents have a peculiar time dependence.

Ultimately the E fields are best shorted out by currents flowing north-south along the field lines and east-west in the E and F1 layers. These current paths begin to participate after a time of the order of the Alfvén wave transit time from the location of the upward-expanding debris to the ionospheric F1 layer along the path which follows the field lines. The transit time is roughly 1 second. It will be shown that the slowing-down time associated with this type of current flow is approximated by

$$\tau_c \equiv \frac{v}{\mathrm{d}v/\mathrm{d}t} \cong \frac{\int_{l_0}^{l_{\max}(t)} \rho_P \,\mathrm{d}l}{B^2 \bar{\Sigma}_y}$$

where $\bar{\Sigma}_y$ is the height-integrated east-west conductivity of the ionosphere following the notation of MAEDA (1953), and $\int_{l_0}^{l_{\max}} \rho_p \mathrm{d}l$ is the integral of farthest point reached by transverse Alfvén waves from the burst at the time t. If Maeda's value of 2×10^{-8} emu-cm is used for $\bar{\Sigma}_y$, and $\int \rho_p \mathrm{d}l$ is taken to be 2×10^{-10} gm-cm (assuming an average plasma density of 5×10^{-18} gm/cc and a distance of 400 km), then we obtain $\tau_c = 0.1$ seconds. The physical basis for this number is as follows:

The polarization charges which accumulate on the east and west faces of the bubble as the result of its unbalanced expansion in the upward direction tend to spread out along the field lines. As the charges spread along the field they create an east-west E field in places where the air plasma is initially at rest. At every point the rising E field results in a polarization current and a corresponding acceleration of the plasma. This polarization current flow regulates the spreading-out of charge along the field so that it proceeds with a well-defined velocity approaching the Alfvén velocity. In this way the perpendicular momentum of the expanding debris is transferred laterally along the field to an ever-increasing mass of ambient plasma, and the velocity of the debris itself must therefore decrease. If the velocity of the wave is a decreasing function of distance along the direction of propagation, then the wave front will tend to remain steep. Behind the front the plasma velocity and the associated E field (both normal to B) follow the variations of velocity and E field in the upward-moving debris. When the wave front reaches the F1 layer of the ionosphere the signal begins to be damped by collisions with neutral air and (equivalently) the E field begins to be drained by conduction currents. A quasi-steady state is approached wherein the flow of conduction currents in the F1 layer and below are balanced by a corresponding flow of polarization currents in the moving collisionless plasma above the F1 layer. The E field which drives the conduction currents is associated with the velocity of the moving plasma via $\mathbf{E} = -\mathbf{v} \times \mathbf{B}$, while the polarization currents are associated with its deceleration via

$$\mathbf{j}_p = -\frac{\rho_P}{B^2}\frac{\mathrm{d}v}{\mathrm{d}t} \times \mathbf{B}.$$

The total polarization current flowing between the east and west faces of a very long tube of 1 square centimeter cross section aligned with the B field is

$$I_P = \int_{l_0}^{l_{\max}(t)} j_P \, \mathrm{d}l$$

where l_0 is a point near the top of the F1 layer and $l_{\max}(t)$ is the farthest point at which the plasma is in motion. We shall assume that B, v, and $\mathrm{d}v/\mathrm{d}t$ are all constant between $l_{\max}$, and that the conductivity along the field lines is infinite. Then the current becomes

$$I_P = -\frac{1}{B}\frac{\mathrm{d}v}{\mathrm{d}t}\int_{l_0}^{l_{\max}} \rho_P \, \mathrm{d}l \, .$$

At equilibrium this current must be balanced by an equal and opposite flow of conduction current. The conduction current is $I_C = E\bar{\Sigma}_y = -vB\bar{\Sigma}_y$. Setting $I_P = -I_C$, we have

$$\frac{\mathrm{d}v}{\mathrm{d}t} = -\frac{B^2 v \bar{\Sigma}_y}{\int_{l_0}^{l_{\max}} \rho_P \, \mathrm{d}l}$$

which is equivalent to the equation for τ_c written previously. Thus through the conduction currents the energy of the upward expansion is eventually transferred to neutral particles in the lower ionosphere.

The strength of the MHD shock perpendicular to the B field is further depleted by the loss of plasma along the field lines. There is as yet no adequate calculation of this effect, as the particle losses depend in a critical way on the shapes of the B field and the MHD shock surface near the ends of the bubble where they are hardest to predict. Some losses are certainly to be expected, and they are definitely indicated by the data. The present calculations neglect these effects.

The applicability of the simple snowplow model is limited to times when the debris velocity is much greater than Alfvén speed. For later times it is possible to do better with a Lagrangian finite difference procedure coded for a computer. This step has been taken, and Figure 11 shows some of the results.

The code is designed to compute the hydrodynamic behavior of the expansion in each of several radial directions separately, ignoring any interactions between them. For each separate direction the calculation is done as though the system remained spherically symmetric – i.e., all quantities are taken to be functions of R and t alone; however, the initial values of plasma density, density of neutrals, field strength, etc., are different functions of R for the different directions. The debris mass and initial radial momentum are the same for all directions. Some details of the code are described in Appendix I.

Figure 11 shows the computed positions of the outermost debris cell boundry at various times in a vertical plane through the burst field line.

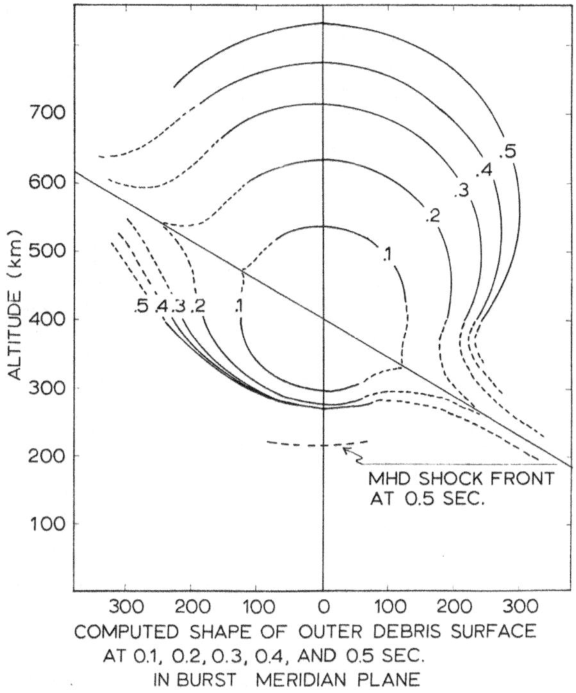

Fig. 11. Computed Shape of Outer Debris Surface at 0.1, 0.2, 0.3, 0.4 and 0.5 seconds in Burst Meridian Plane.

Since, as we have already pointed out, the various directions of expansion are all interdependent, we make no strong claims for the applicability of this code which treats them as independent. The intent of the code is to provide a picture of the early development of asymmetries in the expansion, so that the two-dimensional effects can be intelligently patched in. Since both the interaction between directions of motion and the particle losses along field lines would tend to reduce the perpendicular velocities, the dimensions and velocities calculated with this code can be regarded as upper limits. By $H+0.5$ seconds the computed velocities are all well below Alfvén speed, so that the coupling effects should have become strong. The perpendicular dimensions should not have increased appreciably beyond this point.

On the basis of these calculations the fact that the horizontal debris dimensions did not exceed 500 kilometers is quite reasonable. It is also to be

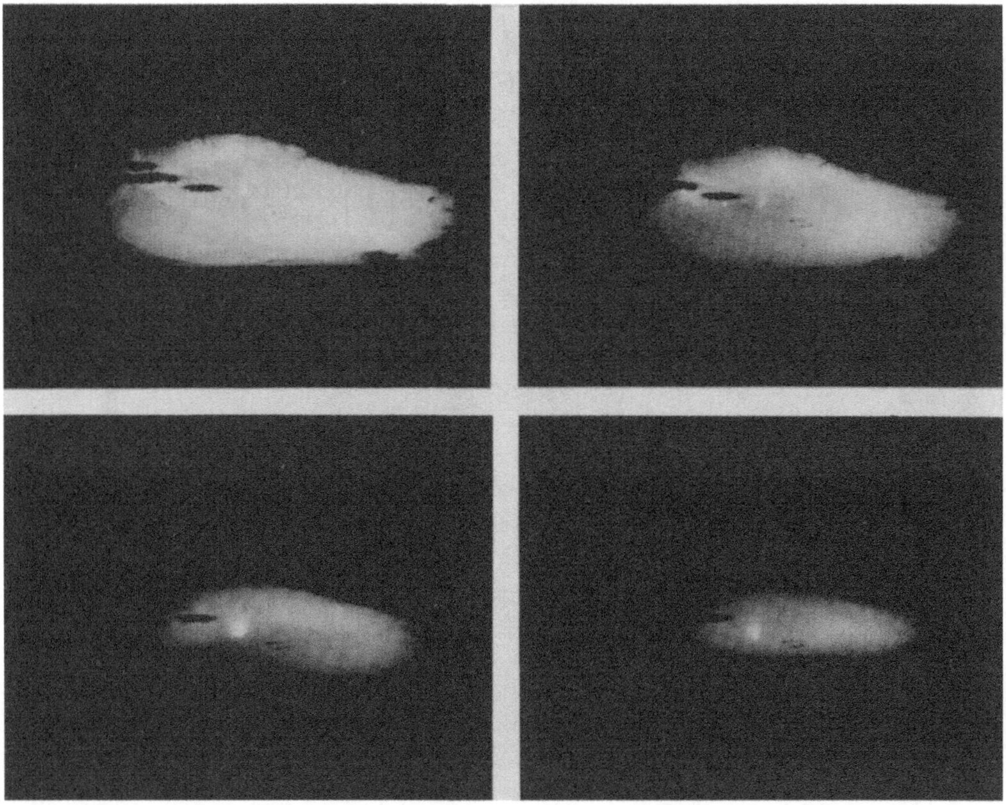

Fig. 12. (a) Maui at 3 seconds, (b) Maui at 8 seconds, (c) Maui at 16 seconds, (d) Maui at 24 seconds.

expected that after about one second a recollapsing phase should have begun.

Let us return to some more of the data.

Figure 12 is a sequence of photographs taken from Maui at 4, 8, 16, and 24 seconds. They clearly show a downward-moving shock which converges on the burst point with a downward velocity of approximately 40 kilometers per second. The location of the burst point is marked by the remains of the booster rocket. The shock arrives at the burst point at about $H+14$ seconds and overshoots. The velocity and timing of this converging shock serves as further evidence that the upward expansion was stopped and turned around at a point less than 500 kilometers above the burst, and moreover that dissipative processes (conduction currents etc.) acted to prevent the debris from returning at a velocity

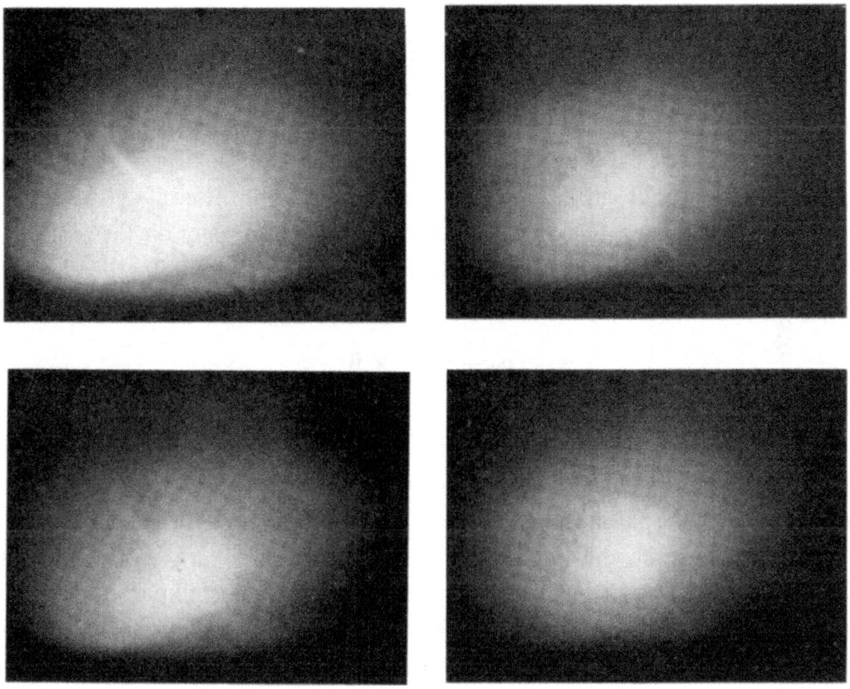

Fig. 13. View from Aircraft; (a) 4 seconds, (b) 8 seconds, (c) 12 seconds, (d) 16 seconds.

at all comparable with the velocity at which it moved out. Figure 13 is a sequence of photographs from the KC-135 aircraft covering the same time span. The downward shock is also visible in these photographs, though less clearly than in the Maui photos.

While both the data and the calculations indicate that the vertically upward-moving debris stopped at a distance of less than 500 km above the burst, it is also a known fact that a large quantity of fission debris went to extremely high altitudes. The nature of the process by which this high altitude injection occurred is shown fairly clearly by the Christmas Island photographs (Figures 8, 9 and

10) together with the Canton Island photographs (Figures 6a, b). The Canton Island photographs show that a jet of debris moved straight along the magnetic meridian. The Christmas Island pictures show the remains of the same jet as viewed from the side. It is apparent that a part of the jet was undeflected by the bending field lines, but coasted in a straight line across the field for several hundred kilometers (until it passed out of the field of view of the camera). As it travelled it lost material into the field at a relatively constant rate, resulting in a high but thin curtain-like formation. Between 10 and 60 seconds after burst the higher southern end of the formation drifted upwards across the field. As viewed from below (Canton Island) it drifted toward the west.

The motion of a jet of plasma in a bending magnetic field is subject to the same sorts of influences as the vertically-moving debris which we have already discussed. As the field bends away from the direction of motion of the plasma a small east-west current flows. The current density controls itself at a level just sufficient to maintain the correct polarization charge density on the east and west sides of the plasma mass, so that $\mathbf{E}_1 = -\mathbf{v}_1 \times \mathbf{B}$. The current flow decelerates the plasma via

$$\rho_P \frac{dv}{dt} = -\mathbf{j} \times \mathbf{B}.$$

If no process exists whereby the surface polarization charge can leak away, then j becomes extremely small. (It becomes zero if the angle between $\mathbf{v}$ and $\mathbf{B}$ does not change.) Then the plasma mass tends to coast indefinitely in its original direction.

In fact, however, the surface charge can leak off in a variety of ways, and a current must flow in the plasma to replenish it. The most important leakage process is one which we have discussed previously; namely, the leakage along field lines leading to generation of transverse Alfvén waves. Other leakage currents arise from gradients of plasma density and field strength produced through the displacement of ambient plasma.

The current required to maintain the E field is independent of the plasma density. From

$$\frac{dv}{dt} = -\frac{1}{\rho_P}\mathbf{j} \times \mathbf{B}$$

it follows that the less dense portions of the plasma mass are decelerated fastest. As the jet of ionized debris moved across the bending B field, its thin surface layers were progressively peeled off – i.e., their motion perpendicular to B was stopped and they became constrained to follow the field lines. The progressive sloughing off of plasma into the field accounts for the high curtain formation seen in the photographs, as well as the elongated distribution of debris in the southern conjugate region reported by D'Arcy and Colgate (1965).

The observed westward drift of the curtain, as seen from Canton and Johnston Islands, was due to gradient B and curvature drifts of energetic positive ions. The drift velocity is proportional to the particle kinetic energy, and it was measured to be about 0.5 km/sec. The electrons which accompany the positive ions drift to the east, but at a negligibly small velocity. The westward drifts of the positive ions constitute an electric current which tends to polarize the plasma mass. The polarization produces an E field directed west to east, and a corresponding $E\times B$ drift of the plasma as a whole, directed upward. If not checked by conduction currents, etc., this upward $E\times B$ drift would become very large; however, it is limited by conduction currents in the lower ionosphere which drain off the polarization charge.

The limiting upward drift velocity can be calculated if the total westward drift current is known, as well as the conductivity of the ionosphere. Since charge densities will be small everywhere, we have $\nabla\cdot j\approx 0$, which says that the westward drift current is exactly balanced by the conduction current in the lower ionosphere. If we consider once again a tube of unit square cross section lying along the B field between the two conjugate points and containing N singly-charged ions which drift to the west with velocity v_w, then the total drift current is $I=Nev_w$. The east-west resistance of the ionosphere is $R=1/\bar{\Sigma}_y$, where $\bar{\Sigma}_y$ is the height-integrated conductivity. Then the E field is $Nev_w/\bar{\Sigma}_y$ and the corresponding upward drift velocity is $v_{up}=E/B=Nev_w/B\bar{\Sigma}_y$. It will be difficult of course to make a realistic guess as to possible values for N. Let us return to the hypothetical bomb which weighs 10^6 grams and is made of iron, and suppose that it produces a jet comprising one tenth of its mass, or 10^{27} ions. Suppose that when the jet is finally stopped the 10^{27} ions have become uniformly distributed over a projected area of 10^4 square kilometers. If the ions now drift west at 0.5 km/sec, and the height integrated conductivity is 2×10^{-8} emu (Maeda, 1953), the corresponding upward velocity is $v_{up}\approx 13$ km/sec. This is of the same order of magnitude as the upward drift observed. The possibility of such an upward drift was discussed in classified literature by C. L. Longmire at the time of the Argus tests. Some interesting laboratory experiments pertinent to these observations have been reported recently by Baker and Hammel (1965).

By 3 minutes after burst an appreciable quantity of debris from the jet had mirrored in the southern conjugate region and returned northward. Figure 14 is a photograph taken from the KC-135 aircraft at $H+3$ minutes. Figure 15 is the corresponding B field projection. At these late times the yellow streaks conformed rather closely to the normal undisturbed B field geometry. The highest visible streak follows a field line 2000 km above Johnston Island. The yellow-emitting region cuts off rather sharply at latitude 18 °N, or about 170 km north of Johnston Island. The red emission appears to be due to the long-lived oxygen 6300 line, excited by collisions produced in the hydromagnetic shock.

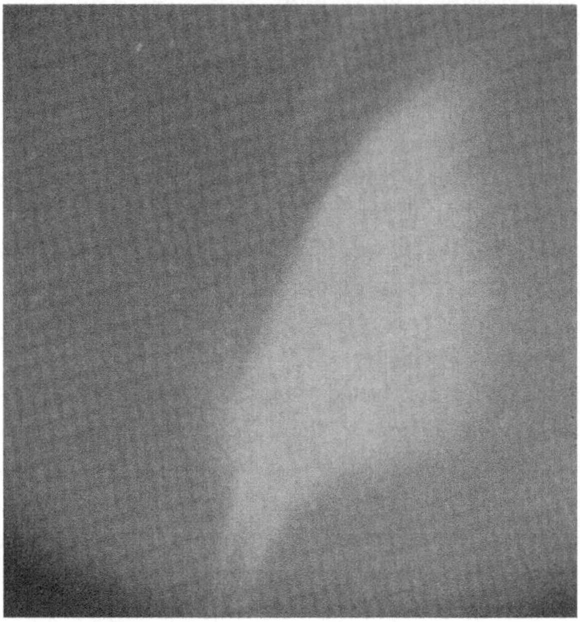

Fig. 14. View from Aircraft at 180 seconds.

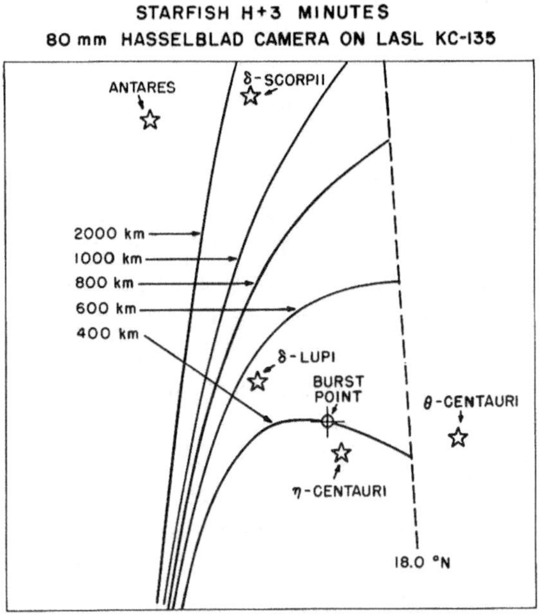

Fig. 15. Aircraft Hasselblad B Field Projection.

Appendix

The computer code is designed to integrate the following simplified MHD equations:

$$\rho_P \frac{dv}{dt} = -\frac{d}{dr}\left(\Delta P_\perp + \frac{\Delta(B_\perp^2)}{8\pi} + Q\right) - KB_\perp^2 v - \left(\rho_P n_n v^2 \frac{m_n}{m_i}\right)(\sigma_x + \sigma_i) \tag{1}$$

$$\frac{d\rho_P}{dt} = -\rho_P \frac{dv}{dr} - 2\frac{\rho_P v}{r} + \left(\frac{\rho_P}{m_i} n_n v\right)[\sigma_x(m_n - m_i) + \sigma_i m_n] \tag{2}$$

$$\frac{dB_\perp}{dt} = -B_\perp \frac{dv}{dr} - 2\frac{B_\perp v}{r} \tag{3}$$

$$\frac{dP_\perp}{dt} = \frac{2P_\perp}{B}\frac{dB_\perp}{dt} + \varepsilon_0\left(\frac{\rho_P}{m_i} n_n v\right)[\sigma_x(m_n - m_i) + \sigma_i m_n] \tag{4}$$

where d/dt is the Lagrangian time derivative. The code assumes spherical symmetry, although the initial profiles of $B_\perp$, ρ_P, n_n, etc. may be chosen to represent any specified direction. $B_\perp$ is the component of **B** perpendicular to the radius vector from the burst point. $P_\perp$ is the part of the plasma pressure arising from velocity components normal to **B**. $\Delta P_\perp$ is the difference between $P_\perp$ and its ambient value, and $\Delta(B_\perp^2)$ is the difference betwene $B_\perp^2$ and the ambient value. K is the local conductivity defined by the Pedersen expression previously written down. Q is a Richtmeyer-Von Neuman viscosity pressure term, necessary for computational reasons. ε_0 is a constant intended to represent the random kinetic energy per gram of ions produced in collision processes. The remaining quantities, ρ_P, n_n, v, m_i, σ_x, and σ_i, have all been defined previously.

The collisional ionization cross section, σ_i, is assumed to be given by a formula due to Firsov (see FEDORENKO, 1959), viz:

$$\sigma = \sigma_z\left[(v/v_z)\frac{1}{5} - 1\right]^2$$

with

$$\sigma_z = \frac{32.7 \times 10^{-16}}{(z_1 + z_2)^{2/3}} \text{ cm}^2$$

and

$$v_z = \frac{23.3 \times 10^6 E_i}{(z_1 + z_2)^{2/3}} \text{ cm/sec}$$

where z_1 and z_2 are the nuclear charges of the ion and atom respectively, and E_i is the first ionization potential of the gas atom in eV. Using values appropriate for 0^+ on 0, $\sigma_z = 5.1 \times 10^{-16}$ cm^2, and $v_z = 5.0 \times 10^7$ cm/sec.

The charge exchange cross section is assumed to be represented by the relation (DALGARNO, 1958)

$$\sigma^{\frac{1}{2}} = - k_1 \ln v + k_2$$

fitted to data for charge exchange of 0^+ with 0 (STEBBINGS *et al.*, 1963; also FITE, 1964). Thus we take $k_1 = 5.2 \times 10^{-9}$, and $k_2 = 1.24 \times 10^{-7}$.

In its present state of refinement the code does not attempt to take account of ionization due to collisions of fast neutrals produced by charge exchange encounters.

The effects of electron collisions and of resonance charge exchange collisions at low energies are lumped together in the conductivity term $kB_\perp^2 v$. k is the Pedersen conductivity previously defined, and the collision frequencies ν_{eN} and ν_{+N} are taken to be $7.2 \times 10^{-10}\, n_n \sqrt{T_e}\,\sec^{-1}$ (NICOLET, 1959) and $8.0 \times 10^{-11}\, n_n$ $\sec^{-1}$ (CHAPMAN, 1956) respectively.

It may be noted that if the collisional terms were dropped from Equations (2) and (4), we would obtain the familiar relationships $\rho / B_\perp = \text{const}$, and $P_\perp / B_\perp^2 = \text{const}$ (6).

The code employs a rather standard Lagrangian forward difference procedure.

Acknowledgement

A very large number of persons have participated in collection of the data presented in this paper. Special mention is deserved by Drs. H. M. Peek, E. B. Bennett, D. Steinhaus, and T. N. K. Godfrey, and Mr. Donald Barnes. We also wish to thank Mr. W. P. Boquist of Edgerton, Germeshausen, and Grier Inc. for furnishing the Canton Island photographs. The authors are indebted to Dr. C. L. Longmire for many helpful discussions.

This paper is based on work performed under the auspices of the U.S. Atomic Energy Commission.

References

BAKER, D. and HAMMEL J.: 1965, *Phys. Fluids* **8,** 713.
CHAPMAN, S.: 1956, *Nuovo Cimento,* **4,** (Suppl.), 1385.
COLGATE, Stirling A.: 1965, *J. Geophys. Res.* **70,** 3161.
DALGARNO, A.: 1958, *Phil. Trans. Roy. Soc.* **A250,** 426.
D'ARCY, Raymond G. and COLGATE, Stirling A.: 1965, *J. Geophys. Res.* **70,** 3147.
FEDORENKO, N. V.: 1959, see *Sov. Phys.-Usp.* **2 (68),** 526–546.
FITE, W. L.: 1964, Ann. Geophys. **20,** 000.
LONGMIRE, C. L.: 1963, *Elementary Plasma Physics,* Interscience Publishers.
MAEDA, H.: 1953, *J. Geomag. Geoelect.* **5,** 94.
NICOLET, M.: 1959, *Phys. Fluids* **2,** 95.
STEBBINGS, SMITH, and EHRHARDT, 1963, Proceedings of the Third International Conference on the Physics of Electronic and Atomic Collisions, London (to be published).
WOBSCHALL, GRAHAM, and MALONE: 1963, *Phys. Rev.* **131,** 1565.

ENERGETIC ELECTRONS FROM SHOCK HEATING IN THE EXOSPHERE

STIRLING A. COLGATE

New Mexico Institute of Mining and Technology, Socorro, New Mexico

Abstract. A significant fraction of the artificial radiation belt formed by the July 9, 1964, Starfish high-altitude nuclear explosion may have been caused by electrons heated in a magnetohydrodynamic shock wave – accelerated in the density gradient of the earth's atmosphere. Calculation of the expected fission beta electron flux from the measured fission debris decaying in transit to the southern conjugate region agrees well with the observed low *L* value (1.22 – 1.32) measurements of Injun I satellite. The measurements (Telstar) of the electron flux at higher *L* do not agree with the observed southern conjugate debris. The shock-injected electrons are predicted to be more than sufficient to account for the discrepancy, but there remains a major uncertainty in shock-dissipation (electron heating) mechanisms.

Introduction

The high-altitude Starfish nuclear explosion of July 9, 1962, was the source of an artificial radiation belt formed in the earth's magnetosphere. Many of the measurements of this radiation belt by the satellites Injun, Telstar, Traac, and Explorer XIV gave evidence for the presence of high energy electrons injected onto the magnetic flux surfaces of the earth's magnetosphere up to several earth's radii (*L* value) from the surface of the earth. The first synthesis of flux contours by HESS and NAKADA (1962) from the Telstar and Injun data has been widely disputed by the original experimental evidence (BROWN and GABBE, 1963; VAN ALLEN *et al.*, 1963). In these discussions, the different sensitivity of the detectors (low energy for Telstar and high energy for Injun) led to possible differences of interpretation. However, with the further calibration of the bremstrahlung sensitivity of the high-energy detector to low-energy electrons, a fairly reasonable reconciliation of the data was made. Nevertheless, the spectrum measured by Telstar was significantly softer (lower energy) than the equilibrium fission spectrum of Beta electrons (BROWN and GABBE, 1963). Theoretical estimates (COLGATE, 1965) would predict that a very small fraction $\leqslant 1$ per cent of neutral fission debris escapes to high *L* value in the neutral state, and so the fission debris deposited at the southern magnetic conjugate point (D'ARCY and COLGATE, 1965) gives a measure of the maximum possible injected beta spectrum on trapped orbits of the magnetosphere. The total debris as a function of *L* value is shown in Figure 1 along with a calculated required deposition distribution to explain the high *L* value contours of the

McCormac (ed.), Radiation Trapped.

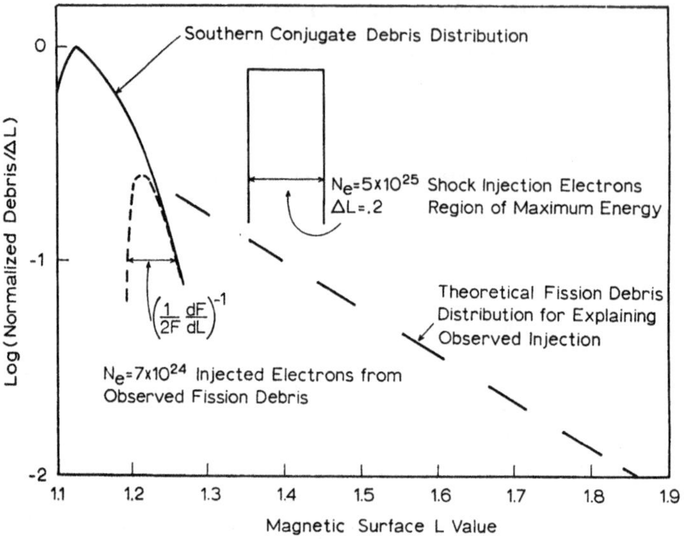

Fig. 1. Observed southern conjugate debris distribution and expected electron injection from debris and shock heating.

Telstar data. It is evident that an extrapolation of the ionized debris injection mechanism cannot explain the observed electrons at high L value. This paper explores the possibility that the electrons observed at high L value are due to a shock hydrodynamic origin rather than a radioactive decay origin.

Fission Debris Injection

The beta injection expected from the observed distribution of debris (D'Arcy and Colgate, 1965) can be calculated in the following manner: Y=fission yield in fission/megaton, F=fraction of debris observed in the southern conjugate region south of the cut-off L value for confinement ($L \geqslant 1.22$), f=the fractional beta decay in transit, V=volume of flux surface, ε=trapping fraction of electrons injected in transit.

Since only those betas injected during transit of the debris will be magnetically confined, the trapped electron density N_e for a nominal one megaton of fission becomes

$$N_e = \frac{YFE\varepsilon}{V} \text{ electrons/cc} \tag{1}$$

and the flux

$$\Phi = 0.7\, cN_e \text{ cm}^{-2} \text{ sec}^{-1} \tag{2}$$

for $E > 200$ keV.

$Y=1.5\times10^{26}$ fission/megaton and from the distribution of southern conjugate debris as a function of L (D'ARCY and COLGATE, 1965), $F=0.33\ F_0$, where F_0 is the total fraction of fission products observed at the southern conjugate region. From D'ARCY and COLGATE (1965), $F_0=0.27$ so that $F=0.09$. The volume of flux surface becomes

$$V=(2\pi r_0 L)\left(\frac{1}{2Fr_0}\frac{\mathrm{d}F}{\mathrm{d}L}\right)^{-1} l, \tag{3}$$

where l is the length of the flux surface $=0.6\ r_0$, and r_0 = radius of earth. The logarithmic gradient of the debris distribution was related theoretically to the plasma instability mixing length (COLGATE, 1965) and agrees with the experimental observation (Figure 1) so that

$$((1/2Fr_0)(\mathrm{d}F/\mathrm{d}L))^{-1}=0.1\ r_0$$

and so that

$$V=0.44\ r_0^3=10^{26}\ \mathrm{cc}. \tag{4}$$

The fractional beta-decay in transit is porportional to the total betas per fission and to the fraction of fission products having decay periods between 2 and 10 seconds; that is, the mean period of debris transit free of the earth's atmosphere.

This becomes

$$f=5(t_1^{-1/5}-t_2^{-1/5})=1.3\ \text{betas/fission}. \tag{5}$$

The fraction trapped of an isotropically emitted distribution becomes (MCILWAIN, 1961)

$$\epsilon=1/l\int_{-l/2}^{l/2}\left(1-\frac{B}{B_0}\right)^{1/}\mathrm{d}l\simeq0.4. \tag{6}$$

Therefore $\Phi=1.4\times10^9\ \mathrm{cm}^{-2}\ \mathrm{sec}^{-1}$. (7)

This agrees satisfactorily with the Injun I observations of O'BRIEN *et al.*, (1962) of $10^9\ \mathrm{cm}^{-2}\ \mathrm{sec}^{-1}$ flux.

VAN ALLEN *et al.* (1963) estimate from a synthesis of Injun I, Telstar and Explorer XIV data that the trapped flux decreased to 10 per cent at $L=1.8$ and to 1 per cent at $L=2.2$. If we note that the confinement per $(\mathrm{d}L)$ volume increases as L^2 per $\mathrm{d}L$, but that the trapping efficiency ε increases to approximately 0.8 at $L=1.8$ and that the decay fraction $f\simeq2.8$ at $L=1.8$, then the debris density expected in the southern conjugate region at $L=1.8$ corresponds to the dashed curve in Figure 1. This curve has a logarithmic slope $(1/F\ \mathrm{d}F/\mathrm{d}L)^{-1}$

$=0.2$ compared to the measured debris distribution of $(1/F\,\mathrm{d}F/\mathrm{d}L)^{-1}{}_{\text{observed}}$ $=0.05$. This difference of a factor of 4 leads one to suspect a different mechanism for the high L value injection.

Shock Injection

It is recognized that a strong shock wave speeds up and increases in strength in a density gradient. CHISNELL (1955) showed that a plane parallel shock wave speeds up as

$$u = u_0(\rho/\rho_0)^{-\alpha} \tag{8}$$

where

$$\alpha = 1/2\left\{1 - \left[1 + 2\left(\frac{\gamma - 1}{2\gamma}\right)^{1/2}\right]^{-1}\right\}, \tag{9}$$

u is the shock or fluid velocity as a function of density ρ, and u_0 and ρ_0 are the corresponding values at an earlier time and higher density. γ is the ratio of specific heats of the medium. Recently GROVER and HARDY (1964) improved the accuracy of Chisnell's approximation for steep density gradients and by using computing machine calculations verified solutions appropriate to exponential atmospheres. An accurate value of α for these conditions is

$$\alpha = 1/2\left\{1 - \left[1 + 1.6\left(\frac{\gamma - 1}{2\gamma}\right)^{1/2}\right]^{-1}\right\}. \tag{10}$$

This same behavior has been observed asymptotically in supernova shock wave calculations (COLGATE and WHITE, 1965), but in these hydrodynamic machine calculations, the spherical divergence was included. An approximate spherical correction that agrees with their calculations and those of Grover and Hardy for large r becomes

$$u = u_0\left(r\frac{\mathrm{d}M}{\mathrm{d}r}\right)^{-\alpha} = u_0\left[\left(\frac{r}{h}\right)^3 e^{-r/h}\right]^{-\alpha}. \tag{11}$$

The Collision-Free Shock

In the upward direction, the coupling mechanism by plasma ionization (COLGATE, 1965) is ineffective at one scale height (60 km) above this explosion; so, if we exclude the electrostatic shock mechanism because of the large electron-to-ion mass ratio, the only remaining mechanism for shock formation is the collision-free magnetic shock. In the present limit of theoretical speculation, such shocks are limited to an Alvin Mach number less than 2, so that for propagation, the magnetic field must be large enough to give the required pressure, namely

$$(3B)^2/8\pi \geqslant \rho_0 (u_0)^2 . \tag{12}$$

Choosing $u_0 = 2 \times 10^8$ cm/sec of escaped debris and $B = 0.3$ gauss gives $\rho_0 \leqslant 10^{-18}$ g/cm^3.

For a nominal ton of debris, the noninteracted debris density becomes less than 10^{-18} g/cm^3 at $r = 600$ km, so a small fraction of noninteracting debris is sufficient to drive a magnetic shock at the initial piston speed of 2×10^8 cm/sec. Since expression 11 indicates a decreasing shock (or fluid) velocity for $r \geqslant 3h$, we use the constant piston speed of the debris to initiate the speed-up at $r = 3h$. At this altitude ($r = 600$ km, $\rho = 2 \times 10^{-16}$ g/cm^3, $N_0 = 8 \times 10^6$, h = 80 km), the probability of debris ionization of neutral atoms is 1 to 2 per cent (Colgate, 1965) so that an ionization density coupled to the magnetic field of $n_i \leqslant 10^5$/cc, $\rho = 1.5 \times 10^{-18}$ g/cc is assumed. The ambient ionization (F layer) is similarly $n_e = 5 \times 10^4$ e/cc (Johnson, 1961) and the initial shock conditions become $u_0 = 2 \times 10^8$ cm/sec and $r = 3h$. The ambient ionization density continues to decrease to $n_e \simeq 500$/cc at 1100 km altitude with scale height $h = 85$ km and mean molecular weight decreasing from 16 to approximately 3. Above this altitude, the density remains roughly constant because of the large scale height associated with the light, ionized hydrogen. Therefore, the number of scale heights above the shock initiation point becomes

$$\ln \left(\frac{N_i}{500} \frac{16}{3} \right) = 7 \text{ scale heights};$$

therefore,

$$u_f = u_0 \left[\left(\frac{7}{3} \right)^3 e^{-7} \right]^{-\alpha} = (1.3 \times 10^{-2})^{-\alpha} . \tag{13}$$

Assuming that the shock dissipation mechanisms scatter electrons into three degrees of freedom, then $5/3 \leqslant \gamma \leqslant 2$, so $\alpha = 0.21$ and

$$u_f = 2.5\, u_0 = 5 \times 10^8 \text{ cm/sec} . \tag{14}$$

The mean energy of an ion of molecular weight 3 becomes 380 keV, and the total number of accelerated ions becomes the number of ambient ions in a layer one scale height thick and subtending approximately one sterradian from the explosion. Therefore, $N = n_e h r^2$. For the conditions

$$n_e = 500/\text{cc}, \quad h = 85 \text{ km}, \quad r = 700 \text{ km}, \quad N_i = 1.5 \times 10^{25} \text{ions}. \tag{15}$$

Since the ambient fractional ionization is small, the charge per ion will be unity, so the total injected electrons will be equal to the accelerated ions giving

$N_e = N_i = 4 \times 10^{25}$. The energy distribution of these electrons will be dependent upon the heating or acceleration that they undergo in the shock transition so that expression (15) is an upper limit to the possible high-energy electron injection. The electron energy spectrum depends upon the shock-dissipation mechanism.

Shock-Dissipation Mechanism

The mechanism of collision-free magnetohydrodynamic shocks (ADLAM and ALLEN 1958; DAVIS *et al.*, 1958; VEDENOV *et al.*, 1962) depends upon the formation of strong ion wave pulses in which the charge separation electric field accelerates the ions and causes a rapid transverse drift of the electrons. This electron current gives rise to the perturbed magnetic field. Self-consistent models have been machine-calculated by AUER *et al.*, (1962) demonstrating many features of such a shock. In particular, the scattering and coalescence of the strong pulses have taken place in a one-dimensional model in a fashion that gives confidence to a picture of an irreversible turbulence. However, in this model, as in previous theoretical speculations, the electrons have been assumed to be adiabatic. It has been recognized that the rapid transverse electron drift should be unstable to electrostatic oscillations; if these grow to large amplitude before passage of the wave, then the electron transverse energy would be scattered and become irreversible. The rapidity of thermalization was first calculated by BUNEMAN (1959) who demonstrated by computational techniques that effective thermalization occurred after only 1 to 2 large amplitude plasma periods. Recently KELLOGG (1965a) extended the theory to derive a dynamic friction as a function of streaming and mass ratio. When this is applied to the magnetic shock (KELLOGG, 1965b), a predominant heating of the electrons results. Recently, impressive experimental evidence supports this view in the measurement of electron energy distribution at the boundary of the magnetosphere. In general, the electron energy measurements of FAN *et al.*, (1964) demonstrate the shock acceleration of electrons to greater than 30 keV with total energy $\simeq 10^{-2}$ of the solar wind. Since the expected electron temperature from the Rankine-Hugoniot conditions is very much lower $\simeq 200$ eV, it is evident that a significant fraction of the shock energy has been used in electron acceleration above thermal energies. The further evidence of magnetic field turbulence also has been interpreted as a collision-free shock (SONETT *et al.*, 1964) in excellent agreement with the Rankine-Hugoniot relation.

Therefore, it is highly likely that the Starfish accelerated shock should produce electrons whose mean energy approximately equals the mean fluid motion energy. As this is 380 keV per ion, we expect a comparable electron energy and a nonthermal tail extending to much higher energies. Since the shock electron dissipation mechanism is just a scattering of the transverse electron motion, the major fraction of the electron energy should be perpendicular to the magnetic field and result in "trapped" electrons.

Figure 1 shows the observed southern conjugate debris distribution and the expected injection from this debris. These results agree with the Injun data at low L value. At higher L, electrons were observed by Telstar. If one assumes that these electrons are due to the beta decay of ionized fission debris, then one expects a distribution of debris deposited at the southern conjugate point described by the dashed curve of Figure 1. Superimposed at $1.35 \leqslant L \leqslant 1.45$ is the expected shock injection, which assumes that the bomb debris escaping upward in one steradian interacts only weakly with the ambient air and generates a magnetohydrodynamic collision free shock whose internal energy is dissipated by collective ion-electron interaction. It is evident that the total number of electrons is more than required to explain the Telstar observations. A slightly less efficient dissipation mechanism or shock speed-up would inject fewer electrons whose energy was above the detection threshold. The shock progressing to higher L would give rise to a comparable number of electrons but of lower mean energy.

Spectrum

Assuming constant density (ionized hydrogen) above the region of maximum shock energy ($L_0 = 1.4$) then the similarity solution of spherical blast waves of G. I. Taylor (1941) gives

$$u^2 \sim u_0^2 \left(\frac{r}{r_0} \right)^{-3} \tag{16}$$

which conserves total energy. Applying this to the shock above $L_0 = 1.4$ gives a shock energy spectrum E

$$E = E_{\max} \left(\frac{L_0 - 1}{L - 1} \right)^3, \qquad L \geqslant L_0 . \tag{17}$$

If the shock dissipation mechanism is not too sensitively dependent upon shock strength, then the shock energy E_s should be proportional to the injected electron energy.

It is a pleasure to acknowledge many fruitful discussions with Ray D'Arcy and John Troutman.

References

Adlam, J. H. and Allen, J. E.: 1958, *Phil. Mag.* **3**, 448.
Auer, D. L., Hurwitz, H., and Kilb, R. W.: 1962, *Phys. Fluids* **5**, 298.
Brown, W. L. and Gabbe, J. D.: 1963, *J. Geophys. Res.* **68**, (3), 607.
Buneman, O.: 1959, *Phys. Rev.* **115**, 503.

Chisnell, R. E.: 1955, *Proc. Roy. Soc. A.* **232,** 350.
Colgate, S. A.: 1965, *J. Geophys. Res.* **70,** 3161.
Colgate, S. A. and White, R. H.: 1965, Lawrence Radiation Laboratory Report UCRL-7777 (to be published in *Astrophys. J.*, 1965).
D'Arcy, R. G. and Colgate, S. A.: 1965, *J. Geophys. Res.* **70,** 3147.
Davis, L., Lüst, R., and Schlüter, A. S.: 1958, *Z. Naturforsch.* **13a,** 916.
Fan, C. Y., Gloechler, G., and Simpson, J. A.: 1964, *Phys. Rev. Letters* **13,** 149.
Grover, R. and Hardy, J. W.: 1964, Lawrence Radiation Laboratory, Report UCRL-7980T, Livermore, California.
Hess, W. H. and Nakada, P.: 1962, *Science* **138** (3536), 53.
Johnson, F. S. (ed): 1961, *Satellite Handbook,* Stanford Univ. Press, Stanford, California.
Kellogg, P.: 1965a, *Phys. Fluids* **8,** 102.
Kellogg, P.: 1965b, Conference on Collision Free Shocks, Ames Research Lab., and Univ. of Minnesota Report CR-64..
McIlwain, C. E.: 1961, *J. Geophys. Res.* **66,** 3681–3691.
O'Brien, B. J., Laughlin, C. D., and Van Allen, J. A.: 1962, *Nature* **195,** 939.
Sonett, C. P., Colburn, D. S., Davis, L., Jr., Smith, E. J., and Coleman, P. J., Jr.: 1964, *Phys. Rev. Letters* **13,** 153.
Van Allen, J. A., Frank, L. A., and O'Brien, B. J.: 1963, *J. Geophys. Res.* **68** (3), 619–626.
Vedenov, Velikov, and Saddeev: 1962, *J. Nucl. Fission, Suppl.* **2,** 465.

SESSION VII

SYNCHROTRON RADIATION

Chairman

PAUL NAKADA

Goddard Space Flight Center

SYNCHROTRON RADIATION MEASUREMENTS NEAR THE MAGNETIC EQUATOR *

GERARD R. OCHS

National Bureau of Standards, Boulder, Colorado

Abstract. The radiation produced in the radio spectrum by relativistic electrons injected into trapped orbits in the earth's magnetic field at the time of the Starfish nuclear explosion of July 9, 1962, has been monitored near Lima, Peru, for one year at 30 Mc/sec and almost three years at 50 Mc/s. After an initial brief transient increase of 4.5×10^4 °K, the excess antenna temperature at 50 Mc/s rapidly decreased to 1×10^4 °K. Later, the temperature T_{50} at a time t days after the explosion approximately followed the law,

$$T_{50} = 4600\, e^{-(t/270)},$$

for $t > 90$ days. At 30 Mc/s, the initial transient increase was not observed, but the subsequent behavior was similar. The decay was more rapid and the early signal was of greater intensity, however. After the first three months, the temperature T_{30} approximately followed the law,

$$T_{30} = 16000\, e^{-(t/180)}.$$

1. Introduction

Most experimental measurements of radiation trapped in the earth's magnetic field have logically been made in situ from artificial satellites. One property, however, which allows detection and measurement under certain conditions from ground-based stations near the equator, is that resulting from the radiation of magnetically-trapped relativistic electrons. This radiation was originally observed in a synchrotron particle accelerator, from which it takes its name, and SCHWINGER (1949) first explained its origin. It was later shown to be the source of much of the cosmic electromagnetic radiation observed by radio astronomers.

DYCE and NAKADA (1959) discussed the possibility of detecting synchrotron radiation from the natural Van Allen belts, but no experimental confirmation of the low predicted level was obtained. The vastly greater radiation from the Starfish high-altitude nuclear explosion of July 9, 1962, was, however, easily detected by DYCE and HOROWITZ (1963), GLOVER *et al.* (1963), GOLDMAN and HOROWITZ (1963), and OCHS *et al.* (1963).

This paper presents measurements made at Jicamarca Radar Observatory near Lima, Peru, on 30 and 50 Mc/s, stressing the radiation intensity and the

* This work has been supported jointly by the U.S. Government through the National Bureau of Standards and by the Government of Peru through the Instituto Geofisico del Peru.

decay observed. The calibration techniques employed are very similar to those used by radio astronomers. As is customary in radio astronomy, brightness temperature has been used to describe the intensity of the radiation, even though the source is non-thermal.

Jicamarca Radar Observatory is located nearly on the magnetic equator (latitude 11°56′S, longitude 76°52′W, magnetic latitude 1°N), in an ideal position to observe the radiation. Primarily the observatory was established to make incoherent scatter studies of the earth's upper atmosphere. The large antenna aperture and the receiving facilities necessary for these studies have been particularly useful for the synchrotron measurements made at 50 Mc/s.

2. Measurements at 30 Mc/s

Figure 1 shows a plot in which the ordinate is plotted in equivalent noise diode current. This measure is directly proportional to the total intensity received per unit bandwidth in a 40 kc/s bandwidth centered at 30 Mc/s. The initial peak is not recorded as the receiver was first connected in August 1962. The antenna consisted of an array of 4 half-wave dipoles having approximately 60° of beam-

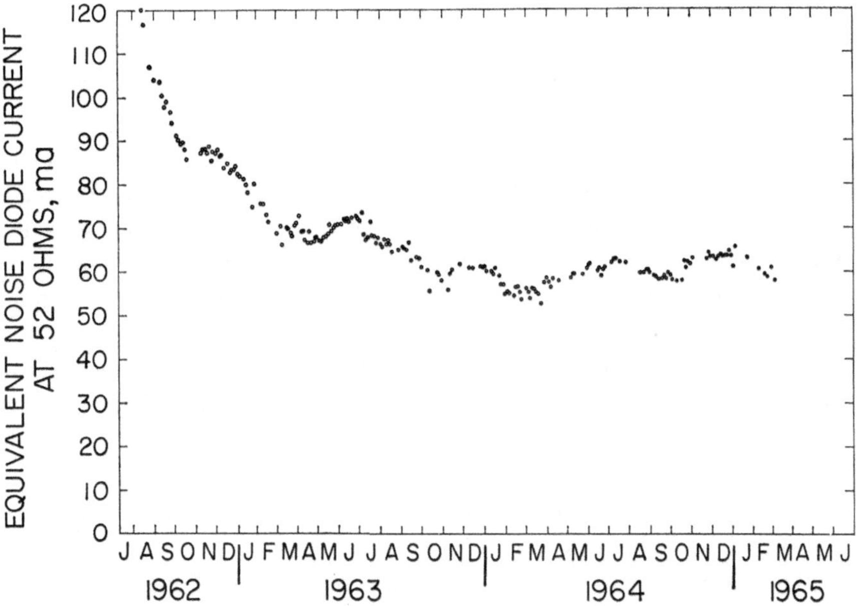

Fig. 1. 30 Mc/s noise at Jicamarca. The ordinate is proportional to total intensity received per unit bandwidth in a 40 kc/s bandwidth.

width between half-power points. A riometer (relative ionospheric opacity meter) was used as a receiver. This is a servo-noise-diode controlled receiver having a long-term stability approaching that of the noise diode. The receiver-transmission line combination was calibrated with an external emission-limited noise diode source. The individual data points have been obtained by averaging sixteen readings per day at one hour intervals, omitting an 8 hour period centered

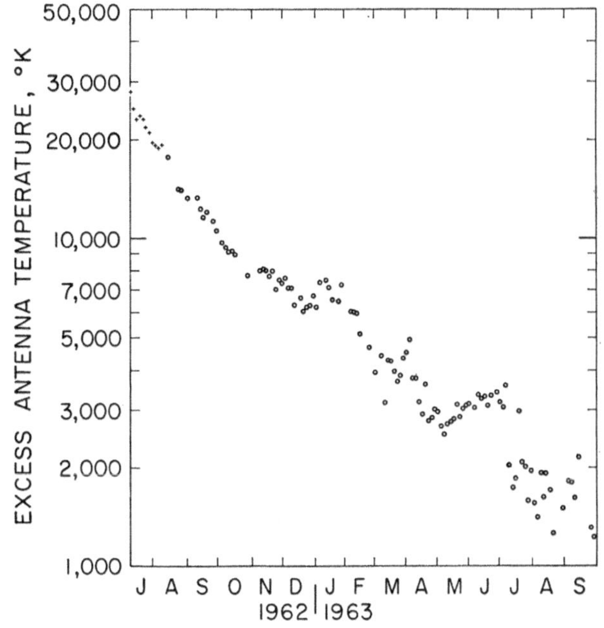

Fig. 2. Excess 30 Mc/s antenna temperature at Jicamarca.

on the principal transit of the galactic plane through the antenna beam. Every third day has been plotted, except for some days when the Jicamarca radar transmitter caused excessive interference. There is a yearly cycle discernible, showing that the method of plotting is not entirely effective in removing the cosmic noise and absorption variations.

Since the synchrotron signal as measured by this method has essentially disappeared into the background noise by the end of 1963, it seemed reasonable, in the absence of a better method, to regard the remainder of the curve as the pre-explosion level. In the logarithmic plot of Figure 2, this has been done. A smooth background curve was formed from the January, 1964, to March, 1965, data and this background subtracted from corresponding days of the year in the July, 1962, to December, 1963, portion of the curve. The roughness still present in the curve is probably due to uncorrected ionospheric absorption changes.

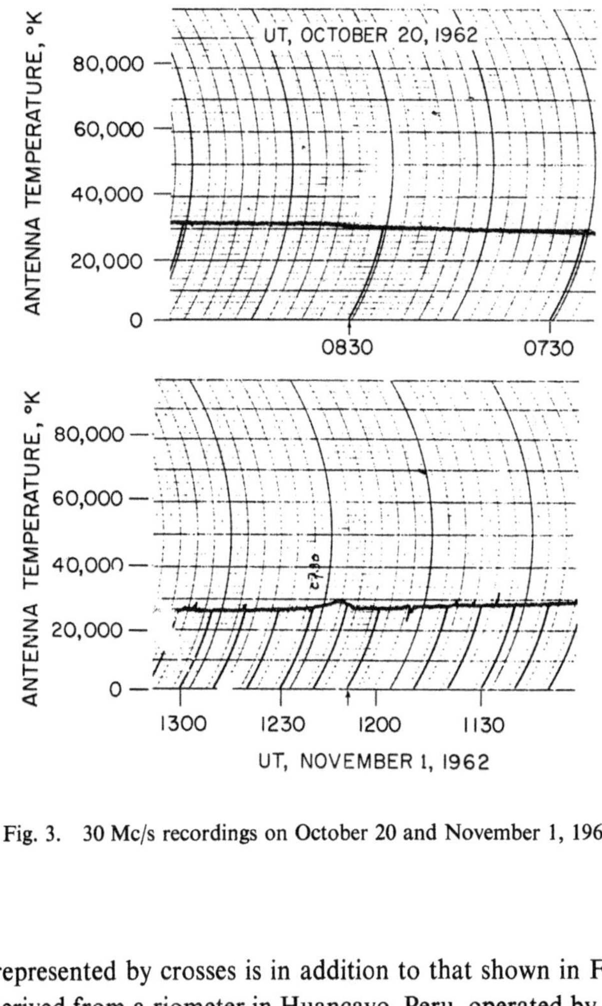

Fig. 3. 30 Mc/s recordings on October 20 and November 1, 1962.

The data represented by crosses is in addition to that shown in Figure 1, and has been derived from a riometer in Huancayo, Peru, operated by the Instituto Geofisico del Peru for the Air Force Cambridge Research Laboratory. The initial peak intensity is unknown because of equipment overload at this signal level.

In Figure 2 the ordinate has been converted to excess antenna temperature. The conversion was made on the assumption that the polarization of the signal as received was random so that one-half of the incident power was accepted by the polarized antenna. For the widebeam 30 Mc/s antenna, this is essentially true because of the differential Faraday effect resulting from the signal passing through the ionosphere over a wide range of angles with the earth's magnetic field, as explained in PETERSON and HOWER (1963).

As expected the early decay is more rapid. If after September, 1962, an exponential decay of the form $A = A_0 e^{-t/\tau}$ is assumed, $\tau = 180$ days approximately. In OCHS (1964) a fit to a decay curve of the form $A = A_0(1 + t/\tau)^{-1}$ was attempted with the justification that this should represent the number of particles of a given energy range remaining in a given orbital shell at time t after injection, according to WELCH and WHITAKER (1959). Attempts to fit this function to the data now available are unsatisfactory. The data appear to be much closer in accord with the theory of WALT (1964) and the time-decay history measured by VAN ALLEN (1964).

On at least one, and possibly two other occasions, high-altitude nuclear explosions produced effects on the 30 Mc/s record, as shown in Figure 3. The trace of November 1, 1962, is unique as far as these records go and seems quite likely to have been caused by the detonation occurring at the time marked by the arrow. The trace of October 20 has a slight rise, somewhat after a reported explosion. This order of change in the record is common, however.

3. Measurements at 50 Mc/s

The 50 Mc/sec measurements have been made using the main array at Jicamarca. This consists of a filled aperture of two superimposed arrays having linear but perpendicular polarizations. The half-power beamwidth is 1.1 degrees. The array feeds are available separately so that any polarization may be synthesized. By changing the phasing of the array, it is possible to change the pointing direction over a range of approximately 6 degrees.

For a month previous to the Starfish explosion, half of this antenna had been used in a search for natural synchrotron radiation from the Van Allen belts. The antenna direction was approximately perpendicular to the earth's magnetic field at F-region heights. The results were negative although the presence of a linearly polarized component equivalent to two to three times the minimum predicted by Dyce and Nakada could have been detected. Shortly after the Starfish explosion, this equipment was driven into a very non-linear operating range, but did indicate large amounts of noise power linearly polarized in the east-west direction.

The other half of the antenna, having approximately a 1.1 by 2.2 degree elliptical beam pattern, was connected in circular polarization during the time of Starfish explosion. The synchrotron radiation received during the first

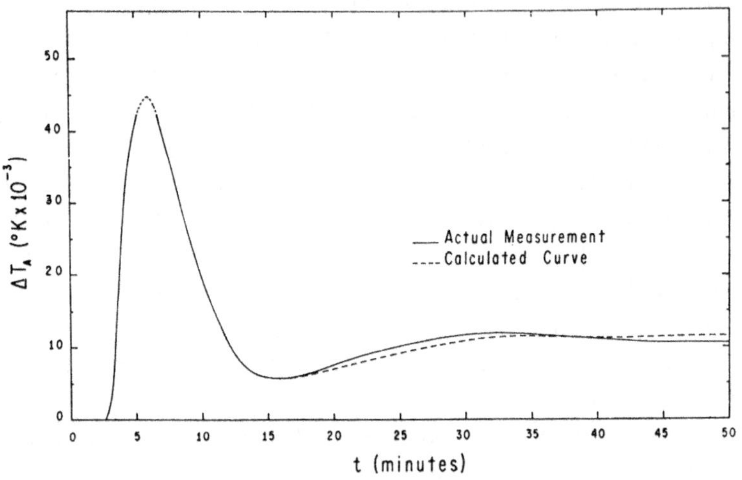

Fig. 4. Total synchrotron radiation received at Jicamarca at 50 Mc/s for the first 50 minutes following the explosion. The half-power beamwidth of the antenna was 1.1 by 2.2 degrees.

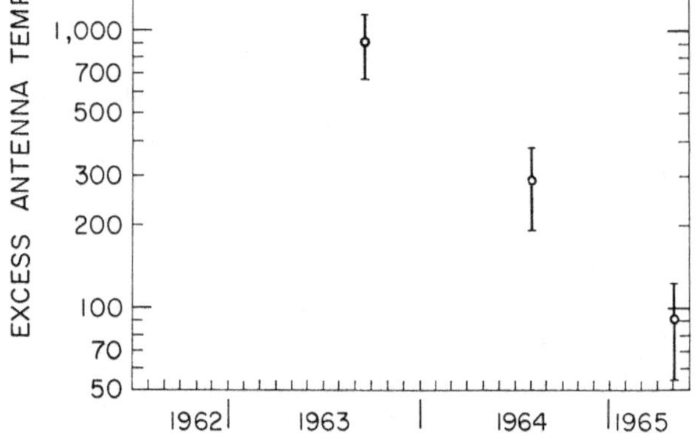

Fig. 5. Excess 50 Mc/s antenna temperature at Jicamarca. The half-power beamwith of the antenna was 1.1 degrees.

50 minutes is shown in Figure 4, taken from OCHS *et al.* (1963). In this paper D. T. Farley, Jr. obtained a description of the energy spectrum of the electrons from this early synchrotron radiation behavior.

The long-term decay taking place after the first hour is shown in Figure 5. Here again the data are not at all compatible with a decay function of the form

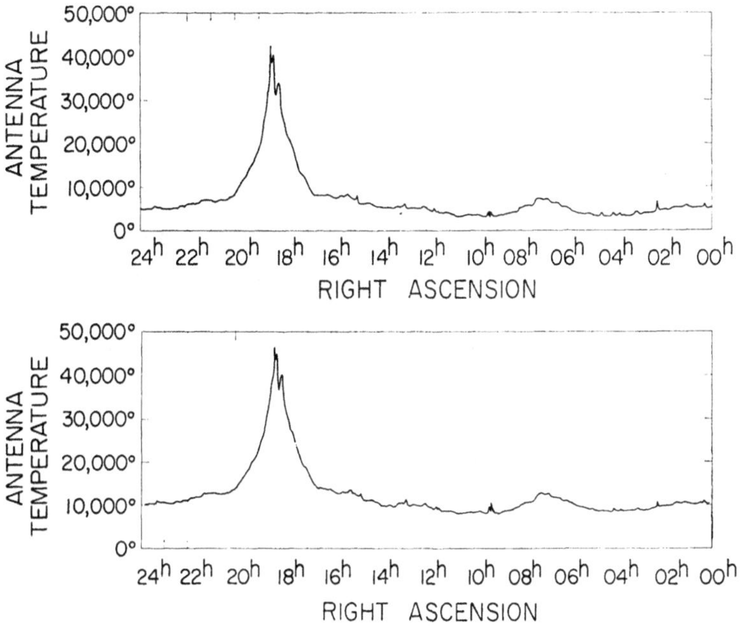

Fig. 6. Comparison of 50 Mc/s antenna temperature at declination − 12.9°. The upper trace was taken during June 30–July 3, 1962, and the lower trace during August 25–26, 1962. The antenna was circularly polarized with half-power beamwidth of 1.1° and directed 1.5 degrees southwest of vertical.

$A = A_0(1 + t/\tau)^{-1}$. Fitting an exponential decay $A = A_0 e^{-t/\tau}$ to the curve after September, 1962, $\tau = 270$ days approximately.

The first nine data points were obtained by subtracting the cosmic noise data obtained before the explosion from those afterwards. Figure 6 shows the records from which one such point was obtained. The lower trace, taken in the latter part of August, 1962, was still nearly double the intensity of the upper trace, which was taken prior to Starfish. All of these records were taken in the same sense of circular polarization, using a riometer with the bandwidth of the receiver portion increased from 40 kc/s to 1 Mc/s.

As the signal decayed to lower values, a second method was employed, which discriminated against the randomly polarized background. This method measured the difference between the maximum and minimum responses obtained by rotation of the direction of linear polarization of the antenna. A third polarization-sensitive technique utilized receiving equipment already present for incoherent scatter measurements. Outputs sensitive to the two rotational directions of circular polarization were obtained from the antenna and the fractional correlation between these two signals measured. This fraction was then multiplied by the total sky brightness at the time of the measurement.

The three techniques all measure different polarization parameters, but these parameters will decay at the same rate, provided that the polarization does not change. In Figure 5, the 1964 data point was obtained by method two and the 1965 point by method three. The different methods were related empirically by making simultaneous comparisons of methods one and two and then two and three. The readings have been compared either during the early morning hours when ionospheric electron density is least, or for the same times of day. It has been necessary to reject certain portions of the data due to irregular ionospheric effects upon the signal. The phenomena of equatorial spread *F* is an outstanding example of such an irregularity.

The uncertainty of the 1965 data point is difficult to assign. Very small amounts of cross coupling between polarizations can cause large increases in the observed fractional polarization as shown by COHEN (1958). The value shown at least places an upper limit on the emission.

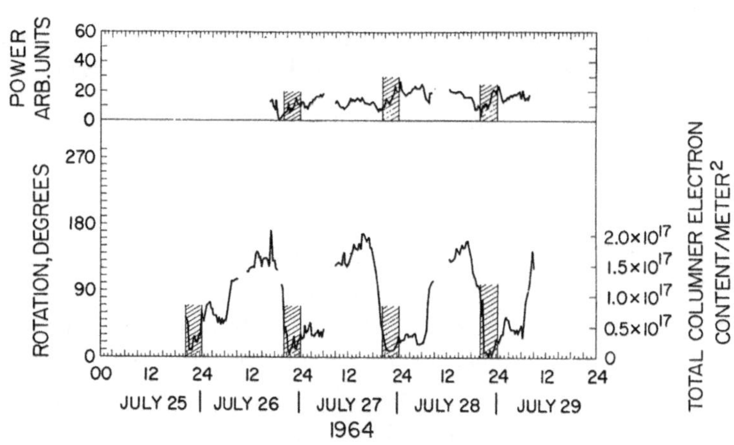

Fig. 7. Diurnal variation of synchrotron signal characteristics at antenna declination −12.9°. Shaded areas indicate unreliable data due to high galactic background noise.

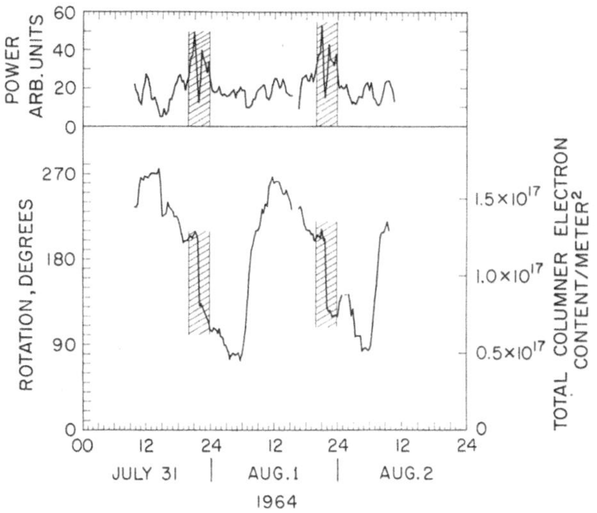

Fig. 8. Diurnal variation of synchrotron signal characteristics at antenna declination − 16.2°. Shaded areas indicate unreliable data due to high galactic background noise.

Some of the measurements made in 1964 included that of the Faraday rotation of the polarized signal in the ionosphere. Figures 7 and 8 show some of the incidental information recorded. The principal reason for observing the Faraday rotation of the synchrotron signal was to be more certain that synchrotron radiation was in fact what was observed. A secondary result gave information on the earth's magnetic field orientation over Jicamarca. The amount of rotation and its direction were close to what would reasonably be expected, knowing that the major linear component of the synchrotron radiation was east-west polarized when generated. By assuming a known magnetic field, the total electron content up to the point of generation of the radiation may be calculated from the Faraday rotation curve, and a scale has been added on the right-hand side of Figure 7 and 8 representing this calculation. A constant value (0.26 gauss) which is appropriate for the magnetic field at the electron density peak was used and no allowance made for the change in intensity of the earth's magnetic field as a function of altitude. These Faraday rotation results are in good quantitative agreement with measurements made using incoherent scattering at the same time of day, although on different days from those on which the synchrotron radiation was observed. The crosshatched areas on the curves indicate unreliable data due to the very high and sharply peaked background cosmic noise level at these times (approximately 40000 degrees at the peak of the galactic pass). The upper curves in Figures 7 and 8 represent the power difference observed between linearly polarized antennas aligned along and perpendicular to the direction of maximum response. The variation in intensity is an indication of the extent to which ionospheric effects disturb the measurement.

Since the Faraday rotation of the signal has been observed at two angles of passage through the magnetic field, the difference of these being known, it is possible to determine approximately these angles at F-region heights. By this method, the normal to the magnetic field at F-region heights is approximately 2.1 degrees north of vertical at Jicamarca Radar Observatory.

The exact polarization of the synchrotron radiation is of theoretical interest but it has been difficult to measure at Jicamarca. No possibility existed with the 30 Mc/s equipment because of the randomizing effect of the wide beamwidth antenna on the signal polarization. Even with the 50 Mc/s antenna, the wide angular extent of the signal presented problems. The differential Faraday rotation across the 1.1 degree beamwidth caused significant depolarization in the daytime. In addition, approximately 25% of the energy absorbed by the antenna arrived from directions outside the main beam. The bandwidth accepted by the receiver (1 Mc/s in most cases) also caused a differential Faraday effect although not as serious as the beamwidth problem.

Blumle (1963) discusses the distortion of the polarization of a signal which is propagated quasi-transverse to the earth's magnetic field. In general, a linearly polarized wave splits into two elliptically polarized magnetoionic modes which propagate with different phase velocities. The polarization of the waveform at any point along the propagation path is that resulting from the combined effect of the mode polarizations. In the quasi-transverse region, the ellipticity of the mode polarizations is such that the resultant waveform, even though originally linear, becomes elliptical at all points along the propagation path except those where the differential phase path is an integral multiple of the wavelength. The elliptical polarization measured by Ochs *et al.* (1963) with circularly polarized antennas can then be explained as resulting from an alteration, upon passage through the ionosphere, of a waveform that was originally linear.

The effects of all of the above problems can be minimized by taking data only in the early morning hours when electron density is lowest. Nevertheless irregular ionospheric effects may still be present. Observations with linear polarization have indicated maximum to minimum power ratios of approximately 4 to 1. Due to the various depolarizing factors mentioned earlier, this should be taken as a lower limit. Westfold (1959) predicts that the synchrotron radiation should be composed of two orthogonal linearly polarized components, having a power ratio of 7 to 1. Thus the polarization measurements, although not definitive enough for proof, are not in disagreement with Westfold.

Acknowledgements

I am indebted to Dr. K. L. Bowles for originally suggesting this experiment, and to both him and Dr. D. T. Farley for their interpretation of the initial results. The assistance of the Staff of Jicamarca Radar Observatory, particularly R. D. Abney, A. Arevalo, J. L. Green, and Dr. W. K. Klemperer, is gratefully acknowledged. G. F. Miller built the polarization-sensitive gear and Dr. P. Bandyopadhyay furnished the Huancayo riometer records.

References

BLUMLE, L. J.: 1963, *J. Geophys. Res.* **68,** 5886.
COHEN, M. H.: 1958, *Proc. IRE* **46,** 172.
DYCE, R. B. and NAKADA, M. P.: 1959, *J. Geophys. Res.* **64,** 1163.
DYCE, R. B. and HOROWITZ, S.: 1963, *J. Geophys. Res.* **68,** 713.
GLOVER, F., STEIGER, W. R., and WARWICK, J. W.: 1963, *J. Geophys. Res.* **68,** 4650.
GOLDMAN, S. C. and HOROWITZ, S.: 1963, *J. Geophys. Res.* **68,** 4865.
OCHS, G. R., FARLEY, Jr., D. T., BOWLES, K. L., and BANDYOPADHYAY, P.: 1963, *J. Geophys. Res.* **68,** 701.
OCHS, G. R.: 1964, *J. Geophys. Res.* **69,** 1014.
PETERSON, A. M., and HOWER, G.: 1963, *J. Geophys. Res.* **68,** 723–734.
SCHWINGER, J.: 1949, *Phys. Rev.* **75,** 1912.
VAN ALLEN, J. A.: 1964, *Nature* **203,** 1006.
WALT, M.: 1964, *J. Geophys. Res.* **69,** 3947.
WELCH, J. A., Jr. and WHITAKER, W. A.: 1959, *J. Geophys. Res.* **64,** 909–922.
WESTFOLD, K. C.: 1959, *Astrophys. J.* **130,** 241.

General References

COHEN, M. H.: 1964, *J. Geophys. Res.* **69,** 1935.
HORNBY, J. M.: 1964, *J. Geophys. Res.* **69,** 2737.
NAKADA, M. P.: 1963, *J. Geophys. Res.* **68,** 4079.
PAWSEY, J. L. and BRACEWELL, R. N.: 1955, *Radio Astronomy*, Clarendon Press, Oxford, England.

Discussion

Prof. VAN ALLEN asked if it was indeed possible to measure 40 °K in a 10000 °K background even with polarization techniques. OCHS replied that since the cosmic radio noise was unpolarized at 50 mc/s the Jicamarca observatory is capable of measuring the polarized synchrotron emission from the radiation belts down to 10 °K in a background of 4000 °K from cosmic noise. These measurements would be assisted by Faraday rotation in the ionosphere.

THEORETICAL MODEL OF SYNCHROTRON RADIATION AND COMPARISON WITH OBSERVATIONS

ALLEN M. PETERSON

Stanford Research Institute and Stanford University, Stanford, California

and

GLEN L. HOWER

Washington State University (formerly Stanford University)

1. Introduction

A. PURPOSE

The process of synchrotron emission from high-energy electrons spiraling in a magnetic field has received considerable attention in recent years. This interest arises since the process offers a possible explanation for such things as the non-thermal component of galactic and extra-galactic radiation, the polarized emission from a planet Jupiter, and the VHF noise generated during severe solar disturbances. Also, the observation of VHF noise following high altitude nuclear explosions has greatly stimulated interest in the process.

In this paper we will review the characteristics of the process, study graphically the effects of the various parameters, and present general techniques for computing numerical results. Although the results are discussed with reference to emission from electrons moving in the earth's magnetic field over the auroral zone and over the geomagnetic equator, they are perfectly general and may be applied to other problems.

B. DEFINITION OF SYMBOLS

The symbols used most frequently are defined below. Other symbols are defined as they occur in the text. MKS units are used throughout the report.

e = electronic charge
m_0 = rest mass of electron
μ_0 = permeability of free space
ε_0 = permittivity of free space
c = velocity of light
v = velocity of particle
B = magnetic flux density
H = magnetic-field intensity

$\beta\ = v/c,\ \gamma = 1/(1-\beta^2)^{\frac{1}{2}}$
$R\ =$ radius of circular orbit $= m_0 v\gamma/eB$
$\omega_0 =$ angular gyro-frequency $= v/R = eB/m_0\gamma$
$E_{\mathrm{T}} =$ total energy of particle $= mc^2$
$E\ =$ kinetic energy $= m_0c^2(\gamma-1) = E_{\mathrm{T}} - m_0c^2$
$\omega\ =$ angular wave frequency, $f = \omega/2\pi =$ frequency.

2. Characteristics of Synchrotron Emission

A. PHYSICAL DESCRIPTION OF THE PROCESS

A charge moving uniformly in free space does not radiate since the Poynting vector corresponding to the fields of the moving charge falls off too rapidly to contribute to energy flow over a distant surface. In the case of an accelerated charge, however, the fields may fall off as $1/r$ instead of $1/r^2$, and the charge thus gives rise to radiation fields in the usual sense. If the acceleration is perpendicular to the velocity – as for a charge moving at right angles to a uniform magnetic field – the motion is circular and the resulting emission is referred to by the terms "synchrotron radiation", "cyclotron radiation", "gyro radiation", or "magnetic bremsstrahlung".

At low electron energies the emission from the revolving charge is equivalent to emission from two dipoles at right angles to each other and to the field. It thus occurs at the angular frequency of the circular motion (the gyro frequency) and is maximum at right angles to the plane of the circular motion (along the magnetic field). This is the case referred to by the terms "cyclotron" or "gyro" radiation.

The term "synchotron" is applied when the electron velocities are relativistic. For any such particles, the emission is strongly concentrated in directions near the direction of the instantaneous velocity. Thus, emission in the direction of an observer located in the orbital plane of a revolving particle

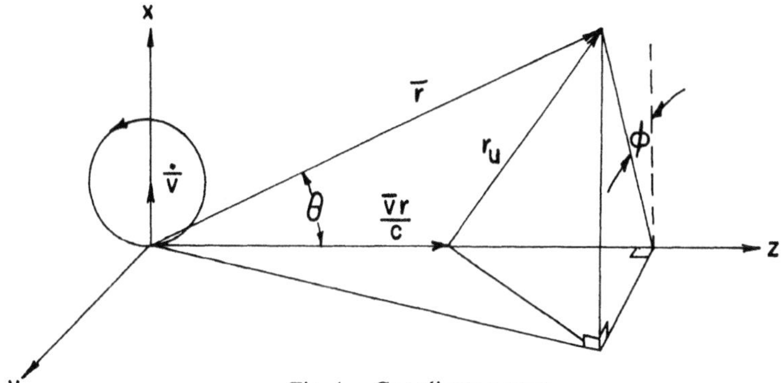

Fig. 1. Coordinate system.

occurs only while the electron is moving very nearly toward that observer. The emission is received in the form of pulses that recur at the gyro frequency, and since radiation toward the observer occurs only for motion along a small portion of the total circular path, the duration of the pulses will be short compared to the fundamental period. Consideration of the Fourier components of such a waveform indicates that a characteristic of the radiation will be its high harmonic content. In the case of these relativistic electrons the fundamental frequency is low, so that the separation between adjacent harmonics is small, and we therefore have essentially a continuous spectrum instead of the single frequency radiation of the low-energy particle.

B. RADIATION FIELDS

The fields of the accelerated charge may be obtained from classical electromagnetic theory using the retarded Lienard-Weichert potentials (PANOFSKY and PHILIPS, 1962). The portions of these fields varying as $1/r$ (the radiation fields) are given by

$$\mathbf{E} = \frac{e}{4\pi\varepsilon_0 s^3 c^2}[\mathbf{r} \times (\mathbf{r}_u \times \dot{\mathbf{v}})], \quad \mathbf{B} = \frac{\mathbf{r} \times \mathbf{E}}{rc}, \tag{1}$$

where $\mathbf{r}$ is the "retarded" radius vector from the charge to the observing point; $\mathbf{v}$ and $\dot{\mathbf{v}}$ are the velocity and acceleration of the charge, $\mathbf{r}_u = \mathbf{r} - \mathbf{v}r/c$; and $s = r - (\mathbf{v} \cdot \mathbf{r})/c$. These quantities are identified in Figure 1 for the case of a charge in uniform circular motion. A polar coordinate system is used in which the instantaneous velocity of the charge is directed along the z-axis and the circular orbit lies in the x-z plane.

From the expressions for the fields, it is noted that $\mathbf{B}$ is always perpendicular to $\mathbf{r}$ and $\mathbf{E}$, while $\mathbf{E}$ is perpendicular to $\mathbf{r}$ and the direction defined by $\mathbf{r}_u \times \dot{\mathbf{v}}$. Examination of these quantities reveals that the emission has a strong linear polarization in the plane of the circular orbit (perpendicular to the magnetic field). This is a very important characteristic and provides a possible means for experimentally identifying the process.

C. RADIATION PATTERN

From the Poynting vector corresponding to the above fields, the angular distribution of the radiated power may be derived. This gives the following result (PANOFSKY and PHILIPS, 1962)

$$P(\theta, \phi) = \frac{e^2 R \omega_0^2}{16\pi^2 \varepsilon_0 c^3} \left[\frac{\left(1 - \frac{v}{c}\cos\theta\right)^2 - \left(1 - \frac{v^2}{c^2}\right)\sin^2\theta\cos^2\phi}{\left(1 - \frac{v}{c}\cos\theta\right)^5} \right] \tag{2}$$

where θ and ϕ are the angles defined in Figure 1.

Examples of the radiation patterns are presented in the rough sketches of Figure 2. It is seen that the dipole pattern corresponding to low electron energies changes into extremely directional patterns for the higher energies. The maximum values of $P(\theta, \phi)$ relative to the low-energy ($\beta=0.01$) case are given on

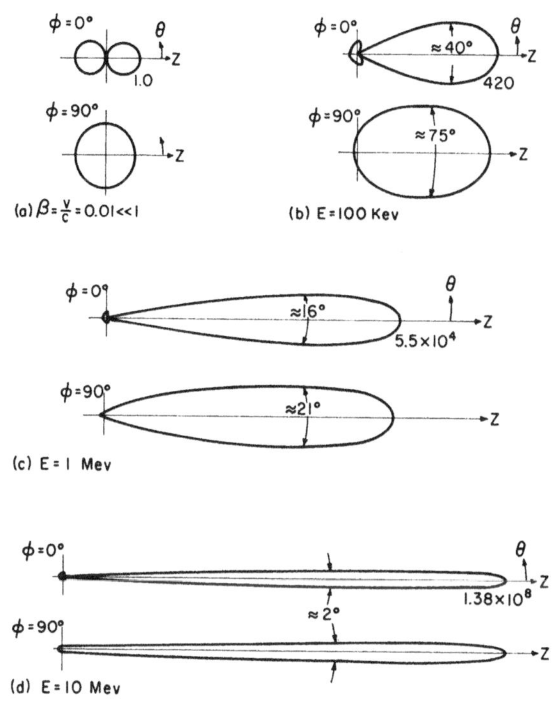

Fig. 2. Examples of radiation patterns (total power).

the diagrams, as are the approximate half-power beamwidths. The patterns are not symmetrical about the z-axis, as can be noted by examining the cross sections in the planes $\phi=0°$ and $\phi=90°$. In the plane $\phi=0°$, there are zeros at $\theta=\pm\arccos(v/c)$, while no zeros occur in the plane $\phi=90°$. Although the difference is great at low energies the distinction disappears in the patterns for higher energies.

At high energies, the important region of the diagrams occurs for $\theta \ll 1$. By approximating the sine and cosine terms in expression (2) it may be shown that the beamwidth of the pattern is on the order of m_0c^2/E_T, where E_T is the total energy of the electron. This leads to the often-quoted result that the emission is essentially confined within a cone of apex angle m_0c^2/E_T whose axis lies along the direction of motion.

With the aid of this result, the duration of the pulses reaching an observer in the orbital plane may be expressed in more definite form. From Figure 3 it is seen that appreciable radiation is launched toward the observer for only that

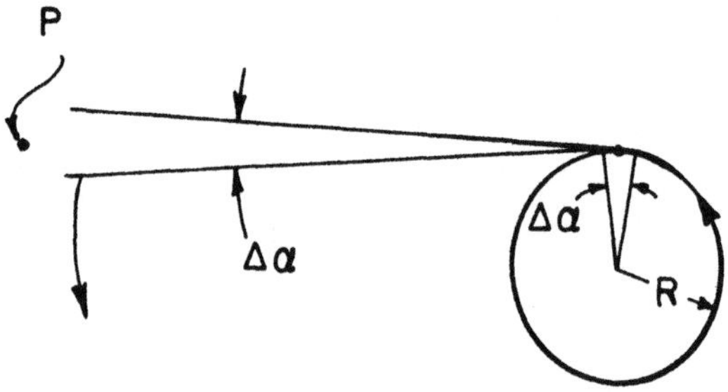

Fig. 3. Effect of the cone of emission in determining pulse duration.

length of time necessary for the particle to rotate through an angle $\Delta\alpha \approx m_0c^2/E_T$. After proper relativistic corrections, the duration of the pulse seen by an observer at point P is approximated by $\Delta t = (m_0/eB)\gamma^{-2} = (m_0/eB)(m_0c^2/E_T)^2$ (PANOFSKY and PHILIPS, 1962; SHKLOVSKY, 1960). The pulses are separated by

$$\tau = \frac{2\pi}{\omega_0} = \frac{2\pi m_0 \gamma}{eB} = \frac{2\pi m_0}{eB}\frac{E_T}{m_0c^2}.$$

D. FREQUENCY SPECTRUM

From the preceding discussion, the distribution of the radiated power with respect to frequency should consist of harmonics of the angular gyro frequency ω_0. Thus,

$$P(\omega) = \sum_{n=1}^{\infty} \delta(\omega - n\omega_0) P_n.$$

This distribution has been derived by a variety of authors. Both PANOFSKY and PHILIPS (1962) and SCHWINGER (1949) present the following expression for the power radiated at the frequency of the nth harmonic into a unit solid angle at an angle ψ from the orbital plane.

$$P_n(\psi) = \frac{e^2\omega_0\beta^3 n^2}{8\pi^2\varepsilon_0 R}\left[J_n'^2(n\beta\cos\psi) + \frac{\tan^2\psi}{\beta^2} J_n^2(n\beta\cos\psi)\right]. \tag{3}$$

In this equation, n is the harmonic number and J signifies a Bessel function in the usual manner.

Following Schwinger's work we may examine some limiting cases of the above expression. If $\beta \cos\psi \ll 1$, as would be the case for either low electron energies or for directions nearly at right angles to the orbital plane, the equation may be approximated as

$$P_n(\psi) = \frac{e^2 \omega_0 \beta}{8\pi^2 \varepsilon_0 R} \frac{(n\beta/2)^{2n}}{[(n-1)!]^2} (\cos\psi)^{2n-2} (1 + \sin^2\psi). \tag{4}$$

The radiated power thus decreases rapidly with increasing harmonic number in the case of low-energy ($\beta \ll 1$) electrons. For example, in this energy range the fundamental (gyro) frequency at auroral altitudes will have its maximum value on the order of 1 Mc. Emission at 12 Mc ($n \approx 12$) for $\beta \approx 0.1$ is then less than emission at the fundamental frequency by a factor on the order of 10^{-18}. It is apparent that emission by electrons whose energies lie in the range of this approximation will not contribute to synchrotron emission at frequencies above 10 Mc.

The total power radiated at a frequency $\omega = n\omega_0$ may be obtained by integrating Equation (3) over all angles. In the relativistic limit ($\gamma \gg 1$), this may be accomplished using various approximations with the following results (PANOFSKY and PHILIPS, 1962; LANDAU and LIFSHITZ, 1951).

$$P_n = 0.52 \left(\frac{e^2}{4\pi\varepsilon_0 c} \right) \omega_0^2 n^{1/3} \qquad \text{for} \quad 1 \ll n \ll \gamma^3 \tag{5}$$

$$P_n = \frac{1}{2\sqrt{\pi}} \left(\frac{e^2}{4\pi\varepsilon_0 c} \right) \omega_0^2 \frac{n^{1/2}}{\gamma^{1/2}} \exp\left(\frac{2n}{3\gamma^3} \right) \quad \text{for} \quad n \gg \gamma^3 . \tag{6}$$

Emission by high-energy electrons thus *increases* with increasing frequency over the range $1 \ll n \ll \gamma^3$ and then decreases exponentially for $n \gg \gamma^3$. This variation is sketched in Figure 4. It is seen that the emission is appreciable up to at least a frequency of the order of $f_0 \gamma^3$.

Schwinger gives a high-energy approximation that is similar to Equation (5) but only for the watts/steradian radiated in the orbital plane. From Equation (3) with $\psi = 0$ and $\beta \approx 1$,

$$P_n(0) = \frac{e^2 \omega_0 n^2}{8\pi^2 \varepsilon_0 R} [J_n'(n)]^2 .$$

Using an approximation given by WATSON (1952), the expression becomes

$$P_n(0) = \frac{6^{1/3}}{16\pi^3 \varepsilon_0} \left[\Gamma\left(\frac{2}{3}\right) \right]^2 \frac{\omega_0 e^2}{R} n^{2/3} \tag{7}$$

for $\gamma^2 \gg 1$ and $1 \ll n \ll 3/2\gamma^3$. Again, the increase in emission with frequency over this range of frequencies is shown. Retaining the angular dependence, the general expression (3) in the region of interest ($\psi^2 \ll 1$) becomes

$$P_n(\psi) = \frac{e^2 \omega_0 n^2}{24\pi^4 \varepsilon_0 R}(1 - \beta^2 + \psi^2)^2 \left[K_{2/3}^2(\xi) + \frac{\psi^2}{1 - \beta^2 + \psi^2} K_{1/3}^2(\xi) \right] \tag{8}$$

where

$$\xi = \frac{n}{3\gamma^3}\left(\frac{1 - \beta^2 + \psi^2}{1 - \beta^2}\right)^{3/2} = \frac{n}{3}(1 - \beta^2 + \psi^2)^{3/2}.$$

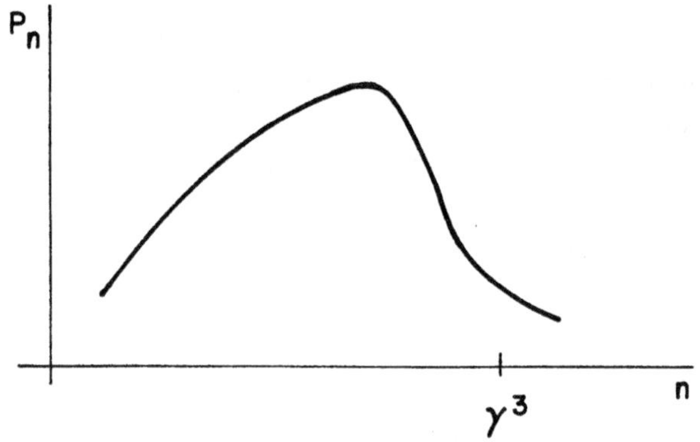

Fig. 4. Radiated power vs n.

This equation is valid for $\gamma \gg 1$, $\psi^2 \ll 1$, and $n \gg 1$. In the above expression the original Bessel functions have been replaced by their asymptotic forms for large order and argument (WATSON, 1952).

Since the spectrum is essentially continuous we may replace n by ω/ω_0 and consider $P_{\omega/\omega_0}(\psi)$ as a continuous function of ω. Also, we note that the number of harmonics in an interval $\Delta\omega$ is the integer value of $\Delta\omega/\omega_0$ so that the number of harmonics in a unit interval ($\Delta\omega = 1$) is $1/\omega_0$. Therefore, the expression $1/\omega_0 P_{\omega/\omega_0}(\psi)$ is the power radiated at frequency ω into a unit solid angle at an angle ψ with the orbital plane *per unit angular frequency* interval (i.e., watts/ steradian – radian/sec). Then $2\pi 1/\omega_0 P_{\omega/\omega_0}(\psi)$ is the emission in watts/steradian-cps.

After inserting the numerical values we finally arrive at the following expression for the instantaneous power per unit solid angle per cycle per second at an angle ψ from the orbital plane.

$$P(\psi, f) = 3.4 \times 10^{-45} \frac{Rf^2}{\beta^2}(1 - \beta^2 + \psi^2)^2 \left[K_{2/3}^2(\xi) + \frac{\psi^2}{1 - \beta^2 + \psi^2} K_{1/3}^2(\xi) \right] \tag{9}$$

where

$$\xi = 0.698 \times 10^{-8} \frac{fR}{\beta} (1 - \beta^2 + \psi^2)^{3/2} .$$

E. THE FUNCTIONS $K_{1/3}(\xi)$ AND $K_{2/3}(\xi)$

Before the expressions for emission from high-energy electrons may be used in calculations, the values of the Bessel functions must be determined. These functions are referred to as modified Bessel functions of the second kind and are not widely tabulated, although WATSON (1952) gives a short table for $K_{1/3}(\xi)$. The values may be determined, however, from the modified Bessel functions of the first kind, $I_\nu(\xi)$, by using the relation (WATSON, 1952)

$$K_\nu(\xi) = \frac{\pi}{2 \sin \nu\pi} [I_{-\nu}(\xi) - I_\nu(\xi)] .$$

Adequate tables of $I_\nu(\xi)$ for $\nu = \pm 1/3$, $\pm 2/3$ are available for example, (see *Computation Lab.*, 1949) and are used in constructing the curves of $K_{1/3}(\xi)$ and $K_{2/3}(\xi)$ given in Figure 5. The relationship between these functions,

$$- K_{2/3}(\xi) = \frac{d}{d\xi} [K_{1/3}(\xi)] - \frac{K_{1/3}(\xi)}{3\xi} ,$$

may also be useful and is included here for reference.

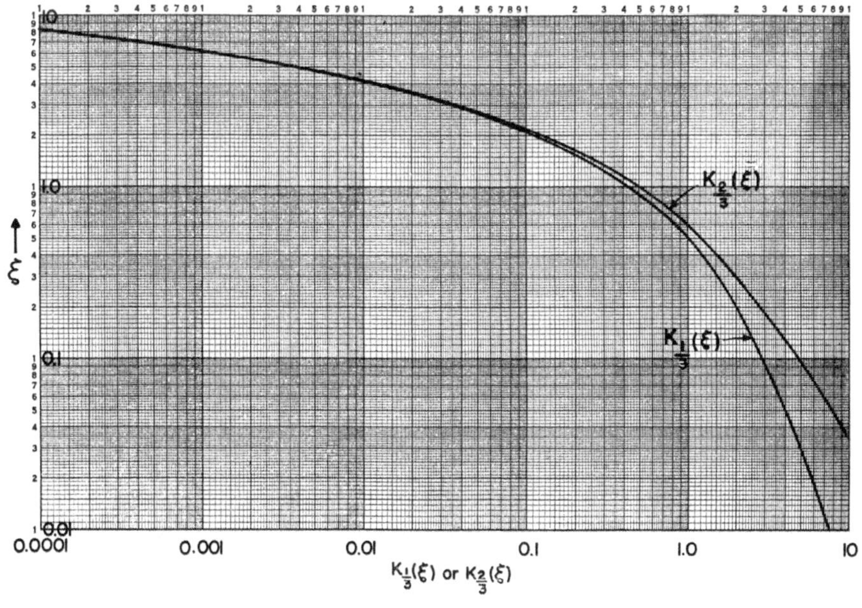

Fig. 5. Plot of the functions $K_{1/3}(\xi)$ and $K_{2/3}(\xi)$.

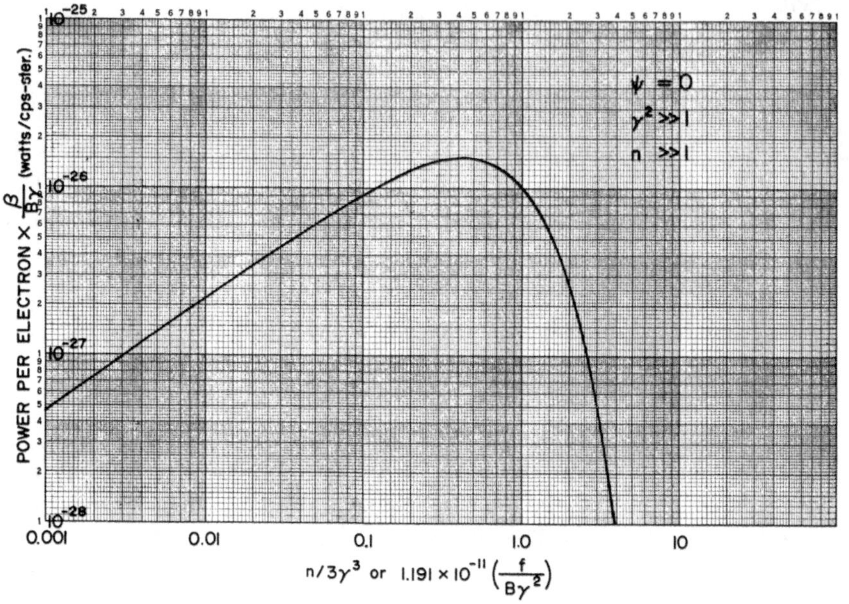

Fig. 6. Normalized spectrum of high-energy particles for the range $n \gg 1$.

F. CALCULATED SPECTRA

With the values for the Bessel functions available, we may now calculate the spectra for high-energy ($\gamma^2 \gg 1$) particles in the range $n \gg 1$ from either Equation (8) or Equation (9). We will first consider only emission in the orbital plane ($\psi = 0$). With $\psi = 0$ and after multiplying by $2\pi/\omega_0$ to obtain the emission in watts/cps-steradian, Equation (8) becomes

$$P(0, n) = 4.08 \times 10^{-26} \frac{B\gamma}{\beta} \left(\frac{n}{3\gamma^3}\right)^2 K_{2/3}^2 \left(\frac{n}{3\gamma^3}\right). \tag{10}$$

Equation (9), which is already in terms of watts/cps – steradian, becomes

$$P(0, f) = 4.08 \times 10^{-26} \frac{B\gamma}{\beta} \left(1.191 \times 10^{-11} \frac{f}{B\gamma^2}\right)^2 K_{2/3}^2 \left(1.191 \times 10^{-11} \frac{f}{B\gamma^2}\right). \tag{11}$$

In Figure 6, a plot of $P(0, n)$ vs $n/3\gamma^3$ is given. This same curve is also a plot of $P(0, f)$ vs $1.191 \times 10^{-11} f/B\gamma^2$. The curve may be regarded as a normalized curve from which the spectrum in the range $n \gg 1$ may be obtained for any high-energy ($\gamma^2 \gg 1$) electron at any desired magnetic-field intensity.

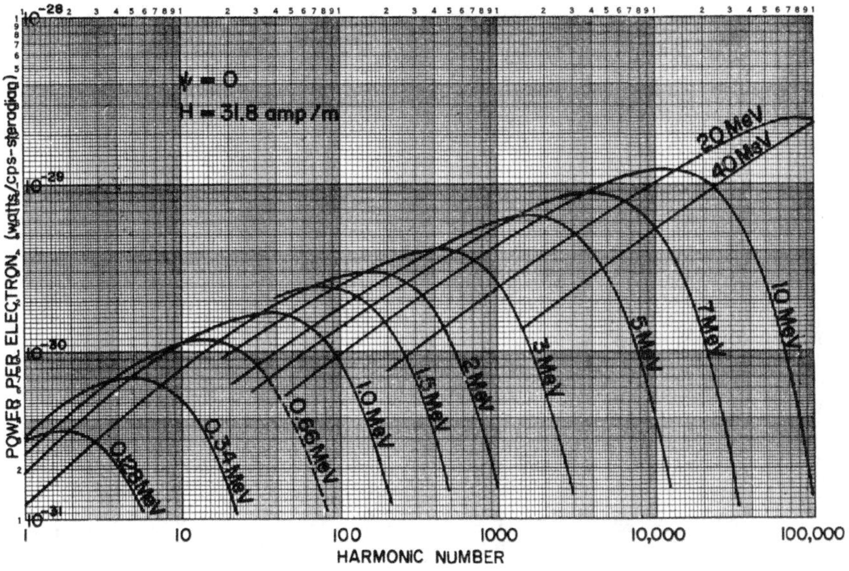

Fig. 7. Spectra of single electrons (harmonic number).

The spectra for electrons in the energy range below about 1 or 1.5 MeV must be obtained by a different method. In this energy range, γ^2 is not large compared to unity nor is β small compared to unity, so that neither the high- nor the low-energy approximations may be employed. Thus the more general expression (Equation (3)) with $\psi = 0$ must be used.

$$P(0, n) = \frac{e^3 B}{4\pi\varepsilon_0 m_0 c} \frac{\beta^2 n^2}{\gamma} [J_n'(n\beta)]^2 . \tag{12}$$

This calculation is straightforward if we employ tables such as those of JAHNKE and EMDE (1945) and the recursion relation

$$J_n'(\xi) = \tfrac{1}{2}[J_{n-1}(\xi) - J_{n+1}(\xi)] . \tag{13}$$

The spectra of electrons in the energy range between 128 keV and 40 MeV are shown in Figure 7 for a field intensity of 31.8 amp/m and the direction $\psi = 0$. This field intensity is chosen since it is typical of about the 750 km level over the auroral zone. Since the curves are plotted in terms of the harmonic number (n), we see from Equation (10) and (11) that a change in B will change the magnitude of the emission for each value of n by the same amount. Thus the effect of changing B will be to shift the entire set of curves vertically. For example, if the field intensity is reduced to one-fourth of the given value, the entire set of curves is shifted downward such that the points on the new curves are one-fourth of their previous values. By plotting the curves on a transparent overlay, we thus have a very simple means of determining these spectra at any given field intensity.

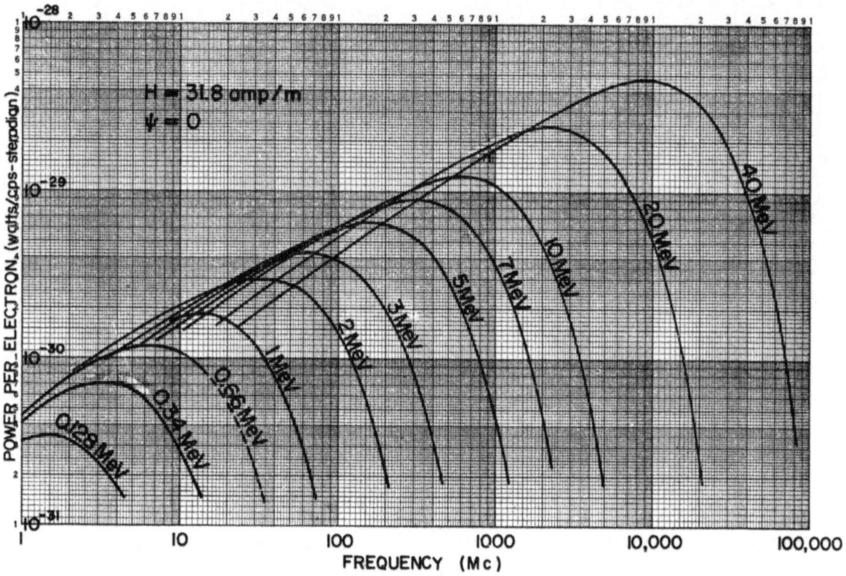

Fig. 8. Spectra of single electrons (frequency).

This same information on the spectra of the individual particles may also be presented in terms of frequency instead of harmonic number. This presentation is shown in Figure 8. From Equation (11) we see that for the high-energy particles, B enters into both the coefficient and argument, so that a change in B affects both the abscissa and ordinate scales. Thus the curves would have to be shifted horizontally as well as vertically. This is illustrated in Figure 9 for the 10 MeV curve. We see that the reduction in field intensity by one-fourth, shifts the curve downward and to the left such that the new peak emission is one-fourth of the original peak and occurs at a frequency which is one-fourth of the previous frequency of peak emission.

By this technique it can easily be shown that the curves of Figure 8 for energies between 2 and 40 MeV are in substantial agreement with the curves published by DYCE and NAKADA (1959).

G. EMISSION PER ELECTRON VS ELECTRON-ENERGY CURVES

Another way of examining the relationships between the various parameters is to consider emission at a single frequency and plot the contribution of each electron as a function of electron energy. For a field intensity of 31.8 amp/m and $\psi = 0$, the values may be read directly from Figure 8. For any other field

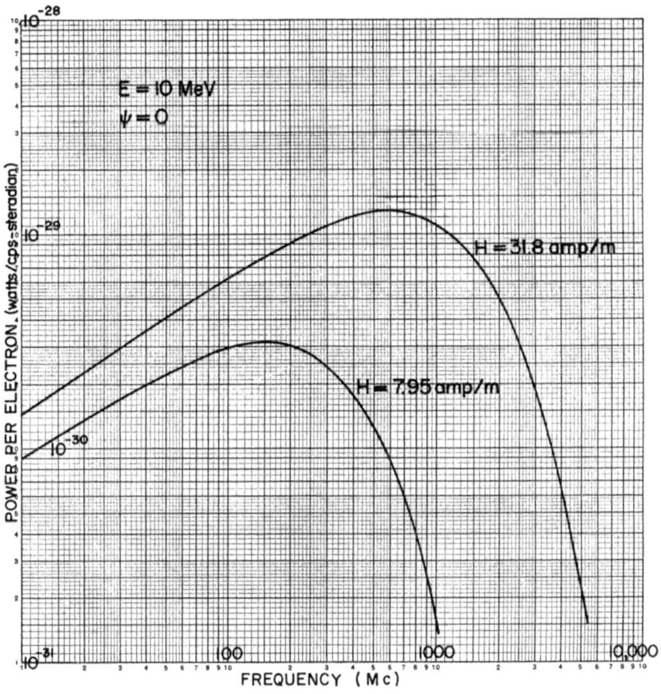

Fig. 9. Effect of change in B on the spectrum of A 10-MeV electron.

intensity the values could be obtained after sliding the curves as previously discussed. Of course, this effect may be accomplished without physically sliding the curves. For example, at a field intensity of $1/4 \times 31.8 = 7.95$ amp/m, we need merely read the values for 30-Mc emission at a frequency of $4 \times 30 = 120$ Mc and then multiply them by 1/4.

The information for these curves may also be obtained from Figure 7 by computing the proper harmonic number $n = f/f_0 = 3.58 \times 10^{-11} f\gamma/B$, reading the emission from the curve corresponding to the particular value of γ, and multiplying this emission by $H/31.8$ or $B/(4 \times 10^{-5})$.

H. EFFECT OF CHANGES IN FIELD INTENSITY

Plots of the 30-Mc emission per electron as a function of electron energy for two different field intensities are shown in Figure 10. This particular frequency is chosen since it is of primary interest in auroral noise studies. In Figure 9 we have seen that an increase in field intensity increases the emission at the low end of the frequency spectrum by a relatively small amount, while greatly en-

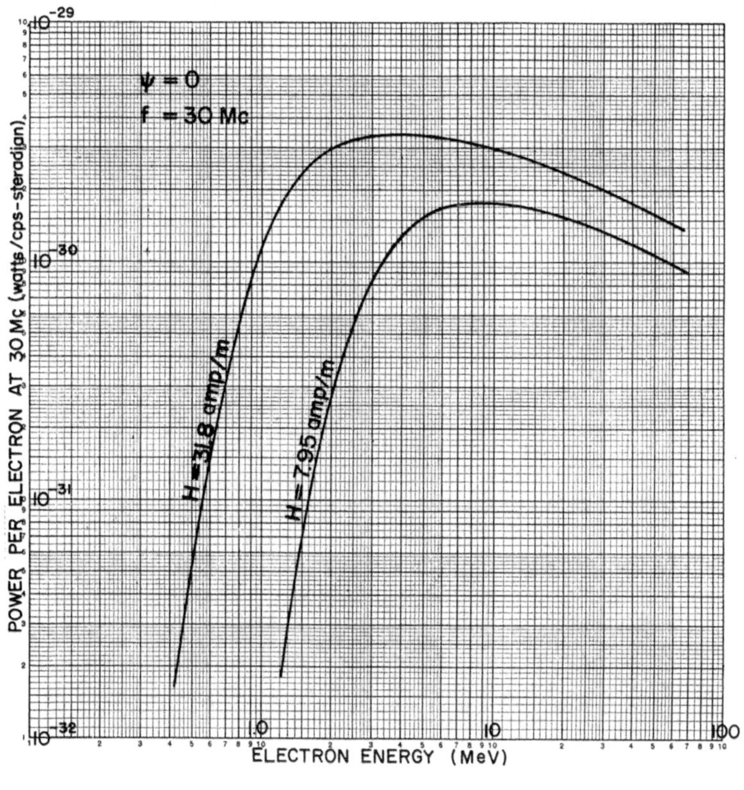

Fig. 10. Plot of 30-Mc emission per electron vs electron energy.

hancing emission at the high-frequency end of the spectrum. This effect is also indicated in Figure 10 where we see that 30 Mc is at the low end of the frequency spectrum for particle energies greater than about 7 MeV but at the high end for energies less than about 2 MeV. Thus, an increase in field intensity extends to lower values the range of electron energies that contribute to emission at a given frequency.

The lower curve in Figure 10 corresponds to about the 3000-km level over the magnetic equator, while, as we have previously mentioned, the upper curve is typical of about the 750-km level over the auroral zone. It is interesting to note that electrons with energies between about 500 keV and 2 MeV may contribute significantly to 30-Mc emission from the auroral zone, while these same particles would give a negligible contribution in the weaker field. Since satellite measurements indicate that the density of particles decreases rapidly with increasing energy, particles in this energy range might actually provide the dominant contribution to VHF auroral noise signals generated by the synchrotron process.

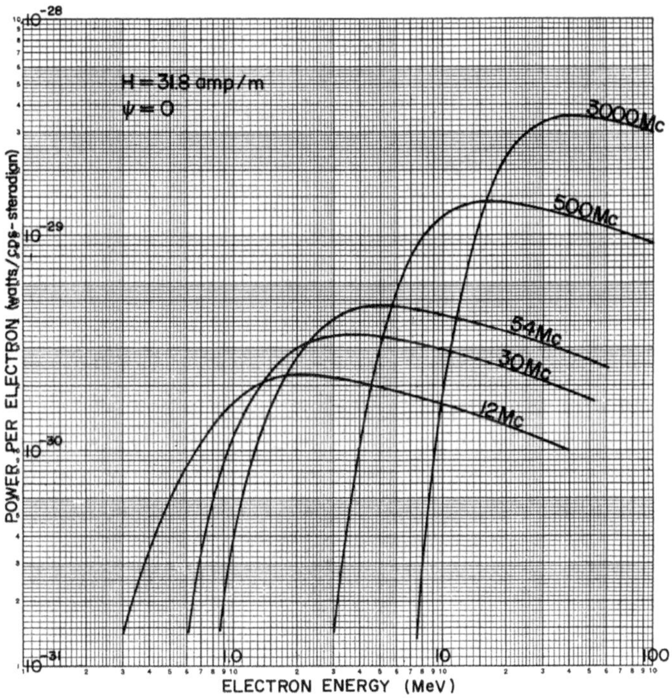

Fig. 11. Power per electron vs electron energy for frequencies of 12, 30, 54, 500, and 3000 Mc.

I. EFFECT OF CHANGES IN THE OBSERVING FREQUENCY

Figure 11 shows the power per electron vs electron-energy curves for a constant field intensity and several different frequencies. It is seen that the higher frequencies require higher energy particles, as we have already noted from the spectra of Figures 7 and 8.

Another conclusion of importance that may be drawn from these curves is that the ratio of signals at different frequencies is dependent on the energy spectrum of the emitting particles. For example, suppose that we observe emission from a group of electrons at 30 and 54 Mc. For the particular field intensity used in the figure, the curves for these two frequencies cross at about 2 MeV. Therefore, if all of the emitting particles have energies below 2 MeV, the 30-Mc emission would exceed the 54-Mc emission; while if all the particles had energies greater than 2 MeV, the opposite would be true. In general, we see that information on the *energy* spectrum of the emitting particles may be obtained from observations of the frequency spectrum of their combined emission.

Valuable additional information can therefore be obtained by making experimental measurements over a range of frequencies. Such a procedure could prove or disprove a postulated synchrotron generation mechanism if the energy spectrum of the emitting particles were known. Conversely, the energy spectrum of a group of particles could be studied by such measurements if the synchrotron mechanism is assumed to be operative. For example, measurements made at lower frequencies from a satellite above the ionosphere might be useful for studies of the energy spectrum in the outer Van Allen belt in the energy range above a few hundred keV.

J. EFFECT OF THE ANGLE ψ

The determination of the effect of viewing the emission away from the orbital plane ($\psi \neq 0$) is somewhat more difficult, as can be seen by examining Equation (9). In this case the total emission must be computed as the sum of two terms whose relative magnitudes are dependent on ψ. For computational purposes it is convenient to define the quantity

$$Z^2 \equiv \frac{1}{1 - \beta^2 + \psi^2} = \frac{\gamma^2}{1 + \psi^2\gamma^2}$$

and rewrite Equation (8) or (9) in the form

$$P(\psi, f) = 4.08 \times 10^{-26} \frac{BZ^2}{\beta\gamma} \left(\frac{n}{3Z^3}\right)^2 K_{2/3}^2\left(\frac{n}{3Z^3}\right)$$
$$+ 4.08 \times 10^{-26} \frac{BZ^4\psi^2}{\beta\gamma} \left(\frac{n}{3Z^3}\right)^2 K_{1/3}^2\left(\frac{n}{3Z^3}\right). \tag{14}$$

The value of the first term in this expression may be obtained from Figure 6 if we first re-label the horizontal axis, "$n/3Z^3$" and the vertical axis, "Power per Electron × $\beta\gamma/BZ^2$". The second term may be obtained by a similar procedure after plotting its "normalized" curve. This result is given in Figure 12. The technique of sliding the curves to account for changes in the various parameters may therefore be applied to each term in Equation (14) separately and the results added to obtain the total emission.

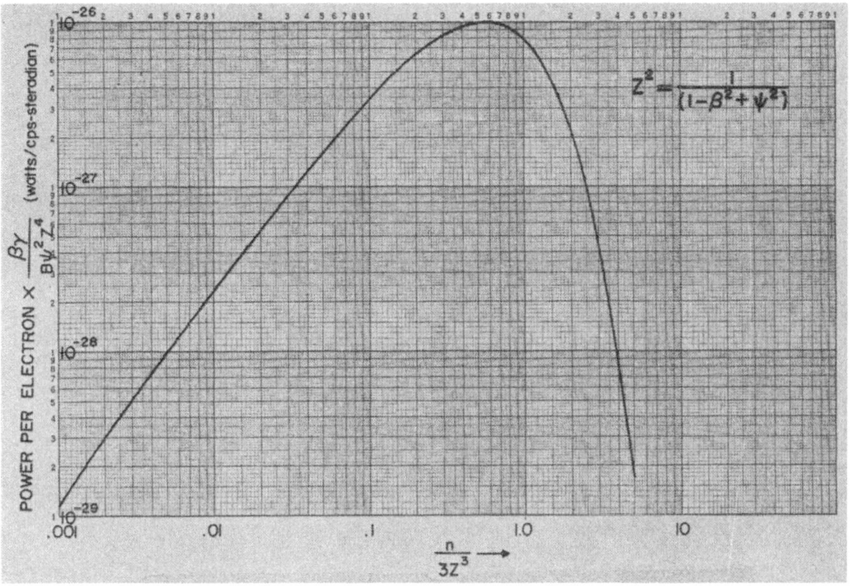

Fig. 12. Plot of second term of Equation (14).

The effect of ψ on the frequency spectrum of a 10-MeV electron is shown in Figure 13. It is seen that an increase in ψ causes a large decrease in the high-frequency components without greatly changing the low-frequency end of the spectrum. In Figure 2 we found that for the *total power* contributed by all frequencies ($\int P(\psi, f)\, df$), the beamwidth to the half-power points for a 10-MeV electron was only about 2 deg (0.035 radian). In practice, however, we are primarily interested in the effect on the signal received within a relatively narrow band of frequencies. Figure 13 indicates that the beamwidth of narrowband emission is strongly dependent on the frequency of observation. For the contribution of the 10-MeV electron at 30 Mc, the beamwidth is about 18 deg, while at 1000 Mc it is less than 4 deg.

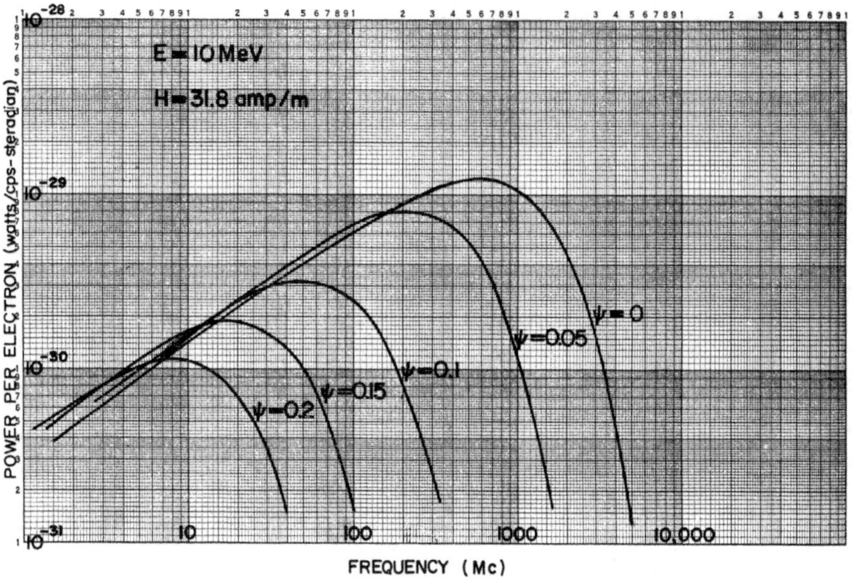

Fig. 13. Effect of ψ on spectrum of a 10-MeV electron.

It is of interest to examine the beam patterns for emission at a particular frequency. If we define

$$K_\psi \equiv \frac{\text{Emission at frequency } f \text{ in the direction } \psi}{\text{Emission at frequency } f \text{ in the direction } \psi = 0},$$

a plot of this quantity as a function of ψ then represents the desired pattern. From Equation (8) we obtain

$$K_\psi = (1 + \psi^2\gamma^2)^2 \left[\frac{K_{2/3}^2\left(\dfrac{n}{3Z^3}\right)}{K_{2/3}^2\left(\dfrac{n}{3\gamma^3}\right)} + \psi^2 Z^2 \frac{K_{1/3}^2\left(\dfrac{n}{3Z^3}\right)}{K_{2/3}^2\left(\dfrac{n}{3\gamma^3}\right)} \right]. \tag{15}$$

Some examples of the patterns obtained from this expression are given in Figure 14. Figure 14a shows the patterns at three different frequencies for a 10-MeV electron. Similar variations are obtained by holding the frequency constant and allowing the magnetic field to change. For example, these same curves also represent the emission patterns at 30 Mc for field intensities of 9.55,

31.8, and 79.5 amp/m (left to right). Figure 14b shows 30-Mc emission patterns for three different electron energies.

These sketches indicate that the emission at a particular frequency may actually increase as we move away from the orbital plane before beginning to

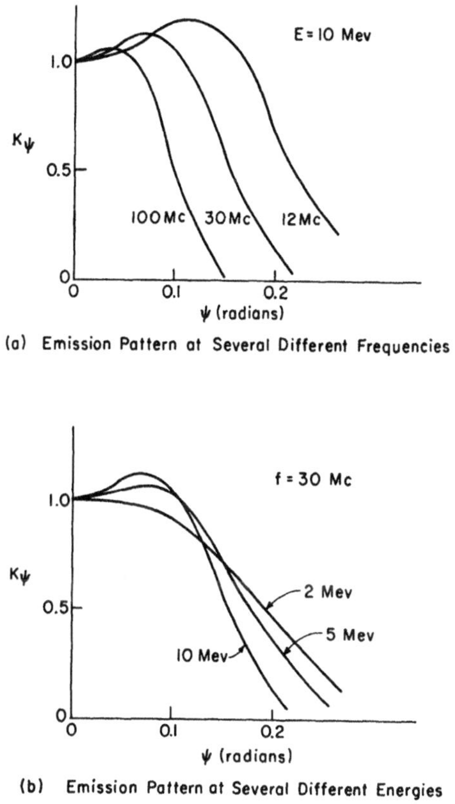

Fig. 14. Examples of emission patterns.

drop off. For the lower energies (which happens to be the energy range giving the primary contribution to 30-Mc emission from electrons moving in the earth's magnetic field), this effect disappears and the beam patterns are approximately gaussian in shape.

Additional information of a more general nature may be obtained by plotting K_ψ as a function of $n/3Z^3$ for various values of $n/3\gamma^3$. Some examples of these curves are shown in Figure 15a. It is seen that for $n/3\gamma^3 < 0.1$ and

$n/3Z^3 > 0.5$, the curves are essentially the same. The value of K_ψ is reduced to 0.5, in this case at approximately $n/3Z^3 = 0.82$. This condition leads to the following expression for the half-power angle which, as noted above, is applicable for $n/3\gamma^3 < 0.1$.

$$\psi_{1/2}^2 \simeq 1.68 \times 10^7 \left(\frac{B}{f\gamma}\right)^{2/3} - \frac{1}{\gamma^2} \text{ radians}. \tag{16}$$

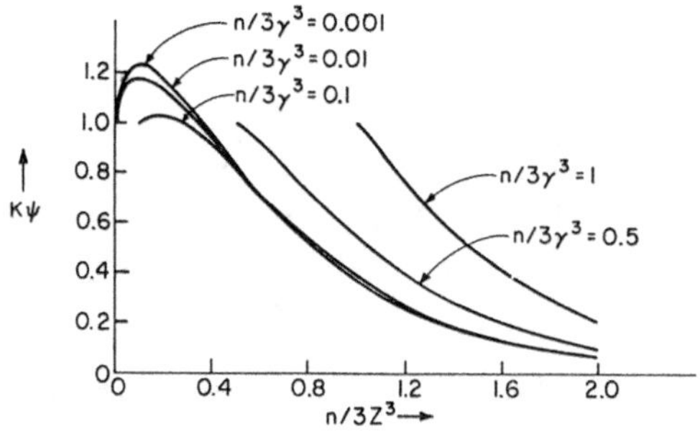

(a) Kψ vs n/3Z³ for Various Values of n/3γ³

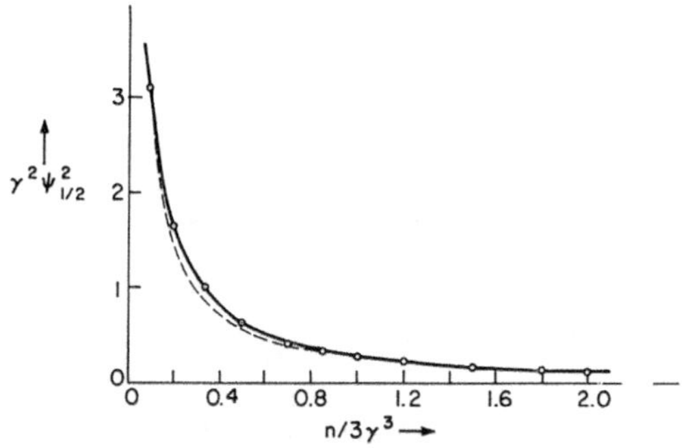

(b) γ²ψ² vs n/3γ³ for Kψ = 0.5

Fig. 15. Information on emission beamwidths.

The emission beamwidths thus decrease with increasing energy in this range.

In Figure 15b the value of the quantity $\gamma^2\psi^2$ at which K_ψ becomes equal to 0.5 is plotted as a function of $n/3\gamma^3$ (solid line). The dotted line is a plot of the expression $\gamma^2\psi^2 = 0.3/(n/3\gamma^3)$, which is a good representation of the desired curve in the range $n/3\gamma^3 > 0.7$ and is a fair representation for all $n/3\gamma^3 > 0.1$.

Therefore,

$$\gamma^2 \psi_{1/2}^2 \approx \frac{0.3}{n/3\gamma^3} = \frac{0.3\gamma^2}{1.191 \times 10^{-11} Bf}$$

and

$$\psi_{1/2} \approx 1.59 \times 10^5 \sqrt{B/f} \text{ radians}. \tag{17}$$

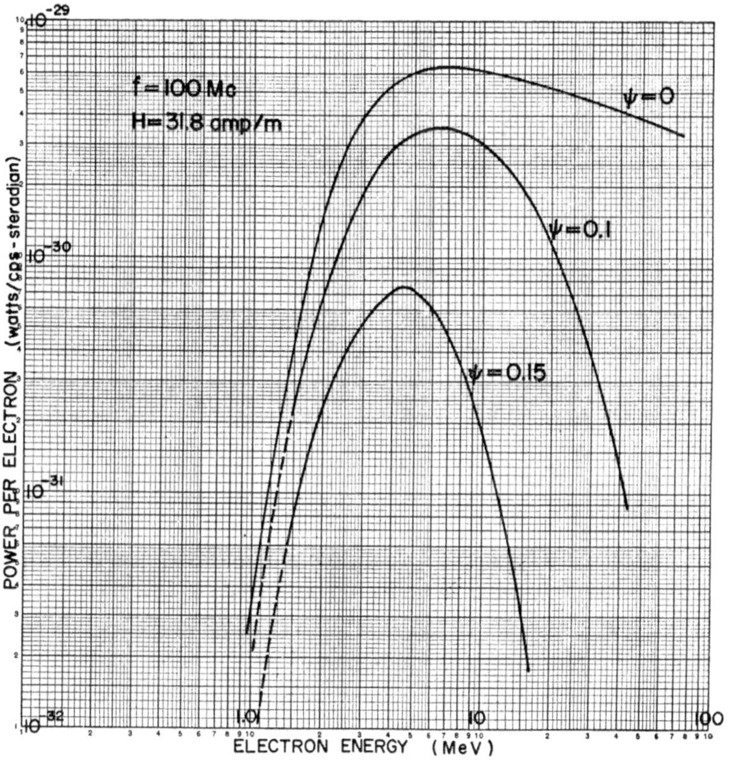

Fig. 16. Power per electron vs electron-energy curves for several different values of ψ.

This latter expression shows that the beamwidth of emission is essentially *independent of energy* in this range of $n/3\gamma^3$. For the case of VHF emission from electrons in the earth's magnetic field, this would be the generally applicable expression. This situation arises since the density of particles falls off rapidly with increasing energy in this range so that the dominant contribution to the received signal comes from the energy range characterized by values of $n/3\gamma^3$ between 0.1 and about 3.

For 100-Mc emission from electrons in a field of 31.8 amp/m the condition $n/3\gamma^3 > 0.1$ implies $\gamma < 17$, or electron energies less than about 8 MeV. From the preceding discussion we would expect the contribution of particles with $E < 8$ MeV to fall to one-half value at approximately the same value of ψ. That this actually occurs is illustrated by the emission per electron vs electron-energy curves for this situation given in Figure 16. For $\psi = 0.1$ radian, the contributions of the high-energy particles have decreased considerably while for energies below 8 or 10 MeV the reduction is much less ($\approx 1/2$) and is essentially independent of energy. Therefore, for 100-Mc emission from electrons having energy less than 8 MeV, we would expect an overall emission beamwidth of about 2×0.1 radian or 11.5 deg. This compares very favorably with the value computed from Equation (17) – that is, $2 \times 5.6 = 11.2$ deg.

Figure 16 also shows that if emission were received from particles with energies above 10 MeV, the emission beamwidth could be considerably less than the value determined above. Thus, the overall emission beamwidth at a particular frequency is, in general, determined by the energy spectrum of the emitting particles. However, we have seen that in many cases of practical importance the emission is received from a range of particle energies over which the individual half-power beamwidths are relatively independent of energy. The overall emission beamwidth for such cases may then be determined from Equation (17).

K. HELICAL MOTION OF ELECTRONS

We have considered the emission from an electron in uniform circular motion such as would occur in the case where the electron's velocity is at right angles to a uniform magnetic field. In the case of an electron in the earth's field, however, the actual motion is much more complex.

Consider a particle of total kinetic energy W_T moving in the earth's magnetic field. This total energy can be broken up as $W_T = W_\perp + W_\parallel$, where $W_\perp$ and $W_\parallel$ are the energies associated with the components of velocity perpendicular and parallel to the field, respectively. The basic principle involved is that the magnetic moment or the ratio of $W_\perp$ to B ($\mu = W_\perp / B$) is a constant. This is true for fields which vary slowly in a distance large compared to the gyro radius (a case of adiabatic invariance). For an electron of total energy W_T moving along a field line toward the earth, the quantity $W_\perp$ must therefore increase as the particle moves into the stronger portions of the field. At some critical value of the field B_R, all of the velocity will be perpendicular to the field. This is called the mirror point or point of reflection. From the expressions above:

$$B_R = \frac{W_T}{\mu} = \frac{W_T B}{W_\perp} = \frac{W_T B}{W_T \sin^2 \alpha} = \frac{B}{\sin^2 \alpha}$$

where B is the magnetic field at a reference point and α is the angle between the field and the velocity of the particle at that point.

Thus a charged particle which is "trapped" by the earth's field moves in a helical path and oscillates back and forth along a field line between opposite hemispheres. There is also a longitudinal drift at right angles to the field which need not be considered here.

In addition to these trapped electrons we may also have particles passing through the antenna beam which are "dumped" from the outer Van Allen belt or which are new arrivals from the sun moving with large velocity components along the field. These particles do not encounter a field intensity sufficient to satisfy the mirroring condition and are thus lost in the lower atmosphere after one passage through the antenna beam.

Radiation from these high-energy electrons moving in helical paths is again confined essentially within a cone-of-apex angle m_0c^2/E_T whose axis lies along the instantaneous direction of motion. The emission from such an electron is thus no longer a maximum at right angles to the field but in a direction making an angle α with the field. SCHWINGER (1949) considers radiation from a high-energy ($\gamma^2 \gg 1$) electron moving in an arbitrary path. He arrives at a result which is identical with Equation (9) if we interpret the R as the *instantaneous* radius of curvature of the path and ψ as the angle between the direction of emission and the *instantaneous* orbital plane. The frequency spectra for an electron moving at an angle α with the field may thus be obtained by replacing the R in Equation (9) by $R/\sin\alpha$ (see DYCE and NAKADA, 1959).

For this situation Equation (11) becomes

$$P(0,f) = 4.08 \times 10^{-26} \frac{B\gamma \sin\alpha}{\beta} \left(\frac{1.191 \times 10^{-11}}{\sin\alpha} \frac{f}{B\gamma^2} \right)^2 K_{2/3}^2 \left(\frac{1.191 \times 10^{-11}}{\sin\alpha} \frac{f}{B\gamma^2} \right). \tag{18}$$

Therefore, the normalized curve of Figure 6 applies if we consider the abscissa scale to represent the quantity $(1.191 \times 10^{-11}/\sin\alpha) \times (f/B\gamma^2)$ and the ordinate scale to represent the power per electron multiplied by $\beta/(B\gamma \sin\alpha)$. After sketching the spectrum of a given particle for a particular α, the spectrum for other values of α may be obtained by the technique of sliding the curve. For example, if the spectrum for $\alpha = 90°$ is given and we wish to obtain the spectrum for $\alpha = 45°$, we must shift the given curve downward and to the left such that the peak emission is reduced to $\sin\alpha = 0.707$ of its former value and the peak occurs at a frequency which is 0.707 times the former frequency of peak emission.

The effect of the angle α on the spectrum of a 10-MeV electron is shown in

Figure 17. Each curve is for emission in the direction $\psi=0$ which now corresponds to the direction making an angle α with the magnetic field. The variations with ψ in each case would be similar to that shown in Figure 13. We see from Figure 17 that for emission in the instantaneous orbital plane, the high-frequency portion of the spectrum for a particular particle drops off much faster as α deviates from 90 deg than the low-frequency portion. Also, it is important to note that the emission spectra are essentially unchanged as long as α does not vary from 90 deg by more than 20 deg.

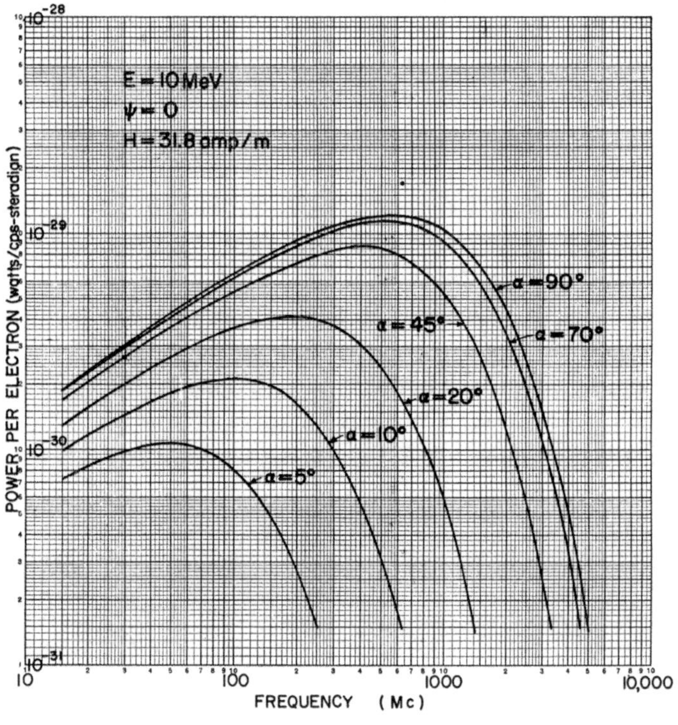

Fig. 17. Effect of angle α on Spectrum of 10-MeV Electron.

3. Calculations of Synchrotron Emission for Starfish Electrons

A. GENERAL DISCUSSION

In the preceding sections the characteristics of synchrotron emission have been examined, and methods for calculating the emission from individual electrons were presented. The factors determining the maximum possible power that might be received from a given electron are summarized below.

(1) The frequency of observation (f).

(2) The energy of the electron (β, γ).

(3) The instantaneous radius of curvature (R).

This quantity is itself a function of the electron energy and also of the magnetic field (B) and the angle between the velocity and the field (α).

(4) The angle between the instantaneous orbital plane and the direction of emission (ψ).

The functional dependence of the emission per electron might therefore be expressed as $P(\psi, f, E, B, \alpha)$ watts/cps ster in terms of the symbols previously defined.

To compare theory and experiment in an actual case it is necessary to calculate the total emission from a large number of electrons. Although a completely accurate determination of this total emission is generally impossible, we can indicate techniques for making reasonable estimates. Some of the factors that must be considered in calculating the noise to be expected from the electrons trapped during the Starfish explosion are listed below.

(1) The geometry of the emitting region, relative to the magnetic field and the position of the receiving station, as it affects the emission per electron.

(2) The extent of the emitting region.

(3) Possible ionospheric effects along the propagation path between emitter and receiver.

(4) The number of electrons in each energy interval and their distribution throughout the emitting region.

(5) The characteristics of the receiving system.

The calculation of the total power from electrons whose individual radiation is known is, in principle, very simple. Since the motions are uncorrelated we need only sum the individual power contributions. We have determined that $P(\psi, f, E, B, \alpha)\, df\, d\Omega$, where $d\Omega$ is an element of solid angle in the direction ψ, is the power emitted by a given electron in the direction of interest. If r is the distance from the receiver to the electron, and A_{eff} is the effective antenna area in this direction, the quantity

$$P(\psi, f, E, B, \alpha)\frac{A_{\text{eff}}}{r^2} \times \frac{1}{A_{\text{eff}}} = P(\psi, f, E, B, \alpha)/r^2$$

is the radiation received per square meter of effective antenna area (watts/cps m^2). Thus, if $\varrho(E, \alpha, \mathbf{r})\, dV\, dE\, d\alpha$ is the number of electrons in the volume element $d\mathbf{V}$ located at vector distance $\mathbf{r}$ having energies in the range dE about E and helix angles in the range $d\alpha$ about α, the total radiation received is given by

$$P_r = \iiint \frac{P(\psi, f, E, B, \alpha)}{r^2} \cdot \varrho(E, \alpha, \mathbf{r}) \, dV \, dE \, d\alpha \, \frac{\text{watts}}{\text{cps m}^2} \tag{19}$$

As the starting point for calculation, a receiver and overhead-looking antenna are assumed to be located on the magnetic equator. The radiation from trapped electrons is assumed to uniformly fill the antenna beam and an equivalent 'sky-temperature' can be calculated for the synchrotron radiation.

In this situation we may therefore express the emission in terms of a 'brightness' of the region viewed, b watts/cps m^2 ster

$$b = \frac{\iiint \frac{P(\psi, f, E, B, \alpha)}{r^2} \varrho(E, \alpha, r) A \, dr \, dE \, d\alpha}{A/r^2} \tag{20}$$

$$b = \iiint P(\psi, f, E, B, \alpha) \varrho(E, \alpha, r) \, dr \, dE \, d\alpha \tag{21}$$

It is interesting to note that the brightness expression used by NAKADA (1963) may be obtained from Equation (1) by expressing the density function ϱ in terms of the directional flux of particles (electrons/cm^2-sec-steradian) and by noting that the pitch angle distribution is generally much wider than the beam pattern of synchrotron emission. It is often convenient to express the 'brightness' in terms of an equivalent 'blackbody' temperature T. When this is desirable the following relationship (PIDDINGTON, 1961) can be used

$$b = 2kT/\lambda^2 \tag{22}$$

where k = Boltzmann's constant, 1.38×10^{-23}; λ = radio wavelength in meters; T = temperature in degrees Kelvin.

Figure 18 shows the omnidirectional electron flux maps prepared by Nakada from the estimates of BROWN and GABBE (1963), and VAN ALLEN *et al.* (1963). Figure 19 shows the variation of electron volume density with height over the magnetic equator corresponding to these contours together with a curve based on an estimate by HESS (1963). Calculations based on the Hess model have been described in some detail previously (PETERSON and HOWER, 1963).

B. NUMERICAL RESULTS

The electrons are assumed to have a fission β-decay spectrum, as given by CARTER *et al.* (1959); (HESS, 1963; VAN ALLEN *et al.*, 1963)

$$N = N_0 \, e^{-(0.575E + 0.55E^2)} \tag{23}$$

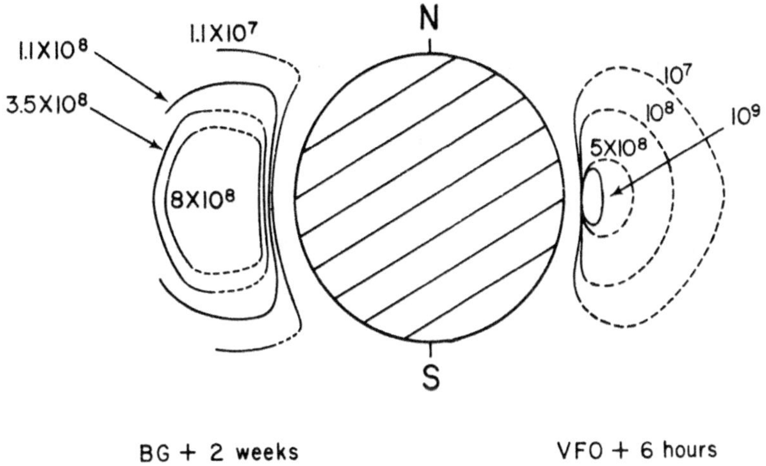

Fig. 18. The BG and VFO flux maps that were used in calculations. Numbers are omnidirectional fluxes in electrons/cm² sec.

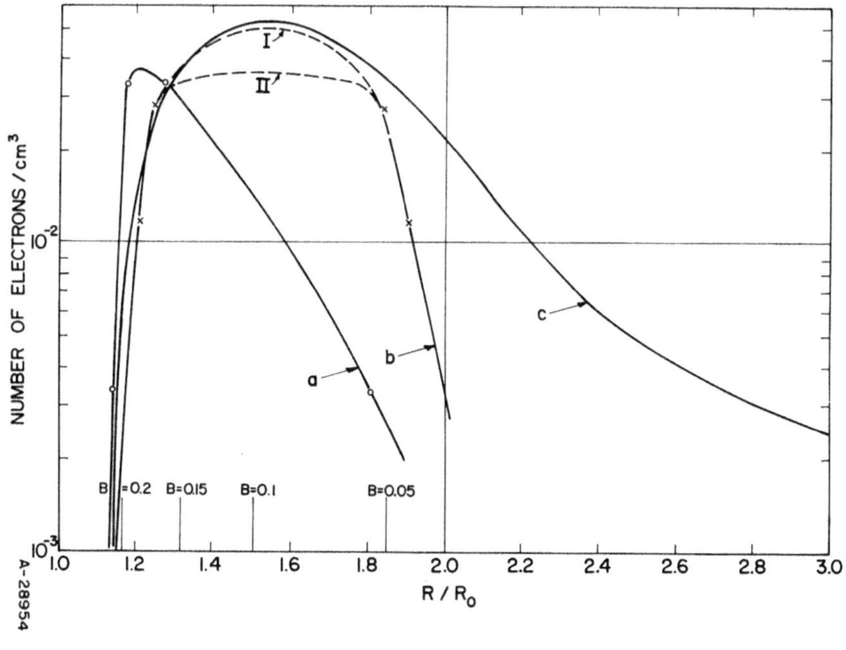

Fig. 19. Density of electrons vs height over magnetic equator in earth radii for the flux contour of: (a) VAN ALLEN, FRANK, and O'BRIEN (1963); (b) BROWN and GABE (1963; (c) HESS (1963).

where E=electron energy in MeV; N_0=total volume density; N=fraction of N_0 at a given energy. We will consider the brightness at a constant frequency f. The effects of variation of ψ and α can be accounted for approximately by assuming that those electrons having pitch angles within $\psi_{1/2}$ (the radiation pattern beamwidth) of perpendicular to the earth's field contribute to the received power and that all others do not. An estimate of this useful fraction can be obtained in the following fashion. Assume that distribution of pitch angles is isotropic. Then the fraction of useful electrons

$$F_{\alpha,\psi} = \frac{\text{solid angle within } \psi_{1/2} \text{ of } 90° \text{ from field}}{\text{total solid angle}}$$

$$= \frac{2\pi(2\psi_{1/2})}{4\pi} = \psi_{1/2}$$

Therefore

$$F_{\alpha,\psi} \cong \psi_{1/2} = 1.59\sqrt{B/f}$$

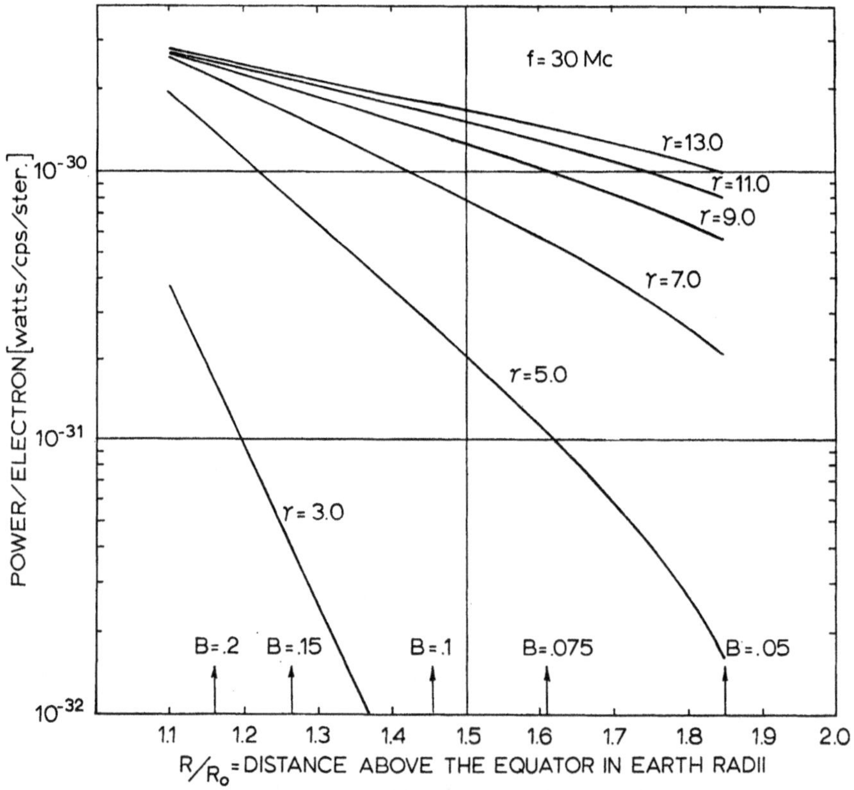

Fig. 20. Power per electron at a frequency of 30 Mc/s as a function of height in earth radii for a number of different electron energies.

Using this approximation, (21) becomes

$$b = \int\int [F_{\alpha,\psi}] P(0^\circ, f_0 E, B, 90^\circ) \varrho(E, r) \, dr \, dE \tag{24}$$

As a means of carrying out the indicated integration, it is convenient to calculate the radiated power per electron in the orbital plane as a function of distance

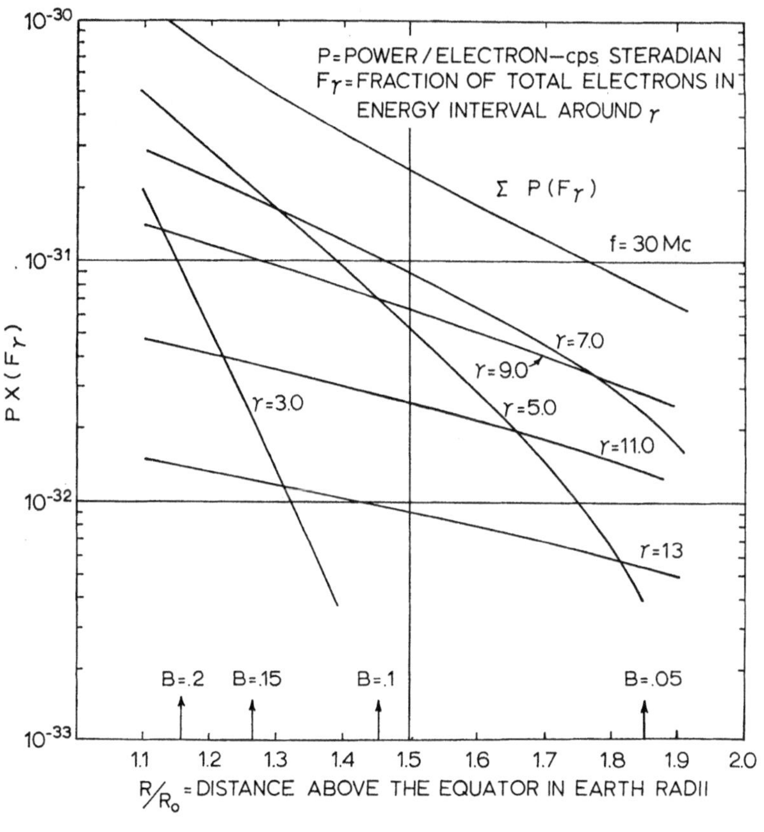

Fig. 21. Power per electron (P) times the fraction ($F\gamma$) of electrons in energy interval around γ. The $\Sigma P(F\gamma)$ curve is the power per electron (watts/cps ster) in the orbital plane ($\psi = 0^\circ$ at 30 Mc/s, assuming that the energy spectrum of the electrons is a fission β spectrum.

above the earth with electron energy (γ) as a parameter. The results of this calculation for a frequency of 30 Mc/s are shown in Figure 20, based on curve c of Figure 19. It is apparent from this calculation that for heights in excess of a few tenths of an earth radius only the higher-energy electrons contribute appreciably to received power at 30 Mc/s.

A useful next step in the calculation is to scale the curves of Figure 20 in accordance with the spectrum of (23) from which the fraction of total electrons in different energy groups $F\gamma$ can be obtained (i.e. fraction near $\gamma=3$, $\gamma=5$, etc.). This has been carried out, and the results are presented in Figure 21. From this figure it can be seen that for $1.1<R/R_0<2.0$ the principle contribution to the radiated power comes from electrons of energy in the range from $\gamma\approx 5$ to $\gamma\approx 9$. For low energies the power per electron radiated decreases rapidly, and for higher energies the number of electrons in the energy group decreases rapidly.

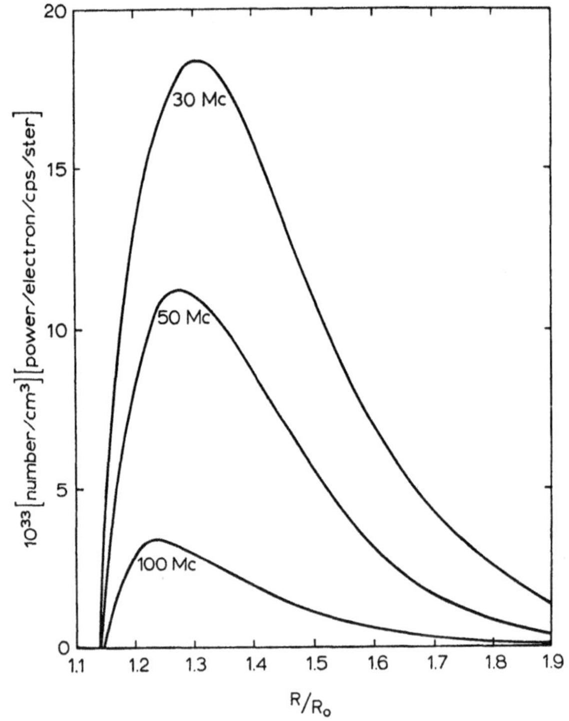

Fig. 22. Synchrotron power versus height over the magnetic equator for frequencies of 30, 50, and 100 Mc/s.

It is next necessary to sum the contributions over the various energy groups to give the average power per electron as a function of height. This is shown in Figure 21 as the curve labeled $\Sigma PF\lambda$.

Next, the number of electrons per unit volume (Figure 19) can be multiplied by $\Sigma PF\lambda$ to give the curves of Figure 22. These curves are a measure of the relative contribution at 30, 50, and 100 Mc/s of electrons at different heights to the total brightness (power/unit area cps ster) of the sky over the equator. In Figure 23 are shown the synchrotron power at 30 Mc/s versus height corresponding to curves *a* and *c* of Figure 19.

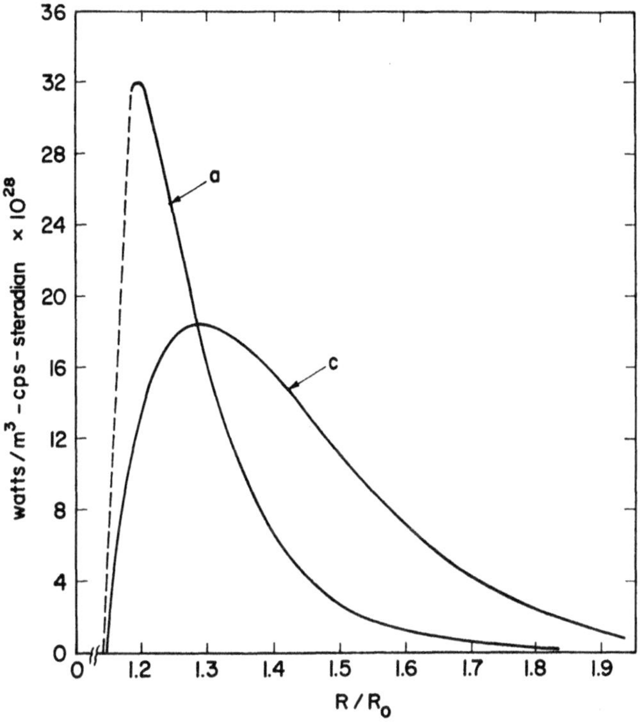

Fig. 23. Synchrotron power at 30 Mc versus height over magnetic equator for the flux contours of: (a) VAN ALLEN, FRANK, and O'BRIEN (1963); (c) HESS (1963).

Numerical integration of these curves over the height region gives the final value of the brightness at a particular frequency for the given distribution of trapped electrons. Thus for 30 Mc/s

$$b = 4.75 \times 10^{-21} \text{ watt/cps m}^2 \text{ ster}$$

and

$$T = (\lambda^2/2k)b = 1.7 \times 10^4 \text{ °K}$$

In Figure 24 is shown the results of calculations of the type just described which have been carried out for a number of frequencies (10, 30, 50, and 100 Mc/s). The corresponding sky temperatures are summarized in Table I.

TABLE I

Frequency, Mc/s	Temperature, °K
10	8.5×10^5
30	1.7×10^4
50	3.2×10^3
100	1.9×10^2

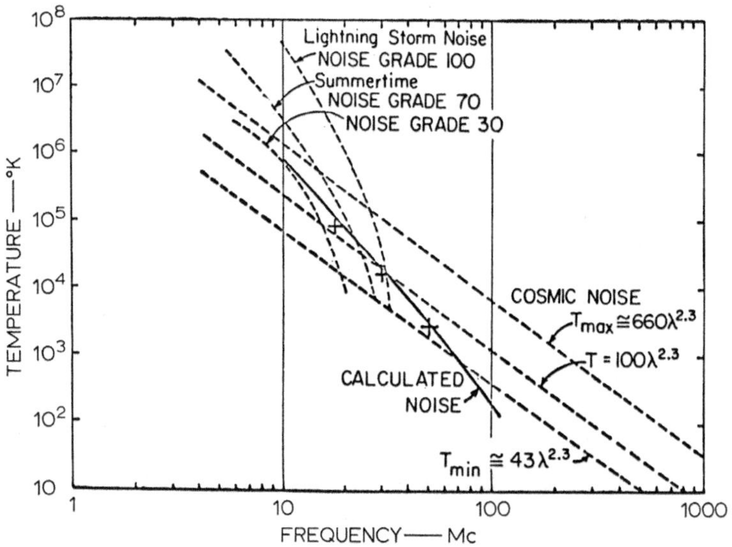

Fig. 24. Comparison of synchrotron noise (solid line is calculated noise; crosses are measured noise) with cosmic noise and lightning storm noise (straight dashed lines are cosmic noise; curved dashed lines are lightning storm noise).

In Figure 24 these results are compared with background sky temperatures (Menzel, 1960), cosmic and lightning storm noise (*International Radio Consultative Committee*, 1957), and with measured values of synchrotron radiation. The heavy solid curve is the calculated synchrotron noise curve, and the large crosses are measured synchrotron noise at 18, 30, and 50 Mc/s at a time about three weeks after the explosion. This time corresponds rather well to the time at which the electron flux contours were measured.

The brightness temperature computed from the Van Allen-Frank-O'Brien contours is found to be approximately 75 percent of the value corresponding to the Hess model. The value obtained using the Brown-Gabbe model depends on how one extrapolates in the region between the points where experimental

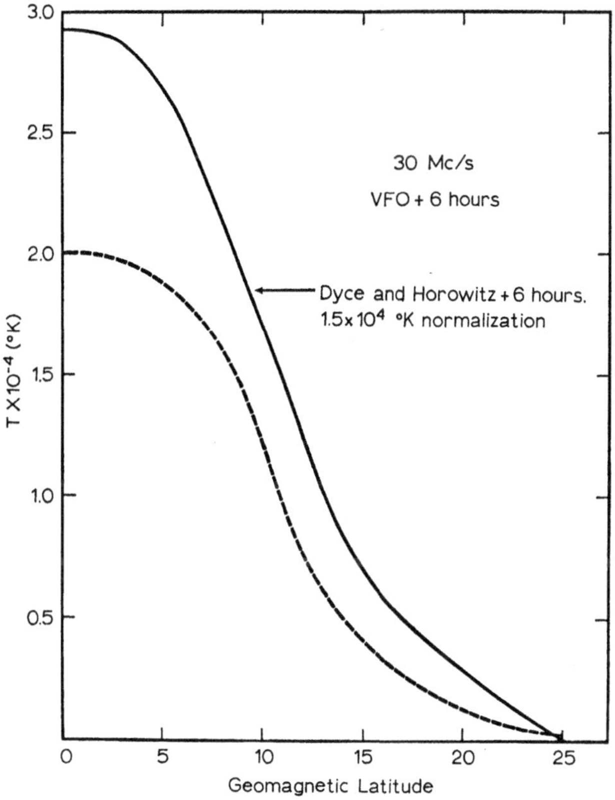

Fig. 25a. Dotted line shows predicted sky temperatures at +6 hours for vertically directed antennas with $\cos^2 Z$ pattern for the VFO flux estimate at different geomagnetic latitudes. Solid curve is the renormalized experimental curve of Dyce and Horowitz.

data are plotted. If the rounded variation labeled (I) in Figure 19 is used, the results are essentially the same as for the Hess model, since the chief differences occur at altitudes above 1.8 earth radii. Because of the lower field intensities, particles in this region do not contribute significantly to the emission received at the equator. If the flatter density variation labeled (II) in Figure 19 is assumed, a value is obtained which is essentially equal to that corresponding to the Van Allen-Frank-O'Brien contours.

It is important to note that the density variation for the Van Allen-Frank-O'Brien model is determined by a very few points which are themselves difficult to read from the published flux contours. Thus, the brightness determined from the curve of Figure 23 may vary considerably depending on how the points are read, especially in the low altitude region which gives the strongest contribution to the total emission.

C. OTHER CALCULATIONS

Theoretical studies of the synchrotron radiation following the Starfish explosion have also been carried out by NAKADA (1963) and by KAUFMAN (1964). Nakada using a simple analytic approach calculated not only the equatorial value of the noise temperature but also estimated the variation with latitude. Figure 25a shows Nakada's calculation based on the Van Allen-Frank-O'Brien contours of the latitude variation of sky temperature (dotted curve) together with an experimental curve (DYCE and HOROWITZ, 1963) as normalized by Nakada to an assumed 1.5×10^4 °K minimum cosmic noise background on the equator at 30 Mc/s. It should be noted that if the original experimental curve of Dyce and Horowitz (see Figure 29) is accepted without changes, then very close agreement exists with Nakada's calculations. Figure 25b is a similar curve calculated by Nakada based on the contours of Brown and Gabbe (see Figure 18) along with a "normalized" experimental curve of Dyce and Horowitz.

KAUFMAN (1964) carried out detailed computer calculations including the effects of electron distributions (to the extent known) and antenna patterns. His valuable results are as yet unpublished.

D. POLARIZATION OF SYNCHROTRON RADIATION

Characteristically, synchrotron radiation, when emitted, is predominantly linearly polarized at right angles to the magnetic field. However, the polarization of the received signals at the ground depends on ionospheric conditions along the propagation path.

A radio wave linearly polarized at right angles to the field corresponds to a quasi-transverse (QT) characteristic wave (RATCLIFFE, 1960). Since a characteristic wave propagates without changing polarization, the same linear (east-west) polarization would be observed at the receiver if the QT approximation were valid all along the propagation path. However, for VHF signals in the earth's ionosphere, this approximation is only valid for directions of propagation within a fraction of a degree of the perpendicular to the field. Thus, the only possibility of receiving this mode at the ground would be with an antenna looking vertically at the magnetic equator and having a very narrow beamwidth

($\approx 1°$). Even in this case, local irregularities in the field would make it improbable that the beam would stay at right angles to the field at all points along the path.

Under these same conditions, the quasi-longitudinal (QL) approximation will be valid for all directions of propagation not within a few (2 or 3) degrees of the perpendicular. Therefore, most of the signal reaching a given antenna

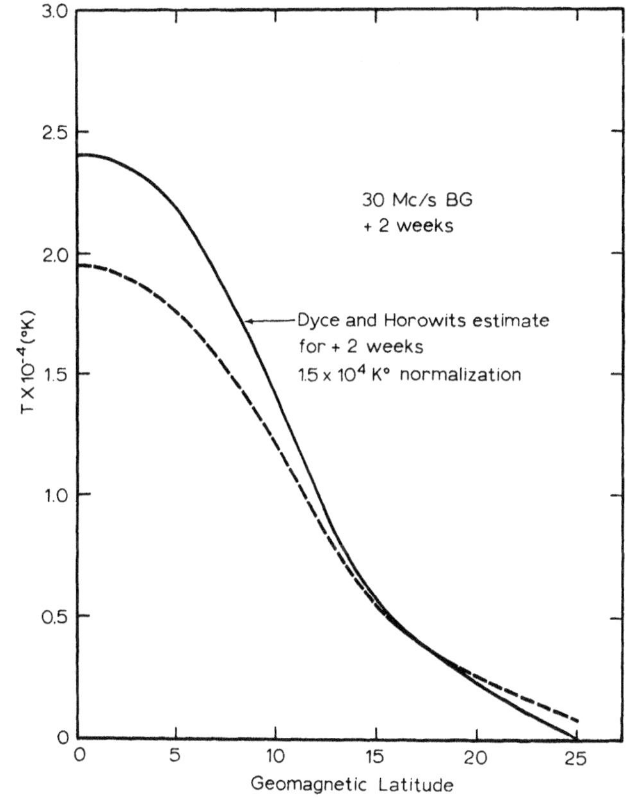

Fig. 25b. Dotted line shows predicted sky temperatures for vertically directed antennas with $\cos^2 Z$ pattern for the BG flux estimates at +2 weeks at different geomagnetic latitudes. Solid curve is the experimental curve of Dyce and Horowitz; this curve has been renormalized and a decay has been included.

will travel via this mode. For an ionized region in which QL conditions exist, a linearly polarized signal is split into two equal and oppositely rotating circularly polarized waves. Since these waves travel with different velocities, the linearly polarized wave obtained at any point by combining the two has a plane of polarization that is rotated with respect to the original plane (Faraday rotation).

For frequencies that are large compared with both the gyrofrequency and the plasma frequency existing in the medium, the amount of the Faraday rotation (Ω) may be expressed as

$$\Omega = \frac{\mu_0 e^3}{8\pi^2 \varepsilon_0 m^2 c} \frac{1}{f^2} \int HN \cos\theta \, \mathrm{d}s \text{ radians}$$

where N is the electron density of the medium, θ is the angle between the magnetic field H and the direction of propagation, and ds is an element of the path. Most of this total rotation thus occurs near the peak of the F region at the lower altitudes where the electron densities are highest and will vary markedly from night to day as the ionosphere density varies. In determining whether the polarization of synchrotron emission will be detectable, however, we are not concerned with the total rotation but with the spread in the planes of polarization (differential rotation) of signals arriving from different parts of the emitting region.

Consider emission from a thin shell of electrons over the magnetic equator for a particular ray along which QL conditions exist. Since H, N, and cos θ are essentially constant along the path through the shell, the amount of rotation occurring within the emitting region is given by

$$\Delta\Omega \approx (0.0297/f^2) NH \cos\theta (\Delta r)$$

where Δr is the thickness of the shell and H, N, and cos θ are the values of these quantities within the shell. The orientation of the axes of polarization of signals originating at various points along this ray would thus differ by $\Delta\Omega$ on emerging from the emitting region. Since the rotation occurring along the rest of the propagation path would be the same for each component of the signal at a given frequency, $\Delta\Omega$ is also the spread in polarization angles that would be observed with a very narrow beam antenna oriented in this direction. As an example, consider emission from a 500-km thick shell, 1500 km over the magnetic equator, and located in an ambient electron density 0.5×10^{10} electrons/m^3. Then the spread in polarization angles occurring along a ray for which $\cos\theta \approx 0.1$ is about 0.1 radian.

For observations made with a wide beam antenna, the signals from different parts of the emitting region would travel through the ionosphere along oblique paths of different lengths. Also, the appropriate values of cos θ would be different along the various paths. Using a simplified model of the ionospheric electron density profile below 500 km with a peak of 1×10^{11} m^{-3} and considering a 40° beam antenna, calculations indicate that for the example of the previous paragraph the differences in path length could produce an additional differential rotation of the order of 1 radian, whereas the differences in cos θ

could add several radians. An additional effect for a vertically directed antenna at the equator is the fact that the sense of the Faraday rotation is opposite for signals propagating from the north and south. It would appear, therefore, that detection of polarization is unlikely unless observations are made with narrow beam antennas, in which case the polarization should be easily detected.

E. SUMMARY OF SYNCHROTRON NOISE OBSERVATIONS

At radio frequencies in the 20–120 Mc range special radio receivers known as "riometers" were used to monitor absorption and also the synchrotron noise associated with the Starfish event. In this frequency range broad-band cosmic radio noise can be regularly monitored for geophysical research purposes. In the absence of ionospheric disturbances, a few days of observation are adequate to determine the normal (or "quiet-day") profile of received power versus sideral time within a few percent. A depression below this normal curve can then be interpreted as attenuation (principally D-region absorption). A rise of the power level above the normal curve implies a new source of radio power in addition to the usual cosmic noise power. This excess can be expressed as the brightness temperature of an equivalent black body radiator which is viewed by the riometer antenna.

In addition to the special network of riometers in the mid-Pacific certain other geophysical research centers were in strategic locations to observe the synchrotron noise produced by the Starfish shot. One of these was located at Huancayo, Peru, which is very close to the magnetic equator, and Figure 26 shows riometer records from this location (Dyce and Horowitz, 1963). Note that time proceeds from right to left and that the ordinate is linearly proportional to power. The top curve shows the shape of the normal curve on the day before the detonation. The maximum near the left edge of the chart is due to the passage of the galactic center overhead as the earth rotates. Synchrotron radiation appears very shortly after shot time, temporarily driving the record off scale. The power level remains above the preshot normal curve for the rest of the records. This general pattern was observed at most of the riometer sites in the mid-Pacific area, though the records were in many cases disturbed by absorption as well as synchrotron noise.

Figure 27 shows the 30 Mc riometer record from Wake Island, about 2500 km west of Johnston Island. Significant synchrotron noise was not observed until ten minutes after the detonation at which time the shell of trapped electrons has migrated eastward and reached overhead at Wake Islands having nearly progressed around the world. The maximum synchrotron noise occurred at about +25 minutes. Shortly thereafter the noise level becomes about constant, implying that the trapped electrons have become approximately uniformly distributed in longitude.

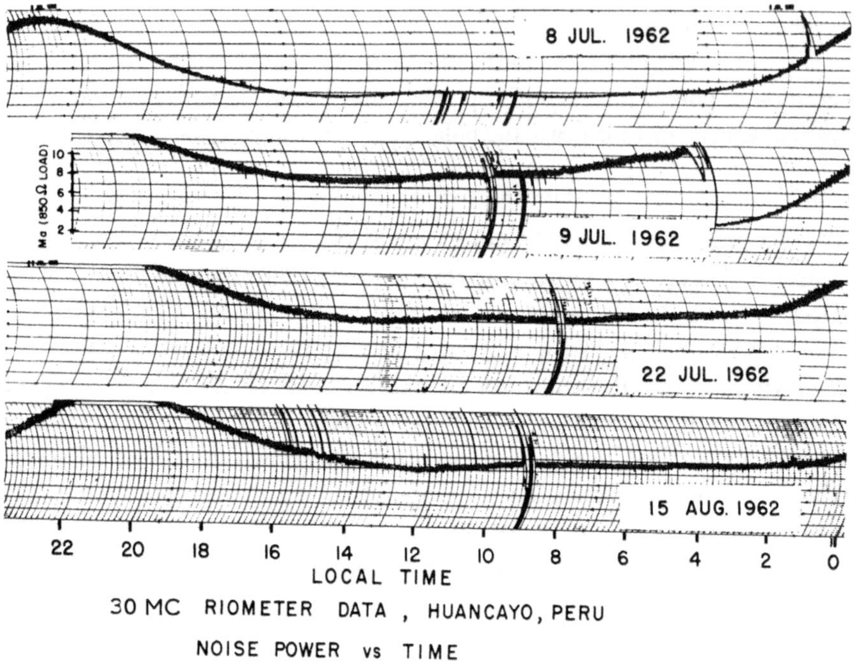

Fig. 26. 30-Mc riometer data, Huancayo, Peru; noise power vs time.

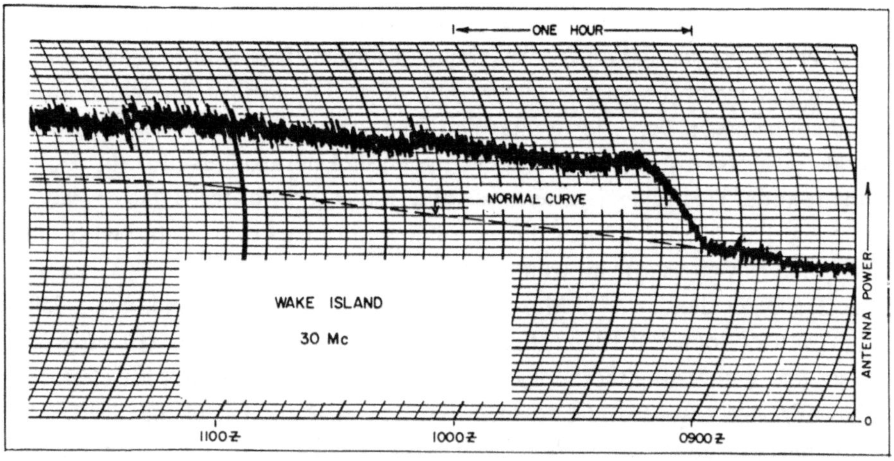

Fig. 27. Effect of wave of synchrotron radiation after the relativistic electrons have migrated once around the earth. Except for brief attenuation in the first 15 sec after shot time, the record is essentially undisturbed until +10 min.

Figure 28 shows a noise record (Ochs *et al.*, 1963) taken at the Jicamarca Observatory in Peru at a frequency of 50 Mc. Two peaks can be seen on this record. The first large peak in the noise record, 6 minutes after the detonation corresponds to the first pass of the electrons overhead. The second smaller peak is caused by the drifting electrons coming overhead for the second time. This peak is very broad because the different drift rates of electrons lead to a rapid dispersion of the initially tightly-bunched group.

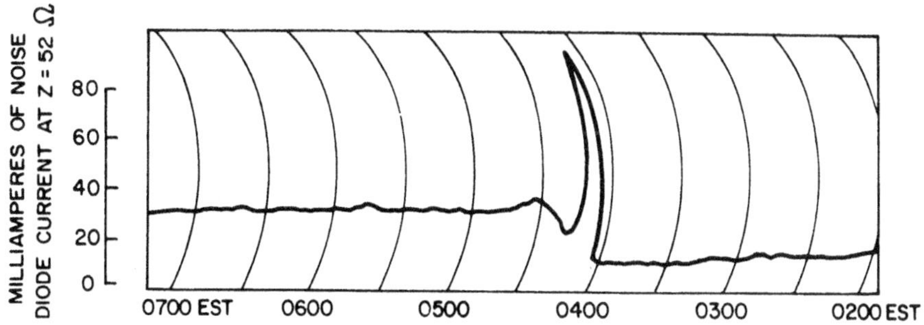

Fig. 28. Total radio noise power at 50 Mc/s (Jicamarca).

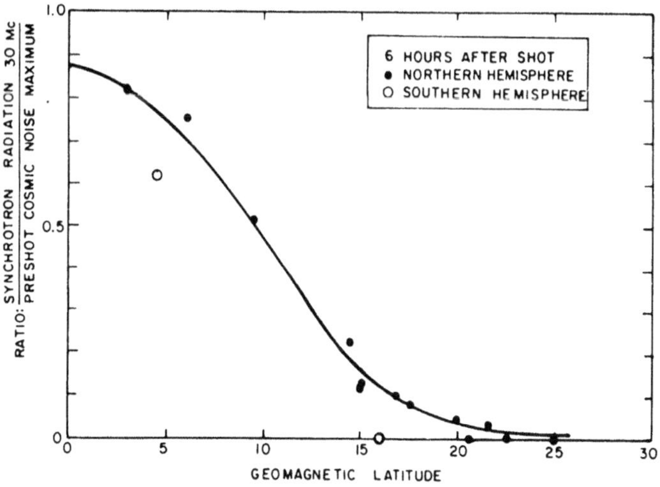

Fig. 29. 6 hours after shot.

Figure 29 shows the dependence of synchrotron noise on magnetic latitude (Dyce and Horowitz, 1963), at a time six hours after the explosion. There was a maximum of synchrotron noise at the magnetic equator and the intensity fell to half maximum at 12° from the equator.

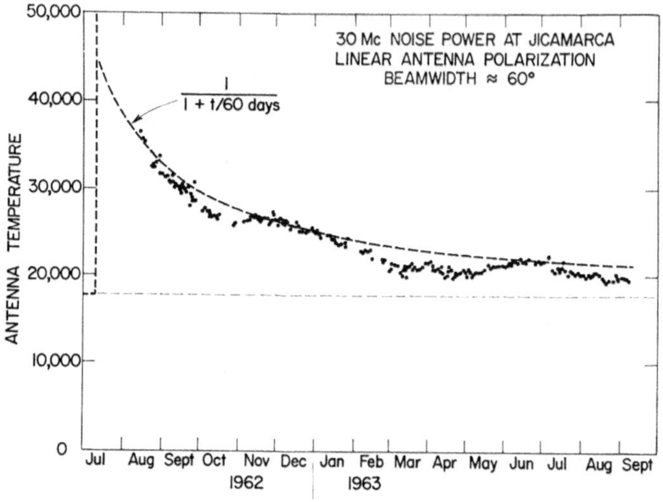

Fig. 30. 30-Mc noise power at Jicamarca, linear antenna polarization, beamwidth 60°.

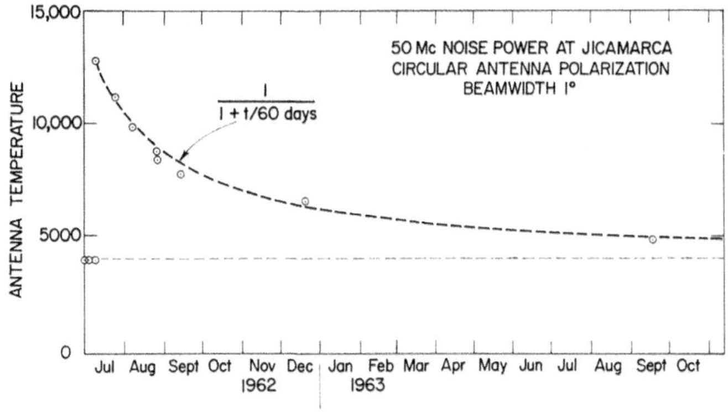

Fig. 31. 50-Mc noise power at Jicamarca, circular antenna polarization, beamwidth 1°.

Figures 30 and 31 show the decay of synchrotron noise at 30 Mc and 50 Mc at Jicamarca, Peru (Ochs, 1964) during a period of fifteen months. The initial noise output following the explosion was about three times the cosmic noise background and has now decreased to within about 10% of the background.

4. Conclusions

Calculations of synchrotron radiation based on the trapped electron model agree reasonably well with measurements made at several radio frequencies at stations near the magnetic equator. These calculations show that the radio noise in the 20–100 Mc/s frequency range is generated primarily at heights above the magnetic equator between 1.2 and 1.6 earth radii and is a very insensitive measure of electrons moving at heights above about 1.6 earth radii because of the weak magnetic field at these heights. Van Allen has suggested that few of the trapped electrons at heights above 1.5 earth radii are high-energy fission β's from the July 9 nuclear explosion but instead are a part of the previously existing natural belt. Since the synchrotron noise measured is essentially independent of the presence or absence of electrons in this outer region, these measurements do not help to resolve the differences. However, measurements at 30 and 50 Mc/s, together with the calculations presented here, do show that the trapped electrons at heights below about 1.6 earth radii must contain at least as many electrons of energy above 1 MeV as occur in the fission electron spectrum.

Electrons trapped on high L shells will not yield observable synchrotron noise at the ground, except when near mirror points which are themselves within a small fraction of an earth's radius of the earth. Noise from these mirroring electrons would, in addition, only be seen by antennas which are directed nearly at right angles to the earth's magnetic field. At temperate latitudes in the northern hemisphere noise would only be observed when looking generally northward. The noise intensity would be expected to be reduced greatly below that which was measured by equatorial stations with vertically directed antennas.

Polarization measurements at equatorial stations using narrow beam antennas provide an extremely sensitive indication of synchrotron radiation from trapped electrons and will be particularly useful for measurements at later times when this component of noise becomes weak in comparison to the cosmic noise background.

References

Brown, W. L. and Gabbe, J.: 1963, *J. Geophys. Res.* **68**, 607.
Carter, R. E., Reines, F., Wagner, J. J., and Wyman, M. E.: 1959, *Phys. Rev.* **113**, 280.
Computation Laboratory of the National Applied Mathematics Laboratories, National Bureau of Standards: 1949, *Tables of Bessel Functions of Fractional Order*, Vol. II, Columbia University Press, New York.
Dyce, R. B., and Nakada, M. P.: 1959, *J. Geophys. Res.* **64**, 1163.
Dyce, R. B. and Horowitz, S.: 1963, *J. Geophys. Res.* **68** 713.
Egan, R. D. and Peterson, A. M.: 1960, *J. Geophys. Res.* **65** 3830.
Hess, W. N.: 1963, *J. Geophys. Res.* **68** 000.
International Radio Consultative Committee: 1957, *Rept. 65*, Revision of atmospheric radio noise data, published by International Telecommunications Union, Geneva.
Jahnke, E. and Emde, F.: 1945, *Tables of Functions*, Dover Publications, New York.

Kaufmann, R. L., 'Synchrotron Radiation from the Starfish Electron Belt', *Technical Report*, Department of Physics, University of New Hampshire, Durham, New Hampshire, 1964 (NASA Contract NAS W-155).

Landau, L., and Lifshitz, E.: 1951, *The Classical Theory of Fields*, Addison-Wesley, Reading, Mass.

Menzel, D. H.: 1960, *The Radio Noise Spectrum*, Harvard University Press, Cambridge, Mass.

Nakada, M. P. 11963, *J. Geophys. Res.* **68**, 4079.

Ochs, G. R., Farley, D. T., Bowles, K. L., and Bandyopadhay, *J. Geophys. Res.* **68**, 701.

Ochs, G. R.: 1964, *J. Geophys. Res.* **69**, 000.

Panofsky, W. K. H. and Philips, M.: 1960, *Classical Electricity* and *Magnetism*, Addison-Wesley Publishing Co., Reading, Mass. (2nd edition 1962).

Peterson, A. M., and Hower, G. L.: 1963, *J. Geophys. Res.* **68** 723.

Piddington, J. H.: 1961, *Radio Astronomy*, Harper and Brothers, New York.

Ratcliffe, J. A.: 1959, *The Magneto-Ionic Theory*, Cambridge University Press, Cambridge, England.

Shklovsky, I. S.: 1960, *Cosmic Radio Waves*, Harvard University Press, Cambridge, Mass.

Schwinger, J.: 1949, *Phys. Rev.* **75**, 1912.

Van Allen, J. A., Frank, L. A. and O'Brien, B. J.: 1963, *J. Geophys. Res.* **68**, 619.

Watson, G. N.: 1952, *Theory of Bessel Functions*, 2d ed., Cambridge University Press, Cambridge, England.

Westfold, K. C.: 1959, *Astrophys. J.* **130**, 241.

Discussion

Dr. Walter Brown wondered if there was any evidence for electron bunching in phase, and of correlative effects in synchrotron radiation. Dr. Peterson replied that small amounts of such effects should change results rather drastically. Baggerly commented that in synchrotrons where phase bunching of electrons does occur, there appears to be no evidence for correlative effects. Peterson wondered whether such correlative effects might not occur in whistler amplification. However, he added that although such effects were likely at the gyrofrequency of electrons, they were unlikely at the higher harmonics which characterize synchrotron radiation.

SESSION VIII

EFFECTS OF TRAPPED RADIATION ON SENSORS, ELECTRONIC DEVICES, ETC.

Chairman

DR. CHARLES HULBURT

Defence Atomic Support Agency

EFFECTS OF TRAPPED RADIATION ON ELECTRONIC DEVICES

W. L. BROWN
Bell Telephone Laboratories, Inc., Murray Hill, New Jersey

1. Introduction

The high-energy electrons and protons trapped in the earth's magnetic field provide a severe environment for the long-time reliable performance of complex satellite systems in space. The environment is extremely dependent upon the orbit of a satellite due to the enormous spatial variations in the trapped particle flux. This flux rises rapidly above the top of the atmosphere and ultimately falls to very low values at the boundary of the magnetosphere. It has a rich variety of structure in between, depending on particle type and energy, and upon time because of variations in the activity of the sun. Other papers at this Institute have dealt in detail with the observed distributions of the trapped particles and with the geophysical mechanisms responsible for them. It has been amply demonstrated that a great deal of our present understanding is still in quite a primitive state. This paper is not concerned directly with geophysics but with the consequences of the geophysics of the magnetically trapped radiation, folded through the solid state physics of materials and devices, to the hardware of unmanned vehicles in space. The paper is organized to discuss the different classes of important radiation effects; to consider in detail the radiation effects on semiconductor devices, often the most sensitive elements of unmanned space systems; and finally, to illustrate with the Telstar Satellite experiments, the observations of radiation damage in space and their comparison with expectations based on laboratory experiments and a knowledge of the space environment.

2. Radiation Damage

A. PRIMARY MECHANISMS

Radiation has two fundamentally different mechanisms through which it interacts with materials. These interactions in some cases are extremely desirable, but the present paper is limited to considerations of their damaging or destructive consequences. The first mechanism is that of ionization. A high-energy charged particle strips electrons off of atoms with which it has close encounters. This is shown in Figure 1 for the atoms of a homopolar, crystalline solid such as silicon. The electrons primarily involved are those in the outermost electronic shells of the atoms, the electrons that form the chemical bonds of the atoms to one another in a solid or even in a single molecule. In many chemical systems, such disruption of chemical bonds provides an opportunity for electronic rearrangements that yield new molecular species or new inter-

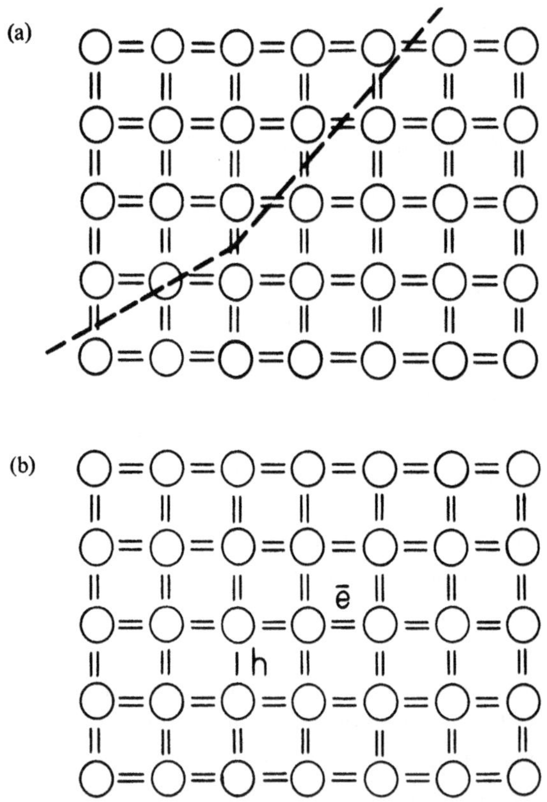

Fig. 1. Ionization by a high energy charged particle in a solid.
(a) Interaction with a valence electron. (b) The resulting hole-electron pair.

connections between those originally present. It may produce very corrosive and volatile molecular fragments. It may shorten long-chain polymers (as in the biological molecule DNA) with a wide variety of subsequent results, or it may cross-link polymer chains to make higher molecular weight materials. In optical materials, it may result in formation of color centers that dramatically alter the emission and absorption characteristics of the materials.

In a semiconductor like silicon, ionization produces holes and electrons whose existence is transient. These mobile electrical carriers may live long enough to provide an external electrical signal indicating the passage of a charged particle, as in a semiconductor particle detector, for example, but in a period of microseconds following the ionization event the original electronic bonds are

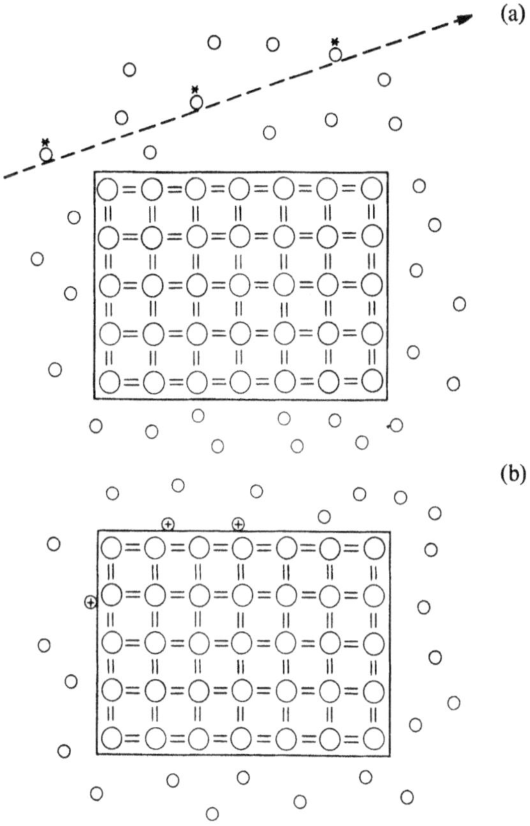

Fig. 2. (a) Ionization of gas adjacent to the surface of a semiconductor device by a high energy charged particle. (b) Deposition of ions on the surface.

restored by recombination. In the presence of massive pulsed ionization sources (such as a nuclear detonation) the transient currents that flow in electronic devices may disrupt normal circuit functions, but rarely in a way immediately destructive to the device, and following the transient the ionization effects in transistors or diodes will have disappeared. The radiation level encountered in space is ordinarily of no consequence from this point of view.

There is an important exception to the lack of concern for semiconductors in response to ionization effects in space, and that is at their surfaces. Figure 2 illustrates a particular case. Ionization in the material surrounding a semiconductor device, either a gas or an inorganic liquid or an oxide coating, can result in the deposition of charged ions at the device surface. These charges can,

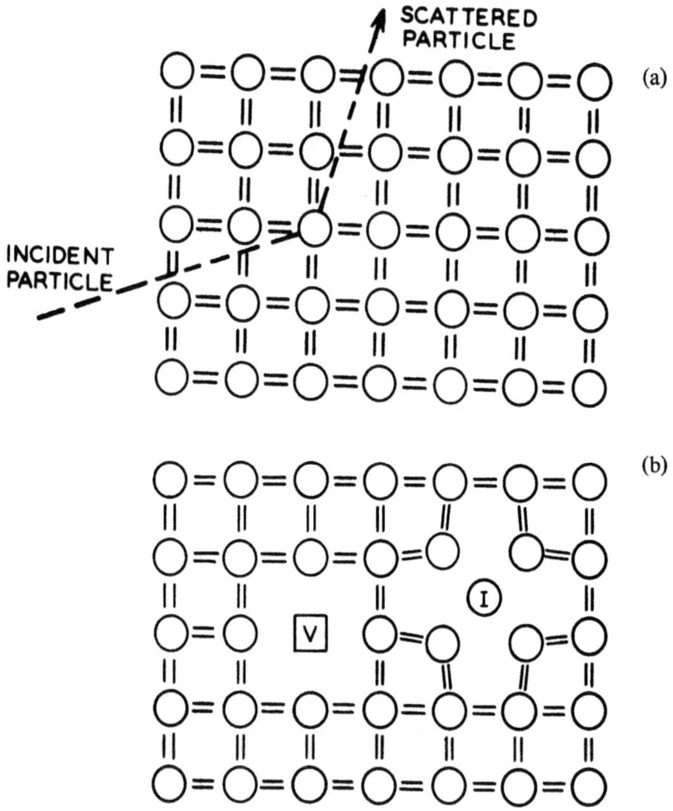

Fig. 3. Atomic displacement by a high energy particle. (a) Collision with the nucleus of an atom. (b) Formation of a vacancy and interstitial by displacement of the struck atom.

even at surface concentrations of a tiny fraction of a monolayer, tremendously alter the electrical properties of a device. Some more detailed aspects of these phenomena will be discussed in Section 2C.

The other primary interaction of radiation with materials is in atomic displacement. Rather than producing ionization by interaction with electrons, a high energy particle actually collides with the nuclei of the material and transfers enough momentum to them to knock them out of their normal positions, Figure 3. In a gas or liquid, this is usually quite unimportant. However, in a crystalline solid, the vacancy (where the atom originally belonged) and the interstitial (the displaced atom in an unnatural position) are defects in an otherwise highly-ordered structure. These defects alter the electrical properties

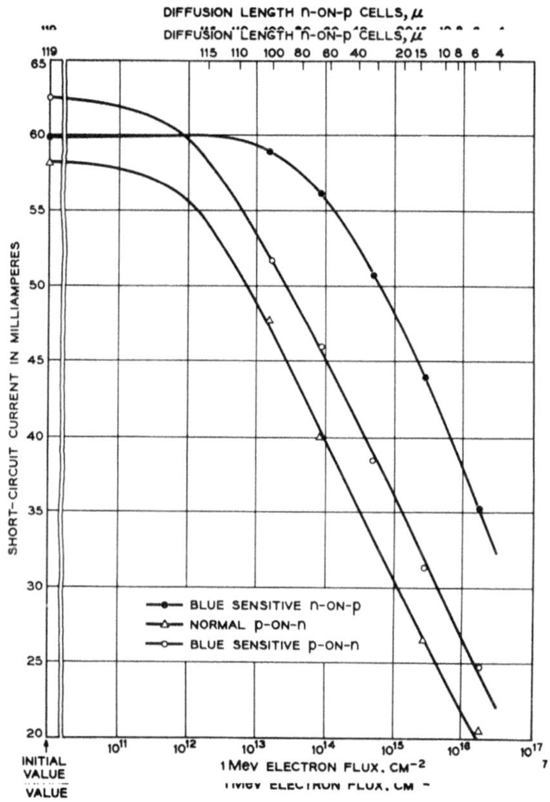

Fig. 4. Decrease in solar cell short-circuit current by 1-MeV electrons.

of the material very much like the alterations produced by chemical impurities. In metals a very high concentration of defects can be tolerated. The electrical properties of these materials normally depend on electron concentrations that are approximately equal to the atomic concentration of the solid, and the defect concentration must be at least some small fraction of the atomic concentration before the damage has a pronounced effect. In semiconductors, however, the whole range of electrical properties depends on a balance of impurities at a level of 1 in 10^6 or 10^8 of the atomic concentration, or even less. Radiation produced defects thus play a highly sensitive role, and often an even more sensitive role than this simple argument would suggest when the detailed nature of the electrical properties of the defects are examined. Some device-oriented aspects of this problem will be discussed in Section 2B.

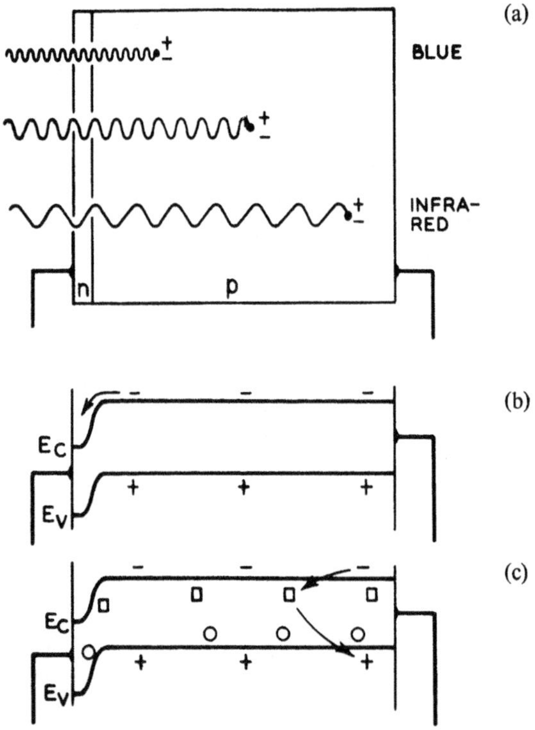

Fig. 5. Solar cells. (a) Production of hole-electron pairs by absorption of light of different wavelengths. (b) Diffusion of carriers to the solar cell junction. (c) Recombination of carriers at radiation produced defect centers.

B. SEMICONDUCTOR DAMAGE BY ATOMIC DISPLACEMENT

Probably the most familiar consequence of radiation damage in space is the degradation of the electrical power output from solar cells. The silicon solar cell case will be used as an illustration here, both because of the importance of the solar cell problem itself, and because the damage effect in solar cells provides information of value to a number of other related cases. Figure 4 shows the decrease in the short-circuit current available from the three different types of solar cells under the solar illumination of space in response to radiation damage by 1 MeV electrons at normal incidence (ROSENZWEIG *et al.*, 1963). The plot is semi-logarithmic, and after the high energy particle flux begins to have a substantial effect, the primary decrease in current occurs approximately logarithmically with flux. This shape is a coincidence which actually depends

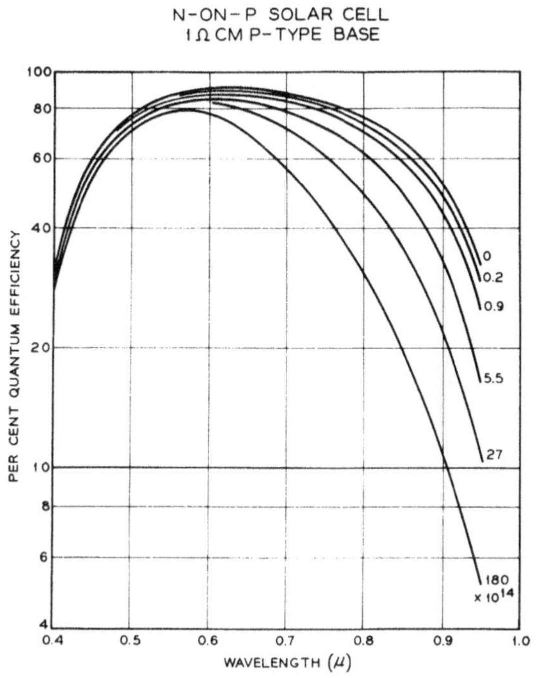

Fig. 6. Quantum efficiency of a solar cell for different concentrations of radiation produced defects. The curves are labeled with the integral damaging flux of 1-MeV electrons/cm^2.

upon the optical properties of silicon and the spectral characteristics of the illumination. It is worthwhile to look briefly at what is happening.

Figure 5 is a highly schematic view of a solar cell. The device might be formed by a shallow diffusion (approximately 0.5 microns) of an n-type impurity (phosphorous) into a relatively thick (approximately 0.5 millimeters) base of approximately 1 ohm-centimeter p-type silicon. This was, in fact, the form of solar cells used on the Telstar satellites (SMITH, 1963). Photons are incident from the left in the figure. For monochromatic light, absorption occurs exponentially with distance into the material, but the characteristic absorption distance varies tremendously with wavelength in the silicon case. Infrared light of about 1 micron wavelength, which is just barely energetic enough to break one of the valence bonds of the crystal, is very weakly absorbed and thus has an absorption which occurs almost uniformly through the material thickness. Blue light with a wavelength of about 0.4 microns on the other hand, is very strongly absorbed,

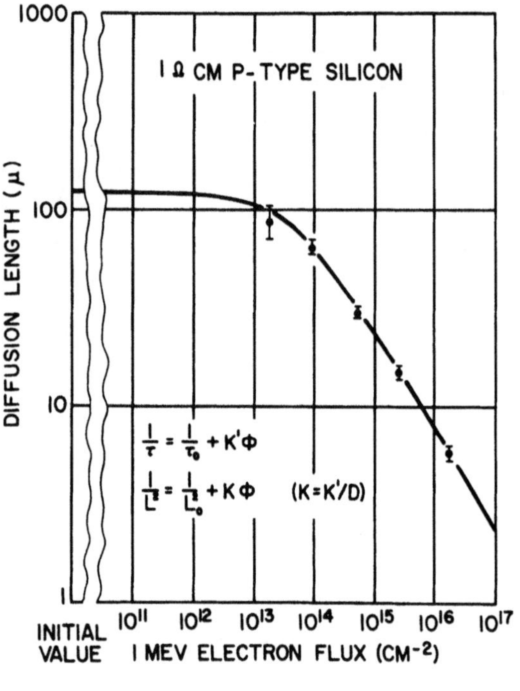

Fig. 7. Diffusion length vs. integral flux of 1-MeV electrons.

within a characteristic distance of the order of 1 micron. Such is the intended indication of Figure 5a. The absorption event associated with a single photon results in production of one hole-electron pair just as was illustrated in one of the multiplicity of ionizing interactions that a high energy charged particle has with a semiconductor (Figure 1). The pairs produced by red light are more deeply distributed in the material than those produced by blue.

In order for these pairs to be available as a current to an external circuit, they must reach the junction between the n and p-type materials. There the holes and electrons will be separated by the junction field, Figure 5b. Pairs reach the junction by diffusion, and they do so while processes of recombination are working to return the electrons to holes in the broken crystalline bonds. The pairs produced farthest from the junction take the longest time to arrive, and are thus subject to much higher probability of recombination. For well-made solar cells the diffusion length, the mean distance which pairs can diffuse before recombination, may be 100 to 200 microns, so that pairs from a relatively broad part of the solar spectrum are efficiently included in the solar cell current.

Radiation damage defects act as recombination centers for holes and electrons, as shown in Figure 5c. At small defect concentrations they primarily influence the collection of pairs produced by red light. This is shown directly in Figure 6, which gives the quantum efficiency, the number of collected pairs per incident photon, as a function of wavelength (ROSENZWEIG *et al.*, 1963). The lower curves correspond to increasing the concentration of radiation damage defects.

To a very good approximation, the rate of recombination is proportional to the concentration of added defects, and the total rate of recombination is the sum of this rate and the rate originally present in the material. This can be written in terms of the inverse lifetimes as

$$\frac{1}{\tau} = \frac{1}{\tau_0} + \frac{1}{\tau_{RD}} = \frac{1}{\tau_0} + K' \Phi \tag{1}$$

where τ_0 is the initial lifetime, and K' is the inverse lifetime associated with defects created per unit particle flux Φ. This can be re-expressed in terms of diffusion length $L = \sqrt{D\tau}$ where D is the diffusion constant as

$$\frac{1}{L^2} = \frac{1}{L_0^2} + K \Phi . \tag{2}$$

The diffusion length can be directly measured in solar cells and Figure 7 shows the experimental values plotted at various integrated fluxes of 1 MeV electrons. The curve is expression (2) above. K for this case is 1.7×10^{-10}. These simple ideas are seen to provide an entirely adequate explanation of the results. Furthermore, when expression (2) is folded together with the spectrum of solar illumination and with the optical properties of silicon (KLEINMAN, 1961), the result is the approximately logarithmic variation of short-circuit current with bombarding flux illustrated in Figure 4.

The constant K of expression (2) depends on a great many detailed variables. Primary among these are the energy and particle type of the bombarding radiation and the impurity type and resistivity of the silicon base material.

Figure 8 gives the energy variation of K for monoenergetic electrons relative to the value of K at 1 MeV and for 1 ohm-centimeter p and n-type silicon (ROSENZWEIG *et al.*, unpublished). K is an increasing function of energy as is expected from the relativistic Rutherford scattering cross section for electrons. The very rapid rise of the damage constant with energy above 0.2 MeV is due to near-threshold effects. At about this electron energy the maximum energy transfer

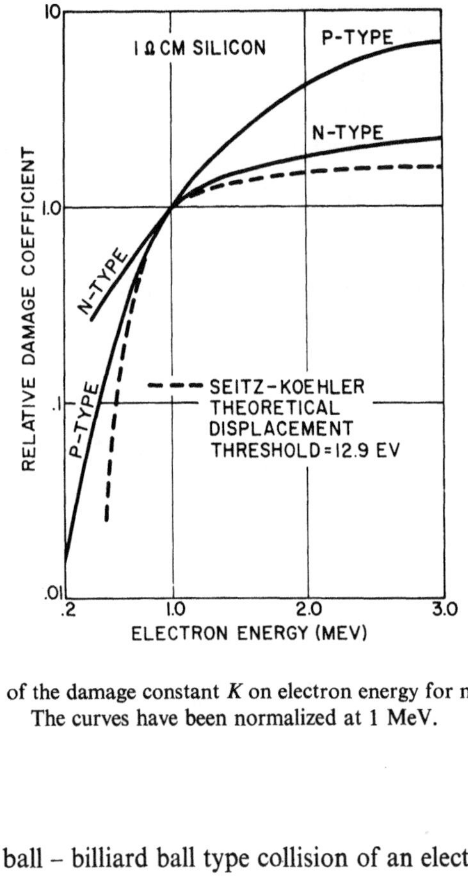

Fig. 8. Dependence of the damage constant K on electron energy for n- and p-type silicon. The curves have been normalized at 1 MeV.

in this ping-pong ball – billiard ball type collision of an electron with a silicon nucleus begins to exceed the minimum required for stable displacement of an atom.

The dashed line plotted in Figure 8 is a theoretical curve expressing the energy variation to be expected if only a simple threshold for formation of only simple vacancies and interstitials were involved (SEITZ and KOEHLER, 1956). The experimental curves do not agree well with this theory, and are also quite different from one another. It is not unexpected that the shapes at low energies are at variance with the theory because of the simplicity of the threshold assumption for formation of vacancies and interstitials. However, the fact that the n and p-type curves are not alike is strong evidence that the radiation defects being formed involve association of the vacancies and the interstitials originally created by the incident radiation with other impurities and imperfections already present in the crystals. The fact that the p-type curve continues to rise rather rapidly at higher energies suggests in addition that formation of di-vacancies may

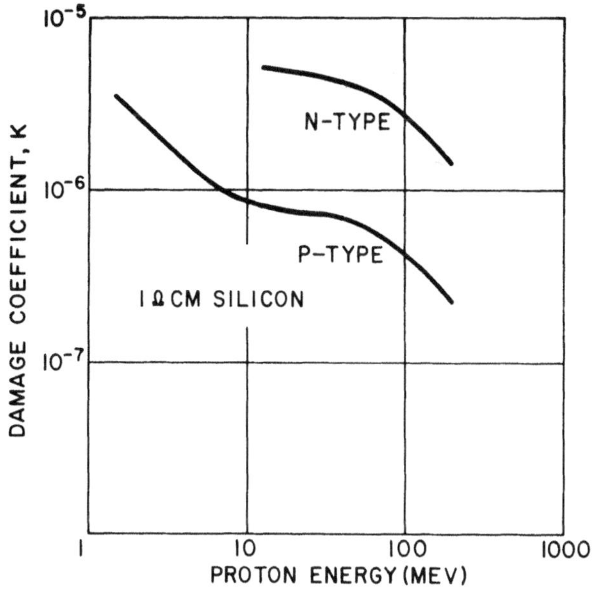

Fig. 9. Dependence of the damage constant K on proton energy for n- and p-type silicon.

well be important. The value of K for 1 MeV electrons in n-type silicon is 1.22×10^{-8}, 72 times greater than for p-type. The recombination properties of a particular defect in n and p-type material could only coincidentally be the same, but this large difference suggests that the stable defects are, themselves, different in the two materials. The physics of radiation damage in silicon has been studied in considerable detail and these effects observed in solar cells, although not unambiguously identified with more fundamental measurements of the defect structure, are clearly reflecting the same complexity that has been previously found (Hulin, 1965).

Figure 9 shows the energy dependence of K for monoenergetic protons (Rosenzweig *et al.*, 1964). Although the p-type material has been less extensively investigated, over the range examined there seems to be a consistent factor of approximately 6 between the two material types. The principle energy dependence is now inverse, decreasing K with increasing energy. For damage by heavy particles rather than by electrons, this is to be expected because energy transfers are no longer limited to anything like the threshold for displacement, and the cross section for interaction is dropping as the speed of the particle increases. An approximate $1/E$ variation with modification by a logarithmic term is anticipated (SEITZ and KOEHLER, 1956), and indeed is observed at the lower and higher energy portions of the curve, but the plateau is puzzling. It has been suggested (BAIKER *et al.*, 1963) that including the nuclear part of the interaction of protons with silicon nuclei, rather than just the coulomb part, provides a correction in the appropriate energy region. However, the quantitative aspects of this theory are still in doubt. There is continuing disagreement as to whether the energy dependence displayed in Figure 9 is experimentally correct, and questions of the minority carrier concentration dependence of the observed values of L, which enter the determination of K, are still at issue. J. M. DENNEY and Coworkers at Space Technology Laboratories disagree with the results of Figure 9. They find a dependence almost inversely proportional to proton energy in the range from 6 to 94 MeV. Under actual conditions of solar cell shielding, and with the types of proton spectra that exist in space, the most important part of the energy dependence is that below 10 MeV, and here the shape is relatively well agreed upon.

The changes of diffusion length produced by recombination centers introduced with radiation damage are important not only to solar cells, the case for which they have been illustrated above and from which damage coefficients have been obtained, but also for other semiconductor diodes and transistors which depend upon transport of minority carriers by diffusion. The forward voltage drop of a p-i-n power diode is another example. In this case the very low forward resistance of the normal device depends upon a high concentration of carriers injected into its base region to effectively increase the base region conductivity from what would otherwise be a relatively low value. The carriers are introduced at the junctions and must diffuse to fill the base. If the lifetime of the carriers in the base is shortened by radiation damage the mean injected carrier concentration in the base will decrease and the voltage drop at fixed forward current will correspondingly increase. Figure 10 shows this increase as a function of the damaging flux for devices with a 2 mil and a 0.6 mil base width in 1000 ohm cm p-type material. The simple theory indicates that the resistance of the base layer (at fixed current) varies inversely as the carrier lifetime and hence linearly with the bombardment flux. The curves are in fairly good agreement with this explanation. The simple theory also predicts a resistance variation with the square of the intrinsic layer width so that a half-decade decrease in base width provides a decade less sensitivity to radiation damaging flux, as shown.

In transistors a decrease in lifetime in the base region decreases the efficiency of transport of minority carriers from emitter to collector and thus results in a loss in common emitter gain. Figure 11 shows the computed behavior of dc common emitter gain for a 2N2222, a 350 megacycle n-p-n device. High fre-

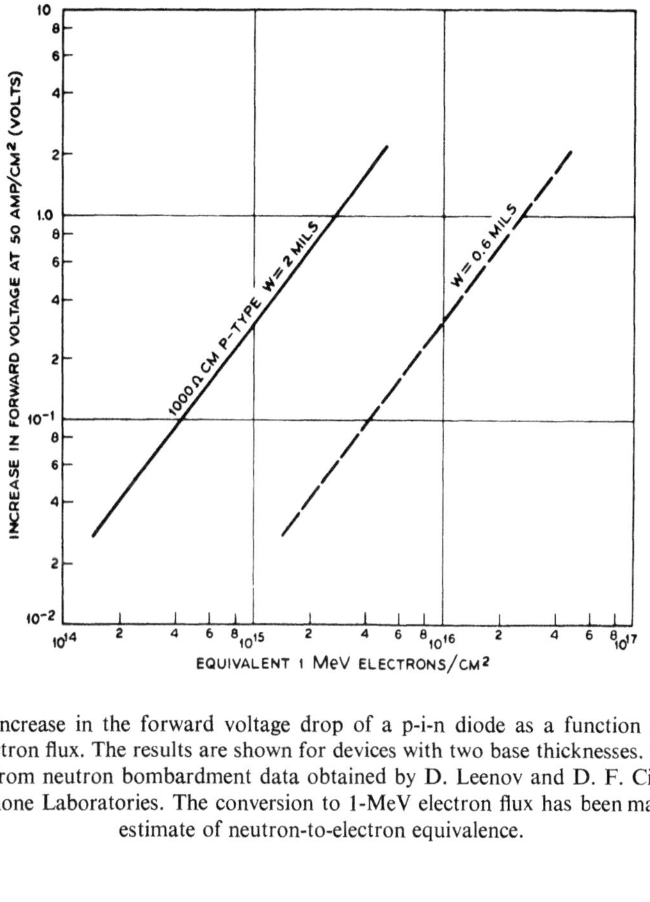

Fig. 10. Increase in the forward voltage drop of a p-i-n diode as a function of integral 1-MeV electron flux. The results are shown for devices with two base thicknesses. The curves are taken from neutron bombardment data obtained by D. Leenov and D. F. Ciccolella of Bell Telephone Laboratories. The conversion to 1-MeV electron flux has been made with an estimate of neutron-to-electron equivalence.

quency transistors such as this have very narrow base widths (the order of 1 micron) and hence require only very short diffusion times for transport of electrons from emitter to collector. As a result, the shortening of lifetime by radiation damage must be quite severe before it affects carrier transport efficiency substantially. For lower frequency devices the damaging effects are expected to be correspondingly more severe as has been well verified experimentally (Easley and Dooley, 1960). The two curves of Figure 11 are for different collector currents and show the normally observed initial gain dependence on injection level. As we will see later in Section 2C, other effects in these devices are likely to arise at lower damaging fluxes than those important for bulk radiation damage.

C. SURFACE EFFECTS ON SEMICONDUCTORS

As was noted in Section 2A, although ionization effects in the bulk of semiconductors are usually unimportant in the radiation environment of space,

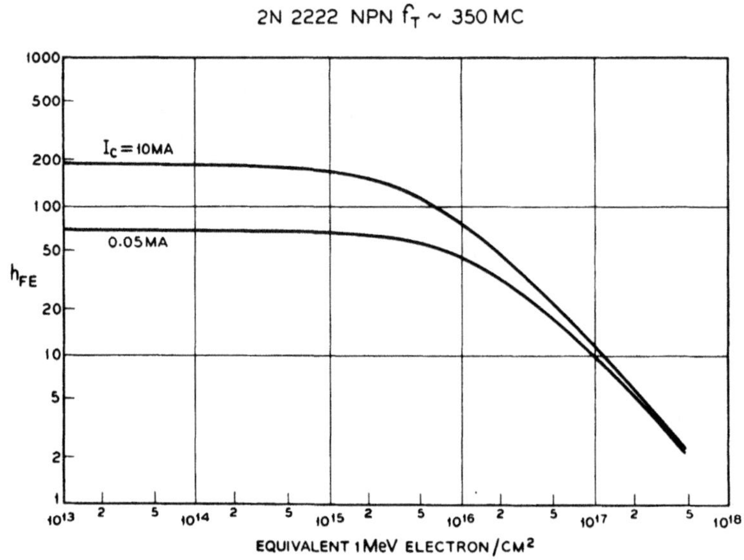

Fig. 11. Decrease in common emitter current gain of a high frequency planar epitaxial silicon n-p-n transistor as a function of integral 1-MeV electron flux.

ionization at the surfaces of semiconductor devices can be extremely important because of the device sensitivity to fractional monolayer changes in charge at their surfaces. It is this same surface sensitivity even in the absence of radiation that has resulted in development of elaborate passivation and encapsulation techniques in order to achieve device reliability. Radiation surface sensitivity of semiconductor devices has been recognized for a number of years (KEISTER and STEWART, 1957). The situation in which the surface sensitivity of transistors to radiation was first interpreted with a model of the process, is shown in Figure 12 (PECK *et al.*, 1963). The device is a diffused n-p-n mesa transitor encapsulated in a can filled with approximately one atmosphere of nitrogen.

When the device is under bias there is an electric field between the transistor can and the silicon because the can is electrically connected to the collector, and there is a fringing field at the device surface near the collector to base junction. Ions produced by radiation in the gas-filling drift to the device surface under influence of these electric fields. Figure 13 shows on a larger scale the consequences of this ionic deposition. The fields are so directed at the reverse

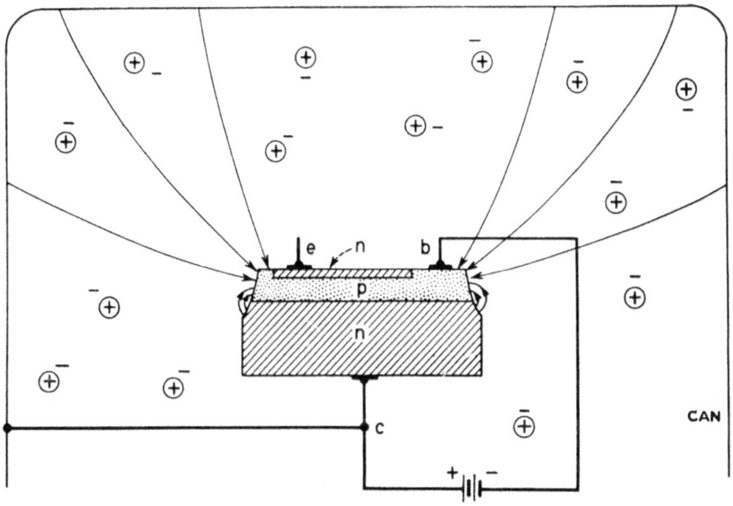

Fig. 12. Ionization of the gas surrounding a reverse biased n-p-n mesa transistor.

biased collector junction that positive ions are urged to the surface of the base and negative ions to the surface of the collector. In each case the electric field the ions create in the underlying material is such as to induce an inversion layer at the surface, a layer dominated by carriers of the type opposite to that of the original material. Such inversion layers are known to be responsible for changes in junction leakage current and junction breakdown properties, and for changes in carrier transport efficiency. Effects of these kinds are observed with radiation in a wide variety of different cases and with widely different sensitivity.

It has been possible to carry out some relatively simple experiments which illustrate at least the rudimentary validity of the simple model of Figures 12 and 13. Devices were fabricated with the can electrically independent of the device so that the electric field between can and silicon could be adjusted independent of device bias. Some of the results of these experiments are shown in Figures 14 and 15 (BLAIR, 1963), Figure 14 shows the collector-to-base reverse current measured in a high intensity gamma radiation field as the potential of the can was varied with respect to the transistor base, keeping the base to collector voltage fixed. It is clearly the case that positive can potentials greatly enhance the collector current. This is in qualitative agreement with the formation of an inversion layer by positive ions on the surface of the base. The can-to-device

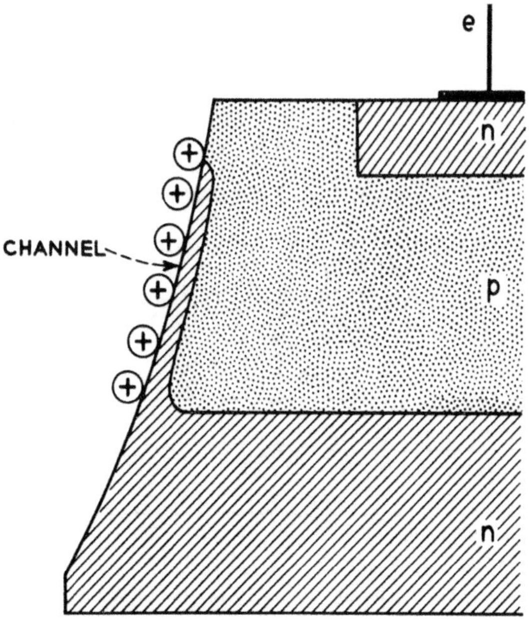

Fig. 13. Formation of a channel on the base of the n-p-n transistor of Figure 12 as a result of ion deposition on the surface.

electric field is encouraging the deposition of such ions when the can is positive. The response of the device to a change in can potential is not immediate: time is required for the ions to reach and build up a layer on the surface.

At the points marked with stars in Figure 14 (approximately steady state conditions), gain measurements were also made. Summaries of I_{CBO} and h_{fe} are shown in Figure 15. The rapid rise of I_{CBO} with positive can potential is accompanied by a precipitous drop in h_{fe} as might be expected in development of an inversion layer on the base region. The effects on both device parameters at large negative can-to-device potential can probably be explained in terms of negative-ion effects on the surface of emitter and collector regions.

The actual chemistry and physics going on at the device surfaces is much more subtle than illustrated in Figure 12 and 13. Supposedly identical encapsulated devices, for example, often exhibit radiation sensitivities differing by an order of magnitude or more. The changes do not always even vary monotonically with radiation exposure. It seems evident from the work that has been done that minor impurities either left on the device surface in fabrication or

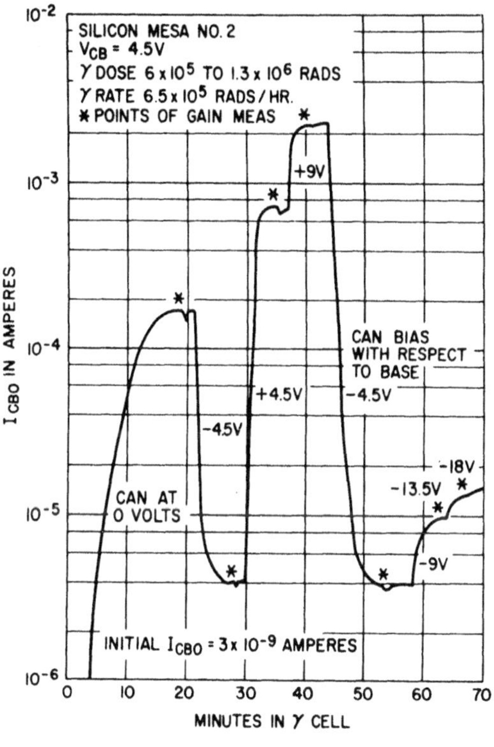

Fig. 14. Variations in collector-to-base reverse current on an n-p-n transistor in a gamma radiation field as the device-to-can potential is changed.

possibly deposited there from the walls of the encapsulation at some later time, serve as active sites on the surface. At such sites it is likely that there is charge exchange with the ions of an encapsulating gas and quite probably a further exchange with the semiconductor itself on a somewhat longer time scale.

The changes due to radiation are not in general permanent and the ionized state of the surface may be allowed or encouraged to recover. As a result there is not in general a dependence on total exposure independent of exposure rate.

On the other hand, the effects observed are not independent of the previous radiation history of a device. These questions are quite complicated and represent a major difficulty in control of device fabrication.

The gas encapsulation of transistors is not essential to their operation, although the gas has in the past been chosen to optimize device stability under

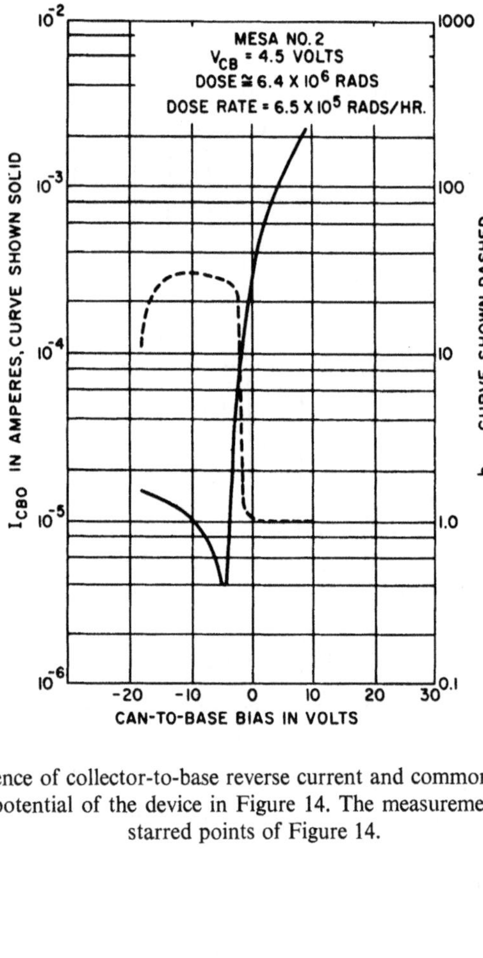

Fig. 15. Dependence of collector-to-base reverse current and common emitter current gain on device-to-can potential of the device in Figure 14. The measurements were made at the starred points of Figure 14.

normal or even severe operating conditions in a normal environment. It is now possible to produce a passivating oxide directly on the device surface. This should prevent ions of the encapsulating gas, if any, from reaching the semiconductor surface and avoid inversion layer formation. Oxide protected devices are in general less sensitive to radiation than mesa devices, but apparently the oxide layer, or impurities within its few thousand ångstrom thickness, or impurities at the interface of the oxide with a semiconductor, are sensitive to radiation and the protection is incomplete.

Figure 16 shows the results of radiation on the planar epitaxial 2N2222 transistor considered in Section 2.2.3 (Figure 11). The curves are based on unpublished data obtained by D. S. Peck and E. R. Schmid (Bell Telephone Laboratories). The previously computed curves due to bulk damage are

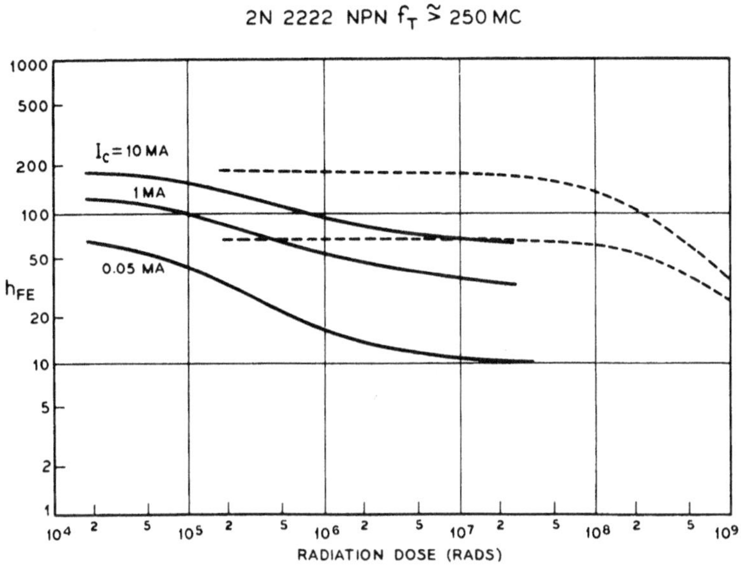

Fig. 16. Dependence of common emitter current gain of a high frequency planar epitaxial silicon n-p-n transistor on radiation dose, based on data obtained by D. S. Peck and E. R. Schmid of Bell Telephone Laboratories with gamma ray bombardment. The calculated bulk damage curves of Figure 11 are reproduced for comparison with the surface dominated effects actually observed, and are placed on the dose scale assuming 1-MeV electron bombardment.

reproduced, but the added solid curves are experimentally determined gain changes due to ionization effects, in this case produced by gamma rays. Perhaps even more serious than the dominance of surface radiation effects over bulk radiation damage is the observation that not all 2N2222's behave alike. One batch of devices may be extremely insensitive to radiation while another may be even much more sensitive than Figure 16 indicates. A great deal more work on the surface chemistry and physics is required before adequate understanding of these empirical observations is obtained.

3. Radiation Damage Results on the Telstar Satellites

Section 2 above has dealt entirely with laboratory measurements and their interpretation. This section will be devoted to radiation damage observed in space and the comparison of this damage with expectations based on these laboratory experiments and the knowledge of the particle distributions that make up the satellite environment.

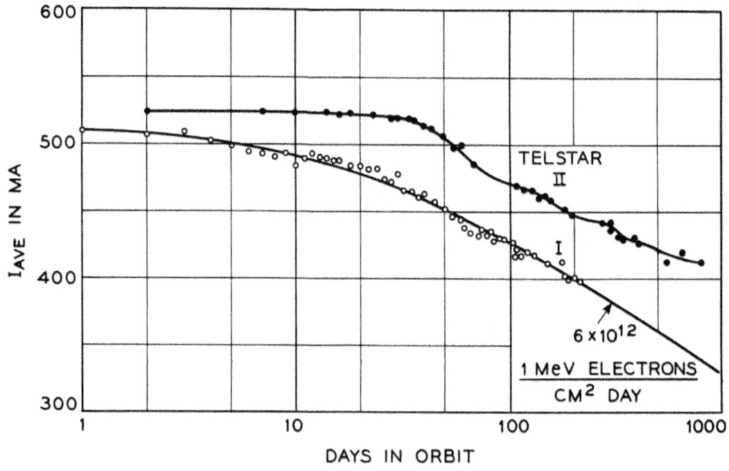

Fig. 17. The solar power plant current for Telstar I and II as a function of their time in orbit. The curve through the Telstar 1 points has the shape of the laboratory determined curve of Figure 4.

A. SOLAR POWER PLANT

Figure 17 shows the time dependence of the current from the solar power plants of Telstar I (Brown *et al.*, 1963) and II (Brown and Gabbe, 1964) as available to a load at 24.8 volts. The currents have been normalized to illumination at mean solar distance. The solid curve through the points of Telstar I has the shape of the laboratory-determined degradation curve. Fitting to such a curve implies that the damaging radiation is essentially constant in time. The curve has been shifted on the time scale to best fit the points. The fit corresponds to the statement that the damage to the power plant is that which would have been expected in the laboratory if the bare solar cells had been irradiated at normal incidence with a flux of 6×10^{12} 1 − MeV electrons/cm^2 day. The curve is a reasonably good fit to the observations. In the Telstar II case however, the curve has no such significance, because the bumps along the curve are clearly inconsistent with the shape expected for a constant damaging flux. We shall try to account for the absolute magnitude of the damage observed in both cases and for the irregularity in the case of Telstar II.

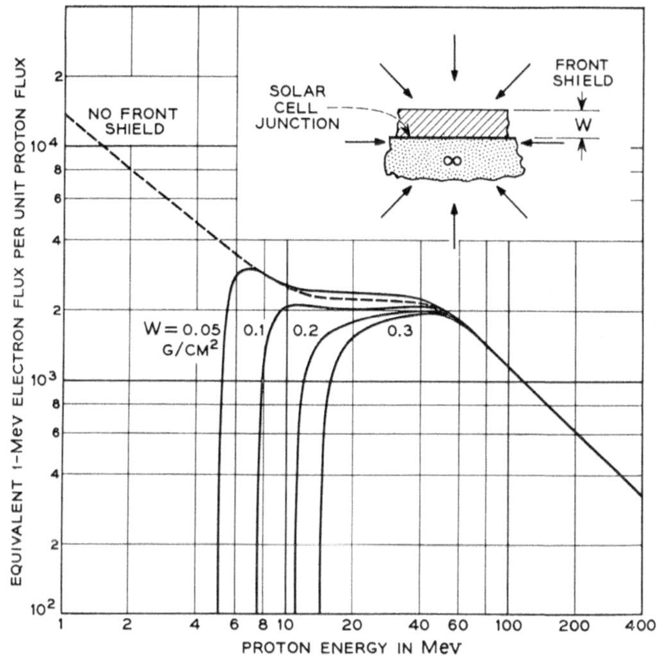

Fig. 18. Equivalent 1-MeV electron flux corresponding to damage by a monoenergetic omnidirectional flux of protons at several shield thicknesses.

B. EQUIVALENT DAMAGING FLUX

The solar cells on the Telstar satellites are shielded with synthetic sapphire covers 0.3 g/cm^2 in thickness on the front and by effectively an infinitely thick layer of material from the rear (SMITH *et al.*, 1963). The cells are furthermore being irradiated from all directions by protons and electrons with a wide spectrum of energy. The effect of the shielding and the omnidirectional character of the incident radiation modify the damage coefficients displayed and discussed in connection with Figures 8 and 9. The energy spectrum requires an appropriate weighting of the responses from monoenergetic particles.

It is convenient to express the damage produced in an omnidirectional flux of mixed particles in terms of the equivalent flux of a single monoenergetic particle which would give the same result under laboratory conditions (SMITS *et al.*, 1962). One-MeV electrons have been chosen in this case as a standard, and all of the damaging conditions are specified in terms of the equivalent 1 MeV

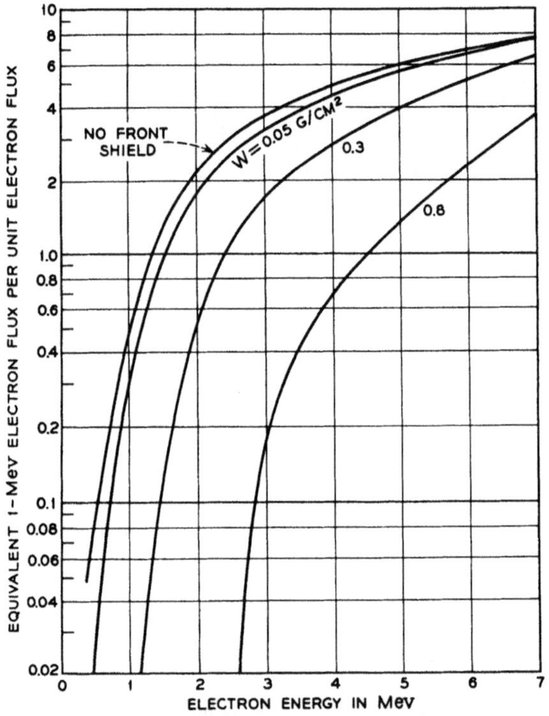

Fig. 19. Equivalent 1-MeV electron flux corresponding to damage by a monoenergetic omnidirectional flux of electrons at several shield thicknesses.

electron flux at normal incidence without shielding which would produce the same solar cell damage, or more exactly, the same change in diffusion length for minority carriers. Implicit in this definition is the assumption that an actual equivalence exists; that is, that the damage produced by different particles and at different energies is nearly enough the same that the effect of a mixture can be expressed in terms of some single component. This is not in detail true, but it is a sufficiently good approximation to be very valuable for this type of calculation.

Figures 18 and 19 give the equivalent 1-MeV electron flux for an omnidirectional flux of protons or electrons incident on 1 ohm-cm p-type silicon under a number of different conditions of front shielding. The proton damage under shielding conditions has been calculated by Brown *et al.* (1963) from results of Figure 9 and the known penetration properties of protons. Because

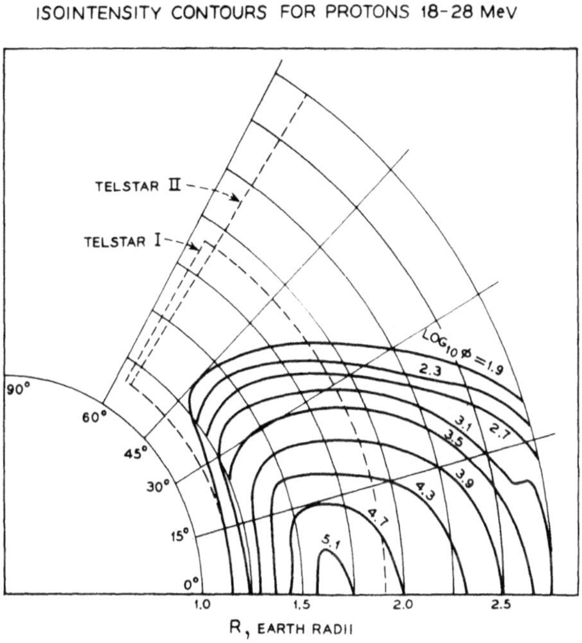

Fig. 20. Isoflux contours in *R*-*λ* space for omnidirectional 18–28 MeV protons as measured by Telstar II.

of the complex penetration characteristics of electrons, Figure 19 is based on a combination of empirical results and theoretical interpolation. Above 3 MeV the curves are based on data obtained by J. Denney and Coworkers at Space Technology Laboratories. In all cases shielding is assumed infinite from the rear. Notice that at 0.3 g/cm^2 front shielding in Figure 18 the equivalent flux goes to 0 for protons below 14 MeV. A proton of 14 MeV is just capable of penetrating the shielding. Because, as discussed in other papers of this Institute, the trapped proton flux in space drops rapidly with increasing energy and because the damage per proton drops at high energies as well, those protons with energies just above 14 MeV will be most effective in damage to the Telstar solar cells. In Figure 19 for electrons, note that the zero-shielding line has the value 0.5 at 1 MeV. This simply expresses the fact that only half the electrons in an omnidirectional flux are creating damage because of the back shielding. The energy threshold for damage by electrons at 0.3 g/cm^2 shielding is less sharply defined than for the proton case as the result of the straggle of electron energy loss in materials, but damage at 0.3 g/cm^2 is primarily produced by

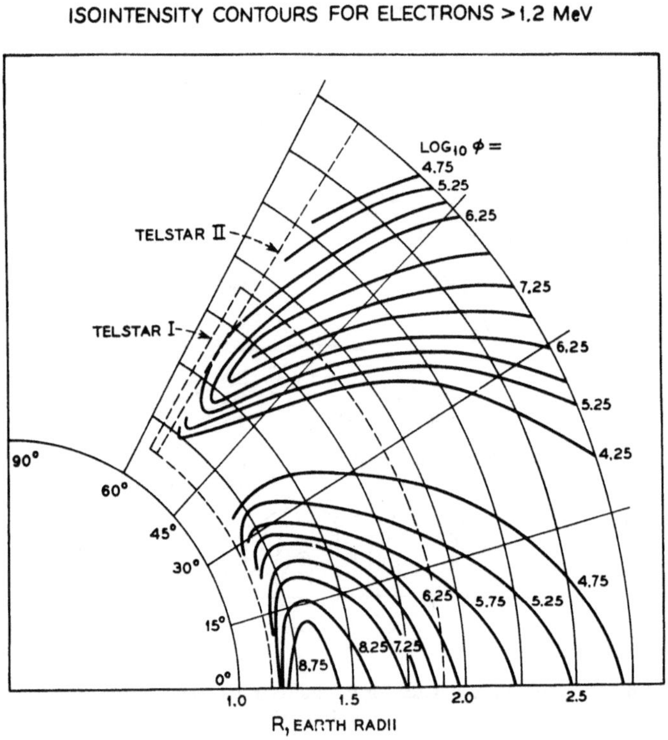

Fig. 21. Isoflux contours in R-λ space for electrons of > 1.2 MeV as measured by Telstar II during the summer of 1963.

electrons of greater than 1.2 MeV in energy. In the electron case because the damage efficiency increases rather rapidly with energy, the higher energy portions of the electron spectrum are particularly important.

C. THE RADIATION ENVIRONMENT OF THE TELSTAR SATELLITES

Table I gives the salient features of the orbits of the two Telstar satellites. The principle difference between them is the much higher apogee of Telstar II. Particle detectors on the satellites provided data for direct mapping of the

TABLE I

Several data on Telstar I and II satellites

	Telstar I	*Telstar* II
Launch date	July 10, 1962	May 7, 1963
Apogee	3507 stat. mi	6713 stat. mi
Pergigee	592 stat. mi	604 stat. mi
Orbital inclination	44.9°	42.8°
Orbital period	158 minutes	225 minutes
Period of precession of line of apsides	182 days	150 days

radiation environment. The results for protons and part of the results for electrons have already been discussed in earlier papers (GABBE and BROWN, 1966; ROBERTS, 1966). Flux maps for protons between 18 and 28 MeV and electrons greater than 1.2 MeV are shown in Figures 20 and 21. Because of the Telstar solar cell shielding these particles are of particular importance for

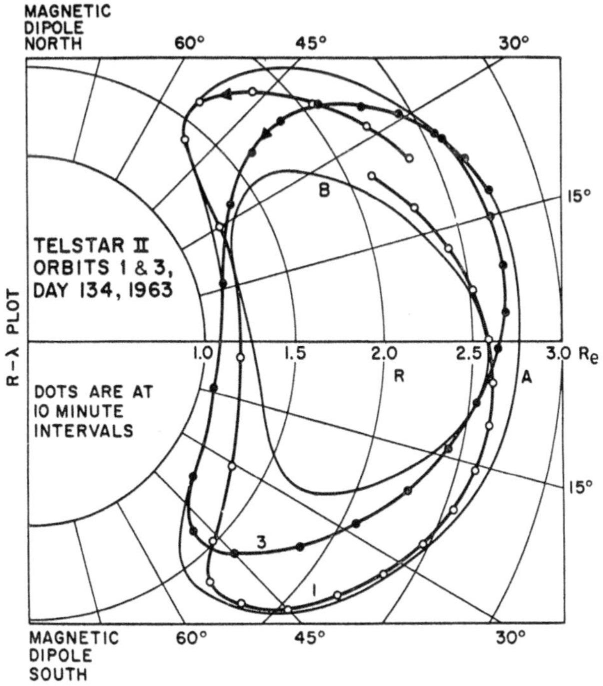

Fig. 22. Representative orbits of Telstar II on Day 134 1963, with apogee of the satellite near the equator. The orbits for the day are marked with points at 10 minute intervals. Solid boundary lines *A* and *B* indicate the extremes of the orbital coverage for the day.

consideration of damage. The boundaries of the regions traversed by the two satellite orbits are indicated in Figures 20 and 21 by the dashed outlines. These regions are completely covered only over a complete period of precession of the line of apsides (see Table I). However it is evident that the satellites spend part of their time in the most intense portion of the proton belt and the inner electron belt and sample the high intensity ridge of the outer electron belt as well.

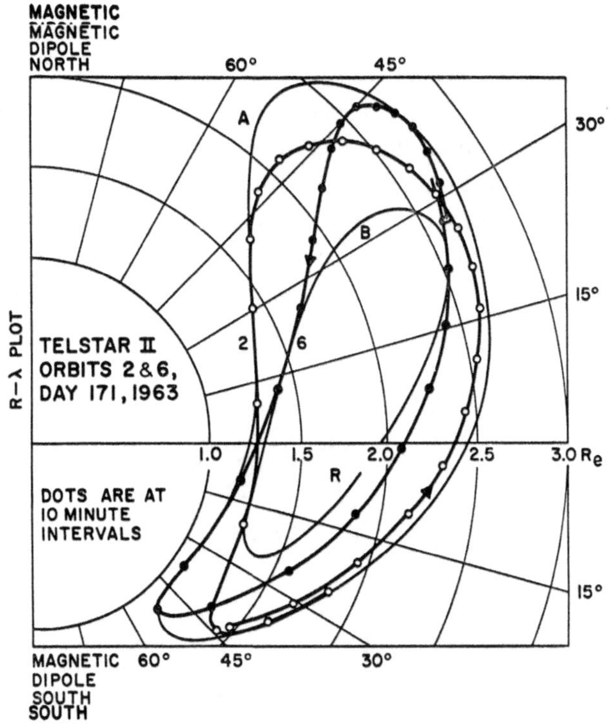

Fig. 23. Orbits of Telstar II on Day 171.

D. ORBITAL VARIATIONS

The actual trace of the satellite in R-λ space is irregular because of the offset in the earth's magnetic dipole and other relatively smaller scale irregularities in the earth's magnetic field. Such traces for Telstar II on three different days in its orbital period of precession are shown in Figures 22, 23 and 24 (Brown and Gabbe, 1964). On day 134, Figure 22, apogee of the satellite is near the equator. The first and third orbits of the day are indicated together with bounding lines A and B which represent the outer and inner extremes of the whole group of orbits for that day. Notice that no time is spent on the equator between $L=1.3$ and 2.5. Figure 23 shows the situation 37 days later when apogee has precessed northward. The equatorial hole is beginning to close. Finally Figure 24 is the situation on Day 208, 1963 when apogee has reached its northern extreme. The crescent-shaped orbits now heavily traverse the equator between $L=1.45$ and $L=1.75$. Examining the flux maps of Figures 20 and 21 it is clear

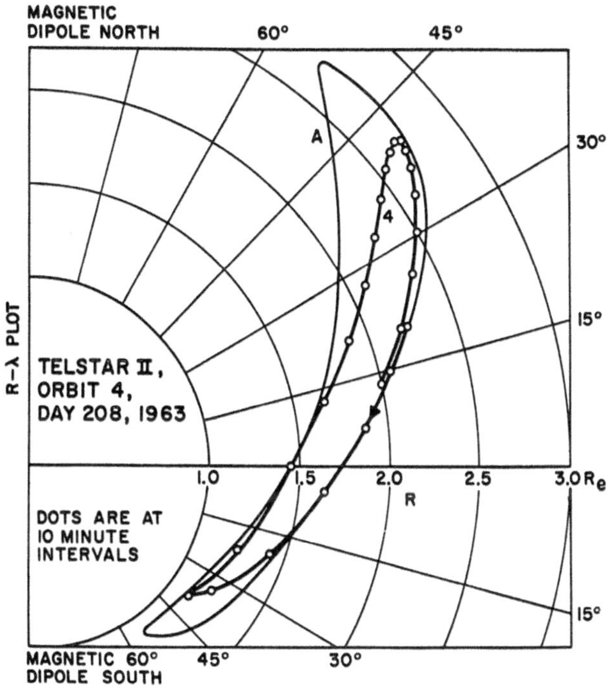

Fig. 24. Orbits of Telstar II on Day 208. The region inside the outer boundary line is substantially fully covered at this time.

that a substantial variation in the satellite exposure should be anticipated during the period of orbital precession.

Figure 25 is the computed daily exposure of the two satellites to protons between 18 and 28 MeV. The difference is striking. The variations in the daily exposure for Telstar I are less than a factor of two, but the variations for Telstar II exceed a factor of ten. The exposures are a maximum when the satellite apogee is full north or south and a minimum when apogee is on the equator.

Telstar I's lower apogee never allows it to avoid the high intensity proton belt as Telstar II is seen to do.

A variation similar to that of Figure 25 exists for the exposure to electrons, though the peak to valley ratio for Telstar II is only about a factor of five. The

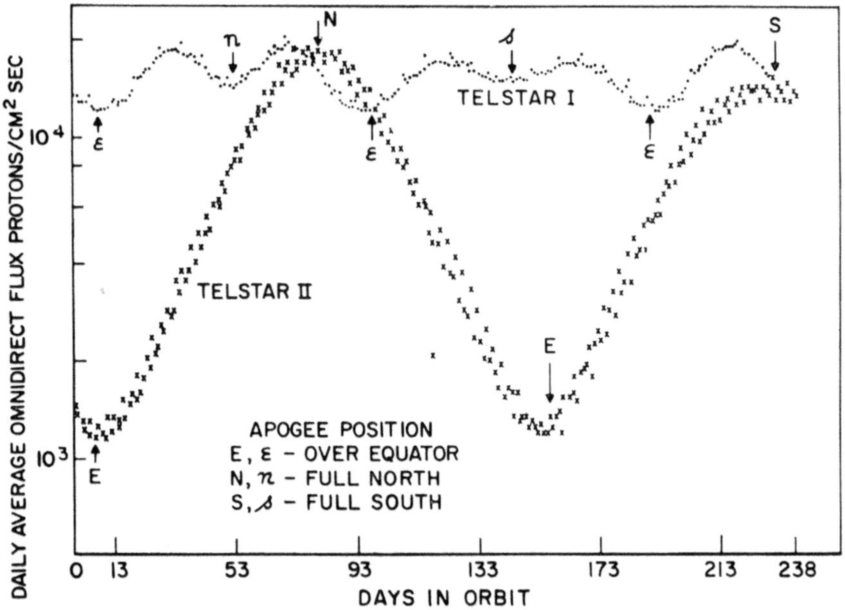

TELSTAR	AVE OMNIDIR FLUX CM^{-2} SEC^{-1}
I	1.5×10^4
II	7.3×10^3

Fig. 25. Average daily exposure of Telstar I and Telstar II orbits to 18–28 MeV protons (measured by Telstar II). The locations of orbital apogee on the equator or at northern or southern extremes are indicated.

Telstar I orbit also has a substantial expected variation in exposure when use is made of the Telstar II electron flux map (Figure 21). The peak of the high energy electron distribution is low enought to be circumscribed even with Telstar I's lower apogee. In actuality, because of a much broader distribution of high energy electrons from the Starfish nuclear test during the first weeks of Telstar I's orbital life, there was less variation in the average daily electron exposure than if this satellite had been launched a year later.

E. ADJUSTED TIME SCALE

Bulk radiation damage is a cumulative effect and as has been seen in Figure 4 it varies for solar cells approximately in proportion to the logarithm of the total exposure. It is thus convenient to indicate the cumulative exposure as a

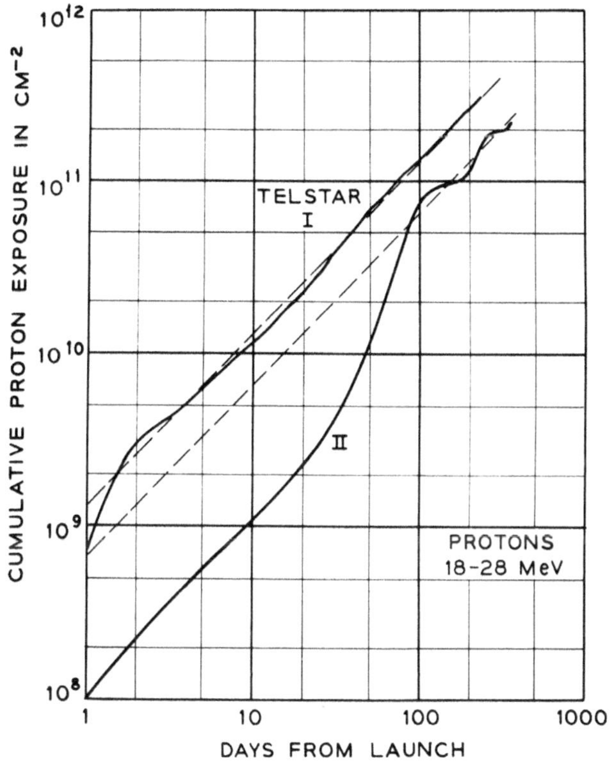

Fig. 26. Cumulative exposure of Telstar I and II orbits to 18–28 MeV protons. The dashed lines show the cumulative exposure corresponding to the average exposures over a period of precession of the line of apsides.

function of time-in-orbit as in Figure 26. Telstar I has small variations around the dashed line which is the cumulative exposure to 18–28 MeV protons the satellite would have received if the exposure were uniform. In Telstar II, however, the cumulative exposure starts far below the mean line and does not reach that line until 90 days after launch. An effective time scale can be defined from such a plot. The effective orbital time is the time at the average exposure rate which would have been required to produce the actual cumulative flux. Ten days after launch, when Telstar II had accumulated just over 10^9 protons/cm^2 in the 18–28 MeV energy region, would correspond to an effective orbital time of about 1.5 days. At 90 days the effective orbital time would equal the actual orbital time.

The effective orbital time scale has been used with the data of Figure 17 to produce the results in Figure 27. The lumps of Figure 17 have now largely disappeared. The remaining scatter may be due to incompleteness of the analysis.

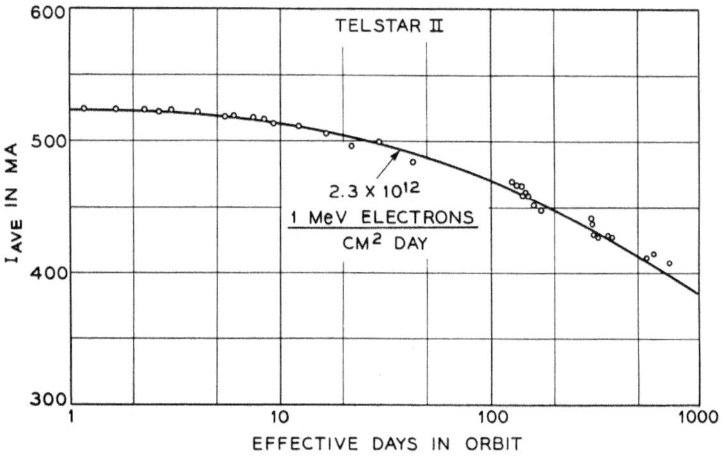

Fig. 27. Solar cell power plant current from Telstar II (Figure 17) plotted on the effective orbital time scale defined by the 18–28 MeV proton exposure (Figure 26).

In principle the effective orbital time should be defined on the basis of the exposure of the satellite to all particles whereas here only the dominant damaging protons have been used. The smooth curve in Figure 27 is a laboratory degradation curve as in Figure 17 for Telstar I, and its position along the time scale corresponds to an equivalent 1-MeV electron exposure of $2.3 \times 10^{12}/cm^2 day$.

F. ABSOLUTE DAMAGE COMPARISONS

Having verified that the time variations in the power plant degradation are clearly those to be expected because of the precession of the satellite orbit, and having obtained average equivalent 1-MeV daily exposures for Telstar I and II from Figures 17 and 27, it is now feasible to see how well the calculated and measured equivalent 1-MeV damages agree. For this, the average proton exposure and its spectrum and the average electron exposure and its spectrum are required in order to fold them through the curves of Figures 18 and 19.

In the high intensity region as previously shown (Gabbe and Brown, 1966, Roberts, 1966), the differential proton energy distribution is well represented by a power law with a power of approximately 4.5. Such a spectrum yields an equivalent 1-MeV electron flux at 0.3 g/cm^2 shielding of 1150 per omnidirection proton above 14 MeV. The electron spectrum in this same region, from which the principle electron exposure is derived, is reasonably approximated by an exponential spectrum with an *e*-folding energy of 1.2 MeV. The equivalent flux in this case for 0.3 g/cm^2 shielding is 0.83 1-MeV electrons for each omnidirectional electron above 1.2 MeV.

From these equivalents and the integrated flux of protons and electrons derived from the flux maps and the satellite orbital positions, the computed equivalent fluxes are obtained as shown in Table II. The table includes the equivalent 1-MeV flux corresponding to the protons, to the electrons and the sum. The sums are to be compared with the equivalent flux determinations from the solar power plant degradation. In both cases the computed equivalence is

TABLE II

Equivalent 1 MeV electron fluxes in cm^{-2} day^{-1} for damage to Telstar solar cell power plants

	Radiation particle counter measurements			*Observed degradation of solar cell power plant*
	Proton equivalent	*Electron equivalent*	*Total equivalent*	
Telstar I	4.9×10^{12}	3×10^{12}	7.9×10^{12}	6×10^{12}
Telstar II	2.4×10^{12}	5×10^{11}	2.9×10^{12}	2.3×10^{12}

higher than the observed, although the agreement is quite good considering the approximations that have been made. The discrepancy may possibly result from long-time annealing of the damage, a subject which has not yet been extensively studied. It also might indicate the nonadditivity of damage from electrons and protons when they are present simultaneously although laboratory experiments have failed to show up such a mechanism. It is most likely to be a reflection of inaccuracy in the knowledge and treatment of the damaging environment.

F. IONIZATION EFFECTS

The direct experimental evidence for ionization produced damage effects in space is rather complicated to obtain and as yet is quite limited. There have been studies of the deterioration of paints and other thermal coatings in response to the combination of micrometeroides, ultraviolet light, and low energy particle radiation. The ultraviolet contribution falls into the class of ionization effects that change the chemistry of the materials. There has been speculation concerning discoloration of transparent solar cell cover materials as having contributed to the deterioration of the output of various solar power plants (FISCHELL *et al.*, 1966). At least a major part of such an effect would be due to ionization, conduction electrons being produced in the material to fill defects originally present, thus producing optically absorbing centers. Many more engineering studies are planned for the future as a necessary verification of the adequacy of laboratory experiments to predict behavior in space.

For ionization-produced surface effects in semiconductor devices very little information exists from space. Largely this is due to the difficulty of designing reasonable experiments at the present state of understanding of the effects even under well controlled laboratory conditions. There is also the strong conviction that the characterization of the relevant space environment for this class of effect is not a very difficult part of the problem. Finally there is the belief that semiconductor device manufacturers will improve their products to a place where these effects can be ignored.

The observation of a failure in some electronic circuit in which there may be hundreds of transistors may of course be blamed on ionization damage, but it is rare that assignment of such a failure can be made clearly. One case of this kind was failure of a command circuit in Telstar I (MAYO *et al.*, 1963). The difficulty was very extensively investigated both in the laboratory on other identical units, through records of the satellite response before the failure was recognized and through observations of the satellite operating parameters, temperature, power supply behavior, transmitter signal levels and so forth. The difficulty was narrowed to a small portion of the command decoder. A way was devised to bypass the faulty part by command pulse reshaping and command control was regained. This failure was with a very high level of certainty due to a gross decrease in transistor gain, in either one of two transistors. This decrease, and the subsequent recovery of the transistor gain to an operating level after the trick command sequence had removed reverse bias from the transistor, was quite certainly due to radiation produced surface effects of the kind discussed in Section 2B. Such a failure would have occurred with very low probability if the ionization dose had been less than 10^5 rads as anticipated before satellite launch.

The radiation exposure was recalculated after observation of the high electron intensity in space following July 9, 1962. Figure 28 gives the ionization dose versus shielding thickness from this calculation for the Telstar I orbit, under assumption of a fission beta spectrum for electrons and an inverse power law for the protons. The transistors in question were located at about 0.7 g/cm^2 shielding depth and were, according to the figure, exposed to a dose of about 600 rads/hour, almost entirely due to penetrating electrons. Over the period from July 9 to November 15, the date of command failure, the total dosage figure was in excess of 10^6 rads. With the transistor screening and selection

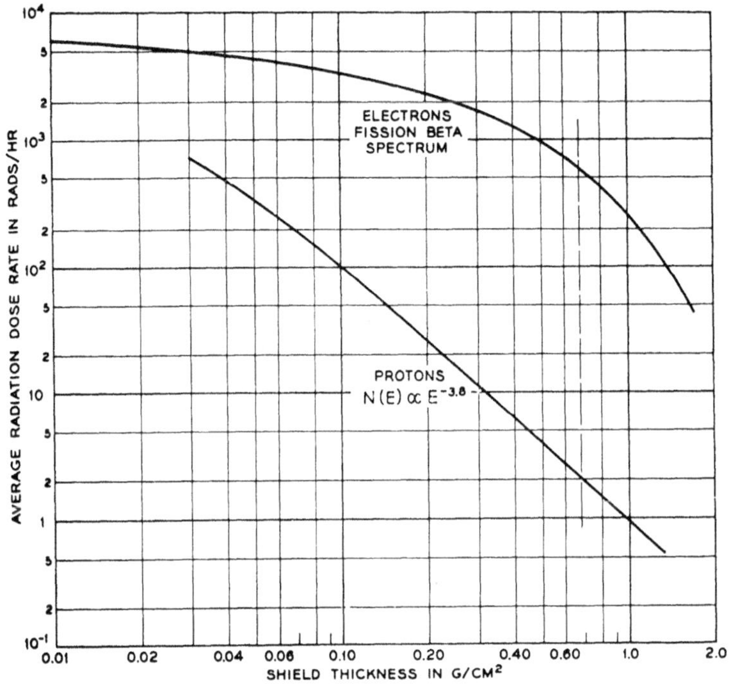

Fig. 28. Ionization dose vs. shielding thickness for the Telstar I orbit assuming all electrons above 0.5 MeV have a fission beta spectrum.

procedures (PECK and WOOLEY, 1963) which had been devised to cope with the ionization surface effects, this excess of actual dose above expected dose was too much and devices were known to fail by drastic gain reduction at the higher dose level. No doubt there are other satellite failures which are due to ionization effects in transistors, but very few satellite orbits have placed them either in space or in time in as severe a radiation environment as that sustained by Telstar I.

4. Conclusion

There is no question that radiation damage in space deserves attention for all satellite and space probe missions, but the level of relevant concern will vary enormously with the characteristics of the orbit and the conservativeness of the system design. There is no need for pessimism in considering prospects for satellites even with very long lives, so long as the radiation response of the system is well enough examined in the laboratory and the levels of exposure in space are well enough understood to permit extrapolation in time. Although the physical and chemical properties of radiation damage are complex and still poorly understood in detail, they are not beyond our ability to interpret nor our ability to place under reproducible control.

5. Acknowledgements

The work that has been sketchily covered in this paper represents the contributions of a large number of people and in particular: W. Rosenzweig, H. K. Gummel and F. M. Smits in studies of radiation damage in solar cells; D. S. Peck, R. R. Blair, and G. L. Miller in work on surface ionization effects, and J. D. Gabbe who has contributed heavily to work on the radiation environment and satellite exposure.

References

Baiker, J. A., Flicker, H., and Vilms, J.: 1963, *Appl. Phys. Letters* **2**, 104.
Blair, R. R.: 1963, *IEEE Trans. Nucl. Sci.*, N.S. **10**, 5.
Brown, W. L. and Gabbe, J. D.: 1964, *Project Telstar Report on Communications and Radiation Experiments with Telstar II* (Bell Tel. Labs., Inc., Murray Hill, N.J.), Section 5.
Brown, W. L., Gabbe, J. D., and Rosenzweig, W.: 1963, *Bell System Tech. J.* **42**, 1505.
Denney, J. M. *et al.*: unpublished.
Easley, J. W. and Dooley, J. A.: 1960, *J. Appl. Phys.* **31**, 1024.
Gabbe, J. D. and Brown, W. L.: 1966, this volume, p. 165.
Hulin, M. (ed.): 1965, *Radiation Damage in Semiconductors* (Vol. 3 of 7th Int. Conf. on the Phys. of Semiconductors, Paris-Royaumont, 1964), Dunod, Paris, 1965.
Keister, G. L. and Stewart, H. V.: 1957, *Proc. IRE* **45**, 931.
Kleinman, D. A.: 1961, *Bell System Techn. J.* **40**, 85.
Leenov, D. and Ciccolella, D. F.: unpublished.
Fischell, R. E., Martin, J., Radford, W. E., and Allen, W. E.: 1966, this volume, p. 793.
Mayo, J. S., Mann, H., Witt F. J., Peck, D. S., Gummel, H. K., and Brown, W. L.: 1963, *Bell System Tech. J.* **42**, 4.
Peck, D. S. and Woolley, M. C.: 1963, *Bell System Tech. J.* **42**, 4.
Peck, D. S. and Schmid, E. R.: unpublished.
Peck, D. S., Blair, R. R., Brown, W. L., and Smits, F. M.: 1963, *Bell System Tech. J.* **42**, 95.
Roberts, C. S.: 1966, this volume, p. 403.
Rosenzweig, W. M., Augustyniak, W. M., and Wright, K. A.: unpublished.
Rosenzweig, W. M., Gummel, H. K., and Smits, F. M.: 1963, *Bell System Tech. J.* **42**, 399.
Rosenzweig, W. M., Smits, F. M., and Brown, W. L.: 1964, *J. Appl. Phys.* **35**, 2707.

SEITZ, F. and KOEHLER, J. S.: 1956, in *Solid State Physics*, Academic Press, New York, p. 328.
SMITH, K. D., GUMMEL, H. K., BODE, J. D., CUTTRISS, D. B., NIELSEN, R. J., and ROSENZWEIG, W. M.: 1963, *Bell System Tech. J.* **42**, 1765.
SMITS, F. M., ROSENZWEIG, W. M., and BROWN, W. L.: 1962, *Radiation Damage to Semiconductor Solar Devices*, Vol. I of *Proceedings of the Solar Working Group Conference*' Washington D. C., Feb. 27 and 28.

Discussion

Leo L. BAGGERLY, LTV, queried as to conditions which should exist so as to promote recovery from damage – in particular whether or not the presence of ions would aid in recovery from damage due to surface effects. The response was to the effect that the recovery and/or self annealing of solid state devices as a function of time after damage has occurred is not well understood. It is known to occur in some cases and not in others. There is no method, of which the presenter was aware, at present of predicting response insofar as self nealing and/or recovery is concerned.

John CLADIS, Lockheed, queried as to the effectiveness of neutrons in producing damage in transistor type devices. The response was that neutrons are very effective in producing damage in solid state devices, i.e. the damage threshold is low; however, they are not sufficient in number, under normal geophysical conditions, to be of concern. Artificial disturbances could be of considerable concern, were the solid state devices within line of sight of the event.

M. J. DOWNEY, ESTC, queried as to current thinking on the effects of flux rate as compared to total accumulated flux. At one time it was thought that the damage criteria was almost solely a function of total flux and relatively insensitive to flux rate. It has since become evident that this earlier conclusion was in error since there have been instances where very high and also very low flux rates have produced effects out of proportion to that expected from the total accumulated flux. The points at which rate starts to have an appreciable effect depends strongly on the type and design of the solid state device and perhaps on the characteristics of the ionization produced by the incident flux.

Leo L. Baggerly queried as to the importance of considering back scattering in connection with particle interaction with solid state devices and the associated shielding materials. Consideration must be given to back scattering, however, in the paper presented the effects of back scatter and the increased path length which results was exactly offset by the cosine squared term in relation to the angle of incidence for unshielded devices. For the ordinary thin shields this is a reasonable approximation because the back scattered electrons are, on the average of much lower energy. For thick shields there may be an appreciable effect since something akin to refraction takes place which shifts the path closer to the vertical.

J. W. KELLER, NASA, queried as to calculations of electron damage on an omnidirectional flux basis. The response was that BTL has not made such calculations since it was quicker and easier to make the measurements. It was indicated that Berger of NBS has made Monte Carlo calculations indicating exit angles and that Boeing has done considerable work in the area of concern. (See papers by Adams and Keller, Session IX.)

RADIATION DAMAGE TO ORBITING SOLAR CELLS AND TRANSISTORS

R. E. FISCHELL, J. H. MARTIN, W. E. RADFORD, and W. E. ALLEN
Applied Physics Laboratory, The Johns Hopkins University, Silver Spring, Md.

Introduction

The Johns Hopkins University, Applied Physics Laboratory began in-flight solar cell degradation studies with the launch of the 1961-01 satellite on June 29, 1961. Similar experiments were flown on the 1961-$\alpha\eta$1 and 1961-$\alpha\eta$2 satellites. Data obtained from these satellites covered a time period both before and after Operation Starfish on July 9, 1962. Subsequently the Applied Physics Laboratory has flown solar cell and electronic component experiments on the Satellite 1962 $\beta\kappa$, ANNA I-B, 1963-38C, and 1964-83C. A brief description and the test results from each of these experiments are presented below. Some interpretations of the radiation belt contents and effects of solar cell shielding are also discussed.

Description of Experiments

Table I lists all satellites on which Applied Physics Laboratory solar cell and transistor experiments have been flown. Launch dates, orbital parameter, present status, and basic circuit descriptions are included in the table. An experimental panel consisting of 20 series connected P-on-N, 1 × 1 cm, solar cells were flown on 1961-01. A similar panel consisting of 20 series connected 1 × 2 cm "blue sensitive" P-on-N solar cells were flown on 1961-$\alpha\eta$1. Four separate solar panels were flown on the 1961-$\alpha\eta$2 satellite. Each of these panels consisted of two series-connected 1 × 2 cm P-on-N solar cells. The solar cells in each of the experiments described above were loaded at approximately 0.2 volt per cell, thereby providing a voltage measurement essentially proportional to short circuit current. Six-mil thick cover glass with blue reflecting filters and anti-reflecting coatings were attached to the solar cells. Furane 15-E adhesive was used to attach cover slides for the 1961-01 and 1961-$\alpha\eta$2 experiments while Spectrolab ES-10 adhesive was employed for the 1961-$\alpha\eta$1 cover slides.

The remainder of the Applied Physics Laboratory's degradation experiments are outlined in Tables II through V. Two by two mm solar cells were employed on satellite 1962-$\beta\kappa$ solar cell experiments while all other experimental solar cells were 1 × 2 cm. In each experiment the short circuit current was monitored.

TABLE I

Satellites on which solar cell and transistor experiments were flown

Satellite	*Date of launch*	*Apogee km*	*Perigee km*	*Incli-nation degrees*	*Telemetry accuracy*	*Attitude control system*	*Circuit for solar cells experiments*	*Circuit for transistor experiments*	*Condition*
1961 01	29 Jun. 1961	1146	983	67	±1%	magnetic	*A* $N = 20$ R 75Ω	none flown	T.M. failed 15 July 1961
1961$\alpha\eta$1	15 Nov. 1961	1111	950	32.4	±1%	magnetic	*A* $N = 20$ $R = 75\Omega$	none flown	ceased transmitting 2 Aug. 1962
1961$\alpha\eta$2	15 Nov. 1961	1111	950	32.4	±1%	none	*A* $N = 2$ $R = 7.5\Omega$	none flown	ceased transmitting 12 Aug. 1962
1962$\beta\kappa$	26 Oct. 1962	6335	220	71.4	±1%	spin	*B* $N = 4$ $R = 909\Omega$	*C*	ceased transmitting 13 Dec. 1962
ANNA 1-B	31 Oct. 1962	1180	1075	50.1	±1%	magnetic	*B* $N = 2$* $R = 4\Omega$*	*C*	still transmitting
1963 38C	28 Sept. 1963	1120	1070	89.9	±1%	magnetic	*B* $N = 4$** $R = 1\Omega$**	See Table IV	still transmitting
1964 83C	12 Dec. 1964	1070	1027	89.99	±1%	magnetic	*B* $N = 4$** $R = 1\Omega$	See Tables IV and V	still transmitting

* The GaAs experiment consisted of one cell operating into a 7-ohm resistor.

** The 125 mil fused silica experiment consisted of one cell operating into A 1-ohm resistor.

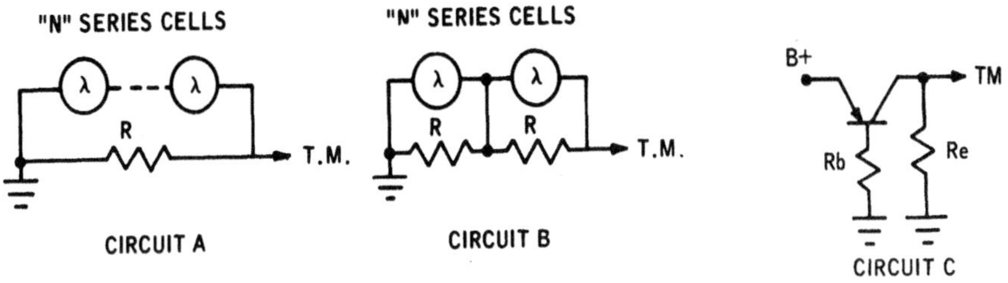

TABLE II

1962$\beta\kappa$ degradation experiments

Experiment	*Device*	*Shielding*	
		Type	*Grams/cm²*
1	P-on-N solar cells (4)	6 mil glass	0.037
2	N-on-P solar cells (4)	6 mil glass	0.037
3	P-on-N solar cells (4)	60 mil quartz	0.336
4	N-on-P solar cells (4)	60 mil quartz	0.336
5	2N327A transistor	bare	
6	2N327A transistor	aluminum	0.4
7	2N327A transistor	aluminum	0.8

TABLE III

ANNA 1-B degradation experiments

Experiment	*Device*	*Shielding*	
		Type	*Grams/cm²*
1	P-on-N solar cells (2)	6 mil glass	0.037
2	N-on-P solar cell (1) (azimuth detector)	6 mil glass	0.037
3	N-on-P pre-irradiated solar cells (2)	6 mil glass	0.037
4	P-on-N solar cells (2)	20 mil quartz	0.112
5	P-on-N solar cells (2)	30 mil sapphire	0.302
6	N-on-P solar cells (2)	30 mil sapphire	0.302
7	gallium arsenide solar cell (1)	6 mil glass	0.037
8	2N327A transistor	10 mil kovar	0.203
9	2N327A transistor	10 mil kovar + 57 mil aluminum	0.609
10	2N327A transistor	10 mil kovar + 250 mil aluminum	1.960

Furane 15-E adhesive was used to attach cover slides on the 1962-$\beta\kappa$ and ANNA experiments with exception of ANNA experiment number 6 which employed the Telstar cover slide attachment technique (SMITH *et al.*, 1963). RTV-602 was used to attach most of the cover slides for the satellites 1963-38C and 1964-83C. For those satellites the 125 mil quartz shielding was mounted just above the solar cell with no adhesive in between.

The transistor experiments on 1962-$\beta\kappa$ and ANNA were mounted on the solar cell test panels. A basic diagram for these experiments is given as part of

TABLE IV

1963 38C degradation experiments

Experiment	*Device*	*Shielding*	
		Type	*Grams/cm²*
1	N-on-P solar cells (4)	none	
2	N-on-P solar cells (4)	1 mil quartz	0.006
3	N-on-P solar cells (4)	3 mil quartz	0.017
4	N-on-P solar cells (4)	6 mil glass	0.037
5	N-on-P solar cells (4)	20 mil quartz	0.112
6	N-on-P solar cells (4)	30 mil sapphire	0.302
7	N-on-P solar cell (1)	125 mil quartz	0.700
8	P-on-N solar cells (4)	6 mil glass	0.037
9	transistors 2N1711(4)	epoglass coated honeycomb	0.280
10	transistors 2N1711(2)	epoglass coated honeycomb	0.280

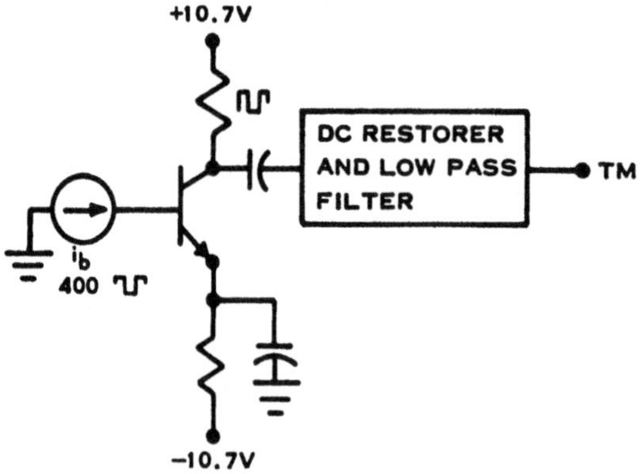

CIRCUIT FOR TRANSISTOR
BETA EXPERIMENTS

TABLE V

1964 83C degradation experiments*

Experiment	Device	Shielding	
		Type	*Grams/cm*[2]
1 (beta)	2N2222 transistors (2)	none	
2 (beta)	2N2586 transistors (2)	none	
3 (beta)	2N2222 transistors (1)	0.1″al	0.685
4 (beta)	2N2586 transistor (1)	0.1″al	0.685
5 (beta)	2N2222 transistor (1)	0.3″al	2.057
6 (beta)	2N2586 transistor (1)	0.3″al	2.057
7 (beta)	2N2907-A transistors (2)	none	
8 (beta)	2N2907-A transistor (1)	0.1″al	0.685
9 (beta)	2N2907-A transistor (1)	0.3″al	2.057
10 (leakage)	2N2222 transistors (2)	none	
11 (leakage)	2N2586 transistors (2)	none	
12 (leakage)	2N2907-A transistors (2)	none	
13 (leakage)	2N2222 transistor (1)	0.1″al	0.685
14 (leakage)	2N2586 transistor (1)	0.1″al	0.685
15 (leakage)	2N2907-A transistor (1)	0.1″al	0.685
16 (leakage)	2N2222 transistor (1)	0.3″al	2.057
17 (leakage)	2N2586 transistor (1)	0.3″al	2.057
18 (leakage)	2N2907-A transistor (1)	0.3″al	2.057
19 (leakage)	100μf, 20 wv, 150D capacitors (2)	none	
20 (short)	39μf, 10 wv, 350D capacitor V=11 volts capacitor V=12 volts	none	
21 (short)	39μf, 10 wv, 350D capacitor V=12 volts	none	
22 (short)	39μf, 10 wv, 350D capacitor V=14 volts	none	
23 (short)	39μf, 10 wv, 350D capacitor V=16 volts	none	

* Solar cell experiments were similar to those on 1963 38C.

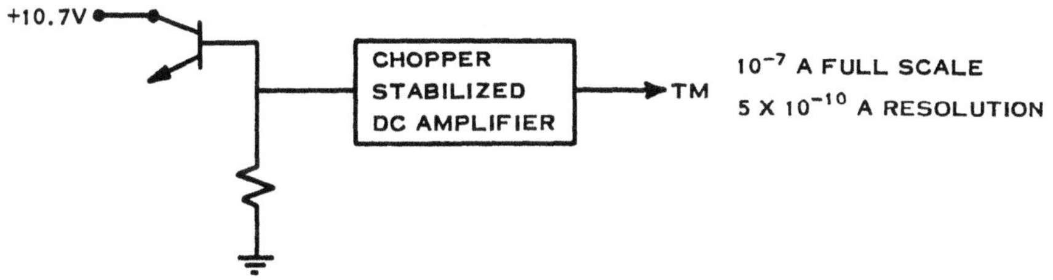

Table I. The test transistors on 1963-38C satellite were placed inside the satellite's outer skin and directly behind an aluminum support brace. There was no other shielding except for the case of the transistors. The circuit diagram for the 400 cps beta (gain) experiment is shown on Table IV. All of the electronic component experiments on the 1963-83C satellite had shielding listed in Table V in addition to transistor cases. Components shielded by 0.3 inch aluminum are reference standards. The circuit diagram for the 400 cycle beta experiments is basically the same as that for the 1963-38C satellite transistor experiment. All beta measurements were made with 0.5 ma bias currents and were therefore not "worse-case" measurements. Transistors which are irradiated while they are biased "off" experience channeling damage. This type degradation does not occur when the transistors are conducting.

Results of Satellite Experiments

Meaningful data was not obtained from several experiments due to various problems; only pertinent data is presented below. The telemetry system of APL Satellites have accuracies better than ± 1 percent. Solar cell data was read when the experiments were near normal incidence to the sun and corrected by dividing

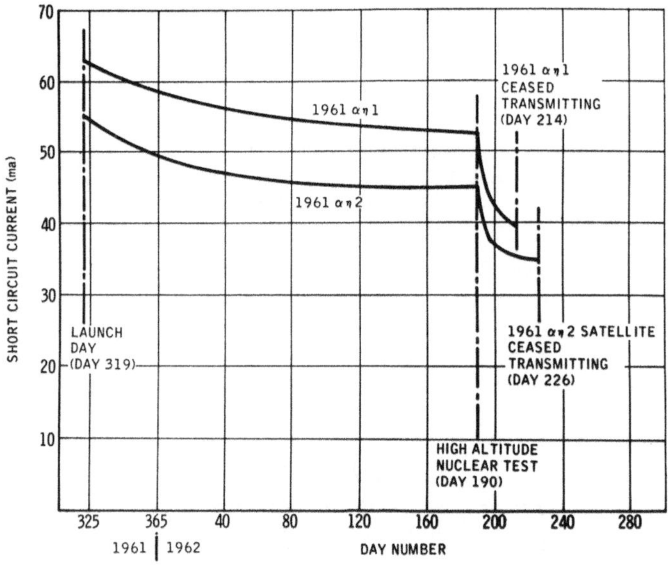

Fig. 1. Solar cell output as a function of time for 1961-$\alpha\eta$1 and 1961-$\alpha\eta$2. Orbit: Apogee 1111 km, Perigee 950 km, Inclination 32.4°.

by the cosine of the angle of incidence. All such data were then normalized for sun intensity at the mean solar distance, i.e. 140 mv/cm^2. Next, these data were plotted as a function of measured temperature for several consecutive passes. The intersection of the straight line fit of these data points with the 50 °F temperature line was considered to be the fully corrected output of the cell. Temperature measurements were not made on the 1961-01, 1961-$\alpha\eta$1, and 1961-$\alpha\eta$2 panels. However, since all readings were at nearly the same angle of incidence, temperature variations for these satellites were assumed to contribute, less than two percent error.

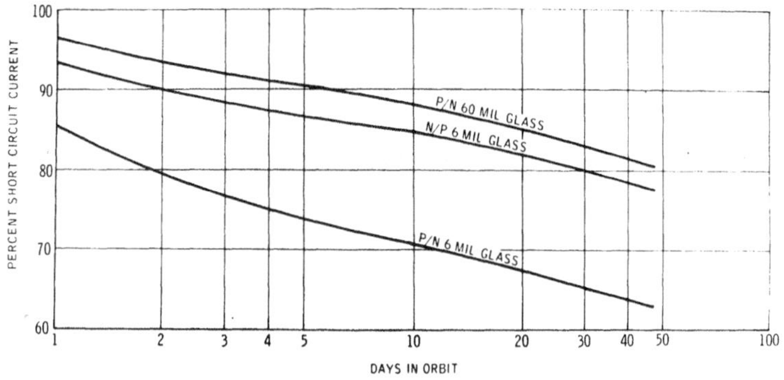

Fig. 2. Results of solar cell experiments on satellite 1962 $\beta\kappa$. Launched 26 October 1962. Orbit: Perigee 220 km, Apogee 6335 km, Inclination 71.4°.

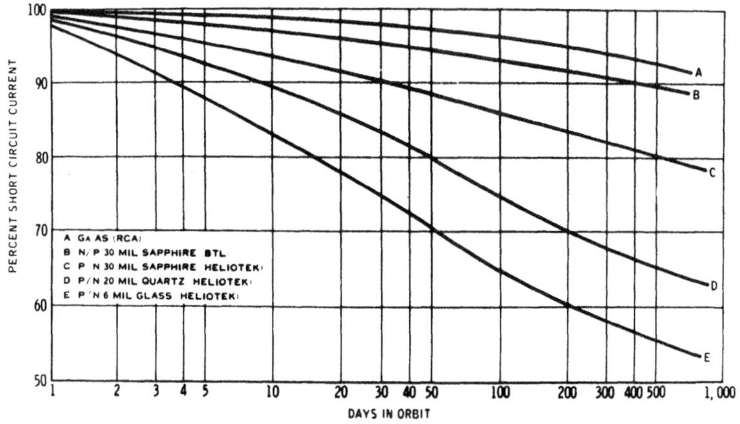

Fig. 3. Results of solar cell experiments on ANNA 1-B Satellite. Launched 31 October 1962. Orbit: Apogee 1180 km, Perigee 1075 km, Inclination 50.1°.

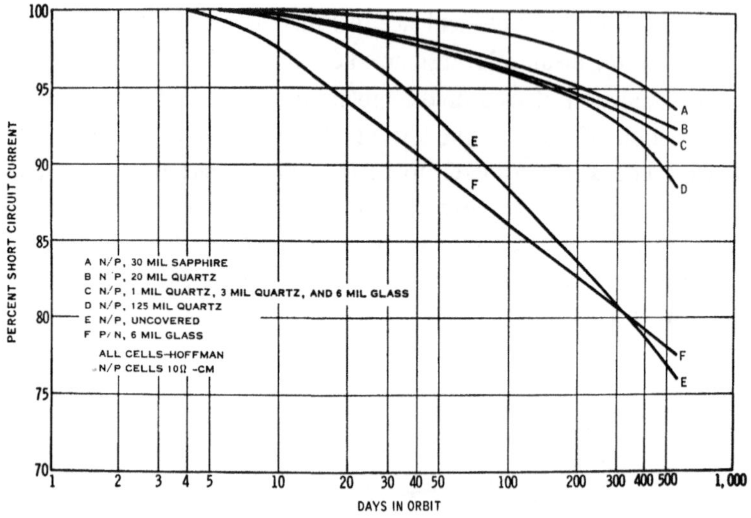

Fig. 4. Results of solar cell experiments on satellite 1963-38C. Launched 28 September 1963. Orbit: Apogee 1120 km, Perigee 1070 km, Inclination 89.9°.

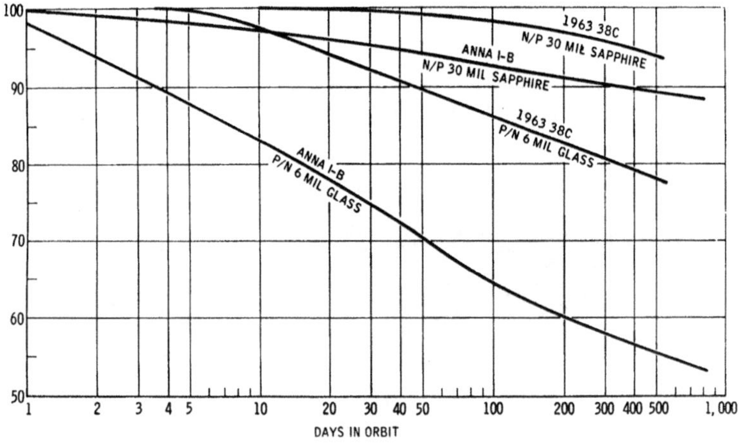

Fig. 5. Comparison of ANNA I-B and satellite 1963-38C solar cell experiment data.

Results of the satellites 1961-$\alpha\eta$1 and 1961-$\alpha\eta$2 experiments are presented in Figure 1. The satellite 1962-$\beta\kappa$ experimental results are shown in Figure 2. Figure 3 is a presentation of the ANNA I-B solar cell data. A decrease in the slope curves *D* and *E* indicate that during the period January through May, 1963 the radiation intensity of the Van Allen Belts had decreased slightly. Results of the 1963-38C solar cell experiments are given in Figure 4.

Figure 5 is a comparison of results of similar experiments on the ANNA I-B and 1963-38C satellites. Rapid initial degradation was noted on the ANNA experiments; however, five days passed before the first noticeable degradation occured on the 1963-38C experiments. This indicates a significant decrease in radiation belt intensity which occurred prior to launch of the 1963-38C satellite in September, 1963.

The ANNA transistor experiments (circuit C, Table I) were devised on the premise that bulk damage would be predominant. Because the output is a function of collector to base leakage as well as gain, telemetered data from these experiments would not be an accurate measure of either. HOGREFE (1963) concluded that since an initial increase in output was followed by a decrease in output, a substantial increase in leakage did occur.

Data received from the satellite 1963-38C transistor experiments were corrected for variations in temperature by normalizing to 90 °F and for variations

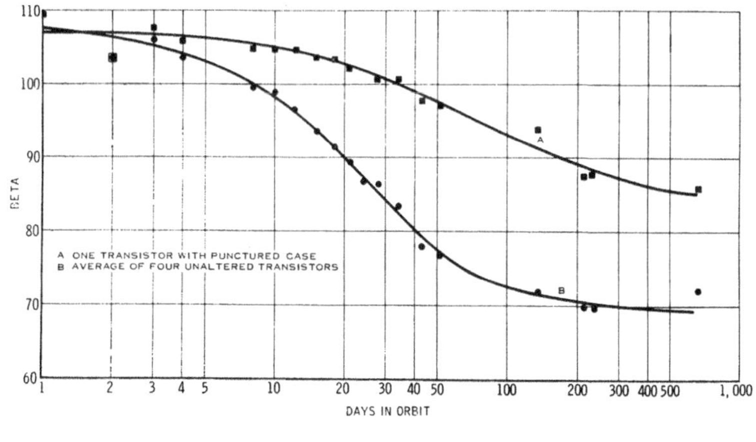

Fig. 6. Beta vs time for fairchild 2N1711 NPN planar transistors flown on 1963-38C satellite.

in power supply voltage to +10.7 volts. Curve A of Figure 6 is a plot of beta as a function of time for one of the transistors with a punctured case. Curve B is a plot of the time variation of the average beta of four unaltered transistors. The gain-bandwidth product of these transistors was 60 mc before flight. Ionization of gases in the case of the unaltered transistors has caused surface damage which, coupled with bulk damage, has resulted in rapid degradation of these components. The single gasless transistor appears to be less effected by surface damage. A decrease in the rate of degradation of both types of transistors is noted to coincide with the decrease in solar cell degradation.

Data received from the electronic component tests on 1964-83C indicates no capacitor degradation had occurred at the time of this writing. Transistor data was corrected as follows: The transistors shielded by 0.3 inch of aluminum, one of each type, are controls and are considered to be undegraded by radiation. The telemetered gain of each control transistor is normalized to its preflight value for each set of readings. The appropriate normalizing factor is then used to correct the test transistors for temperature and line voltage variations. Thus far, base-to-collector leakage has been found to be less than one nanoampere above preflight values. The beta experiments presented in Figure 7 show that

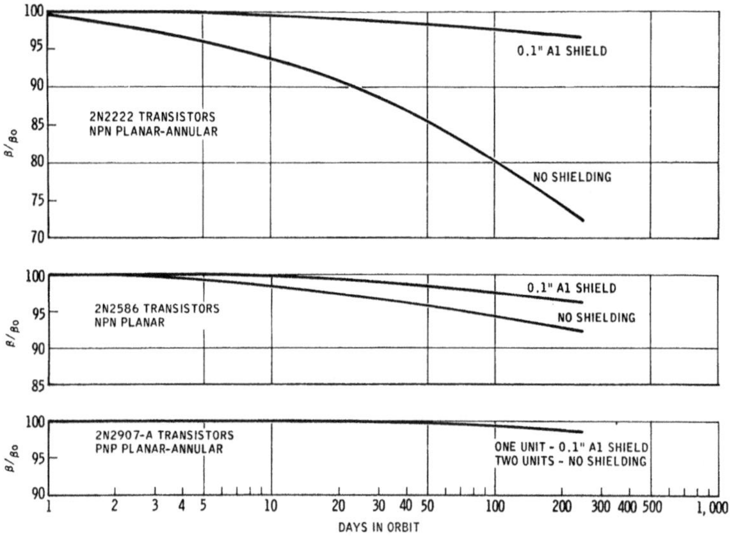

Fig. 7. Results of transistor reliability experiment flown on satellite 1964-83C. Orbit: Apogee 1070 km, Perigee 1027 km, Inclination 89.9°.

both the shielded and unshielded 2N2907-A-PNP bandguard transistors degraded approximately one percent in 250 days, while the unshielded NPN transistors degraded approximately 27 percent in the same period. Preflight gain bandwidth products are as follows: 2N2222, 200 mc; 2N2586, 45 mc; 2N2907-A, 200 mc.

Analysis of Results

Figure 1 shows that during the 236 day period prior to Operation Starfish, the 1961-$\alpha\eta$1 solar cells degraded by 16 percent and the 1961-$\alpha\eta$2 solar cells degraded by 18 percent. In a 20-day period immediately following the high altitude nuclear test, the 1961-$\alpha\eta$1 solar cells were degraded by an additional 22 percent. The 1961-$\alpha\eta$2 solar cells were degraded by an additional 22 percent in 28 days after the test.

The Applied Physics Laboratory conducted an investigation of the degradation of "blue-sensitive" P-on-N and 10 ohm-cm N-on-P solar cells in a normal incident 1 MeV electron environment (MARTIN, 1964). Data obtained from this investigation was plotted as relative output of the solar cells as a function of time integrated 1 MeV flux in electrons/cm^2. Rosenzweig and Brown computed the damage to N-on-P and P-on-N solar cells protected by various shield thickness in a radiation environment consisting of an omnidirectional U^{235} fission electron flux. These results were related to equivalent damage by 1 MeV normal incident electron flux, and were presented as equivalent 1 MeV electron flux as a function of shield thickness for omnidirectional fission electrons incident on the solar cells with infinite back shielding (BROWN *et al.*, 1963).

TABLE VI

Comparison of P-on-N solar cell degradation for periods before and after the "Operation Starfish" high altitude nuclear test*

	1961$\alpha\eta$1		
Experiment	*Period*	*Percent degradation*	*Equivalent* 1 MeV *electrons per* cm^2 *per day incident on bare* P-*on*-N *solar cells*
P-on-N solar cell degradation experiment	236 days prior to "Operation Starfish"	16	39.4g
Same	20 days after "Operation Starfish"	22	8550g
	Satellite 1962$\beta\kappa$		
P-on-N solar cells 60 mil glass covers	47 days – beginning 109 days after "Starfish"	19.4	320g
P-on-N solar cells 6 mil glass covers	47 days – beginning 109 days after "Starfish"	37.0	3340g

* A solar cell attitude detector on the 1961$\alpha\eta$2 satellite degraded less than 7.8 percent from day 95 to day 195. This detector was shielded by a 125 mil quartz window.

By measuring solar cell degradation in orbit and using the laboratory determined degradation rate as a function of 1 MeV electron flux, one can compute the equivalent 1 MeV electron flux encountered by an orbiting solar cell. This was done using the experimental data from the satellites 1961-$\alpha\eta$1 and 1962-$\beta\kappa$ and is presented in Table VI. 1961-$\alpha\eta$1 satellite results indicate that particle flux effective in damaging solar cells protected by six-mil thick glass increased by a factor of 217 after Starfish. Results of the satellite 1962-$\beta\kappa$ experiments indicate that during the period from late October through early December, 1962 the P-on-N cells protected by six-mil glass were subjected to 10.5 times as much damaging particle flux as were cells protected by 60 mil thick glass.

Further, a solar attitude detector, with 125 mils of quartz shielding, on satellite 1961-$\alpha\eta$2 degraded by only 6 percent in the first five days after Starfish. The experimental solar cells degraded by 16 percent during that period.

Table VII shows a comparison of the P-on-N solar cell degradation on ANNA I-B during the first 120 days in orbit. The equivalent one MeV electron flux values were computed by relating the in orbit degradations to laboratory data. Correlation of these values with Rosenzweig's curves yielded the omnidirectional integrated fission electron flux. Had the damaging particles been fission electrons, the fission electron flux would have been approximately equal for all three experiments. However, the solar cells with six-mil thick glass had nearly seven times the fission flux values as compared to the solar cells with 30 mil thick sapphire. It can be similarly shown that the difference in degradation of the P-on-N solar cell experiments on satellite 1962-$\beta\kappa$ may not be

TABLE VII

Comparison of P-on-N solar cell degradation on ANNA 1-B satellite; 120 day period beginning 114 days after "Operation Starfish"

*Experiment**	*Percent degradation*	*Equivalent* 1 MeV *electrons per* cm^2 *per day incident on bare* P-on-N *solar cells*	*Omnidirectional integrated fission electrons per* cm^2 *per day on covered cells*
P-on-N solar cells 30 mil sapphire covers	14.8	63 g	0.42 *T*
P-on-N solar cells 20 mil quartz covers	26.9	358 g	1.1 *T*
P-on-N solar cells 6 mil glass covers	36.7	1250 g	2.8 *T*

* All experiments had infinite back shielding.

explained by an omnidirectional fission electron spectrum. This analysis is also applicable to the results observed from the satellite 1961-$\alpha\eta$2 experiment and solar attitude detector. The 1962-$\beta\kappa$ results may be fully explained by normal inner belt protons.

FILZ and YAGODA (1963) measured an increase in 55 MeV protons at altitudes from 367 to 514 km after Starfish and after the Soviet test of 1962. After July 9, 1962 they observed a factor of six increase in 55 MeV protons. This was followed by a decay to approximately a factor of four above the pre-Starfish value just prior to the Soviet tests when the proton flux was increased to a factor of five over the normal level. As late as March, 1963 they measured 55 MeV proton levels about 3.5 times the level before the high altitude nuclear tests. It may be assumed that protons at other energy levels were also added to the radiation belts at these comparatively low altitudes as a result of these tests. Protons with $E>4.5$ MeV will penetrate six-mil thick glass while those with $E>9.3$ MeV will penetrate 20 mils of quartz. Thirty mil sapphire covers will shield against protons with $E<14$ MeV (ROSS, 1963). Therefore, it is postulated that the high altitude nuclear test caused a significant increase in the number of protons $E>4.5$ MeV to be redistributed at least to an altitude range between 400 and 1200 km.

The ANNA I-B was magnetically oriented. Therefore, assumption of an omnidirectional electron flux is not valid for this satellite. A directional flux results in twice the damage to solar cells as does an omnidirectional flux. Further, the electron bombardment of the ANNA solar cell experiments was at angles between 70° to 90° from the normal. Thus, effective increase of shield thickness and scattering of particles were considered. A comparison of integrated omnidirectional and integrated directional fission electrons is given in Table VIII. Since the fission electron fluxes are in close agreement for the directional case, it is probable that the ANNA I-B degradation was produced largely by a fission electron spectrum.

TABLE VIII

Comparison of integrated fission electrons per cm^2 per day on covered P-on-N solar cells during 120 days of ANNA I-B orbit.

Experiment	*Omnidirectional Electrons*	*Directional Electrons Incident 70° from normal*
30 Mil Sapphire Covers	0.42 T	1.5 T
20 Mil Quartz Covers	1.1 T	1.2 T
6 Mil Glass Covers	2.8 T	1.9 T

Earlier mention was made that the radiation intensity decreased during the period January to May, 1963. This is indicated on curve *D* and *E* of Figure 3. By use of curve *E*, P-on-N solar cells with six-mil cover slides, one finds that after 50 days in orbit those cells have degraded by 29.7 percent. Had the radiation intensity remained constant, the cells under question would have degraded by 46.4 percent at the end of 500 days. However, the cells degraded by 44.6 percent during that period. By the same analysis, the P-on-N solar cells with 20 mil quartz cover slides degraded 34.8 percent at the end of 500 days while they would have degraded by 36.5 percent in the same period had the intensity level not decreased.

An interesting comparison of the effects of radiation environment is given in Figure 5. The P-on-N solar cells with six-mil glass covers on ANNA I-B had experienced 1 MeV radiation flux equivalent to 850 G electrons/cm^2/day during the first 500 days in orbit while the equivalent experimental solar cells on satellite 1963-38C had experienced an equivalent 1 MeV radiation flux of only 44 G electrons/cm^2/day, a factor of 19.3 less. This is explained by the fact that ANNA I-B was launched while the radiation intensity was higher than when the satellite 1963-38C was launched.

Figure 4 shows that heavy cover slides may not, in themselves, be the solution to the problem of solar cell degradation in the radiation environment. The N-on-P solar cell with 125 mil quartz cover window, curve *D*, degraded by 10.5 percent in 500 days while the N-on-P solar cells with six-mil glass degraded by only 8 percent in the same time. This difference is attributed to discoloration of the thick cover slide.

To determine the power-to-weight ratio of a solar array after one year in an orbit similar to that of the 1963-38C satellite, one must consider the number of solar cells required to deliver the required power after one year and the cover slide thickness to be used. Thus, if *n* solar cells with six-mil glass covers are required, reference to curves *C*, *B*, and *A* of Figure 4 show that 0.92 *n* cells with 20 mil quartz covers or 0.69 *n* cells with 30 mil sapphire will be required to deliver the same power after one year. However, when one considers the relative mass of cover slides compared to the weight of (15 mil thick) solar cells, the power-to-weight ratio is significantly increased by using six-mil cover slides. To use Figures 3 and 4 for power systems analysis one must assume operation toward the short circuit side of the solar array IV curve.

Conclusions

A review of the data analysis presented above indicates the damage to solar cells in a 1000 km orbit during the early months after Starfish may not necessarily have been mostly a result of fission spectrum electrons. Although many energetic electrons were introduced into the inner belt, it is concluded that some high energy (>4.5 MeV) protons were redistributed to altitudes including 1000 km. The results of flight experiments indicate optimum power-to-weight ratio for a solar cell power supply and one year life expectancy may be obtained by use of N-on-P solar cells with six-mil glass covers. By projection of curve *C*, Figure 4, one may predict the degradation of optimum solar power supply for several years in a 1000 km polar orbit. Thus, the optimum power to weight design should degrade by 7 percent in one year and by 11 percent in three years. Orbital data obtained thus far indicates high reliability (against radiation damage) electronic circuits may be designed for operation in the radiation belts by use of the 2N2907-A transistors with minimal shielding or by use of 2N2586 and 2N2222 transistors with 0.60 grams/cm^2 of shielding.

Acknowledgements

The authors wish to acknowledge the significant contributions of Mr. A. F. Hogrefe without which this paper would have been incomplete. His efforts made possible presentation of all the transistor and capacitor data included and his technical advice in this area was most valuable. The technical discussions with Dr. A. J. Zmuda were most helpful.

References

Brown, W. L., Gabbe, J. D., and Rosenzweig, W.: 1963, *Bell System Tech. J.* **42**, 1505.
Filz, R. and Yagoda, H.: 1963, *Space Res.* **4**.
Hogrefe, A. F.: 1963, The Johns Hopkins University, Applied Physics Laboratory, Report TG-510, July 1963.
Martin, J. H.: 1964, Fourth Annual Photovoltaic Specialists Conference, Cleveland, Ohio, June 1964.
Ross, B.: 1963, ASME Aviation and Space Hydraulic and Gas Turbine Conference, Los Angeles, California, March 1963.
Smith, K. D., Gummel, H. K., Bode, J. D., Cuttriss, D. B., Nielsen, R. J. and Rosenzweig, W.: 1963, *Bell System Tech. J.* **42**, 1765.

Discussion

W. Brown, BTL, commented that changes in the Starad apogee may have caused the observed shift in source characteristics and that calculation of the effects of non-omnidirectional vs omnidirectional fluxes for the oriented satellites would be very difficult since pitch angles and pitch angle distributions would have to be determined and utilized as an independent input for each point in space which the satellites traversed.

A METHOD FOR CALCULATION OF AVERAGE RADIATION FLUXES ON SATELLITES

D. H. SOWLE and R. W. LOWEN

Space Science Laboratory, General Dynamics/Convair, San Diego, Calif.

Abstract. A method of calculating the time averaged flux on a satellite in orbit in the lower radiation belt (altitude $\lesssim$ 9000 km) is described. The flux encountered by a satellite is shown to converge rapidly to the average value. The deviation from the calculated average for a satellite whose period is commensurate with that of the earth is shown to be unimportant for design applications.

1. Introduction

The designer of a vehicle which will orbit in the radiation belt needs to know the time integrated flux of charged particles which will be encountered by the vehicle. For many purposes an average of the flux rather than a detailed time history will suffice. We will describe briefly a method of calculation of average radiation flux on an orbiting satellite and discuss the accuracy of the method. The description of the method will be intuitive as the detailed mathematics have been discussed elsewhere (Sowle and Lowen, 1965).

2. Description of the Method

We suppose the space which surrounds the earth and contains the radiation belts to be divided into cells of width Δl in longitude, $\Delta\lambda$ in latitude, and Δz in altitude. The dimensions of the cells are to be chosen small enough that for practical purposes the charged particle flux can be considered constant within a cell. A satellite will pass through each cell it visits many times in its useful lifetime. Rather than follow the satellite in its trajectory, summing contributions from cells as it passes, and thus repeating time consuming operations many times, we choose to find the time averaged flux by summing the product of the flux in each cell with the fraction of time spent there,

$$\langle F \rangle = \sum F_{abc} W_{abc} . \tag{1}$$

where F_{abc} is the flux of interest in cell abc and W_{abc} is a weighting function equal to the fraction of time the satellite spends there.

The utility of the method depends upon finding a simple expression for the weight factors, W. We inquire into conditions for which we can write

$$W_{abc} = W_a W_b(i) W_c(A, P) \tag{2}$$

where W_a is the fraction of time spent in the ath longitude interval, W_b is the fraction of time spent in the bth latitude interval, and W_c is the fraction of time spent in the cth altitude interval. W_b is expected to depend on the inclination, i, and W_c is expected to depend on the apogee altitude, A, and the perigee altitude, P.

Consider first a circular orbit, so $W_c(A, P)=1$ at altitude Z. The distribution of longitudes at which the satellite crosses a given latitude will tend to become uniformly spaced as the earth rotates under the satellite orbit (see Figure 1).

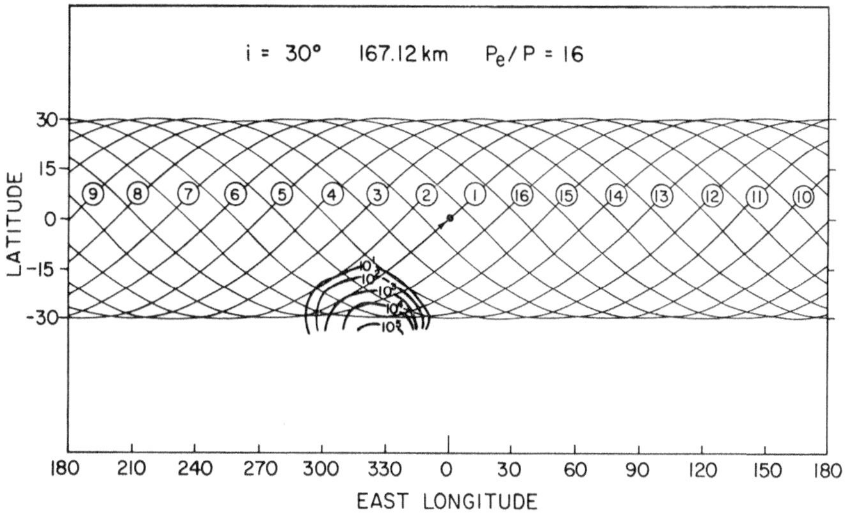

Fig. 1. Locus of longitude-latitude path traced by a low altitude satellite whose period is commensurate with that of the earth. Successive orbits are numbered assuming injection at $l=0°$, $\lambda=0°$. Omnidirectional flux contours for electrons with $E>0.5$ MeV are shown.

If the period of the satellite, P, is not commensurate with the period of the earth, P_e, the average longitude spacing between crossings becomes small and we can write

$$W_a \approx \frac{\Delta l}{2\pi}. \tag{3}$$

We may hope to avoid commensurability problems for the region of space where the radiation belts are fairly stable since in this region $P \ll P_e$ thus the set of longitudes at which the satellite crosses a given latitude will be fairly closely spaced even for commensurate periods. For the moment we assume the commensurability question is answered favorably so (3) holds. We therefore find that for circular orbits (2) holds. It is straightforward to obtain the fraction of time spent in $\Delta\lambda$, between λ_1 and λ_2 as

$$W_b(i) = \pi^{-1}\left[\sin^{-1}(\sin\lambda_2/\sin i) - \sin^{-1}(\sin\lambda_1/\sin i)\right]. \tag{4}$$

Consider now an elliptical orbit. We direct attention to some increment of the orbit lying between z and $z+\mathrm{d}z$, (Figure 2). As the perigee precesses, the orbit rotates as a rigid body, and that segment of the orbit lying in $\mathrm{d}z$ executes

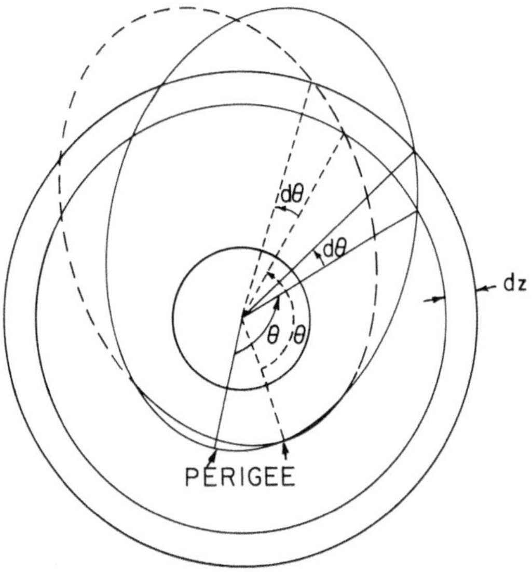

Fig. 2. The angular increment dθ of an elliptical orbit undergoes circular motion in the orbital plane as the perigee precesses.

circular motion at inclination i. If the precession rate is such that the orbital segment of interest remains in a latitude cell long enough for the satellite to obtain a good longitude sampling, then (3) will hold for elliptical orbits as well as circular. We expect (3) to hold since the satellite, earth, and perigee periods are respectively hours, one day, and many days.

The fraction of time that the element $\mathrm{d}z$ spends in the latitude increment $\Delta\lambda$

is given by Equation (4) after an integral number of perigee cycles. Since Equation (4) has no altitude dependence we see that Equation (2) holds for orbits where there are no commensurability problems and where the perigee precession rate is sufficiently slow. The fraction of time spent in altitude increment dz between z_1 and z_2 is

$$W_c(A,P) = 2[\pi(A+P+2R)]^{-1}\{[(A-z_1)(z_1-P)]^{\frac{1}{2}} - [(A-z_2)(z_2-P)]^{\frac{1}{2}}\} + \pi^{-1}\{\sin^{-1}[(A+P-2z_1)/(A-P)] - \sin^{-1}[(A+P-2z_2)/(A-P)]\}. \quad (5)$$

It is thus possible to construct average orbital flux tables for particles of interest as a function of inclination and altitude of circular orbits. Elliptical orbit averages may be easily synthesized by hand by applying Equation (5) to the circular orbit averages. A set of tables ranging from 200 km altitude to 8800 km altitude may be calculated using Equations (1) through (4) with less effort than one elliptical orbit average when the satellite trajectory is followed. Higher altitudes than about 8800 km become less interesting because the radiation belts vary relatively rapidly as altitude increases, thus accurate prediction becomes impossible.

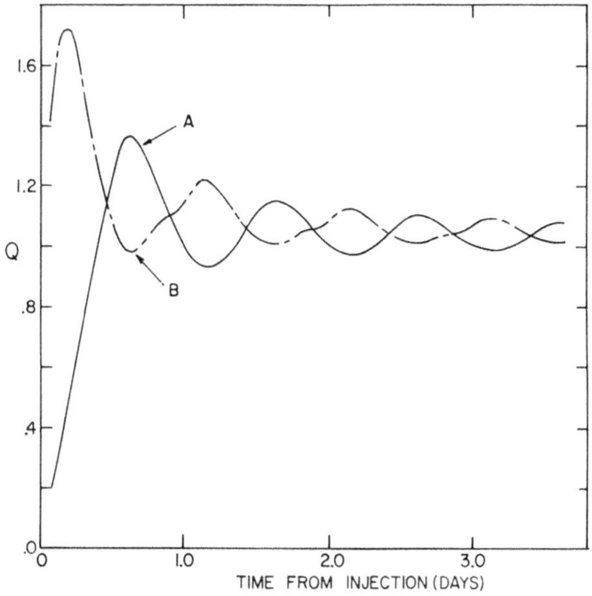

Fig. 3. Comparison of long-term and current average proton fluxes for a circular orbit at 1000 km and 30° inclination. Initial positions: A, 30 °N, 305 °E; B, 30 °S, 326 °E.

Examples of the rate of convergence to the tabular average for a noncommensurate circular orbit are given in Figure 3. We define Q as the ratio of the "current" average to the tabular average

$$Q(t) = \langle F \rangle^{-1} t^{-1} \int_0^t F(t')\,\mathrm{d}t' \tag{6}$$

Figure 4 shows Q for two noncommensurate elliptical orbits, we see that the method works well for this type orbit.

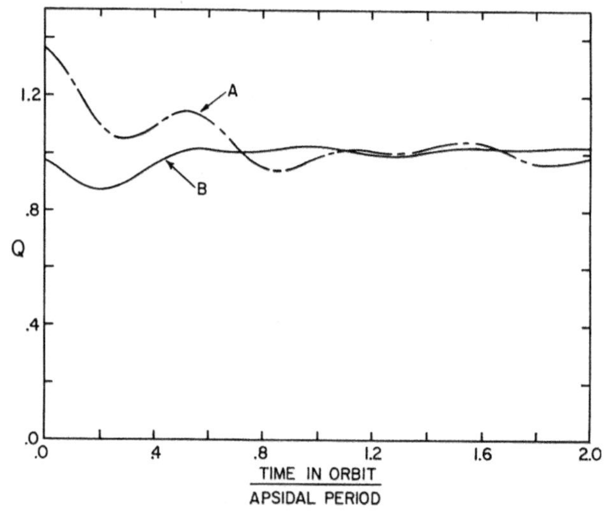

Fig. 4. Comparison of long-term and current average proton fluxes on two elliptic orbits; orbit A: $P = 350$ km, $A = 1200$ km, $i = 98°$, initial position 34 °N, 61 °E; orbit B: $P = 648$ km, $A = 3426$ km, $i = 30°$, initial position 14 °N, 178 °E.

We now turn to the question of commensurate orbits.

The largest spacing in longitude for a commensurate orbit with apogee under 8800 km is 72°. For a circular orbit with $i = 90°$ this occurs at altitude 8041.828 km, $P_e/P = 5$. Figure 5 shows the ratio of the orbital average to the tabular for this orbit as a function of initial satellite longitude. We see the deviation from the average is insignificant for satellite design purposes.

At lower altitude the ratio P_e/P becomes larger and the longitude spacing smaller but the radiation belt becomes more variable in l. We therefore have made calculations for several other orbits, with ratios of $P_e/P=10$, 15, and 16. The results are shown in Figure 5. We see that in none of these cases does

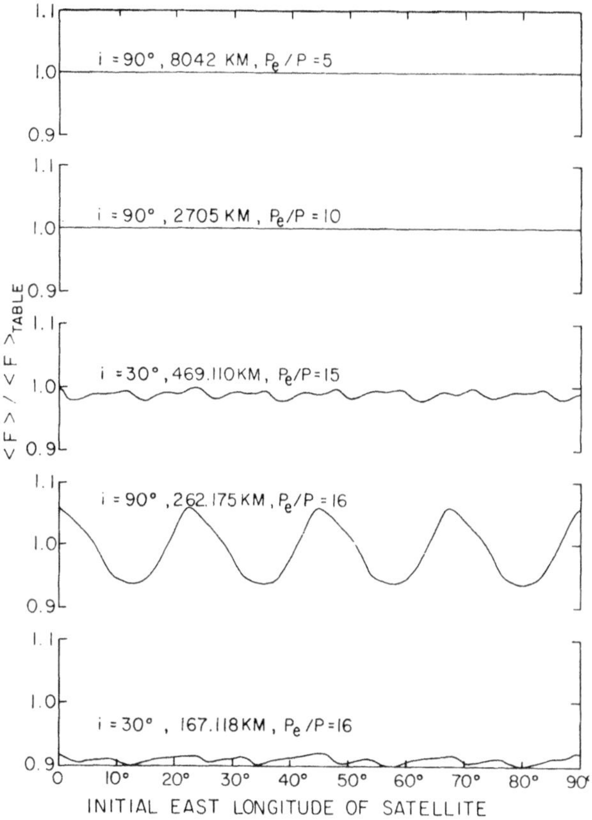

Fig. 5. Ratio of the average omnidirectional electron flux ($E>0.5$ MeV) to the tabular value versus initial satellite longitude, l_2 for commensurate orbits covering the altitude range from 167 km to 8042 km.

commensurability cause difficulties. The method may therefore be applied to any circular orbit which lies in the range of the tables.

Extension of the method to elliptical orbits imposes the additional requirement that the perigee not precess too rapidly to allow a good longitudinal average to be obtained in each cell $\Delta\lambda$. We have found $\Delta\lambda=5°$ to be adequate. No orbit in the range covered by the table precesses more rapidly than a few degrees per day, so there is always adequate time to obtain a good average and the method may be applied to elliptical as well as circular orbits.

Reference

Sowle, D. H. and Lowen, R. W.: 1965, *J. Spacecraft Rockets* **2**, 153.

SESSION IX

RADIATION DOSES RECEIVED BY MANNED FLIGHT IN THE TRAPPED RADIATION BELTS

Chairman

DR. JOHN B. CLADIS

Lockheed Missile and Space Company

PREDICTIONS AND OBSERVATIONS FOR SPACE FLIGHT *

DUANE A. ADAMS

Air Force Weapons Laboratory, Kirtland AFB, N.Mex.

and

BRIAN W. MAR

The Boeing Company, Aero-Space Division, Seattle, Wash.

Abstract. Comprehensive shielding methods are available to predict the dose from radiation which penetrates a space vehicle. These methods are integrated with trajectory and environment definitions to predict the dose to humans in space. The contribution of solar-particle event and cosmic radiation to the total dose is predicted. Measurements of dose on the Gemini GT-4 flight and on unmanned satellites are reported. The human response to radiation must be better defined before the radiation hazard to manned space flight can be fully assessed.

1. Introduction

The radiation trapped in the earth's magnetic field has been the subject of intense analytical and experimental examination, as is evidenced by the comprehensive data reported in the *Proceedings* of this symposium. This paper summarizes the prediction methods and results and the experimental observations of the doses to manned space flights resulting from space radiation. In addition to the radiation trapped in the earth's magnetic field, the radiation dose from solar-particle events and from galactic cosmic radiation is considered in order to give a more complete picture of the radiation hazards in space. The methods of analyzing trapped radiation are easily extended to untrapped radiation from solar-particle events and from galactic cosmic rays.

The evaluation of space radiation hazards began in the late 1950's and was stimulated by the discovery of geomagnetically trapped particles in 1958. Emphasis was placed on protons and the secondary radiation produced by their interaction with matter. The Symposium on the Protection Against Radiation Hazards in Space ** contained much of the early work on protons and indicated a requirement for analysis of electrons injected by the Starfish high-altitude nuclear

* The technical contributions on prediction methods were prepared by W. R. Doherty, W. R. Sheldon, and M. C. Wilkinson of The Boeing Company. M. Schneider, J. Janni, and G. Radke of the Air Force Weapons Laboratory furnished the unpublished Gemini data, and R. Stovall provided prediction results.

** Gatlinburg, Tennessee, November 5–7, 1962.

detonation of July 9, 1962. Refinements of analytical methods for proton shield analysis and development of electron shielding methods were reported at the Second Symposium on Protection Against Radiation in Space.* The prediction methods reported in this paper were initiated in 1961. They have been continually refined to meet changing mission requirements, and they have been improved by the incorporation of recent experimental and theoretical interaction data.

The predictions of space radiation doses encountered by space missions are presented for various dose points in the human body for a specific space vehicle in selected trajectories. These predictions indicate the importance of the energy spectrum of the incident radiation to the dose deposition and the protection that can be achieved by partial body shielding. Experimental data obtained on manned and unmanned space flights are reported to indicate the correlation between experimental and predicted values.

2. Methodology

The prediction methods used by the Air Force Weapons Laboratory and reported here were developed by personnel of The Boeing Company, who are combining these methods into a single computer system. The prediction method consists of (1) the environmental data stored in the computer memory, which permits interpolation for the intensity, spectrum, and angular distribution for any specified point; (2) the trajectory data, the matrix of points in space where the environmental data are required; (3) the sector analysis, which describes the material distribution of the space vehicle and crew as a function of solid angle about the dose point; and (4) the shielding analysis, which provides a matrix of transmitted dose versus incident particle energy. This shield matrix is multiplied by the incident spectrum to obtain the penetrating flux and dose rate. Complete flexibility is available to evaluate many vehicle types for a specific trajectory or many trajectories for a specific vehicle.

2.1. Radiation environment

The ionizing radiation environment in cislunar space may be classified as geomagnetically trapped or untrapped radiation. Galactic cosmic and solar-particle radiation will be treated as untrapped radiation; the neutral components, chiefly photons and neutrons, make such a small contribution to the dose that they are neglected.

2.1.1. *Trapped Radiation*

The trapped-particle flux is recorded in *B-L* coordinates for this analysis, and

* Gatlinburg, Tennessee, October 12–14, 1964.

all particles adiabatically trapped for at least several drift periods are considered. The omnidirectional flux, the spectral distribution, and the angular distribution at any point are required to describe this environment. Then a flux table is prepared from maps of isoflux contours in *B-L* space with an accuracy of 10 to 20 percent. Currently, the proton maps constructed by W. Hess from 1958 measurements, and the proton and electron maps compiled by J. Vette from 1962 measurements are used in this program. Other tabulations of flux data can also be inserted.

Direct spectral measurements of the protons and electrons have established that the spectrum is strongly *L*-dependent and at least slightly *B*-dependent. The energy spectrum for each particle type is expressed in a simple function of location and particle energy. This expression allows the calculation of the dose rate at any point in space, and by integrating these spectra over time one can obtain the differential energy spectrum of the fluence for any mission.

The angular distribution of the incident flux can be important in dose calculations when the space vehicle is oriented with respect to the earth's magnetic field. (Data from the Gemini GT-4 flight will be presented to demonstrate this effect.) Usually the space vehicle orientation with respect to the field lines is constantly changing and so the fluence is approximately isotropic. The option to compute the angular distribution at a point is seldom used in dose considerations, but it is required for analysis of experimental data obtained at fixed orientation in the earth's magnetic field.

2.1.2. *Untrapped Radiation*

Galactic cosmic rays and high-energy particles ejected from the sun during large solar flares make up the untrapped radiation environment in the vicinity of the earth. Experimental results indicate that both types of radiation are approximately isotropic outside the magnetosphere. The computation of untrapped radiation dose within the magnetosphere requires the specification of an effective spectrum, discussed below, for each of a set of *L* shells and a knowledge of the fluence that reaches the vehicle in each *L* shell. The dose transmission factor for each spectrum is computed. The total dose is the summation of the product of the dose transmission factor multiplied by the fluence received in the corresponding *L* shell. The dose from galactic cosmic radiation is computed separately since it constitutes a small contribution to the dose encountered by an orbiting vehicle.

2.1.2.1. *Solar-Particle Events.* Except for the first few minutes of a solar-particle event, the velocity distribution outside the magnetosphere is roughly isotropic. In most events the particles reaching the earth are predominantly protons, although in a few events appreciable numbers of alpha particles have been observed. In the results presented here, only the proton contribution has been considered. Modifications have been made in the computer program to include the effects of heavier particles. At any time during the event, the particle spectrum can be adequately described by the expression of the form

$$J(>P) = J_0(t) \exp\left[-P/P_0(t)\right] \tag{1}$$

where P is the particle rigidity (momentum/unit charge), $J_0(t)$ and $P_0(t)$ are functions of the time only, and $J(>P)$ is the flux of particles with rigidity $>P$. The rigidity is related to the kinetic energy E of the particle as follows:

$$P(E) = \frac{1}{z}\sqrt{E^2 + 2mc^2E} \tag{2}$$

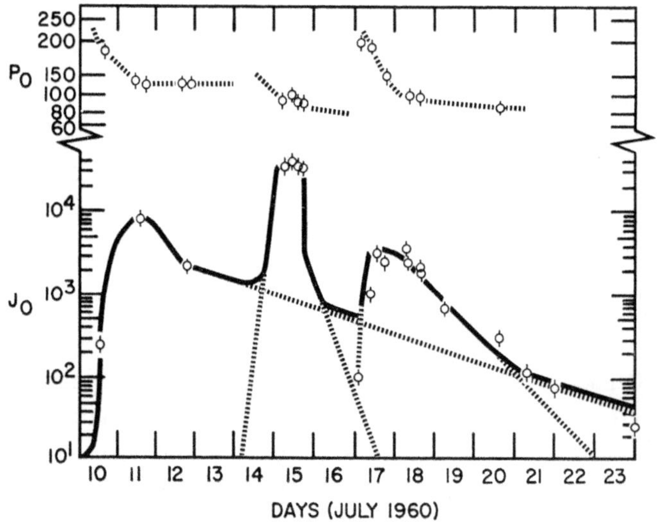

Fig. 1. Solar-particle event characteristics.

A convenient set of units gives P in MV, z the number of electronic charges, mc^2 is the particle rest mass in MeV, and E is in MeV.

The available experimental data indicate that solar-particle events have the general features shown in Figure 1. A few minutes after the optical maximum of the solar flare, the first high-energy particles reach the earth. The particle intensity increases as the lower-energy particles arrive until a maximum intensity is reached from 8 to 30 hours after the beginning of the event. After the maximum, the intensity decreases slowly to the background level. The total duration of the event ranges from 2 to 5 days.

The earth's magnetic field influences the path of a charged particle approaching the vicinity of the earth such that there is a threshold energy for the particle to reach a point (R, L) from outside the magnetosphere. The theory of the allowed cone developed by Störmer (1955) was used by Barton *et al.* (1964) to establish the threshold energy for protons to reach the point (R, L). The dose from solar-particle events at a point in the earth's magnetic field can be computed from the time behavior of the particle intensity, the modification of the energy spectrum with position, and the variation of the spacecraft position with time. In the earth's magnetic field we approximate the spectrum by a sharp cutoff at the threshold energy for vertical incidence and determine the hazard for geomagnetically quiet conditions as well as for magnetically distrubed conditions. The geomagnetic storm that accompanies the larger solar-particle events usually produces a reduction in the magnetic field strength such that the threshold energy for arriving particles at any given point in smaller than the normal value.

Results of solar-flare hazard evaluation are presented in Section 3. This vertical cutoff approximation is a first attempt to compute the effect of the geomagnetic field on untrapped radiation and has the advantage of fitting easily into the calculational procedures established for the trapped radiation.

2.1.2.2. *Galactic Cosmic Rays.* The radiation dose to space vehicles from the galactic cosmic radiation can be calculated from the measurements of Neher and Anderson (1962, 1964), which indicated that 21 percent of the dose to the balloon-borne instruments at polar latitudes was due to albedo arising from the interaction of the primary radiation with the earth's atmosphere and magnetic field. The values presented include the albedo contribution and are appropriate for vehicles in near-earth orbits; for vehicles a large distance from the earth these values should be reduced by 20 percent and then corrected for spherical geometry. Since the measured cosmic-ray doses change very little with atmospheric depth near the top of the atmosphere, the shielding on current space vehicles will not attenuate this dose. The dose rate is observed to change by a factor of two during the sunspot cycle, the maximum dose rate from galactic cosmic radiation coming at the same time as the minimum in solar activity. The measurements of Lin *et al.* (1963) indicate that the counting rate from galactic cosmic rays, as measured by the Explorer VII satellite, is constant above $L=2.5$ and is approximately a linear function of L between values of 1.1 and 2.5. This approximation is conservative, and the dose is slightly overestimated in equatorial regions.

Normalizing the dose rate measurements of NEHER and ANDERSON (1962, 1963) to the L variation, we have the dose rate D as a function of L

$$D = \begin{cases} [0.192(L-1) + 0.072] \text{ mrad/hr} & \text{for } 1 < L < 2.5 \\ 0.36 \text{ mrad/hr} & \text{for } L > 2.5 \end{cases} \tag{3}$$

and at the time of maximum in the solar cycle. For the time of solar minimum

$$D = \begin{cases} [0.54(L-1) + 0.072] \text{ mrad/hr} & \text{for } 1 < L < 2.5 \\ 0.88 \text{ mrad/hr} & \text{for } L > 2.5\,. \end{cases} \tag{4}$$

The resulting dose rates will be much smaller than those arising from the trapped particles for most cases of interest. The maximum galactic dose-rate is 21 millirads/day during solar minimum and 9 millirads/day at the time of solar maximum. For cases where the dose from trapped particles exceeds these values significantly, it is recommended that the values for $L>2.5$ be used to estimate the galactic radiation dose.

2.1.3. *Spacecraft Trajectories*

The space-radiation flux is a sensitive function of position and time; hence, the radiation flux encountered during a mission depends critically on the accuracy with which the spacecraft trajectory is specified. Experience indicates that the required order of accuracy, compatible with the variability of the radiation environment, is generally $\pm$ tens of kilometers in altitude, ± 1 degree in latitude and longitude, and ± 1 minute in time; these values are also compatible with present uncertainties of flux values in the radiation belts.

For low-altitude orbits the flux of trapped radiation is zero for more than half the geocentric spherical surface. A survey of the low-altitude flux data indicates that the flux of trapped particles for altitudes under 800 km is confined to the Atlantic between 110 °W and 80 °E. For altitudes under 400 km the region shrinks somewhat so that it is contained between 75 °W and 60 °E. Therefore, if calculations of flux outside this region are bypassed, a 50-percent reduction in computer time will be realized. Since its shape and location are only approximately known, this region is made generously large, and the flux of trapped particles is set equal to zero for longitudes west of 120 °W and east of 90 °E for altitudes below 800 km. An option to bypass this geographic test is provided for studies of untrapped radiation.

Some commonly used methods of minimizing computer time include the use of interpolated B-L coordinates, variable step integration methods, and tape storage of results of trajectory calculations for use with subsequent environments or vehicles.

2.2. Shielding

The penetration of the incident radiation in a spacecraft is highly dependent on the energy and type of radiation and the physical characteristics of the spacecraft. In addition the bodies of the astronauts provide significant self-shielding. Penetrating protons are generally the primary dose component for shield thickness of less than 30 g/cm^2 of aluminum, and the penetrating electron dose is only significant for shields thinner than 5 g/cm^2. Bremsstrahlung produced by the electrons can be significant and large amounts of additional shielding are required to significantly reduce the dose.

The basic method used for space radiation shielding calculations is the summation of the dose contribution from individual "sectors". A sector, i, is defined as a portion of space, subtended by a solid angle ω_i about any dose point. Sectors are selected on the basis of regions of uniform shielding mass (where an average mass or thickness, $\bar{t}_i$, can be used to adequately represent the shielding characteristics of each sector). For an isotropic incident flux distribution, the dose contribution from any sector of solid angle ω_i, and thickness $\bar{t}_i$, is

$$D_i = \frac{\omega_i}{4\pi} T(\bar{t}_i) \tag{5}$$

where $T(\bar{t}_i)$ is the dose penetrating a sphere a uniform thickness $\bar{t}_i$ per incident unit flux. The function $T(\bar{t}_i)$ is usually an integral over the incident particle energy distribution

$$T(\bar{t}_i) = \int_E N(E) T(E, \bar{t}_i) K(E, \bar{t}_i) \mathrm{d}E \tag{6}$$

where $N(E)$ is the differential number spectrum, No./cm^2-sec-MeV; $T(E, \bar{t}_i)$ is the fraction of particles of initial energy E penetrating shield of $\bar{t}_i$; and $K(E, \bar{t}_i)$ is the dose conversion factor, rad/No./cm^2-sec. These basic relations become complex when multilayered shields, nonisotropic flux distributions, and secondary particles are considered.

The ultimate goal of radiation analysis is to provide a quantitative measure of the radiation effect on a manned or unmanned system. Before the analytical methods are discussed, a brief discussion is presented on the units used to measure the energy deposition. The rad, defined as 100 ergs deposited per gram of material, is the unit of dose commonly used by workers in the field of radiation damage. It is defined independently of absorber material, type of particle, or energy spectrum of the incident radiation, and it can be easily calculated or measured. For biological work a Relative Biological Effectiveness (RBE) is defined as the ratio of the dose of X or gamma radiation to produce a given biological response to the dose for the type of radiation of interest to produce the same response. This concept does not lend itself to any generalization, since an RBE must be measured for every radiation type for each biological response. A new term, Quality Factor (QF), has been introduced by the International Commission on Radiological Protection. This factor defines the ratio of a dose of X or gamma radiation to produce whole-body biological damage to the dose from any given type of radiation. The QF should be used for hazard analysis while the RBE is a term that has been reserved for experimenters to report actual measurements of specific radiobiological effects.

2.2.1. *Principles of Sectoring*

The geometrical analysis of vehicle shielding, commonly called sectoring, consists of first placing a spherical coordinate system at the dose point; the vehicle is then divided into N sectors, or solid angles, each subtending a certain solid angle, ω_i. Each sector is selected in such a way that the optical depth and the material composition in the sector can be considered constant for shielding calculations. A large number of sectors may be required to meet this criterion in a complex space vehicle or the human body. The Apollo spacecraft in a particular analysis has been divided into 108 sectors, while Gemini has been divided into 720 sectors for a particular analysis. These sector analyses have been largely intuitive and have not been verified experimentally.

To determine the radiation shielding properties provided by the space vehicle and the astronaut's body at a particular dose point, the shielding calculations require the total number of sectors about the dose point, the solid angle subtended by each sector, and the material composition and optical depth of each material. Future shielding codes may require additional data, such as the linear distance between the dose point and the materials (slant path), the angle between the normal to the materials and the slant path, and the angle between the axis of symmetry of the radiation flux and the normal to the sector to account for scattering and nonisotropic flux distributions.

The human body is a significant shield for most space radiations and it must be considered when computing doses at actual body points, such as the eye or blood-forming organs. An attempt to realistically define the selfshielding of a human body was made by Dye (1962a). Dye chose 12 body points for analysis, using the seated, 75-percentile, U.S. Air Force man as a model. These points are representative of the various points for which radiation tolerances are most often given. He then determined, with the methods of descriptive geometry, the tissue thickness for each of these body points as a function of solid angle.

The absorbed dose in rads can be determined using these isothickness plots for each radiation type and spectrum. For these calculations the density is assumed to be 1 g/cm^3, and the average Z is assumed to be that of H_2O.

There are two basic assumptions in the Dye model astronaut: (1) it is a homogeneous model, and (2) it is a seated astronaut. The error incurred by assuming homogeneity was checked by including bone tissue in the femur calculation and by including the lung cavity in the chest calculation. It was shown that the former was in error by less than 10 percent and that the latter was in error by less than a few percent. The single posture restriction can be corrected by determining the isothickness contours for other postures.

Alternative astronaut models have been proposed and constructed by using combinations of geometrical shapes such as spheres, truncated cones, cylinders, and ellipsoids. Although these combinations may achieve a model whose position and posture may be changed and, with considerable sophistication, may achieve a heterogeneous model, the errors inherent in the methodology would seem larger than those of the Dye model.

When the sectoring is completed, it may be used directly as input data for the shielding codes. In general the number of sectors the shielding codes must consider can be reduced significantly since (1) Sectors of the same thickness can be combined to form one large sector with greater angular weight; (2) Electrons encountered in space have no dose contribution for shields thicker than 5 g/cm^2. These sectors need not be considered by the electron penetration codes; (3) The more penetrating radiations, protons and bremsstrahlung, can be treated with little error by combining sectors with similar optical depth; (4) Materials with similar shielding characteristics can be combined, thus reducing the total amount of sectoring data that must be processed by the shielding codes.

2.2.2. *Electrons*

A pseudo Monte Carlo analysis of electron penetration was made to develop prediction methods for shielding calculations. Details of the Monte Carlo methods are described by BERGER (1963) and MAR (1963). The significant results of this analysis were that (1) a dimensionless equation could be used to predict the number flux penetrating a shield, (2) the angular distribution of the electron population achieves an equilibrium condition after a certain thickness, and (3) the differential energy spectra of the penetrating electrons can be approximated by a simple exponential relationship.

The Monte Carlo scheme used in this analysis employed a series of straight-line segments to represent the tortuous path that an electron follows when slowing down in a material. The electron is degraded in energy by ionization loss and bremsstrahlung loss in each segment. The scattering of the electron after each segment is computed from the Bethe-modified Moliere scattering relations (BETHE, 1953). This procedure is illustrated in Figure 2. The agreement with existing theoretical and experimental data was better than 10 percent.

The Monte Carlo results indicated that the angular distribution of penetrating electrons attain an equilibrium distribution after penetrating a diffusion length (observed to be about one-fifth the electron range by FRANK (1959)) and was independent of the incident angle of the electron. Monte Carlo data were correlated to obtain a simple representation of the number transmission and energy spectrum of the penetrating electrons. The number transmission is influenced by the initial direction of the electron, as shown in Figure 3. The

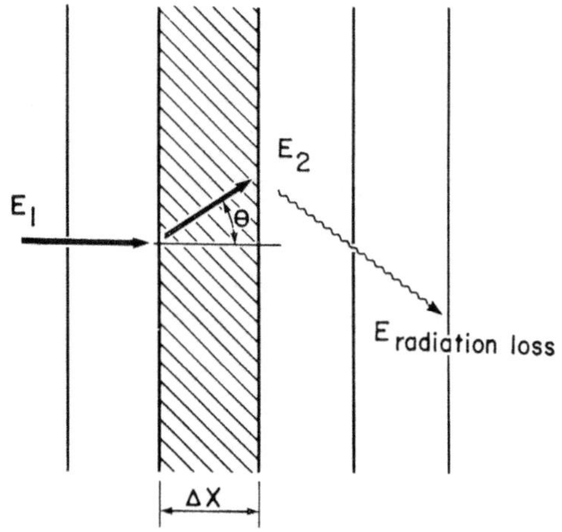

$$E_2 = E_1 - E_{\text{ionization loss}} - E_{\text{radiation loss}}$$

$$E_{\text{ionization loss}} = \frac{\Delta X}{\cos\theta} \cdot \frac{dE}{dX}(E_1)$$

Fig. 2. Monte-Carlo analysis.

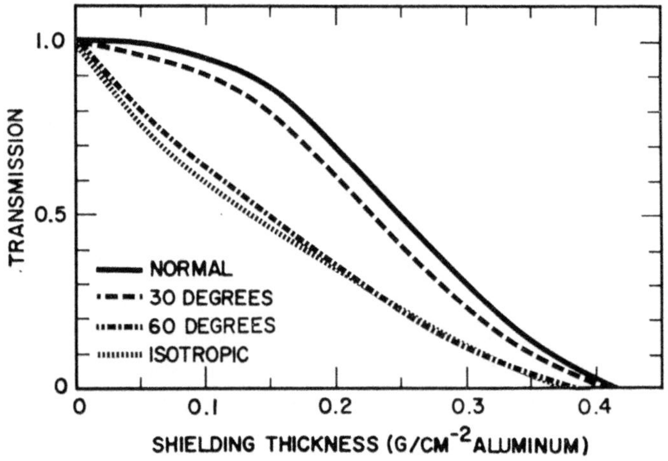

Fig. 3. The effect of incident angle on 1 MeV electron transmission.

normal incident electron transmission is observed to be higher than electrons incident from any other direction and the transmission of normal incident electrons was used to provide a conservative model for electron penetration.

The normal transmission data were analyzed using the dimensionless length and energy parameters of SELIGER (1955) and MAKHOV (1960). This procedure resulted in the expression

$$T(X,E,Z) = \exp\{-[E(0.585Z^{-0.271}/X)^{0.848}]^{-7(Z-3.25)-0.24}\} \tag{7}$$

and the penetrating spectrum was fit by the expression

$$\begin{aligned} \Phi(E) &= A \exp[-b_0(E_0 - E)] \quad & E < E_{\max} \\ &= 0 & E > E_{\max} \end{aligned} \tag{8}$$

where A and b are functions of Z, X, E_0.

These expressions form the basis for a simple electron penetration calculation. A nomograph (shown in Figure 4) has been prepared based on these equations for rapid hand calculations.

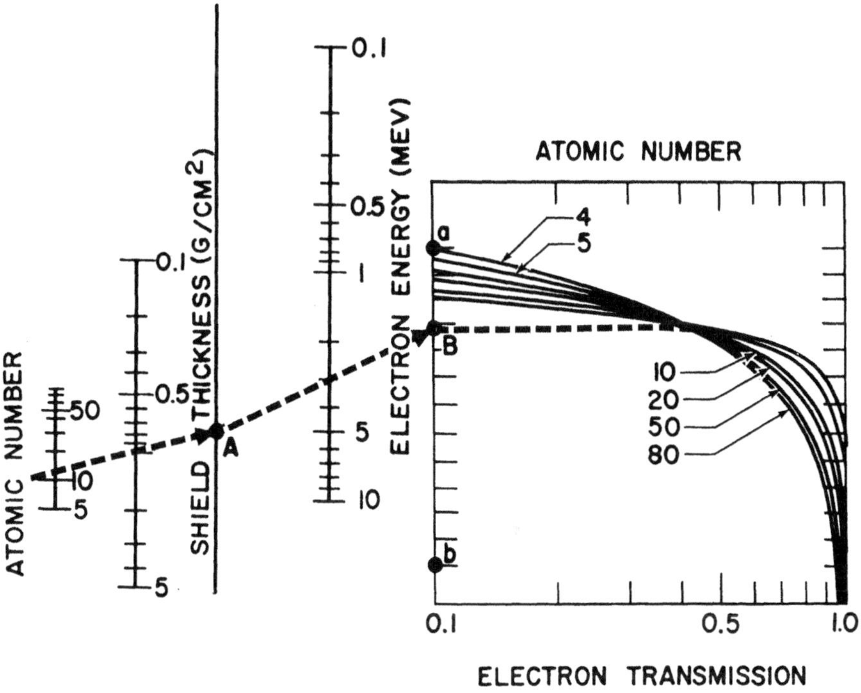

Fig. 4. Transmission of normally incident electrons.

The electron dose transmission factor is a form of Equation (6) that (1) computes the number of electrons penetrating the shield from incident electrons of energy E_0, (2) integrates over all incident energies to obtain the total number of electrons of energy E penetrating the sector, (3) computes the energy deposited by these electrons, and (4) integrates over all final electron energies to compute the total dose.

The total dose transmission from all sectors is expressed as

$$D_e = \sum_{i=1}^{n} D(i)\, S_e F_e(i)\, \Omega(i)\, \omega(i) \tag{9}$$

which is the sum of the dose contribution from each sector i, defined as the free space flux S_e, times the fraction $F_e(i)$ of this flux entering the sector (shown by EVANS (1961) to be 0.25 for an isotropic flux), times the dose transmissions through a shield as thick as sector $D(i)$, times the fractional solid angle $\omega(i)$, subtended by sector (i), times the fraction $\Omega(i)$ scattering to the dose point (3/2 for cosine squared distribution).

A multislab calculation is made assuming that an equivalent thickness of any material that produces an equal attenuation of electrons will also produce an identical energy and angle distribution of electrons at that point. Obviously this is not exact since the ratio of radiation to ionization loss is dependent upon atomic number. There is no advantage in using a multislab approximation when the atomic numbers of the shields are similar, since both single transmission curves for these shields are close to the multislab case. For shield thicknesses less than one-half the electron range, the multislab approximation provides 15-percent accuracy in the calculated number transmission. Maximum errors occur at large percentages of the electron range where large differences exist in the atomic numbers of the shielding materials.

The bremsstrahlung calculations are based on the expression

$$D_B = \int K(h\nu) \prod_{i=1}^{N} B(\mu_i(h\nu) X_i) \exp - \mu_i(h\nu) X_i \times \int 0.0007\, ZE\Phi_e(E) [4(1 - h\nu/E) - 3h\nu/E \ln (E/h\nu)] \,\mathrm{d}E \,\mathrm{d}h\nu \tag{10}$$

where D_B is the bremsstrahlung dose created in the n-layered shield by an incident spectrum of electrons $\Phi_e(E)$; $K(h\nu)$ is the flux-to-dose conversion factor for photons of energy $h\nu$; $B(\mu_i(h\nu) X_i)$ is the buildup factor for photons of energy $h\nu$ for $\mu_i X_i$ mean free paths in shield i; μ_i is the mass absorption coefficient; and X_i is the slab thickness. The expression 0.0007 ZE is the average amount of energy radiated by bremsstrahlung in a thick target as shown by

EVANS (1955). The term in the brackets is the energy radiation per unit energy interval derived by WYARD (1952):

$$[4(1 - h\nu/E) - 3h\nu/E \ln (E/h\nu)].$$

The dose from a truncated cone shield composed of *n* layers has been derived by FODERARO and OBENSHEIN (1955) and used for this calculation. An option assuming that all bremsstrahlung is produced on the surface of the shield is used when the first shield layer is thicker than the incident electron range. The agreement between the results of the surface source bremsstrahlung model and the volume distributed source is within 5 percent for shields greater than 5 g/cm^2.

2.2.3. *Protons*

Standard transport equations have been used by ALSMILER *et al.* (1963), ALLEN *et al.* (1961), and DYE (1962b) to describe the propagation of an incident proton spectrum, and the resulting cascade and evaporation products, through various shield configurations. The approach generally used in the solution of these equations (some form of finite difference approximation) has the disadvantage of requiring considerable computer time to achieve results of reasonable accuracy. By formulating the transport equations as a series of integral equations (using the first collision and straight-ahead approximations) it was found possible to achieve a high degree of accuracy with a minimum of computer time.

The proton spectrum at point x is related (DYE, 1962b) to the incident spectrum by

$$\Phi_p(E_{out}, X) = \Phi_p(E_{in}, 0)\frac{S(E_{in})}{S(E_{out})} \tag{11}$$

where $\Phi_p(E_{out}, X)$ is the proton flux of energy E_{out} at a distance X in the shield resulting from a proton flux of energy E_{in} incident on the surface of the shield, and $S(E)$ is the stopping power for protons of energy E. For a multilayer absorber of different material types the procedure is to repeat the calculation as the proton flux passes through each layer. The total number of protons penetrating a shield of thickness X is $\Phi_p(X) = \int \Phi_p(E, X)\, dE$.

For thin absorbers, that is, X much less than the mean free path of protons in the absorber between inelastic collisions, this procedure gives proton fluxes of adequate accuracy. As absorbers of increasing thickness are considered, it becomes necessary to attenuate the incident proton flux to account for those protons removed by inelastic collisions, using the relationship

$$\Phi_p(E_{out}, X) = \Phi_p(E_{in}, 0)\frac{S(E_{in})}{S(E_{out})}\exp - \int_{E_{in}}^{E_{out}} \frac{\mu(E)}{S(E)} dE \tag{12}$$

where $\mu(E)$ is $N_0/A\sigma_{inelastic}$ (N_0 is Avogadro's number, A is the atomic weight of absorber material M, and $\sigma_{inelastic}$ is the inelastic cross section of material M).

The incident proton flux generates secondary protons in the absorber through inelastic collisions with absorber nuclei. While these secondary protons may be emitted with various angular distributions, ranging from isotropic to highly anisotropic, they are assumed to be traveling in the direction of the incident proton.

The flux of secondary protons of energy E is obtained by the integration of the production rate of secondaries over all possible primary proton energies that can produce secondaries of energy E at a point X, and then integrating this integral over all volume space in the shield. Secondary neutrons and gamma rays are computed in a similar method substituting the proper production rate for each neutron or gamma (BARTON *et al.* (1964)).

While this flux of secondary protons in turn is further attenuated by inelastic collisions and produces third-generation particles, the process will be terminated at this point. The results indicate that, for the proton spectra and shield materials of interest in space vehicle shielding, no significant error is made in terminating the cascade at the second order.

The proton dose is computed by Equation 6 using $S(E_{out})$ as the dose conversion factor. The neutrons generated by the incident protons present a more complex dose-deposition problem. Because the neutron is not a charged particle, it deposits energy only by collisions with the absorber's nuclei. These collisions in turn result in a transfer of energy to the struck nucleus, either by absorption of the incident neutron, or by elastic or inelastic scattering. In any case, the initial energy of the neutron can be dissipated in a variety of ways, ranging from emission of γ-ray, neutron, proton, or heavier particles by the excited nucleus to the ionization in the absorber resulting from the kinetic energy imparted to the nucleus in the collision. An exact calculation of the absorbed dose would follow these tertiary particles through their respective histories as energy depositors. As this would take a great deal of time, an approximate dose conversion factor developed by GIBSON (1962) is used that gives an upper-limit estimation of the absorbed dose.

For space vehicle shielding analysis, where thin shields and low secondary doses are the rule, this method provides sufficient accuracy. The neutron dose at a point X is computed in the same manner as protons, that is by integrating the product of the neutron fluence of a given energy and at a given position, and of the dose conversion factor for neutrons of the same energy over all neutron energies.

The gamma rays present a problem similar to, although less complex than, the neutrons. The primary absorption mechanisms of high-energy gamma rays (Compton scattering and pair production) result in energetic electrons and photons of degraded energies. The range of the electrons is small, and their energy can be considered as deposited locally. The secondary gamma rays, however, have mean absorption ranges that can be significant in comparison to the human body. Due to their relative lack of importance in comparison to the other radiations, as will be shown in the following discussion, a simple upper-limit dose conversion factor will be used. The expression for the gamma dose then has the same form as the other dose expressions.

2.3. Summary of Methodology

A summary of the existing computer programs and their incorporation is shown in Figure 5. The work being done to integrate the codes into a unified program

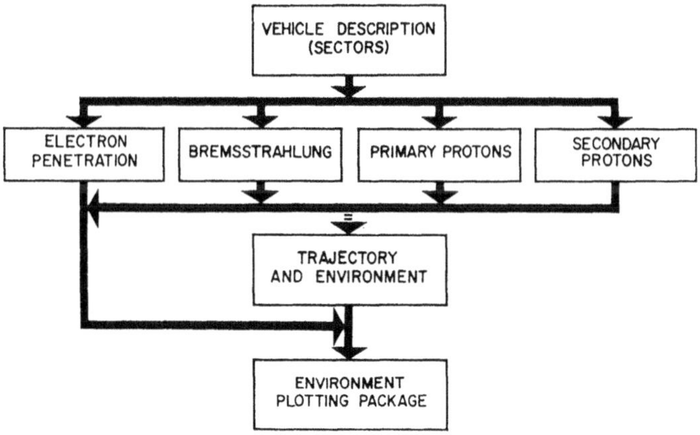

Fig. 5. Computer program for mission analysis.

system may be of interest. The electron Monte Carlo program is not shown since it is not normally used in connection with the others. Using predetermined energy spectra with the electron penetration, bremsstrahlung, primary proton, and secondary proton programs, dose-rate per unit particle flux number can be calculated. Thus flux-to-dose conversion numbers can then be used with the "trajectory and environment" program to obtain dose-rates as well as fluxes as a function of time along a trajectory.

The essential input data to the shielding codes consist of the shield material distribution of a vehicle, the basic shielding properties of the materials involved, and the spectra specifications. Shield distributions are given in terms of sectoring data; that is, average thicknesses and material types for a closed set of solid angular sections of a vehicle. Shielding properties include cross-section data and other similar quantities for each material. Having once defined the radiation environment by flux maps in *B-L* space, the essential input to make predictions for a given trajectory consists of only the trajectory specification. The program contains subroutines that generate trajectory points for circular and elliptical trajectories.

3. Predictions

In this section we apply the methods which have been discussed above to the problem of calculating the doses from space radiation. It should first be pointed

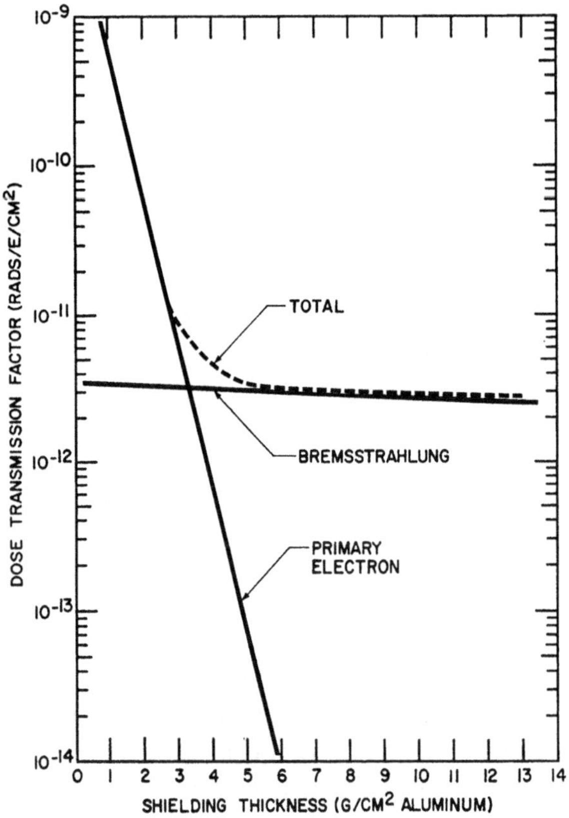

Fig. 6. Dose transmission factor for fission spectrum electrons penetrating a spherical aluminum shell.

out that in general it is not possible to simply speak of the "dose" that an astronaut would receive on a given mission. Rather, it is necessary to specify the dose as a function of depth in the human body, usually referred to as the depth dose distribution. For the galactic cosmic rays this distribution is rather uniform across the body, while for electrons and protons the depth dose distribution is strongly dependent on the energy spectrum of the incident radiation.

Figure 6 illustrates the variation of dose as a function of shield thickness for fission spectrum electrons. A very steep depth-dose distribution is observed for thin shields; however, once the primary electrons are stopped, the secondary bremsstrahlung dose is fairly constant with increasing shield thickness.

For protons trapped in the radiation belts and also for solar flare protons there can be large variations of dose with depth. This is strongly dependent on the spectral shape. For the 14 July 1959 solar flare ($P_0 = 80$) the ratio of skin dose to the midline dose has been calculated to be 40. For the inner belt protons this same ratio is slightly over 3. Table I presents a summary of the depth dose distribution for several values of P_0 and for the inner belt protons. These calculations do not take into account the dose due to secondaries, but as can be seen from Figure 7, this is a very minor contribution for thin shields.

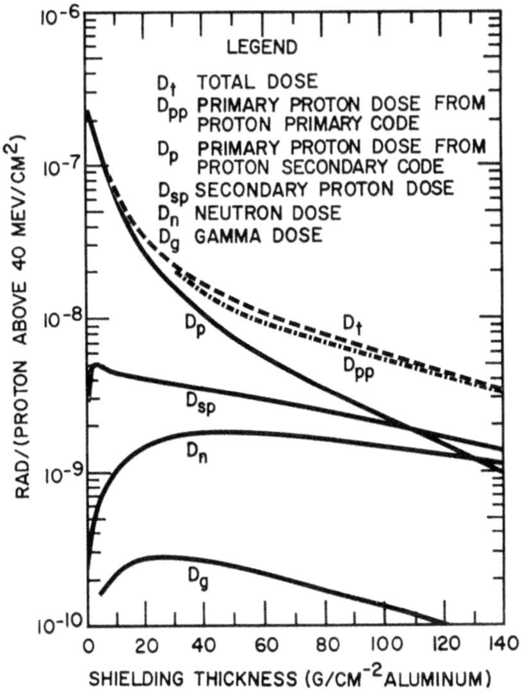

Fig. 7. Freden-White spectrum – protons incident.

For a spacecraft such as the Gemini, portions of the capsule are less than 1 g/cm^2 thick. Considering the chest as the dose point, there is a region of the Gemini above the astronauts subtending about 15 percent of the total solid angle where the shielding averages only 1.6 g/cm^2 thick. This compares with a region behind the astronauts subtending 25 percent of the total solid angle where the shielding averages nearly 7 g/cm^2. Thus it turns out that for a solar flare with a rather soft spectrum the skin dose may be as much as a factor of 10 higher at the chest than at the back.

TABLE I*

Dose in rads from 1 proton/cm^2 incident on 2 g/cm^2 aluminum shielding

Symbol	$P_0 = 80$	$P_0 = 124$	$P_0 = 195$	*Inner belt***
Chest 0	1.212×10^{-7}	1.134×10^{-7}	1.077×10^{-7}	9.097×10^{-8}
Waist 0	9.642×10^{-8}	9.228×10^{-8}	9.031×10^{-8}	7.869×10^{-8}
Waist 1	1.662×10^{-8}	2.786×10^{-8}	4.042×10^{-8}	4.661×10^{-8}
Waist 4	6.831×10^{-9}	1.554×10^{-8}	2.760×10^{-8}	3.489×10^{-8}
Midline	3.003×10^{-9}	9.514×10^{-9}	2.058×10^{-8}	2.792×10^{-8}

* Note: All solar-flare spectra are normalized to one particle greater than 30 MeV.
** Composite spectrum of Vette representing a region for $L \leqslant 1.2$. This spectrum is normalized to one particle greater than 20 MeV.

Although at the present time it is difficult to assess the effect of a particular depth-dose distribution on mission performance or what the long-term effects might be on the individual concerned, the few experiments that have been performed indicate that depth-dose distribution is of major importance. JACKSON (1962) has shown that by simulating the depth-dose profile of the June 16, 1959, solar flare in the body of a rat, he was able to increase the skin dose required to produce LD 50 by a factor of 4 over the uniform whole-body dose required to produce the same effect. Experiments at the Air Force Weapons Laboratory using sheep as the irradiated specimen have shown that another form of depth-dose distribution – that caused by partial body shielding – can also considerably increase the dose required to produce LD 50. As was seen above, a typical space vehicle induces partial body shielding with its heterogeneous array of equipment and components.

Figures 8 and 9 show the daily dose rates that might be expected at various altitudes and with different shielding configurations. The electron doses are based on flux maps obtained in early 1963 and do not represent the decay of the Starfish electrons which has occurred since then. For low altitude orbits, that is for orbits below about 500 km, the majority of the dose will be received in the region of the South Atlantic Anomaly. Figure 10 shows those periods of time that are spent passing through the anomaly in a typical low-altitude orbit.

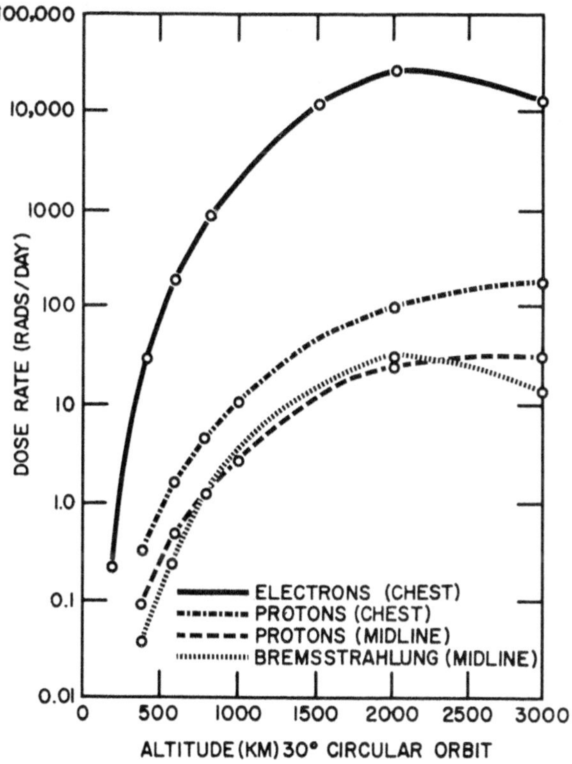

Fig. 8. Dose behind 0.4 g/cm^2 aluminum shield.

In general, one can say that behind 2 g/cm^2 of aluminum the total skin dose from trapped radiation in any orbit below 400 km will not exceed 1 rad/day. This, of course, is barring any further nuclear detonations in space.

Besides altitude, another orbital parameter affecting the dose, but to much less extent, is the inclination of the orbit. In comparing the 30-, 45-, and 60-degree orbits below about 400 km, one finds that a slightly smaller dose will be received in the 30-degree orbit than in the 45-degree orbit. The smallest dose is received in the 60-degree orbit, because a smaller portion of time is spent in

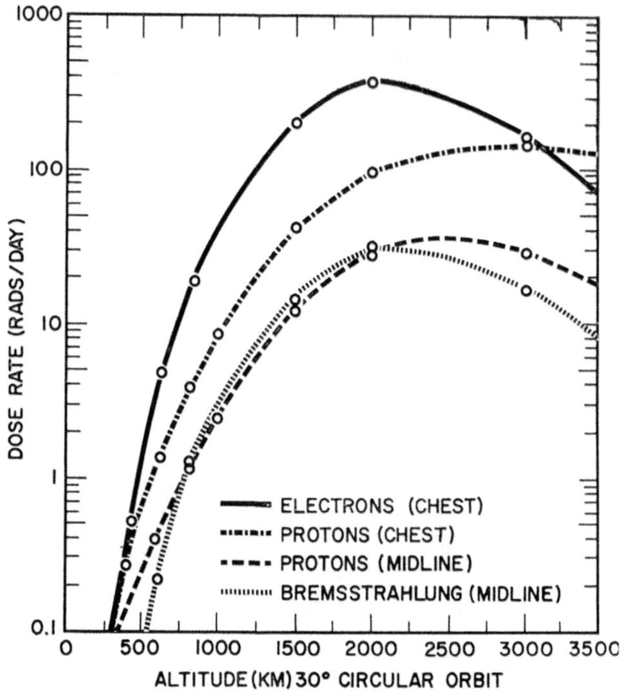

Fig. 9. Dose behind 2 g/cm² aluminum shield.

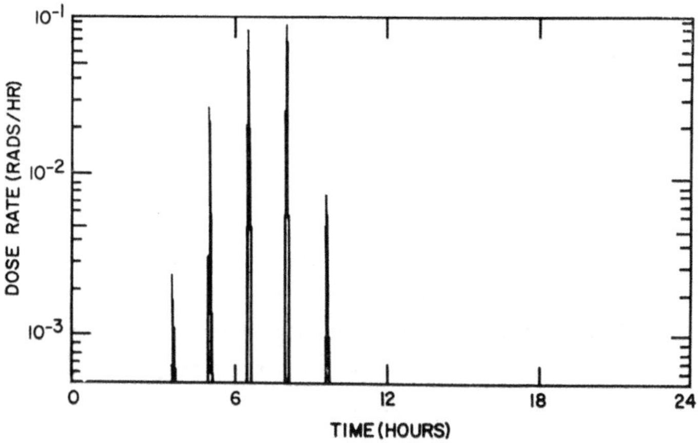

Fig. 10. Time variation of dose in low-altitude orbit.

the anomaly region. In comparing the 45- and 60-degree circular orbits with a 30-degree orbit at 1000 km, we find the following: The dose received from protons in a 45-degree orbit will be about 25 percent lower than in a 30-degree orbit; while in a 60-degree orbit it will be about a factor of 2 lower; for electrons the doses will be about 40 percent and a factor of 3 lower in the 45- and 60-degree orbits respectively.

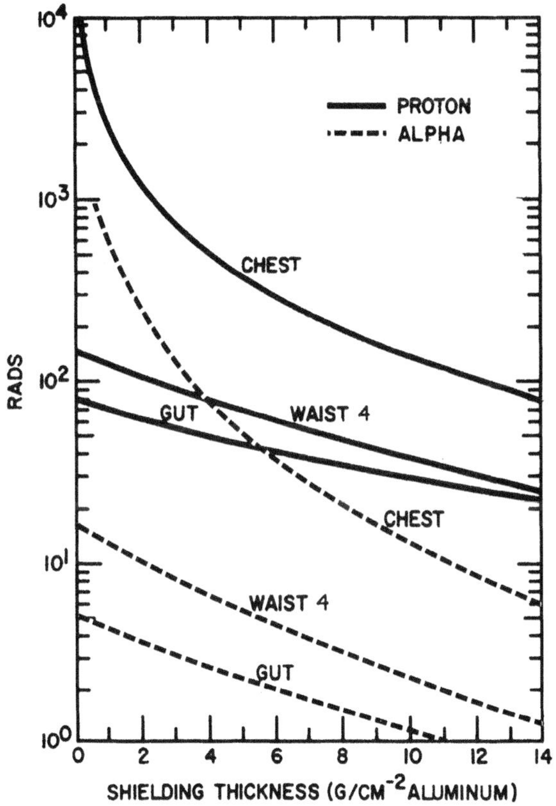

Fig. 11. Total proton and alpha-particle doses for solar cycle 19 at various body points.

When one is traveling in free space (defined as being away from the influences of the earth's magnetic field) the only radiations of biological concern encountered will be solar-flare particles and galactic cosmic rays. To determine the dose from these particles in free space it is only necessary to multiply the total flux by the proper dose transmission factors. Figure 11 shows the total dose that would have been received by an astronaut spending the entire solar cycle 19 (8 years) outside the earth's magnetic field. When a vehicle is in a near-earth orbit, however, a portion of each revolution is shielded from these charged particles by the earth's magnetic field and by the earth's mass. For each point along a trajectory the magnetic field will deflect all particles below a given energy known as the cutoff energy. Thus, the largest dose from solar-flare particles for a near-earth orbiting vehicle will be received in a polar orbit, with the dose decreasing as the inclination of the orbit decreases. This effect is illustrated in Table II. The numbers in this table do not account for the shielding

TABLE II

Solar flare doses in radians

Location	23 Feb 1956		14 July 1959		12 Nov 1960		*Shielding*
	Chest	*Midline*	*Chest*	*Midline*	*Chest*	*Midline*	
Free space	92.0	16.0	145.0	2.5	128.0	8.9	Gemini
	108.0	20.6	158.0	4.0	148.0	12.0	2 g/cm^2
	57.0	19.2	40.0	3.2	58.0	9.6	5 g/cm^2
400 km 90°	31.0	6.0	45.0	0.9	41.0	3.3	Gemini
	36.0	7.6	48.0	1.5	47.0	4.5	2 g/cm^2
	20.0	7.0	14.0	1.1	20.0	3.5	5 g/cm^2
400 km 60°	11.0	2.9	17.0	0.4	17.0	1.7	Gemini
	12.8	3.6	18.0	0.7	19.0	2.3	2 g/cm^2
	7.9	3.3	5.6	0.4	9.0	1.8	5 g/cm^2

by the earth's mass, and consequently are too high by about a factor of 2 for near earth orbiting vehicles. For orbits similar to the Mercury or Gemini series – low altitude and inclination around 30 degrees – even the worst solar flare occurring during the last solar cycle would have delivered a dose in the order of millirads.

4. Experimental Measurements

Shortly after the discovery of the earth's radiation belts, dosimetry was designed and flown to obtain a direct measurement of the dose rate. The flights have ranged from probes containing a single dosimeter to a complete biophysical satellite and also instrumentation aboard manned flights. The basic instrument used in these flights is a tissue-equivalent ionization chamber (referred to as TEIC). The chamber walls are constructed from a special conducting plastic that closely resembles human muscle tissue in its radiation interaction characteristics (BAILEY, 1963 and SHONKA *et al.*, 1958). The sensitive volume of the chamber is filled with either dry air or a tissue-equivalent gas. The current from the ion chamber is fed through a diode-connected electrometer tube, and the circuit is designed to yield a logarithmic response for several decades.

Not until late 1962, after the Starfish high-altitude nuclear detonation of July 1962, was a comprehensive mapping of the inner-belt region made with a dosimeter. At that time two TEIC's were flown aboard an Air Force satellite; one was shielded only by 0.4 g/cm^2 to respond to electrons, while the other was shielded by 4.7 g/cm^2 to stop the electrons and measure the secondary bremsstrahlung dose.

The satellite was launched on October 26, 1962, between the first and second USSR high-altitude nuclear detonations. Thus it was possible to measure dose rates from the Soviet tests as well as from Starfish. The principal results of that experiment (Clark and Adams) are summarized as follows:

(1) Dosimetry measurements in 1960 have shown that for $L \leqslant 1.6$ and $B \geqslant 0.2$ the isodose contours are in general agreement with the Explorer IV proton contours. Discrepencies can be explained by a softening spectrum with increasing L. At $B=0.2$ and $L=1.25$ a dose-rate of 0.15 rad/hr was measured behind about 5 g/cm^2 shielding. But data obtained in September 1962 showed that the dose-rate at the same point in B, L space was a factor of 20 higher than for the pre-Starfish environment; and in November 1962 the dose-rate was still a factor of 6 higher at this point. Both measurements after Starfish were made with a chamber shielded by 4.7 g/cm^2 brass, and hence represent a bremsstrahlung dose.

(2) The artificial radiation belt created by the three separate USSR high-altitude nuclear detonations in October 1962 showed a peak dose-rate with the shielded chamber of 12 rads/hr located at an L of 1.82. This measurement was recorded following the second detonation and represents a dose due to bremsstrahlung. The Soviet belts decayed rapidly, although they were still in evidence a month after the last detonation. This is to be contrasted with the Starfish detonation, which occurred in July 1962. Four months after the Starfish event, the heavily shielded chamber gave a peak reading of 30 rads/hr, while the lightly shielded chamber read 23000 rads/hr. These readings were due to bremsstrahlung and penetrating electrons respectively, and they occurred at $B=0.16$ and $L=1.25$.

(3) When the isodose contours from both the heavily shielded and lightly shielded chambers were compared with existing electron and proton maps, no one-to-one correlation was found. The lightly shielded chamber more nearly followed electron contours in all regions of space covered by this satellite; but because of spectral variations and the different minimum energy cutoff used by various experimenters, it was not possible to obtain a one-to-one correspondence for all of the measured B-L space with any map.

The shielded chamber presented an even more complex picture. For a region of space bounded roughly by $L<1.5$, most of the dose was due to bremsstrahlung. Once again, however, the dose varied as did the electron spectrum. With increasing L a larger percentage of the dose was due to protons, particularly for $L \geqslant 1.5$.

Operationally, the most critical need for accurate predictions of radiation dosages is in connection with manned flights, which is also the most difficult task and the area in which most uncertainties reside. All the U.S. space flights so far have been low-altitude, low-angle-of-inclination flights of rather short

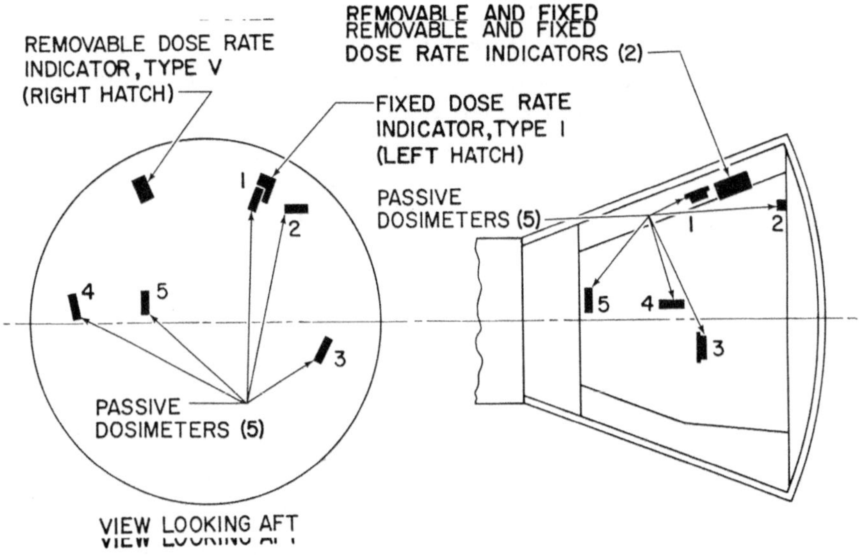

Fig. 12. Experimental equipment spacecraft installation locations.

duration; consequently, no radiation hazard has been involved. However, experiments such as those performed on the recent Gemini-Titan (GT-4) flight allow one the opportunity to check computer predictions in an actual situation. As of this writing the data analysis from those experiments is still being conducted; hence, only selected results of a qualitative nature are presented.

The experiment flown by the Biophysics Branch of the Air Force Weapons Laboratory consisted of 2 tissue equivalent ionization chambers (as described earlier in this section), one mounted on each hatch, and five small hermetically sealed units each containing a variety of passive dosimetry. (Figure 12 shows the general location of the experiments within the capsule.) The TEIC located on the left hand hatch is referred to as the type I instrument and was permanently attached to that location. The instrument on the right hand hatch, however, had its sensor head on a coiled lead so that it could be moved about the spacecraft cabin. It is referred to as the type V instrument. The TEICs flown on this flight incorporated many improvements over previous versions and allowed measurements as low as 1 mrad/hr to be made with 20% accuracy. The TEICs each had a spherical sensor head which made them essentially an omnidirectional detector. This is important to keep in mind when examining the following data.

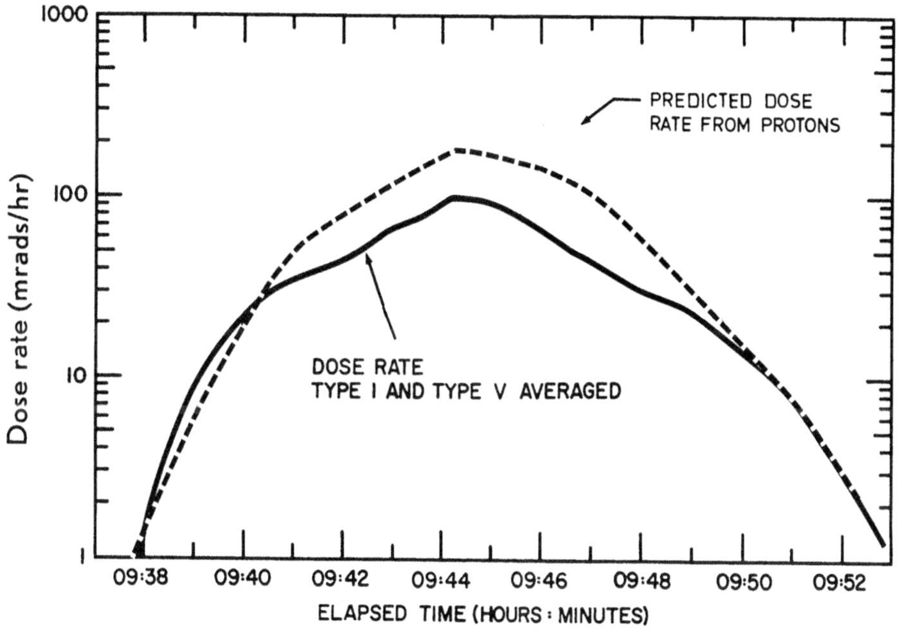

Fig. 13. Dose received by Gemini-Titan 4 flight in seventh revolution.

Figure 13 shows a graph of the dose-rate as a function of time on the seventh revolution as it passed through the anomaly region. The dose-rate plotted is an average of the values obtained from the two instruments, each located in their respective hatch positions. Also plotted on this same graph is the predicted proton dose-rate for that same pass. Here one sees that the predicted proton dose-rate is nearly a factor of 2 higher than the measured dose-rate, which includes the dose from all types of particles. The calculated electron dose-rate based on a map generated by VETTE (1963) gave a peak dose-rate of 0.7 rads/hr, using a fission spectrum. Clearly, the measurements from the TEICs indicate that there was no such contribution to the dose from penetrating electrons. This can be explained in part by the decay which has taken place in the more than two years between the time when the flux map measurements were made and the GT-4 flight. Also, a softening of the spectrum could alter the dose within a spacecraft considerably.

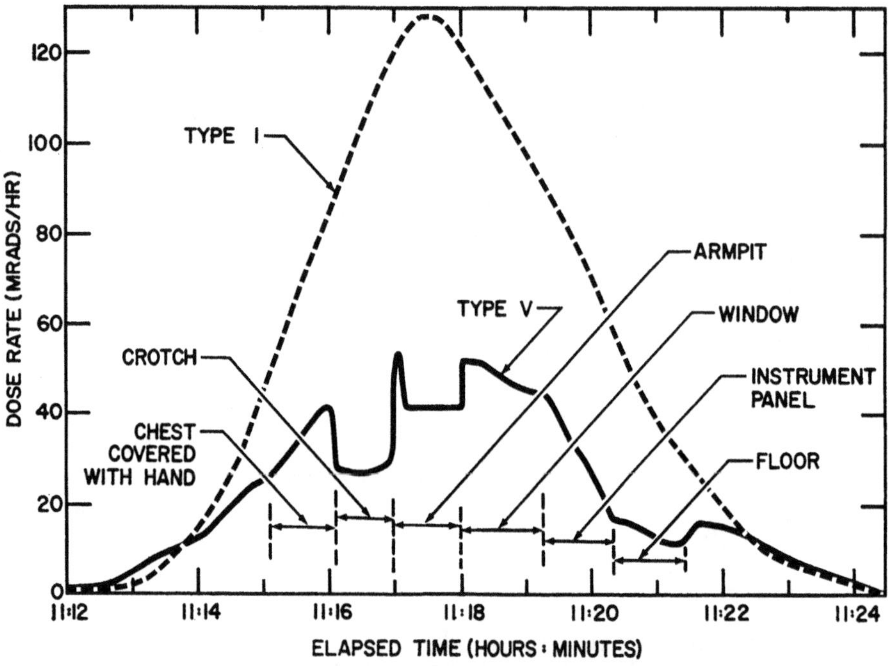

Fig. 14. Effect of detector location on penetrating dose on Gemini-Titan 4 flight in eight revolution.

Figure 14 is a graph of the eighth revolution for that portion of the time which it spent in the anomaly. On this orbit one of the astronauts performed a series of experiments by placing the movable type V unit at various positions throughout the capsule. It can be seen that as the instrument is moved away from its position near the hatch, the dose decreases. This is because the hatch, which is the thinnest portion of the vehicle, subtends a smaller solid angle as the instrument is moved away from it. The six positions in which the instrument was placed are fairly self-explanatory. The positions chosen were necessarily limited because of the immobility of the astronauts within the capsule.

Figure 15 demonstrates the pitch angle distribution as it is modulated by the spacecraft tumbling. The dosimeters are omnidirectional instruments, hence the modulation that is seen is due to nonuniform vehicle shielding about the dose point. At present there has been no attempt to correlate the particle measurements on GT-4 with the dosimetric measurements.

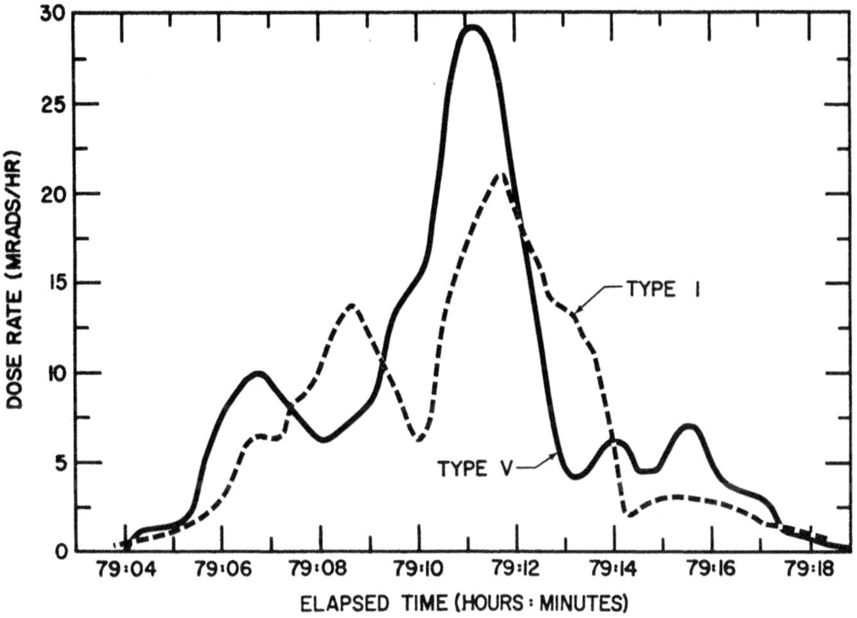

Fig. 15. Dose received by Gemini-Titan 4 flight in fifty-first revolution measuring the anisotropy of the incident flux.

Each of the hermetically sealed passive dosimeter packs contained Lithium Fluoride, Calcium Fluoride, pocket ionization chambers, and Toshiba glass. The pack located near the Type I TEIC gave an integrated dose of about 55 mrads. Of this it has been calculated that 15 mrads are due to the galactic cosmic ray background and the remainder due to passes thru the anomaly. By integrating the dose-rate measured by the TEICs a somewhat higher total dose was recorded. This would indicate the presence of electrons which were measured by the TEICs but which couldn't penetrate the somewhat heavier shielding around the passive units.

5. Conclusions

The prediction of the radiation doses received by manned flights in the trapped radiation belts is based on experimental measurements of the intensity, energy spectra, and angular distribution of the radiation in space, and the transport theory of these radiation penetrating the space vehicle. The measurement of the radiation environment is a continuing effort; however, the verification of shielding calculations for space missions awaits further manned space flights. Additional experiments are required to verify shielding procedures in spacecraft prior to flight and to obtain greater confidence in predicting the anticipated doses as manned flights extend into higher-radiation fields. The biological response to the radiation doses received in manned flight must be better defined before the radiation hazard to manned space flight can be fully assessed.

References

Allen, R. I., Bly, F., Dessler, A., Douglass, C., Perkins, J., Price, H., Smith, E., and Schofield, W.: 1961, *Lockheed Report* NR-140.

Alsmiller, R. G., Jr. and Murphy, F. E.: 1963, *Oak Ridge National Laboratory Report* ORNL-3317.

Bailey, N. A. and Hoalst, K. M.: 1963, *Nucleonics* **21**, 68.

Barton, J. A. *et al.*: 1964, *Air Force Weapons Laboratory Document* WL-TDR 64-71.

Berger, M. J.: 1963, *Methods in Computational Physics,* Vol. 1, Academic Press, New York.

Bethe, H. A.: 1953, *Phys. Rev.* **89**, 1256.

Clark, B. C. and Adams, D. A.: 1965, *Life Sciences and Space Research* **3**.

Dye, D. L.: 1962a, *Boeing Document* D2-90107.

Dye, D. L.: 1962b, *USAEC* TID-7652, 633–661.

Evans, R. D.: 1955, *The Atomic Nucleus,* McGraw-Hill, New York.

Evans, R. D.: 1961, *NASA Report* 63270-05-01.

Foderaro, A. and Obenshain, F.: 1955, *Westinghouse Electric Corporation Document* WAPD-TN-508.

Frank, H.: 1959, *Z. Naturforsch* **14a**, 247–261.

Gibson, W. A.: 1962, *AEC Report* ORNL-3260.

Jackson, K. L.: 1962, Paper C-6 of the *Proc. of Symp. on Protection Against Radiation Hazards in Space,* Gatlinburg, Tennessee, TID-7652.

Lin, W. C., Venkatesan, D., and Van Allen, J. A.: 1963, *J. Geophys. Res.* **68**, 4885, 4897.

Makhov, A. F.: 1960, *Fizika Tverd. Tela* **2**, 2161.

Mar, B. W.: 1963, *Boeing Document* D2-90414.

Neher, H. V. and Anderson, H. R.: 1962, *J. Geophys. Res.* **67**, 1309–1313.

Neher, H. V. and Anderson, H. R.: 1964, *J. Geophys. Res.* **69**, 807–814.

Seliger, H. H.: 1955, *Phys. Rev.* **100**, 1029–1037.

Shonka, F. R., Rose, J. E., and Faella, G.: 1958, *Proc. of the 2nd U. N. Int. Conf. on the Peaceful Uses of Atomic Energy* **21**, 184.

Störmer, C.: 1955, *Polar Aurora,* Oxford University Press, London.

Vette, J. I.: 1964, presented at the *2nd Symp. on Protection Against Radiation in Space,* Gatlinburg, Tennessee.

Wyard, S. J.: 1952, *Proc. Phys. Soc.* (London) **A65**, 377.

Discussion

J. VETTE commented on the low Gemini dose measurements relative to the predicted values in the region of the Brazilian geomagnetic anomaly between L values of 1.2 and 1.6. He said that his model of the proton flux, which Adams used in the calculation, probably overestimated the flux existing during the Gemini flight because of the decay of the protons which appear to have been redistributed in the magnetosphere by the Starfish detonation.

SHIELDING TECHNIQUES

J. WARREN KELLER

Headquarters, National Aeronautics and Space Administration Washington, D.C.

Introduction

The geomagnetically trapped radiation constitutes only one of several components of the total environment of charged particles which are found in space. Other major components include the galactic cosmic radiation, the solar cosmic radiation originating in solar flares and the solar wind.

The presence of these radiation fields constitutes a major concern for all space missions as a result of the harmful effects of such radiations upon both man and equipment. It is desirable to alleviate this problem by increasing the radiation resistance of affected items. However, since this is not possible in all cases, especially for man, one must resort to shielding the sensitive items from the harmful radiations. Consequently, it has been necessary to develop a comprehensive charged particle shielding technology. This paper will review the physics of the interaction of charged particles with matter, describe the techniques currently used for calculating the penetration of such particles through shields and discuss a portion of the data currently being generated to assist in such calculations.

Passage of Radiation Through Matter

The geomagnetically trapped radiation consists primarily of protons and electrons. Of interest is the manner in which these particles pass through matter including interactions with the atoms of the penetrated material and the subsequent production of secondary radiations.

PROTONS

Protons lose energy chiefly through inelastic collisions with bound electrons in the stopping material, causing excitation and ionization. Since the mass of an electron is much less than that of a proton, the maximum fraction of the proton energy that may be lost in a collision is $\sim 1/460$. This process results in an almost continuous reduction of energy as the proton penetrates through the medium, with the rate of energy loss decreasing when the proton energy becomes low enough (<0.1 MeV) for it to capture electrons. Protons are also slowed down or removed from the beam through collisions with nuclei of the stopping material and by the emission of electromagnetic radiation, but the contribution of these processes is very small compared to the losses through ionization and excitation. Collisions with nuclei can be important, however, from the viewpoint of secondary radiation production as will be discussed later.

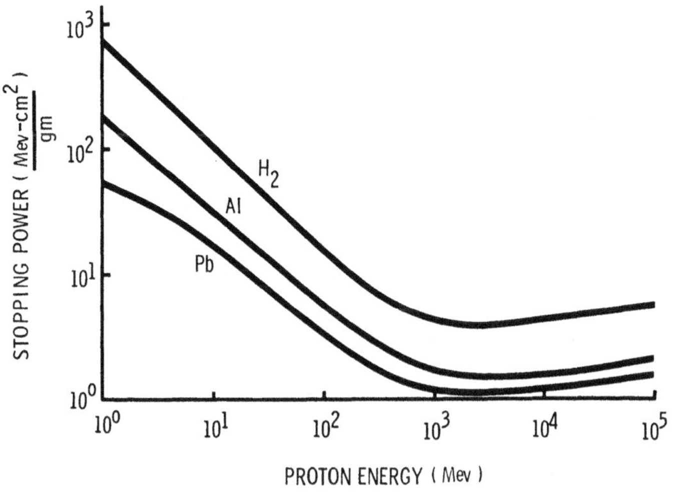

Fig. 1. Stopping power as a function of proton energy.

The first part of the slowing down process – during the time that the particle is charged – can be adequately treated from a theoretical point of view. When the proton is slow enough to capture electrons, the theoretical problem becomes extremely difficult. However, reasonably adequate experimental information is available in this energy region which is not very important for shielding because of the small range involved.

The average rate of proton energy loss by ionization and excitation in passing through material is given by the Bethe-Bloch formula as:

$$-\frac{\mathrm{d}E}{\mathrm{d}X}=\frac{4\pi e^4 N}{mv^2}\left\{Z\left[\ln\frac{2mv^2}{I}-\ln(1-\beta^2)-\beta^2\right]-C_k\right\} \tag{1}$$

where E=kinetic energy of the proton; X=material thickness (gm/cm^2); N= number of atoms per gram of material; $v=\beta c$=proton velocity; m=electron mass; Z=atomic number of the material; e=electron charge; I=average ionization potential of atoms of material (roughly proportional to Z); C_k=correction term for k shell binding. The value $-\mathrm{d}E/\mathrm{d}X$ is referred to as the stopping power of the material and is usually denoted as $S(E)$. Curves of the stopping power versus energy for several materials are shown in Figure 1. It is seen from Equation (1) that the stopping power is generally proportional to NZ which is just the electron density of the material. Since in general, the electron density decreases with increasing Z, the low-Z materials on a weight basis are more effective than high-Z materials in slowing down protons with hydrogen forming a singular point with by far the highest number of electrons per unit weight since the hydrogen atom has only one nucleon for each electron.

Since the mass of the proton is ~2000 times that of an electron, the proton undergoes at most a minute deflection in an ionizing collision and, therefore, travels in virtually a straight line through the material until stopped. Also, since only a small fraction of the proton energy can be lost in a single collision, a very large number of collisions are required to stop the particle. Therefore, the statistical nature of the problem results in all particles of equal energy traveling an almost equal distance through a given material before being stopped. The distance traveled is referred to as the range of the particle and can be obtained by integration of the stopping power formula as

$$R = \int_0^E \frac{1}{-\mathrm{d}E/\mathrm{d}X} \mathrm{d}E$$

or, in a more general sense the distance traversed by the particle in going from energy E_1, to energy E_2 is given by

$$R_{E_1 \to E_2} = \int_{E_2}^{E_1} \frac{1}{-\mathrm{d}E/\mathrm{d}X} \mathrm{d}E. \tag{2}$$

Proton range-energy relationships have been calculated by a number of groups. However, the latest tabulation is that of BARKAS and BERGER (1964). Curves of range versus energy for a variety of materials are shown in Figure 2.

It was mentioned earlier that collisions with nuclei can be important because of secondary radiation production. Inelastic collisions of high energy nucleons with nuclei are generally interpreted as taking place in two phases. The first phase is the rapid development of a nucleonic cascade; the second is the de-excitation of a thermally excited nucleus by the process of evaporation.

The cascade develops when the incident nucleon collides with nucleons in the nucleus which, in turn, either escape from the nucleus or collide with still other nucleons, etc. The nucleons that are involved in the collision cascade but do not escape from the nucleus are assumed to share their energy with the entire residual nucleus leading to an excited nucleus from which lower energy nucleons "boil-off" in the comparatively slower second phase or evaporation process.

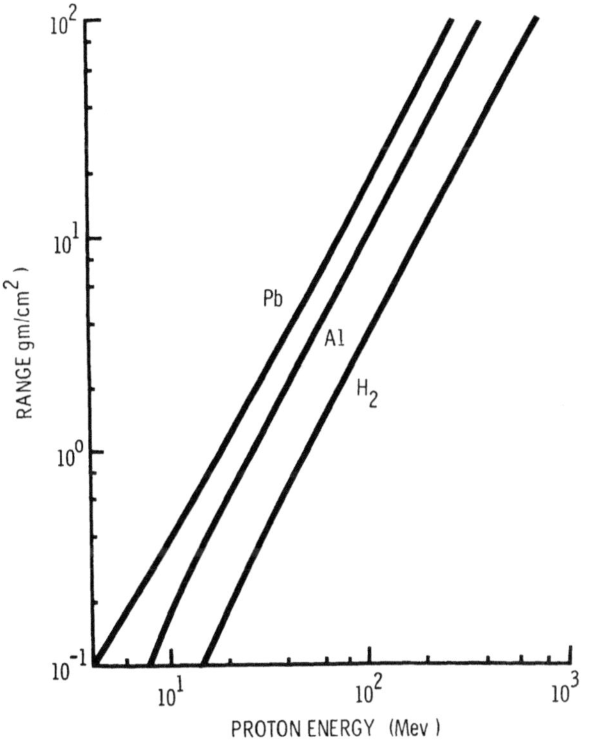

Fig. 2. Proton range as a function of energy.

Detailed experimental data yields and the energy and angular distributions of secondary particles resulting from nuclear collisions are sparse. However, calculations of these distributions are available. If the cascade phase of the reaction is considered to consist of a sequence of collisions between free nucleons for which cross-sections are known, the cascade may be treated by the Monte Carlo method. Several groups have performed such calculations with the most detailed and complete to date being carried out by BERTINI (1963). These data compare reasonably well with the small amount of existing experimental data. Several relatively high-energy nucleons (both neutrons and protons) may be produced by single collisions by particles in the energy range of geomagnetically trapped protons. Such secondary nucleons are found to be concentrated in the forward hemisphere with respect to the direction of the incident proton. The evaporation process has also been treated using a statistical

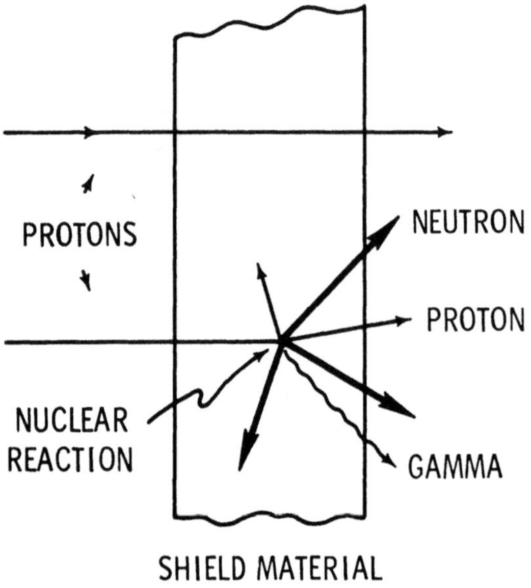

Fig. 3. Schematic diagram illustrating proton path through shield material for proton undergoing ionization losses only (upper) and for proton undergoing a nuclear reaction (lower).

Monte Carlo process (DOSTROVSKY *et al.*, 1958). Here, the nucleons are of much lower energy and are emitted isotropically. While the mean free path for inelastic nuclear collisions is of the order of 50 to 100 g/cm^2, the resulting secondary radiation can become significant for thick shields of a few tens of grams per square centimeter. Gamma rays produced in these collisions have not been found to be important for shielding.

Protons penetrating through matter, then, fall generally into the two classes illustrated in Figure 3. The upper example is for the proton which penetrates the shield undergoing continuous energy degradation but causing no nuclear reactions. The lower example is for the proton which has a nuclear collision so that the incident proton is stopped with secondary radiations being initiated.

ELECTRONS

Low energy electrons as in the case of protons lose energy in passing through matter principally through inelastic collisions with bound electrons resulting in excitation and ionization. The theory of energy loss by electrons through such collisions is parallel to that for protons. For low velocities the energy loss per unit path length for protons and electrons is almost the same for particles of the same velocity. The average rate of energy loss through such ionization collisions for electrons is given by

$$-\left(\frac{\mathrm{d}E}{\mathrm{d}X}\right)_{\text{ion}} = \frac{2\pi e^4 NZ}{mv^2}$$

$$\left\{\ln\frac{mv^2 E}{2I^2(1-\beta^2)} - (2\sqrt{1-\beta^2} - 1 + \beta^2)\ln\pi\, 2 + 1 - \beta^2 + \tfrac{1}{8}(1-\sqrt{1-\beta^2})^2\right\}$$

where E is the kinetic energy of the incident electron and the other symbols are as in Equation (1).

Electrons also lose energy in the form of electromagnetic radiation or bremsstrahlung (braking radiation). Bremsstrahlung is produced when the electrons are decelerated in the electric fields of nuclei of the stopping material. The average rate of energy loss through electromagnetic radiation may be expressed as

$$-\left(\frac{dE}{dX}\right)_{\text{rad}} = N\int_0^E E_\gamma \sigma(E, E_\gamma, Z)\,\mathrm{d}E_\gamma$$

where N is the number of atoms per unit mass, E_γ is the energy of the emitted photon and $\sigma(E, E_\gamma, Z)\,\mathrm{d}E_\gamma$ is the cross-section for the production of a photon in the energy range between E_γ and $E_\gamma + \mathrm{d}E_\gamma$ by electrons of kinetic energy E in a stopping material having atomic number Z. The cross-section, $\sigma(E, E_\gamma, Z)$, is found to be roughly proportional to Z^2/E_γ.

Both $(\mathrm{d}E/\mathrm{d}x)_{\text{ion}}$ and $(\mathrm{d}E/\mathrm{d}x)_{\text{rad}}$ are plotted in Figure 4 as a function of electron energy for several materials. It is seen that for low-energy electrons the radiation energy losses are much smaller than the ionization losses, however, the fractional energy loss from radiation increases with both electron energy and the atomic number of the stopping material.

To obtain the average path length for electrons in going from energy E_1 to E_2, one must consider both ionization and radiation losses so that the parallel equation to Equation (2) for protons is

$$R_{E_1 \to E_2} = \int_{E_2}^{E_1} \frac{1}{-\,\mathrm{d}E/\mathrm{d}X}\,\mathrm{d}E \tag{3}$$

where

$$\frac{\mathrm{d}E}{\mathrm{d}X} = \left(\frac{dE}{\mathrm{d}X}\right)_{\mathrm{ion}} + \left(\frac{\mathrm{d}E}{\mathrm{d}X}\right)_{\mathrm{rad}}$$

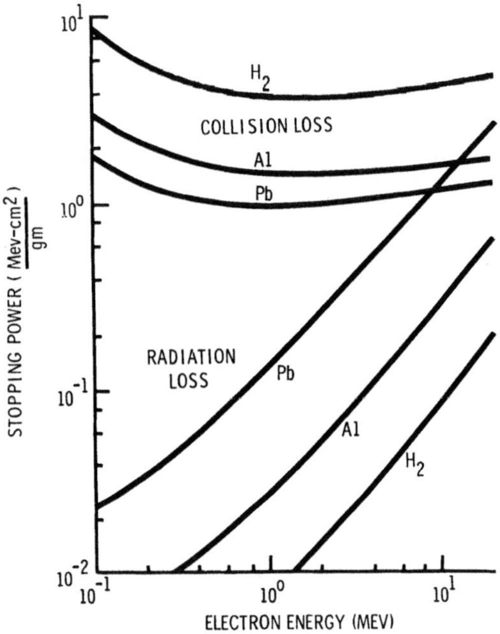

Fig. 4. Stopping power as a function of electron energy for both collision loss and radiation (bremsstrahlung) loss.

Values of $(\mathrm{d}E/\mathrm{d}x)$, $(\mathrm{d}E/\mathrm{d}x)_{\mathrm{ion}}$, and $(\mathrm{d}E/\mathrm{d}x)_{\mathrm{rad}}$ have been newly calculated by BERGER and SELTZER (1964) for a wide range of materials and electron energies. It is seen from Figure 4 that for materials of primary interest for spacecraft and for the energy range of interest with respect to geomagnetically trapped electrons, radiative losses are of little importance in determining electron penetration.

The total electron path length as defined by Equation (3) is of only limited usefulness in electron shielding considerations for the following reasons. For electrons, unlike protons, large fractions of the particle energy may be transferred in a single electron-electron collision resulting in a large variation or "straggling" of the energy loss as well as significant deflections in the direction of travel. Also, elastic Coulomb collisions with the atoms of the stopping material result in many large angle deflections of the particles. Consequently, the path of an electron as it passes through an absorber is very tortuous as

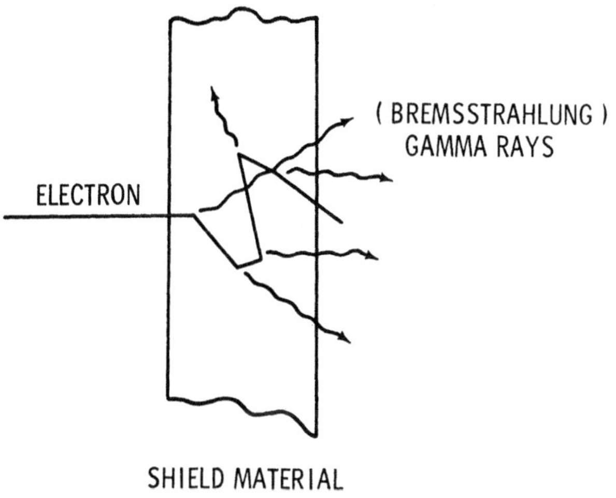

Fig. 5. Schematic diagram illustrating electron path through shield material showing effect of multiple scattering and bremsstrahlung radiation.

illustrated in Figure 5. The multiple Coulomb scattering has the effect of considerably increasing the actual distance traveled by an electron in passing through a given thickness of material and, therefore, increasing the energy loss. In practice, the actual path length is found to be from 1.2 to 4 times the thickness of material traversed with this factor being highest for low energy electrons and high Z materials.

The average path length as defined by Equation (3) and the "range" or thickness of material which an electron can just penetrate are two quite different quantities. Figure 6 is a typical transmission curve for monoenergetic electrons showing the effect of range straggling and illustrating three different concepts of electron range which are commonly (often erroneously) used. These are the "average range", the "extrapolated range" and the "maximum range". One must be sure that he knows which concept of range he is using since these quantities can differ significantly.

Although radiative losses may be neglected in determining penetration of geomagnetically trapped electrons, the high penetrating power of electromagnetic radiation compared to the primary electrons requires that the bremsstrahlung radiation be considered in determining the radiation field behind a shield.

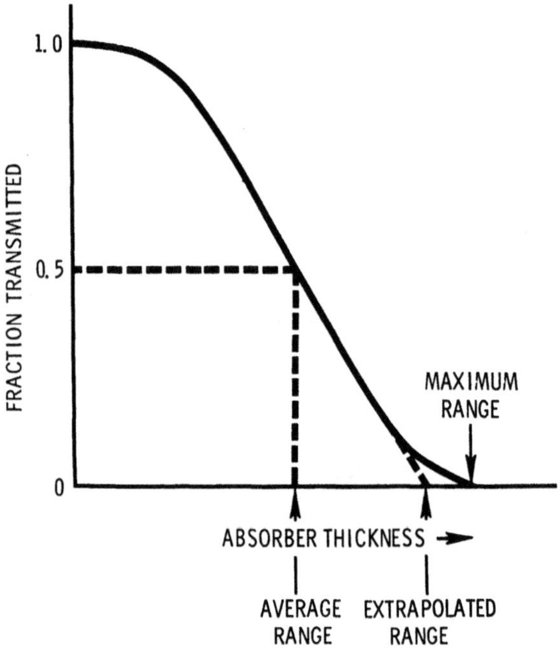

Fig. 6. Example of transmission curve for monoenergetic electrons normally incident on an absorber, illustrating three concepts of electron "range".

A useful quantity in dealing with bremsstrahlung production is the average fraction of the electron kinetic energy, E, which is converted to electromagnetic radiation as the electron is being completely stopped in a thick absorber. Wu (1941) has shown this fraction $f(E)$ to be represented by

$$f(E) = \frac{1.98 \times 10^{-4}[1.96E + 2]}{[1 + 0.35 \log_{10}(82/Z)]} \cdot Z \tag{4}$$

where Z is the atomic number of the stopping material. Incident electrons of energy E can produce bremsstrahlung photons having any energy between 0 and E. The spectral distribution of the emitted bremsstrahlung has been shown by WYARD (1952) to be independent of the atomic number of the stopping material. This distribution is seen to be a function only of the ratio of the photon energy E_γ to the incident energy, E, of the electron. The spectral intensity may be expressed as

$$I(E_\gamma, E) = \frac{f(E)}{1.25}\left\{4\left(1 - \frac{E_\gamma}{E}\right) - 3\frac{E_\gamma}{E}\ln\left(\frac{E}{E_\gamma}\right)\right\} \tag{5}$$

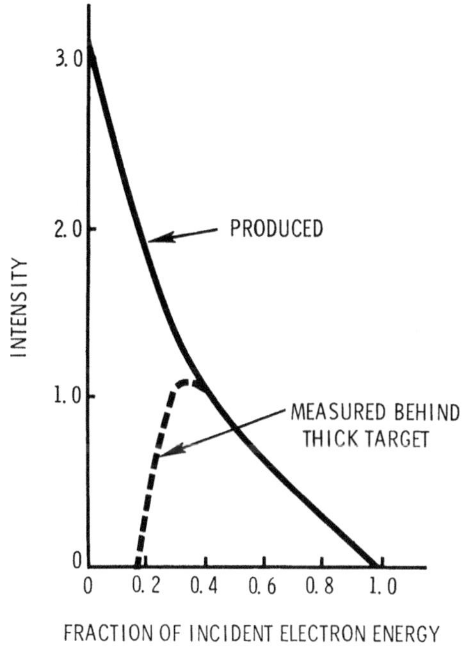

Fig. 7. Typical spectral distribution of bremsstrahlung produced by electron which is completely stopped in the absorber. Dotted curve is representative of spectrum which might be measured behind high-Z target.

The distribution is plotted in Figure 7 along with a typical spectrum which might be measured behind a thick target of high-Z material where the low energy bremsstrahlung have been absorbed. It is interesting to note from Equation (4) that the fraction of energy which goes into thick target bremsstrahlung is proportional to Z whereas the cross-section for bremsstrahlung production is proportional to Z^2.

Much simpler first order approximations for the expressions given in Equations (4) and (5) have been derived (EVANS, 1955) which are useful where great accuracy is not required. These are expressed as:

$$f(E) \approx kZE$$

and

$$I(E_\gamma, E) \approx 2kZ(E - E_\gamma)$$

where k is a semiempirical constant having a value $\approx 7 \times 10^{-4}$ MeV^{-1}.

The angular distribution of the emitted bremsstrahlung is of interest. For electrons having energies less than several hundred keV, the bremsstrahlung produced in a thick target are quite isotropic. As the energy is increased, however, the distribution tends to peak in the forward direction (with respect to the incident electron). On the other hand, as one goes to higher-Z materials where more scattering takes place the distribution becomes less peaked. The angular distribution of the thick target bremsstrahlung can range anywhere from essentially isotropic for low energy electrons stopped by high-Z materials to an almost unidirectional distribution for very high energy electrons stopped by low-Z materials.

Secondary radiations other than bremsstrahlung are produced by electron interactions with the stopping material, however, they have been found to be relatively unimportant with respect to shielding. These include X-rays from electronic transitions by excited atom and the secondary electrons or delta rays produced in ionization collisions.

Shielding Calculations

The technology commonly referred to as shielding consists mostly of techniques for calculating the radiation field behind a shield. The involved nature of the interactions of nuclear particles with matter make such calculations quite difficult so that; in many instances, approximations are required. It was found at the time of the discovery of the geomagnetically trapped radiation belts that much of the charged particle interaction data needed for even crude approximations were not in hand. Since that time, a continuing effort has been devoted to the generation of the needed data.

GEOMETRY

The very complex nature of an actual spacecraft structure makes it almost impossible to perform radiation transport calculations which take into consideration all different materials and geometrical irregularities. To do so would require a Monte Carlo calculation of such fantastic proportions as to be completely impractical. Consequently, techniques for approximating the geometry must be employed.

The most commonly employed approach is that referred to as "sectoring". In this technique which is illustrated in Figure 8, the vehicle structure is considered to consist of a number of laminated slab sectors which are representative of both the composition and thicknesses over the vehicle region covered by the sector. The dose, then, at the point of interest is the sum of the doses obtained from slab geometry calculations for each sector weighted by the solid angle subtended by the sector. For example, referring to Figure 8, the dose, D, at a point internal to a vehicle in an isotropic flux is given by

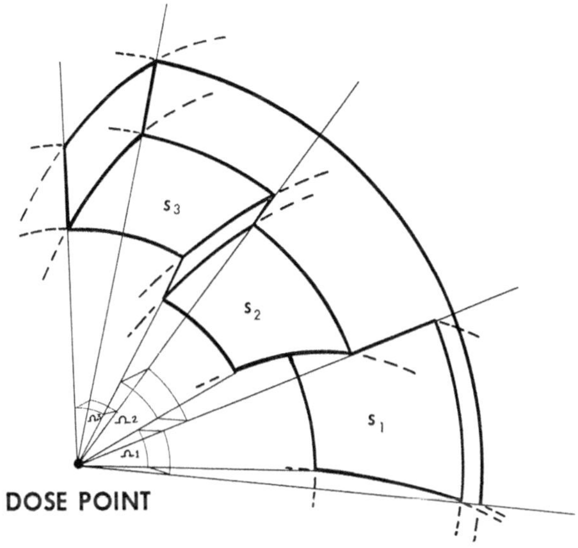

Fig. 8. Schematic diagram illustrating sectoring technique used in shielding calculations.

$$D = \phi \sum_i \frac{\Omega_i}{4\pi} D(S_i)$$

where Φ is the total omnidirection flux incident on the spacecraft, Ω_i is the solid angle subtended at the dose point by the sector, S_i, and $D(S_i)$ is the dose at the dose point per unit flux incident on the ith sector. With this approach, the dose calculation reduces to that of determining the spectrum and angular distribution of both the primary and secondary radiations which penetrate a number of slab shields and, in turn, the dose contribution from each sector at the dose point. The number of sectors required to describe a given vehicle is dictated by the complexity of the vehicle. Since homogenization is involved in any reasonable sectoring of a vehicle geometry care must be taken so that thin regions which permit the major portion of the dose are not averaged in with thick regions. The sectors should include all material between the radiation field and the dose point in question including self shielding of the body in the case foman.

A special, simplifying case of the sectoring technique is often utilized in calculations where detailed accuracy is not required. This is the single sector or spherical shield geometry. For the spherical shield, the dose calculation is reduced to a single slab calculation. This type of calculation is found to be quite adequate for parametric studies where detailed vehicle designs are not available. Also, the shield effectiveness of a spacecraft is often defined in terms of the equivalent spherical shield thickness.

PROTONS

As was discussed earlier, protons in passing through matter travel essentially straight paths while they more or less continually lose energy. Therefore, if nuclear collisions are neglected, a proton incident on a shield can be assumed to emerge from the shield traveling in the same direction as it entered but with a reduced energy. If the path length through the shield in the direction of travel is greater than the range of the proton, the particle is stopped in the shield. It may be shown (ALLEN *et al.*, 1961) that for a beam of protons passing through a thickness, X, of material the emerging spectrum is determined by the equation

$$\phi(E,X)\mathrm{d}E = \phi(E',0)\frac{S(E')}{S(E)}\mathrm{d}E \tag{6}$$

where $\Phi(E, X)\,\mathrm{d}E$=flux of protons having energies between E and $E+\mathrm{d}E$ at a distance X into the shield; $S(E)$=the stopping power of the shield material for protons of energy E (as defined earlier); E'=the incident energy of protons degraded to energy E in traversing a distance X in the shield material. The use of Equation (6) with existing stopping power and range data along with the fact that protons are not significantly deflected in passing through matter make the determination of the energy and angular distributions of primary protons and, in turn, their dose contribution at a point behind a shield a rather straight forward matter. If one takes into consideration removal from the beam through nuclear reactions as must be done for very thick shields Equation (6) becomes

$$\phi(E,X)\mathrm{d}E = \phi(E',0)\frac{S(E')}{S(E)}\exp\left[-\int_0^x \sum(E_S')\,\mathrm{d}S\right]\mathrm{d}E$$

where $\Sigma(E'_S)$ is the reaction cross-section at a distance, S, into the shield for protons which enter the shield with energy E'.

The dose is obtained by folding the rate of energy loss into the flux-spectrum of particles at the desired point and integrating over energy to determine the energy deposition per unit mass with 100 ergs/cm equaling one rad.

Calculations for the secondary radiation component produced in nuclear collisions of protons is much more difficult than those for the primary com-

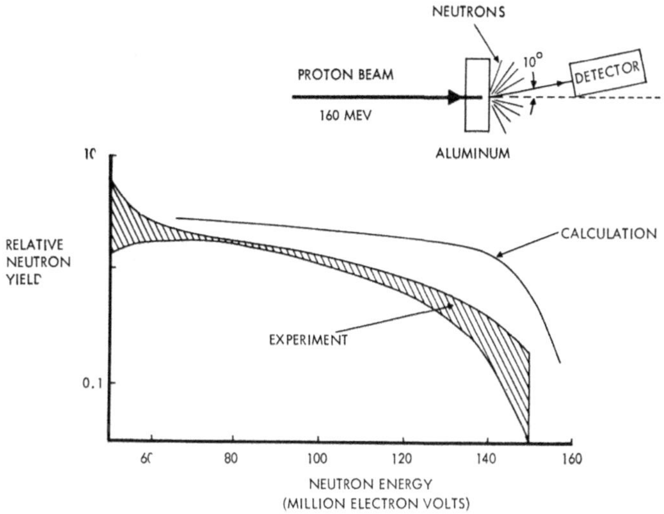

Fig. 9. Comparison of calculated and measured spectra of secondary neutrons produced by 160 MeV protons incident on a thick aluminum target. Data are for an angle of 10° from direction of incident protons.

ponent. The fact that several nucleons may be emitted at various angles and energies following a single nonelastic collision, and the fact that the energies of the cascade particles may be high enough to produce tertiary effects again suggest the use of Monte Carlo techniques. A very detailed Monte Carlo code (KINNEY, 1964) has been produced to handle this problem utilizing the particle yield data produced by BERTINI (1963). Also, experiments are being conducted (MAIENSCHEIN *et al.*, 1962) to check the accuracy of this code along with the cross-sections of Bertini. A typical comparison of the (GIBSON *et al.*, 1965) experimental data with calculations made for the same thick aluminum target bombarded with 160 MeV protons are shown in Figure 9. The measured quantity is the energy distribution for neutrons escaping the target at an angle of 10 degrees to the direction of the incident protons. While the results are not in excellent agreement, it must be remembered that the calculations are based on theoretically determined cross-sections and yields.

To use such a time consuming calculational procedure on a routine basis for shielding calculations is not practical. Therefore, certain approximations to the actual case are in general use. The approximation most commonly used to

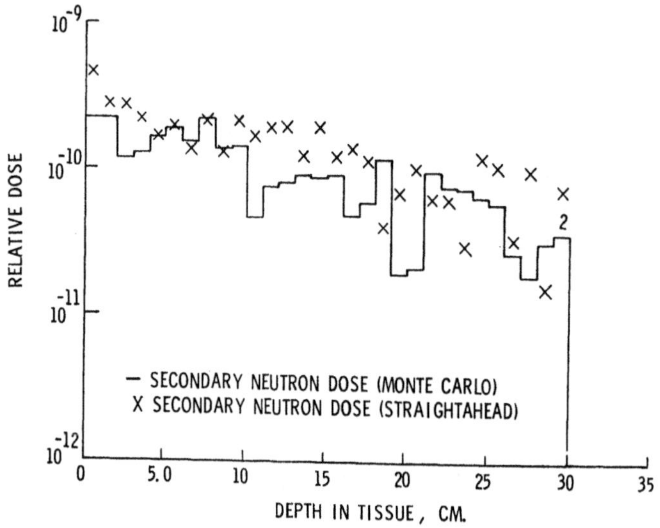

Fig. 10. Comparison of dose calculated using "straight ahead" approximation with that obtained using Monte Carlo technique for secondary neutrons produced by 100 MeV protons incident on 10 gm/cm² of aluminum followed by 30 cm of tissue.

simplify calculations of the secondary radiation component originating from proton interactions is the so called "straight ahead" approximation. The cascade nucleons produced in the interaction are all assumed to be directed in the same direction as that of the incident particle or "straight ahead". The logic in the approximation is that the overestimate in the secondaries produced by particles directed at the point of interest will compensate for the secondaries that would have reached the point from interactions of primary particles not so directed. The Monte Carlo code discussed above has provided a basis for evaluation of the "straight ahead" approximation. Calculations have been made (ALSMILLER *et al.*, 1965) for identical cases with both the Monte Carlo code and a "straight ahead" code using the same cross-section data. The results for the secondary neutron component for one case is shown in Figure 10. The agreement is seen to be quite good. Similar calculations have been made for a variety of energies, angular distributions and materials. The results shown are typical. The approximation usually overestimates the dose and appears to have the same validity for elements between carbon and copper.

ELECTRONS

Shield penetration calculations are much more difficult for electrons than for primary protons. The scattering and straggling (discussed earlier) so complicate the problem that, in general, the use of Monte Carlo techniques are required.

Several Monte Carlo computer codes have been written for electron slab penetration with one of the most advanced being that of BERGER and SELTZER (1965). In this code, the electron trajectory is subdivided into a number of sufficiently short, straight-line segments, so that total angular deflection along each is small. Using the results of analytical treatments for the probability distribution of the net angular deflections, the problem is treated as if the electrons suffer discrete energy losses and angular deflections at the end of each segment as the histories are traced through the slab. This electron penetration portion of the code is used to determine the source distribution for the bremsstrahlung photons which are then traced through the shield.

As in the case of the proton codes, the calculational procedures are being checked by comparison with experimental results. Figure 11 shows a comparison of thick target bremsstrahlung results obtained using the above discussed Monte Carlo code with those obtained experimentally (DANCE and BAGGERLY, 1965). The agreement is seen to be quite excellent.

Again, however, it is not practical to perform detailed Monte Carlo calculations on a routine basis for every problem of interest and, therefore, less accurate approximate procedures are generally employed. In many cases for space vehicle shielding considerations, the penetration of the electrons themselves need not be considered since the range of the incident particles are not sufficient to penetrate the shield. In these cases, bremsstrahlung constitutes the only significant radiation component behind the shield. Where electron penetration is of importance, however, one procedure which has been followed has been to perform detailed parametric calculations using a Monte Carlo code as described above for monoenergetic electrons normally incident on a large number of slab materials and, in turn, to generate analytic expressions which fit the data (transmission as a function of energy, material thickness, and atomic number) and give a basis for interpolation. The problem then reduces to one of determining an appropriate equivalent flux normally incident on the slab sector in question.

For a shield exposed to an isotropic flux of particles, the number of particles incident on one square centimeter is just one-fourth of the omnidirectional flux, Φ. After an electron loses about 20 percent of its energy in passing through matter, its direction (because of Coulomb scattering) is virtually independent of its initial direction. Therefore, the assumption of a normal incident flux equal to $\Phi/4$ is not unreasonable. The use of such a flux, however, does clearly result in an overestimate of the radiation penetrating the slab since backscattering

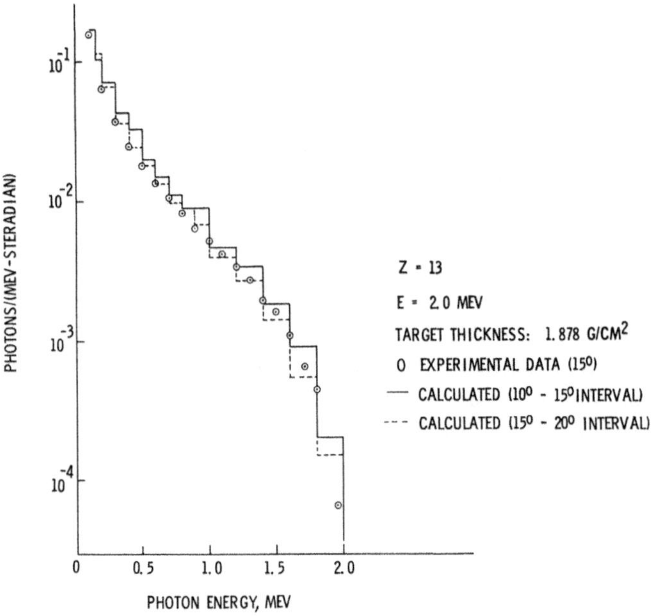

Fig. 11. Comparison of calculated and measured spectra of bremsstrahlung produced by 2 MeV electrons normally incident on thick aluminum target. Measurements are for an angle of 15° from direction of incident electrons.

increases with increasing angle of incidence from the normal. To correct for this a normal flux of $\Phi/8$ has been used in some cases. In line with the above discussion, the electrons are frequently assumed to emerge from the inside of the shield with an isotropic distribution over the forward hemisphere.

In calculating electron bremsstrahlung for geomagnetically trapped electrons, the approximations normally used involve the simplifying assumption that the bremsstrahlung is produced on the outside surface of the shield. This is not a poor assumption since with the steep energy spectra which are encountered, a great fraction of the total energy loss occurs in the first few grams per square centimeter of the stopping material.

Expressions such as those in Equations (4) and (5) or results obtained from the Monte Carlo codes may be used to determine the spectrum of the emitted photons with the photons then attenuated through the shield. Again, an incident electron flux of one-fourth the omnidirectional isotropic flux should be used in determining the intensity. In the earlier section on electron penetration, it was pointed out that depending upon the electron energy and the atomic number of the stopping material, the angular distribution of the bremsstrahlung emitted by each electron lay somewhere between an isotropic distribution and one peaked in the forward direction. Therefore, the distribution of the total bremsstrahlung considered to be produced at the shield surface is usually considered to be isotropic over the forward hemisphere which represents a compromise between the two extremes. The photons are attenuated through the shield using exponential attenuation and buildup factors. For most thicknesses involved in space vehicles, the attenuation of the photons is not large so that the bremsstrahlung dose tends to set a lower limit for the reduction of dose through added shielding. Obviously, the bremsstrahlung dose can be minimized if a low-Z material is used to stop the electrons (minimizing bremsstrahlung production) followed by a high-Z material to attenuate the photons. One must be careful in using the procedure outlined above to make bremsstrahlung calculations for such a configuration since the Z dependence of bremsstrahlung production could result in more bremsstrahlung being produced by those electrons reaching the high-Z layer than were produced in the first layer. In such cases, a second source would have to be considered at the surface of the high-Z layer.

Summary

It is obvious from the foregoing section that a significant effort is being devoted to development of a charged particle shielding technology adequate for the design needs for all future space vehicles. Quite accurate calculational techniques exist for all the radiation components involved in shielding against geomagnetically trapped radiations. However, these procedures are limited by their complexity when applied to realistic geometries and in some cases by the lack of adequate basic particle interaction input data. The first of these problems can be alleviated by simplifying geometrical approximations while the basic input data are needed for even the simplest geometries. The approach which is currently being taken is to develop improved theoretical means for developing the needed data. These theoretical techniques are being carefully checked experimentally for accuracy before being used to generate parametric data. The present effort in this area appears to be adequate to cope with the problems as they arise with regard to spacecraft shielding.

References

ALLEN, R. I. *et al.*: 1960 *Lockheed Report* NR-104.
ALSMILLER, R. G. *et al.*: 1965, *Proceedings of Second Symposium on Protection Against Radiations in Space,* NASA SP-71, 177–181.
BARKAS, W. H. and BERGER, M. J.: 1964, NASA SP-3013.
BERGER, M. J. and SELTZER, S. M.: 1964, NASA SP-3012.
BERGER, M. J. and SELTZER, S. M.: 1965, *Proceedings of Second Symposium on Protection Against Radiations in Space,* NASA SP-71, 437–448.
BERTINI, H. W.: 1963, *Oak Ridge National Laboratory,* ORNL-3383.
DANCE, W. E. and BAGGERLY, L. L.: 1965, NASA CR-334, Part 1.
DOSTROVSKY, I. *et al.*: 1959 *Phys. Rev.* **111**,.
EVANS, R. D.: 1955, *The Atomic Nucleus,* McGraw Hill Book Co.
GIBSON, W. A. *et al.*: 1965, *Proceedings of Second Symposium on Protection Against Radiations in Space,* NASA SP-71, 337–339.
KINNEY, W. E.: 1964, *Oak Ridge National Laboratory,* ORNL-3610.
MAIENSCHEIN, F. C. *et al.*: 1962, *Proceedings of the Symposium on the Protection Against Radiation Hazards in Space,* USAEC, TID-7652, 523–594.
WU, C. S.: 1941, *Phys. Rev.* **59**, 481.
WYARD, S. J.: 1952, *Proc. Phys. Soc.* **A65**, 377.

Discussion

W. BROWN asked if laminated shielding is useful in attenuating fission betas and the attendant bremsstrahlung. J. KELLER replied that the best approach is to stop the betas in a material of low atomic number. In this way the bremsstrahlung is reduced. J. CLADIS commented that if the bremsstrahlung becomes a problem despite the use of low Z shielding materials, the use of superconducting magnets to deflect the betas away from the vehicle surface should be investigated.

A MODEL PROTON ENVIRONMENT ABOVE 4 MeV

JAMES I. VETTE

Aerospace Corporation, El Segundo, Calif.

Abstract. The procedure for constructing a proton environment from existing satellite data is described. Four different flux maps are given to describe the 4–15, 15–30, 30–50 and above 50 MeV energy ranges. In each of these energy ranges an exponential energy spectrum is used and the variation of the spectrum with position is taken into account. A comparison of the model environment with experimental data is presented and the results of orbital integrations are shown.

Introduction

We have been engaged in a program to produce trapped radiation model environments that will be of general use to space engineers, system planners and space physicists. An inner zone electron environment was completed earlier and has been described (VETTE, 1964). An updated version is presently being made. The proton environment above 4 MeV has recently been finished. This paper will discuss the construction of that proton model and the comparison of it with existing satellite data. Lower energy proton maps are being developed and will be presented at a later date.

We have used measurements from many satellites to obtain a fairly complete spatial and spectral coverage. Because the protons are quite stable over periods of several years in most of the inner zone, it is possible to use both old and new data. The measurements that have mainly been used in determining this environment are given in Table I. We will use the code letter and number shown in the last column in referring to each measurement hereafter.

Procedure

Many of the measurements give the omnidirectional flux but Relay I experiments provide fluxes perpendicular to the magnetic field line. These have been converted to omnidirectional fluxes by performing the integration

$$J(B,L) = 2\pi B \int_{B_m = B}^{B_m = B_c} \frac{j_\perp (B_m, L)\, \mathrm{d}B_m}{B_m^{3/2} (B_m - B)^{1/2}}$$

where B is the magnetic field strength, L is the McIlwain shell parameter and

TABLE I

Data used in proton environments

Satellite	Data Time Period	Nominal Energy Range (Mev)	Reference	Text Code for Data
Explorer 4 1958 E 1	July–Oct. 1958	>31 >43	McIlwain, 1961	P1 P2
Injun 1 1961 02	July–Dec. 1961	>40	Bostrom *et al.*, 1965	P3
1962 Aγ1 1962 BE1	Sept.–Oct. 1962	60–120	Freden and Paulikas, 1964	P4
Explorer 15 1962 BΛ1	Oct.–Jan. 1962–63	40–110	McIlwain, 1963	P5
Relay 1 1962 Bγ1	May–Sept. 1963	> 34 18.2–35	McIlwain *et al.*, 1964	P6
Injun 3 1962 Bγ2	Dec.–Sept. 1962–63	40–110	Valerio, 1964	P8
Telstar 1 1962 AE1	July–Feb. 1962–63	50–75	Brown *et al.*, 1963	P9
1962 K1	April 1962	> 59	Imhof and Smith, 1964	P10
1962 Aγ1 1962 BE1	Sept.–Oct. 1962	5–20	Freden and Paulikas, 1964	P11
Relay 1 1962 Bγ1	Dec.–Jan. 1962–63	5–8.6	Brown *et al.*, 1964	P12
Explorer 15 1962 BΛ1	Oct.–Jan. 1962–63	4–13	Private Communication From W. L. Brown and J. D. Gabbe	P13
Relay 1 1962 Bγ1	May–Sept. 1963	>5.2	McIlwain *et al.*, 1964	P14

B_c is the cutoff value where $j_\perp$ vanishes. Once the data are all in the omnidirectional form it is necessary to obtain a self consistent energy spectrum in order to compare the various experiments. This is usually the most difficult step.

For certain energy intervals and for large regions of space the integral spectrum can be represented by an exponential spectrum. To illustrate this point we show some of the equatorial spectrum given by Fillius and McIlwain (1964) in Figure 1. The curves in the figure are the sum of two exponentials; this gives a reasonable fit above a few MeV for all their equatorial data. The data of Naugle and Kniffen (1963) can also be fit in this manner. The representation in this form is too restrictive to describe the whole spatial and energy domain we desire to include. However, the use of a simple energy form such as an exponential has advantages. From a model environment standpoint this form is easy to use in computation, and allows one to account for spatial variations of the energy spectrum without reverting to general functions of three variables.

We have broken the environment into four energy bands: 4–15, 15–80, 30–50, >50 MeV and have used an exponential spectrum within each band.

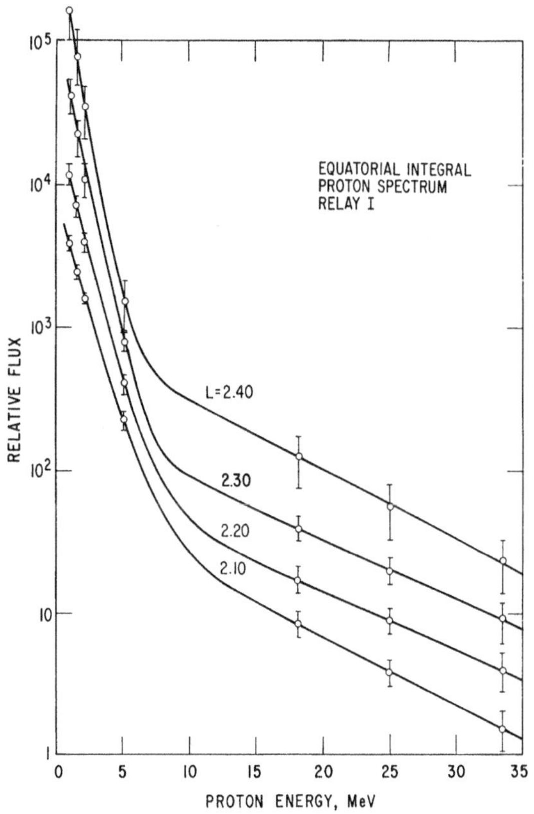

Fig. 1. Integral proton spectra from Relay I, $L = 2.1, 2.2, 2.3, 2.4$ at the equator. The curves drawn through the points are the sum of two exponential spectra.

The energy parameter, E_0, is a function of both B and L so we are representing the omnidirectional integral flux as

$$J(>E, B, L) = F_i(B, L) \exp \frac{E_i - E}{E_{0i}(B, L)}$$

in each energy band. The distribution function F_i represents the integral flux for energies greater than E_i. We have chosen 4, 15, 34 and 50 MeV for the values of E_i.

Proton Map AP1 30–50 MeV

The data above 30 MeV was studied first. Since there were not many efficiency curves available for the various detectors, the nominal thresholds or energy bands given in Table I were used. Using the ratios of various measurements an $E_0(B, L)$ function was obtained and is shown in Figure 2. Because the P6

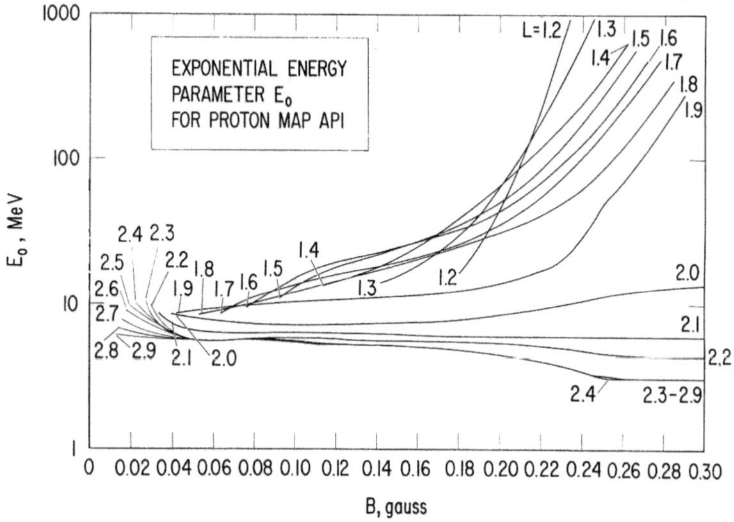

Fig. 2. The exponential energy parameter, E_0, for proton map AP1. This function is to be used over the energy range 30–50 MeV.

measurement was the most extensive in the regions where there was not enough data to obtain a good spectrum we have selected $E_1 = 34$ MeV for this proton map. The various data in the energy band above 30 MeV have been converted to the flux above 34 MeV in the following manner. If an experimental measurement of flux above energy E_a is denoted by

$$J_m(> E_a, B, L)$$

the quantity

$$F_m(B, L) = \frac{J_m(E_a, B, L)}{\exp\left[(34 - E_a)/E_0(B, L)\right]}$$

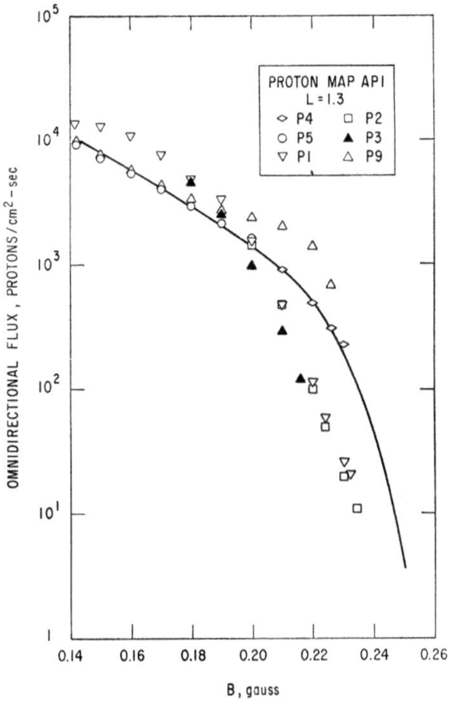

Fig. 3. Comparison of AP1 with experimertal data at $L = 1.3$. The P1, P2, and P3 data were obtained near solar maximum.

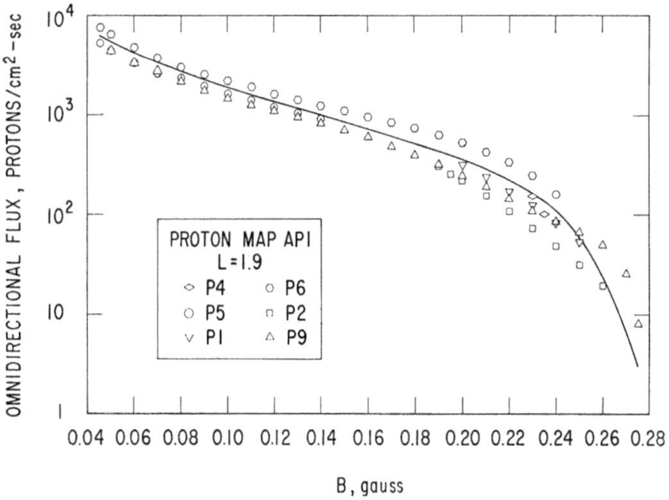

Fig. 4. Comparison of AP1 with experimental data at $L = 1.9$.

represents the data converted to the flux above 34 MeV by the model spectrum. If a measurement covers an energy band $E_a - E_b$, then the conversion is made by

$$F_m(B, L) = \frac{J_m(E_a - E_b, B, L)}{\exp\left[(34 - E_a)/E_0\right] - \exp\left[(34 - E_b)/E_0\right]}$$

The results of this conversion for two L values are given in Figures 3–4. We have used points at fixed B values to plot the converted data. An eyeball fit is made from these plots to form the $F(B, L)$ for the map. It has been found necessary to draw the B-L flux map representation to show some of the inconsistencies that developed in selecting the B dependence. An iteration between the B-L map and the F-B plot produced a smooth representation in both systems that is consistent with the data. At some time in the future it may be feasible to obtain analytical fits but at present we feel the eyeball method is adequate and consistent with the accuracy of the input data.

At the low L and high B values the P1, P2, P3 data cluster together and fall below the later data as can be seen in Figure 3. This is either the solar cycle change suggested by BLANCHARD and HESS (1964) or the redistribution of high energy protons following Starfish as determined by FILZ (1964). The P9 data converted to >34 MeV seems too large at high B values when the spectrum is hard. The energy window of this detector probably depends on the energy spectrum. We have selected the higher fluxes to reflect 1963 conditions for AP1. In general the data and the model are in reasonable agreement and the situation shown in Figure 4 is typical. The radical decrease of protons >34 MeV observed by MCILWAIN for $L>2.0$ following the large magnetic storm of September 23, 1963 (MCILWAIN, 1965a), has not been included. However the fluxes seem to be returning slowly to prestorm values (MCILWAIN, 1965b).

Proton Map AP2 15–30 MeV

It is apparent from the data that the slope of the integral spectrum will change radically somewhere in the 6–20 MeV region and that the specific energy at which this occurs is a function of B and L. The only data available to us in this region was P7 which could not give any detail to clarify this point. Using AP1 to calculate fluxes >30 MeV and using P7, an energy spectral function E_0 was determined. The map obtained by this procedure is designated AP2 and the distribution function for $E>15$ MeV is shown in Figure 5.

Proton Map AP3 > 50 MeV

Although there was insufficient data available above 50 MeV to carry out a

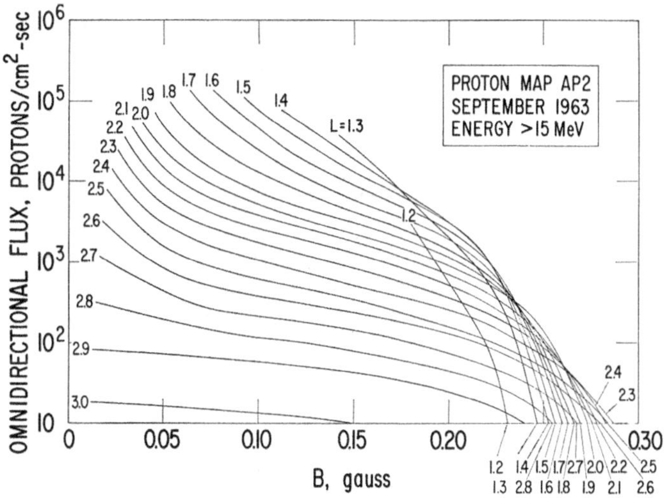

Fig. 5. The flux versus *B* plot of proton map AP2, $E > 15$ MeV.

similar procedure to that used in constructing AP1, it was apparent that higher energy protons have a different spectral behavior as a function of position. One notices in Figure 2 that the spectrum gets softer on a given L shell ($L < 2.0$) as the equator is approached. At higher energies the opposite effect occurs on the basis of the data of IMHOF and SMITH (1964) and the recent measurements of FREDEN, BLAKE and PAULIKAS (1965). Using the spectral behavior indicated by these measurements and the data at 350 MeV of GARMIRE (1963) a spectral function for high energy protons was obtained. The distribution function for this high energy map was obtained by using AP1 to calculate $J(>50 \text{ MeV})$.

Proton Map AP4 4–15 MeV

Although there is data down to 1 MeV in the inner zone, the spectrum in the 1–5 MeV range does not conform well to an exponential description for $L < 1.6$ and more detail is needed to describe the spectrum properly. The additional measurements made by McIlwain on Relay I should clarify this picture when they are available. In the interests of obtaining a model down to lower energies that maintains the same spectral description as the other maps, a lower limit of 4 MeV was selected. Using P11, P12, P13 and P14 along with the distribution function for AP2, a spectral function E_0 for the 4–15 MeV range was obtained. The conversion of the various measurements to 4 MeV at selected L values by means of this spectrum produce graphs similar to those shown in Figures 3–4. This AP4 model fits the data reasonably well except for $L<1.5$ where the spectrum causes difficulty. As more data in the 1–20 MeV range become available a better model can be constructed. The distribution function of AP4 is for $E>4$ MeV.

Orbital Integrations

Another useful way of presenting a trapped radiation environment is by giving the average flux encountered in various circular orbits. The results obtained with the AP4 map are given in Figure 6. Although the maps are most accurate over the energy bands described earlier, it is possible to obtain reasonable answers outside these bands. Since it is sometimes cumbersome to use more than one map for a specific purpose, we have computed the orbital integrations for 23 energy bands with each map. It is hoped that these results will be useful

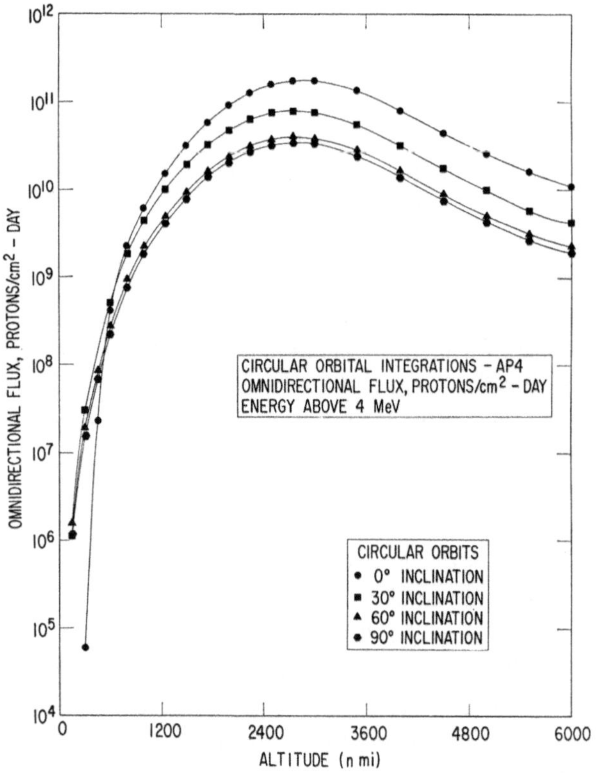

Fig. 6. Orbital integrations with AP4, $E>4$ MeV.

in selecting a single map best suited for a given purpose. The complete results of this calculation are too extensive to be given here but are available.

Summary

A proton model environment has been constructed from existing satellite data. In order to obtain good agreement with this data, four different flux maps for energies greater than 4, 15, 34, and 50 MeV were used with an exponential energy spectrum that varies with B and L. These maps have been used to obtain a detailed energy spectrum for the fluxes accumulated by a satellite in circular orbits. A more detailed presentation of this environment will be available.

Acknowledgements

The author wishes to thank Mrs. J. A. Lane and Messrs. A. B. Lucero, E. F. Martina and R. G. Pruett for their efforts in plotting, calculating and analyzing computer outputs as well as assistance in constructing the model environment. Messrs. K. W. Hubbard and J. K. Okazaki are commended for their extensive computer programming contributions. The author is grateful to all the experimenters listed in Table I who have contributed their data and provided valuable discussions. In particular Drs. W. L. Brown, J. D. Gabbe, and C. S. Roberts of Bell Telephone Laboratories, Dr. C. E. McIlwain of the University of California at San Diego and Dr. G. A. Paulikas of the Aerospace Corporation have provided unpublished data and valuable discussions during this effort.

This work has been supported by the National Aeronautics and Space Administration under Contract W-11, 683 and the U.S. Air Force under contract AF 04(695)-469.

References

Blanchard, R. C. and Hess, W. N.: 1964, *J. Geophys. Res.* **69**, 3927.
Boström, C. O., Zmuda, A. J., and Pieper, G. F.: 1965, *J. Geophys. Res.* **70**, 2035.
Brown, W. L., Buck, T. M., Medford, L. V., Thomas, E. W., Gummel, H. K., Miller, G. L., and Smits, F. M.: 1963, *Bell System Tech. J.* **42**, 899.
Brown, W. L., Davidson, L. W., and Medford, L. V.: 1964, Preprint from Bell Telephone Laboratories, 'The Energetic Particle Environment of Relay 1', Bell Telephone Laboratories Experiments.
Fillius, R. W. and McIlwain, C. E.: 1964, *Phys. Rev. Letters* **12**, 609.
Filz, R., Yagoda, H., and Holeman, E.: 1964, *Space Research* **4**, 642.
Freden, S. C., Blake, J. B., and Paulikas, G. A.: 1965, *Trans. Am. Geophys. Union* **46**, 141.
Freden, S. C. and Paulikas, G. A.: 1964, J. *Geophys. Res.* **69**, 1259.
Garmire, G.: 1963, J. *Geophys. Res.* **68**, 2627.
Imhof, W. L. and Smith, R. V.: 1964, J. *Geophys. Res.* **69**, 91.
McIlwain, C. E.: 1961, *J. Geophys. Res.* **66**, 3681.
McIlwain, C. E.: 1963, *Science* **142**, 355.
McIlwain, C. E.: 1965a, *Space Research* **5**, 374.

McIlwain, C. E.: 1965b, *Trans. Am. Geophys. Union* **46**, 141.
McIlwain, C. E., Fillius, R. W., Valerio, J., and Dave, A.: 1964, Preprint from University of California at San Diego, 'Relay I Trapped Radiation Measurements'.
Naugle, J. E. and Kniffen, D. A.: 1963, *J. Geophys. Res.* **68**, 4065.
Valerio, J.: 1964, *J. Geophys. Res.* **69**, 4949.
Vette, J. I.: 1964, 'The Updating and Dissemination of the Knowledge of Trapped Radiation – Model Environments', Second Symposium on Protection Against Radiation in Space, Gatlinburg, Tennessee, October 12–14, 1964.

SOME NEW METHODS FOR SPACE SHIELDING CALCULATIONS

R. W. LOWEN and D. H. SOWLE

Space Science Laboratory, General Dynamics/Convair, San Diego, California

Abstract Some new, simple methods of limited but quite useful accuracy for use in space shielding calculations are described. The energy dissipation by protons isotropically incident on a plane slab is solved analytically. Energy and angular distributions of the bremsstrahlung emitted by electrons incident either normally or isotropically on thick targets are calculated using a semi-classical cross-section and electron distributions for an infinite medium. The results are compared with experiment and with calculations by Berger. The energy dissipation of electrons from a plane perpendicular source is calculated by an age equation model and compared with the accurate calculations of Spencer.

1. Introduction

A year and a half ago, Dave Sowle and I found ourselves with a responsibility for developing improved methods for doing space shielding calculations, as part of a company-sponsored program at General Dynamics/Convair. Because our objectives were purely practical rather than practically pure, and because of our budgetary boundary conditions, the methods we chose to pursue were different from the usual ones. Briefly, we sought methods which were physically sound but as simple and inexpensive as possible; we were willing to make some sacrifice of accuracy in return. It seemed unnecessary for our purposes to strive for accuracy much better than that to which the external radiation environment can be predicted, somewhat poorer than ten per cent in our experience.

I should like to describe briefly the methods we have used and to show some results for proton penetration of slabs, thick-target bremsstrahlung, and electron penetration.

2. Proton Penetration of a Plane Slab

Protons are relatively nice, because both range straggling and elastic scattering can be neglected; the main effect is just ordinary energy loss by inelastic collisions. This lets one tackle complex geometries, and many groups have developed elaborate computer codes for this purpose. We have stayed with the simple slab geometry, and have done the calculation analytically.

The stopping powers of several low- to intermediate-Z materials are shown in Figure 1 (BARKAS *et al.*, 1964). It is apparent that over the energy range of greatest interest, a few MeV to a few hundred MeV, the curves can be fitted very well by a power law. Furthermore, the parameters (especially the exponent) do not vary much for the different materials. Thus a laminate can be approximated from results for a homogeneous slab, and in particular we may consider an astronaut as just another layer of the shield!

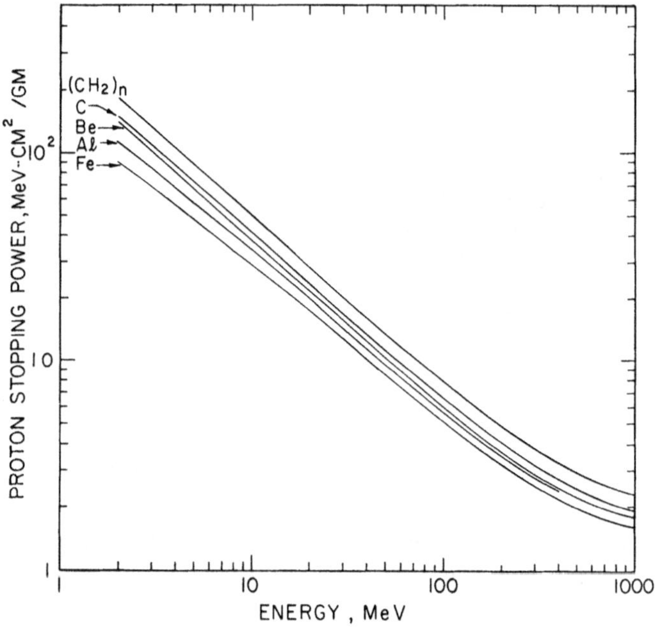

Fig. 1. Stopping power for protons of several low- and intermediate-Z materials (BARKAS *et al.*, 1964).

The assumptions of our calculation are: (1) The proton flux in free space is isotropic; (2) The stopping power can be represented by

$$dE/d\xi = -kE^{-p}(\text{MeV cm}^2/\text{g}),$$

where E=current proton energy (MeV), k, p=empirical constants, (3) The energy spectrum of incident protons is a power law,

$$\phi(E_0) = (n-1)\varepsilon^{n-1}E_0^{-n}, E_0 > \varepsilon \quad \text{and} \quad n > 1,$$

where E_0 = incident proton energy (MeV), ε = low energy cut-off, n = a constant.

Under these assumptions, the energy dissipation at depth ξ (g/cm^2) in a slab can be calculated without further approximation. The result, for unit omnidirectional flux in free space, is

$$\text{Dose} = 1.6 \times 10^{-8} \frac{k\Gamma\left(\dfrac{p+2}{p+1}\right)\Gamma\left(1+\dfrac{n-2}{p+1}\right)(n-1)\varepsilon^{n-1}}{\left[k(p+1)\xi\right]^{\frac{p+n-1}{p+1}}} \text{ rad} \tag{1}$$

A somewhat similar result for a spherical geometry has since been found independently by NAKACHE (1964).

Note that if the power-law energy spectrum is not currently in vogue, one can often approximate other spectra by a few terms of this form.

3. Thick-Target Bremsstrahlung

Let me first write an expression for the flux of X-ray energy from a slab exposed on one side to mono-energetic electrons:

$$\frac{\mathrm{d}^2F}{\mathrm{d}k\,\mathrm{d}\Omega} = \frac{N_0 k \xi_0}{A} \int \mathrm{d}x \int \mathrm{d}t \int \mathrm{d}\Omega_1 I(t,x,\theta_1)\, T(x,k,\theta) \cdot \frac{\mathrm{d}^2\sigma(\chi,t,k)}{\mathrm{d}k\,\mathrm{d}\Omega} \left(\frac{\text{MeV}}{\text{cm}^2 \text{ sec MeV sterad}}\right) \tag{2}$$

where $\cos\chi = \cos\theta \cos\theta_1 + \sin\theta \sin\theta_1 \cos(\varphi_1 - \varphi)$. (φ_1, θ_1) are azimuth and polar angles with respect to the normal for electrons; (φ, θ) are the corresponding angles for photons. N_0, ξ_0, and A are, respectively, Avogadro's number, the electron range in the material (g/cm^2), and the gram atomic weight. The depth (g/cm^2)/ξ_0 is x, and k is the photon energy in units of the electron rest energy. Finally, I, T, and $\mathrm{d}^2\sigma/\mathrm{d}k\,\mathrm{d}\Omega$ are the electron flux, the X-ray transmission, and the thin-target bremsstrahlung cross-section, respectively. The problem consists in finding suitable formulas for these functions and carrying out the four-fold integration in (2).

It is something of a shock when one learns that the elaborate formulas for the thin-target bremsstrahlung cross-section, differential in energy and angle, which adorn textbooks of quantum electrodynamics do not agree well with experiment for electron energies of the order of the rest energy. This is presumably due to failure of the Born approximation. One may make empirical corrections to these complex formulas (BERGER *et al.*, 1964); we have preferred to use a simple semi-classical formula which seems to do quite well. We adopted

$$k\frac{d^2\sigma}{dk\,d\Omega} = \frac{Z^2\alpha r_0^2\eta}{\pi\beta^2(\tau+1)^2}\ln\left[\frac{\lambda}{4k}(\sqrt{\tau}+\sqrt{\tau-k})^2\right]\left[\frac{(1-\beta\cos\chi)^2+(\cos\chi-\beta)^2}{(1-\beta\cos\chi)^4}\right] \tag{3}$$

where Z, α, and r_0 are the atomic number, the fine-structure constant, and the classical electron radius, respectively. β is v/c, and τ is the electron kinetic energy in units of rest energy. η and λ are two adjustable parameters. The angular dependence of (3) is just what one gets by transforming a dipole radiation pattern in the rest frame of the incident electron to the laboratory frame of coordinates. The logarithmic factor resembles the non-relativistic limit of the Bethe-Heitler (HEITLER, 1944) formula.

To choose η and λ we compared (3) with data from STARFELT and KOCH (1956) and KOCH and MOTZ (1959). The "natural" values $\eta=1$ and $\lambda=4$ turned out to give a fairly good fit to the angular distributions. In Figure 2 we show

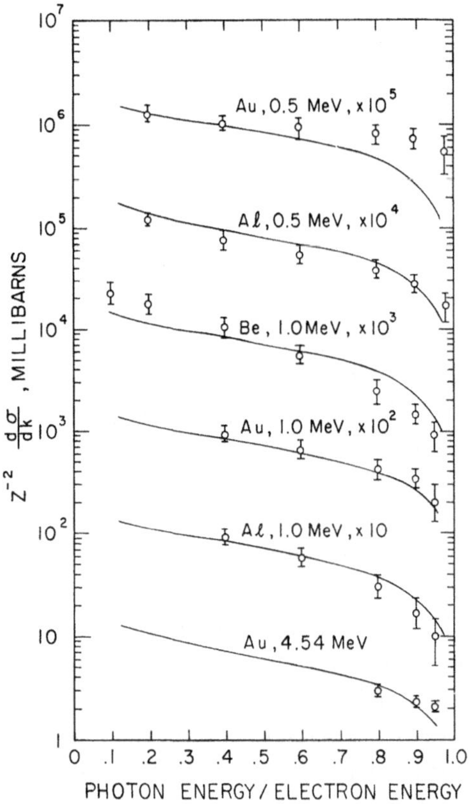

Fig. 2. Comparison of a semi-classical thin-target bremsstrahlung cross-section with data from STARFELT and KOCH (1956) and KOCH and MOTZ (1959).

some comparisons of data with (3) integrated over angle,

$$k\frac{d\sigma}{dk} = (16/3)Z^2\alpha r_0^2\beta^{-2}\eta \ln\left[\frac{\lambda}{4k}(\sqrt{\tau} + \sqrt{\tau - k})^2\right] \tag{4}$$

for $\eta = 1$ and $\lambda = 4$.

The transmission factor which appears in (2) is, of course, just

$$T(x, k, \theta) = \exp\left[-\mu(k)(D - \xi_0 x)\sec\theta\right] \tag{5}$$

For photon energies above 100 keV or so in light materials the mass absorption coefficient $\mu \sim 0.03$ cm^2/g. Because very few radiation belt electrons have ranges greater than a few g/cm^2, the x-dependence of T is not very important except very near $\theta = 90°$. We therefore evaluate T at $x = \bar{x}$, some convenient mean value, and remove it from the integrals in (2).

This permits a great simplification; the x-integration can now be done first. We write

$$I_1(t, \theta_1) = \int I(t, \theta_1, x)\,dx \tag{6}$$

The point of this is, that while one does not know even the simpler function I_1 for a slab, it is easy to find the I_1 appropriate to an infinite medium, and we expect this to be a fair approximation for the low- to intermediate-Z materials of prime interest.

The electron flux, I_1, is taken to be the solution to the Fokker-Planck approximation to the transport equation,

$$\frac{\partial I_1}{\partial t} + f_1(t)\frac{\partial}{\partial(\cos\theta_1)}\left[\sin^2\theta_1 \frac{\partial I_1}{\partial(\cos\theta_1)}\right] = Q \tag{7}$$

where Q is a source term and

$$f_1(t) \equiv \frac{\pi\xi_0 N_0}{A}\int_{-1}^{1} d(\cos\theta)(1 - \cos\theta)\sigma(t, \theta) \tag{8}$$

with $\sigma(t, \theta)$ the elastic scattering cross-section. The solution is evidently

$$I_1(t, \theta_1) = \sum_{n=0}^{\infty} A_n P_n(\cos\theta_1)\exp\left[-n(n+1)v(t)\right], \tag{9}$$

where $v(t) = \int_t^1 f_1(t')\,dt'$. The A_n depend on the type of source considered; for a beam of unit flux,

$$A_n = (n + \tfrac{1}{2}) \tag{10}$$

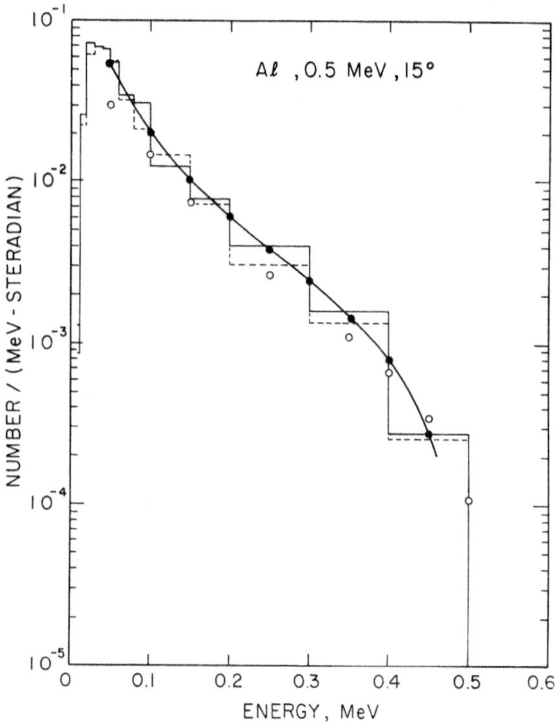

Fig. 3a

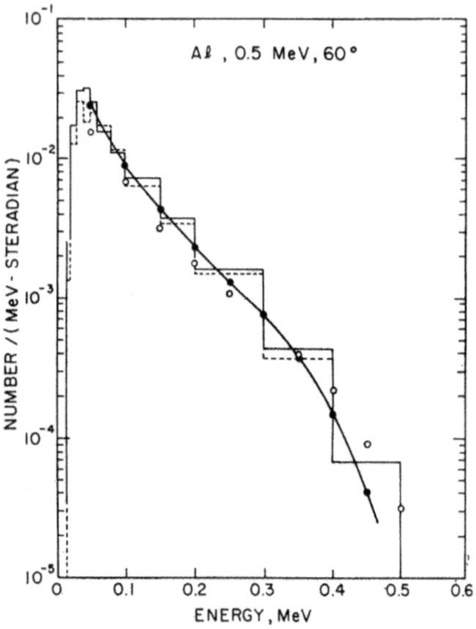

Fig. 3b

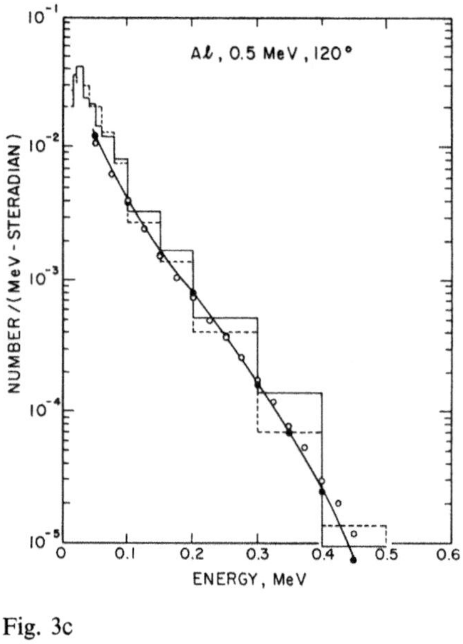

Fig. 3c

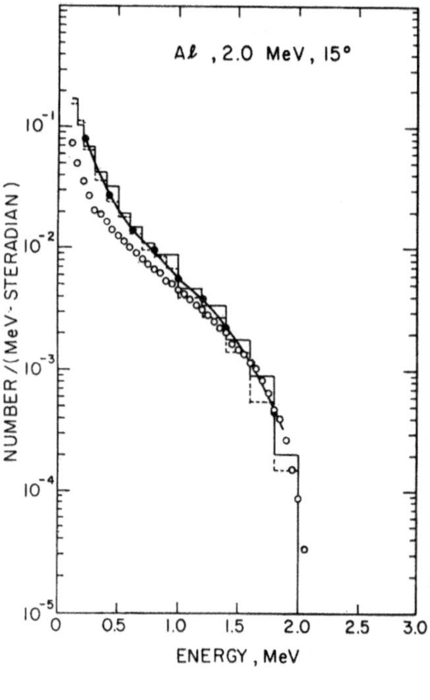

Fig. 3d

For isotropic free-space flux normalized to unit omni-directional flux,

$$A_0 = (16\pi)^{-1}$$
$$A_1 = (8\pi)^{-1}$$
$$A_2 = 5/64\pi$$
$$A_n = 0, \quad n \text{ odd and} > 1,$$
$$A_n = \frac{(-)^{n/2}(n-3)(n-5)(\dots)3\cdot 1}{(n+2)n(n-2)(\dots)4\cdot 2}, \quad n \text{ even and} \geqq 4.$$

For the beam case, the series can be summed in the limit $t \to 1\,(v \to 0)$, the result is

$$I_1(t, \theta_1) \underset{t\to 1}{\sim} \exp(v/4)(2v)^{-1} \exp(-\theta^2/4v), \tag{12}$$

which is just the usual Gaussian multiple scattering approximation.

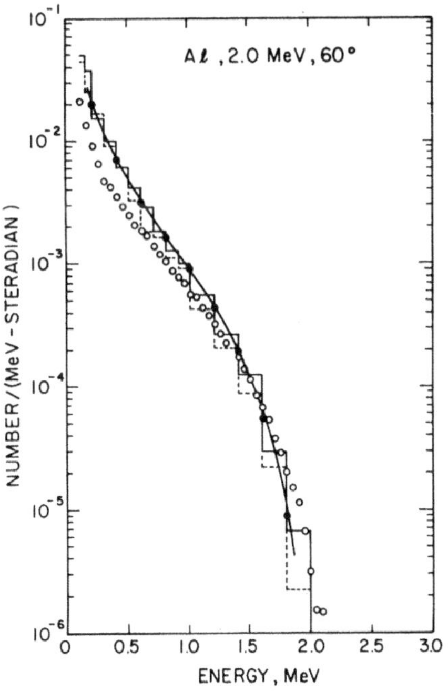

Fig. 3e

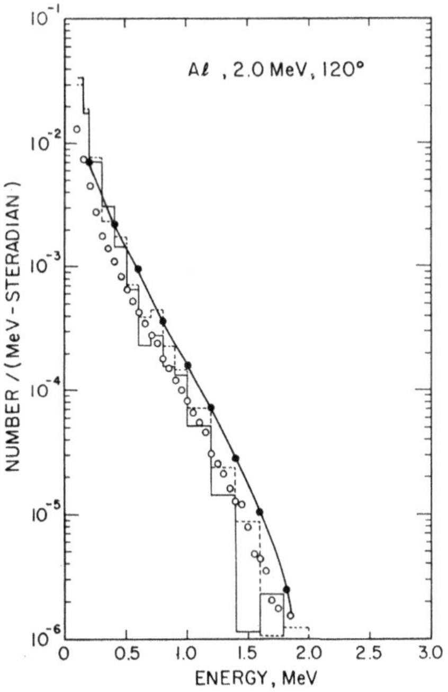

Fig. 3f

Fig. 3. Comparison of thick-target bremsstrahlung calculations with calculations by Berger and Seltzer (1964) (histogram) and data by Baggerly *et al.* (1964) (open circles).
(a) Aluminum, 0.5 MeV electrons, 15° exit angle; (b) Aluminum, 0.5 MeV electrons, 60° exit angle; (c) Aluminum, 0.5 MeV electrons, 120° exit angle; (d) Aluminum, 2.0 MeV electrons, 15° exit angle; (e) Aluminum, 2.0 MeV electrons, 60° exit angle; (f) Aluminum, 2.0 MeV electrons, 120° exit angle.

We have built a small computer code to finish the evaluation of (2) along these lines, for both a normal electron beam and for an isotropic electron flux. We actually took account of the angular dependence of the thin-target bremsstrahlung only when the electron angular distribution was narrower; otherwise we treated the bremsstrahlung as straight-ahead. A few of our results for a

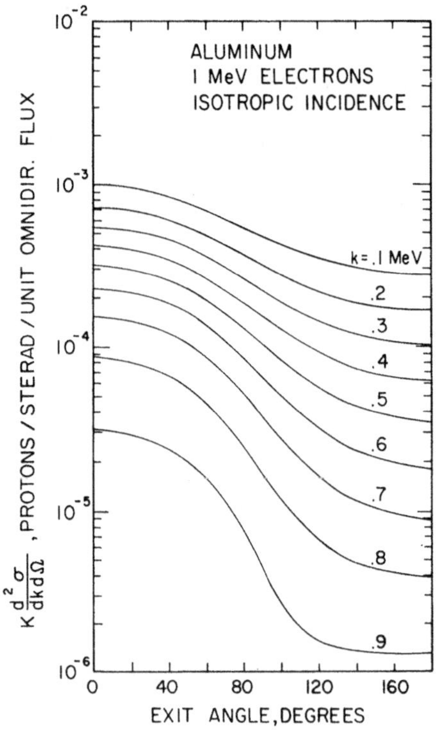

Fig. 4. Thick-target bremsstrahlung from 1 MeV electrons isotropically incident on aluminum.

normally-incident beam are shown in Figure 3, compared with Monte Carlo calculations by Berger and Seltzer (1964) and experimental data from Baggerly *et al.* (1964).

A typical result for the case of isotropic electrons is shown in Figure 4. I should like to say that we realize that our rough approximation for the electron flux function is surely poorer for this case. Nevertheless, it is certainly conservative where it is wrong, and it enables us to get estimates like this very readily. We have not seen published results of better calculations for this practical case.

One comment should be made regarding Figure 4: all of the curves are flatter than a $\cos\theta$ dependence. Because Gauss's theorem tells us that the interior of a "nicely-shaped" convex space vehicle would be uniformly irradiated if the bremsstrahlung flux followed a $\cos\theta$ distribution, these results indicate that an astronaut would be unwise to spend too much of his time near the walls.

4. Electron penetration

Intrigued by a footnote remark by L. V. SPENCER (1955) to the effect that the old electron penetration model of BETHE *et al.* (1938) gave results better than order-of-magnitude while being right nowhere, we decided to have a look at it for possible use in bremsstrahlung calculations. These authors constructed a simple model which patched together two extremes of electron behavior, the straight-ahead and diffusion approximations. Electrons were treated as straight-ahead until the r.m.s. Gaussian multiple scattering angle (neglecting change of energy) became 1 radian; then they were injected to diffuse.

We complicated the BRS model somewhat, allowing the beam electrons to spread into the angular distribution given by (12) but modified to incorporate a sink at $\theta=\theta_c$; the electrons spreading over this boundary were injected to diffuse. Because we used simplified stopping-power and elastic-scattering formulas which treated logarithmic factors cavalierly, we allowed ourselves the luxury of a second free parameter in the spatial age, weakly dependent on Z. We applied this model to the calculation of the energy dissipation of electron beams in infinite media, because the results could be compared with the extensive calculations of SPENCER (1959); no comparable body of experimental data is known to us.

Some results are shown in Figure 5. These do not necessarily represent the best choices of our parameters. We consider that the results are really rather encouraging. This age-diffusion equation model at least offers the advantage of being extremely flexible with regard to boundary conditions. A computer code to do electron penetration of a slab by this age equation model is being developed, but unfortunately has not yet been completed.

References

BARKAS, Walter H. and Berger, Martin J.: 1964, NASA SP-3013.

BAGGERLY, L. L., DANCE, W. E., FARMER, B. J., and JOHNSON, J. H.: 1964, paper given at the Second Symposium on Protection against Radiations in Space, Gatlinburg, Tenn., Oct. 12–14, 1964.

BERGER, Martin J., and SELTZER, Stephen M.: 1964, NASA SP-3012.

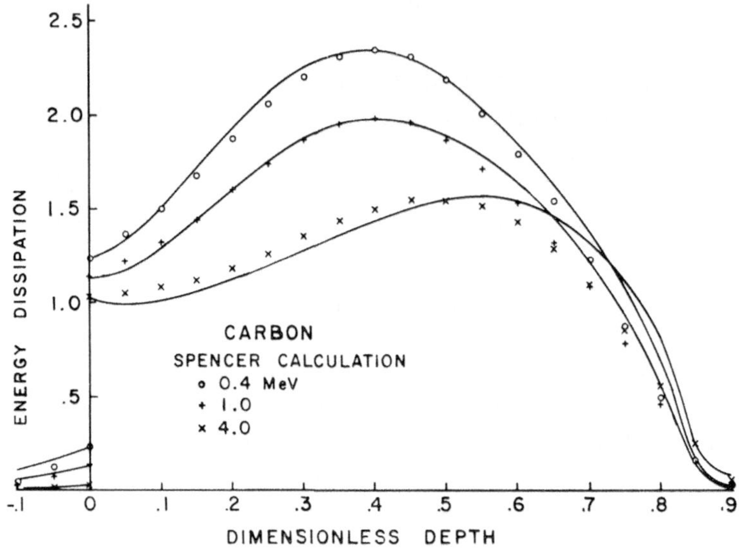
2.5
2.0
1.5
1.0
.5
ENERGY DISSIPATION
CARBON
SPENCER CALCULATION
o 0.4 MeV
+ 1.0
× 4.0
-.1
0
.1
.2
.3
.4
.5
.6
.7
.8
.9
DIMENSIONLESS DEPTH

Fig. 5a

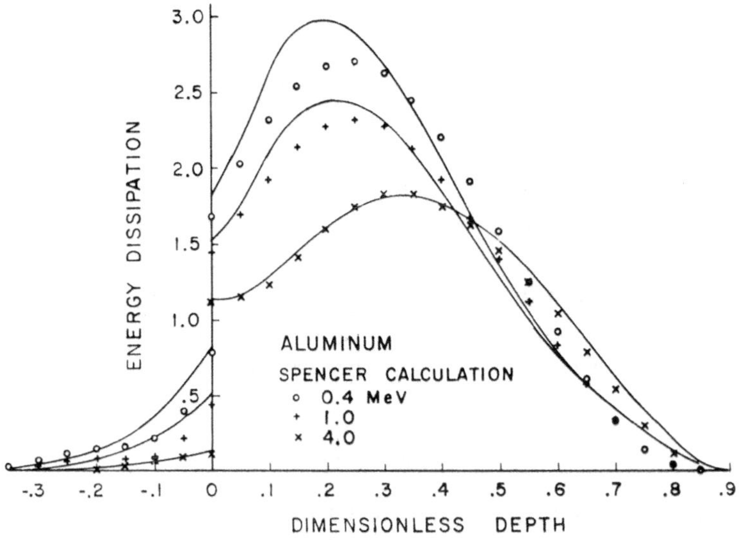
3.0
2.5
2.0
1.5
1.0
.5
ENERGY DISSIPATION
ALUMINUM
SPENCER CALCULATION
o 0.4 MeV
+ 1.0
× 4.0
-.3
-.2
-.1
0
.1
.2
.3
.4
.5
.6
.7
.8
.9
DIMENSIONLESS DEPTH

Fig. 5b

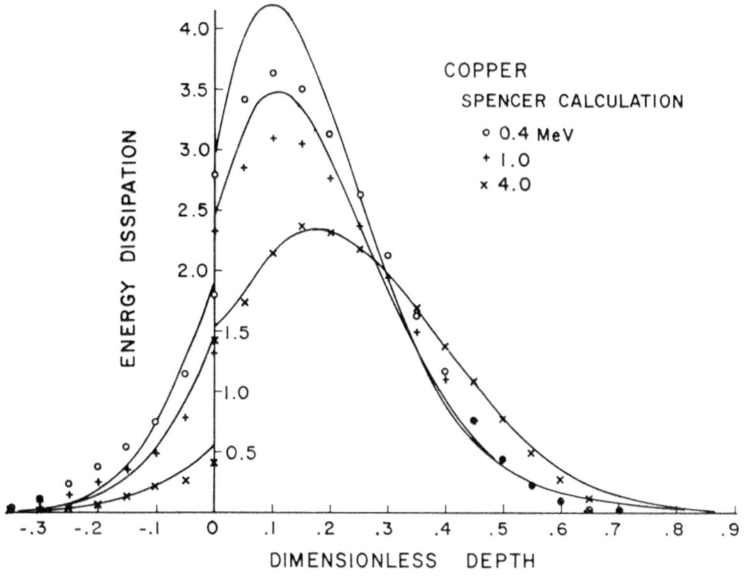

Fig. 5c

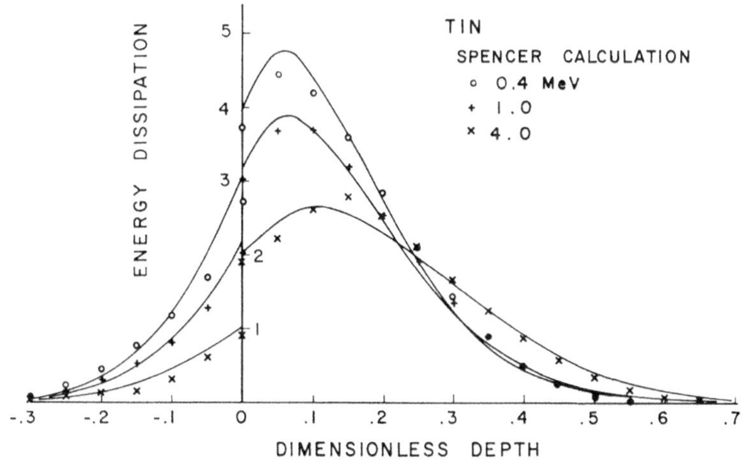

Fig. 5d

Fig. 5. Energy dissipation by electrons from a plane-perpendicular source in various infinite media. Comparison of age equation calculations with the accurate calculations of Spencer (a) Carbon, (b) Aluminum, (c) Copper, (d) Tin.

BERGER, M. J. and SELTZER, S. M.: 1964, paper given at the Second Symposium on Protection against Radiations in Space, Gatlinburg, Tenn., Oct. 12–14, 1964.
BETHE, H. A., ROSE, M. E., and SMITH, L. P.: 1938, *Proc. Am. Phil. Soc.* **78**, 573.
KOCH, H. W. and MOTZ, J. W.: 1959, *Rev. Mod. Phys.* **31**, 920.
HEITLER, W., *The Quantum Theory of Radiation,* 2nd ed., Oxford Univ. Press, London, 1944, p. 166.
NAKACHE, F. R.: 1964, Second Symposium on Protection against Radiations in Space, Gatlinburg, Tenn., Oct. 12–14, 1964 (Abstract)
SPENCER, L. V.: 1955, *Phys. Rev.* **98**, 1597.
SPENCER, L. V.: *Energy Dissipation by Fast Electrons,* NBS Monograph 1, U. S. Department of Commerce, National Bureau of Standards, Sept. 10, 1959.
STARFELT, N. and KOCH, H. W.: 1956, *Phys. Rev.* **102**, 1598.

SUMMARY

BILLY M. McCORMAC

After over sixty lectures and a vast amount of discussion on the various aspects of radiation trapped in the earth's magnetic field, it is important to attempt to present a summary. The Institute did not attempt to reach agreements or to determine popular opinion but rather to see that the experimental and theoretical results were presented and adequately discussed. Each individual has to make his own evaluation as to the importance and credibility of the various information. My personal summary of the Institute is presented below from the point of view of future research requirements.

The McIlwain B-L space coordinate system is as good as the knowledge of the field, B. The currently used Jensen-Cain-1962 field is inadequate, especially for low altitude mirror point work. However, POGO I data (launch date October, 1965) is essential to reduce the inaccuracies in the field to the same size as those for satellite location, atmospheric densities, etc. Some individuals may desire to use the 1965 field for low altitude mirror point work, but it is expected that most users will not change their coordinate system until McIlwain revises his code on the basis of a magnetic field, using POGO data, which may be 1966 or 1967. The B-L space coordinate system should continue to be used for the stable trapping region. There does not seem to be a satisfactory coordinate system for the transition region.

The stable trapping region has not been adequately measured. This is especially true for the outer belt. Many more measurements are needed to look for secular variations, diurnal and 27-day solar cycle fluctuations, etc. The inner proton belt seems very stable; however, the outer belt is effected by magnetic storms and other natural phenomena. Currently, one cannot unravel spatial and temporal variations. This may never be possible with one satellite. Diffusion and acceleration have not been separated. During catastrophic events, large changes in the B-L coordinate system may be occurring at high L values. Observed particle fluctuation have some unknown dependence on B, L, and particle energy. There is some evidence of acceleration, both at high and low L values, when the magnetic field fluctuation has a period equal to the drift period. Many more experimental observations are required. Matched satellite pairs (separated by a short distance) are needed to try to resolve temporal and spatial variations, and more theoretical studies are required. For example, more attention could be given to the theoretical aspects of producing E fields perpendicular to B.

Greatly increased theoretical effort on basic processes is needed for the interaction of the solar wind and the magnetosphere. Two general types of intuitive models have been advocated, but they are not detailed enough to employ. Several basic theoretical problems have been identified for study: (1) collisionless shock, (2) magnetopause viscous force, (3) dynamics of the geomagnetic tail, (4) neutral sheath, and, (5) particle acceleration.

Only very fragmentary data are available on the solar wind, the transition region, and the tail. These data show very large variations in particle flux and energy. It is not possible with current satellites to separate spatial and temporal variations. Any instabilities have not been measured. The plasma and magnetic field in the transition region need to be measured to determine if the plasma is sufficient to carry off any transport currents. If so, there will be no E fields parallel to B for acceleration. It is likely that instabilities exist but since they cannot be investigated in the laboratory, they have to be measured. Two satellites separated by a short distance are absolutely essential for measuring instabilities and making spatial and temporal variation separations. The ratio of the energies of the protons and electrons are needed at many locations throughout these regions.

The solar wind is the source of all electrons and protons of energy less than 100 MeV by some diffusion and acceleration processes which are currently not understood. There does seem to be some local acceleration at injection. Neutron albedo decay only satisfies protons greater than 100 MeV. Unless the neutron albedo decay spectrum above 50 MeV is measured, which is unlikely, and found to be much different than estimated, further theoretical study seems useless.

Precipitation of charged particles produces auroras. The phenomena associated with night aurora seem different from those for day aurora. It is not now known if the aurora particles have been trapped, if the aurora zone is inside or outside the stable trapping region or even the magnetosphere. There is a need for coordinated ground based and satellite measurements to determine the energy spectrum, pitch angle distribution, back scattering, location in the magnetic field, etc. for auroras.

Artificial radiation sources are needed to study the belt. To resolve many problems the source must be known which seems impossible from natural phenomena. Suggestions to use positrons or alpha particles were not enthusiastically received. Some International endeavor to employ small nuclear detonations at different L shells may be more practical. The U.S. test, Star Fish, on July 9, 1962, trapped about 3 per cent of the electrons below L of 1.5 and by some mechanism accelerated natural electrons above L of 1.5. It is not possible to separate diffusion from acceleration with the data available. Atmospheric loss explains the data for $L \leqslant 1.25$. Important problems remaining are (1) to determine the decay rate for electrons at $L > 1.25$, (2) to explain the transfer of electrons accross L shells, (3) to explain the non-fission spectrum above L of 1.5.

Since synchrotron radiation depends on so many variables, its measurement does not provide any unique constraints on the trapped electrons. To date, measurements have only been made on the electrons trapped from Star Fish. It may be possible in the next few years to make synchrotron measurements on the natural electrons.

The nuclear test, Star Fish, focused attention on the degradation of sensitive electronics by trapped electrons and protons. Solar cells and transistors both suffer surface as well as bulk effects. Enough experimental data has been obtained for the derivation of empirical procedures which provide for the design of reliable systems with predictable degradation. However, the basic processes occurring in the solar cells and transistors by the electrons and protons are poorly understood.

Current theoretical and experimental efforts for determining space vehicle shielding by mass and the dose inside these vehicles seems to be quite adequate; however, the response of human beings to radiation must be refined before the radiation hazard to manned space flight can be fully assessed. There is no hazard for low altitude orbits up to 500 to 800 km altitude, providing there are no more high altitude nuclear tests; however, the center of the inner belt is very hazardous for manned satellites.

This Institute did much to identify important areas of agreement and disagreement and to focus attention on key problems for the future. The attendees concluded that a sequel of two weeks duration should be held in 1967. The following areas will be importants (1) Experimental and Theoretical Research on the Outer Magnetosphere, (2) Injection of Particles, (3) Acceleration of Particles, (4) Life Times above L of 1.25, (5) Experimental Observations in the Belts, (6) Secular Variations, (7) Solar Wind, (8) Artificial Sources, (9) New Representation of the Magnetic Field, and (10) Diffusion and Precipitation of Particles.

INDEX OF PERSONS*

* Italic numbers indicate that the person mentioned is denoted there as an author of a paper in this proceedings or as a participant in the discussion.

INDEX OF SUBJECTS

Batch number: 10824463

Printed by Printforce, the Netherlands